Introduction to Energy and Climate

Introduction to Energy and Climate
Developing a Sustainable Environment

by
Julie A. Kerr

CRC Press
Taylor & Francis Group
Boca Raton London New York

CRC Press is an imprint of the
Taylor & Francis Group, an **informa** business

CRC Press
Taylor & Francis Group
6000 Broken Sound Parkway NW, Suite 300
Boca Raton, FL 33487-2742

First issued in paperback 2019

ISBN-13: 978-1-4987-7439-0 (hbk)
ISBN-13: 978-0-367-87888-7 (pbk)

Library of Congress Cataloging-in-Publication Data

Names: Gines, Julie K., author.
Title: Introduction to energy and climate : developing a sustainable
environment / Julie Kerr.
Description: Boca Raton : Taylor & Francis, [2017] | Includes bibliographical
references.
Identifiers: LCCN 2016057705 | ISBN 9781498774390 (hardback : alk. paper)
Subjects: LCSH: Power resources--Environmental aspects. | Climatic changes. |
Sustainability. | Sustainable living. | Renewable energy sources.
Classification: LCC TD195.E49 G36 2017 | DDC 363.738/74--dc23
LC record available at https://lccn.loc.gov/2016057705

Visit the Taylor & Francis Web site at
http://www.taylorandfrancis.com

and the CRC Press Web site at
http://www.crcpress.com

Contents

SECTION II Energy Sources, Their Future, and Sustainable Living

Introduction

Climate change is a challenging problem, with an overwhelming body of scientific evidence supporting it. Its effects are now being felt on a global scale. Although the issue has manifested itself on a worldwide basis, it still carries with it its share of contention—generated largely from resistance in the political arena and traditional energy sector, combined with the general public being largely unwilling to adopt a more environmentally friendly lifestyle. Because climate change is not just confined to the realm of the scientific community, nor does it have one simple, predefined solution, it is still largely seen as volatile and controversial—especially in highly developed countries, where significant lifestyle changes may be required. This is a multifaceted issue that must be addressed and solved, however, if we are to continue to enjoy an equitable and comfortable lifestyle. Climate change's multiple dimensions are addressed in this book. They involve scientific, economic, sociological, political, psychological, and personal issues, making it a topic that affects every person on the earth now and in the future.

Gaining a firm understanding of the climate issue and the effects it is having on our environment is only the first step, however. The next phase—and equally important—is gaining an understanding of what is involved in fixing the problem and how to do it—in other words, how to successfully create a sustainable environment that is equitable and fair globally. The book will analyze those issues by first looking at the concept of energy: what it is, how we use it, and how much we currently need and will need in the future. It then analyzes the systems we have been used to for decades and why our current system simply will not work as a mechanism to attain future sustainability.

Similar to the climate change issue, we must look at multiple aspects, not just different energy types and their individual feasibility. As responsible land stewards, we must find a way to successfully balance and mesh these multifaceted issues, such as types of energy, amount of available resources, the geographic feasibility to make it practical, political attitudes, and personal willingness to become educated and make appropriate lifestyle choices. If necessary, we must also implement improved urban and transportation planning systems, as well as find creative, yet fair, ways to fund them. We must develop policies to assist third-world countries in order to ensure they also have an equitable opportunity to sustainable food and energy, and develop necessary geopolitical frameworks enabling sustainable systems to operate. These issues will require education, integration, and cooperation on a global scale.

This book offers a unique approach by proposing a truly comprehensive solution to deal with the issues of both problems in the short and long terms. It reaches across diverse sectors of the society and global culture to link the topics and offer new insights for practical solutions.

Beginning with a detailed discussion of the scientific topics involved, it endows the reader with a firm foundation of the concepts necessary to understand and be able to maturely reason through a myriad of diverse, difficult issues. This timely volume puts on center stage those crucial ideas that are usually overlooked, misunderstood, or lost in media sensationalism. Enlightening and empowering readers, it looks at the decisions that must be made to achieve true global sustainability. This volume is a perfect fit for anyone currently employed or entering into a career related to climate change, energy resources, sustainability, or green living.

The first part of the book is dedicated toward understanding the climate and the science of climate change. Chapter 1 begins by providing a thorough overview of the environment—and how natural resources, climate, energy resources, and the concept of sustainability all work together and how humans fit into the picture. It looks at key environmental issues, population, ecosystems, and the concept of conservation in order to lay a solid groundwork for the entire book. Following this, the book is divided into two major sections: Section I discusses the earth's climate balance and the consequences of climate change and Section II discusses energy sources, their future, and sustainable living.

Chapters 2, 3, and 4 provide an overview of the physical science aspects of climate change, divided into three central categories: the climate system, greenhouse gases, and the effects on ecosystems. It is imperative to have a basic, working knowledge of these areas so that you can then connect the importance of them to the remainder of the climate change issues. Next, we discuss the sociological and psychological issues that surround climate change in Chapter 5, including what makes us act the way we do, our ability to cope, the power of the media, and the effect of disinformation.

Following this, Chapter 6 explores geopolitical considerations, such as the mind-set of countries and how climate change is perceived, how changes in legislation are needed to make a difference, and what the Environmental Protection Agency (EPA) is doing about the problem. We explore the U.S. Climate Action Plan, the Clean Power Plan, what other countries are also involved in to deal with cutting carbon emissions, and how the world is working together to solve the growing problem. Then, in Chapter 7, we discuss public perception and the economic implications of a warming world. We take a good look at the current threats to several sectors of the economy and where we are headed if emissions are not controlled, including the impacts on our food sources, natural resources, recreational areas, and our very health. We will look at the all-to-real challenges we face now from climate impacts and those that our children will be facing in the near future.

In Chapter 8, we shift to mitigation options and look at ways that industry, manufacturing, other businesses, and government can fight climate change through various efforts that are now in place and what projected outcomes are possible if efforts are made now to clean up and cut back carbon output.

This begins Section II on energy. Chapter 9 begins with an overview of energy: its basics, concepts, and quantification. We learn about the laws of energy, how it behaves, how it is converted and compared, and how it moves and interacts with other matter. Once you have a solid base for understanding what energy actually is—and is not—we move into nonrenewable energy in Chapter 10 and discuss all the petroleum products, coal, nuclear, and other fossil energy—their characteristics, uses, and prospects for the future. Following that, Chapter 11 moves to the renewable forms of energy—solar, geothermal, wind, hydropower, bioenergy, and secondary energy sources, such as electricity and hydrogen. Like the previous chapter, we discuss their characteristics, uses, and prospects for the future. Following this, we explore the consequences of energy consumption on the society in Chapter 12—both nonrenewable and renewable. Each form of energy has its pros and cons, and we discuss each in detail so that the consumer can make the most educated choices possible.

Once you have the basics—and then some—mastered, in Chapter 13 we move to the geopolitics of energy, and you will see how energy affects different parts of the world and their economies in fundamentally different ways. You see what a key role energy issues can play in a national, or even international, setting and ultimately determine policies that span multiple countries and often set the tone for how several countries get along.

Next, we move to the applications in sustainability, and in the next two chapters, we examine the applications of technology. In Chapter 14, we look at the importance of urban planning and green buildings and the new directions these fields have taken in the past few years. In Chapter 15, we examine smart transportation, alternative fuels, and new fuel technology. These chapters present several exciting new concepts that are sure to shape urban areas for the next few decades. Chapter 16 addresses agriculture and industry and you will see how these major industries are going green and the huge differences it is making in cleaning up the environment. After that, we hit some exciting topics in Chapter 17: Green careers and ecotourism. This is the section where you can get all kinds of ideas if you would like a "green collar" job—that is, a job in sustainability. You may also want to take a green vacation some time, and you can learn all about that in this chapter on ecotourism: what it is, how you do it, what it involves, and if it is for you.

In Chapter 18, we bring you back to your own home. You can take a sustainability test to find out how green you really are, and then get loaded with ideas on how to become even greener than you have ever been, both inside your home and in your backyard with all the ideas we provide you with. Chapter 19 illustrates how to live a sustainable lifestyle if that is something that interests you—from making your own cleaning agents to building your own compost pile. We can teach you that and much, much more!

Finally, when we arrive at the end of the book, Chapter 20 reviews all the benefits that are afforded you and all future generations if you choose to make just the few simple changes that are recommended in your life—either through small steps or in a few big leaps. Whatever you choose, by the end, there will be no doubt: you will know how to live a green sustainable life and do your part to make the right changes—and choices—for the benefit of future generations.

Author

Julie A. Kerr has been an earth scientist for nearly 40 years, promoting a healthier and better managed environment. She holds a PhD degree in earth science with an emphasis on remote sensing satellite technology from the University of Utah, Salt Lake City, Utah, and has dealt with the hands-on applications of resource and climate change management, focusing on the delicate balance of the multifaceted issues involving the political, economic, and sociological components in both short- and long-term applications with real-world ramifications. She has been involved specifically with projects such as forest management, classification and monitoring of desertification, rangeland change detection and management, vegetation inventory and health assessment, and land-use change.

Dr. Kerr approaches the subject with a background uniquely blended with conservation and land-use management expertise with the U.S. Bureau of Land Management, military applications with the U.S. Air Force, and university-based research. She has also had environmental experience serving in the professional community as the president of the Intermountain Chapter of the American Society of Photogrammetry and Remote Sensing (ASPRS) and spent years dedicated to the promotion of a healthier environment through participation with various environmental groups. The prolific author has published more than 25 books, many of which are focused on environmental topics such as global warming, climate change management, conservation of natural resources, green living, and the earth sciences.

1 Understanding the Environment
Resources, Climate, Energy, and Sustainability

OVERVIEW

In order to begin our journey in striving to understand the role of climate and energy in building a sustainable future, we must start with a foundation. In order to do this, before diving right into the complex issue of climate change, we will first take a look at some environmental basics. Because all these issues are so closely intertwined, this chapter will deal with the issue of understanding our environment. We will begin with a look at how humans affect the environment, including the issue of overpopulation; how our desire for a higher standard of living is taking a toll on the natural world; why our current lifestyle is polluting and exhausting precious resources, such as the water, land, and the atmosphere; and just what we are doing— informed and less-informed—to contribute to climate change. Next we will examine the most critically important environmental issues of today and look at the reasons why they are important, as well as why those reasons tend to change over time. Coupled with that will be a discussion of environmental ethics and why—although it is not a science per se—it has its place in the scientific community.

From that, we will shift gears and look at the world's rapidly growing population and what the projections currently see our future looking like as far as the worldwide production and distribution of food, its availability, and scarcity; and why those concepts may not be closely related. We will also examine the world's water resources and you will learn that it not only is one of earth's most precious resources, but one of its most scarce. From there, we will shift gears a bit and take a look at the concept of ecosystems—what they are and how they work. We will also explore the importance of biodiversity and why the earth depends on it, as well as what the consequences would be without its benefits. In conclusion, we will focus on the concept of conservation and why that is the way of the future.

It is important to understand how all these topics are interwoven and affect us every day in order to make sense of why the issues of climate change and energy resources are two of the most fundamental components of developing a sustainable environment. Therefore, we will address the basics as a springboard to launching into the first half of the book on climate change.

INTRODUCTION

In order to understand the significance of where climate change and energy choices fit in with the importance of developing a sustainable environment and lifestyle, it is necessary first to understand some fundamental concepts about why sustainability is so important. Sustainability is important for many reasons. Its primary purpose is to protect our natural environment, and human and ecological health, while driving innovation and not compromising our way of life. It is important because we only have a finite amount of natural resources and people are living longer. The United Nations projects that there will be more than 10 billion people living on the earth by the year 2100, which—without sustainable practices— will seriously impact the environment and our quality of life (Gillia and Dugger, 2011).

Sustainability is also important in order to protect our technological resources. It requires a wide range of minerals and other materials to manufacture technological components. If these are not conserved wisely, production of future technology could be seriously hampered. Sustainability is important to provide for basic human needs, such as food, water, and shelter. If we continue down the path of a fossil fuel diet for our energy sources rather than sustainable options, the cost and environmental penalty will eventually become prohibitive. Our food supply is another critical area of concern. With a rapidly growing population, we must find better ways to grow food. Current methods simply will not suffice. Sustainable agricultural practices, however, such as crop rotation, effective seeding practices, contour farming will provide higher yields while protecting the health of the soil. They will enable larger amounts of food to be produced without the impact the environment is currently experiencing. In addition, urban growth is expected to increase. If we continue to develop urban areas in a nonsustainable manner, pollution will increase with the use of fossil fuels. Conversely, if sustainable development practices are employed, new housing and business development can be designed to coincide harmoniously with the environment, enabling urban expansion.

Sustainable development is also critical in the battle to deal with climate change. According to the United Nations, "Climate change is one of the greatest challenges of our time and its adverse impacts undermine the ability of all countries to achieve sustainable development." Sustainable development practices would enforce a lower use of fossil fuels, which are not sustainable and produce enormous amounts of greenhouse gases (GHGs). As the population increases and energy demands increase, sustainable energy sources must be used in order to mitigate the consequences of climate change.

Sustainable practices also promote financial stability. This would enable resource-poor economies to have access to renewable energy as well as resultant access to new jobs and related markets. Sustainable living also promotes healthy biodiversity, resulting in a healthier environment for everyone. So let us take a look at how understanding the environment and our role in it directly affects sustainability.

DEVELOPMENT

One of the more frequently used words today seems to be environmentalism, such as: "Is it environmentally friendly?" "What will that do to the environment?" "What kind of environment will we be leaving for our children?" and "Those plastic bags will not degrade; they will sit in the environment forever." Sound familiar? And the good thing: it seems more people are thinking about the environment these days. People are becoming more environmentally aware and willing to do something positive in its behalf. They are willing to make sacrifices to protect it by changing their personal habits. It has been a long time coming, but it is beginning to happen. The not so good thing: population numbers continue to rise at accelerated rates, which means a bigger demand for natural resources, more use of—and stress on—the land, more pollution, and more problems to solve.

So what is the status of our environment, and what is our role in it? Why do we continue to hear increasingly more about saving our planet—especially when we also seem to be inundated with other important issues, such as economic crises, wars and conflicts, new diseases, and a myriad of rising social problems? Does the environment take a back seat? Should it be front and center? Or could all of these problems be connected and need to be looked at from a systems viewpoint?

Understanding the Environment and Our Role in It

As complicated as it might seem at times, these problems are all connected, and the rest of the answer is quite simple: our environment matters so much because it—the earth—is our home and the only one we will ever have. The pressures that the collective population is putting on it are enormous, and we cannot continue as we have in the past. We must now make intelligent decisions if we are to have enough and provide for those who follow.

Our environment is everything around us—the land we live on, the air we breathe in, the water we drink, food we grow, soil we grow our food in, and the animals we share the planet with. The unity of all this, called the biosphere, is a global ecological system—each component dependent on, and affected by, another. The concept of the biosphere is shown in Figure 1.1 and was coined by Austrian geologist Eduard Suess in 1875 as "the place on earth's surface where life dwells." Based on this, the concept of an ecosystem was introduced in 1935 by English botanist Sir Arthur Tansley.

FIGURE 1.1 Conceptual diagram of the Earth System showing the four subcomponents of the geosphere (lithosphere, atmosphere, hydrosphere, and cryosphere), the biosphere (terrestrial and marine), climatic processes, hydrologic cycle, biogeochemical cycles, and solar energy. (Designed by Jim Tomberlin, U.S. Geological Survey.)

The earth's biosphere covers every part of the planet—from the equatorial to polar regions, from the deepest depths of the oceans to the upper reaches of the atmosphere, and into the earth's substrate. Life exists everywhere, extending roughly from 5900 feet (1800 m) above the earth's surface (flying altitude of birds) to 27,467 feet (8372 m) below the ocean's surface. There are also extreme exceptions, such as yaks that live at elevations as high as 17,700 feet (5400 m), bar-headed geese that migrate at altitudes of 27,200 feet (8300 m), and microscopic organisms that live at ocean depths seven miles deep (11.3 km), 40 miles (64.4 km) high in the atmosphere, and deep inside rocks found in the earth's crust (Science Daily, 1998).

The environment is critical to every human and living organism on earth. It is where we live, breathe, eat, and survive, and it functions in a way similar to a well-oiled machine—when all the components are working as they are designed, it runs harmoniously. If one part of the system becomes out of sync, however, it can cause problems that quickly spread from one component to another, rapidly degrading the system; possibly leading to failure. Think of a car being run without oil. How long could that car run before it became vulnerable? Before it became completely nonfunctional? Now think of an ecosystem that had polluted water introduced into it. The plants in that system require clean water in order to grow. Other life relies on it as a drinking source for survival. Before long, the quality of life would decline, and then eventually become unsustainable.

As we will see in Chapter 2, there are key cycles—such as water, carbon, and nitrogen—that function in the earth's various environments, enabling the ecosystems to exist and thrive.

If, however, a component to that system becomes out of sync with the ecosystem in which it is contained (such as marine, forest, arctic, and desert), it still impacts the entire planet. Although environmental problems may be diverse, they can all have far-reaching implications toward the health of the planet and the life on it.

As an illustration, consider air pollution and the impact it can have on global weather. It may seem obvious that air pollution can have negative impacts on humans living nearby by causing health issues and local environmentally related issues, such as respiratory illness—but new research at Texas A&M University, College Station, Texas, has discovered a direct link between increased air pollution and changing weather patterns, particularly the formation of powerful, destructive storms (Texas A&M Today, 2014). What the researchers have discovered is that pollutants, in the form of particulates, are being released into the atmosphere in great amounts in Asia due to their fast-growing economic base. The particulates—also called aerosols—were interfering with and altering the solar radiation balance by scattering and absorbing the solar radiation, which was, in turn, altering the formation of clouds. Through monitoring the activity going on today and comparing it with the conditions of the area from 1850 (preindustrial era), they have concluded that the pollutants are affecting the storm formations over the Pacific Ocean. This area, which is the source of the weather patterns that feed North America and elsewhere, is now experiencing stronger, more intense storms. These strong storms, in turn, impact the resulting weather at both regional and global levels, causing damage, destruction, and other negative effects.

Another reason why taking care of the environment is important is to keep it from deteriorating. This condition—called environmental degradation—threatens the earth's natural resources. Because some of these natural resources are important to health and life, such as clean water and air, fertile soil, and food supply—it is imperative to understand the environment and our role in it. Also important are the aesthetic qualities the natural environment has to offer. This was recognized >25 years ago. According to the World Health Organization "good health and well-being require a clean and harmonious environment in which physical, psychological, social, and aesthetic factors are all given their due importance" (WHO, 1989). This also applies to urban settings. In what the United Kingdom's Forest Research Department calls quality green infrastructure in urban settings, it not only improves environmental quality but makes a positive contribution to landscape character and aesthetic appeal. They define local environmental quality as tangible elements, such as cleanliness and personal security. They also include less tangible elements being equally important—such as visual quality and environmental pollution. Environmental pollution includes the contamination of the physical and biological components of both the earth and the atmosphere to the extent that natural environmental processes are negatively affected. In some ecosystems—such as the fragile Arctic—this does not take much.

According to Forest Research, the public thinks of environmental quality holistically, often describing it as "aesthetically pleasing," "clean," and "well maintained." In addition, a high level of aesthetic quality is seen as a sign of care within the urban environment that leads ultimately to a sense of community (Forest Research, 2010).

Although much evidence exists concerning the connection between human health and well-being and a healthy environment, the challenge is still getting enough people to take drastic enough action so that a significant difference can be made before it is too late.

HOW HUMANS AFFECT THE ENVIRONMENT

Everything living has an impact on the environment. For this reason, no matter what we do, there will be an inevitable impact of some degree. The act of simply existing creates an impact. The thing that differentiates humans from other species, however, is that we have very few limits placed upon us and we have the ability to place a huge burden on our environment. Due to our quest for a comfortable lifestyle, humans have used and altered their environment to fit their needs and wants. This has especially been evident during the past 250 years since the Industrial Revolution began. During this time period, various environments on earth have taken a beating—some more than others. Some of the more prominent effects include the following:

- Overpopulation
- Our desire for a higher standard of living
- Environmental pollution
  - Water pollution
  - Land pollution
  - Air pollution
- Climate change

The following sections titled Overpopulation, Our desire for a higher standard of living, Environmental pollution, and Climate change will discuss these impacts.

Overpopulation

There are currently 7.4 billion people on earth today. Increased populations make excessive demands on the earth's natural resources. Many of the impacts involve demands on the agricultural and livestock sectors as the requirement for food continues to increase. These impacts include the following:

- The use of chemical fertilizers, insecticides, and herbicides increases production. They pollute the soil, water, and air with toxic chemicals. For example, when fertilizers run off into streams and other water bodies, they create algal blooms that kill the aquatic animals living there.
- Large-scale farming of animals increases their chances of contracting diseases, such as mad-cow disease or avian flu. According to a study by

Grace et al. (2012), researchers have discovered 13 animal-to-human diseases (called zoonoses) that are responsible for 7.2 million deaths a year. The study, called "Mapping of Poverty and Likely Zoonoses Hotspots," revealed that most of the deaths were in low- and middle-income countries, the highest incidences occurring in Ethiopia, Nigeria, Tanzania, and India. In addition, the waste generated by animals can be detrimental to local water quality.

- Cutting down forests and other native vegetation to create new areas to grow crops has a negative effect on the environment because it promotes habitat loss and threatens the survival of the indigenous plant and animal species.
- Creating monocultures is a drawback. Monocultures are used in order to keep production costs low, but their use harms the soil's ability to produce over time and reduces biodiversity.
- Transportation distance from farm to market is an issue. The greater the distance food and other market items have to travel to reach the consumer, the greater the impact that its required transportation has on the environment.

Our Desire for a Higher Standard of Living

In the most developed countries, our demand on the environment's natural resources has exceeded the earth's supply capacity. The mind-set has been that the earth can generate more to meet our needs. The truth, however, is that the turning point of consumption and related mind-sets occurred along with the advent of the Industrial Revolution, which began in the eighteenth century. This marked the end of sustainable living. Human nature being what it is, as we became used to the newly introduced comforts of life, we wanted more.

The transportation sector is another area affected by the desire for affluent living. The burning of fossil fuels greatly contributes not only to the pollution problem but also to climate change. This includes all modes of transportation: land, air, and water. Nowhere has it been harder to try and change the mind-set of people to conserve and cut back on gas-guzzling vehicles than in developed countries, the United States being one of the prime examples.

Another area of environmental stress comes from our desire to control our artificial (indoor) environments. Opting for personal comfort over environmental diligence, many have become used to the comforts of overcooling indoor air in the summer and overwarming it in the winter to stay more comfortable. The downside of this is that it requires large amounts of energy to achieve this.

Environmental Pollution

The earth has always been efficient at being able to recycle waste through the ages. Since the Industrial Revolution, however, humans—by far the most polluting species—are now generating much more than the planet can handle. The result is pollution. Environmental pollution occurs today worldwide. It occurs at different levels and in different forms; and it does not just affect the planet as a whole. It impacts all species—from humans down to the tiniest organisms.

Water Pollution

Without water, there can be no sustainable life on this planet. Yet, there is very little of it. Surprisingly, of all the water on earth, only 2.5 percent is freshwater (drinkable). Of that 2.5 percent, less than 1 percent is actually available. With that limited amount, all 7.4 billion people on earth require up to 13 gallons (50 L) a day of freshwater for drinking, cooking, and basic living. This is not including the water required for agriculture or for animals. Furthermore, of that 1 percent that is available, 70 percent of that goes to irrigation.

There are several types of water pollution, including nutrient, surface water, oxygen depleting, ground water, microbiological, suspended matter, chemical, oil spillage, sewage and wastewater, septic tank, ocean and marine dumping, underground storage leakage, and atmospheric pollution.

Nutrient pollution occurs when wastewater, fertilizer, or sewage contain high levels of nutrients. If the nutrients get washed into water sources, excess algae and weed growth occur. This makes water undrinkable and depletes the oxygen content. Surface water pollution occurs in rivers, lakes, ponds, and oceans. This happens when hazardous substances dissolve and mix with the water. Oxygen-depleting pollution occurs when biodegradable matter ends up in the water and causes microorganisms to grow, which in turn use up the oxygen in the water, depleting it.

Groundwater pollution occurs when pesticides are used on plants and soils, and then are washed down to the groundwater level by rain. Microbiological contamination is found in untreated stream water that can contain natural pollution caused by microorganisms such as protozoa, bacteria, and viruses. This type of pollution can kill fish and make humans sick. Suspended—or particulate—matter is the contaminant that can kill aquatic life that lives at the bottom of the water body. Chemical pollution usually originates from a point source (such as a specific industry or farm) and can include poisons, toxins, metals, solvents, and chemicals. Oil spills are more localized in occurrence, but their harmful effects are spread for miles. One of the most well known of these is the BP oil spill that occurred in 2010 in the Gulf of Mexico. More than 1000 animals died—many of them endangered.

Sewage and wastewater is especially dangerous because of the deadly diseases associated with them. In some countries, untreated sewage is dumped into rivers or the ocean. Ocean and marine dumping consists of paper, food, plastics, rubber, and metals. Although some items may decompose within weeks (like paper), others take years; and all act as pollutants (Figure 1.2).

The effects of water pollution are varied and depend on which chemicals are dumped and in which locations. Unfortunately, many bodies of water near urban areas are extremely polluted. One of the biggest effects is that it kills

FIGURE 1.2 The Oshiwara River in Mumbai—severely polluted with solid and liquid waste generated by Mumbai. MMRDA studying 5 Rivers. (Courtesy of Jan Jörg; https://en.wikipedia.org/wiki/Oshiwara_River.)

the life that depends on these water bodies, which in turn disrupts the natural food chain. Polluted water also causes the spread of disease. It can severely affect the health of ecosystems.

The Great Pacific Garbage Patch An extreme example of ocean pollution is the Great Pacific Garbage Patch (also referred to as the Pacific trash vortex). This is an area in the North Pacific Ocean where an extremely large collection of marine debris has collected over time, caused by people tossing their litter into the water. The marine debris that it is composed of is a collection of litter that people have thoughtlessly thrown in the oceans, seas, and other significant bodies of water for years, eventually pulled to this location due to the predominant flow of the ocean's major current patterns. The entire area extends from North America's West Coast eastward to Japan. Within the entire region, there are actually two principal areas of trash collection—referred to as the Western Garbage Patch (located near Japan) and the Eastern Garbage Patch (located between Hawaii and California).

The North Pacific Subtropical Gyre helps confine the Great Pacific Garbage Patch. By definition, an ocean gyre is a system of circular ocean currents formed by the earth's wind patterns and the forces created by the earth's rotation. In this case, the North Pacific Subtropical Gyre is created by the proximal interaction of the California, North Equatorial, Kuroshiro, and North Pacific currents. All four currents move in a clockwise direction around an area of approximately 7.7 million square miles (20 million km^2) in size. In a gyre, the currents on the outside are actively moving, but the area inside the borders of the currents is very stable. This is where the garbage is currently being trapped and deposited. For example, if someone throws a plastic water bottle into the ocean from the coast of California, it will join the California Current and travel southward toward Mexico. From there, it could enter the North Equatorial Current and head across the Pacific Ocean. If the bottle reaches Japan and then gets swept up in the Kuroshiro Current, it can be carried to the north, to eventually be intercepted by the westward North Pacific Current. The two vortexes—the Eastern and Western Garbage Patches—will eventually capture the bottle (Figure 1.3).

The reason for the accumulation of garbage is that the majority of it is not biodegradable. Much of it is composed of plastics that do not wear down, but instead break into smaller

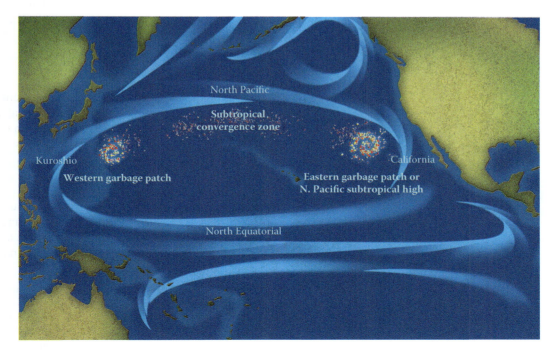

FIGURE 1.3 The Great Pacific Garbage Patch. (Courtesy of NOAA, Silver Spring, MD.)

and smaller pieces. In fact, the majority of the collection is composed of tiny bits of plastics, called microplastics. The microplastics are formed through a process called photo-degradation, where the sun breaks down the plastics into tinier pieces. Scientists have been able to collect as much as 750,000 bits of microplastics in 0.6 mile (1 km²), which equates to about 1.9 million bits per square mile (National Geographic, 2012). Nested within this mass are permanent island-like features composed of larger plastic objects. Some extend more than 50 feet (15 m) in length. The majority of the debris originates from plastic bags, bottle caps, plastic water bottles, and Styrofoam cups.

Much of the microplastics cannot even be seen by the naked eye, but collectively they make the water look murky and opaque. Found floating within this thick, soupy mass are also other, larger pieces of refuse, such as fishing gear or spilled items from container ships. Both oceanographers and ecologists have found that almost three-quarters of the marine debris actually sink to the bottom of the ocean, making this refuse deposit very thick and massive. Just how massive? No one has been able to calculate it. The North Pacific Subtropical Gyre is simply too massive for scientists to trawl it. In addition, because all the garbage is not floating—but suspended within the ocean column—it is impossible to calculate the actual mass from the surface (Figure 1.4).

Garbage from the North American coast takes about 6 years to reach the Great Pacific Garbage Patch, whereas that from Japan and other Asian countries takes around a year. Besides the aesthetic problems, it is very harmful to marine life. Some species, such as the loggerhead sea turtle, often mistake plastic bags for jellyfish—one of their most favorite foods. Other species feed resin pellets in the water to their young, mistaking them for fish eggs. Seals and other marine mammals get entangled in abandoned plastic fishing nets

and drowned—an unfortunate event termed *ghost fishing*. The refuse also blocks the incoming sunlight. If sunlight cannot reach plankton and algae further beneath the ocean's surface, it disrupts the food chain and subsequent availability of fish.

The problem of cleanup is also problematic. The National Ocean and Atmospheric Administration's Marine Debris Program has estimated that it would take 67 ships per year to clean up less than 1 percent of the North Pacific Ocean. The most realistic way to control this problem is to eliminate our use of disposable plastics and increase our use of biodegradable resources. Organizations such as the Plastic Pollution Coalition and the Plastic Oceans Foundation are now using social media and direct action campaigns in an attempt to get manufacturers' and businesses' attention in the matter.

Land Pollution

Land pollution is the degradation of the earth's surface and soil. The land is being polluted and abused constantly. The degradation occurs directly or indirectly due to human activities, which lessen the quality and/or productivity of the land. Sometimes we allow the land to get to the polluted condition it is in before we realize just how degraded we have allowed it to become. Caused by human activity, there are several practices that degrade the land. Some include poor agricultural practices, deforestation, urban development, transportation expansion, industrial and mining practices, recreational activities and development, overuse of resources, poor conservation practices, and inappropriate activities.

Depending on the ecosystem involved, misuse of the land can cause degradation that can leave long-term scars. Some areas may never recover. For example, Arctic ecosystems are so fragile that misuse can permanently damage the ecosystem. In desert ecosystems, just driving over fragile soils can leave scars on the landscape that last for centuries.

When cities expand and urban sprawl occurs, natural habitats are built over and eliminated leaving the native species with nowhere to go. This can be seen in the mountain foothills of the western United States with mule deer (*Odocoileus hemionus*) and moose (*Alces alces*) (Figure 1.5). Urban spread often intrudes in their natural environment. Then when the herd migrate to lower elevations in the winter for food and gravitate to the landscaped yards that replace their previous habitat, landowners get angry.

Another major issue is that proper land use planning is not done and other issues arise, such as the following:

- Urban areas are built in flood zones.
- Areas are overbuilt in regions where there is not enough water so that drought years force mandatory water rationing.

FIGURE 1.4 An example of extreme amounts of ocean debris—mostly plastics. (Courtesy of Ray Boland, NOAA.)

FIGURE 1.5 A moose (*Alces alces*) visiting a resident's yard in Park City, Utah. (Courtesy of Julie A. Kerr.)

- Areas that are not well planned out with sustainability in mind often result in wasted land.
- Areas developed without being public transportation, or other green transportation, friendly, encourage residents to commute with private cars, contributing to climate change.

Poor agricultural practices include allowing manure to collect to the level where rainwater washes it as runoff into water sources. Clearing the land of natural vegetation can leave it vulnerable to erosion as well or negatively alter the local ecosystem. In addition, the use of fertilizers and herbicides can cause harmful pollution, and if the land is not farmed with conservation processes in mind, such as crop rotation and alternating with fallow fields, it can deplete the soil of necessary nutrients.

Once land is cleared of its natural vegetation and converted to dry land, it will never go back to its original fertility again regardless of what is done to it. Changing the natural use of land destroys it. With overpopulation and the human need to keep expanding and building, land can be ruined and wasted.

Along these same lines of overcrowding are landfills. Every year each household is responsible for producing tons of garbage. Those items that can be recycled—plastic, paper, and metal—reduce the amount that end up in the landfills, but the remainder that cannot be recycled adds to the collective problem, not only causing land—but visual—pollution. In matters of urbanization and construction, wasted materials such as wood, bricks, tile, metal, cement, and plastic can be

seen at construction sites and scattered during high winds. Sewage treatment areas can also be a problem. Large amounts of solid waste can be left over from sewage treatment, which can subsequently end up in landfills, adding to that pollution problem. In areas where nuclear power is used, the disposal of nuclear waste becomes an environmental issue because left-over radioactive material contains harmful chemicals that are detrimental to human health. If they are buried underground, and later leak into the soil, it presents additional environmental issues. Unpaved roads are another form of land pollution. They can erode and deteriorate into adjacent ditches, which can then flood the roads during heavy storms, causing further erosion and land degradation.

Industry is another key contributor to land pollution. As the population grows and increases the demand for housing, food, services, and goods, the industrial level of an area increases in order to support the population. The spread of industry in some cases, however, has led to deforestation and emissions of highly toxic chemicals into the soil, atmosphere, and water. In the United States, which is similar to many industrialized nations, the greatest source of pollution originates from industrial sources. For this reason, the U.S. Environmental Protection Agency (EPA) keeps track of the management of certain toxic chemicals that may pose a threat to human health and the environment in their annual Toxics Release Inventory (TRI). The TRI tracks the management of certain toxic chemicals that may pose a threat to human health and the environment. Industrial sectors must report annually on how much of each chemical is released to the environment and/or managed through recycling, energy recovery, and treatment. Information is submitted by thousands of U.S. facilities on more than 650 chemicals and chemical categories. This information then helps support informed decision-making by industry, government, nongovernmental organizations, and the public.

The latest TRI report released for 2014 performance showed the following about U.S. industry in 2014:

- Of the 25.45 billion pounds of TRI chemicals that were reported as managed as waste
  - Thirty-seven percent was recycled.
  - Fourteen percent was used for energy recovery.
  - Thirty-four percent was treated.
  - Sixteen percent was disposed of or released.
- Of the 3.89 billion pounds of TRI chemicals that were disposed of or released
  - Nineteen percent was to the atmosphere (on-site).
  - Six percent was to water sources (on-site).
  - Sixty-five percent was to land (on-site).
  - Eleven percent was to off-site disposal.

Compared to the prior year's amounts, production-related waste had decreased 13 percent, and disposal and releases had decreased 6 percent. TRI data can be used in combination with other data sources to provide a more complete picture of what is going on with chemical use, management, and releases (EPA, 2014).

These facts make the industrial sector an important one to consider and monitor for environmental impacts. The most common industrial polluters include the oil refining, petrochemical, chemical, pesticide, food processing, iron and steel, and metal smelting industries. Each of these industries is a major energy producer and can contribute significantly to environmental degradation.

The good news is that environmental progress has been made during the past 150 years regarding industrial practices. In the United States, because industries must now comply with regulations, emissions have been reduced in many places. Cleanup programs have also been implemented, as have sustainable development initiatives. Nongovernment or special interest groups, such as the Sierra Club and Friends of the Earth, have done a lot to bring pressure onto industry that pollutes the environment, in an effort to force cleanup and better practices. Although work still needs to be done, steps are moving in the right direction.

Mining plays a significant role in land degradation. In the process of extracting metals and minerals from the earth, it requires the clearing of large areas of land. The chemicals used in mining can be released into the environment and cause large-scale problems. Some of the ill effects associated with mining include deforestation, water pollution, land pollution, and loss of biodiversity. When large areas are deforested, it permanently degrades and changes the character of the land, leading to erosion and infertility. Harmful elements can be introduced into the area such as sulfuric acid and metal oxides. When these compounds are introduced into the local waterways, they contaminate local rivers with heavy metals. Chemicals, such as mercury, cyanide, sulfuric acid, arsenic, and methyl mercury, are used in various stages of mining. When these are released into water bodies, they pollute the water. If they are introduced into the ground and slowly percolate downward, they contaminate the groundwater. Chemicals also leak onto the land, and this can change the chemical composition of the area. If the chemicals poison the soils, the soils become infertile. When large areas are cleared for mining activities, this leads to loss of biodiversity. This is especially a problem in rainforest areas where the cutting down of trees is in itself a significant threat to the many plants, trees, birds, and animals that live within it.

The recreational industry is another area that has seen tremendous growth recently. With a surge of more people than ever before spending time outdoors—275 million visitors to the U.S. national parks each year—care must be taken to avoid land degradation (NPS, 2016). Hikers, mountain bikers, and off-road vehicle users must take care to stay on designated paths; camping enthusiasts have rules of etiquette to follow; and others who enjoy the outdoor experience need to adopt a leave-no-trace philosophy in order to avoid harming the landscape.

Air Pollution

Air pollution is a serious environmental concern. Environmental awareness has seemed to make a difference in certain areas

of the world. According to the EPA (2016a), air quality has increased over the past 15 years. From 2000 to 2014,

- Ozone decreased 18 percent.
- Lead decreased 87 percent.
- Nitrogen dioxide decreased 43 percent.
- Carbon monoxide decreased 60 percent.
- Sulfur dioxide decreased 62 percent.

The EPA warns, however, that 127 million people still live in areas that exceed air quality standards, and this causes serious problems to the young, the elderly, and those with existing respiratory problems. Many factors contribute to the problem, such as the size of the population; climate, weather patterns, and air flow; local geography; transportation patterns; and industrial activity, making the issue unique to each location. In this regard, one variable that must be closely examined is the energy choice used in the industrial, residential, and transportation sectors (Figure 1.6). As we will see further on in this book, there are many choices available to us on several levels that allow us to take action for a healthier environment, such as smarter land planning for new development, smarter transportation choices, retrofitting less efficient structures with green technology, and eliminating poor agricultural and other land use practices. The common denominator is that it will take the joint effort of all of the society.

THE COUNTRIES WITH THE CLEANEST AIR IN THE WORLD

- Luxembourg
- Ireland
- Sweden
- New Zealand
- Denmark
- Estonia
- Brunei Darussalam
- Finland
- Canada
- Iceland

FIGURE 1.6 Smog over Almaty city, Kazakhstan, January 12, 2014. (Courtesy of Igors Jefimovs.)

Climate Change

The earth is getting increasingly warmer—summers are growing hotter, glaciers are melting, sea levels are rising, and weather events are becoming more unpredictable. Human-induced climate change has only emerged as a serious issue in the past few decades, but climatologists have found undisputable evidence that humans are slowly changing the earth's climate and environment. The CO_2 levels in the atmosphere are slowly climbing—404.16 ppm as of February 2016—we have long since passed the threshold level of 350 ppm that climate scientists warned was the tipping point. Figure 1.7 illustrates the presence of CO_2, and its key sources and sinks between the lithosphere and troposphere. Although the change so far is not rapid, it will begin accelerating, and we will see temperatures rise higher than they have in the past 420,000 years.

Natural climate change occurs over time due to factors such as the relationship between the earth's rotation, axis, position, and revolution around the sun, as well as a result of major volcanic eruptions (what has traditionally been small increments of change over thousands of years). These types of gradual changes allow most species to survive by migration or adaptation. However, warming has increased dramatically during the last century at an unnatural rate, focusing attention on the human factor. Many activities, such as the burning of fossil fuels for energy and massive deforestation, are contributing to the atmospheric warming at an alarming rate. As this continues, it will create problems in the distribution of species and their habitats increase the occurrence of severe weather events, contribute to sea-level rise, and trigger a host of health and quality-of-life issues that will affect everyone on earth. No ecosystem will escape the impact of human-induced global warming. According to the Intergovernmental Panel on Climate Change (IPCC), the impact of human activities is approximately 10 times that of natural and solar factors.

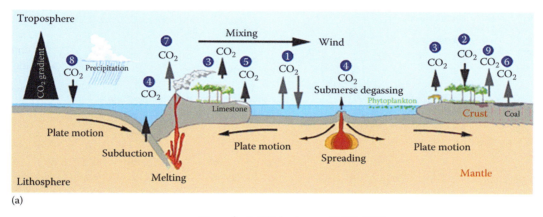

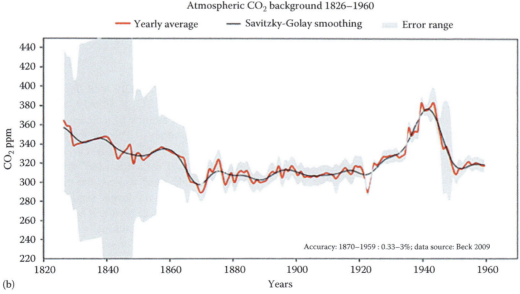

FIGURE 1.7 (a) CO_2 sources and sinks in the boundary layer of the lithosphere-troposphere: (1) ocean degassing/absorption, (2) photosynthesis, (3) respiration, (4) submerse geological degassing, (5) limestone weathering, (6) surface coal oxidation, (7) volcanic degassing and subduction degassing, (8) precipitation absorption, (9) soil respiration. Main globally effective controllers for CO_2 flux (IPCC). (From http://www.biomind.de/realCO2/.) (b) Atmospheric CO_2 background 1826–1960. (From http://www.biomind.de/realCO2/.) *(Continued)*

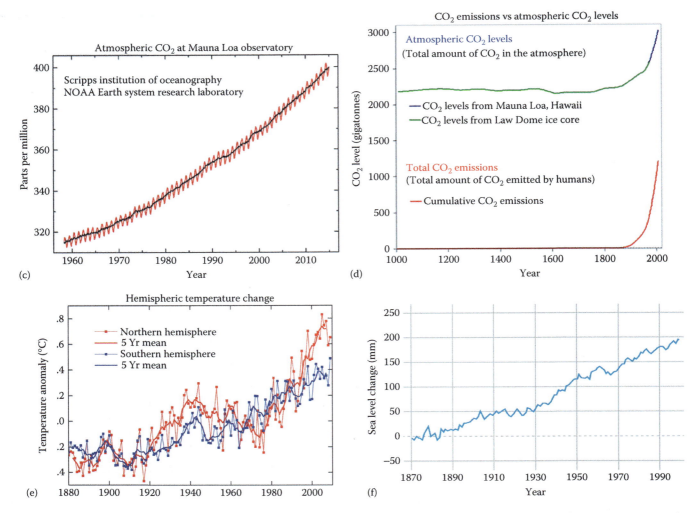

FIGURE 1.7 (Continued) (c) Atmospheric CO_2 1960-present. Dashed red line indicates monthly mean values, black line represents the average seasonal cycle. (From National Oceanic and Atmospheric Administration [NOAA], Silver Spring, MD.) (d) CO_2 emissions versus atmospheric CO_2 levels. (From Department of Energy.) (e) Hemispheric temperature change, 1880-present. (From NASA Goddard Institute for Space, New York, NY.) (f) Sea-level change, 1870–2000. Blue line indicates ground data collected from coastal tide guage records. (From CSIRO Marine and Atmospheric Research.)

The fear is that our influence will cause irreversible catastrophes (Didyouknow.org, 2011).

The effects on sea-level rise alone are chilling. A new study published in *Nature Climate Change* (Clark et al., 2016) reports finding that even if countries could meet carbon reduction targets to stay below the 2°C global warming threshold, sea-level rise will eventually inundate many major coastal cities around the world. This means that 20 percent of the world's population will eventually have to migrate away from coasts swamped by rising oceans. Cities—including New York, London, Rio de Janeiro, Cairo, Calcutta, Jakarta, and Shanghai—would all be submerged. Low-elevation islands would be completely submerged. This conclusion was drawn by examining past climate change events and modeling simulations of the future. It was discovered that there is a strong relationship between the total amount of carbon pollution humans emit, and how far global sea levels will rise. Even though ice sheets melt slowly, carbon dioxide stays in the atmosphere a long time, enabling the

resident CO_2 to eventually melt the ice and raise sea level. In other words, what CO_2 is already in the atmosphere is locked in for the future melting of ice caps and glaciers. So what does this mean measurably? This means we have already committed the planet to an eventual sea-level rise of 5.5 feet (1.7 m). If we stay within the 1 trillion ton (907 trillion kg) carbon budget—an amount calculated to keep the earth below the 2°C warming above preindustrial levels, sea levels will *still rise* a total of approximately 30 feet (9 m). If we continue on the same fossil fuel diet we have traditionally been on, we could eventually trigger 165 feet (50 m) of sea-level rise. In a recent study published in the *Proceedings of the National Academy of Sciences*, scientists are also concerned that the Antarctic ice sheet could melt more quickly than previously thought (Kopp et al., 2016; Rietbroek et al., 2016). Over the past century, global sea level has risen much faster than at any time in the past 2000 years, and most of the recent sea-level rise is due to human-caused climate change. They believe that several feet (m) of sea-level rise this century will occur—with a possibility of 5 feet (1.5 m) or more. In short, the earth is warming about 50 times more rapidly than when it comes out of an ice age.

WHAT IS SO SPECIAL ABOUT 2°C?

The magic number—the number to avoid seems to be 2°C when it comes to climate change. As long as we do not pass that threshold of "additional warming," there is still hope. By the end of 2015, however, the planet had warmed 1°C (1.8°F)—we are already half way to that magical "threshold of doom." So what, exactly, is this threshold, where does it come from, and what does it mean?

Put simply, the 2°C threshold is a target—a way to express how much warming humanity can tolerate before experiencing the most destructive and dangerous effects of climate change on the environment. In order to come up with a realistic, obtainable goal that participants could wrap their minds around and commit to, the UN Climate Panel agreed upon the 2°C as a collective target that countries could agree to. Although there is still uncertainty of the absolute effects it will cause, climate scientists do agree that there will likely be more drought and intense heat waves, which will cause significant disruptions to the world's food supply, along with sea-level rise significant enough to flood many cities and potentially cause huge migrations from countries such as Bangladesh, India, and Vietnam.

To date, efforts made have not been enough to keep carbon emissions below what they need to be to keep within the 2°C limit. Figure 1.8 illustrates what the carbon emissions must drop by from now until 2050 if we are to keep within the 2°C constraint. The problem is that the three largest culprits—the United States, Europe, and China—will account for most of the budgeted CO_2 in 2030, leaving no allowance for other countries, such as Africa and India, making it an unrealistic goal (Figure 1.8) (Peters et al., 2015).

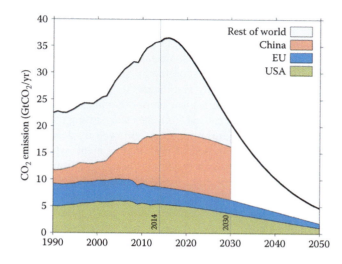

FIGURE 1.8 Cumulative emissions, global carbon budgets, and the implications for climate mitigation targets. (From Peters, G. P. et al., *Measuring a Fair and Ambitious Climate Agreement Using Cumulative Emissions*, IOP Publishing, Bristol, 2015; http://www.wri. org/blog/2013/12/global-food-challenge-explained-18-graphics.)

The effects of climate change are becoming more evident every day. From changing weather patterns and reduced water availability to deforestation and melting ice caps—the examples are all around us. So what are the next steps? What are governments and businesses doing to address the impact of climate change? What are individuals doing?

Some governments are putting in place strategies to develop green industries such as renewable energy, and they are enacting regulations to reduce carbon emissions. On the business side, some companies are minimizing their own footprint on the environments and communities they work in by adopting corporate responsibility agendas and by reviewing their carbon emissions and supply chains. Some individuals are doing their part by participating in the green movement. But is it currently enough?

No.

The answer: all governments, all businesses, and all individuals need to do their part. The clarification: sustainable development and living. Burning fossil fuels is harmful, as are many other practices, including deforestation. The first half of this book will discuss in depth the issues of climate change, its causes, and its consequences. This will lead into the second half of the book on energy choices and creating a sustainable society. In order to accomplish these goals—what society so far has failed to do, even in light of compelling evidence for the need to do so—there are basic milestones we much reach. We must strike the right balance with the general population, and in order to do this, education is critical. Doom and gloom has no place here; neither does placid acceptance. Education and implementation of constructive solutions are what are needed to find answers to serious challenges. We must face the truth: even implementing sustainable practices now—we will have wins, but we will also have some losses. Also of supreme importance, the political contingency must be educated, made to understand, and forced to stop looking at the issue through fossil fuel-colored glasses. Those who are being swayed by misinformation need to be replaced by those without blinders on. The time for skeptics is over if effectual change is to happen. The political arena must be supportive if any long-term change is to happen, because in order to be effective, it will require policy change, and governments will have to work together. In addition, it is important for each of us to be on board with the issue and educated, because that is the first step for community action, and it will take community action—all of us working together—to make lasting changes. Sound like a big undertaking?

Yes.

Is it possible?

For all of our sakes—now and future generations—it has to be.

ENVIRONMENTAL ISSUES AND WHY THEY ARE IMPORTANT

Environmental issues put in the most straightforward way are simply those issues that, in some manner, either directly or indirectly have an impact upon the environment. That effect can be miniscule or hugely significant. It can range from soil erosion from a single rainstorm on poorly cared for soil to a catastrophic landslide caused from the clear-cutting of forests, subsequent loss of natural nutrients, followed by devastating erosion on a soil composed of high clay content, which has been heavily built up with residential housing, sitting seemingly dormant until the right rainstorm saturates the hillside and causes massive destruction and death in a community. It comes in all sizes and shapes, various geographic settings and conditions. No place on the earth's surface is immune from human destruction. Environmental issues are also as much the result of what we do to the environment as what we do not do. For example, when we do not value the biodiversity of an area, certain species can be forced into extinction. Coral reefs are a case in point. Highly essential to the environment in a number of ways, today they face numerous threats from human activity. In recent years, many have seen large areas of die-off due to human interference and rising water temperatures (Figure 1.9). Because they provide such rich ecosystems for multitudes of aquatic species, this has become a serious environmental impact in recent years.

When environmental issues face us every day—both on the news and around us—it may be puzzling as to why they continue. At the core level, the importance of environmental issues is based on several factors, including (but not limited to) the following:

- Needs.
- Wants.
- Ideals.
- Economics.
- Historic status.
- Culture and traditional values.
- Political values.
- Popularity.
- Incorrect, uninformed, or biased information.
- Who is presenting the information?

FIGURE 1.9 This image shows a bleached brain coral. (From NOAA.)

According to a study by Lyytimäkia et al. (2012), although we will face many new environmental problems, the environmental scientists and policy makers may not recognize them. In fact, which environmental issues they choose to address will most likely depend on things such as media attention and socio-economic dynamics—a topic discussed in detail in Chapter 5. It may be a purely personal choice of who is in charge. And because environmental issues do not remain static over time, coupled with new scientific findings continuously coming to light, issues that do gain momentum toward reaching solutions may be abandoned as new information comes along or an entirely new issue eclipses it (or is perceived to eclipse its importance). During the study, the researchers focused on trying to understand what factors cause an issue to fail to receive attention and the implications of not prioritizing certain types of environmental information. They also focused on potential solutions, such as (1) acknowledging our limitations in understanding, (2) thinking we have all the information we need when we do not and making false assumptions, (3) not caring enough about the issue to find a meaningful solution, or (4) having a potential solution but being unwilling to share it with others. All of these issues hamper solving environmental problems.

TODAY'S MOST CRITICAL ENVIRONMENTAL ISSUES

Although we have ascertained that some environmental issues tend to change in importance over time and evolve as more information is gathered and solutions are implemented, the following list represents some of the currently recognized more critical environmental concerns:

- *Acid rain*—Destruction from its increase due to climate change and GHGs
- *Acidifying oceans*—Ocean die-off, threatened marine ecosystems
- *Agricultural pollution*—Overuse of fertilizers and herbicides
- *Air pollution*—Both indoor and outdoor air quality, carbon emissions, ozone, particulates, and other harmful emissions
- *Aligning governance to global sustainability*— Moving toward a global perspective
- *Biological pollutants*—Bacteria, viruses, molds, mildew, dander, pollen, ventilation, and infections
- *Bridging science and policy*—Achieving a better understanding between the scientific, economic, and political arenas
- *Carbon footprint*—Responsibility of individuals to reduce their effect on the environment
- *Clean water sources*—Preventing contaminants from entering water sources
- *Climate change*—Issues related to the greenhouse effect, global, dimming, and sea-level rise
- *Consumerism*—Overconsumption and its effect on the planet
- *Consumption*—Creating more equitable distribution/ consumption designs

- *Contaminated drinking water*—Contamination of freshwater used for household needs
- *Dams*—The impact of dams on the environment
- *Deforestation*—Land degradation, erosion, loss of soil fertility, and productivity
- *Degradation of inland waters in developing countries*—Water pollution control
- *Ecosystems and endangered species*—Destruction of environments and habitats
- *Effect on marine life*—Climate change and human exploitation
- *Energy sources and conservation*—Finding feasible renewable energy sources
- *Food safety*—The effects of hormones, antibiotics, preservatives, toxic contamination, and lack of proper health controls
- *Food supplies*—Finding more efficient ways to feed a growing world population
- *Food security*—Regulation of health issues and development of efficient delivery and supply agreements
- *Food waste*—Inefficient use of food and poor distribution
- *Genetic modification/engineering*—Genetically modified foods and genetic pollution
- *Increasing pressures on coastal ecosystems*—Land-use patterns and fishing practices
- *Integrating biodiversity and economic agendas*—Better communication between scientific, economic, and ethics components
- *Integrated ocean governance*—Better ocean management systems
- *Intensive farming*—Irrigation methods, overgrazing, monoculture, methane emissions, and deforestation
- *Invasive species*—Overgrazing, wildfire, non-rehabilitated land
- *Land management*—Developing a working system in order to make efficient, timely decisions
- *Land use*—Urban sprawl, lack of free space, habitat destruction, and fragmentation
- *Light and noise pollution*—Effects on human health and behavior
- *Logging*—Deforestation, clear-cutting, destruction of wildlife habitat, and GHG emissions
- *Loss of biodiversity*—Impacts of land use, human practices, and climate change
- *Loss of endangered species*—Destruction of habitats, ecosystems, and species
- *Medical waste*—Toxic and biohazardous items
- *Mining*—Acid mine drainage, soil and air pollution from toxic emissions, and heavy metals
- *Moving toward a green economy and transforming human capabilities*—Transferring to a green infrastructure and mind-set
- *Nanotechnology*—Future effects of nano-pollution and nano-toxicology

- *Natural disasters*—Their impact on all aspects of the environment
- *Natural resource depletion*—Overexploitation of the earth's natural resources
- *Nuclear issues*—Nuclear fallout, nuclear meltdown, and radioactive waste issues
- *Overexploitation of natural resources*—Destruction and depletion of natural resources
- *Overfishing/fishing practices*—Negative effects on marine ecosystems, blast fishing, cyanide fishing, bottom trawling, whaling, and other forms of exploitation
- *Overpopulation*—Effects on resources and land use
- *Ozone layer depletion*—Damage to the earth's ozone layer caused by CFCs
- *Poaching*—Decimation of wildlife and harm to ecosystems
- *Public health issues*—Food safety, nutrition, physical activity and obesity, drug abuse
- *Responding to new national and international pressures*—Negotiation issues
- *Soil contamination*—Soil erosion, conservation, salinization, and contamination by waste, pesticides, and lead
- *Sustainable communities*—Reducing reliance on nonrenewable energy, supporting local suppliers, and going green
- *Toxins*—Household and industrial waste, chlorofluorocarbons, heavy metals, pesticides, herbicides, toxic waste, and so on
- *Water–land interactions*—Better monitoring and understanding between ecosystem interfaces
- *Water pollution*—Contamination of the earth's limited freshwater; related issues, such as acid rain, ocean dumping, urban runoff, oil spills, ocean acidification, and wastewater
- *Waste management*—Littering, landfills, marine debris, and toxins
- *Wildlife conservation*—Care of wildlife habitat

ENVIRONMENTAL ETHICS

Ethics are a broad way of thinking about what constitutes a good life and how to live one. They address questions of right and wrong—making the right decisions. Applied ethics approaches these issues with a special emphasis on how they can be lived out in a practical manner. Environmental ethics apply ethical thinking to the natural world and the relationship between humans and the earth. They are a key feature of environmental studies, but they also have application in many other fields as society attempts to deal in meaningful ways with pollution, resource degradation, the threat of extinction, and global climate disruption.

Environmental ethics has a specific focus: moral focus and decision-making efforts toward the way we look at, care for, and interact with our environment. True environmental ethics results in an in-depth understanding of the significance of nature and an awareness of the ethical dimension of sustainability.

The concept of what we care about becomes rooted in our moral vision—we protect and care for our environment because we value it for itself and as an important extension of our society. When we instinctively act ethically, we are intrinsically aware of it subconsciously. For instance, if you are consuming resources (degrading the earth's ability to provide the services that humans need) at a faster rate than they can naturally be replenished; and if you believe this is wrong, you are basing this on some kind of moral principle that you hold. This is the moral process of believing that the earth has ethical significance.

Although moral reasoning is not a substitute for science, it does provide a powerful companion to scientific knowledge about the earth. Because science does not teach us to care, it stands to reason that scientific knowledge does not, by itself, provide reasons for environmental protection. Science provides data, information, and knowledge. Environmental ethics then takes this information and asks moral questions, such as the following: How then, should we live? Why should we care? Why is it important to me? Environmental ethics builds on scientific understanding by bringing human values, moral principles, and improved decision making in tandem with science. For this reason, environmental ethics must be interdisciplinary, because it draws on other fields of academic inquiry, rather than standing by itself. Due to its unique nature, usually by just asking the question "What is the right thing to do?" can open up new insight for the solution to environmental problems. Thinking ethically about the environment has the potential to help everyone discover environmental solutions.

In summary, the concept of environmental ethics is to protect and sustain environmental diversity and ecological systems (Nature Education, 2014). It also builds upon scientific understanding by bringing human values, moral principles, and improved decision-making into the scientific arena.

THE EFFECTS OF HUMAN POPULATION GROWTH

Human overpopulation is one of the most serious environmental issues. Overpopulation leads to many of the planet's most taxing problems today: environmental pollution of the air, land, and water; desertification; climate change; food shortages; health issues; habitat loss and species extinction; deforestation; overgrazing; poor agricultural practices; irresponsible land-use practices; natural resource depletion; lower life expectancies in the fastest growing countries; and increased emergence of new epidemics and pandemics.

SPECIES EXTINCTION

Human beings are currently causing the greatest mass extinction of species since the extinction of the dinosaurs 65 million years ago with rates 1000–10,000 times more rapid than normal. According to the International Union for Conservation of Nature Red List of Threatened Species, in 2015 there are 79,837 species on the Red List

and 23,250 of them are endangered species threatened with extinction—that is nearly one-third of the total. This means that if present trends continue, within a few decades, at least half of all plant and animal species on earth will become extinct as a result of habitat loss, pollution, climate change, acidifying oceans, invasive species, overexploitation of natural resources, overfishing, poaching, and human overpopulation. The trend is alarming, as illustrated in Table 1.1.

With world population currently at 7.3 billion people, the current growth rate is one billion people every 12 years; which equates to about 220,000 per day. Because of this, we are facing resource shortages, social pressures (including wars), and a multitude of other challenges like we have never encountered before.

According to howmany.org (2016), food resources are being negatively impacted by the world's current population. Every day one billion people (one out of every seven) go to bed hungry. Each day, 25,000 people die due to malnutrition and hunger-related diseases. Of these 72 percent are children that are less than 5 years old. In addition, approximately one billion people do not have access to drinking water, or

TABLE 1.1

Rises in Threatened Species

Year	Number of Threatened Species
1998	10,533
2000	11,046
2002	11,167
2004	15,503
2006	16,117
2008	16,928
2010	18,351
2012	20,219
2014	22,413
2015	23,250

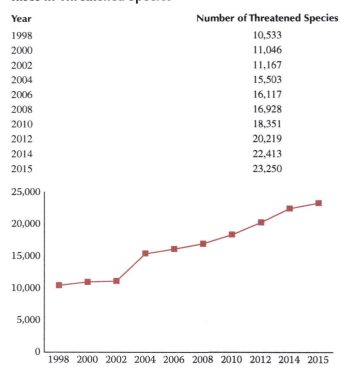

Source: International Union for Conservation of Nature; http://cmsdocs. s3.amazonaws.com/summarystats/2015-4_Summary_Stats_Page_ Documents/2015_4_RL_Stats_Table_1.pdf.

water for agriculture and sanitation. Existing aquifers are being used at a faster rate than they can be replenished, and other areas that rely on glacial water are finding diminishing resources due to their increased melting in response to climate change.

Overpopulated areas are also seeing higher air pollution levels in both industrialized and nonindustrialized areas—where traditional fossil fuels are not the main source of fuel, wood, and dung are. Unfortunately, all sources pollute the air.

Other negative effects of overpopulation include habitat destruction and fragmentation as urbanized areas expand. Human activities are pushing thousands of plant and animal species into extinction. According to Hinrichsen and Robey (2016), two of every three species are estimated to be in decline.

The types of actions that people take today will determine the type of environment available for the populations of tomorrow. If population and irresponsible use of natural resources continue rather than transitioning to sustainable development, humanity faces a bleak environment and future. The time to take action is now. The impacts to the environment have not taken place overnight and transitioning to a sustainable solution will not happen overnight either; but steps can start to be taken in the right direction Positive steps include managing our cities better, using energy more efficiently, phasing out subsidies that encourage waste, slowing population growth, and protecting and conserving natural resources.

If every country today would commit to these measures, we would be able to implement sustainable development, but it will require a mixture of community involvement, effective natural resource management, cleaner technologies, and cleaner living standards.

WORLDWIDE PRODUCTION AND DISTRIBUTION OF FOOD

The reality of food production and distribution for those who are forward thinking is that by 2050, we will need to be able to feed 9.6 billion people—the projected future population of the world. The question that begs to be answered is this: how can we do it sustainably?

Sustainable here means how to feed the world population while also advancing rural development using appropriate conservation methods, reducing GHG emissions through the use of smarter green technologies, and protecting valuable ecosystems from further degradation, while rehabilitating those we can. We are, indeed, facing a global food challenge today. There are several reasons this crisis has become so urgent. According to the World Resources Institute (WRI), the biggest hurdle is trying to feed an exploding population—two billion more people expected in just over 30 years. Added to that, more than half of that growth is expected to occur in sub-Saharan Africa—the same region where one-fourth of its population is *already* undernourished.

Shifting diets is a principal factor in the problem. The consumption of meat and milk—already high in the wealthier developed countries such as the United States, Canada, Western Europe, and Australia, is currently escalating in China and India—who have some of the largest populations on earth. It is expected to also remain high in Brazil and Russia. Table 1.2 shows global consumption of meat and milk products from 2006 projected to 2050 (Ranganathan, 2013). You may wonder what the drawbacks of the meat and dairy industry are. These commodities are very resource intensive to produce, much more so than plant-based diets are, making them a nonsustainable, impractical solution for the long term.

TABLE 1.2

Global Consumption of Meat and Milk Products

Region	Livestock (KCAL/Person/Day)			Beef and mutton (KCAL/Person/Day)		
	2006	2050	Percent change	2006	2050	Percent change
European Union	864	925	7	80	75	–6
Canada and USA	907	887	–2	117	95	–19
China	561	820	46	41	89	116
Brazil	606	803	33	151	173	15
Former Soviet Union	601	768	28	118	156	32
Other OECD	529	674	27	64	84	31
Latin America (ex. Brazil)	475	628	32	59	86	45
Middle East and North Africa	303	416	37	59	86	45
Asia (ex. China, India)	233	400	72	24	43	79
India	184	357	94	8	19	138
Sub-Saharan Africa	144	185	29	41	51	26
World	413	506	23	50	65	30

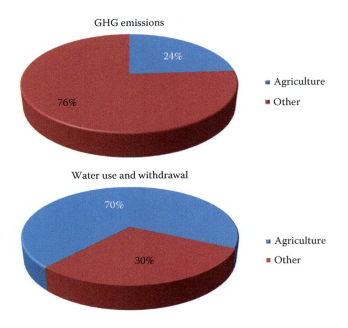

GRAPH 1.1 Agriculture's share of global environmental impact. (Courtesy of Julie A. Kerr.)

Now let us add to increasing population and shifting diets. Before 2050, the world will need to produce ~69 percent more food calories than was produced a decade ago. The rub is that we *cannot* do it by producing more food if we are going to become more sustainable in our agricultural production methods and leave a smaller environmental footprint. Currently, the agricultural industry is responsible for about one-fourth of global greenhouse emissions, uses more than one-third of the earth's landmass (excluding Antarctica), and is responsible for almost three-fourths of freshwater use (Graph 1.1). If that is not a big enough challenge, climate change—as it accelerates—is expected to negatively impact what arable land we are using through drought, resulting in *lower crop yields*. And the area expected to be impacted the most: sub-Saharan Africa—the area with the fastest-growing population. India and portions of China are also expected to suffer extreme negative effects—other areas with dense populations.

Now let us look at biofuels and how they relate to food supply. Although a very promising concept for more sustainable transportation energy, biofuels compete for land and crops against a hungry world. According to WRI, governments that plan to produce 10 percent of all transport fuels from biofuel sources by 2050 will require 32 percent of all global crop production—yet supply only 2 percent of global energy in return, arguably not economically, sustainably, or socially wise. Conversely, WRI reports that through the cessation of biofuel production as an alternative energy source, it would close the food gap by as much as 14 percent. Another argument that could be made in favor of closing the food gap is that it would provide more jobs for farmers—especially those that are poor and could benefit the most.

TABLE 1.3
Potential Solutions to the Global Food Crisis

Solution	Rationale
Reduce food loss and waste	Today one-fourth of the food calories produced are lost or wasted. Cutting this in half could close the food gap by 20 percent by 2050.
Shift to healthier diets	Lower the consumption of beef (it generates 6 times more GHG emissions per unit of protein than pork, chicken, and eggs. Shifting 20 percent could spare hundreds of millions of acres (hectares) of forest and rangeland.
Slow population growth	The goal would be to achieve *replacement-level fertility.* Sub-Sahara's population is expected to more than double by 2050.
Boost crop yields	This must be accomplished in a sustainable manner.
Improve land and water management	This includes sustainable farming techniques such as crop rotation, reduced tillage, mulching, contour farming, and agroforestry.
Shift agriculture to less prime lands	It helps to avoid deforestation and protects natural resources from being exploited.
Increase aquaculture's productivity	This will reduce the impact on wild-caught fish.

Source: World Resources Institute, Creating a sustainable food future: A menu of solutions to sustainably feed more than 9 billion people by 2050, http://www.wri.org/sites/default/files/wri13_report_4c_wrr_online.pdf. (accessed March 15, 2016), 2013–2014.

What it comes down to is finding the best mix between narrowing the large food gap, supporting sustainable economic development, and finding a feasible way to reduce agriculture's already-substantial impact on the environment. In this day of emerging revolutionary technologies, we are at the threshold for young, great minds to discover ways to make that happen. Table 1.3 lists some potential solutions.

With some serious effort and creative thinking, it is imperative that countries—both suppliers and consumers—step up and produce viable results now.

FOOD: SCARCITY VERSUS DISTRIBUTION

Another thing that is important to consider, especially with regard to climate change, energy resources, and sustainable living, is the issue of food scarcity and distribution. According to the organization A Well-Fed World (2016), both factors are important to consider, and both are complex and interconnected issues. Scarcity is a problem because of rapidly increasing population and resources that are just as quickly being depleted. In addition, it is also necessary to consider that governments do not often provide enough food to those who need it the most.

In theory, although there should be enough food to feed the global population, when vast amounts are fed to animals to support a meat and other animal-based foods (dairy and eggs) industry, it diminishes that supply. It is known that animals are very inefficient converters of food (they eat more than they produce). As a result, animal-based foods are highly resource intensive and they require many more resources (land, water, food, and energy) than if we were to eat plant-based foods directly.

In addition, much of the "extra" food is produced in areas far away from those who need it most. It is also used as animal feed to produce meat for those who can most afford it. When looked at this way, animal-based foods can be considered a form of overconsumption. Its redistribution reduces the amount of available food and it increases the price of basic food staples. Put bluntly, those with the financial resources are getting the benefit at the expense of the poor, whose hunger needs are not being met. In the concept of achieving a global sustainable community, this is counterproductive.

Therefore, global hunger results are due to a plethora of complex problems, including both scarcity and distribution. Food scarcity at the global level is currently a growing issue because past food stores are being drawn down (awfw.org). Although this is occurring, and population is increasing, available land, water, and other finite resources are decreasing. We are being forced to do more with less, and it is only going to intensify. Now it is the time to find more sustainable, resource-preserving solutions, because we not only have to do more with less but also have to do it better.

Water Resources

Clean water is the lifeblood of healthy communities. We rely on clean water to survive. Yet, as environmental pressures mount, we are heading toward a water crisis in many parts of the world. These growing pressures on water resources—from population and economic growth, climate change, pollution, poor management, and waste—have serious impacts on our social, economic, and environmental well-being. Today, many of the most important aquifers are being overpumped, causing widespread lowering of groundwater levels. Some major rivers, such as the Colorado River in the western United States and the Yellow River in China, do not even flow all the way to the ocean in most years. Major droughts, like those seen in California recently, are enlarging the gap between the region's water use and available water—demand is exceeding supply. Today, half of the world's wetlands have been destroyed for development, eliminating that valuable resource. And what water is remaining is becoming degraded, which does not bode well for community or ecosystem health. It is estimated that 780 million people worldwide still lack access to clean water, and for this reason, thousands die each day (Pacific Institute, 2016). What we need to focus on is better meeting basic human needs, improving our water management skills, and better balancing human needs with the needs of the natural world. If these areas were better developed and efficient, there would

be less wasteful use of water, better water management strategies, more attractive economic incentives, better financial support, active application of existing technologies, and well-thought-out and enacted conservation plans.

Dirty water is the world's biggest health risk, threatening both the quality of life and public health. When water from rain and melting snow runs off roads, roofs, gutters, and other contaminated surfaces, it picks up toxic chemicals, dirt, trash, and disease-carrying organisms. Water resources can also become contaminated from pollution due to factory farms, industrial plants, and mining. This, in turn, can lead to contamination of our drinking water, degrade and destroy habitats, and cause beaches to be closed. Protecting clean water requires ensuring that proper legislation is in place to ensure that pollution control programs are being applied to important waterways and wetlands, which provide much of the critical habitat and drinking water. Increasing protection levels in order to reduce contaminants such as bacteria and viruses is also key toward the public's health and well-being. For problem areas, such as runoff and areas prone to sewage overflow, more stringent monitoring and controls need to be put in place.

Many of the most drastic and directly felt impacts of climate change will relate to water—from more severe and frequent droughts to excessive flooding. Climate change is currently threatening the condition of many of the world's lakes, rivers, and streams; and sources of drinking water are being drawn down and not replenished. Effects, such as severe water shortages, sea-level rise, saltwater intrusion, harm to fisheries, and more frequent and intense storm events, will continue. Now it is the time for officials, planners, and citizens to plan for these issues and develop contingencies to deal with the effects a changing climate will bring.

Because water is such a critical resource, it is one that must be managed carefully with clear, well-planned-out guidelines. The Natural Resources Defense Council recommends the following objectives:

- To promote water efficiency strategies to eliminate waste
- To protect water from contamination (pollution) by enforcing clean water acts and promoting green infrastructure
- To prepare all geographical water use areas (towns, cities, regions) for the water-related challenges they will face due to climate change
- To ensure all waterways have enough water to support aquatic ecosystems

The importance of protecting our global water resources now and into the future will be one of the key factors in environmental sustainability.

AIR POLLUTION AND CLIMATE CHANGE

Another important resource is the earth's atmosphere. Good air quality refers to the degree which the air is clean and free from pollutants, such as smoke, dust, and smog. Several pollution indicators are used to determine air quality. In order for life on earth to remain healthy, air quality must be good. Conversely, when air quality drops due to high levels of pollutants, human health, plants, animals, and natural resources are all threatened.

Air quality can be degraded by natural or man-made sources. Natural sources include things such as volcanic episodes and dust storms. Man-made sources include pollution from cars, toxic gases from industrial processes, coal-powered plants, burning wood or other material, and landfills. When air quality is affected enough, the quality becomes so bad that it can lead to severe health problems for humans.

Pollution can come from both stationary sources and moving sources. Stationary sources include power plants, smelters, motor machines, manufacturing facilities, and burning wood and coal. Moving sources include trucks, buses, planes, cars, and trains. Once these pollutants are released into the atmosphere, the quality of the air is dependent on three factors:

- The amount of pollutants
- The rate at which they are released in the atmosphere
- How long they are trapped in an area

Pollutants include sulfur dioxide (SO_2), particulate matter, hydrocarbons and volatile organic compounds (VOCs), lead, carbon dioxide (CO_2), carbon monoxide (CO), nitrogen oxides (NO_x), and smoke/dust/smog. Many pollutants are gases, but others are very tiny solid particles (dust, smoke, soot). These can even affect the indoor air quality. Other indoor contaminants include cigarette smoke, mold, dust mites, pet dander, formaldehyde, VOCs, and radon gas.

If the flow of air is fairly good, the pollutants will mix with the air and quickly disperse. If the air flow is slow or mountains stand in the way as obstacles, then pollutants tend to remain in the air. When this happens, pollutants tend to build up rapidly. This is also common with some weather patterns. For example, in the winter in areas that have restricted air flow due to mountains, when high pressure occurs and a layer of cold sits above a warmer layer of air, a strong inversion will be created, trapping the pollutants close to the ground.

Because of the heavy increase in the number of vehicles and industries in the past few decades, air quality in many places has gone from bad to worse. Every year, millions of people across the world die due to the inhalation of toxic gases present in the atmosphere. According to Conserve Energy Future, if we do not clean up our atmosphere soon and effectively, the worsening air quality can cause severe harm to the entire planetary ecosystem (Conserve Energy Future, 2016).

The first place to start curbing emissions is with automobile traffic. If the number of cars was reduced on the planet and public transportation was used instead, it would help tremendously. One good thing is that today urban planning can help reduce traffic numbers. With the increase of faster electric trains and other forms of public transport, this is also helping move the society in the right direction. Through the use of cleaner vehicles, public transportation, and green fuels, we can improve air quality even further.

Industry is another major source of air pollution. It emits large amounts of pollutants into the air, which lead to acid rain, ozone layer depletion, and climate change. It is imperative that regulation policies be enforced to regulate industrial emissions. There are control devices called "pollution-eating nanoparticles" that can provide effective ways to reduce industrial air pollution by absorbing or destroying toxic emissions and contaminants from industries. These are effective when control devices are placed in exhaust streams. Two examples are titanium dioxide and ultraviolet light. When they are installed in an industrial facility, titanium dioxide reduces pollution by absorbing toxic emissions, although ultraviolet light breaks up NO_x gases and VOCs. Other control devices are employed as well, such as electrostatic precipitators, dust cyclones, particulate scrubbers, and baghouses. Switching to renewable energy, however, is the most effective approach for improving air quality standards in industrial manufacturing.

The other critical component necessary to ensure clean air is the establishment of adequate policies and clean air plans. Local, national, and international agencies must work together to set adequate air quality standards and to make sure these levels are reached and sustained. In order to create effective policies, the underlying pollution challenges must be identified and addressed. Examples of these plans in action include the Clean Air for Europe and the EPA. They have both set standards for restricting air pollution and have continued to work on directives for clean air. For a universally clean environment, all countries must participate in a clean air program.

On a different, but complimentary scale, air pollution reduction efforts can also involve land-use, urban, and transportation planning strategies. By using these approaches, they can assess the root causes of air pollution and adopt clean air action plans and support programs. In order to be a part of the solution, society, the public, businesses, manufacturing facilities, and government authorities need to work as a team to realize positive outcomes of having a healthier and more sustainable environment. Also of note is that the sources, causes, and impacts of air quality are often interlinked. That means dealing with one air purity concern (such as cutting back on fossil fuels) not only helps to reduce the emission of particulate matter but also lessens CO_2, CO, and NO_x. When efforts are made to improve air quality, they contribute to improved human well-being and health. They also increase environmental productivity.

ECOSYSTEMS AND THE IMPORTANCE OF BIODIVERSITY

Humans rely on a healthy, functioning natural world every day. We rely on it for adequate food, clean water, breathable air, and resources to make the useful products we use every day. In fact, we rely on it for both tangible and intangible benefits beyond what we probably think about—until it is gone. Because of our rapidly expanding population, however, people are becoming more aware of the impact we—as a society—are having as a whole on the environment.

We are becoming collectively aware of how the impact of a few can have a global impact and how finding replacements for depleted systems can be cost-prohibitive.

Because the current global population is consuming natural resources at a much faster rate than the planet can replenish them, we are putting current and future generations at risk and exacerbating the already unequal access to the goods and services fundamental to life.

One measure of a healthy, productive natural world is biodiversity. Biodiversity is the variety of all forms of life, and it is necessary for the existence and healthy functioning of all ecosystems. The number of species of plants, animals, and microorganisms; the huge diversity of genes in these species; and the different ecosystems on earth—such as rainforests, coral reefs, and deserts—are all part of the biological diversity that exists on this planet. Biodiversity enhances an ecosystem's productivity where each species—regardless of how small—all have an important role to play. For instance, if there is greater species diversity, there is increased sustainability for all life forms. Biodiversity supports habitats for all species, by providing several unique environments in which species can thrive. Ecosystems exist at all scales, in multiple types, can be nested in larger ones, or exist as corridors between habitats. With this in mind, well-thought-out conservation and sustainable development strategies strive to recognize this as being an integral part to any approach in preserving biodiversity.

According to the U.S. EPA, biodiversity provides the core benefits that humans derive from their environment; that it is fundamental for the provision of ecosystem services which we depend on for food, water security, air, and many other natural benefits, as shown in Figure 1.10. Because biodiversity promotes ecosystem productivity, where each species (no matter how small) all have an important role to play, this makes biodiversity—and conserving it—extremely important. Healthy biodiversity provides many benefits. For example, consider the following:

- Healthy ecosystems have a better chance of withstanding and recovering from disasters.
- A larger number of plant species allow for a greater variety of crops to be produced.
- Greater species diversity promotes natural sustainability for all life forms.

FIGURE 1.10 An eco-wheel that shows natural resources provided by biodiversity, the benefits and beneficiaries, and the drivers of change. (From EPA.)

TABLE 1.4
Services Provided by Healthy Biodiversity

Ecosystem Services
- Protection of water resources
- Soils formation and protection
- Storage and recycling of nutrients
- Pollution breakdown and absorption
- Contribution to climate stability
- Maintenance of ecosystems
- Recovery from unpredictable events

Biological Resources
- Food
- Medicinal resources and pharmaceutical drugs
- Wood products
- Ornamental plants
- Breeding stocks, population reservoirs
- Future resources
- Diversity in genes, species, and ecosystems

Social Benefits
- Research, education, and monitoring
- Recreation and tourism
- Cultural values

Source: Shah, A., Why is biodiversity important? Who cares? *Global Issues,* http://www.globalissues.org/article/170/why-is-biodiversity-important-who-cares#WhatisBiodiversity (accessed March 11, 2016), 2014.

According to Global Issues Organization, a healthy biodiversity provides several natural services. A few of these are listed in Table 1.4. Genetic diversity can also help fight against chances of extinction in the wild. In addition, species depend on each other. Each species depends on the services provided by other species for its survival. They have a type of cooperation tied to mutual survival that is often referred to as a "balanced ecosystem."

At least 40 percent of the world's economy and 80 percent of the needs of the poor are derived from biological resources. In addition, the richer the diversity of life, the greater the opportunity for medical discoveries, economic development, and adaptive responses to such new challenge as climate change.

—The Convention about Life on Earth

There are many stressors and drivers of change that can have a consequential effect on biodiversity, as several human activities can have a negative effect on biodiversity within ecosystems. One of the most profound is the destruction of habitat due to a rapidly growing human population. Overexploitation of a species is also detrimental. This includes activities such as overfishing and overhunting species, occasionally to the point of extinction (Rosser et al., 2002). For example, in 2008, the Food and Agriculture Organization (FAO) estimated that approximately 32 percent of fish stocks were overexploited, depleted, or recovering from depletion (FAO, 2010).

Overuse of the land is another major stressor on biodiversity. This can happen through many activities, such as outdoor recreation (damaging plant life and stressing local animal populations), overgrazing and introducing invasive species onto the land to compete with indigenous species, and many other uses and impacts. The problem with invasive species is that they can outcompete native species to the point of extinction. Examples of invasive species include kudzu, cheat grass, and the Emerald Ash Borer Beetle found in the United States, which can completely alter ecosystems and negatively affect the overall biodiversity. Pollution is another serious driver. Both terrestrial and aquatic species can be harmed from this.

There are many recognized benefits for conserving biodiversity, such as enabling species to cope with environmental stressors more efficiently and preserving genetic diversity so that a wide range of species can thrive. If there is a loss in genetic diversity, it will decrease the species' coping ability to survive. As a result, humans lose the benefits these species have to offer—from food sources, medicinal value, aesthetic value, pollination and pest control services, recreational value, and so forth.

Figure 1.11 illustrates some of the services that ecosystems provide, such as clean air and water; natural hazard mitigation; climate stabilization; recreation, culture, and aesthetic values; food, fuel, and materials; and biodiversity conservation.

Many of the decisions we make—from how to develop community infrastructure to managing the land—have an impact on the provision of ecosystem services. What we need to be more aware of when making decisions is that ecosystems play a role in everything we do. Therefore, by considering the true value of ecosystem services in our policy and decision-making processes, we can better manage our resources.

Ecosystems also help maintain a stable climate. Figure 1.12 illustrates some examples of natural resources that help stabilize climate, the benefits, and drivers of change. Healthy ecosystems, and natural resources such as soils, oceans, and vegetation, can reduce or stabilize the rate of flow of GHGs through the long-term capture and storage of carbon—a process called carbon sequestration. By sequestering atmospheric CO_2 in natural resources called *sinks*, it reduces the amount of CO_2 that is available to heat the atmosphere and increase climate instability.

The IPCC estimates that the earth's oceans have taken up 500 gigatons (Gt) of atmospheric CO_2 from the 1300 Gt of total CO_2 human-caused emissions in the past 200 years (IPCC, 2005). In addition to that, plant roots and leaves store carbon at an average of 53.5 tons of carbon per acre (hectare) for U.S. forests. It also works on smaller scales: vegetation can mitigate climate variability and extremes by creating local micro-climates through shading and increased humidity, which can produce a more favorable environment for humans and wildlife. Stressors, such as transportation, energy production, and industrial process, emit GHGs into the atmosphere, which contribute to changing climate conditions. Table 1.5 lists the health impacts and benefits of climate stabilization.

There are also several drivers and stressors of change. Human activities such as transportation, energy production, and industrial processes emit GHGs into the atmosphere, which contribute to climate change. Various land-use patterns and practices can also have an impact. For example,

FIGURE 1.11 Some of the natural resources providing ecosystem services, their benefits, and the drivers of change. (From EPA.)

urbanization and its resulting sprawl increase the use of private cars and the input of GHGs into the atmosphere. The EPA estimates that about 27 percent of GHG emissions come from this source.

A growing worldwide population has resulted in increased demands on food and biofuels, more land under use, and more intensive improper land management. Increased agricultural activities, such as the heavy use of fertilizers, poor cultivation and plowing practices, emission of nitrous oxide, and the fact that livestock produces large amounts of methane, are leading us to destructive drivers of change. Furthermore, cutting down trees removes valuable carbon storage potential. Forests left intact have the ability to sequester CO_2. According to the EPA, over the past several decades in the United States, land-use and forestry activities have resulted in more removal of CO_2 from the atmosphere than emissions, offsetting –15 percent of total U.S. GHG emissions in 2010 (EPA, 2016d).

There are also health impacts and benefits, which are as follows (EPA, 2015):

- Climate stabilization is important for the safety and security of all species.

- Ecosystem protection, management, and carbon sequestration strategies help slow the current rate of changing climate averages and variability, buffering humans from the negative aspects of such change.
- In the near term, trees and other vegetation can help mitigate heat-related hazards by providing local cooling and reducing indoor temperatures through shading.
- Local temperature reduction may decrease hospital admissions and mortality due to heat stroke and other heat-related illnesses.
- Careful use and management of vegetation can also regulate the flow of water.
- Sea-level change in response to climate warming is a potential threat along the Mid-Atlantic and portions of the Gulf Coast. Slowing the rate of sea-level rise could help maintain the quality of drinking water supplies by reducing the rate of saltwater intrusion into aquifers.
- Slowing sea-level rise through carbon sequestration could also provide additional time to move water supply infrastructure and treatment facilities, or to develop alternative treatment technologies.

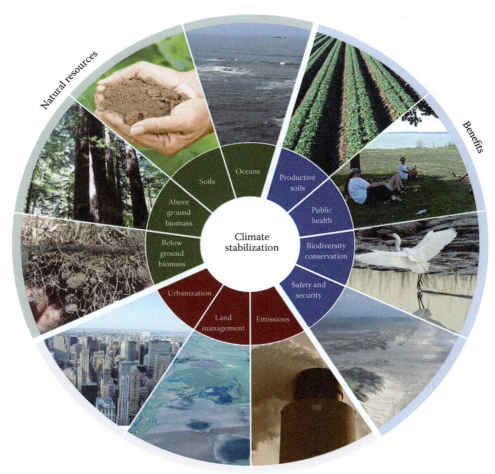

FIGURE 1.12   Natural resources that help stabilize climate, their benefits, and drivers of change. (From EPA.)

TABLE 1.5

Effects of Climate Stabilization

- Carbon sequestration reduces the amount of CO_2 that is available to heat the atmosphere.
- Ocean storage of carbon. The IPCC estimates that the world's oceans have taken up 500 Gt of atmospheric CO_2 from the 1300 Gt total CO_2 human-caused emissions in the past 200 years.
- Vegetation and roots store carbon—an average of 53.5 tons (48,534 kg) per hectare for U.S. forests) (Nowak and Crane, 2002).
- Natural plant species have evolved and adapted to a slowly changing climate, but they can be severely challenged by more rapidly changing environmental conditions.
- Vegetation can counteract climate variability and extremes by creating local micro-climates through shading and increased humidity, producing a more favorable environment for humans and wildlife.

- In the longer term, increasing the stores of carbon in soil is one way to increase soil fertility and productivity. By slowing the rate of change, plant breeders have the opportunity to develop new commercial crop varieties that can flourish under these new climate patterns.
- Encouraging a less rapidly changing climate helps maintain stability in biodiversity, which underpins all services that the earth's ecosystems provide.

As illustrated, well-managed ecosystems deliver many benefits. They contribute significantly to food, water, income security, and climate resilience. Ultimately, healthy ecosystems provide the basis for sustainable development and a functioning society. For this reason, conservation is critical.

SUSTAINABLE DEVELOPMENT, CONSERVATION, AND ENERGY RESOURCES

Sustainable development is defined as the development that meets the needs of the present population without compromising the ability of future generations to meet their own needs. According to the Brundtland Report (Report of the World Commission on Environment and Development: Our Common Future; Brundtland and World Commission on Environment and Development, 1987), it consists of two important key concepts:

1. The concept of *needs*, in particular the essential needs of the world's poor, to which overriding priority should be given.
2. The idea of *limitations* imposed by the state of technology and social organization on the environment's ability to meet the present and future needs (IISD, 2014).

The concept of sustainable development can be interpreted in many different ways, but at its most basic, it is an approach toward development where the goal is to balance different (often competing) needs against an awareness of the environmental, social, and economic limitations the society faces on a daily basis. Rather than focusing on singular needs, it requires we consider the bigger picture—a multidimensional view. Usually, when people use the term sustainability, however, they have a much higher level of achievement in mind at the basic level. This mind-set for some can also include a sustainable environment for all species. The concept of sustainability contains the following components:

- Conservation of biodiversity and ecological integrity (including stopping the unnatural [nonevolutionary] loss of biodiversity)
- Ensuring intragenerational (within generations) and intergenerational (across generations) equity
- Social equity and community participation
- Integration of environmental and economic goals in policies and activities
- Ensuring appropriate valuation of environmental assets
- Dealing cautiously with risk, uncertainty, and irreversibility

Sustainability is equated with renewable fuel sources, reducing carbon emissions, protecting the environment, and keeping ecosystems in balance. Its focus is to protect the natural environment, promote human and ecological health, and employ progressive innovation, while maintaining/improving our standard of living.

The concept of conservation goes hand in hand in achieving sustainability. Preservation of the natural environment is essential for maintaining community sustainability. Conservation implies protection, or keeping something safe. As we have become aware of our impact on the environment over recent years, the definition of conservation has grown to also encompass the effects of human consumption. For this reason, we are now encouraged by conservationists to "reduce, reuse, recycle, buy local, and lessen our carbon footprint...." The accepted concept of conservation now is "to conserve the resources required for the human race to survive." The natural resources the society focuses on as important to conserve include biodiversity, water, air and climate, land, forests, ecosystems, and energy, among others. Biodiversity is specifically important for creating sustainability because of the specialized role each species plays in maintaining an ecological balance. In addition, a high-quality water supply is critical. It is important for the local system as well as the community.

Both natural ecosystems and human health can be negatively impacted by worsening air quality and climate change. Communities can preserve air quality by limiting—or eliminating—the release of harmful chemicals into the air and minimizing other sources of air pollution. Forests are also a critical resource that must be conserved. They provide a bounty of goods and services for us, as well as act as a carbon store, slowing down the warming process we are undergoing during global climate change. In order to protect this valuable resource, we must develop innovative approaches for managing, protecting, and enhancing wildlife populations and habitats.

We must also conserve energy. Energy is required to accomplish most things. Because nonrenewable energy sources currently serve as the dominant force for power generation, home and workplace energy sources, and transportation—which causes an enormous amount of pollution—we must learn to conserve energy and make the transition to renewable fuels, which provide cost-effective and more sustainable alternatives. With the implementation of a successful energy conservation program, we should be able to expect the following:

- Buildings that are designed for maximum energy efficiency
- Smart meters that monitor every aspect of energy use
- Carbon taxes implemented to dramatically reduce GHG production
- Renewable energy public transportation systems commonplace in urban areas
- Energy-efficient products available in local stores

These are some of the topics that will be discussed in the second half of this book. We have got a challenging road ahead of us for sure, but the good thing is that we know in which direction we need to be headed.

CONCLUSIONS

About now, hopefully you are asking yourself some very key questions, such as the following:

- How can we best address the issue of overpopulation?
- Is it possible for each country worldwide to implement some form of sustainable lifestyle?
- How can countries worldwide work together to achieve global sustainability?
- How can we develop sustainable communities at home?
- What would be the most efficient and painless way to shift to a renewable energy lifestyle?
- What environmental issues are important to me?
- Which environmental issues should be important to me?
- What is my carbon footprint and how can I lower it?
- Where do I go from here? What are the next steps?

If any of these questions have crossed your mind, that is good. It shows you are thinking about the important issues. You are thinking about solutions. You are thinking about where you fit into the picture. If these questions have not crossed your mind yet, the goal is that they will as you continue to read. We have just touched on some basics so far. Now we are going to venture into "Part I: The Earth's Climate Balance and Consequences of Climate Change," where we will discuss the role of climate change—what it is, what is currently happening with the climate, what the current climate direction means to you, and what the future holds. As Albert Einstein put so clearly in 1954,

> We shall need a substantially new way of thinking if humanity is to survive.

Or said by Robert Redford,

> I think the environment should be put in the category of our national security. Defense of our resources is just as important as defense abroad. Otherwise, what is there to defend?

Or said by Thomas Edison in 1931,

> I'd put my money on the sun and solar energy, what a source of power. I hope we don't have to wait until oil and coal run out, before we tackle that.

REFERENCES

A Well-Fed World. 2016. Scarcity vs. distribution. http://awfw.org/scarcity-vs-distribution/ (accessed January 29, 2016).

Clark, P. U., J. D. Shakun, S. A. Marcott, A. C. Mix, M. Eby, S. Kulp, A. Levermann et al. February 8, 2016. Consequences of twenty-first-century policy for multi-millennial climate and sea-level change. *Nature Climate Change.* http://www.nature.com/articles/nclimate2923.epdf (accessed March 13, 2016).

Conserve Energy Future. 2016. What is air quality? http://www.conserve-energy-future.com/what is air quality.php (accessed March 8, 2016).

Didyouknow.org. 2011. The great global climate change quick guide. http://didyouknow.org/the-great-global-climate-change-quick-guide/ (accessed March 14, 2016).

Environmental Protection Agency. 2014. TRI national analysis. www.epa.gov/trinationalanalysis/releases-industry-2014-tri-national-analysis (accessed February 29, 2016).

Environmental Protection Agency. 2015. EnviroAtlas benefit category: Biodiversity conservation. https://www.epa.gov/enviroatlas/enviroatlas-benefit-category-biodiversity-conservation (accessed March 9, 2016).

Environmental Protection Agency. 2016a. Air quality trends. www3.epa.gov/airtrends/aqtrends.html (accessed March 3, 2016).

Environmental Protection Agency. February 23, 2016b. GHG emissions: Transportation. http://www3.epa.gov/climatechange/ghgemissions/sources/transportation.html (accessed March 3, 2016).

Environmental Protection Agency. February 23, 2016c. GHG emission: Agriculture. http://www3.epa.gov/climatechange/ghgemissions/sources/agriculture.html (accessed March 2016).

Environmental Protection Agency. February 23, 2016d. GHG emissions: Land use change. http://www3.epa.gov/climatechange/ghgemissions/sources/lulucf.html (accessed March 2016).

Environmental Protection Agency. February 24, 2016e. Climate change: Indicators. http://www3.epa.gov/climatechange/science/indicators/ (accessed June 2016).

Food and Agricultural Organization of the United Nations. 2010. The state of world fisheries and aquaculture: 2010. http://www.fao.org/fishery/sofia/en (accessed March 4, 2013).

Forest Research. 2010. Increasing environmental quality and aesthetics. http://www.forestry.gov.uk/pdf/urgp_evidence_note_003_Environmental_quality_aesthetics.pdf/$FILE/urgp_evidence_note_003_Environmental_quality_aesthetics.pdf (accessed February 21, 2016).

Brundtland, G. H. and World Commission on Environment and Development. 1987. *Our Common Future: Report of the World Commission on Environment and Development.* Oxford, UK:Oxford University.

Gillia, J. and C. W. Dugger. May 3, 2011. U.N. forecasts 10.1 billion people by century's end. http://www.nytimes.com/2011/05/04/world/04population.html?_r=0 (accessed March 20, 2016).

Grace, D., F. Mutua, P. Ochungo, R. Kruska, K. Jones, L. Brierley, L. Lapar et al. 2012. Mapping of poverty and likely zoonoses hotspots. Zoonoses Project 4. Report to the UK Department for International Development. Nairobi, Kenya: ILRI.

Hinrichsen, D. and B. Robey. 2016. Population and the environment: The global challenge. http://www.actionbioscience.org/environment/hinrichsen_robey.html (accessed February 29, 2016).

howmany.org. 2016. Overpopulation: Environmental and social problems. http://www.howmany.org/environmental_and_social_ills.php (accessed January 19, 2016).

IISD. 2014. Smart solutions for a small planet. International Institute for Sustainable Development. http://www.iisd.org/sites/default/files/publications/iisd-2015-brochure.pdf (accessed March 13, 2016).

IPCC. 2005. Special report on carbon dioxide capture and storage. https://www.ipcc.ch/pdf/special-reports/srccs/srccs_wholereport.pdf (accessed March 2016).

Kopp, R. E., A. C. Kemp, K. Bittermann, B. P. Horton, J. P. Donnelly, W. R. Gehrels, C. C. Hay et al. 2016. Temperature-driven global sea-level variability in the common era. *Proceedings of the National Academy of Sciences* 201517056. doi:10.1073/pnas.1517056113.

Lyytimäkia, J., P. Tapiob, and T. Assmutha. 2012. Unawareness in environmental protection: The case of light pollution from traffic. *Land Use Policy* 29(3):598–604.

National Geographic. 2012. Great pacific garbage patch. http://education.nationalgeographic.org/encyclopedia/great-pacific-garbage-patch/ (accessed March 22, 2016).

National Park Service. 2016. National Park Service: About us. www.nps.gov/aboutus/index.htm (accessed February 27, 2016).

Nature Education. 2014. Knowledge project, Ben A. Minteer, Editor. *Environmental Ethics*. http://www.nature.com/scitable/knowledge/environmental-ethics-96467512 (accessed March 4, 2016).

Nowak, D. J. and D. E. Crane. 2002. Carbon storage and sequestration by urban trees in the USA. *Environmental Pollution* 116(3):381–389.

Pacific Institute. 2016. Sustainable water management—Local to global. http://pacinst.org/issues/sustainable-water-management-local-to-global/ (accessed March 6, 2016).

Peters, G. P., R. M. Andrew, S. Solomon, and P. Friedlingstein. 2015. *Measuring a Fair and Ambitious Climate Agreement Using Cumulative Emissions*. Bristol: IOP Publishing.

Ranganathan, J. 2013. The global food challenge explained in 18 graphics. http://www.wri.org/blog/2013/12/global-food-challenge-explained-18-graphics (accessed January 19, 2016).

Rietbroek, R., S. Brunnabend, J. Kusche, J. Schröter, and C. Dahle. 2016. Revisiting the contemporary sea-level budget on global and regional scales. *Proceedings of the National Academy of Sciences* 113:1504–1509. doi:10.1073/pnas.1519132113.

Rosser, A. and S. A. Mainka. 2002. Overexploitation and species extinctions. *Conservation Biology* 16(3):584–586.

Science Daily. August 25, 1998. First-ever scientific estimate of total bacteria on earth shows far greater numbers than ever known before. Athens, GA:University of Georgia. (accessed March 15, 2016).

Shah, A. 2014. Why is biodiversity important? Who cares? *Global Issues*. http://www.globalissues.org/article/170/why-is-biodiversity-important-who-cares#WhatisBiodiversity (accessed March 11, 2016).

Texas A&M Today. January 21, 2014. Asian air pollution affecting world's weather. http://today.tamu.edu/2014/01/21/asian-air-pollution-affecting-worlds-weather/ (accessed February 21, 2016).

United Nations Environment Program. 2005. Millennium ecosystem assessment. Ecosystems and human well-being: Current state and trends. CH 4: Biodiversity. (accessed March 10, 2016).

World Health Organization. 1989. European charter on environment and health; WHO Regional Office for Europe, ICP/RUD/113/conf. Doc.1/ rev. 2 2803r, December 7, 1989, 7pp. http://www.euro.who.int/__data/assets/pdf_file/0017/136250/ICP_RUD_113.pdf (accessed February 22, 2016).

World Health Organization. 2005. WHO air quality guidelines global update 2005: Report on a working group meeting, Bonn, Germany, October 18–20, 2005. WHO Regional Office for Europe. http://www.euro.who.int/__data/assets/pdf_file/0008/147851/E87950.pdf (accessed March 15, 2016).

World Resources Institute. 2013–2014. Creating a sustainable food future: A menu of solutions to sustainably feed more than 9 billion people by 2050. http://www.wri.org/sites/default/files/wri13_report_4c_wrr_online.pdf (accessed March 15, 2016).

SUGGESTED READING

Brown, L. R. and E. Adams. 2015. *The Great Transition: Shifting from Fossil Fuels to Solar and Wind Energy*. New York: W.W. Norton & Company.

Callicott, J. B. 1997. *Earth's Insights: A Multicultural Survey of Ecological Ethics from the Mediterranean Basin to the Australian Outback*. Berkeley, CA: University of California Press.

DesJardins, J. R. 2006. *Environmental Ethics: An Introduction to Environmental Philosophy*. Belmont, CA: Wadsworth.

Dunlap, R. A. 2014. *Sustainable Energy*. CL Engineering.

Groom, M. J., G. K. Meffe, and C. R. Carroll. 2006. *Principles of Conservation Biology*, 3rd ed. Sunderland, MA: Sinauer Associates.

IISD. 2014. Smart solutions for a small planet. International Institute for Sustainable Development. http://www.iisd.org/sites/default/files/publications/iisd-2015-brochure.pdf (accessed March 13, 2016).

Kerr Gines, J. 2012. *The Smart Guide to Green Living*. Norman, OK: Smart Guide Publications.

Martin-Schramm, J. B. and R. L. Stivers. 2003. *Christian Environmental Ethics: A Case Method Approach*. Maryknoll, NY: Orbis.

Weathers, K. C. and D. L. Strayer. 2012. *Fundamentals of Ecosystem Science*. San Diego, CA: Academic Press.

World Resources Institute. 2013–2014. Creating a sustainable food future: A menu of solutions to sustainably feed more than 9 billion people by 2050. http://www.wri.org/sites/default/files/wri13_report_4c_wrr_online.pdf (accessed February 15, 2016).

Young, A. 2014. *Sustainable Energy: How to Save Money Using Renewable Energy, Living Green and Living Sustainably*. Amazon Digital Services, LLC. London, UK: Bloomberg.

Section I

The Earth's Climate Balance and
Consequences of Climate Change

2 The Climate System

OVERVIEW

In order to understand the problem and then be able to manage it effectively, it is imperative to have an understanding of the complex theory and science behind climate change. Although much of the existing literature refers to the issue as *global warming*—the term used predominantly in the past—scientists have largely transitioned to the term *climate change* because of the multifaceted nature of the phenomenon. This book utilizes the current terminology. As background, the term global warming tends to imply that the only consequence is that the earth's atmosphere is slowly getting warmer, unfortunately causing many to minimize the significance of the issue. When the media reports that scientists predict that the earth's atmosphere is warming, some people expect to hear that the temperature will get tens of degrees warmer. Some may have visions of sitting in a hot, steamy sauna—a vision that for those who live in extremely cold climates, such as Siberia or the Yukon, may be very welcome indeed! Therefore, when climatologists predict a temperature rise of 1.1°C–6.4°C, the public may be inclined to ask what the urgency about the issue really is. After all, it is just a few degrees, right?

Wrong.

Look at it this way: During the earth's last ice age, the earth was only about 4°C–6°C *cooler* than it is today. Although that may seem like an insignificant temperature difference, it was enough to blanket huge areas of the earth with thick layers of ice. It had such an enormous impact on ecosystems that it even rendered some species extinct, such as the mammoth and mastodon.

Thus, although a few degrees may seem trivial, the earth's climate is so sensitive that those few degrees can make a significant difference. In addition, although some climate change is natural, scientists have now proven that humans are causing the bulk of the recent changes and rises in temperature, and although those few degrees may not seem like much, it is unfortunately enough to serve as a tipping point—a big enough influence on the climate that, once reached, it will set into motion permanent changes with global ramifications. When the climate changes, it affects the entire earth systems. Many components of the earth's physical system operate on a global scale, such as the hydrologic, carbon, nitrogen, and phosphorus cycles. The earth's various biogeochemical cycles are always changing and represent the continual interactions between the biosphere (life), the lithosphere (land), the hydrosphere (water), and the atmosphere (air). Various substances move endlessly throughout these four spheres. Of the four spheres, the atmosphere transports elements the most rapidly, and climate change will negatively affect these spheres if left unchecked.

This chapter is the first of the three parts that lays the foundation needed to understand how the climate system works as a whole and what climate change is capable of and why. It looks at the climate as a global system and the fact that harmful environmental practices at one location can negatively affect other locations. It then examines the effect that human activities have on the earth's natural carbon cycle—commonly referred to as the human-enhanced carbon cycle. From there, it focuses on global circulation patterns of the atmosphere and oceans and their role in climate change, specifically the consequences of the disruption of the Great Ocean Conveyor Belt—a major circulation pattern responsible for Western Europe's mild climate and possibly for abrupt climate change. Finally, it presents current research information about the effects of sea-level rise and the consequences of a warming world.

INTRODUCTION

Climate change is one of today's most urgent topics. Despite all the controversy and hype the entire climate change topic has generated, there now exists an overwhelming body of scientific evidence that the problem is real, that its effects are being felt right now on a global scale, that some geographic areas are more vulnerable than others, and that it will take the concerted effort of every person working toward the same goals to put a halt to the rise in both temperature and atmospheric greenhouse gases. The fact is that even if all necessary steps are taken right now to stop the increase in greenhouse gases entering the earth's atmosphere, the effects of climate change will still be present for centuries to come. The complicated process has already been set into irreversible motion. The critical key to understanding the effects of climate change is to be knowledgeable about how to control and mitigate the causes of climate change now in order to minimize future damage to the environment and the life living in it. An unhealthy environment affects all forms of life on earth, and through understanding first the causes and effects—the science behind climate change—it will be possible to empower the world's populations to solve the problem through political leadership, realistic sociological mind-sets, and proper economic measures. The scientific consensus is that people's behavior is largely responsible for today's climate change problem. Another factor that makes this a volatile and controversial issue is that it is not just confined to the realms of the scientific community, nor does it have just one simple, predefined solution—it has multifaceted dimensions involving economic, sociological, political, psychological, and personal issues, making this a topic affecting the future outcome of every person's life on earth now and in the future. To make the problem even more sensitive, it is projected that those who will suffer the worst effects are not even those who have caused the bulk of the problem—the undeveloped nations will take the brunt of what the developed nations have largely caused.

The scientific community has many theories about climate change. The topic is extremely challenging in nature because there are so many factors involved in it, making it difficult to pinpoint its exact causes and solutions across the board. The earth's climate is extremely complicated, and climatologists are conducting daily research in order to improve their understanding of all the interrelated components.

In 2014, about 40 billion tons of carbon was pumped into the atmosphere (Plait, 2014). Studies show that concentrations of carbon dioxide (CO_2) have increased by about one-third since 1900. During this same time period, experts say the earth had warmed rapidly. For this reason, a connection has been made that humans are contributing significantly to climate change. Even scientists who are skeptical about the climate change issue recognize that there is much more CO_2 in the atmosphere than ever before.

Although it is certainly true that the atmosphere is warming up, that is only one part of what is going on. As the earth's atmosphere continues to warm, it is setting off an avalanche of other mechanisms that will do even greater harm to the earth's natural ecosystems. Glaciers and ice caps are melting, sea levels are rising, and ocean circulation patterns are changing, which then change the traditional heat distributions around the globe. Seasons are shifting and storms are becoming more intense, leading to severe weather events. Droughts are causing desertification, crops are dying, and disease is spreading. Some ecosystems are shifting where they still can; others are beginning to fail. In short, humans are changing the earth's climate—and not for the better.

DEVELOPMENT

In order to understand climate change management issues, it is necessary to have a good understanding and working knowledge of the basic scientific theory behind climate change: how the earth's atmosphere works, how it interacts with the earth's surface, what causes climate change, how it affects the various earth systems, and why it is an issue that must be addressed today. In practical land management, principally overseen by the world's government agencies and appropriated entities, much of the knowledge relied on in order to manage the land and make appropriate policy decisions to properly provide for present and future management capabilities and expertise is current research, largely provided by the world's principal research institutions—chiefly governmental, academic, and private—such as the National Aeronautics and Space Administration (NASA), numerous universities worldwide, and esteemed organizations such as the Pew Center on Global Climate Change, respectively. Therefore, Chapters 2 through 4, while merely touching on some of the most prominent key issues, provide an overview of the recent scientific research and work that has been done on the climate change issue,

laying the basic groundwork, to promote an understanding of the complex, interwoven issues and key concepts of the climate change problem, thereby creating a firm foundation for why this issue is critical to both the society and the environment's future, as well as why politics, sociology, and economics play such a critical role in its management. Consulting the references listed at the conclusions of each chapter will also provide a greater depth of detail in each subissue presented. This research has been crucial toward enabling managers to gain an understanding of all the individual facets of the problem in order to be able to put proper and effective policy in place to provide efficient land stewardship practices for both the present and the future. It is crucial that you, the reader, understand the scientific concepts behind the complex climate change issue in order to gain an appreciation for the immense challenge that land managers, urban planners, financial analysts, engineers, and other professionals concerned with sustainability face today. In planning for future climate change situations based on a plethora of interactive scenarios where each has the ability to change the long-term outcome, making effective management decisions one of the most challenging tasks of this century—a task that will affect many future generations. Chapters 2, 3, and 4 provide a brief but comprehensive introduction to climate change science and present the key scientific concepts behind it. Only through understanding the depth of the scientific principles and processes involved is it possible to grasp the interwoven complexity of the issue and gain an appreciation for why informed climate management practices are critical for the future of both humanity and the environment when dealing with climate, energy, and sustainability.

WHAT IS CLIMATE CHANGE?

Climate change is a term used to refer to the increase of the earth's average surface temperature, due largely to a buildup of greenhouse gases in the earth's atmosphere, primarily from the burning of fossil fuels and the destruction of the world's rain forests. The term was coined by Svante Arrhenius, a Swedish scientist and Nobel Laureate, in 1913. The term is often used to convey the concept that there is actually more going on than just rising temperatures. It is a misnomer that the issue involves just the atmosphere and temperature rising a few degrees; hence, the major reason for the name change from global warming to climate change in order to more accurately convey the multifaceted nature of the phenomenon. Climate change encompasses long-term changes in climate, which include temperature, precipitation amounts, and types of precipitation, humidity, and other factors.

Today, climate change has become one of the most controversial issues in the public eye, appearing frequently in print and televised news reports, documentaries, scientific and political debates, and other venues and economic issues, and the messages can be contradictory and confusing. One goal of this book is to set that record straight by presenting various points of view and clarifying them with the facts. Climate change also receives a lot of attention because it is more than just a scientific issue—it also affects economics, sociology, and people's personal lifestyles and standards of living. It is one of the hottest current political issues, not only in the United States but worldwide. Political positions on climate change have become major platforms and debating issues as public demands toward a solution have intensified in recent years.

More than 2500 of the world's most renowned climatologists, represented by the Intergovernmental Panel on Climate Change (IPCC), support the concept of climate change and agree that there is absolutely no scientific doubt that the atmosphere is warming. They also believe that human activities—especially burning fossil fuels (oil, gas, and coal), deforestation, and environmentally unfriendly farming practices—are playing a significant role in the problem. The IPCC is an organized group of more than 2500 climate experts from around the world that consolidates their most recent scientific findings every 5–7 years into a single report, which is then presented to the world's political leaders. The World Meteorological Organization (WMO) and the United Nations Environment Programme established the IPCC in 1988 to specifically address the issue of climate change. As a result of their comprehensive analysis, the IPCC determined that this steady warming has had a significant impact on at least 420 animal and plant species in addition to natural processes. Furthermore, this has not just occurred in one geographical location, but worldwide.

Unfortunately, the science of climate change does not come with a crystal ball. Scientists do not know exactly what will happen, such as what the specific impacts to specific areas will be, nor can they say with certainty when or where the impacts will hit the hardest, making it all the more difficult for land resource managers, planners, politicians, and economists to do the best job possible in long-range planning. But they are certain that the effects will be serious and globally far-reaching. According to the National Oceanic and Atmospheric Administration (NOAA)/NASA/EPA Climate Change Partnership, potential impacts include increased human mortality, extinction of plants and animal species, increased severe weather, drought, and dangerous rises in sea levels.

Although climatologists still argue about how quickly the earth is warming and how much it will ultimately warm, they do agree that climate change is currently happening, and that the earth will continue to keep warming if something is not done soon to stop it.

DID YOU KNOW?

There are several interesting facts concerning climate change. For instance, did you know that as of January 2016 (Ppcorn, 2016):

- The year 2015 was the world's hottest year on record. Every year of the past decade has been in the running for the world's hottest year on record. Before 2014 was declared the hottest, it was 2010.

- Every year the CO_2 concentration level in the atmosphere steadily climbs higher and that the earth will reach a "tipping point" of no return if humans do not reverse the trend. In order to avoid a rise of 6°C by the end of this century—which is the critical level identified by climate change research scientists—total global emissions need to have peaked by 2015 and reduced by at least 80 percent by 2050 to prevent nonreversible damage. The peak limit will not be met.

- Although every person plays a role in the problem and has a responsibility for its solution, the lion's share of the problem can be attributed to 23 of the world's most wealthy nations—which constitute only 14 percent of the world's population. These "developed" countries are guilty of producing 60 percent of the world's carbon emissions since 1850 when industrialization became an integral part of their lifestyles. Even though many of these nations pledged to reduce CO_2 below 1990 levels by 2012 through the ratification of the Kyoto Protocol, their emissions have risen instead.

- Climate change costs the United States more than $100 billion each year. This is due to costs spent combating the effects of crop loss, rising sea levels, foot and water shortages, and higher temperatures—all because of climate change.

- Thirty-seven percent of Americans still believe climate change is a hoax. Ignoring numerous studies that show climate change is real, nearly 4 of 10 Americans do not believe it. And these are some of the biggest offenders.

- The United States is responsible for 80 percent of the world's fossil fuel use annually. Behind other progressive countries on the problem, the United States still obtains 84 percent of its total energy from coal, oil, and gas (fossil fuels).

- The earth's human inhabitants release a collective 37 billion metric tons of CO_2 annually. These emissions need to decrease more than 5 percent each year just to give us a 50/50 chance of keeping under the 2°C threshold leading to disastrous effects.

- By 2050, we could see nearly 40 percent of the plant and animal species on earth become extinct. That is only 33 years from now.

- Up until 2007, the United States emitted the highest levels of greenhouse gases. China is now the largest emitter of CO_2, surpassing the United States. As of 2015, China emitted nearly 30 percent of the global total.

- Climate change is a problem that affects every nation sociologically, politically, and economically. In fact, the World Health Organization estimates that climate change was causing 150,000 deaths worldwide from malaria, malnutrition, diarrhea, and flooding.

- The low-lying island of Tuvalu in the Pacific has already evacuated 3000 of its inhabitants to New Zealand.
- By 2020, up to 250 million people in Africa and 77 million in South America will be under increased water stress because supplies will no longer meet demand.
- By 2025, tens of millions more will go hungry due to low crop yields and rising global food prices.
- A rise in sea level of just 1 meter would displace 10 million people in Vietnam and 8–10 million in Egypt.
- The number of people in Africa at risk of coastal flooding will rise from 1 million in 1990 to 70 million by 2080 according to the Stern Enquiry into Climate Change and Developing Countries.
- Nearly two-thirds of Americans do NOT believe that climate will affect their lives.

Let us take a brief look at the status of climate change at the end of 2015 before we continue. These are some highlights from the year and where we stand on getting climate change voluntarily under control (Hood, 2016):

- While registering as the hottest year on record for the planet as a whole, 195 states pledged to curb the carbon pollution that drove climate change at the December 12 Paris Agreement talks (United Nations Framework Convention on Climate Change) (Figure 2.1).
- Experts believe it is now up to each country to take appropriate action in order to make a difference. The Paris Agreement is the key to our salvation or too-little-too-late will depend on what happens now: who takes what action, when, and how much.
- Experts have identified the year 2015 as the tipping point for climate change. It is sink or swim now.

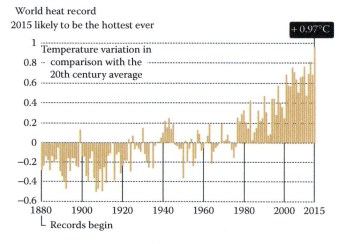

World heat record
2015 likely to be the hottest ever

FIGURE 2.1   Trend in the average temperature of the earth's surface 1880–2015. (Courtesy of NOAA, Silver Spring, MD.)

- The Paris talks was seen as the most serious gathering where officials seem to be coming to grips with the real issues. The seriousness was prompted by the deadly extreme weather seen in 2015, as well as recent technological and scientific advancements in climate science.
- Specific climate events served as powerful illustrations from 2015: the most powerful hurricane ever registered; freakish, above-freezing temperatures at the North Pole in December; and life-threatening droughts in eastern and southern Africa.
- The elevated CO_2 levels have likely pushed back the next Ice Age by 50,000 years—a shocking result as to the extent to which human activity has destroyed the earth's natural cycles. In tandem with that idea, the earth's biochemical systems have also been permanently impacted an altered.
- Just added to the list of global risks and concerns is "climate change mitigation and adaptation."
- On the brighter side, the balance of investment is shifting away from fossil fuels and toward renewables, as reported by Bloomberg New Energy Finance.

THE GLOBAL SYSTEM CONCEPT

Although different types of climate exist in different parts of the world, the climate does work as a complete unified global system—conditions and actions in one area can impact conditions and actions in others. For example, because the earth shares one atmosphere, what goes on in China will affect the United States. Similarly, what goes on in South America eventually affects Africa, and so on. NASA has reported that soot (black carbon) originating from industrial practices is being deposited on Arctic snow and ice, causing incoming *electromagnetic radiation* to heat up the now-darker surfaces that the soot is covering, causing more snow and ice to melt, creating a cycle of increased melting.

Because of these interactions, climate change is also a global issue—it is affecting the earth in different ways, such as destabilizing major ice sheets, melting the world's glaciers, raising sea levels, contributing to extreme weather, and shifting biological species northward and higher in elevation, ensuring that no area of the earth is immune from its effects. In fact, NASA scientists have determined that the atmospheric concentration of carbon dioxide, the principal greenhouse gas, is now higher than it has been for the past 650,000 years.

A remark often heard is that the earth's temperature and CO_2 levels have risen in the past so these cyclic variations are normal. This is true, as can be seen in the illustration (Figure 2.2). As paleoclimatologists have studied ice age cycles they have determined that the atmospheric CO_2 closely follows the earth's surface temperature. It is a known fact that as biological activity rises (with increases in temperature), CO_2 levels also rise because more CO_2 is

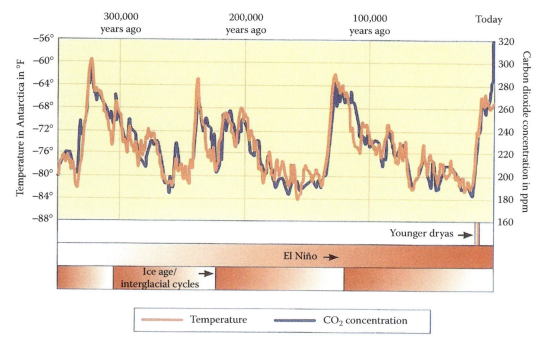

FIGURE 2.2 Fluctuations in temperature (orange line) and in the amount of carbon dioxide concentrations in the atmosphere (blue line) over the past 350,000 years. The temperature and carbon dioxide concentrations at the South Pole run roughly parallel to each other, showing the strong correlation between the two. (From Casper, J. K., *Global Warming Trends: Ecological Footprints*, Facts on File, New York, 2009.)

produced and is entered into the atmosphere. This increases warming and encourages more CO_2 production. It is critical to note, however, that under natural processes (non-anthropogenic), CO_2 concentrations never exceeded 300 ppm (parts per million) until the Industrial Revolution occurred in the late eighteenth century. It was at that point that CO_2 levels began to continually rise. In 1958, when CO_2 concentrations began to be recorded, the atmosphere was at 315 ppm. In March 2016, CO_2 concentrations were at 404.16 ppm and are still steadily rising (co2.earth, 2016). In addition, NASA's scientists from the Goddard Institute for Space Studies in New York released a report in 2014 outlining that their scientific studies provided clear evidence that an enhanced greenhouse effect is ongoing. Their study involved satellites, data from buoys, and computer models to study the earth's oceans. Scientists were able to conclude that more energy from the sun is being absorbed by the atmosphere than is emitted back to space, which is causing the earth's energy to become unbalanced and is now significantly warming the earth. NASA scientists claim that the world is entering "largely uncharted territory as atmospheric levels of greenhouse gases continue to rise." Today, scientists have no doubt that the recent spike in the graph showing the rise of CO_2 concentration is due to anthropogenic activity. In summary, because the earth is a global system, climate change must be approached as a global issue with far-reaching cause-and-effect issues, and it will take all nations working toward a common solution to effectively mitigate the problem (ScienceDaily, 2010).

THE ATMOSPHERE'S STRUCTURE

The atmosphere can be thought of as a thin layer of gases that surround the earth. The two major elements, nitrogen and oxygen, make up 99 percent of the volume of the atmosphere. The remaining 1 percent is composed of what is referred to as "trace" gases. The trace gases include water vapor (H_2O), methane (CH_4), argon (Ar), carbon dioxide (CO_2), and ozone (O_3). Although they only make up a small portion of the atmosphere, the trace gases are very important. Water vapor in the atmosphere is variable: Arid regions may have less than 1 percent, the tropics may have 3 percent, and over the ocean, there may be 4 percent.

The atmosphere is not uniform from the earth's surface to the top; it is divided vertically into four distinct layers: (1) the troposphere; (2) the stratosphere; (3) the mesosphere; and (4) the thermosphere, designated by height and temperature. The lowest level, closest to the surface of the earth, is the troposphere. It extends upward to an average height of 12 kilometers. This level is critical to humans because all of the earth's weather occurs in this layer. In this level, the temperature gets cooler with increasing height. In fact, the temperature at the surface of the earth averages 15°C, and at the tropopause (the top of the troposphere) the temperature is just −57°C. In addition, the moisture content decreases with the altitude in this layer; above the troposphere, there is not enough oxygen in the atmosphere to sustain life; and winds increase with height. To illustrate this, one of the most pronounced wind systems, the jet stream, is located at the top of the troposphere.

The climate of an area is the result of both natural and *anthropogenic* factors. Natural elements come from the following four principal inputs:

- Atmosphere
- Lithosphere
- Biosphere
- Hydrosphere

The human factor influences climate when it alters land and resource uses—a principal concern for land, wildlife, and natural resource managers worldwide. For example, when a natural forested area is converted into a city, it has a direct effect on climate—one of the most notable, the *urban heat island effect*.

Even though climate does change naturally, those types of changes occur so slowly (such as over millennia) that they are not readily detectable by humans. Climate changes caused by humans, however, are occurring much faster (such as in centuries or even decades) and are becoming noticeable within a few generations. Often, a change in one part of the climate will produce changes in other areas of climate as well. Since the Industrial Revolution began in the late eighteenth century—and especially since the introduction and use of fossil fuels involved in the rapid modernization of the twentieth and twenty-first centuries in developed countries—the global average temperature and atmospheric CO_2 concentrations have increased notably. Because CO_2 levels are higher now than they have been in the past 650,000 years and surface temperatures on earth have risen significantly during the same time, this has led scientists to the conclusion that humans are responsible for some of the unusual warmth that exists today. In support of this notion, climatologists at both NOAA and NASA have run two types of computer models: one of climate systems with natural climate processes alone and another with natural climatic processes combined with human activities. The models that include the human activities more accurately resemble the actual climate measurements of the twentieth century, giving scientists further conviction that human activity does play a significant role in the climate change issue that is apparent today.

Since the 1970s, instruments on satellites have also monitored the earth's climate. In addition, measurements of the atmospheric CO_2 have been obtained since 1958, when the world's first monitoring station was built on top of Mauna Loa, the highest mountain on Hawaii, to enable scientists to monitor the atmospheric CO_2 levels in order to document any increases. They were able to determine a documentable increase—one that is still occurring today—known worldwide as the famous "Keeling Curve." Currently, the CO_2 monitoring network has expanded to more than 100 stations globally in order to track the concentrations of carbon dioxide, methane, and other greenhouse gases.

Even though scientists know natural climate variability and cycles will continue to occur, they also expect CO_2 levels to rise and climate change to increase because of human interactions with climate. How much variability actually occurs will ultimately depend on the choices humans make in future population growth, energy choices, technological developments, and global policy decisions.

Dr. Rajendra Pachauri, the chairman of the IPCC, said at an international conference attended by 114 governments in Mauritius in January 2005 that the world has "already reached the level of dangerous concentrations of carbon dioxide in the atmosphere." He recommended immediate and "very deep" cuts in the pollution levels if humanity is to survive. He also said, "Climate change is for real. We have just a small window of opportunity and it is closing rather rapidly. There is not a moment to lose" (Lean, 2005).

In the most recent IPCC Synthesis Report (2014), the following observations were made: According to Ban Ki-Moon, UN Secretary-General, "With this latest report, science has spoken yet again and with much more clarity. There is no ambiguity in their message. Leaders must act. Time is not on our side." In fact, the unrestricted use of fossil fuels must be phased out by 2100 if the world is to avoid the dangerous effects of climate change. The IPCC reported that most of the world's electricity *must* be produced from low-carbon sources by 2050. If it was not, the world would be facing "severe, pervasive, and irreversible" damage.

According to Thomas Stocker, a Swiss Climatologist and cochair of the IPCC Working Group I, "Our assessment finds that the atmosphere and oceans have warmed, the amount of snow and ice has diminished, sea level has risen and the concentration of carbon dioxide has increased to a level unprecedented in at least the last 800,000 years" (IPCC Synthesis Report, 2014). U.S. Secretary of State, John Kerry, reported from the 2014 conference that climate change was "another canary in the coal mine. Those who choose to ignore or dispute the science so clearly laid out in this report do so at great risk for all of us and for our kids and grandkids."

Chris Rapley, head of the British Antarctic Survey, commented that the huge west Antarctic ice sheet may be starting to disintegrate, an event that would raise sea levels around the world by 5 m. He said, "The IPCC report characterized Antarctica as a slumbering giant in terms of climate change. I would say it is now an awakened giant. There is real concern." (CICERO, 2006). According to the American Geophysical Union, "Natural influences cannot explain the rapid increase in the global near-surface temperatures observed during the second half of the 20th century" (American Institute of Physics, 2010).

THE CARBON CYCLE: NATURAL VERSUS HUMAN AMPLIFICATION

The carbon cycle is an extremely important earth cycle and plays a critical role in climate change. Carbon dioxide enters the atmosphere during the carbon cycle. Because it is so plentiful, it originates from several sources. Vast amounts of carbon are stored in the earth's soils, oceans, and sediments at the bottoms of oceans. Carbon is stored in the earth's rocks, which is released when the rocks are eroded. It exists in all living matter. Every time animals and plants breathe, they exhale CO_2.

When examining the earth's natural carbon cycle, it is important to understand that the earth maintains a natural carbon balance. Throughout geologic time, when concentrations of CO_2 have been disturbed, the system has always gradually returned to its natural (balanced) state. This natural readjustment operates very slowly.

Through a process called diffusion, various gases that contain carbon move between the ocean's surface and the atmosphere. For this reason, plants in the ocean use CO_2 from the water for photosynthesis, which means that ocean plants store carbon, just as land plants do. When ocean animals consume these plants, they then store the carbon. Then when they die, they sink to the bottom and decompose, and their remains become incorporated in the sediments on the bottom of the ocean.

Once in the ocean, the carbon can go through various processes: It can form rocks, then weather; and it can also be used in the formation of shells. Carbon can move to and from different depths of the ocean, and also exchange with the atmosphere. As carbon moves through the system, different components can move at different speeds. Their reaction times can be broken down into two categories: short-term cycles and long-term cycles.

In short-term cycles, carbon is exchanged rapidly. One example of this is the gas exchange between the oceans and the atmosphere (evaporation). Long-term cycles can take years to millions of years to complete. Examples of this would be carbon stored for years in trees; or carbon weathered from a rock being carried to an ocean, being buried, being incorporated into plate tectonic systems, and then later being released into the atmosphere through a volcanic eruption.

Throughout geologic time, the earth has been able to maintain a balanced carbon cycle. Unfortunately, this natural balance has been upset by recent human activity. Over the past 200 years, fossil fuel emissions, land-use change, and other human activities have increased atmospheric carbon dioxide by 30 percent (and methane, another greenhouse gas, by 150 percent) to concentrations not seen in the past 420,000 years (which is the time span of the longest fully documented ice core record).

Humans are adding CO_2 to the atmosphere much faster than the earth's natural system can remove it. Figure 2.3 illustrates the greenhouse effect. Prior to the Industrial Revolution, atmospheric carbon levels remained constant at approximately 280 ppm. This meant that the natural carbon sinks

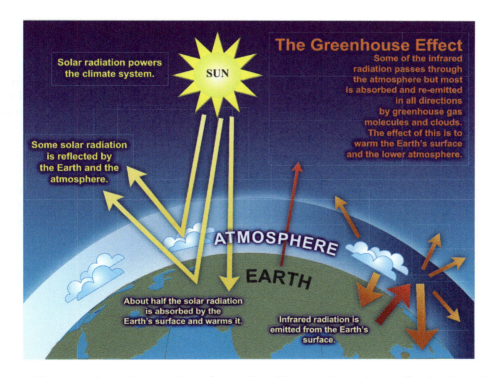

FIGURE 2.3 Diagram illustrating the earth's natural greenhouse effect. The natural greenhouse effect is what makes earth a habitable planet—if the greenhouse effect did not exist, the earth would be too cold, and life could not exist. As more greenhouse gases are added to the atmosphere through the burning of fossil fuels, deforestation, and other measures, however, less heat is able to escape to space, warming the atmosphere unnaturally—causing a situation called the enhanced greenhouse effect. This is what is contributing to climate change today. (From Gines, J. K., *Climate Management Issues: Economics, Sociology, and Politics*, CRC Press, New York, 2012.)

were balanced between what was being emitted and what was being stored. After the Industrial Revolution began and carbon dioxide levels began to increase—from 315 ppm in 1958 to 391 ppm in 2010 to 404 ppm in 2016—the "balancing act" became unbalanced, and the natural sinks could no longer store as much carbon as was being introduced into the atmosphere by human activities, such as transportation and industrial processes. In addition, according to Dr. Pep Canadell of the National Academy of Sciences, 50 years ago for every ton of CO_2 emitted, natural sinks removed 600 kilograms (Global Carbon Project, 2007). In 2006, only 550 kilograms was removed per ton, and the amount continues to fall today, which indicates the natural sinks are losing their carbon storage efficiency. In a 2013 BBC News Report, they reported the evidence of European forests that were nearing their "carbon saturation point," which means that the forests' ability to sequester carbon was dramatically slowing. This means that, although the world's oceans and land plants are absorbing great amounts of carbon, they simply cannot keep up with what humans are adding. The natural processes work much slower than the human ones do. The earth's natural cycling usually takes millions of years to move large amounts from one system to another. The problem with human interference is that the impacts are happening in only centuries or decades—and the earth cannot keep up with the fast pace. The result is that each year the measured CO_2 concentration of the atmosphere gets higher, making the earth's atmosphere warmer.

Another way that humans are contributing to climate change is through deforestation. By burning, or cutting down, the rain forests, two things happen: (1) the forest can no longer store carbon and (2) burning the trees releases the carbon that had been stored long term in the wood back to the atmosphere, further exacerbating the problem.

THE HYDROLOGIC CYCLE AND THE RELATIONSHIPS BETWEEN THE LAND, THE OCEAN, AND THE ATMOSPHERE

The hydrologic cycle is also important because it plays a direct role in the function of healthy ecosystems. It describes the movement of all the water on earth. It has no starting point, and involves the existence and movement of water on, in, and above the earth. The earth's water is always moving and changing states—from liquid to vapor to ice and back again. This cycle, illustrated in Figure 2.4, has been in operation for billions of years, and all life on earth depends on its continued existence. What have scientists concerned is that climate change is affecting the major components of the water cycle and it is having a negative impact.

When the cycle is in equilibrium, water is stored as a liquid in oceans, lakes, rivers, the soil, and underground aquifers in the rocks. Frozen water is stored in glaciers, ice caps, and snow. It is also stored in the atmosphere as water vapor, droplets, and ice crystals in clouds. Water can change states and

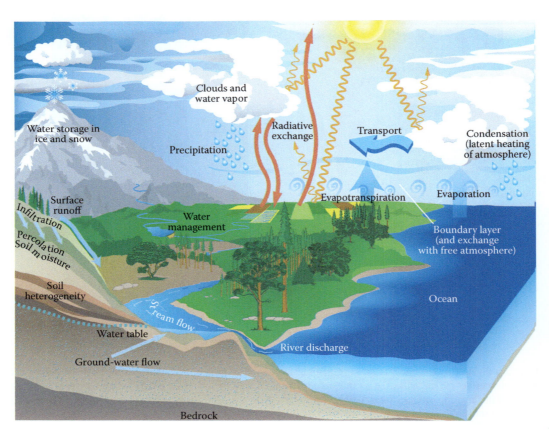

FIGURE 2.4   The earth's hydrologic cycle. (Courtesy of NASA, Washington, DC.)

move to different locations. For instance, water can move from the ocean to the atmosphere when it evaporates and turns from a liquid to a gas (vapor). Plants release water as a gas through transpiration. Water in the atmosphere condenses to form clouds, which can then form rain, snow, or hail, and return to the earth's surface. Water that comes back to earth can be stored where it lands (in an ocean) or can flow above the ground (river), or can infiltrate and move under the ground (as groundwater).

This is a recount of how the earth's natural water cycle works. When climate change becomes an issue, however, it enhances the water cycle, making it more "extreme." Because climate change causes the earth's atmosphere to be warmer, it develops a higher saturation point, enabling the atmosphere to evaporate and hold more water from the earth's surface. This may have a twofold effect. First, in areas where more water vapor exists, more clouds will form, causing more rain and snow. In other areas, especially those further away from water sources, more evaporation and transpiration (together referred to as evapotranspiration) could dry out the soil and vegetation. The result would be fewer clouds and less precipitation, which could cause drought problems for farmers, ranchers, cities, and wildlife habitat. All ecosystems in these areas on an international scale could be negatively impacted.

In areas that are receiving increasing amounts of precipitation, such as portions of Japan, Russia, China, and Indonesia, they will also experience more water infiltrating into the ground and over the surface. It could increase the levels of lakes and rivers, causing serious flooding, and may even form new lakes. Wetter conditions will also affect the plants and animals in the area.

The drier areas, such as Africa and the southwestern United States, will also experience serious impacts. As the ground dries out from evapotranspiration, the atmosphere loses an important source of moisture. This, in turn, creates fewer clouds, which means there is less rain, making the area more arid. As less water is available to infiltrate the ground, less will be able to live off, on, or in the soil. Rivers and lakes would dry up, vegetation would die off, and the land would no longer be able to support humans, animals, and other life.

There are many cycles operating on earth, with elements continually passing in and out of them. All natural cycles on earth must work together in a state of dynamic equilibrium. This means that as substances move and change at different times and places, they do it in a way that does not negatively impact the entire working system—all the components in the system work together and complement each other. This dynamic equilibrium changes as the seasons change, because different needs must be met at different times. In the spring and summer when plants grow, they need carbon so they take it from the atmosphere and the soil. Then, when the growing season is over in the fall and winter, plants release carbon back to the soil and atmosphere. This plays a significant role especially in the earth's Northern Hemisphere because most of the landmasses are located there, creating a global seasonal change of carbon dioxide in the atmosphere. The oceans and atmosphere also interact extensively. Oceans are more than a moisture source for the atmosphere. They also act as a heat source and a heat sink (storage), as well as a carbon sink.

GLOBAL ENERGY BALANCE

Interactions between energy from the sun, the earth, and the atmosphere all have an effect on the earth. This is called the global energy balance, an energy balance that also plays a role in climate. If the energy balance changed and the atmosphere began retaining more heat, this could trigger climate change. It is this scenario that climatologists are concerned about now and in the future.

The global energy balance regulates the state of the earth's climate. Modifications to it—a concept called "forcings"—can be either natural sources or human-made sources and could cause global climate to change. Natural forcings might include variations in the sun's intensity, a shift in the earth's orbit around the sun, a shift in the earth's tilt, or an increase in volcanic activity. Human-caused forcings could include burning fossil fuels, changing land-use patterns, or deforestation.

Greenhouse gases in the atmosphere do have an effect on the global energy balance. Without the natural amount of greenhouse gases in the atmosphere, the earth would not be an inhabitable planet because it would be too cold. The natural amounts of carbon dioxide, water vapor, and other greenhouse gases make life possible on earth because they keep the atmosphere warm.

Several things can have an effect on the energy balance, such as clouds and atmospheric aerosols. Clouds can interact in several ways with energy. They can block much of the incoming sunlight and reflect it back to space. In this way, they have a cooling effect. Clouds also act like greenhouse gases, and they block the emission of heat to space and keep the earth from releasing its absorbed solar energy. The altitude of the cloud in the atmosphere can affect the energy budget. High clouds are colder and can absorb more surface-emitted heat in the atmosphere; yet they do not emit much heat to space because they are so cold. Clouds can cool or warm the earth, depending on how many clouds there are, how thick they are, and how high they are. For this reason, it is not yet fully understood what effect clouds will have on surface temperatures if climate change continues into the next century and beyond.

Climatologists have proposed different opinions. Some think that clouds may help to decrease the effects of climate change by increasing cloud cover, increasing thickness, or decreasing in altitude. Others think that clouds could act to increase the warming on earth if the opposite conditions occurred. According to Anthony Del Genio of NASA's Goddard Institute for Space Studies, when air temperatures are higher, clouds are thinner and less capable of reflecting sunlight, which increases temperatures on earth, exacerbating climate change. Del Genio, in an interview with *CNN*, said that "in the larger context of the climate change debate I'd say we should not look for clouds to get us out of this mess. This is just one aspect of clouds, but this is the part people assumed would make climate change less severe" (Environmental News Network, 2000). After 15 years of climate modeling, the overall consensus in 2015 based partly on a study conducted just prior to that was that clouds could not be counted on to counteract climate change, but could possibly contribute to additional warming (Clement et al., 2009).

One way that scientists try to predict the future climate change and the effects of climate change is through the analysis and interpretation of mathematical climate models. In these computer models, climatologists attempt to account for all items that affect climate. Cloud cover is one of those variables. Today, this is still one of the most difficult variables to control and interpret. The climate is so sensitive as to how clouds might change, that even the most complicated, precise models developed today often vary in their climate change prediction under all the different methods available for cloud modeling.

The main reason clouds are so difficult to model is that they are so unpredictable. They can form rapidly and complete their life cycle in a matter of hours. Other climate variables work on a much slower timescale. Clouds also occur in a relatively small geographic area. Other climate variables operate on a much larger—regional or global—scale. According to climate research scientists at NASA, the world's fastest supercomputers can only track a single column of the surface and the atmosphere every 80–322 kilometers. By comparison, a massive thunderstorm system might cover only 32 kilometers. Features that are small, fast, and short-lived are hard to predict. This is one of the reasons why predicting specific individual weather events is more difficult than predicting long-term climate changes over broad areas.

Clouds are just one thing that can change the global energy balance; snow and ice can also do it. If the earth becomes cold enough, allowing large amounts of snow and ice to form, then more of the sun's energy will be directly reflected back to space because snow and ice have a high albedo. Over a period of time, this will change the global energy balance and the global temperature. Conversely, if the earth warms, the snow and ice will melt. This lowers the surface albedo, allowing more sunlight to be absorbed, which will warm the earth more.

Deforestation can also upset the global energy balance if forested area is removed and land is left bare; the ground can then reflect more sunlight back to space, causing a net cooling effect. However, if the forest material is burned, then the CO_2 stored in the trees is released into the atmosphere, contributing to climate change. Also, forests are good reservoirs of existing CO_2. The plants store and hold the CO_2, keeping it out of the air. If the forest is burned, not only does the already stored CO_2 now enter the atmosphere, but any future storage potential of CO_2 in that forest is now destroyed, creating two negative conditions of CO_2 toward climate change.

Atmospheric aerosols (tiny smoke particles) can be added to the atmosphere by sources such as fossil fuels, biomass burning, and industrial pollution. Aerosols can either cool or warm the atmospheric temperature depending on how much solar radiation they absorb versus how much they scatter back to space. Fossil fuel aerosols can also pollute clouds. Scientists need to do a lot more research on aerosols before they fully understand the full impacts of aerosols on climate change. The composition of the aerosol, its absorptive properties, the size of the aerosol particles, the number of particles, and how high in the atmosphere all have an effect on whether they cool or warm the atmospheric temperature and by how much.

Another effect aerosols have on clouds is that as aerosols increase, the water in the clouds gets spread over more particles; and smaller particles fall more slowly, resulting in a decrease in the amount of rainfall. Scientists believe aerosols have the potential to change the frequency of cloud occurrences, cloud thickness, and amount of rainfall in a region. Like clouds, aerosols are also a challenge to accommodate in climate models.

RATES OF CHANGE

Climate change "drivers" (causes) often trigger additional changes (feedbacks) within the climate system that can amplify or subdue the climate's initial response to them. For instance, if changes in the earth's orbit trigger an interglacial (warm) period, increasing CO_2 may amplify the warming by enhancing the greenhouse effect. When temperatures get cooler, CO_2 enters the oceans, and the atmosphere becomes cooler. Sometimes, the earth's climate seems to be quite stable; other times, it seems to have periods of rapid change. According to the U.S. Environmental Protection Agency (EPA), interglacial climates (such as like the climate today) tend to be more stable than cooler, glacial climates. Abrupt or rapid climate changes often occur between glacial and interglacial periods.

There are many components in a climate system, such as the atmosphere, the earth's surface, the ocean surface, vegetation, sea ice, mountain glaciers, deep ocean, and ice sheets. All of these components affect, and are affected by, the climate. They all have different response times, however, as shown in Table 2.1.

TABLE 2.1
Climate System Components and Response Times

Climate System Component	Response Time	Example
Fast Responses		
Land surface	Hours to months	Heating of the earth's surface
Ocean surface	Days to months	Afternoon heating of the water's surface
Atmosphere	Hours to weeks	Daily heating; winter inversions
Sea ice	Weeks to years	Early summer breakup
Vegetation	Hours to centuries	Growth of trees in a rain forest
Slow Responses		
Ice sheets	100–10,000 years	Advances of ice sheets over Greenland
Mountain glaciers	10–100 years	Loss of glaciers in Glacier National Park
Deep ocean	100–1500 years	Deepwater replacement

The amount of change applied, as well as the component's ability to naturally respond determines largely what the climate actually ends up doing. For instance, if there is a slow climate change, but the system component reacts quickly (in a short period of time), then the response will be visible. If the climate change is rapid, but then reverts back to its previous condition and the component's response time is naturally slow, then there will be no response. If the climate change alternates from one extreme to another at a rate that the components can keep up with, these changes will be seen as visible adaptations. It is these types of rates of change that are most enlightening for climatologists because it allows them to more efficiently model all the subtle components of the climate system.

ATMOSPHERIC CIRCULATION AND CLIMATE CHANGE

When pressure differences alone are responsible for moving air, the air—or wind—will be pushed in a straight path. Winds do follow curved paths across the earth, however, as illustrated in Figure 2.5. Named after the scientist who discovered this effect, Gustave-Gaspard Coriolis, this phenomenon is called the Coriolis force, which is an apparent drifting sideways (a property called deflection) of a freely moving object as seen by an observer on earth.

Given in nonvector terms, the property can be described as follows: at a given rate of rotation of the observer, the magnitude of the Coriolis acceleration of the object is proportional to the velocity of the object and also to the sine of the angle between the direction of movement of the object and the axis of rotation.

In mathematical terms, the vector formula for the magnitude and direction of the Coriolis acceleration is

$$a_c = -2\Omega \times v$$

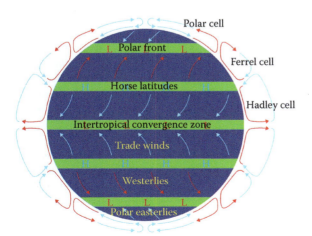

Polar cell
Polar front
Ferrel cell
Horse latitudes
Hadley cell
Intertropical convergence zone
Trade winds
Westerlies
Polar easterlies

FIGURE 2.5 Diagram representing the earth's major atmospheric circulation patterns. Major wind systems, such as the trade winds and Westerlies, lie between permanent bands of high or low pressure located at specific latitudes. (Courtesy of NASA, Washington, DC.)

where:

a_c is the acceleration of the particle in the rotating system
v is the velocity of the particle in the rotating system
Ω is the angular velocity vector which has a magnitude equal to the rotation rate w and is directed along the axis of rotation of the rotating reference frame

The equation may be multiplied by the mass of the object in question to produce the Coriolis force as follows:

$$F_c = -2m\Omega \times v$$

The *Coriolis effect* is the behavior added by the *Coriolis acceleration*. The formula implies that the Coriolis acceleration is perpendicular to both the direction of the velocity of the moving mass and the frame's rotation axis. Therefore, the following conditions apply:

- If the velocity is parallel to the rotation axis, the Coriolis acceleration is *zero*.
- If the velocity is straight inward to the axis, the acceleration is *in the direction* of local rotation.
- If the velocity is straight outward from the axis, the acceleration is *against the direction* of local rotation.
- If the velocity is in the direction of local rotation, the acceleration is *outward from the axis*.
- If the velocity is against the direction of local rotation, the acceleration is *inward to the axis*.

Therefore, the Coriolis force is the tendency for any moving body on or above the earth's surface, such as an ocean current, an air mass, or a ballistic missile—to drift sideways from its course because of the earth's rotation underneath. In other words, a moving object appears to veer from its original path. In the Northern Hemisphere, the deflection is to the right of the motion, whereas in the Southern Hemisphere, it is to the left.

The Coriolis deflection of a body moving toward the north or south results from the fact that the earth's surface is rotating eastward at greater speed near the equator than near the poles, because a point on the equator traces out a larger circle per day than a point on another latitude nearer either the North or the South Pole (the equator is a great circle and other latitudes are smaller). A body traveling toward the equator with the slower rotational speed of higher latitudes tends to fall behind or veer to the west relative to the more rapidly rotating earth below it at lower latitudes. Similarly, a body traveling toward either the North or the South Pole veers eastward because it retains the greater eastward rotational speed of the lower latitudes as it passes over the more slowly rotating earth closer to the pole.

The practical applications of the Coriolis force are important when calculating terrestrial wind systems and ocean currents. Scientists studying the weather, ocean dynamics, and other related earth phenomena must take the Coriolis force into account. The Coriolis also affects regional and global weather patterns because it interacts with the jet streams.

According to a study conducted in 2006 by Gabriel Vecchi of the University Corporation for Atmospheric Research, the trade winds in the Pacific Ocean are weakening as a result of climate change (Vecchi, 2006). This conclusion is based on the findings of a study that showed the biology in the area may be changing, which could be harmful to marine life and have a long-term effect of disrupting the marine food chain. Researchers predict that it could also reduce the biological productivity of the Pacific Ocean, which would impact not only the natural ecosystem and balance, but would affect the food supply for millions of people.

The study used climate data consisting of sea-level atmospheric pressure over the past 150 years and combined that with computer modeling to conclude that the wind has weakened by about 3.5 percent since the mid-1800s. The researchers predict another 10 percent decrease is possible by the end of the twenty-first century.

Some of the computer modeling simulations included variables such as the effects of human greenhouse gas emissions, whereas other simulations included only natural factors that affect climate such as volcanic eruptions and solar variations. Vecchi concluded that the only way they could account for the observed weakening of the trade winds was through the model that included human activity—specifically from greenhouse gases and the burning of fossil fuels.

Vecchi believes climate change is to blame because in order for the ocean and atmosphere to maintain an energy balance, the rate that the atmosphere absorbs water from the ocean must equal the amount that it loses to rainfall. As climate change increases the air temperature, more water evaporates from the ocean into the air. The atmosphere cannot convert it into rainfall and return it back fast enough. Because the air is gaining water faster than it can release it, it gets overloaded and the natural system compensates by slowing the trade wind down, decreasing the amount of water being drawn up into the atmosphere in order to maintain the energy balance. The drop in winds reduces the strength for both the surface and subsurface ocean currents and interferes with the cold water upwelling at the equator that is responsible for supplying ocean ecosystems with valuable nutrients, which are the lifeblood of the fishing industry.

This was later substantiated by Dr. Larry Mayer Unh of the University of New Hampshire in 2015, who also believes that this is one of the major contributing factors that contributes to El Niño climate patterns, and a continued weakening could lead to an increase in both the severity and the frequency of them. Other detrimental effects can be expected to occur as well, such as disrupted food chains, reduced fish harvests, and variations in episodes of droughts and floods. He emphasized that a continued weakening puts the entire ocean ecosystem at risk (Unh, 2015).

This is one example of how climate change can affect atmospheric circulation. It is also affecting oceanic circulation. In a recent study conducted in 2015 by the Potsdam Institute for Climate Impact Research, a slowdown of the Great Ocean Conveyor Belt has been observed. This is the major current that helps drive the Gulf Stream, which greatly modifies the climate in Western Europe, bringing warmer temperatures to their higher latitudes. Although a slowdown will not trigger another Ice Age, it could cause other negative effects, such as an increase in sea level, which could result in disastrous flooding of cities such as London, Boston, and New York. The researchers discovered that the overturning circulation began weakening in the North Atlantic around 1970, which they believe was caused by an unusual amount of sea ice traveling out of the Arctic Ocean, melting, and lowering the local salinity. The circulation began to recover in the 1990s, but was only temporary. According to Stefan Rahmstorf of the Potsdam Institute for Climate Impact Research, the circulation has further weakened due to the massive Greenland ice sheet losing enough freshwater due to melting, which is weakening the circulation by lowering the water's salinity. This is substantiated by the fact that the ocean region south of Greenland and between Canada and Britain is becoming colder, which indicates that there is now less northward heat transport. Rahmstorf also points out that, although the winter of December 2014 through February 2015 was the warmest on record for the globe as a whole, there was a major anomaly—there were also record cold temperatures in the middle of the North Atlantic (where the Great Ocean Conveyor Belt flows). This equates to a circulation that is about 15–20 percent weaker—something that has not happened since the year 900. Adding to the scientific unease, this was not predicted in climate models. Although the IPCC, in 2013, projected a slowdown of the current from 1 to 54 percent (most likely 11–34 percent) over the course of this century, actual data collected by Rahmstorf and his team showed that we were already within that window of change as of 2015 (Rahmstorf, 2015).

Extreme Weather

A serious concern about climate change is the potential damage that can be done to humans, property, and the environment as a result of extreme weather events, such as severe drought and storms. The EPA studies the aspects of change in extreme weather and climate events. Although extreme weather events are typically rare, they have noted that climate change is increasing the odds of more extreme weather events taking place. Although establishing the most likely causes behind an extreme weather event can be challenging, because each event is due to a combination of multiple factors (including natural variability), scientists have been able to make a connection between extreme climate patterns (and individual events) and climate change by focusing on whether an extreme weather event was made *more likely* by climate change. What they have determined is that there have been changes in some types of extreme weather events in the United States over the past several decades—including more intense and frequent heat waves, less frequent and intense cold waves, and regional changes in floods, droughts, and wildfires. This rise in extreme weather events fits a pattern that is expected with a warming climate. Scientists project that climate change will make some of these extreme weather events more likely to occur

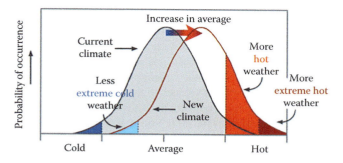

FIGURE 2.6 As the planet warms, more extreme weather events are expected to take place. (Courtesy EPA, Washington, DC.)

and/or more likely to be severe (Environmental Protection Agency, 2016) (Figure 2.6).

Another recent example of climate change increasing the intensity of hurricanes is evident from climate-fueled Hurricane Sandy that surged up and over retaining walls and destroyed countless homes of New York residents in 2012. Because of the extreme destruction wrought from this storm, the NOAA announced it would now employ a new practice of keeping hurricane and tropical storm watches and warnings in effect even after storms lose their tropical characteristics if they pose significant danger to life and property. The intensity of Sandy's post-tropical phase caught residents unprepared and its disastrous storm surge caused damage of approximately $50 billion, making it the second costliest cyclone to hit the United States since 1900. At least 147 people died in the Atlantic basin. According to Michael Oppenheimer, professor of geosciences and international affairs at Princeton University, and a member of the IPCC, "Hurricanes could become more intense as the earth warms. They are frightening, destructive, and extremely costly, and we expect future hurricanes to leave an even greater trail of damage in their wake" (Livescience, 2013) (Figure 2.7).

FIGURE 2.7 Coastal damage caused by Hurricane Sandy in Mantoloking, New Jersey. (Courtesy of U.S. Air Force by Master Sgt. Mark C. Olsen—http://www.defense.gov/photoessays/PhotoEssaySS.aspx?ID=3316, http://www.defense.gov/dodcmsshare/photoessay/2012-10/hires_121030-F-AL508-081c.jpg, Public Domain, https://commons.wikimedia.org/w/index.php?curid=22549477.)

The WMO has warned that extreme weather events, such as drought, hurricanes, and heavy rainfall, may very likely increase because of climate change. It is an organization of meteorologists from 189 countries. It is a specialized agency of the United Nations and serves as its voice on the state and behavior of the earth's atmosphere, its interaction with the oceans, the resulting climate as a product from the interaction, and the resulting distribution of water resources. There are not many forces in nature that can compare to the destructive capability of a hurricane. These storms can have winds blow for long periods of time at 249 kilometers per hour or higher. Not only the wind destructive but also the rainfall and storm surges can cause significant damage and loss of life (Climateaction, 2010).

Hurricane Katrina, which formed on August 23, 2005 and dissipated on August 30, 2005, affecting the Bahamas, South Florida, Cuba, Louisiana, Mississippi, Alabama, and the Florida Panhandle, was one of the deadliest hurricanes in the history of the United States, killing more than 1800 people and destroying more than 200,000 homes. There were more than 900,000 evacuees, many relocated to states in the western United States. It was also the costliest hurricane in U.S. history at more than $75 billion in estimated damages. By 2010, only about 40 percent of the New Orleans pre-Katrina residents had returned to the city. The demographics, however, are partly determined by race, age, and income level. As of 2013, only 30 percent of residents of predominantly lower income, black areas in one precinct had returned. Younger residents were also less likely to return.

A hurricane—or tropical cyclone—forms over tropical waters—between latitudes 8° and 20° in areas of high humidity, light winds, and areas where the sea surface is warm. Typically, temperatures must be 26.5°C or warmer to start a hurricane, which is why climate change and the heating of the ocean is such a concern.

Protecting life and the environment from severe weather triggered by climate change currently has many research scientists at the National Hurricane Center at NOAA and elsewhere engaged in theoretical studies. For example, computer modeling and the collection and analysis of field data are providing an avenue to gain a better understanding of the mechanics of climate change. It also sheds light on climate change's interaction with the environment, which helps improve forecasting methods, rapid response procedures, and safety protocols.

In an article published in Switzerland by the Environment News Service (2010), extreme weather events, such as wildfire outbreaks across Russia, record monsoonal flooding in Pakistan, rain-induced landslides in China, and the calving of a huge chunk of ice off the Petermann Glacier in Greenland, fall in line with what the WMO are warning of when they project "more frequent and intense extreme weather events due to global warming" (WMO/UNFCCC, 2011).

These identified unprecedented events that are occurring concurrently around the world are causing loss of human life and property. The WMO has brought to public attention the fact that the similar timing of all these incidences closely coincides with the IPCC's Fourth Assessment Report published back in 2007, suggesting a scientific connection in support of climate change (IPCC, 2007). In the IPCC's Summary for Policy Makers, it states "that the frequency and intensity of extreme events are expected to change as earth's climate changes, and these changes could occur even with relatively small mean climate changes." The summary points out that some of these changes have already occurred, specifically the examples previously mentioned. In Moscow, they experienced 30 daily record temperature rises in June 2010, leading to massive forest and peat fires. In Pakistan, the monsoonal flooding was so intense that they received 300 millimeters over a 36-hour period, and the Indus River in the northern portion of the country reached its highest water levels in 110 years. Southern and central areas in the country were also flooded, killing more than 1600 people. As a result, more than six million people have been driven from their homes. The Pakistan government reports more than 40 million of their citizens have been adversely affected by flooding thus far. China is also being negatively impacted by the worst flooding in decades. A resulting mudslide in the Gansu province on August 7, 2010, killed more than 700 and left more than 1000 others unaccounted for. In China, their government has reported that 12 million people have lost their homes to unexpected flooding.

NASA's Aqua satellite, via the moderate resolution imaging spectroradiometer (MODIS) sensor, discovered the calving of a major iceberg—the size of those typically found in Antarctica—on August 5, 2010. It is the largest chunk of ice to calve from the Greenland ice sheet in the past 50 years, dwarfing the tens of thousands of much smaller icebergs that normally calve from the glaciers of Greenland.

What is significant about these extreme weather events, according to the WMO, is that "all these events compare with, or exceed in intensity, duration or geographical extent, the previous largest historical events," leaving human interference as the critical link (WMO, 2011). Figure 2.8 shows the calving of the Petermann Glacier in Greenland on August 5, 2010.

FIGURE 2.8 On August 5, 2010, an enormous chunk of ice, ~251 km² in size, broke off the Petermann Glacier along the northwestern coast of Greenland. According to climate experts at the University of Delaware, the Petermann Glacier lost about one-fourth of its 70-km-long floating ice shelf. The recently calved iceberg is the largest to form in the Arctic in 50 years. (Courtesy of NASA, JPL, La Cañada Flintridge, CA.)

It is a chunk of ice, roughly 251 square kilometers in size (Environment News Service, 2010).

In a report issued by NOAA on November 5, 2015, human activities, such as greenhouse gas emissions and land use, influenced specific extreme weather and climate events in 2014, including tropical cyclones in the central Pacific, heavy rainfall in Europe, drought in East Africa, and stifling heat waves in Australia, Asia, and South America. Figure 2.9 illustrates the location and type of events analyzed in the study. Key findings of the report are shown in Table 2.2.

According to Stephanie C. Herring, PhD, lead editor for the report, "Understanding our influence on specific extreme weather events is ground-breaking science that will help us adapt to climate change. As the field of climate attribution science grows, resource managers, the insurance industry, and many others can use the information more effectively for improved decision making and to help communities better prepare for future extreme events" (NOAA, 2015).

The Role of Ocean Circulation in Climate Change

There are two factors that make water more dense (which causes it to sink) or less dense (which keeps it on the surface): (1) salt content and (2) temperature. Atmospheric flow and ocean currents are the mechanisms that carry heat from the equator to the poles. There are many processes that can alter the circulation patterns, and when this happens, it can change the weather of an area. If the ocean did not distribute heat throughout the world, the equator would be much warmer and the poles would be much colder.

The role of the oceans is equally as important in transporting heat from the equator to the polar regions as the atmosphere, as shown in Figure 2.10. In terms of how much heat and water it can hold, its capacity is much greater than that of the atmosphere. In fact, the world's oceans can store approximately 1100 times more heat than the atmosphere. The oceans also contain 90,000 times more water than the atmosphere does.

As more knowledge is gained about climate change, also gained is a better appreciation of the role the oceans play in shaping the earth's climate. For this reason, much more research has been done on the oceans in the past 15 years, leading to the discovery that the ocean's depths have warmed considerably since 1950. According to scientists at the Woods Hole Oceanic Institution, up until recently, scientific models predicting climate change could not account for where the projected warmth had gone; it was unaccounted for in the atmosphere. This discrepancy in the model had caused much confusion until researchers finally figured out that the world's oceans were storing the "missing" greenhouse warming. Water has a tremendous capacity to hold heat. The warming had occurred, but up until then, no one had thought to look toward the oceans for the answer (Schmitt, 2010).

Now that scientists understand this relationship, those that study climate change agree that including the ocean system in climate change studies is critical. Not only do the oceans have an enormous thermal capacity, the constant movement of slow and fast water can affect the weather for months at a time.

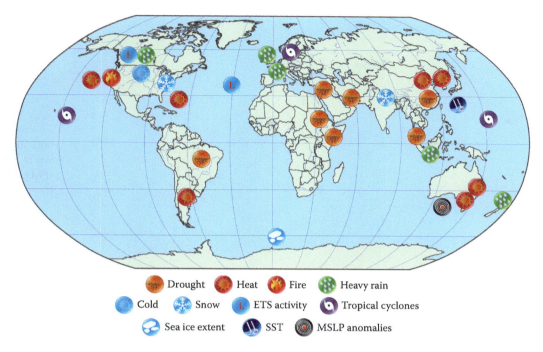

FIGURE 2.9 Location and type of events analyzed in "Explaining Extreme Events of 2014 from a Climate Perspective." (Courtesy of NOAA.)

TABLE 2.2

Key Findings of NOAA Extreme Events of 2014 Report

North America:
- The overall probability of California wildfires has increased due to human-induced climate change.
- Though cold winters still occur in the upper Midwest, they are less likely due to climate change.
- Cold temperatures along the eastern United States were not influenced by climate change, and eastern U.S. winter temperatures are becoming less variable.
- Tropical cyclones that hit Hawaii were substantially more likely because of human-induced climate change.
- Human-induced climate change and land use both played a role in the flooding that occurred in the southeastern Canadian Prairies.

South America:
- The Argentinean heat wave of December 2013 was made 5 times more likely because of human-induced climate change.
- Water shortages in southeast Brazil were not found to be largely influenced by climate change, but increasing population and water consumption raised vulnerability.

Europe:
- All-time record number of storms over the British Isles in winter 2013–2014 cannot be linked directly to human-induced warming of the tropical west Pacific.
- Extreme rainfall in the United Kingdom during the winter of 2013–2014 as not linked to human-caused climate change.
- Extreme rainfall in the Cévennes Mountains in southern France was 3 times more likely than in 1950 due to climate change.
- Human influence increased the probability of record annual mean warmth over Europe, Northeast Pacific, and Northwest Atlantic.

Middle East and Africa:
- Two studies showed that the drought in East Africa was made more severe because of climate change.

Asia:
- Extreme heat events in Korea and China were linked to human-caused climate change.
- Devastating 2014 floods in Jakarta are becoming more likely due to climate change and other human influences.
- Meteorological drivers that led to the extreme Himalayan snowstorm of 2014 have increased in likelihood due to climate change.
- Human influence increased the probability of regional high sea surface temperature extremes over the western tropical and northeast Pacific Ocean during 2014.

Australia:
- Four independent studies all pointed toward human influence causing a substantial increase in the likelihood and severity of heat waves across Australia in 2014.
- It is likely that human influences on climate increased the odds of the extreme high-pressure anomalies south of Australia in August 2014 that were associated with frosts, lowland snowfalls, and reduced rainfall.
- The risk of an extreme 5-day July rainfall event over Northland, New Zealand, such as was observed in early July 2014, has likely increased due to human influences on climate.

Antarctica:
- All-time maximum of Antarctic sea ice in 2014 resulted chiefly from anomalous winds that transported cold air masses away from the Antarctic continent, enhancing thermodynamic sea ice production far offshore. This type of event is becoming less likely because of climate change.

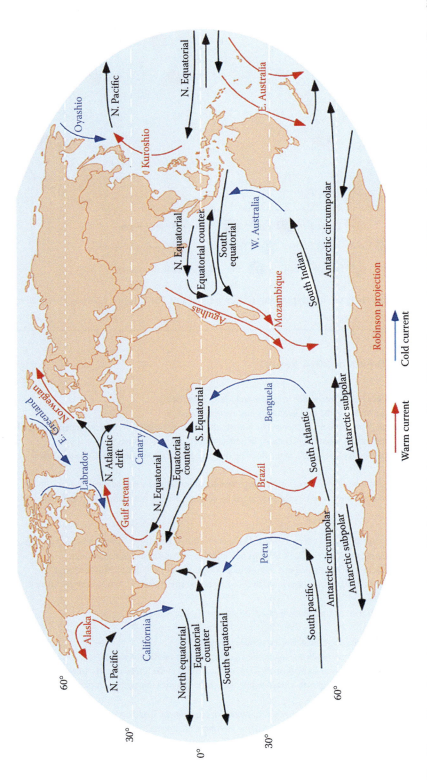

FIGURE 2.10 The earth's major ocean currents are responsible for the global transport of heat. Without them, many areas would be much cooler than they currently are. (From Gines, J. K., *Climate Management Issues: Economics, Sociology, and Politics*, CRC Press, New York, 2012.)

It is important for climatologists to understand these interactions in the ocean in order to be able to predict regional trends in climate. It is also important to understand the deepwater processes as well as processes that occur near the surface in order to understand what mechanisms drive climate. The oceans also play a critical role in balancing the CO_2 levels. The amount of CO_2 stored in the atmosphere and that dissolved in the ocean maintain an equilibrium. If something happens to upset this balance—such as changes in chemistry—then sudden shifts in the CO_2 levels can affect the climate. This is one of the concerns about the steadily increasing levels of greenhouse gases. If the oceans reach the point where they cannot continue to absorb any more CO_2, it could upset the balance of ocean currents and climate patterns on a global scale.

In the world's oceans, the properties of density, temperature, and salinity all work together and result in distinct characteristics that ultimately relate to climate change. Solar energy is absorbed by seawater and stored as heat in the oceans. Some of the energy that is absorbed may evaporate seawater, which increases its temperature and salinity. When a substance is heated, it expands and its density is lowered. Conversely, when a substance is cooled, its density increases. The addition or subtraction of salts also causes seawater density to change. Water that has higher salinity will be denser than lower salinity water.

Pressure is another factor that affects density. It increases with depth, as does the density of water mass. Because high-density water sinks below the average density seawater, and low-density seawater rises above the average density seawater, this distinct change in density generates water motion. This concept is extremely important in the world's oceans because it is a chief mechanism controlling the movement of major currents and ocean circulation patterns.

Oceanographers and climatologists are interested in the distributions of both the temperature and the salinity in the world's oceans because they are two factors that determine the vertical thermohaline circulation in the oceans. The term *thermohaline* comes from two words: *thermo* for heat and *haline* for salt. Of the three factors—temperature, salinity, and pressure—that have an effect on water density, temperature changes have the greatest effect. In the ocean, the thermocline (a water layer within which the temperature decreases rapidly with depth) acts as a density barrier to vertical circulation. This layer lies at the bottom of the low-density, warm surface layer and the top of the cold, dense bottom waters. The thermocline keeps most of the ocean water from being able to vertically mix because these two layers are so drastically different. In the polar regions, the surface waters are much colder than they are anywhere else on earth. This means they are denser, so that little temperature variation exists between the surface waters and the deeper waters— basically eliminating the thermocline. Because there is no thermocline barrier, vertical circulation can take place as the surface waters sink (a process called downwelling), where they replenish deepwaters in the major oceans.

Water surface temperatures have significant effects on coastal climates. Because seawater can absorb large amounts of heat, it enables coastal locations to have cooler temperatures in the summer than inland areas. Coastal currents also affect local climate. For example, Los Angeles, California, and Phoenix, Arizona, are at similar latitudes, yet Los Angeles has a much more moderate summer climate because of the effect of the ocean.

THE GREAT OCEAN CONVEYOR BELT AND CONSEQUENCES OF DESTABILIZATION—ABRUPT CLIMATE CHANGE

One of the most significant features in the ocean is the thermohaline circulation—more commonly referred to as the Great Ocean Conveyor Belt. This massive, continuous loop of flow, shown in the illustration, plays a critical role in determining the world's climate. The two mechanisms that make this conveyor belt work are heat and salt content (Figure 2.11).

The Great Ocean Conveyor Belt plays a major part in distributing the sun's thermal energy around the globe after the ocean has absorbed it. In fact, if it was not for this flow, the equator would be much hotter, the poles would be much colder, and Western Europe would not enjoy as warm a climate as they currently do.

The Great Ocean Conveyor Belt does not move fast, but it is enormous. It carries as much water within it as 100 Amazon Rivers could hold. The mechanism that drives it is the differences in density that range throughout it. It is the ratio of salt and temperature that determines the density. The colder and saltier the water is, the denser it is and tends to sink. When it is warm and fresh, it is less dense and rises to the surface.

The Conveyor Belt literally travels the world. In general, in a continuous loop, it transfers warm water from the Pacific Ocean to the Atlantic as a shallow current, then returns cold water from the Atlantic to the Pacific as a deep current that flows further south. Specifically, as it travels past the north of Australia, it is a warm current. It travels around the southern tip of Africa, then moves up into the Atlantic Ocean. At this point, it turns into the Gulf Stream, which is a very warm, north-flowing current critical for providing warmth to Western Europe and the northeast coast of the United States. After it passes Western Europe and heads to the Arctic, the surface water evaporates and the water cools down, releasing its heat into the atmosphere. It is this released heat that Western Europe enjoys as part of their moderate climate.

At this point, the water becomes very cold and increases in salinity, becoming very dense. As its density increases, it begins to sink. The cold, dense water descends hundreds of meters below the surface of the ocean. It now travels slowly southward through the deep ocean abyss in the Atlantic and flows to the Southern Hemisphere south of Australia and heads north again, until it eventually mixes upward to the surface of the ocean and starts the process again. This entire conveyor belt cycle takes about 1000 years to complete.

Scientists at Argonne National Laboratory (ANL) have been actively researching how long it takes water to move through this conveyor system by tracking the levels of the radioactive argon-39 isotope in the earth's ocean circulation.

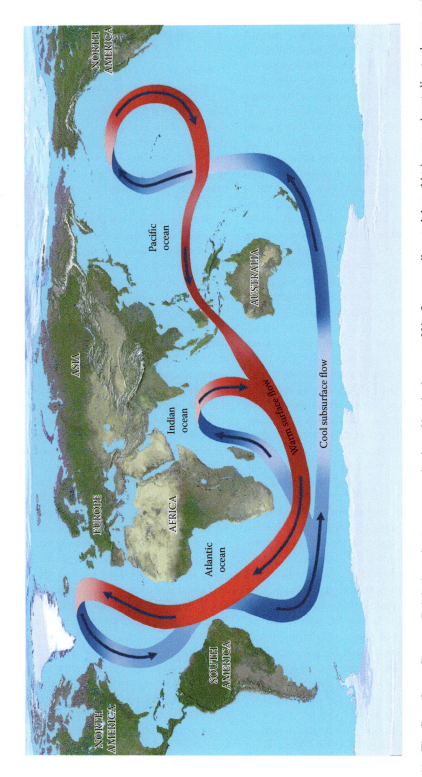

FIGURE 2.11 The Great Ocean Conveyor Belt is the major transport mechanism of heat in the ocean. If its flow were disrupted, it could trigger an abrupt climate change, such as an ice age in Western Europe. (Courtesy of NOAA, Silver Spring, MD.)

They are interested in tracking changes in ocean circulation because scientists have long believed that this system plays an important role in climate moderation. According to Ernst Rehm, a physicist at ANL, "We have some idea that if the 'conveyor belt' stops, then the warm water that is brought to Europe will stop. We have some idea that this may cause an Ice Age in Europe" (Climate Policy Watcher, 2012). In addition, scientists believe that climate change could modify the ocean's circulation if water temperatures were to rise (the heat component of thermohaline) and glaciers melt and release more of their locked up freshwater into the oceans, lowering the salinity levels (the salt component, or haline). If climate change caused this to happen—and some scientists claim it already is—it could slow, or stop, the thermohaline circulation (Argonne National Laboratory, 2003).

In a recent update, research shows that Arctic sea ice is melting faster than expected. As the earth continues to warm and Arctic sea ice melts, the influx of freshwater from the melting ice is making seawater at high latitudes less dense. Data show that the North Atlantic has become fresher over the past several decades. The less dense water will not be able to sink and circulate through the deep ocean as it does currently. This could disrupt or stop the Global Ocean Conveyor. Scientists estimate that, given the current rate of change, the Global Ocean Conveyor may slow, or even come to a halt, within the next few decades. The ramifications of this could cause cooling in Western Europe and North America. If it were to stop completely, the average temperature of Europe would cool 5°C–10°C (Windows to the Universe, 2011).

The Great Ocean Conveyor Belt plays an extremely important role in shaping the earth's climate; a slight disruption in it could destabilize the current and trigger an abrupt climate change. Climatologists at NOAA and NASA believe that as the earth's atmosphere continues to heat from the effects of climate change, there could be an increase in precipitation as well as an influx of freshwater added to the polar oceans as a result of the rapid melting of glaciers and ice sheets in the Arctic Ocean. They believe that large amounts of freshwater could dilute the Atlantic Gulf Stream to the point where it would no longer be saline—and, hence, dense enough—to sink to the ocean depths to begin its return from the polar latitudes back to the equator.

Measurements taken over the past 40 years have shown that salinity levels within the North Atlantic region are slowly decreasing. What makes this situation so serious is that if cold water stopped sinking—which means the Gulf Stream would slow and stop—there would be nothing left to push the deep, cold current at the bottom of the Atlantic along, which is what ultimately drives the worldwide ocean current system today. If this were to happen, the results would be dramatic. Western Europe and the eastern part of North America would cool off. Temperatures could plummet up to 5°C. This is about the same temperature difference as the average global temperature during the last Ice Age and today.

Effects of Sea-Level Rise

There are several impacts associated with rising sea levels making the world's coastlines vulnerable. As climate change continues and sea levels rise, storm surges will increase in intensity, destroying land further inland from the coastal regions. Flooding will become one of the major problems, and associated with flooding are several other negative impacts. As ocean waters move inland, freshwater areas will become contaminated with saltwater. As saline water intrudes rivers, bays, estuaries, and coastal aquifers, they will become contaminated and unusable. Wildlife that depends on freshwater will have their habitat negatively impacted and drinking water will become unusable. Erosion will increase along coastlines, causing disaster for many of the world's population that currently reside along the coasts. It will leave many people homeless and be economically devastating, especially in undeveloped countries.

As wetlands, mangroves, and estuaries are impacted, fragile habitats will be lost worldwide. Species will become threatened, endangered, and extinct. Other marine ecosystems will also be harmed, such as coral reefs. Reef habitats are extremely fragile and significant physical changes in their environment can quickly destroy them. The most vulnerable areas are the low-lying countries of the world with extremely large coastal populations, such as Bangladesh, the Maldives, Vietnam, China, Indonesia, Senegal, Tuvalu, Mozambique, Egypt, the Marshall Islands, Pakistan, and Thailand. Affected developing countries do not have the economic resources to implement adaptation measures, such as building sea walls to hold back rising waters. If sea levels rise, the inhabitants of the coastal areas will have no other choice but to move inland to higher ground, if possible, losing what they have at lower levels. If mass migrations result, this could lead to a host of other negative issues, such as hunger, disease, and civil unrest.

Island states are particularly vulnerable. One of the nations that are most at risk is the Maldives. This nation lies in the Indian Ocean and is composed of nearly 1200 individual islands. Their elevation above the sea level is only 2 meters. With a population of more than 200,000 people, if sea levels were to rise significantly, the entire country could become uninhabitable, leaving the entire countries' population homeless. The Marshall Islands and Tuvalu, in the Pacific, face a similar situation. Rising sea levels would first contaminate drinking water supplies, then drown the landmasses, leaving the population homeless. Other vulnerable locations include London, Amsterdam, Shanghai, New York City, many of the Caribbean islands, and Jakarta.

In a study conducted by Sugata Hazra, an oceanographer at Jadavpur University in Kolkata, India, over the past 30 years, 80 square kilometers of the Sundarbans has disappeared because of rising sea level, displacing more than 600 families. Another area, Ghoramara, has had all but 5 square kilometers of its land submerged, which is now half the size it was in 1969. The Sundarbans represents some of the world's biggest collection of river delta islands that lie between India and Bangladesh. Sea-level rise has contaminated their drinking water and destroyed the forested areas in the ecosystem. It has also threatened the existence of the wildlife, including the Bengal tiger. More than four million people live on the tiny island state, and hundreds of families have already been pushed out of their homes and been forced to move to refugee camps on neighboring islands. This is just one example of how rising sea levels are presently impacting developing countries (George, 2010). In the past few years, satellite images show that 9990 hectares of the Sundarbans forests has disappeared and several islands have completely vanished. The coastlines are retreating by as much as 200 meters a year, equating to ecosystem degradation and biodiversity loss amounting to about $107 million annually (Krishnan, 2015).

The impacts will also be felt in the United States. Both the Atlantic and Gulf Coasts face serious impacts in the face of encroaching ocean levels and saline waters. Washington DC is one of the more vulnerable areas. Higher sea levels would flood the Potomac and encroach on many famous, historical landmarks. Baltimore and Annapolis are in a similar situation.

In the Mississippi delta, the loss of wetlands is a serious issue. Changes in sea level can cause wetlands to migrate landward. The Atlantic coast is one of the more sensitive areas to wetland vulnerability. Not only is this a problem for natural habitats, but historically, these areas have been one of the most rich commercial fisheries in the world. If wetlands are endangered or destroyed, it would also have significant economic ramifications. These issues make the monitoring and control of rising sea levels a critical concern. Areas particularly in danger include Florida, Mississippi, Louisiana, North Carolina, South Carolina, Alabama, Georgia, and Texas.

Future Projections

One of the key issues is that of future sea-level rise. Because the ocean's thermal inertia is so great, it will take decades for the oceans to adjust their levels to the heat absorbed. In fact, for the heating caused by greenhouse gas emissions already released into the atmosphere, sea levels are still trying to find a point of equilibrium. Therefore, even if all greenhouse emissions stopped today, there would still be a lag time for the oceans to stop rising. During this lag time, the oceans will likely rise another 13–30 centimeters by 2100. In the 2007 Fourth Assessment Report of the IPCC, a sea-level rise of 18–58 centimeters by 2100 was projected. In the 2013 Fifth Assessment Report of the IPCC, the map of regional sea-level rise is shown in Figure 2.12. A summary of their key findings is shown in Table 2.3.

TABLE 2.3

Key Findings of the IPCC Fifth Assessment Report on Sea Level Rise

- It is *virtually certain* (99%–100% probability) that global mean sea-level rise will continue for many centuries beyond 2100, with the amount of rise dependent on future emissions.
- Medium confidence that global mean sea-level (GMSL) rise by 2300 will be <1 m for a radiative forcing corresponding to CO_2 concentrations <500 ppm but ≥1 m for 700–1500 ppm.
- Larger sea-level rise could result from sustained mass loss by ice sheets, and some part of the mass loss might be irreversible.
- Sustained warming greater than a certain threshold above preindustrial would lead to the near-complete loss of the Greenland ice sheet (high confidence).
- The threshold is estimated to be >1°C (*low confidence*) but <4°C (medium confidence) global mean warming with respect to preindustrial.
- It is *very likely* (90%–100% probability) that sea level will rise in about >95% of the ocean area.
- About 70% of the coastlines worldwide are projected to experience sea-level change within 20% of the global mean sea-level change.
- It is *very likely* that the twenty-first-century mean rate of GMSL under all RCPs will exceed that of 1971–2010, due to the same processes.
- It is *very likely* that sea level will rise in about >95% of the ocean area.
- It is *very likely* that there will be a significant increase in the occurrence of future sea-level extremes.
- It is *virtually certain* that global mean sea-level rise will continue for many centuries beyond 2100, with the amount of rise dependent on future emissions.

Based on the extensive work that has been done to date, scientists have a clear idea of where conditions are going in the future. According to the United States Geological Survey (USGS), based on the information obtained from both tidal gauges and satellite measurements worldwide, scientists can say with confidence that sea-level rise has increased during the twentieth century. Based on the data acquired from Australia's Commonwealth Scientific and Industrial Research Organization, data gathered from January 1993 to December 2015 shows that sea level has risen on average 3.3 mm/year (Figure 2.13). Increased scientific knowledge has also clarified some issues that were not well understood previously, such as that the large polar ice sheets are far more sensitive to surface warming that initially thought, with significant changes currently being observed on the Greenland and West Antarctic ice sheets. Scientists now realize that these melting ice sheets can add water mass much more quickly to the oceans than previously assumed and play a significant part in overall global sea-level rise. A notable consensus among specialists in climate change at USGS is also marked today. It is largely recognized and accepted that there could be a rapid collapse of the polar ice sheets, and scientists have keyed in on the fact that anthropogenic actions, such as burning fossil fuels, could result in triggering an abrupt sea-level rise before the end of this century. They stress public education and political policy be brought to the forefront in order to deal

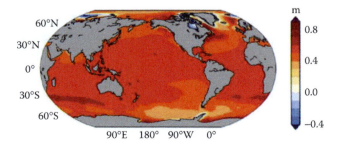

FIGURE 2.12   Regional sea-level rise by the end of the twenty-first century. (Courtesy of IPCC, Geneva, Switzerland.)

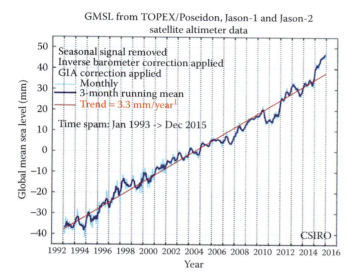

FIGURE 2.13 High-quality measurements of (near)-global sea level have been made since late 1992 by satellite altimeters, in particular, TOPEX/Poseidon (launched August 1992), Jason-1 (launched December 2001), and Jason-2 (launched June 2008). These data have shown a more-or-less steady increase in GMSL rate of ~3.2 ± 0.4 mm/year over that period. This is more than 50% larger than the average value over the twentieth century. Whether or not this represents a further increase in the rate of sea-level rise is not yet certain. (Courtesy of CSIRO, Canberra, Australia.)

most effectively with a situation that affects every person living on earth now and in the future. Figure 2.14a shows the drastic decrease in ice coverage in the Arctic over the past 30 years—approximately half now of what it was in 1979. Today the ice extent is below 4 million square kilometers. Ice volume shows a comparable rapid decrease. Figure 2.14b shows the annual average temperature in the Arctic region from 1979 to 2012. Figure 2.14c illustrates the average sea ice extent from 1979 to 2012. Satellites have measured a warming trend of 0.53°C per decade; which is considerably higher than the 0.16°C per decade global temperature increase. Since 1979, the Arctic has warmed about 3.3 times faster than the earth in general. Referred to as the "Arctic amplification," it is partially caused by the disappearance of sea ice and the effect

this has on regional albedo (referred to as the *ice-albedo feedback*). Scientists recognize that greenhouse gases play a major role in the sea ice decline of recent decades (Hagelaars, 2013). Because of these relationships and effects, it is important that the scientific basis of climate change be well understood when dealing with the sustainable ramifications caused by climate change.

CONCLUSIONS

The information presented in this chapter was geared to provide insight into the physical processes of climate change as well as the anthropogenic component and its overall significance to the problem. Based on the nature of the earth's natural processes, there is already substantial evidence that climate change is well underway, and the society as a whole can no longer ignore the issue and sit on the sidelines. In addition, the days of passing the buck are over—it is no longer "somebody else's problem." We all have a vested interest in its outcome. In order to become empowered and contribute meaningfully to the solution, however, it is necessary to understand the pertinent multifaceted issues—the scientific, political, economic, social, and technological components. Although each different in nature, they are all important pieces to the complete issue—a phenomenon so immense it touches nearly every fabric of society from broad, international scales to personal ones. Similar to just reading a language out loud without truly understanding the words and meanings, without a working knowledge of the scientific foundation it rests upon, it is not possible to comprehend its vastness and relevance to other critical components of our lives—namely, those that encompass the economic, political, and social aspects, in addition to future energy choices and sustainable lifestyle decisions. For example, without a working understanding of the Great Ocean Conveyor Belt and the possibility of abrupt climate change, it is not possible to understand the significance and urgency of economic issues and the loss of homes, food, natural resources, incomes, and basic security; the social aspects of personal loss, fear, riot, unrest, and migration; or the political aspects of security, defense, order, peace, and leadership. Understanding first the scientific basis is critical to finding workable, meaningful long-term solutions that will find an effective fit in all cultures and societies.

Not often have problems had such far-reaching consequences. Being the first world society to be faced with this dilemma, there is, unfortunately, no prior historical experience to fall back on. This is a modern problem that not only will our decisions and actions affect lifestyles and opportunities today, but will also affect those of generations long into the future. For that, we owe it to ourselves and future generations to obtain a good grasp of the basic scientific theory behind climate change. Without this crucial background, it is not possible to take a convincing stand backed by facts, examples, and viable suggestions for solutions. Knowledge is power and personal, and community education is critical to successful solutions. The correct scientific concepts need to be taught and understood so that the misinformation also being distributed by procrastinators and

September 16, 2012

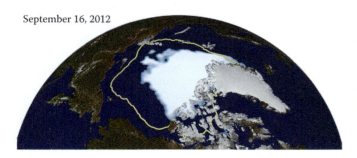

FIGURE 2.14 (a) Line indicates the average minimum ice extent of the past 30 years compared to the minimum ice extent on September 16, 2012. (b) Average annual temperature in the Arctic region from 1979 to 2012. (c) Average sea ice extent from 1979 to 2012. (Courtesy of NSIDC, Boulder, CO.)

those in denial is put to rest in order to stop muddling the issue and delaying action. Perhaps one of the best ways to look at, and approach the issue, is from the standpoint of a global community where it is necessary to reach across borders, boundaries, personal differences, and comfort levels to solve this immense problem together—for this is a war unlike any fought before and we are all on the frontlines.

REFERENCES

American Institute of Physics. 2010. Statement on human impacts on climate change. *American Institute of Physics Policy Statements.* http://www.aip.org/gov/policy12.html (accessed October 3, 2010).

Argonne National Laboratory. 2003. Physicists track great ocean conveyor belt. *Frontiers* (Argonne National Laboratory). http://www.anl.gov/Media_Center/Frontiers/2003/d8ee.html (accessed November 2, 2010).

Casper, J. K. 2009. *Global Warming Trends: Ecological Footprints.* New York: Facts on File.

CICERO. January 31, 2006. UK report warns of global catastrophes. *KLIMA Climate Magazine.* http://www.cicero.uio.no/webnews/index_e.aspx?id=10601 (accessed November 2, 2010).

Clement, A. C., R. Burgman, and J. R. Norris. July 24, 2009. Observational and model evidence for positive low-level cloud feedback. *Science* 325(5939):460–464. http://science.sciencemag.org/content/325/5939/460.abstract (accessed March 23, 2016).

Climateaction. August 19, 2010. Extreme weather events evidence of global warming, research suggests. Climateaction/UNEP. http://www.climateactionprogramme.org/news/extreme_weather_events_evidence_of_global_warming_research_suggests/ (accessed April 3, 2016).

Climate Policy Watcher. June 15, 2012. The Great Ocean Conveyor Belt. https://www.climate-policy-watcher.org/earth-surface-2/the-great-ocean-conveyor-belt.html (accessed May 13, 2017).

Co2 earth. 2016. Daily atmospheric CO2 of the earth. https://www.co2.earth (accessed May 10, 2017).

ENN (Environmental News Network). October 9, 2000. Clouds' role in global warming studied. *CNN.com.nature.* http://archives.cnn.com/2000/NATURE/10/09/clouds.warming.enn/ (accessed March 31, 2016).

Environment News Service. August 17, 2010. Extreme weather events signal global warming to world's meteorologists. *Environmental News Service.* http://www.ens-newsire.com/ens/aug2010/2010-08-17-01.html (accessed March 15, 2016).

Environmental Protection Agency. 2016. Understanding the link between climate change and extreme weather. https://www3.epa.gov/clihttps://www3.epa.gov/climatechange/science/extreme-weather.htmlmatechange/science/extreme-weather.html (accessed April 3, 2016).

George, N. March 24, 2010. Disputed Isle in Bay of Bengal Disappears Into Sea. *ABC News Technology.* http://abcnews.go.com/Technology/wireStory?id=10188225 (accessed March 15, 2016).

Gines, J. K. 2012. *Climate Management Issues: Economics, Sociology, and Politics.* New York: CRC Press.

Global Carbon Project. October 22, 2007. Contributions to accelerating atmospheric CO_2 growth from economic activity, carbon intensity, and efficiency of natural sinks. *CSIRO.* http://www.globalcarbonproject.org/global/doc/Press_GCP_Canadelletal2007.doc (accessed November 8, 2015).

Hagelaars, J. March 25, 2013. Melting of the Arctic sea ice. https://ourchangingclimate.wordpress.com/2013/03/25/melting-of-the-arctic-sea-ice/ (accessed April 3, 2016).

Hood, M. January 18, 2016. 2015 a 'tipping point' for climate change. http://phys.org/news/2016-01-climate-experts.html (accessed March 23, 2016).

Intergovernmental Panel on Climate Change. 2007. WG1—The physical science basis. http://www.ipcc.ch/ (accessed April 6, 2016).

IPCC Synthesis Report. 2014. Selected quotes. http://www.indiaenvironmentportal.org.in/media/iep/infographics/climate%20quotes/index.html (accessed March 23, 2016).

Krishnan, M. March 26, 2015. Rising sea levels threaten Sundarbans forests. D.W. http://www.dw.com/en/rising-sea-levels-threaten-sundarbans-forests/a-18342772 (accessed April 3, 2016).

Lean, G. January 23, 2005. Global warming approaching point of no return, warns leading climate expert. *Independent/UK.* http://www.commondreams.org/headlines05/0123-01.htm (accessed October 15, 2015).

Livescience. April 5, 2013. Perfect storm: Climate change and hurricanes. *Livescience.* http://www.livescience.com/28489-sandy-after-six-months.html (accessed April 3, 2016).

Plait, P. August 20, 2014. Did I say 30 billion tons of CO_2 a year? I meant 40. http://www.slate.com/blogs/bad_astronomy/2014/08/20/atmospheric_co2_humans_put_40_billion_tons_into_the_air_annually.html (accessed March 23, 2016).

Ppcorn. January 10, 2016. 10 shocking facts about global warming. http://ppcorn.com/us/2016/01/10/10-shocking-facts-about-global-warming/ (accessed March 23, 2016).

Rahmstorf, S., J. E. Box, G. Feulner, M. E. Mann, A. Robinson, S. Rutherford, and E. J. Schaffernicht. March 23, 2015. Exceptional twentieth-century slowdown in Atlantic Ocean overturning circulation. *Nature Climate Change* 5. www.nature.com/natureclmatechange.

Schmitt, R. November 19, 2010. The ocean's role in climate. WHOI website. http://www.whoi.edu/page.do?pid=12455&tid=282&cid=10146 (accessed December 1, 2015).

ScienceDaily. 2010. Carbon dioxide controls earth's temperature, new modeling study shows. *Science Daily*, October 15. http://www.sciencedaily.com/releases/2010/10/101014171146.htm (accessed May 31, 2017).

Unh, L. M. 2015. How will climate change affect the trade winds? *Quora.* https://www.quora.com/How-will-climate-change-affect-the-trade-winds (accessed April 2, 2016).

Vecchi, G. 2006. Trade winds weaken with global warming. University Corporation for Atmospheric Research. http://www.vsp.ucar.edu/about/stories/gVecchi.html (accessed December 7, 2010).

Windows to the Universe. January 26, 2011. Melting Arctic sea ice and the global ocean conveyor. http://www.windows2universe.org/earth/polar/icemelt_oceancirc.html (accessed April 3, 2016).

World Meteorological Organization/United Nations Framework Convention on Climate Change. 2011. Fact sheet: Climate change science: The status of climate change science today. http://unfccc.int/press/fact_sheets/items/4987.php (accessed May 12, 2017).

World Meteorological Organization. 2011. Weather extremes in a changing climate: Hindsight on foresight. WMO-No.1075. https://issuu.com/climateandhealth/docs/weather_extremes_in_a_changing_clim (accessed May 11, 2017).

SUGGESTED READING

Abbot, J. and J. S. Armstrong. 2015. *Climate Change: The Facts.* Berlin, Germany: Stockade Books.

Gore, Al. 2009. *Our Choice: A Plan to Solve the Climate Crisis.* Emmaus, PA: Rodale Books.

Kinver, M. August 18, 2013. European forests near 'carbon saturation point.' *BBC News.* http://www.bbc.com/news/science-environment-23712464 (accessed March 23, 2016).

Mathez, E. A. 2009. *Climate Change: The Science of Global Warming and Our Energy Future.* New York: Columbia University Press.

NOAA. November 5, 2015. New report finds human-caused climate change increased the severity of many extreme events in 2014. http://www.noaanews.noaa.gov/stories2015/110515-new-report-human-caused-climate-change-increased-the-severity-of-many-extreme-events-in-2014.html (accessed April 3, 2016).

Pearson, R. 2011. *Driven to Extinction: The Impact of Climate Change on Biodiversity.* New York: Sterling Publishing.

Schmidt, G. and J. Wolfe. 2009. *Climate Change: Picturing the Science.* New York: W. W. Norton & Company.

Volk, T. 2010. *CO$_2$ Rising: The World's Greatest Environmental Challenge.* Cambridge, MA: MIT Press.

3 Greenhouse Gases

OVERVIEW

Climate change is undoubtedly one of the biggest challenges humankind has ever faced. The bad news is that a large share of it is our own doing. The good news is that there is still a window of opportunity left to fix it if we act now. That is right. This is a significant, global problem, but with a transition of habits and some dedicated work, there is still a window of hope. But the clock is ticking and we are quickly running out of time.

This chapter discusses the greenhouse gases—the chief contributor and reason for the rise in temperature and trigger for the negative global impacts the earth is now experiencing and will experience for thousands of years to come. The chapter begins by presenting the basics of solar radiation and the processes the Sun's energy takes to get to the earth. Once it reaches the planet, it explains the various pathways it takes and the complications that then occur because of the greenhouse gases that humans have overloaded the atmosphere with. This introduces the important concept of radiative forcing and how that has now permanently impacted the earth's energy balance.

Next, the various greenhouse gases are introduced, their properties, and why their collection and buildup in the atmosphere is so detrimental. The concept of climate change potential of each of the greenhouse gases is presented as an illustration of why this whole process has such long and far-reaching effects—why some of the greenhouse gases can remain in the atmosphere and cause damage for thousands of years. The chapter then presents the concept of carbon sequestration and why this process is so important, as well as various carbon sinks and sources and why these must be managed in order to help mitigate the rate of greenhouse gas influx into the atmosphere. Next, it examines the impacts of deforestation, the burning of fossil fuels, and coal emissions, and lays the foundation as to why these activities are so environmentally detrimental. In conclusion, it discusses various health concerns and the key impacts to the earth's ecosystems as a result of the increasing concentration of greenhouse gases in the atmosphere and why it is imperative that the world's nations come together now in order to effectively mitigate the problem before it is too late.

INTRODUCTION

According to Dr. James E. Hansen of the National Aeronautics and Space Administration (NASA), one of the world's foremost climate experts, there is still a small window of time left where actions can be taken to slow down the climate change processes that have been put in motion in order to keep temperatures from climbing over the 2°C mark. But everyone has to play a part and make a commitment—large or small. As Canadian philosopher Marshall McLuhan said, "There are no passengers on Spaceship Earth. We are all crew."

One of the areas where humans can have the biggest impact in mitigating the rise in atmospheric temperature is through the control of greenhouse gas emissions. The emission of greenhouse gases is the most significant factor in climate change and the area where attention is most needed. Most greenhouse gases are emitted through the burning of fossil fuels—coal, oil, and gas. Fossil fuels are burned not only for public and private transportation, but also in the manufacture of most commodities, the generation of most electricity, most manufacturing and industrial processes, and commerce and commercial transportation. Developed countries are pumping greenhouse gases into the atmosphere at alarming rates. In order to halt the destruction being caused by greenhouse gases, action needs to be taken immediately. In order to make this happen, however, public education is critical. This chapter lays the foundation as to why greenhouse gases are so detrimental and why their mitigation is critical.

DEVELOPMENT

As we saw in Chapter 2, the earth needs the natural greenhouse effect in order for life to survive. It is the process in which the emission of infrared radiation by the atmosphere warms the planet's surface. The atmosphere naturally acts as an insulating blanket, which is able to trap enough solar energy to keep the global average temperature in a comfortable range in which to support life. This insulating blanket is actually a collection of several atmospheric gases, some of them in such small amounts that they are referred to as trace gases.

The framework in which this system works is often referred to as the greenhouse effect because this global system of insulation is similar to that which occurs in a greenhouse nursery for plants. The gases are relatively transparent to incoming visible light from the Sun, yet opaque to the energy radiated from the earth.

These gases are the reason why the earth's atmosphere does not scorch during the day and freeze at night. Instead, the atmosphere contains molecules that absorb the heat and reradiates the heat in all directions, which reduces the heat lost back to space. It is the greenhouse gas molecules that keep the earth's temperature ranges within comfortable limits. Without the natural greenhouse effect, life would not be possible on earth. In fact, without the greenhouse effect to regulate the atmospheric temperature, the Sun's heat would escape and the average temperature would drop from 14°C to −18°C, a temperature much too cold to support the diversity of life that exists today on the planet.

RADIATION TRANSMISSION

Understanding radiation transmission is important to understanding how and why specific gases in the atmosphere are affected during the greenhouse gas process. The earth reflects 30 percent of the incoming solar radiation back out into space. The remaining 70 percent is absorbed and warms the atmosphere, land, and oceans. In order to maintain an energy balance, keeping the earth from getting too hot or cold, the amount of incoming energy must roughly equal the amount of outgoing energy (see Figure 3.1).

The majority of insolation is in the short and medium wavelength regions of the spectrum. The highest energy, short wavelengths, such as the gamma rays, X-rays, and ultraviolet (UV) light is absorbed by the mid to high levels of the atmosphere. This is desirable because if these wavelengths traveled to the surface, they could cause harm to life on earth. For example, exposure to UV light can lead to cancer. The medium wavelengths—referred to as visible light—travel to the earth's surface. These waves can be absorbed or reflected at the earth's surface as well as by the CO_2 and water vapor in the atmosphere.

When wavelengths from the electromagnetic spectrum reach the earth, several things can happen. They can get reflected; the earth's surface, the ocean, the clouds, or the atmosphere can absorb them; and they can be scattered. Of that which does enter the earth's atmosphere, nearly one-quarter of that is reflected by clouds and particulates (tiny suspended particles) in the atmosphere. Of the remaining visible light that does reach the earth's surface, approximately one-tenth is reflected upward by snow and ice, due to their extremely high albedo (high reflectivity). As the infrared radiation (heat) is emitted upward, it does not simply escape to space. Water vapor and other gases in the atmosphere absorb it and trap it in the atmosphere. This trapped heat then radiates in all directions: upward, down toward the earth's surface again, and sideways. This trapped heat energy is what constitutes the greenhouse effect.

THE NATURAL AND ENHANCED GREENHOUSE EFFECT

In this "greenhouse environment," it is the combination of water vapor and trace gases that are responsible for trapping the heat radiated from the Sun. As stated, this natural amount keeps the earth habitable instead of at −18°C, in which case the earth would look much different—there would most likely be very little, if any, liquid water available. The entire earth would have an ecosystem similar to that of the harshest areas in Antarctica.

The earth's natural greenhouse effect is critical for the survival and diversity of life. Since the Industrial Revolution (over the past 250 years or so), the natural greenhouse effect has been augmented by human interference. Carbon dioxide (CO_2), one of the atmosphere's principal greenhouse gases, has been altered to such an extent by human activity that the natural greenhouse effect is no longer in balance. The earth's energy balance now must contend with what is referred to as the enhanced greenhouse effect or anthropogenic greenhouse warming. CO_2 is being added in voluminous amounts as a result of human activity, deforestation, agricultural practices, the burning of fossil fuels for transportation, urban development, the heating and cooling of homes, and industrial processes. In fact, CO_2 in the atmosphere has increased 31 percent since 1895. Concentration of other greenhouse gases, such as methane and nitrous oxides (also related to human activity), have increased 151 percent and 17 percent, respectively.

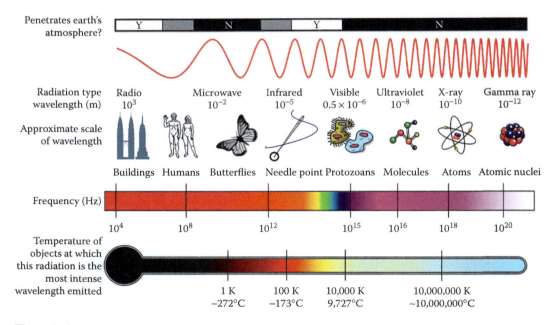

FIGURE 3.1   The sun's electromagnetic spectrum ranges from short wavelengths, such as X-rays, to long wavelengths, such as radio waves. The majority of the sun's energy is concentrated in the visible and nearly visible portions of the spectrum—the wavelengths located between 400 and 700 nm. (From Gines, J. K., *Climate Management Issues: Economics, Sociology, and Politics*, CRC Press, New York, 2012.)

At the beginning of the Industrial Revolution, the CO_2 content in the atmosphere was 280 parts per million (ppm). By 1958, it had increased to 315 ppm; by 2004, 378 ppm; and by 2011, 392.4 ppm. By February 2016, it had climbed to a record of 404.16 ppm. According to Dr. James E. Hansen, a world-renowned expert on climate change at NASA's Goddard Institute for Space Studies (GISS) in New York, "Climate is nearing dangerous tipping points. We have already gone too far. We must draw down atmospheric CO_2 to preserve the planet we know. A level of no more than 350 ppm is still feasible, with the help of reforestation and improved agricultural practices, but just barely—time is running out" (Hansen, 2008).

Although not the only greenhouse gas, CO_2 does seem to be the one most focused on because it is one of the most important and prevalent. It makes up about 25 percent of the natural greenhouse effect because there are several natural processes that put it in the atmosphere, such as the following:

- *Forest fires*: When trees are burned, the CO_2 stored within them is released into the atmosphere.
- *Volcanic eruptions*: CO_2 is one of the gases released in abundance by volcanoes.
- *Ocean equilibrium*: Oceans both absorb and release enormous amounts of CO_2. Historically they have served as a major storage facility of CO_2 put into the atmosphere from human sources. Recently, however, scientists at NASA and National Oceanic and Atmospheric Administration (NOAA) have determined that the oceans have become nearly saturated and are approaching their limits as a carbon store.
- *Vegetation storage*: Trees, plants, grasses, and other vegetation serve as significant stores of carbon. When they die and decompose, half of their stored carbon is released into the atmosphere in the form of CO_2.
- *Soil processes*: When vegetation dies, the other half of their stored carbon is absorbed by the soil. Over time, some of this carbon is released into the atmosphere as CO_2.
- *Biological factors*: Any life form that consumes plants that contain carbon also emits CO_2 into the atmosphere through breathing. This includes animals, insects, and even humans.

With this amount of CO_2 entering the atmosphere, fortunately there are some processes that help keep it in check. Plants and trees remove CO_2 from the atmosphere through photosynthesis, the oceans absorb large amounts of CO_2, and phytoplankton take in CO_2 through photosynthesis. Ideally, keeping CO_2 in balance is desirable, but throughout the earth's history, this has not always been possible. When CO_2 levels have dropped, the earth has consistently experienced an ice age. Since the last ice age, CO_2 levels remained fairly constant, however, until the Industrial Revolution, when billions of tons of extra CO_2 began to be added to the earth's atmosphere.

The turning point was the use of fossil fuels—coal, oil, and gas—made from the carbon of plants and animals that decomposed millions of years ago and were buried deep under the earth's surface, subjected to enormous amounts of heat and pressure. When this carbon is converted into usable energy forms and is burned as fuel, the carbon combines with oxygen and releases enormous amounts of CO_2 into the atmosphere.

According to NASA/GISS, CO_2 levels have not been as high as they are today over the past 400,000 years (Hansen, 2007). As James Hansen of NASA explained, "The total energy imbalance now is about six-tenths of a watt per square meter. That may not sound like much, but when added up over the whole world, it's enormous. It's about 20 times greater than the rate of energy use by all of humanity. It is equivalent to exploding 400,000 Hiroshima atomic bombs per day 365 days per year. That's how much extra energy Earth is gaining each day. This imbalance, if we want to stabilize climate, means that we must reduce CO_2 from 391 ppm, parts per million, back to 350 ppm. That is the change needed to restore energy balance and prevent further warming" (Hansen, 2012).

CO$_2$ FAST FACTS

- Global carbon (C) emissions from fossil fuel use were 9.795 gigatons (Gt) in 2014.
- Fossil fuel emissions were 0.6 percent above emissions in 2013 and 60 percent above emissions in 1990 (the reference year in the Kyoto Protocol).
- Fossil fuel emissions accounted for about 91 percent of total CO_2 emissions from human sources in 2014. This portion of emissions originates from coal (42%), oil (33%), gas (19%), cement (6%), and gas flaring (1%).
- Changes in land use are responsible for about 9 percent of all global CO_2 emissions.
- In 2013, the largest national contributions to the net growth in total global emissions were China (58%), the United States (20%), India (17%), and European Union-28 (EU28) (a *decrease* of 11%).
- The 2014 level of CO_2 in the atmosphere was 43 percent above the level when the Industrial Revolution started in 1750.
- February 2016 was the warmest February in the past 137 years. The temperature deviated from the twentieth-century average by 1.21°C. This is the tenth consecutive month that the monthly temperature record has been broken.
- NOAA's global analysis for 2015 lists 2015 as the warmest year on record since 1880 at 0.90°C above the twentieth-century average. The year 2014 is the second warmest at 0.74°C above the average. The year 2013 was the fourth warmest at 0.66°C above the average.

Source: NOAA Global Analysis.

The enhanced greenhouse effect has not come as a surprise to climate scientists, other experts, and decision makers, however. Climate scientist, Charles David Keeling of the Scripps Institution of Oceanography, made enormous strides in establishing the rising levels of CO_2 in the earth's atmosphere and the subsequent issue of the enhanced greenhouse effect and climate change. Keeling set up a CO_2 monitoring station at Mauna Loa, Hawaii, in 1957. Beginning in 1958, Keeling made continuous measurements of CO_2, which still continue today. He chose the remote Mauna Loa site so that the proximity of large cities and industrial areas would not compromise the atmospheric readings he was collecting. Air samples at Mauna Loa are collected continuously from air intakes at the top of four 7-meter towers and one 27-meter tower. Four air samples are collected each hour to determine the CO_2 level.

The measurements he gathered show a steady increase in mean atmospheric concentration from 315 ppm by volume (ppmv) in 1958 to more than 392 ppmv by 2011. The increase is considered to be largely due to the enhanced greenhouse effect and the burning of fossil fuels and deforestation. This same upward trend also matches the paleoclimatic data shown in ice cores obtained from Antarctica and Greenland. The ice cores also show that the CO_2 concentration remained at approximately 280 ppmv for several thousand years, but began to rise sharply at the beginning of the 1800s (Figure 3.2).

The Keeling Curve also shows a cyclic variation of about 5 ppmv each year. This is what gives it the sawtooth appearance. It represents the seasonal uptake of CO_2 by the earth's vegetation (principally in the Northern Hemisphere because that is where most of the earth's vegetation is located). The CO_2 lowers in the spring because new plant growth takes CO_2

out of the atmosphere through photosynthesis, and it rises in autumn as plants and leaves die off and decay, releasing CO_2 gas back into the atmosphere.

Mauna Loa is considered the premier long-term atmospheric monitoring facility on earth. It is also one of the world's most favorable locations for measuring undisturbed air because vegetation and human influences are minimal, and any influence from volcanic vents has been calculated and excluded from the measurement. The methods and equipment used to create the continuous monitoring record have been the same for the past 58 years, rendering the monitoring program highly controlled and consistent.

Keeling was a true pioneer in establishing the existence of climate change. His work was instrumental in showing scientists worldwide that the CO_2 level in the atmosphere was indeed continually rising. This was the first strongly compelling evidence that climate change not only existed but was accelerating. Because of the worthwhile merits and significance of his work, NOAA began monitoring CO_2 levels worldwide in the 1970s. There are currently about 100 sites continuously monitored today because of what was started on Mauna Loa.

Ralph Keeling, Charles David Keeling's son, says, "The Keeling Curve has become an icon of the human imprint on the planet and a continuing resource for the study of the changing global carbon cycle. Mauna Loa provides a valuable lesson on the importance of continuous earth observations in a time of accelerating global change" (Keeling, 2008).

The Intergovernmental Panel on Climate Change (IPCC) has projected that with continued climate change, temperatures will rise an average of 1.4°C–5.8°C in the next 100 years (IPCC, 2001). The polar regions, however, are expected to rise more than the average. Currently, with much of the polar areas being covered by snow and ice, they have a high albedo, so much of the insolation is reflected back to space. As temperatures warm, however, the snow and ice will melt, exposing darker surfaces (land and water), which will naturally absorb more heat, melt more ice, and uncover more dark surface area. This could continue in a cycle until all the snow and ice are melted, greatly changing the heat balance in the polar regions, resulting in a rapid warming (Riebeek, 2010).

RADIATIVE FORCING

When discussing the enhanced greenhouse effect and climate change, climate scientists often refer to a concept called "radiative forcing." This is a measure of the influence that an independent factor (ice albedo, aerosols, land use, carbon dioxide) has in altering the balance of incoming and outgoing energy in the earth–atmosphere system. It is also used as an index of the influence a factor has as a potential climate change mechanism. Radiative forcings can be positive or negative. A positive forcing (corresponding to more incoming energy) warms the climate system. A negative forcing (corresponding to more outgoing energy) cools the climate system.

As shown in the illustration, examples of positive forcings (those which warm the climate) include greenhouse gases,

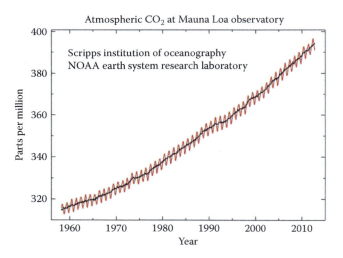

FIGURE 3.2 Collected since 1958, the data in the Keeling Curve have been instrumental in providing convincing evidence that CO_2 levels are rising worldwide. The jagged line depicts the readings collected from the Mauna Loa observation station, and the smoother line depicts the data collected from the south pole. Without these continuous data, it would be much more difficult to determine the existence of climate change. (Courtesy of NOAA.)

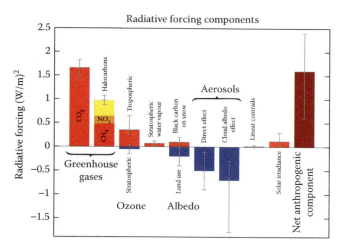

FIGURE 3.3 Graph illustrating the concept of radiative forcing. In order to understand the climate change, it is necessary to understand the components in the atmosphere that control the overall warming and cooling on short- and long-term bases. (From NASA/GISS, Forcings in GISS climate model, http://data.giss.nasa.gov/modelforce/.)

tropospheric ozone, water vapor, solar irradiance, and anthropogenic effects (see Figure 3.3). Those that have a negative rating (they cool the climate) include stratospheric ozone and certain types of land use. Climatologists often use radiative forcing data to compare various cause-and-effect scenarios in climate systems where some type of change (warmer or cooler) has taken place. Radiative forcings and their effects on climate are also built into climate models in order to enable climatologists to determine the effects from multiple input factors on climate change. The greenhouse gases that remain in the atmosphere the longest will have the greatest impact on climate change, making it imperative that climatologists not only be able to identify them but also understand them and how they react with the atmosphere.

The Earth's Energy Balance

Even if greenhouse gas emissions were stabilized at today's rates, climate change would not stop. According to scientists at NASA, CO_2 levels are rising because there is more CO_2 being emitted into the atmosphere than natural processes such as photosynthesis and absorption into the oceans can presently remove. There is no longer a carbon balance, and there is a net gain that continues to increase each year. As the CO_2 levels increase, the temperatures continue to rise.

Because the earth's carbon system is already so out of balance, in order to stop climate change, emissions must actually be reduced over the coming years, not just stabilized. Even with the best of intentions, the earth's temperature would not react immediately—there is a delayed reaction. This occurs because there is already an abundance of excess energy stored in the world's oceans. This "lag" time responding to the reaction is referred to as thermal inertia. Because of thermal inertia, scientists at NASA have determined that the

0.6°C–0.9°C of climate change that has already occurred this past century is not the full amount of warming the environment will eventually reach from the greenhouse gases that have already been emitted into the atmosphere. Even in the extreme case that all greenhouse emissions stopped immediately, the earth's average surface temperature would continue to climb another 0.6°C or more over the next several decades before temperatures leveled off.

The half-life of CO_2 is about 100 years. Therefore, most of the CO_2 being released today will still be in the atmosphere in 2116. Although there are natural factors that control the CO_2 levels and climate, NASA supports the notion that human influence is changing the climate on earth—many agree today's documented climate change trend is at least partly anthropogenic in origin. The IPCC (2013) says, "The balance of evidence suggests that there is a discernible human influence on global climate."

According to Dr. James E. Hansen, this "lag time" is a key reason why it is risky to hold off any longer trying to control greenhouse gas emissions (Climate Policy Watcher, 2012). The longer the society waits to take positive action to stop climate change, the more severe and long-lasting the consequences will become. He stresses that the time to act is now, not some time in the future.

If climate change is allowed to continue, there will be many negative effects, such as the disappearance of ecosystems, leading to the extinction of many species; there will be a greater number of heat-related deaths; there will be a greater spread of infectious diseases, such as encephalitis, malaria, and dengue fever through the proliferation of disease-carrying mosquitoes; malnutrition and starvation as a result of droughts; loss of coastal areas due to sea-level rise from melting ice caps and thermal expansion of the oceans; unpredictable agricultural production; and threatened food supply and contaminated water. Impacts will be health related, social, political, ecological, environmental, and economic.

Greenhouse Gases

It is the existence of trace gases in the atmosphere that act like the glass in a greenhouse. The trace gases serve to trap the heat energy from the Sun close to the earth. Most greenhouse gases occur naturally and are cycled through the global biogeochemical system. It is the greenhouse gases added by human activity that are trapping too much heat today and causing the atmosphere to overheat. There are several different types of greenhouse gases; some exist in greater quantities than others. They include water vapor, CO, methane, nitrous oxide, and halocarbons. They capture 70–85 percent of the energy in upgoing thermal radiation emitted from the earth's surface.

Water vapor—the most common greenhouse gas—accounts for roughly 65 percent of the natural greenhouse effect. When water heats up, it evaporates into vapor and rises from the earth's surface into the atmosphere, where it forms clouds and acts as an insulating blanket to help keep the earth warm or it can reflect and scatter incoming sunlight. This is why cloudy nights are warmer than clear nights. As water vapor condenses and cools, it then comes back to the earth as snow or rain and continues on its way through the water cycle.

The second most prevalent greenhouse gas, CO_2 comprises one-fourth of the natural greenhouse effect. Humans and animals exhale CO_2, vegetation releases CO_2 when it dies and decomposes, burning trees in a forest fire or burning during deforestation releases it, and burning fossil fuels (such as exhaust from cars and industrial processes) are all common sources of CO_2.

Methane is a colorless, odorless, flammable gas, formed when plants decay in an environment of restricted air. It is the third most common greenhouse gas and is created when organic matter decomposes without the presence of oxygen—a process called anaerobic decomposition. One of the most common sources is from "ruminants"—grazing animals that have multiple stomachs in which to digest their food. These include cattle, sheep, goats, camels, bison, and musk ox. In their digestive system, their large fore-stomach hosts tiny microbes that break down their food. This process creates methane gas, which is released as flatulence. Livestock also emit methane when they belch. In fact, in one day a single cow can emit one-half pound of methane into the air. Each day 1.3 billion cattle burp methane several times per minute.

Methane is also a by-product of natural gas and decomposing organic matter, such as food and vegetation. Also present in wetlands, it is commonly referred to as "swamp gas." Since 1750, methane has doubled its concentration in the atmosphere, and it is projected to double again by 2050. According to Nick Hopwood and Jordan Cohen at the University of Michigan, every year 350–500 million tons of methane is added to the atmosphere through various activities, such as the raising of livestock, coal mining, drilling for oil and natural gas, garbage sitting in landfills, and rice cultivation (Hopwood and Cohen, 2008). Because rice is grown in waterlogged soils, such as swamps, they release methane as a by-product.

Nitrous oxide (N_2O), another greenhouse gas, is released from manure and chemical fertilizers that are nitrogen based. As the fertilizer breaks down, N_2O is released into the atmosphere. Nitrous oxide is also contained in soil by bacteria. When farmers plow the soil and disturb the surface layer, N_2O is released into the atmosphere. It is also released from catalytic converters in cars and also from the ocean. According to Hopwood and Cohen, nitrous oxide has risen more than 15 percent since 1750. Each year 6–12 million metric tons is added to the atmosphere principally through the use of nitrogen-based fertilizers, the disposal of human and animal waste in sewage treatment plants, automobile exhaust, and other sources that have not yet been identified.

Halocarbons include the fluorocarbons, methyl halides, carbon tetrachloride (CCl_4), carbon tetrafluoride (CF_4), and halons. They are all considered to be powerful greenhouse gases because they strongly absorb terrestrial infrared radiation and stay in the atmosphere for many decades.

Fluorocarbons are a group of synthetic organic compounds that contain fluorine and carbon. A common compound is chlorofluorocarbon (CFC). This class contains chlorine atoms and has been used in industry as refrigerants, cleaning solvents, and propellants in spray cans. These fluorocarbons are harmful to the atmosphere, however, because they deplete the ozone layer, and their use has been banned in most areas of the world, including the United States.

Hydrofluorocarbons (HFCs) contain fluorine and do not damage the ozone layer. Fluorocarbon polymers are chemically inert and electrically insulating. They are used in place of CFCs because they do not harm or break down ozone molecules, but they do trap heat in the atmosphere. HFCs are used in air conditioning and refrigerators. The best way to keep them out of the atmosphere is to recycle the coolant from the equipment they are used in.

Fluorocarbons have several practical uses. They are used in anesthetics in surgery, as coolants in refrigerators, as industrial solvents, as lubricants, as water repellants, as stain repellants, and as chemical reagents. They are used to manufacture fishing line and are contained in products such as Gore-Tex and Teflon.

Climate Change Potential

Although there are several types of greenhouse gases, they are not the same—they all have different properties associated with them and the differences are critical, especially their half-life's—the duration of time they remain active in the atmosphere. They also differ significantly in the amount of heat they can trap.

It is important to note that these gases and their effects will continue to increase in the atmosphere as long as they continue to be emitted and remain there. Even though these gases represent a very small proportion of the atmosphere—less than 2 percent of the total—because of their enormous heat holding potential, they are a significant component to the atmosphere and represent a serious problem for climate change.

The U.S. Environmental Protection Agency (EPA) has identified three major groups of high global warming potential (GWP) gases: (1) HFCs, (2) perfluorocarbons (PFCs), and (3) sulfur hexafluoride (SF_6). These represent the most potent greenhouse gases; PFCs and SF_6 also have extremely long atmospheric lifetimes—up to 23,900 years. Because their lifetime is so incredibly long, for practical management purposes, once they are emitted into the atmosphere, they are considered to be there permanently. According to the EPA (2011), once present in the atmosphere, it results in "an essentially irreversible accumulation."

HFCs are man-made chemicals; most of them developed as replacements for the prior used ozone-depleting substances that were common in industrial, commercial, and consumer products. The GWP index for HFCs ranges from 140 to 11,700, depending on which one is used. Their lifetime in the atmosphere ranges from 1 to 260 years. Most of the commercially used HFCs have a lifetime of less than 15 years, such as HFC-134a, which is used in automobile air conditioning and refrigeration.

PFCs generally originate from the production of aluminum and semiconductors. They have very stable molecular structures and usually do not get broken down in the lower atmosphere. When they reach the mesosphere 60 kilometers above the earth's surface, high-energy UV electromagnetic energy destroys them, but it is a very slow process, which enables them to accumulate in the atmosphere for up to 50,000 years.

Sulfur hexafluoride has a GWP of 23,900, making it the most potent greenhouse gas. It is used in insulation, electric power transmission equipment, magnesium industry, and semiconductor manufacturing to create circuitry patterns on silicon wafers, and also as a tracer gas for leak detection. Its accumulation in the atmosphere shows the global average concentration has increased by 7 percent per year during the 1980s and the 1990s—according to the IPCC from less than 1 part per trillion (ppt) in 1980 to almost 4 ppt in the late 1990s (IPCC, 2007).

In order to understand the potential impact from specific greenhouse gases, they are rated as to their GWP. The GWP of a greenhouse gas is the ratio of climate change—or radiative forcing—from one unit mass of a greenhouse gas to that of one unit mass of CO_2 over a period of time, making the GWP a measure of the *potential for climate change per unit mass relative to CO_2*. In other words, greenhouse gases are rated on how potent they are compared to CO_2.

GWPs take into account the absorption strength of a molecule and its atmospheric lifetime. Therefore, if methane has a GWP of 23 and carbon has a GWP of 1 (the standard), this means that methane is 23 times more powerful than CO_2 as a greenhouse gas. The IPCC has published reference values for GWPs of several greenhouse gases. Reference standards are also issued and supported by the United Nations Framework Convention on Climate Change, as shown in Table 3.1. The higher the GWP value, the larger the infrared absorption and the longer the atmospheric lifetime. Based on this table, even small amounts of SF_6 and HFC-23 can contribute a significant amount to climate change.

In response to climate change, the U.S. EPA is working to reduce the emission of high GWP gases because of their extreme potency and long atmospheric lifetimes. High GWP gases are emitted from several different sources. Major emission sources of these today are from industries such as electric power generation, magnesium production, semiconductor manufacturing, and aluminum production.

In electric power generation, SF_6 is used in circuit breakers, gas-insulated substations, and switchgear. During magnesium metal production and casting, SF_6 serves as a protective cover gas during the processing. It improves safety and metal quality by preventing the oxidation and potential burning of molten magnesium in the presence of air. It replaced sulfur dioxide (SO_2), which was more environmentally toxic. The semiconductor industry uses many high GWP gases in plasma etching and in cleaning chemical vapor deposition tool chambers. They are used to create circuitry patterns. During primary aluminum production, GWP gases are emitted as by-products of the smelting process.

The best solution found to date to solve the negative impact to the environment and combat climate change is by the EPA working with private industry in business partnerships that involve developing and implementing new processes that are environmentally friendly. In addition, EPA is also working to limit high GWP gases through mandatory recycling programs and restrictions. If a greenhouse gas can remain in the atmosphere for several hundred years, even though it may be in small amounts, it can do a substantial amount of damage. Some of the greenhouse effect today is due to greenhouse gases put in the atmosphere decades ago. Even trace amounts can add up significantly.

TABLE 3.1

Climate Change Potential of Greenhouse Gases

Greenhouse Gas	Lifetime in the Atmosphere (years)	GWP over 100 years (Compared to CO_2)
Carbon dioxide	50–200	1
Methane	12	23
Nitrous oxide	120	296
CFC 115	550	7,000
HFC-23	264	11,700
HFC-32	5.6	650
HFC-41	3.7	150
HFC-43-10mee	17.1	1,300
HFC-125	32.6	2,800
HFC-134	10.6	1,000
HFC-134a	14.6	1,300
HFC-152a	1.5	140
HFC-143	3.8	300
HFC-143a	48.3	3,800
HFC-227ea	36.5	2,900
HFC-236fa	209	6,300
HFC-245ca	6.6	560
Sulfur hexafluoride	3,200	23,900
Perfluoromethane	50,000	6,500
Perfluoroethane	10,000	9,200
Perfluoropropane	2,600	7,000
Perfluorobutane	2,600	7,000
Perfluorocyclobutane	3,200	8,700
Perfluoropentane	4,100	7,500
Perfluorohexane	3,200	7,400

Source: United Nations Framework Convention on Climate Change.

CARBON SEQUESTRATION

Because CO_2 is the most prevalent greenhouse gas in the atmosphere, it is important to understand carbon: what it is used for, where it comes from, where it goes, and how it interacts with other elements. As it relates to climate change, the concept of carbon sequestration—or carbon storage—is especially important. If carbon can be stored, it is removed from the environment, and therefore removed from being a potential source of greenhouse gas. It becomes a sink—a storage place for carbon. In effective management issues, it is important to understand which places store carbon (carbon sink) and which places release carbon (carbon source). Carbon sinks play a direct role in climate change—the more that can be sequestered, the less available to contribute to climate change.

SINKS AND SOURCES

In terms of climate change, storing CO_2 in reservoirs—or sinks—is desirable in order to keep the CO_2 out of the atmosphere so that it does not contribute to ever-increasing atmospheric temperatures. There are several mechanisms whereby carbon can be put into storage.

During the process of photosynthesis, plants convert CO_2 from the atmosphere into carbohydrates and release oxygen during the process. Trees in forests over a period of years are able to store large amounts of carbon, which is why promoting the regrowth of old forests and encouraging the growth of new forests are positive methods of combating climate change. Vegetation sequesters 544 billion metric tons of carbon each year. In fact, the regrowth of forests in the Northern Hemisphere is the most significant anthropogenic sink of atmospheric CO_2. When forests are destroyed during the process of deforestation and the land is cleared for other uses such as grazing, it removes a valuable carbon sink from the ecosystem. In areas where the forests are cleared through burning, this process serves to release the CO_2 back into the atmosphere.

In the oceans, as living organisms with shells die, pieces of the shells break apart and fall toward the bottom of the ocean, slowly accumulating as a sediment layer on the bottom. Small amounts of plankton settle through the deepwater and come to rest of the floors of the ocean basins. Even though this source does not represent huge sources of carbon, over thousands or millions of years it does add up, becoming another useful long-term carbon store. The world's oceans are currently holding huge amounts of CO_2, helping to minimize the effects of climate change. As the oceans become saturated, however, they will not be able to store as much CO_2 as they have previously. Currently, about half of all the anthropogenic carbon emissions are absorbed by the ocean and the land.

A sink can also become a source if the situation changes. Carbon can enter the cycle from several different sources. One of the most common is through the respiration process of plants and animals. Once a plant or animal dies, carbon enters the system. Through the decaying process, the bacteria and fungi break down the carbon compounds. If oxygen is present, the carbon is converted into CO_2. If oxygen is not present, then methane is produced instead. Either way, both are greenhouse gases. A major source into the carbon cycle is through the combustion of biomass. This oxidizes the carbon, which produces CO_2. The most prevalent source of this is through the use of fossil fuels—coal, petroleum products, and natural gas (Figure 3.4). Fossil fuels are large

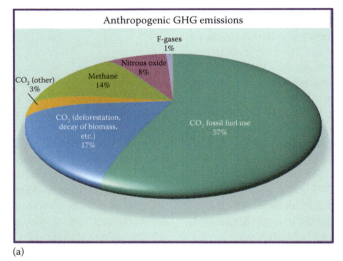

(a)

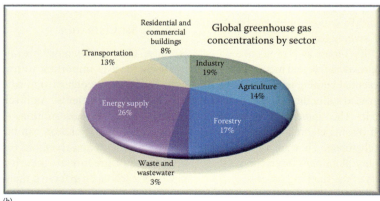

(b)

FIGURE 3.4 (a) Anthropogenic greenhouse gas emissions by type and (b) global greenhouse gas concentrations by sector. (Courtesy of IPCC, Geneva, Switzerland.)

deposits of biomass that have commonly been preserved inside geologic rock formations in the earth's crust for millions of years. Although they are preserved in rock formations, there is no release of carbon to the atmosphere; but as soon as they are processed into fuel products and used for transportation and industrial processes, millions of tons are released into the atmosphere. According to the EPA, the burning of fossil fuels and deforestation releases nearly eight billion metric tons of carbon into the atmosphere each year (FAO, 2014).

Another major source of CO_2 to the atmosphere is from forest fires. When trees burn, the stored carbon is released into the atmosphere (see Figure 3.5). Other gases, such as carbon monoxide and methane, are also released. After a fire has occurred and deadfall is left behind, it will further decompose and release additional carbon into the atmosphere. One of the predictions with climate change is that as temperatures rise, certain areas will experience increased

FIGURE 3.5 Increasing temperatures will cause mountain environments to become drier, increasing the highly flammable tinder in forest areas. Both natural lightning strikes and human carelessness will increase the risk of forest fires. The drier conditions will make the fires spread more quickly and be harder to extinguish, putting both the natural environment and bordering urban areas at extreme risk and causing millions of dollars in damage. (Courtesy of Kelly Rigby, Bureau of Land Management, Salt Lake City, Utah.)

drought conditions. When this occurs, vegetated areas will become extremely dry; then, when lightning strikes, wildfires can easily start, destroying vegetation, increasing the concentration of CO_2 gases in the atmosphere, and warming it further.

Another source of carbon occurs near urban settings. When cement is produced, one of its principal components—lime (produced from the heating of limestone)—produces CO_2 as a by-product. Therefore, in urban areas, this is another factor to consider; areas of increasing urbanization are areas of increasing CO_2 concentrations.

Volcanic eruptions can be a significant source of CO_2 during a major event. Enormous amounts of gases can be released into the atmosphere, including greenhouse gases such as water vapor, CO_2, and sulfur dioxide. It can also be released from the ocean's surface into the atmosphere from bursting bubbles. Phytoplankton, little organisms floating in the world's oceans, also use some of the CO_2 to make their food through photosynthesis.

The earth's zones of permafrost are another potential source of CO_2. According to the Oak Ridge National Laboratory in Tennessee, if climate temperatures continue to rise and cause the permafrost soils in the boreal (northern) forests to thaw, the carbon released from them could dramatically increase atmospheric CO_2 levels and ultimately affect the global carbon balance.

According to Mike Goulden, a scientists working on a NASA project to determine the net exchange of carbon in a Canadian forest, "The soils of boreal forests store an enormous amount of carbon" (McDermott, 2009).

He says the amount of carbon stored in the soil is greater than both the carbon stored in the moss layer and the wood of the boreal forests. Boreal forests cover about 10 percent of the earth's land surface and contain 15–30 percent of the carbon stored in the terrestrial biosphere. Most of the carbon is located in deep layers 40–80 centimeters below the surface, where it remains frozen—the zone called permafrost.

The reason that the permafrost layer is so high in carbon is that the decomposing plant material accumulates on the forest floor and becomes buried before it decomposes. With the steady increase in atmospheric temperatures over recent years, some of the permafrost has been experiencing deepsoil warming and been melting in mid-summer. This process releases CO_2 through increased respiration.

According to Goulden, "Assessing the carbon balance of forest soil is tricky, but if you've got several pieces of evidence all showing the same thing, then you can really start to believe that the site is losing soil carbon. Our findings on the importance of soil thaw and the general importance of the soil carbon balance is already influencing the development of models for carbon exchange." What Goulden is hoping to accomplish in the future is to find a way to use satellite data to improve carbon exchange models (McDermott, 2009).

There are several human-related sources of CO_2. By far, the largest human source is from the combustion of fossil fuels, such as coal, oil, and gas. This occurs in electricity generation, in industrial facilities, from automobiles in the transportation sector, and in heating homes and commercial buildings. According to the Energy Information Administration (EIA), in 2014, petroleum supplied the largest share of domestic energy demands in the United States. This equates to 43 percent of the total fossil fuel-based energy consumption during that year. Coal and natural gas followed, at 22 and 35 percent, respectively, all heavy producers of CO_2. Electricity generation is the single largest source of CO_2 emissions in the United States, according to the EPA. This is one reason why the EPA began their Energy Star® program in 1992, geared toward the production and use of energy efficient appliances.

Industrial processes that add significant amounts of CO_2 to the atmosphere include manufacturing, construction, and mining activities. The EPA has identified six industries in particular that use the majority of energy sources, thereby contributing the largest share of CO_2: (1) chemical production, (2) petroleum refining, (3) primary metal production, (4) mineral production, (5) paper production, and (6) the food processing industry.

In the residential and commercial sectors, the main source of direct CO_2 emissions is due to the burning of natural gas and oil for the heating and cooling of buildings. The transportation sector is the second largest source of CO_2 emissions, primarily due to the use of petroleum products such as gasoline, diesel, and jet fuel. Of this, 65 percent of the CO_2 emissions are from personal automobiles and light-duty trucks.

Specific commodities in the industrial sector that produce CO_2 emissions include cement and other raw materials that contain calcium carbonate that is chemically transformed during the industrial process, which produces CO_2 as a by-product. Industrial processes that use petroleum products also contribute CO_2 emissions to the atmosphere. This includes the production of solvents, plastics, and lubricants that have a tendency to evaporate or dissolve over a period of time. The EPA has identified four principal industrial processes that produce significant CO_2 emissions. They are as follows:

- Production of metals, such as aluminum, zinc, lead, iron, and steel
- Production of chemicals, such as petrochemicals, ammonia, and titanium dioxide
- Consumption of petroleum products in feedstocks and other end uses
- Production and consumption of mineral products, such as cement, lime, and soda ash

Also important are industrial processes that produce other harmful greenhouse gases as by-products, such as methane, nitrous oxide, and fluorinated gases. These gases have a far greater GWP than CO_2. The GWP of methane is 23, that of nitrous oxide is 296, and that of fluorinated gases ranges from 6500 to 8700, posing an even more critical negative effect.

TYPES OF CARBON SEQUESTRATION

Carbon sequestration is the process through which CO_2 from the atmosphere is absorbed by various carbon sinks. Principal carbon sinks include agricultural sinks, forests, geologic formations, and oceanic sinks. Carbon sequestration, or storage, occurs when CO_2 is absorbed by trees, plants, and crops through photosynthesis and stored as carbon in biomass, such as tree trunks, branches, foliage, and roots, as well as in the soil. In terms of climate change and impacts to the environment, sequestration is very important because it has a large influence on the atmospheric levels of CO_2. According to the IPCC, carbon sequestration by forestry and agriculture alone significantly helps offset CO_2 emissions that contribute to climate change. The amount of carbon that can be sequestered varies geographically and is determined by tree species, soil type, regional climate, type of topography, and even the type of land management practice used in the area. For example, in agricultural areas, if conservation tillage practices are used instead of conventional tillage, this limits the introduction of CO_2 into the atmosphere by sequestering larger amounts of CO_2 in the soil. According to the EPA, switching from conventional to conservation tillage can sequester 0.1–0.3 metric tons of carbon per acre per year.

Carbon sequestration does reach a limit, however. The amount of carbon that accumulates in forests and soils will eventually reach a saturation point where no additional carbon will be able to be stored. This typically occurs when trees reach full maturity or when the organic matter contained in soils builds up. According to the EPA, the U.S. landscape currently functions as an efficient carbon sink, sequestering more than it emits. They do warn, however, that the overall sequestration amounts are currently declining because of increased harvests, land use changes, and maturing forests. As far as global sequestration, the IPCC has estimated that 100 billion metric tons of carbon over the next 50 years could be sequestered through forest preservation, tree planting, and improved conservation-oriented agricultural management. In the United States, Bruce McCarl (professor of Agricultural Economics at Texas A&M University) and Uwe Schneider (assistant professor of the Research Unit Sustainability and Global Change Department of Geosciences and Economics at Hamburg University in Germany) have determined that an additional 50–150 million metric tons of carbon could be sequestered through changes in both soil and forest management, new tree planting, and biofuel substitution (Schneider and McCarl, 2003).

Another positive aspect supporting carbon sequestration is that it also affects other greenhouse gases. In particular, methane (CH_4) and nitrous oxide (N_2O) can also be sequestered in agricultural activities such as grazing and growing of crops. Nitrous oxide can be introduced via fertilizers. Instead of using these fertilizers, which can have a negative environmental effect, other practices could be used instead, such as rotational grazing. In addition, if forage quality is improved, livestock methane emissions should be significantly reduced. Nitrous oxide emissions could be avoided by eliminating the need for fertilizer. The EPA stresses that finding the right sequestration practices will help lessen the negative effects of all the greenhouse gases.

Other environmental benefits of carbon sequestration include the enhancement of the quality of soil, water, air, and wildlife habitat. For instance, when trees are planted, they not only sequester carbon but also provide wildlife habitat. When the rain forests are preserved, it keeps both plant and animal species from becoming endangered and helps control soil erosion. When forests are maintained, it also cuts down on overland water flow, soil erosion, loss of nutrients, and improved water quality.

The continuation of climate change, however, can have an impact on carbon sequestration. According to the EPA (2011), "In terms of global warming impact, one unit of CO_2 released from a car's tailpipe has the same effect as one unit of CO_2 released from a burning forest. Likewise, CO_2 removed from the atmosphere through tree planting can have the same benefit as avoiding an equivalent amount of CO_2 released from a power plant."

The experts at the EPA also caution, however, that even though forests, agriculture, and other sinks can store carbon, the process can also become saturated and slow down or stop the storage process (such as traditional agricultural cultivation), or the sink can be destroyed and completely stop the process (such as complete deforestation). Carbon sequestration processes can naturally slow down and stop on their own as they age. In addition, carbon sequestration can be a natural or man-made process. Research is currently in progress to perfect the methodologies that enhance the natural terrestrial cycles of carbon storage that remove CO_2 from the atmosphere by vegetation and store CO_2 in biomass and soils. In order to accomplish this, research of biological and ecological processes is underway by the EPA. Specific technical areas that are currently being researched include the following:

- Increasing the net fixation of atmospheric CO_2 by terrestrial vegetation with an emphasis on physiology and rates of photosynthesis of vascular plants.
- Retaining carbon and enhancing the transformation of carbon to soil organic matter.
- Reducing the emission of CO_2 from soils caused by heterotrophic oxidation of soil organic carbon.
- Increasing the capacity of deserts and degraded lands to sequester carbon. Man-made processes include technologies such as geologic, mineral, and ocean sequestration.

In carbon sequestration, the main goal is to prevent CO_2 emissions from power plants and industrial facilities from entering the atmosphere by separating and capturing the emissions and then securing and storing the CO_2 on a long-term basis. Currently, EPA is involved in research in an attempt to separate and capture the CO_2 from fossil fuels and from flue gases—these are both pre- and postcombustion processes. Underground storage facilities have been also receiving large amounts of attention recently, and their potential is enormous. In 2012, the Department of Energy published a *Carbon Utilization and Storage Atlas*, outlining up to 5700 years of CO_2 storage potential in the United States and portions of Canada. Currently, more than 2.8 trillion liters of both hazardous and nonhazardous fluids are disposed of through a process called underground injection.

IMPACTS OF DEFORESTATION

Deforestation is occurring today at alarming rates, determined through the analysis of satellite imagery. It accounts for about 20 percent of the heat-trapping gas emissions worldwide. The Food and Agriculture Organization (FAO) estimates current tropical deforestation at 15,400,000 hectares per year. This equates to an area roughly the size of North Carolina being deforested annually (Figure 3.6). Tropical forests hold enormous amounts of carbon. In fact, the plants and soil of tropical forests hold 460–575 billion metric tons of carbon worldwide. This equates to each acre of tropical forest storing roughly 180 metric tons of carbon.

When a forest is cut and replaced by pastures or cropland, the carbon that was stored in the tree trunks joins with the oxygen and is released into the atmosphere as CO_2. Because the wood in a tree is about 50 percent carbon, deforestation has a significant effect on climate change and the global carbon cycle. In fact, according to the Tropical Rainforest Information Center, from the mid-1800s to 1990, worldwide deforestation released 122 billion metric tons of carbon into the atmosphere. Currently 1.5 billion metric tons is released into the atmosphere each year (Climate and Weather, 2012). In fact, deforestation in tropical rainforests adds more CO_2 to the atmosphere than the total amount of cars and trucks on the world's roads. According to the World Carfree Network, cars and trucks account for

FIGURE 3.6 Deforestation, shown here in Madagascar, is a serious problem and has become so widespread that its connection to global warming cannot be overlooked. These huge resultant erosional features are called lavaka. The world's rain forests must be managed properly to avoid this kind of environmental damage. (Courtesy of Rhett Butler, Mongabay.com.)

about 14 percent of global carbon emissions, whereas most analysts attribute upward of 15 percent to deforestation. The Environmental Defense Fund states that 32 million acres of tropical rainforest were cut down each year between 2000 and 2009—and the pace of deforestation continues to increase. They warn that unless we change the present system that rewards forest destruction, forest clearing will put another 200 billion tons of carbon into the atmosphere in the coming decades (Scientific American, 2012). Tropical deforestation is the largest source of emissions for many developing countries (Tropical Rainforest Information Centre, 2011).

According to Peter Frumhoff of the Union of Concerned Scientists (UCS), he and an international team of 11 research scientists found that if deforestation rates were cut in half by 2050, it would amount to 12 percent of the emissions reductions needed to keep the concentrations of heat-trapping gases in the atmosphere at relatively safe levels (Global Climate Notes, 2015).

There are multiple impacts of deforestation. Many of the most severe impacts will be to the tropical rain forests. Even though they cover only about 7 percent of the earth's land surface, they provide habitat for about 50 percent of all the known species on earth. Some of these endemic species have become so specialized in their respective habitat niches that if the climate changes, which then causes the ecosystem to also change, this will threaten the health and existence of multiple species, even to the point of forcing them to become extinct. In addition to the species that are destroyed, the ones that remain behind in the isolated forest fragments become vulnerable and sometimes are threatened with extinction themselves. The outer margins of the remaining forests become dried out and are also subjected to die off.

Two other major impacts in addition to the loss of biodiversity are the loss of natural resources, such as timber, fruit, nuts, medicine, oils, resins, latex, and spices, and the economic impact that causes along with the extreme reduction of genetic diversity. The loss of genetic diversity could mean a huge loss for the future health of humans. Hidden in the genes of plants, animals, fungi, and bacteria that may not have even been discovered yet could be the cures for diseases such as cancer, diabetes, muscular dystrophy, and Alzheimer's disease. The FAO of the United Nations says, "The keys to improving the yield and nutritional quality of foods may also be found inside the rain forest and it will be crucial for feeding the nearly ten billion people the earth will likely need to support in the coming decades" (Lindsey, 2009).

Two of the largest climate impacts will revolve around rainfall and temperature. One-third of the rain that falls in the tropical rain forest is rain that was generated in the water cycle by the rain forest itself. Water is recycled locally as it is evaporated from the soil and vegetation, condenses into clouds, and falls back to earth again as rain in a repetitive cycle. The evaporation of the water from the earth's surface also acts to cool the earth. As climatologists continue to learn about the earth's climate and the effects of climate change, they are able to build better models. When the tropical rain forests are replaced by agriculture or grazing, many climate models predict that these types of land-use changes will perpetuate a hotter, drier climate. Models also predict that tropical deforestation will disrupt the rainfall patterns outside the tropics, causing a decrease in precipitation—even to far-reaching destinations such as China, northern Mexico, and the South Central United States.

Predictions involving deforestation can get complicated in models, however. For instance, if deforestation is done in a "patchwork" pattern, then local isolated areas may actually experience an increase in rainfall by creating "heat islands," which increase the rising and convection of air that causes clouds and rainstorms. If rainstorms are concentrated over cleared areas, the ground can be vulnerable and susceptible to erosion.

The carbon cycle plays an important role in the rain forests. According to NASA, in the Amazon alone, the trees contain more carbon than 10 years worth of human-produced greenhouse gases. When the forests are cleared and burned, the carbon is returned to the atmosphere, enhancing the greenhouse effect. If the land is utilized for grazing, it can also be a continual source of additional carbon. It is not certain today whether or not the tropical rain forests are a net source or sink of carbon. Although the vegetation canopies hold enormous amounts of carbon, trees, plants, and microorganisms in the soil also respire and release CO_2 as they break down carbohydrates for energy. In the Amazon alone, enormous amounts of CO_2 escape from decaying organic matter in rivers and streams that flood huge areas. When tropical forests remain undisturbed, they remain essentially carbon neutral, but when deforestation occurs, it contributes significantly to the atmosphere.

Rainforest countries need to give serious consideration about what decisions need to be made for the future. Currently, Papua New Guinea, Costa Rica, and several other forest-rich developing countries are seeking financing from the global carbon market (such as the United States) to create attractive financial incentives for tropical rain forest conservation. Developed nations helping developing nations is one approach to help get a handle on the problem, but developed countries also need to cut back on their own emissions.

ANTHROPOGENIC CAUSES AND EFFECTS—CARBON FOOTPRINTS

When looking for the causes of climate change, it is easy to point fingers and put the blame on industry, other nations, transportation, deforestation, and other sources and activities. But the truth of the matter is that every person on the earth plays a part and contributes to climate change. Even simple daily activities—such as using an electric appliance, heating or cooling a home, or taking a quick drive to the grocery store—contribute CO_2 to the atmosphere. Scientists refer to this input as a "carbon footprint." A carbon footprint is simply a measure of how much CO_2 people produce just by going about their daily lives. For every activity that involves the combustion of fossil fuels (coal, oil, and gas), such as the generation of electricity, the manufacture of products, or any type of transportation, the user of the intermediate or end product is leaving a carbon footprint. Of all the CO_2 found in the atmosphere, 98 percent originates from the burning of fossil fuels. Simply put, it is one measure of the impact people make individually on the earth by the lifestyle choices they make. In order to combat climate change, every person on the earth can play an active role by consciously reducing the impact of his/her personal carbon footprint. The two most common ways of achieving this is by increasing his/her home's energy efficiency and driving less. A carbon footprint is calculated (carbon footprint calculators are available on the Internet), and a monthly, or annual, output of total CO_2 in tons is derived based on the specific daily activities of that person. The goal then is to reduce or eliminate his/her carbon footprint. Some people attempt to achieve "carbon neutrality," which means they cut their emissions as much as possible and offset the rest. Carbon offsets allow one to "pay" to reduce the global greenhouse total instead of making personal reductions. An offset is bought, for example, by funding projects that reduce emissions through restoring forests, updating power plants and factories, reducing the energy consumption in buildings, or investing in more energy-efficient transportation. In order to educate and make people more environmentally conscious, some companies are even advertising today what their carbon footprint is, drawing in additional business support because of their positive environmental commitments. Some commercial products today contain "carbon labels" estimating the carbon emissions that were involved in the creation of the product's production, packaging, transportation, and future disposal.

Carbon footprints are helpful because they allow individuals to become more environmentally aware of the implication of their own choices and actions and enable them to adopt behaviors that are more environmentally friendly—what is referred to as "going green." For example, transportation in the United States accounts for 27 percent of CO_2 emissions. Ways to make a difference include driving less, using public transportation, carpooling, driving a fuel efficient car such as a hybrid, or bicycling. According to the EPA (2013), home energy use accounts for 12 percent of CO_2 emissions in the United States. Therefore, cutting down in these areas by increasing energy efficiency helps lessen the carbon footprint. There are several practical ways to do this, such as lowering the thermostat, installing double-paned windows, and installing good insulation, to name a few. Using compact fluorescent lamps and using energy efficient appliances (such as those listed on the Energy Star program) also increases efficiency and lowers the carbon footprint. Carbon footprints can be a helpful measurement for those who want to take personal initiative and do their part to fight climate change.

Timing is right for individuals to become aware of their personal behavior and how it impacts the environment. The atmospheric concentration of CO_2 has risen by more than 30 percent in the past 250 years. Based on a study conducted by Michael Raupach of the Commonwealth Scientific and Industrial Research Organisation in Australia, worldwide carbon emissions of anthropogenic CO_2 are rising faster than previously predicted. From 1990 to 1999, the increase in CO_2 levels averaged about 1.1 percent per year, but from 2000 to 2004, levels increased to 3 percent per year. For this research, the world was divided into nine separate regions for analysis of specific factors such as economic factors, population trends, and energy consumption. The result of the study showed that the developed countries, which only account for 20 percent of the world's population, accounted for 59 percent of the anthropogenic global emissions in 2004. The developing nations were responsible for 41 percent of the total emissions in 2004 but contributed 73 percent of the emissions growth that year. The developing countries, such as India, are expected to become the major CO_2 contributors in the future. Today, the largest CO_2 emitter is China.

This study is significant because even the IPCC's most extreme predictions underestimate the rapid increase in CO_2 levels seen since 2000. The scientists involved in the study believe this shows that no countries are decarbonizing their energy supply and that CO_2 emissions are accelerating worldwide.

Also from Australia's Commonwealth Scientific and Industrial Research Organisation, Josep G. Canadell calculated that CO_2 emissions were 35 percent higher in 2006 than in 1990, a faster growth rate than expected. Canadell attributed this to an increased industrial use of fossil fuels and a decline in the amount of CO_2 being absorbed by the oceans or sequestered on land.

According to Canadell, "In addition to the growth of global population and wealth, we now know that significant contributions to the growth of atmospheric CO_2 arise from the slowdown of nature's ability to take the chemical out of the air. The changes characterize a carbon cycle that is generating stronger-than-expected and sooner-than-expected climate forcing" (Schmid, 2007c).

In response to the study, Kevin Trenberth from the National Center for Atmospheric Research said, "The paper raises some very important issues that the public should be aware of: namely that concentrations of CO_2 are increasing at much higher rates than previously expected and this is in spite of the Kyoto Protocol that is designed to hold them down in western countries" (MSNBC, 2007c).

According to the Brookings Institution, a research organization in Washington, DC, the carbon footprint of the United States is expanding. As the population of the United States grows and cities expand, they are building more, driving more, and consuming more energy, which means they are emitting more CO_2 than ever before. The Brookings Institution believes that existing federal policies are currently limiting. They believe federal policy should play a more powerful role in helping metropolitan areas so that the country as a whole can collectively shrink their carbon footprint. They believe that besides economy-wide policies to motivate action, five targeted policies should be put in place that are extremely important within metro areas (Brown et al., 2008):

- Promote a wider variety of transportation choices.
- Design more energy-efficient freight operations.
- Require home energy cost disclosure when selling a home in order to encourage more energy efficient appliances in homes.

- Use federal housing policies to create incentives to build with both energy and location conservation in mind.
- Challenge/reward metropolitan areas to develop innovative solutions toward reducing carbon footprints.

AREAS MOST AT RISK

About 634 million people—10 percent of the global population—live in coastal areas that are within just 10 meters above sea level, where they are most vulnerable to sea-level rise and severe storms associated with climate change. Three-quarters of these are in the low-elevation coastal zones in Asian nations on densely populated river deltas, such as India. The other areas most at risk are the small island nations. According to the Earth Institute at Columbia University, the 10 countries with the largest number of people living within 10 meters of the average sea level in descending order are as follows:

China
India
Bangladesh
Vietnam
Indonesia
Japan
Egypt
The United States
Thailand
The Philippines

The 10 countries with the largest share of their population living within 33 feet (10 m) of the average sea level are as follows:

The Bahamas (88%)
Suriname (76%)
The Netherlands (74%)
Vietnam (55%)
Guyana (55%)
Bangladesh (46%)
Djibouti (41%)
Belize (40%)
Egypt (38%)
Gambia (38%)

FOSSIL FUELS AND CLIMATE CHANGE

The burning of fossil fuels (oil, gas, and coal) is one of the leading contributors to climate change, as they are comprised almost entirely of carbon and release CO_2 into the atmosphere when they are burned. In the case of oil, they also contain toxic materials that when burned, or when the fumes are inhaled, are known to cause cancer in humans.

In developed countries, such as the United States, fossil fuels are the principal sources of energy that are used to produce the majority of the fuel, electricity, heat, and air conditioning. In fact, more than 86 percent of the energy used worldwide originates from fossil fuel combustion. Although

for years fossil fuels have been readily available and convenient, they have also played a major role in climate change over the past few decades. According to the Center for Biological Diversity, fossil fuel use in the United States causes more than 80 percent of the greenhouse gas emissions and 98 percent of just the CO_2 emissions. This adds approximately 4.1 billion metric tons of CO_2 to the atmosphere each year and would even be greater if some of the earth's natural carbon sequestration processes such as carbon storage in the world's oceans, vegetation, and soils did not occur (CBD, 2008).

Even though scientists warn that climate change is already under way and will likely continue for the next several centuries due to the long natural processes involved, such as the long lifetimes of many of the greenhouse gases, there are ways the potential effects can be reduced. Because everyone uses energy sources every day, the biggest way to reduce the negative effects of climate change is by using less energy. By cutting back on the amount of electricity used, the use of fossil fuels can be reduced because most power plants burn fossil fuels to generate power. The most effective way to half the emission of greenhouse gases into the atmosphere is through the adoption of nonfossil fuel energy sources, such as hydroelectric power, solar power, hydrogen engines, and fuel cells.

Former Vice President Al Gore, author of *An Inconvenient Truth*, and recipient of the 2007 Nobel Peace Prize (received jointly with the IPCC) said, "It is the most dangerous challenge we've ever faced, but it is also the greatest opportunity we have had to make changes" (MSNBC, 2007b).

Because the IPCC's last two reports were released identifying fuels as a principal cause of climate change with a virtual certainty (99% certainty) (IPCC, 2007, 2013), upgraded from its 2001 report of likely (66%), several achievements and advancements have been made. The levels of research have grown, public awareness has increased, the subject has been incorporated into many school curriculums worldwide, and legislation—local, national, and international—has been passed and is being currently introduced into governments around the world. In addition to the 2007 Nobel Peace Prize shared by Al Gore and the IPCC, the film *An Inconvenient Truth* received two Oscar nominations at the 2007 Academy Awards in Los Angeles, California. Climate change issues are finally receiving the media attention they deserve, making the public more aware of the real issues and the reasons why they need to be addressed now.

Growing public education and awareness has not solved all the problems, however. Even though the public is becoming more educated, the opposition and efforts of skeptics are also raising their voices in protest, continuing to cloud the issues, making it important for the public to pay attention to the facts. In light of this, however, many cities worldwide, foreign countries, and individual states in the United States are taking action to curb fossil fuel emissions. Arnold Schwarzenegger, former governor of California, for example, ordered the world's first low-carbon limits on passenger car fuels. The new standard reduces the carbon content of transportation fuels at least 10 percent by the year 2020.

Climatologists at Lawrence Livermore National Laboratory in California have created a climate and carbon cycle model to examine global climate and carbon cycle changes. What they concluded was that if humans continued with the same lifestyles and habits they are accustomed to today (commonly referred to as a business-as-usual approach), the earth's atmosphere would warm by 8°C if humans use all of the earth's available fossil fuels by the year 2300.

Their model predicted several alarming results: In the next few centuries the polar ice caps will have vanished, ocean levels will rise by 7 meters, and the polar temperatures will climb higher than the predicted 8°C–20°C, transforming the delicate ecosystems from polar and tundra to boreal forest.

Govindasamy Bala, the Laboratory's Energy and Environment Directorate and lead author of the project, said, "The temperature estimate was actually conservative because the model didn't take into consideration changing land use such as deforestation and build-out of cities into outlying wilderness areas" (Williams, 2015).

Their model projected that by 2300, the CO_2 level will have risen to 1423 ppm—roughly a 400 percent increase. The model identified the soil and biomass as significant carbon sinks. However, according to Bala, the land ecosystem would not take up as much carbon dioxide as the model assumes, because it did not take land-use change into account.

The results of the model showed that ocean uptake of CO_2 starts to decrease in the twenty-second and twenty-third centuries as the ocean surface warms. It takes longer for the ocean to absorb CO_2 than it does for the vegetation and soil. By 2300, the land will absorb 38 percent of the CO_2 released from the burning of fossil fuels, and 17 percent will be absorbed by the oceans. The remaining 45 percent will stay in the atmosphere. Over time, roughly 80 percent of all CO_2 will end up in the oceans via physical processes, increasing its acidity and harming aquatic life.

According to Bala, the most drastic changes during the 300-year period will occur during the twenty-second century—when precipitation patterns change, when an increase in the amount of atmospheric precipitable water and a decrease in the size of sea ice are the largest, and when emission rates are the highest. Based on the model's results, all sea ice in the Northern Hemisphere summer will have vanished by 2150.

When referring to climate change skeptics, Bala says, "Even if people don't believe in it today, the evidence will be there in 20 years. These are long-term problems. We definitely know we are going to warm over the next 300 years. In reality, we may be worse off than we predict" (Williams, 2015).

The New Carbon Balance—Summing It All Up

Carbon dioxide enters the air during the carbon cycle. Because it is so plentiful, it comes from several sources. Vast amounts of carbon are stored naturally in the earth's soils, oceans, and sediments at the bottom of oceans. Carbon is stored in the earth's rocks, which is released when the rocks are eroded. It exists in all living matter. Every time animals and plants breathe, they exhale carbon dioxide. The earth maintains a natural carbon balance. Throughout geologic time, when concentrations of CO_2 have been disturbed, the system had always gradually returned to its natural (balanced) state. This natural readjustment works very slowly.

Through a process called diffusion, various gasses that contain carbon move between the ocean's surface and the atmosphere. For this reason, plants in the ocean use CO_2 from the water for photosynthesis, which means that ocean plants store carbon, just as land plants do. When ocean animals eat these plants, they then store the carbon. Then when they die and decompose, they sink to the bottom and decompose, and their remains become incorporated in the sediments on the bottom of the ocean. Once in the ocean, the carbon can go through various processes: it can form rocks, then weather, and it can also be used in the formation of shells. Carbon can move to and from different depths of the ocean, and also exchange with the atmosphere.

As carbon moves through the system, different components can move at different speeds. Scientists break these reaction times down into two categories: (1) short-term cycles and (2) long-term cycles. In short-term cycles, carbon is exchanged quickly. Examples of this include the gas exchange between the oceans and the atmosphere (evaporation). Long-term cycles can take years to millions of years to occur. Examples of this would be carbon stored for years in trees or carbon weathered from a rock being carried to an ocean, being buried, being incorporated into plate tectonic systems, then later being released into the atmosphere through a volcanic eruption.

Throughout geologic time, the earth has been able to maintain a balanced carbon cycle. Unfortunately, this natural balance has been upset by recent human activity. Over the past 150–200 years, fossil fuel emissions, land-use change, and other human activities have increased atmospheric carbon dioxide by 30 percent (and methane, another greenhouse gas, by 150 percent) to concentrations not seen in the past 420,000 years.

Humans are adding CO_2 to the atmosphere much faster than the earth's natural system can remove it. Prior to the Industrial Revolution, atmospheric carbon levels remained constant at around 280 ppm. This meant that the natural carbon sinks were balanced between what was being emitted and what was being stored. After the Industrial Revolution began and carbon dioxide levels began to increase—315 ppm in 1958, to 383 ppm in 2007, to 414 ppm in 2016—the "balancing act" became unbalanced and the natural sinks could no longer store as much carbon as was being introduced into the atmosphere by human activities, such as driving and industry. In addition, according to Dr. Joseph Canadell of the National Academy of Sciences, 50 years ago for every ton of CO_2 emitted, 600 kilograms was removed by natural sinks (Center for Climate Change and Environmental Studies, 2011). In 2006, only 550 kilograms was removed per ton, and the amount continues to fall today, which indicates the natural sinks are losing their carbon storage efficiency. This means that, although the world's oceans and land plants are absorbing great amounts of carbon, they cannot keep up with what humans are adding. The natural processes work much slower than the human ones do. The earth's natural cycling usually takes millions of years to move large amounts from one system to another.

TABLE 3.2

U.S. Energy-Related Carbon Dioxide Emissions by Source and Sector, 2015 (million metric tons)

Sector	Residential	Commercial	Industrial	Transportation	Electric Power	Source Total
Coal	0	6	129	0	1364	1499
Natural gas	252	175	474	49	530	1480
Petroleum	67	40	338	1816	24	2285
Other					7	7
Electricity	723	702	495	4		
Sector total	1043	922	1437	1869	1925	5271

Source: Energy Information Administration, 2015.

TABLE 3.3

Sources of Energy-Related Carbon Dioxide Emissions by Type of Fuel for the United States and the World, 2012 (million metric tons)

	The United States		The World	
	Amount	Share of Total (%)	Amount	Share of Total (%)
Total from fossil fuels	5,270		32,723	
Coal	1,656	31	14,229	43
Natural gas	1,374	26	6,799	21
Petroleum	2,240	43	11,695	36

Source: Energy Information Administration, 2015.

The problem with human interference is that the impacts are happening in only centuries or decades—and the earth cannot keep up with the fast pace. The result is that each year the measured CO_2 concentration of the atmosphere gets higher, making the earth's atmosphere get warmer.

Levels of several greenhouse gases have increased by about 25 percent because large-scale industrialization began about 150 years ago. According to the EIA's National Energy Information Center, 75 percent of the anthropogenic CO_2 emissions added to the atmosphere over the past 20 years is due to the burning of fossil fuels (Casper, 2009 and Gines, 2012).

According to the EIA, natural earth processes can absorb approximately 3.2 billion metric tons of anthropogenic CO_2 emissions annually. An estimated 6.1 billion metric tons is added each year, however, creating a large imbalance, which is why there is a steady, continual growth of greenhouse gases in the atmosphere. In computer models, an increase in greenhouse gases results in an increase in average temperature on earth. The warming that has occurred over the past century is largely attributed to human activity. According to a study conducted by the National Research Council in May 2001, "Greenhouse gases are accumulating in earth's atmosphere as a result of human activities, causing surface air temperatures and sub-surface ocean temperatures to rise. Temperatures are, in fact, rising. The changes observed over the last several decades are likely mostly due to human activities" (Energy Information Administration, 2004; Scientific American, 2009).

Table 3.2 shows the U.S. energy-related carbon dioxide emissions by source and sector, and Table 3.3 shows the

sources of energy-related carbon dioxide emissions by type of fuel for the United States and the world. The EIA has determined that greenhouse gas emissions originate principally from energy use, driven mainly by economic growth, fuel used for electricity generation, and weather patterns affecting heating and cooling needs. Table 3.4 illustrates the world's top CO_2 emitters by country in 2015, and Figure 3.7 illustrates the annual and cumulative emissions by country (1751–2014).

Developing new technologies that create energy by using fossil fuels more efficiently alone is not enough to control the emissions of greenhouse gases being ejected into the

TABLE 3.4

The World's Top Carbon Dioxide Emitters, 2015

Country	Total Emission (%)
China (mainland)	28.03
The United States	15.9
India	5.81
Russian Federation	4.79
Japan	3.84
Germany	2.23
Korea	1.78
Canada	1.67
Iran	1.63
Brazil	1.41
Indonesia	1.32

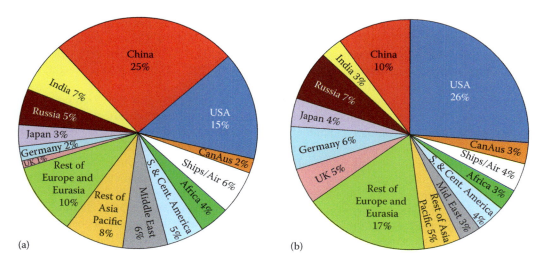

FIGURE 3.7 Annual and cumulative emissions by country, 1751–2014. Current and cumulative fossil fuel carbon dioxide emissions: (a) 2014 annual emissions (9.6 GtC/yr) and (b) 1751–2014 cumulative emissions (396 GtC). (Courtesy of Carbon Dioxide Information Analysis Center, Oak Ridge National Laboratory, Oak Ridge, Tennessee.)

atmosphere. Without the increased use of renewable energy resources, and the weaning off of the dependence on fossil fuels, it will not be possible to bring climate change under control in time to prevent irreversible damage to the environment and life on the planet.

The Concern about Coal Emissions

According to an article in *The New York Times* back on December 4, 2007, titled "Stuck on Coal, Stuck for Words in a High-Tech World," although the society of the twenty-first century has progressed technologically in many ways over the past century, it is still stuck in the old-fashioned, outdated mode of popular energy choices (Revkin, 2007). In particular, even with new technology such as the growing cadre of renewable forms of energy, societies worldwide are still heavily dependent on coal and plan to remain that way despite the repeated pleas and warnings from climatologists about the future, life-threatening consequences of climate change if changes are not made now. It still remains an energy source in demand in 2016.

Using coal-fired plants to generate electricity produces more greenhouse gases for each resulting watt than using oil or natural gas, but coal—often referred to as a "dirty fuel"—is attractive because it is relatively inexpensive. In countries where there are no emission controls (such as China and India), the coal industry is booming today. The International Energy Agency projects that the demand for coal will grow by 2.2 percent a year until 2030, which is a rate faster than the demand for oil or natural gas. Yet, even with warnings to cut back on coal use and greenhouse gas emissions, according to the UCS, the United States currently has plans to increase its emissions by building many more coal-fired plants, with the majority of them lacking the carbon capture and storage technology—equipment that allows a plant to capture a certain amount of CO_2 before it is released and then store it underground (UCS, 2008).

One of the most serious contenders of coal use now and in the future is China. New figures from the Chinese government reveal that coal use has been climbing faster in China than anywhere else in the world. According to a report in *The Economist*, China opens a new coal-fired plant each week (*The Economist*, 2007). Their rising energy consumption is making it more difficult to effectively slow the climate change process. The International Energy Agency in Paris predicts that the increase in greenhouse gas emissions from 2000 to 2030 in just China will be comparable to the increase from the entire industrialized world (Bradsher, 2003).

China is currently the world's largest consumer of coal, and its power plants are burning it faster than the trains can deliver it from the mines in China. As a result, it is also importing coal from Australia to meet its rising demands. Concerning energy products, it has also become the world's fastest growing importer of oil. The Chinese are using more energy in their homes than ever before, and with a population of one and a quarter billion people, energy usage is expected to skyrocket. And this increase in energy usage will affect other energy-related sectors. China, for example, is now the world's largest market for television sets and one of the largest for other electrical appliances as well. It also has the world's fastest growing automobile market. All of these commodities require the use of fossil fuels to manufacture and operate their industries, either directly or indirectly. Energy generated from coal and oil is expected to have a significant impact on climate change.

China is not the only country with a growing demand for energy, however. India, Brazil, and Indonesia are other countries and regions that are also showing a surging demand for energy. Power plants are burning increasing amounts of coal to meet the exploding new demands for electricity to serve both industry and private households. According to the New China News Agency, they have predicted that China's capacity to generate electricity from coal will be almost 3 times as high in 2020 as it was in 2000. China currently uses more coal than the United States, the European Union, and Japan combined" (Bradsher et al., 2006).

In a recent report from the Global Warming Policy Foundation in 2015, China will present a favorable presence at the UN Climate Conference in Paris but will not make any binding commitments, which is summarized in their new report *The Truth About China*. According to economist Patricia Adams, the study's author, "China's Communist Party has as its highest priority its own self-preservation, and that self-preservation depends overwhelmingly on its ability to continue raising the standard of living of its citizens" (Global Warming Policy Foundation, 2015). The study also revealed that more than 2400 coal-fired power stations are under construction or being planned around the world. The new plants will emit 6.5 billion tons of carbon dioxide a year and undermine the efforts at the Paris climate conference to limit global warming to 2°C. China is building 368 plants and planning a further 803. India is building 297 plants and planning to build 149. Japan has planned to build 40 plants and 5 currently under construction (*The Times*, 2015) (Figure 3.8).

If China's carbon usage keeps up with its current economic growth, its CO_2 emissions are projected to reach 8 GT a year by 2030, an amount equal to the entire world's CO_2 production today. In 2000, steel production in China was reported at 127 million metric tons, and in 2006, they produced 380 million metric tons. According to the International Iron and Steel Institute, in 2008, China's production now heads the world at 444 million metric tons, which is twice as much as the production of the United States and Japan combined (Tang, 2010). In 2014, they produced 822 million metric tons. In addition to new construction, the steel is also being used for the manufacture of cars. In 1999, Chinese consumers bought 1.2 million cars. In 2006, 7.2 million cars were sold—an increase of 600 percent. In 2013, 20 million were bought, and in 2015, it was nearly 25 million.

In a report in *The New York Times*, pollution from China's coal-fired plants is already affecting the world. In April 2006, a thick cloud of pollutants originating from Northern China drifted airborne to Seoul, Korea, then across the Pacific Ocean to the West Coast of the United States. Scientists were able to track the progression and route of this "brown cloud" via real-time satellite imagery. According to researchers in California, Oregon, and Washington, DC, a coating of sulfur compounds, carbon, and other by-products of coal combustion were found on mountaintop detectors in the Pacific Northwest.

Steven S. Cliff, an atmospheric scientist at the University of California at Davis, said, "The filters near Lake Tahoe in the mountains of eastern California are the darkest we've ever seen outside smoggy urban areas" (Bradsher and Barboza, 2006).

The sulfur dioxide produced during coal combustion poses an immediate threat to China's population, contributing to roughly 400,000 premature deaths each year. In addition, it causes acid rain that poisons rivers, lakes, wetland ecosystems, agricultural areas, and forest ecosystems. The CO_2 coming from China will exist in the atmosphere for decades. According to the report, concerning China's newly growing economy, "Coal is China's double-edged sword—the new economy's black gold and the fragile environment's dark cloud" (Bradsher et al., 2006).

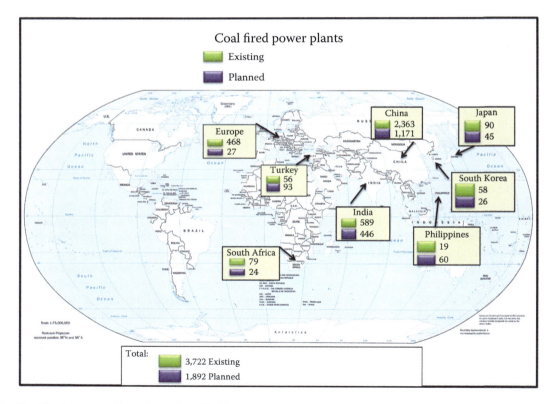

FIGURE 3.8 Coal-fired power stations planned worldwide.

HEALTH ISSUES ASSOCIATED WITH CLIMATE CHANGE

One component of preparing for climate change involves planning for new or increasing threats to human health. As ecosystems change through intense weather events, shifting habitats, wildfires, heat waves, and other effects, humans must be prepared for inevitable changes. There is an overwhelming amount of evidence that rising levels of CO_2 in the earth's atmosphere are having a serious impact on the climate with secondary effects on the earth's physical systems and ecosystems, such as an increase in severe weather events, rising sea levels, migration and extinction of both plant and animal species, shifts in climate patterns, and melting of glaciers and permafrost. Another serious impact that has been clearly identified related to climate change and rising amounts of CO_2 in the atmosphere are the negative health effects being experienced by populations worldwide.

The segments of the population that are at the greatest risk of health problems from the atmospheric changes associated with climate change, such as increased air particulates, greenhouse gases, and pollution, are young children and the elderly of 65 years of age or older. Because young children are still developing physically and breathe faster than adults, they are more at risk to the adverse effects of air quality and extreme temperatures. According to the Environmental Defense Fund, there are four key health-related factors associated with climate change: heat waves, smog and soot pollution, food- and water-borne diseases, and stress from post-traumatic stress disorder (PTSD).

Infants and children of 4 years old and younger are extremely sensitive to heat. When heat waves occur and when children are subjected to the urban heat island effect—the situation in which urban areas are warmer than rural areas because asphalt pavements, buildings, and other human-made structures absorb incoming solar radiation and reemit the energy as long-wave (heat) radiation—they face the risk of becoming rapidly dehydrated and can suffer the negative—sometimes fatal—effects of heat exhaustion or stroke. In addition, because young children's lungs are still developing, they can also suffer irreversible lung damage from being exposed to smog and soot pollution and breathing it into their lungs.

As climate change increases in intensity and food- and water-borne diseases spread into areas where they have never existed before, children are much more susceptible to becoming ill—especially those living in poverty. As extreme weather events occur—such as hurricanes and flooding—and families are left homeless, children are especially prone to complications of PTSD as they try to cope with the upheaval of their lives and often loss of their homes.

The elderly, aged 65 years and older, are also at greater risk than the general population. By the year 2030, one-fifth of the U.S. population is projected to be older than 65 years. Because the elderly often have more frail health and less mobility, they are at greater risk to the negative effects of heat waves. If the income of an elderly person is limited and they cannot afford air conditioning in their home, their health and safety can be in jeopardy because the elderly are more sensitive to changes in temperature and cannot physically adjust to extremes as younger people can.

Another sector of the population at risk is those with chronic health conditions. People, for example, with heart problems, respiratory illnesses, diabetes, or compromised immune systems are more likely to suffer serious health complications because of the effects of climate change than healthy people. Those that suffer from diabetes are at greater risk of death from heat waves. Stagnant hot air masses and areas with high ozone and soot concentrations pay a heavy toll on those with heat and respiratory illnesses. According to the American Lung Association (ALA), their offices in California recognize climate change as one of the greatest threats to public health we will face this century. As testament to it, Californians are already experiencing climate change effects, including increasing air pollution, more frequent and severe heat waves, longer wildfire seasons, severe drought conditions, and other threats to lung health.

In California's favor, they are leading in efforts to reduce climate pollution from cars, trucks, and energy sources in an attempt to reduce impacts to the air, while enhancing multiple public health benefits by reducing the reliance on fossil fuels. Their goal toward a zero carbon economy is a positive movement toward a clean air future. The ALA in California has determined in their report *Driving California Forward* that California's Low Carbon Fuel Standard and Cap and Trade fuels programs will result in cumulative benefits of $10.4 billion by 2020 and $23.1 billion by 2025 from avoided health costs, energy insecurity, and climate change impacts. In addition, the transition to cleaner fuels and more diverse vehicles will result in cleaner air and better health for Californians, increased energy security, as well as a healthier climate (ALA, 2016).

Smog forms when sunlight, heat, and relatively stagnant air meet up with nitrogen oxides and various volatile organize compounds. Exposure to smog can do serious damage to someone's lungs and respiratory systems. In the case of inflammation and irritation to the lungs, it can cause shortness of breath, throat irritation, chest pains, and coughing, and lead to asthma attacks, hospital admissions, and emergency room visits. These consequences are even more severe if people are exposed while being active. More hot days mean better conditions for creating smog that can trigger asthma and other breathing problems.

Those with preexisting conditions of weakened immune systems, such as those suffering from AIDS or cancer are also highly susceptible to catching diseases spread by mosquitoes or other vectors.

The Environmental Defense Fund has identified other groups as being potentially more vulnerable to the negative effects of pollution, illness, and climate change. They include the following:

1. *Pregnant women and their unborn children*: They may be unable to take specific medications or get to properly air conditioned locations.

2. *People living in poverty*: They may not have access to air conditioning or ready access to medical assistance.

3. *People living in areas of chronic pollution*: Consistent exposure to unhealthy air conditions may be compromising; they may have increased exposure to infectious diseases.

4. *Geographic areas prone to harmful climate change*: People in these areas may experience violent storms, coastal flooding and erosion, damage to buildings, and contamination of drinking water.

Table 3.5 shows the risks associated with potential climate change in different regions in the United States.

It is becoming more common today for individual cities to prepare emergency plans outlining actions that need to be taken if the negative effects of climate change and global warming affect their specific geographic area. According to the Environmental Defense Fund, several cities in the United States have already put together working Heat Health Watch/Warning System plans to make their populations prepared in order to deal with various climate issues when necessary. Cities already taking proactive action include Dallas/Fort Worth, Texas; Cincinnati/Dayton, Ohio; Chicago, Illinois; Jackson, Mississippi; Lake Charles, Louisiana; Little Rock, Arkansas; Memphis, Tennessee; New Orleans, Louisiana; Philadelphia, Pennsylvania; Phoenix, Arizona; Portland, Oregon; St. Louis, Missouri; Shreveport, Louisiana; Seattle, Washington; and Yuma, Arizona (Trust for Americas Health, 2009).

Local topography also plays a role in the pollution that an area experiences. Cities located in close proximity to mountain ranges experience unique patterns of recurrent pollution. Large cities that commonly experience photochemical smog—the brown air that often results from car, bus, and truck exhaust (composed of nitrogen oxide, oxygen, hydrocarbons, and sunlight to produce ozone, which can be deadly at lower elevations), as well as sulfurous fog (created from coal-burning plants)—are often found with the negative consequences of air pollution and ill effects on health. Currently, there are some controls implemented on automobiles and coal-fired plants. Catalytic converters are used in some areas. They convert carbon monoxide, nitrous oxide, and hydrocarbons into CO_2, nitrogen, and water. Some cars recirculate exhaust gas along with a catalytic converter to reduce emissions.

In areas where the existence of mountain ranges is a factor, coal-fired plants may burn only low-sulfur content coal. They may also use a process that converts the coal directly to gas (gasification process) or use scrubbing technologies.

In areas where mountain ranges act as physical barriers and trap pollution over cities, air inversions are common, especially during the winter months. In valleys or on the lee side of mountains, if a warmer air mass moves above cooler air, it traps the cooler, denser air underneath and increases the severity of the air pollution. Los Angeles is an example of this, where the warm desert air from the east comes over the mountains to the east of Los Angeles and lies over the cooler Pacific Ocean air. The cooler air becomes trapped because it cannot rise through the less dense warm air above it, and the pollution in the cold air accumulates. In mountain valleys, a similar situation occurs where warm air overlies the colder air that accumulates in the valleys. In cities, heat island effects are common. Warm air filled with pollutants collects and then spreads out over the nearby suburbs. The greenhouse gases contributing the most to the anthropogenic greenhouse effect are listed in Table 3.6.

Based on the research conducted at the Scripps Institution of Oceanography in San Diego, California, an analysis of the pollution-filled "brown clouds" over southern Asia offers hope that the region may be able to slow or stop some of the

---

TABLE 3.5

Potential Regional Effects of Climate Change in the United States

Geographic Region	Potential Negative Effects
Southeast Atlantic and Gulf Coast	Violent storms, strong storm surges and flooding, coastal erosion, more damage to buildings and roads, contamination of drinking water
Southwest	Higher temperatures, less rainfall, hot, arid climate, increased wildfires, worsened air quality
Northwest	Heavy rainfall, flooding, sewage overflow, increased illness, and spread of disease
The Great Plains	Milder winters, scorching summers, decreased agricultural production, intense heat waves
Northeast	Higher temperatures, more allergies, spread of disease by insects and animals
Alaska	Melting permafrost, retreating sea ice, disturbed ecosystems, reduced subsistence hunting and fishing, milder temperatures, increase in insects and forest pests such as the spruce bark beetle

Source: Environmental Defense Fund.

TABLE 3.6

Gases Contributing to the Anthropogenic Greenhouse Effect

Gas	Rate of Increase (% per year)	Relative Contribution (%)
CO_2 (carbon dioxide)	0.5	60
CH_4 (methane)	1	15
N_2O (nitrous oxide)	0.2	5
O_3 (ozone)	0.5	8
CFC-11 (trichlorofluoromethane)	4	4
CFC-12 (trichlorofluoromethane)	4	8

alarming retreat of glaciers in the region by reducing the existing air pollution.

Leading the research team, Dr. V. Ramanathan, a chemistry professor at Scripps, concluded from the research, "Concerning the rapid melting of these glaciers, the third-largest ice mass on the planet, if it becomes widespread and continues for several more decades, will have unprecedented downstream effects on southern and eastern Asia" (*Scripps News*, 2007, UNEP, 2008).

According to Achim Steiner, the United Nations under-secretary general and executive director of the UN Environment Programme, "The main cause of climate change is the buildup of greenhouse gases from the burning of fossil fuels. But brown clouds, whose environmental and economic impacts are beginning to be unraveled by scientists, are complicating, and in some cases aggravating, their effects. The new findings should spur the international community to even greater action. For it is likely that in curbing greenhouse gases we can tackle the twin challenges of climate change and brown clouds and in doing so, reap wider benefits from reduced air pollution to improved agricultural yields" (MSNBC, 2007a).

Jay Fein, program director in the National Science Foundation's Division of Atmospheric Sciences, remarked, "In order to understand the processes that can throw the climate out of balance, Ramanathan and colleagues, for the first time ever, used small and inexpensive unmanned aircraft and their miniaturized instruments as a creative means of simultaneously sampling clouds, aerosols, and radiative fluxes in polluted environments, from within and from all sides of the clouds. These measurements, combined with routine environmental observations and a state-of-the-science model, led to these remarkable results" (NSF, 2006).

What the study was successfully able to reveal was that the effect of the brown cloud was necessary to explain temperature changes that have been observed in the region over the last 50 years. It also clarified that southern Asia's warming trend is more pronounced at higher altitudes than closer to sea level.

Ramanathan concluded, "This study reveals that over southern and eastern Asia, the soot particles in the brown clouds are intensifying the atmospheric warming trend caused by greenhouse gases by as much as 50 percent."

In a subsequent article published in *Science* in 2012, China and India are releasing two million metric tons of carbon soot and other dark pollutants into the air annually. As a result, it is affecting climate thousands of kilometers away, warming some areas and cooling others. For example, it is projected that under current rates, the United States will be up to 0.4°C warmer by 2024 just from the brown clouds alone. These dark aerosols are primarily soot from diesel engines and power plants, which absorb more sunlight than they scatter, warming the atmosphere around them. Concerns from this include the triggering of significant changes in long-term weather patterns, similar to a human-caused El Niño which could alter temperature and precipitation patterns (Perkins, 2012).

His current research includes Project Surya, a cook-stove project that is aimed at eliminating pollutants from traditional biomass cooking that are warming the atmosphere. He is also involved with the California Air Resources Board and R. K. Pachauri (chairman of the IPCC) to initiate a project called ICAMP—a World Bank-sponsored project designed to reduce soot emissions from the transportation sector in India, seen as a practical application in climate change mitigation (Ramanathan, 2015).

It was not until the late 1940s when air pollution disasters occurred on two separate continents that public awareness began to grow concerning outdoor air quality and its effects on human health. Both the 1948 "killer fog" in Denora, Pennsylvania, that killed 50 people and the London "fog" in 1952, where roughly 4000 people died, prompted an investigation into the cause, and it was determined that the widespread use of dirty fuels were to blame. This began the concerted effort for governments to begin taking the problem of urban air pollution seriously (Figure 3.9).

Since this time, many contaminants in the atmosphere have been identified as harmful, and serious efforts have been undertaken to clean up the atmosphere from harmful components. The most common and damaging pollutants include sulfur dioxide, suspended particulate matter (PM), ground-level ozone, nitrogen dioxide, carbon monoxide, and lead. All of these pollutants are tied either directly or indirectly to the combustion of fossil fuels. Even though major efforts are under way today to clean polluted air over cities, many

FIGURE 3.9 A man guiding a bus with a flaming torch through thick fog during the London smog of 1952. (Courtesy of http://www. nickelinthemachine.com.)

cities worldwide still lack a healthy air quality. An inventory completed by the European Environment Agency determined that 70–80 percent of 105 European cities surveyed exceeded World Health Organization (WHO) air quality standards for at least one pollutant. In the United States, an estimated 80 million people live in areas that do not meet U.S. air quality standards, which are comparable to WHO standards. Other areas that do not meet WHO standards include Beijing, Delhi, Jakarta, and Mexico City. In these cities, pollutant levels sometimes exceed WHO air quality standards by a factor of 3 or more. Some of the cities in China exceed WHO standards by a factor of 6. Worldwide, the WHO estimates that up to 1.4 billion urban residents breathe air exceeding the WHO air guidelines and that the health consequences are considerable, with a mortality rate of 200,000–570,000 annually. In addition, the World Bank has estimated that exposure to particulate levels exceeding the WHO health standard accounts for roughly 2–5 percent of all deaths in urban areas in the developing world.

Addressing the issue of public policies that can reduce the health impacts of ambient air pollution, the WHO strongly recommends tackling the issue of inefficient fossil fuel combustion from motor vehicles, power generation, and improving energy efficiency in buildings, homes, and manufacturing processes. In addition, reducing health impacts will require cooperation from authorities at all levels: local, regional, national, and even international. Individuals also have an important role to play. Each of us has the responsibility to choose more efficient and cleaner energy sources and solutions. The key is to get multilevel and cross-sector cooperation, such as the transport, housing, industrial, and energy sectors of various geographic levels all working toward like goals to reduce air pollution and improve health conditions (WRI, 2002, World Health Organization, 2014).

It is stressed, however, that these mortality estimates do not reflect the huge toll of illness and disability that exposure to air pollution brings on a global level. Health effects span a wide range of severity from coughing to bronchitis to heart disease and lung cancer. In developing cities alone, air pollution is responsible for approximately 50 million cases per year of chronic coughing in children younger than 14 years of age. Taking the health effects into consideration, concerning climate change is critical because it affects the future of society.

CONTRIBUTORS TO CLIMATE CHANGE AND POLLUTION

The transportation sector is the largest single source of air pollution in the United States today. It causes almost 67 percent of the carbon monoxide (CO), a third of the nitrogen oxides (NO_x), and a fourth of the hydrocarbons in the atmosphere.

Cars and trucks pollute the air during manufacturing, oil refining and distribution, refueling and, most of all, vehicle use. Motor vehicles cause both primary and secondary pollution. Primary pollution is that which is emitted directly into the atmosphere; secondary pollution is from chemical reactions between pollutants in the atmosphere.

There are six major pollutants: ozone (O_3), PM, NO_x, CO, SO_2, and hazardous air pollutants (toxics). The primary ingredient in smog is O_3. PM refers to particles of soot, metals, and pollen. The finest, smallest textured PM does the most damage, as it travels into the lungs easily. NO_x tend to weaken the body's defense against respiratory infections. CO is formed by the combustion of fossil fuels such as gasoline and is emitted by cars and trucks. When inhaled, it blocks the transport of oxygen to the brain, heart, and other vital organs, making it deadly. SO_2 is created via the burning of sulfur-containing fuels, especially diesel. It forms fine particles in the atmosphere and is harmful to children and those with asthma. Toxic compounds are chemical compounds emitted by cars, trucks, refineries, and gas pumps, and have been related to birth defects, cancer, and other serious illnesses. The EPA estimates that the air toxics emitted from cars and trucks account for half of all cancers caused by air pollution.

Pollution from light trucks is growing quickly. This class of vehicles includes minivans, pickups, and sport utility vehicles. Today, these vehicles represent one of every two new vehicles purchased. For this reason, California regulators and the EPA have recently created new rules requiring light trucks to become as clean as cars over the next 7–9 years.

In 2009, the ALA released a report that identified six out of ten Americans—that is, 186.1 million people—live in areas where air pollution levels endanger lives. This means that almost every major city in the United States is affected by significant amounts of pollution. What was shocking was that despite the growing "green movement," air quality in many cities has become dirtier, not cleaner over the past decade. The three most widespread types of pollution are ozone (or smog), annual particle pollution, and 24-hour particle pollution levels. "This should be a wakeup call," said Stephen J. Nolan, ALA National Board Chair. "When 60 percent of Americans are left breathing air dirty enough to send people to the emergency room, to shape how kids' lungs develop, and to kill, air pollution remains a serious problem." Some of the most significant sources of air pollution include dirty power plants, dirty diesel engines, and ocean-going vessels. All three of these worsen climate change (ALA, 2009).

In 2015, the ALA released a list of the smoggiest cities in the United States. They are in order from least to worst (ALA, 2015), as shown in Table 3.7.

KEY IMPACTS

As the climate change issue has escalated in recent years, scientists have been able to observe the related impacts worldwide—across all continents and oceans. As illustrated in several examples, impacts have been described and future impacts have been predicted if current activity continues.

Due to the lifetimes of the various greenhouse gases, even if all activity were stopped immediately, the effects of climate change would continue to a certain extent, because

TABLE 3.7

Worst Smog Cities in the United States, 2015

Ranking	City
10	Phoenix–Mesa–Scottsdale, Arizona
9	Las Vegas–Henderson, Nevada
8	Modesto–Merced, California
7	Dallas–Fort Worth, Texas
6	Houston–The Woodlands, Texas
5	Sacramento–Roseville, California
4	Fresno–Madera, California
3	Bakersfield, California
2	Visalia, California
1	Los Angeles, California

Source: ALA, State of the air 2015, http://www.stateoftheair.org/ (accessed April 11, 2016), 2015.

past damage has already ensured those changes. There is still a time window, however, to avoid the most catastrophic of changes, if efforts, policies, and appropriate technologies are put into effect now. The chart illustrates the key impacts as a function of increasing global average temperature change (Figure 3.10). It details the changes that would be expected with water sources, ecosystems, food, the status of the coasts, and major health issues and trends of an incremental temperature increase up to 5°C. Resultant impacts expected to a region will vary based on the extent they are able to adapt, the rate of temperature change, and their socioeconomic structure. For example, locations better equipped to handle emergencies and crisis management will fare better than locations that are not. However, even being better equipped will not spare an area from feeling the ramifications of a changing climate.

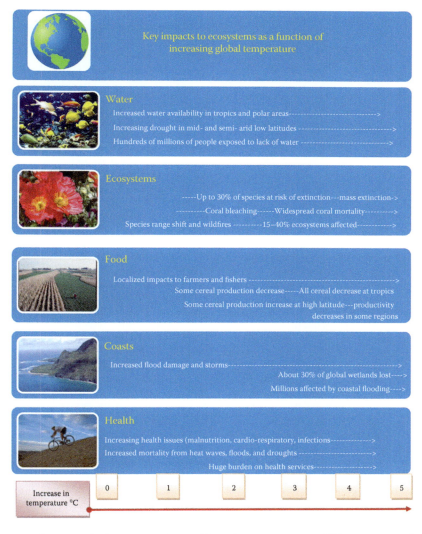

FIGURE 3.10 As temperatures increase, the impacts on the earth's various ecosystems will become more intensified, stressing natural resources to their limits and resulting in possible productive failure, loss of habitat, health disasters, disruption of food and water supplies, and extinction of life. (Courtesy of NOAA.)

CONCLUSION

This chapter has served to illustrate the components and seriousness of the various greenhouse gases, their production, life cycles, and various contributions to climate change. Human activity has played a heavy hand in determining the atmosphere's composition and behavior since the onslaught of the Industrial Revolution, especially from 1850 until the present. Fortunately, Keeling had the foresight to set up a monitoring station in Hawaii and diligently tracked the atmospheric concentrations of CO_2. The results of his graph played an enormous hand in convincing the scientific community years later that climate change was real and deserved the world's attention.

As climate science entered its infancy, scientists began to see the critical role CO_2 and other greenhouse gases were playing in the larger picture. The bigger question was—could they convince the general public of what was going on and trigger what was necessary to solve the problem? Not just regulatory constraints but a major change in mind-set and the way people looked at the environment were needed. Through their subsequent research, they were able to unravel many mysteries and answer several questions, such as where carbon is sequestered (stored), how it is produced, the process by which it is released into the atmosphere, and the worldwide effects it was having.

Much progress has been made in solving and answering this question. And finally, the public at last seems to be embracing this knowledge and choosing to take action. As new discoveries are made, it is hoped that the public will stay informed and adapt to a warming world with appropriate green responses, as those decisions and responses will shape the future landscape.

REFERENCES

ALA (American Lung Association). 2009. 60 Percent of Americans live in areas where air is dirty enough to endanger lives. *ScienceDaily.* https://www.sciencedaily.com/releases/2009/04/090429131158.htm (accessed April 11, 2016).

ALA. 2015. State of the air 2015. http://www.stateoftheair.org/ (accessed April 11, 2016).

ALA. 2016. Climate change in California: Fighting climate change in California. http://www.lung.org/local-content/_content-items/our-initiatives/current-initiatives/climate-change-in-california.html?referrer=https://www.google.com/ (accessed April 10, 2016).

Bradsher, K. October 22, 2003. China's boom adds to global warming problem. *New York Times.* http://www.nytimes.com/2003/10/22/world/china-s-boom-adds-to-global-warming-problem.html (accessed April 10, 2016).

Bradsher, K. and D. Barboza. 2006. Pollution from Chinese coal casts a global shadow. *New York Times,* June 11. http://www.nytimes.com/2006/06/11/business/worldbusiness/11chinacoal.html (accessed May 15, 2017).

Brown, M. A., F. Southworth, and A. Sarzynski. May 2008. Shrinking the carbon footprint of metropolitan America. Brookings Institution, 16 p. http://www.brookings.edu/~/media/Files/rc/papers/2008/05_carbon_footprint_sarzynski/carbonfootprint_brief.pdf (accessed April 6, 2016).

Casper, J. K. 2009. *Fossil Fuels and Pollution: The Future of Air Quality.* New York: Facts on File.

CBD (Center for Biological Diversity). 2008. Climate change solutions: Political and personal. Center for Biological Diversity. http://www.biologicaldiversity.org/programs/climate_law_institute/solutions/index.html (accessed March 30, 2016).

Center for Climate Change and Environmental Studies. 2011. The global carbon cycle. http://www.center4climatechange.com/carboncycle.php (accessed May 15, 2017).

Climate and Weather. 2012. Deforestation. http://www.climateandweather.net/global-warming/deforestation.html (accessed April 4, 2016).

Climate Policy Watcher. June 14, 2012. The Earth's Energy Balance. https://www.climate-policy-watcher.org/greenhouse-gases-2/the-earths-energy-balance.html (accessed May 14, 2017).

The Economist. November 15, 2007. Coal power still going strong: Efforts to curb greenhouse gas emissions have yet to dent enthusiasm for coal. http://www.economist.com/node/10145492 (accessed April 10, 2016).

EIA (Energy Information Center). 2004. Greenhouse gases, climate change, and energy. U.S. Department of Energy. http://www.eia.doe.gov/oiaf/1605/ggccebro/chapter1.html (accessed March 30, 2016).

Environmental Defense Fund. 2017. Potential Regional Effects of Climate Change in the United States. https://www.edf.org/climate/climate-changes-effects-plunder-planet (accessed May 15, 2017).

EPA (Environmental Protection Agency). 2011. Carbon sequestration in agriculture and forestry. U.S. Environmental Protection Agency, Climate Change Home. http://www.epa.gov/sequestration/faq.html (accessed May 2, 2011).

EPA. 2013. Sources of greenhouse gas emissions. https://www3.epa.gov/climatechange/ghgemissions/sources.html (accessed April 4, 2016).

FAO (Food and Agriculture Organization). 2014. Agriculture, forestry and other land use emissions by sources and removals by sinks. Climate, Energy, and Tenure Division, FAO. http://www.fao.org/docrep/019/i3671e/i3671e.pdf (accessed April 4, 2016).

Gines, J. K. 2012. *Climate Management Issues: Economics, Sociology, and Politics.* New York: CRC Press.

Global Climate Notes. December 6, 2015. Impacts of deforestation. http://www.globalclimatenotes.us/greenhouse-gases-2/impacts-of-deforestation.html (accessed April 4, 2016).

Global Warming Policy Foundation. February 12, 2015. New report: The truth about China. http://www.thegwpf.org/new-report-the-truth-about-china/ (accessed April 5, 2016).

Hansen, J. February 2007. Trial of the century, Act II. NASA Science Briefs. http://www.giss.nasa.gov/research/briefs/hansen_12/ (accessed April 12, 2016).

Hansen, J. June 23, 2008. Twenty years later: Tipping points near on global warming. *The Huffpost Green.* http://www.huffingtonpost.com/dr-james-hansen/twenty-years-later-tippin_b_108766.html (accessed November 2, 2011).

Hansen, J. March 2012. Why I must speak out about climate change. https://www.ted.com/talks/james_hansen_why_i_must_speak_out_about_climate_change/transcript?language=en (accessed April 4, 2016).

Hopwood, N. and J. Cohen. 2008. Greenhouse gases and society. http://www.bokashicycle.com/library/GreenhouseGases.pdf (accessed May 15, 2017).

IPCC (Intergovernmental Panel on Climate Change). 2001. Climate change 2001: Synthesis report: Summary for policymakers. https://www.ipcc.ch/pdf/climate-changes-2001/synthesis-syr/english/summary-policymakers.pdf (accessed April 12, 2016).

IPCC. 2007. WG1—The physical science basis. http://www.ipcc.ch/ (accessed April 12, 2016).

IPCC. 2013. WG1 contribution to the IPCC fifth assessment report climate change 2013: The physical science basis. https://www.ipcc.ch/pdf/assessment-report/ar5/wg1/drafts/fgd/WGIAR5_WGI-12Doc2b_FinalDraft_Chapter10.pdf (accessed April 4, 2016).

Keeling, R. F. 2008. Recording earth's vital signs. *Science* 319(5871):1771–1772. http://www.sciencemag.org/content/319/5871/1771.citation (accessed April 12, 2016).

Lindsey, R. March 18, 2009. Tropical deforestation. *Safe Rainforest for Our Children*. http://tropical-rainforest-information.blogspot.com/2009/03/tropical-deforestation.html (accessed April 12, 2016).

McDermott, M. November 12, 2009. Boreal forests store twice as much carbon as tropical—So why aren't we doing more to protect them? *Treehugger*. http://www.treehugger.com/natural-sciences/boreal-forests-store-twice-as-much-carbon-as-tropical-so-why-arent-we-doing-more-to-protect-them.html (accessed April 4, 2016).

MSNBC. August 6, 2007a. Haze added to warming in Asia, study finds: Particulates tied to energy usage—Himalayan glaciers tied to impacts. *MSNBC News*.

MSNBC. October 12, 2007b. Gore, U.N. climate panel win nobel peace prize. *MSNBC News*. http://www.msnbc.msn.com/id/21262661/ns/us_news-environment/ (accessed April 12, 2016).

MSNBC. October 22, 2007c. Study: Warming is stronger. Happening sooner: Higher CO_2 emissions from fossil fuels, and weaker earth, cited as reasons. http://www.msnbc.msn.com/id/21423872/ns/us_news-environment/ (accessed April 12, 2016).

National Oceanic and Atmospheric Administration. 2015. NOAA/NASA Annual Global Analysis for 2015. https://www.giss.nasa.gov/research/news/20160120/noaa_nasa_global_analysis_2015.pdf. (accessed May 14, 2017).

NSF (National Science Foundation). April 17, 2006. Autonomous unmanned aerial vehicles take to the skies to track pollutants. National Science Foundation. http://www.msnbc.msn.com/id/20145043/ (accessed April 12, 2016).

Perkins, S. May 25, 2012. 'Asian Brown Cloud' threatens U.S. *Science*. http://www.sciencemag.org/news/2012/05/asian-brown-cloud-threatens-us (accessed April 11, 2016).

Ramanathan, V. December 2015. Science. Ramanathan in COP-21. http://www-ramanathan.ucsd.edu/ (accessed April 11, 2016).

Revkin, A.C. 2007. Stuck on coal, stuck for words in a high-tech world. *New York Times*, December 4. http://www.nytimes.com/2007/12/04/science/earth/04comm.html (accessed May 15, 2017).

Riebeek, H. 2010. How much more will earth warm? *NASA Observatory*. http://earthobservatory.nasa.gov/Features/GlobalWarming/page5.php (accessed April 4, 2016).

Schmid, R. E. October 23, 2007. Carbon dioxide in atmosphere increasing. *Book Rags*. http://www.bookrags.com/news/carbon-dioxide-in-atmosphere-moc/ (accessed April 5, 2016).

Schneider, U. and B. McCarl. 2003. The potential of U.S. agriculture and forestry to mitigate greenhouse gas emissions: An agricultural sector analysis. Center for Agricultural and Rural Development. 02-WP 300. http://www.card.iastate.edu/publications/DBS/PDFFiles/02wp300.pdf (accessed April 6, 2016).

Scientific American. April 8, 2009. Is global warming a myth? *Scientific American*. http://www.scientificamerican.com/article/is-global-warming-a-myth/ (accessed April 5, 2016).

Scientific American. November 13, 2012. Deforestation and its extreme effect on global warming. http://www.scientificameri-can.com/article/deforestation-and-global-warming/ (accessed April 4, 2016).

Scripps News. August 1, 2007. Pollution amplifies greenhouse gas warming trends to Jeopardize Asian Water Supplies, Scripps Institute of Oceanography, University of California San Diego, San Diego, California. http://healthyamericans.org/assets/files/ClimateChangeandHealth.pdf (accessed April 2, 2016).

Tang, R. 2010. China's steel industry and its impact on the United States: Issues for Congress. CRS Report for Congress, 7-5700, R-41421. http://www.fas.org/sgp/crs/row/R41421.pdf (accessed May 15, 2017).

The Times. December 2, 2015. 2,500 new coal plants will thwart any Paris pledges. *The Times Environment*. http://www.thetimes.co.uk/tto/environment/article4629455.ece (accessed April 4, 2016).

Trust for America's Health. 2009. How can we prevent and prepare for health issues in a changing climate? http://healthyameri-cans.org/assets/files/ClimateChangeandHealth.pdf (accessed May 15, 2017).

UNEP (UN Environment Programme). 2008. Átmospheric Brown Clouds regional assessment report with focus on Asia. http://www-abc-asia.ucsd.edu/ (accessed April 11, 2016).

Union of Concerned Scientists. 2008. Coal power in a warming world: A sensible transition to cleaner energy options, October. http://www.ucsusa.org/sites/default/files/legacy/assets/documents/clean_energy/Coal-power-in-a-warming-world.pdf (accessed May 15, 2017).

United Nations Framework Convention on Climate Change. 2014. Global Warming Potentials. http://unfccc.int/ghg_data/items/3825.php (accessed May 15, 2017).

U.S. Energy Information Administration. March 2017, U.S. Energy-Related Carbon Dioxide Emissions, 2015. https://www.eia.gov/environment/emissions/carbon/pdf/2015_co2analysis.pdf (accessed May 15, 2017).

Williams, B. December 6, 2015. Fossil fuels and global warming. http://www.briangwilliams.us/fossil-fuels-pollution/fossil-fuels-and-global-warming.html (accessed April 5, 2016).

World Health Organization. 2014. Frequently asked questions ambient and household air pollution and health update 2014. http://www.who.int/phe/health_topics/outdoorair/databases/faqs_air_pollution.pdf (accessed April 11, 2016).

World Resources Institute. 2002. Health effects of air pollution. World Resources Institute. http://www.airimpacts.org/documents/local/HealthEffectsAirPollution.pdf (accessed June 10, 2010).

SUGGESTED READING

Alley, R. B. 2004. Abrupt climate change. *Scientific American* 291(5):62–69.

Amos, J. December 12, 2007. Arctic Summers Ice-free by 2013. *BBC News*.

Boyd, R. S. August 21, 2002. Glaciers melting worldwide, study finds. *National Geographic News*. http://news.nationalgeographic.com/news/2002/08/0821_020821_wireglaciers.html (accessed October 18, 2010).

Bridges, A. February 16, 2006. Greenland dumps ice into sea at faster pace. *Live Science*.

Casper, J. K. 2009. *Greenhouse Gases: Worldwide Impacts*. New York: Facts on File, Inc.

Gore, Al. September 18, 2006. Finding solutions to the climate crisis. *Speech presented to New York University School of Law*. http://www.astrosurf.com/luxorion/climate-crisis-al-gore.htm (accessed March 25, 2016).

Tan, Z. 2014. *Air Pollution and Greenhouse Gases: From Basic Concepts to Engineering Applications for Air Emission Control*. Singapore: Springer.

4 Effects on Ecosystems

OVERVIEW

As previously stated, one of the most misunderstood aspects of climate change is not the rise in air temperature by a few degrees—it is the sheer havoc that those "few degrees" will wreak not only on human society but on the natural environment. One aspect of human nature is that if something is not directly affecting us, it probably will not make a lasting impression. For instance, consider this: Every time something upsets the balance in the Middle East and the price of gas skyrockets at the pumps as a result, people complain, drive less, maybe put their Sport Utility Vehicle's (SUV's) up for sale, seriously consider buying a hybrid, adopting a greener lifestyle, and talking about the environment and recycling. Then, right after the prices fall again at the pump, people get comfortable in their "business as usual" ways, and that is the end of it until the next panic at the pumps. What percentage of the general public do you suppose really takes it seriously and follows through for the environment?

If we had to trade places with someone or something that had to live with the permanent consequences of climate change every day, perhaps that would generate the interest and dedication necessary to make real and permanent changes. Knowledge is power, and that is what this chapter's purpose is: to give you a basic, fundamental understanding of exactly how climate change is currently affecting—and will affect—the earth's varied ecosystems.

This chapter will discuss six unique ecosystems and the effect that climate change is currently having, what will happen as temperatures continue to rise, and what the future will look like for the inhabitants of these areas. First, it will take a look at the world's forests—the temperate, boreal (northern), and tropical so that you can get an idea of the destruction affecting these areas. Next it will focus on the world's grasslands and prairies. Following that, we will focus on the earth's most fragile ecosystem—the polar areas. Most of these areas are already feeling the disastrous effects of the warming temperatures. From there, we will travel to the world's great deserts, then mountain areas, and finally to the diverse aquatic regions—lakes, oceans, and coastal areas—where many of the world's most heavily populated urban areas are located, so that you will see why the problem is much more than just a few degrees' rise in temperature.

INTRODUCTION

The earth's ecosystems are each diverse and unique. They vary, sometimes drastically, from one another, and each offers its own blend of life forms, services, and purpose to this complicated planet. They are each rich in their own resources—irreplaceable if they are lost.

Currently, one of the biggest threats to their existence is climate change, caused principally by the action of human activity: chiefly the burning of fossil fuels but also other activities such as deforestation. Changes to ecosystems are being felt worldwide as climates are beginning to shift—some more subtle than others. Documented changes include the dying off of endemic vegetation, migration of vegetation northward (in the Northern Hemisphere), migration of animal species to cooler environments (northward in latitude and higher in elevation in mountainous areas), migration of grasslands and food belts, melting of ice caps and glaciers, loss of sea ice, loss of crucial animal habitat, increased incidences of drought and desertification, increased occurrence of heat waves, an upsurge in wildfire incidences, changes in mountain regimes, an increased lack of mountain water storage, sea-level rise and flooding in coastal regions, die-off of tropical coral reefs, and saltwater contamination in fragile wetland areas. Each one of these changes to the ecosystem they affect causes a negative impact of some type on the life that lives in them. For the birds that migrate, it might mean a 2-week delay in flowering times of local plants and a subsequent lack of food supply. For a dry region suffering from the effects of drought or desertification, it may mean the lack of a food crop during a growing season desperately needed to support several villages. For the polar bear, it may mean the permanent melting of hunting and breeding grounds, restricting where they can travel, thrive—and to some it may spell death.

This chapter will illustrate the effects climate change is having, and will have, on the earth's major ecosystems, what to expect as temperatures steadily climb, and why it is so crucial that this process humans have set in motion be halted now before any more damage is done to the fragile ecosystems that will have no choices once series of events have been set in motion.

DEVELOPMENT

The Intergovernmental Panel on Climate Change (IPCC)'s Climate Change 2014 Synthesis Report: Summary for Policymakers, which is based on the reports of the three Working Groups of the IPCC, including relevant special reports, provided observed changes and their causes, supplying an integrated view of climate change as the final part of the IPCC's Fifth Assessment Report in 2014. A summary of their findings is shown in Tables 4.1 and 4.2.

TABLE 4.1

Habitats Threatened and Endangered by Climate Change

Coral Reefs	Mountain Ecosystems
Coastal wetlands	Prairie wetlands
Mangroves	Ice edge ecosystems
Permafrost ecosystems	

Estimates by the IPCC in their Fifth Assessment Report indicate that 25 percent of the earth's mammals and 12 percent of birds are at significant *global* risk of extinction. Although there are also other stresses on animal habitats, such as exploitation, pollution, extreme climatic events, and diseases, climate change is a significant issue affecting their health and survival, and continuing climate change will increase species' vulnerabilities to rarity and extinction (high confidence level) (IPCC, 2016).

TABLE 4.2

Summary of the IPCC's Findings on Climate Change, 2014

Observed changes and their causes	Human influence on the climate system is clear, and recent anthropogenic emissions of greenhouse gases are the highest in history. Recent climate changes have had widespread impacts on human and natural systems.
Observed changes in the climate system	Warming of the climate system is unequivocal, and since the 1950s, many of the observed changes have been unprecedented over decades to millennia. The atmosphere and ocean have warmed, the amounts of snow and ice have diminished, and sea level has risen.
Causes of climate change	Anthropogenic greenhouse gas emissions have increased since the preindustrial era, driven largely by economic and population growth, and are now higher than ever. This has led to atmospheric concentrations of carbon dioxide, methane, and nitrous oxide that are unprecedented in at least the last 800,000 years. Their effects, together with those of other anthropogenic drivers have been detected throughout the climate system and are extremely likely to have been the dominant cause of the observed warming since the mid-twentieth century.
Impacts of climate change	In recent decades, changes in climate have caused impacts on natural and human systems on all continents and across the oceans. Impacts are due to observed climate change, irrespective of its cause, indicating the sensitivity of natural and human systems to changing climate.
Extreme events	Changes in many extreme weather and climate events have been observed since about 1950. Some of these changes have been linked to human influences, including a decrease in cold temperature extremes, an increase in warm temperature extremes, an increase in extreme high sea levels, and an increase in the number of heavy precipitation events in a number of regions.
Future climate changes, risks, and impacts	Continued emission of greenhouse gases will cause further warming and long-lasting changes in all components of the climate system, increasing the likelihood of severe, pervasive and irreversible impacts for people and ecosystems. Limiting climate change would require substantial and sustained reductions in greenhouse gas emissions which, together with adaptation, can limit climate change risks.
Key drivers of future climate	Cumulative emissions of CO_2 largely determine global mean surface warming by the late twenty-first century and beyond. Projections of greenhouse gas emissions vary over a wide range, depending on both socioeconomic development and climate policy.
Projected changes in the climate system	Surface temperature is projected to rise over the twenty-first century under all assessed emission scenarios. It is very likely that heat waves will occur more often and last longer, and that extreme precipitation events will become more intense and frequent in many regions. The oceans will continue to warm and acidify, and global mean sea level to rise.
Future risks and impacts caused by a changing climate	Climate change will amplify existing risks and create new risks for natural and human systems. Risks are unevenly distributed and are generally greater for disadvantaged people and communities in countries at all levels of development.
Climate change beyond 2100, irreversibility, and abrupt changes	Many aspects of climate change and associated impacts will continue for centuries, even if anthropogenic emissions of greenhouse gases are stopped. The risks of abrupt or irreversible changes increase as the magnitude of the warming increases.
Future pathways for adaptation, mitigation, and sustainable development	Adaptation and mitigation are complementary strategies for reducing and managing the risks of climate change. Substantial emission reductions over the next few decades can reduce climate risks in the twenty-first century and beyond, increase prospects for effective adaptation, reduce the costs and challenges of mitigation in the longer term, and contribute to climate-resilient pathways for sustainable development.
Foundations of decision-making about climate change	Effective decision-making to limit climate change and its effects can be informed by a wide range of analytical approaches for evaluating expected risks and benefits, recognizing the importance of governance, ethical dimensions, equity, value judgments, economic assessments, and diverse perceptions and responses to risk and uncertainty.
Climate change risks reduced by mitigation and adaptation	Without additional mitigation efforts beyond those in place today, and even with adaptation, warming by the end of the twenty-first century will lead to high to very high risk of severe, widespread, and irreversible impacts globally (high confidence). Mitigation involves some level of co-benefits and of risks due to adverse side effects, but these risks do not involve the same possibility of severe, widespread, and irreversible impacts as risks from climate change, increasing the benefits from near-term mitigation efforts.

Source: (IPCC. 2014a).

In the IPCC's Fifth Assessment Report of 2014, they concluded that

> Human influence on the climate system is clear. This is evident from the increasing greenhouse gas concentrations in the atmosphere, positive radiative forcing, observed warming, and understanding of the climate system. (IPCC, 2014b)

In these studies, scientists have been able to break down the natural and human-caused components in order to see how much of an effect humans have had. Human effects can include activities such as burning fossil fuels, certain agricultural practices (such as heavy use of fertilizers and other chemicals, tillage, mismanagement of livestock waste, irrigation erosion, introduction of invasive species, and soil compaction), deforestation, and local pollution from certain industrial processes (such as uncontrolled emissions, chemical use, and noncompliance of specific regulations), the introduction of invasive plant species, and various types of land-use change (such as urbanization, industrialization, and mismanaged recreational activities).

In many cases, scientists do not need to look very far to see the effects a warming world is having on the environment and the earth's ecosystems. Glaciers worldwide are melting at an accelerated rate never seen before (Figure 4.1). The cap of ice on top of Kilimanjaro is rapidly disappearing; the glaciers that have made Glacier National Park in the United States world renowned, and also extend across the border into Canada, are melting and projected to be gone within the next few decades; and the glaciers in the European Alps are experiencing a similar fate.

In the world's tropical oceans, vast expanses of beautiful, brilliantly colored coral reefs are dying off as oceans slowly become too warm. Unable to survive the higher temperatures, the corals are undergoing a process called "bleaching" and are turning white and dying off at rapid rates. In the Arctic, as temperatures climb, ice is melting at accelerated rates, leaving polar bears stranded, destroying their feeding and breeding grounds, causing them to starve and drown (Figure 4.2). Permafrost is melting at accelerated rates. As the ground thaws, it is disrupting the physical and chemical components of the ecosystem by causing the ground to shift and settle, toppling buildings and twisting roads and railroad tracks, as well as releasing methane gas into the atmosphere (another potent greenhouse gas responsible for climate change).

Weather patterns are also changing. El Niño events are triggering destructive weather in the eastern Pacific (in North and South America). Extreme weather events and droughts have become more prevalent in some geographical areas, such as parts of Asia, Africa, and the American Southwest. Animal and plant habitats have been disrupted, and as temperatures continue to climb, there have been several documented migrations of individual species moving northward (toward the poles) or to higher elevations on individual mountain ranges. Migration patterns are also being impacted, such as those already documented of beluga whales, butterflies, and polar bears. Spring, arriving earlier in some areas, is now influencing the timing of bird and fish migration, egg laying, leaf unfolding, and spring planting for agriculture. In fact, based on satellite imagery documentation of the Northern Hemisphere, growing seasons have steadily become longer since 1980.

(a)

(b)

FIGURE 4.1 The Muir Glacier located in Alaska. Photo (a) was taken in 1941, and the photo (b) was taken in 2004. The massive melting that has taken place is attributed largely to anthropogenic warming of the atmosphere. (Courtesy of National Snow and Ice Data Center, Boulder, Colorado; photo (a) by William O. Field, glaciologist for the National Park Service; photo (b) by Bruce F. Molnia, geologist for U.S. Geological Survey.)

Although species have been faced with changing environments in the past and have been able to adapt in many cases, the IPCC scientists view this current rate of change with alarm. They fully expect the magnitude of these changes to increase with the temperatures over the next century and beyond. The concern is that many species and ecosystems will not be able to adapt as rapidly as the effects of climate change cause the environment to change. In addition, there will also be other disturbances along with climate change, such as floods, insect infestations, and spread of disease, wildfire, and drought. Any of these additional challenges themselves can destroy a species or habitat. In particular, alpine (high mountain top) and polar species are especially vulnerable to the effects of climate change because as species move northward (poleward) or higher in elevation on mountains, these species' habitats will shrink, leaving them with nowhere else to go.

FIGURE 4.2 Polar bears are in a precarious situation right now. As the Arctic ice retreats, polar bears are struggling to survive. Lack of ice takes away valuable hunting and breeding grounds as well as migration corridors. (Courtesy of Publitek, Inc., Waukesha, WI.)

Climate zones could shift, completely disrupting land-use practices. For instance, the current agricultural region of the Great Plains in the United States could be shifted instead toward Canada. The southern portion of the United States could become more like Central or South America. Siberia would no longer be a frozen, desolate landscape. Parts of Africa could become dry, desolate wastelands. If this were to happen, it would severely impact the production of agriculture and the agricultural-related industries. Areas where it's currently economically feasible to farm would no longer have the climatic conditions to do so; and areas that may suddenly have a favorable climate for farming may not have fertile soils and would have no farming infrastructure to make it economical. The ripple effect of these disruptions would be felt worldwide. Millions of people would be forced to migrate from newly uninhabitable regions to new areas where they could survive. There would be enough of a disruption that ecosystems worldwide would be thrown out of balance and altered. If, however, the temperature rise falls toward the higher end of the estimate, the results on ecosystems worldwide would be disastrous. There would also

be impacts of climate change on public health as a result. Rising seas would contaminate freshwater with saltwater; there would be more heat-related illnesses and deaths; and disease-carrying rodents and insects, such as mice, rats, mosquitoes, and ticks, would spread diseases such as malaria, encephalitis, Lyme disease, and dengue fever.

Scientists of the IPCC agree that one of the most serious aspects of all this drastic change is that it is happening so rapidly. These changes are happening at a faster pace than the earth has seen in the past 100 million years; although humans may be able to pick up and move to a new location, animals and their associated ecosystems cannot. Unfortunately, the choices humans make and actions they take today are determining the fate of other life and their ecosystems tomorrow.

RESULTS OF CLIMATE CHANGE ON ECOSYSTEMS

World Wildlife Fund scientists believe that climate change could begin causing extinction of animal species in the near future because the heating caused by accelerated climate change severely impacts the earth's many delicate ecosystems—both the land and the species that live within them. Worldwide, there are several species and habitats that have now been identified as being threatened and endangered due to the effects of climate change, which are listed in Table 4.3. Because ecosystems can be altered to the point where the damage becomes irreversible and species must adapt to survive or they will die, it is critical that the issue of climate change be addressed and acted upon now before it is too late.

TABLE 4.3

Some Species Threatened and Endangered by Climate Change

Polar bear (*Ursus maritimus*)	Sea turtles
	Loggerhead: *Caretta caretta*
	Leatherback: *Dermochelys coriacea*
	Hawksbill: *Eretmochelys imbricata*
	Kemp's ridley: *Lepidochelys kempii*
	Olive ridley: *Lepidochelys olivacea*
	Flatback: *Natator depressa*
North Atlantic right whale (*Eubalaena glacialis*)	Giant panda (*Ailuropoda melanoleuca*)
Marine turtles (multiple *chelonian* species)	Pika (*Ochotona princeps*)
Multiple bird species (mountain, island, wetland, Arctic, Antarctic, seabirds, migratory birds) (*Avian*)	Wetland flora and fauna
Snowy owls (*Nyctea scandiaca*)	Salt wetland flora and fauna
Mountain gorilla (Africa) (*Gorilla beringei beringei*)	Cloud forest amphibians
Andes spectacled bear (*Tremarctos ornatus* or Andean bear)	Bengal tiger (*Panthera tigris tigris*)
Staghorn corals	Arctic fox

Source: (World Wildlife Fund, 2015).

One of the most serious constraints on animal and plant adaptation is interrupted migration. Migration for many species, such as geese, elk, salmon, leatherback turtles, wildebeest, and monarch butterflies, is a natural annual act. As climate change impacts ecosystems worldwide, many species unable to survive in the new climatic conditions of the geographic areas where they have always lived will need to migrate to areas with new climates they can survive in. There are several potential problems with this, however. Under the effects of climate change, the environment may change faster than a species can adapt. In other cases, existing land use—such as heavily urbanized areas—may negatively impact wildlife habitat. Wide-ranging wildlife species need secure core habitat where human activity is limited, ecosystem functions are still intact, and wildlife populations are able to flourish. If they do not have this, their long-term health and survival will be negatively impacted. In this situation, corridors connecting core areas are important to have established before the effects of climate change are felt. According to the U.S. Fish and Wildlife Service, it is imperative to save species. Endangered species are nature's "911" because they often serve as an early warning sign for pollution and environmental degradation that can affect human health.

IMPACTS TO FORESTS

Forests are products of hundreds of millions of years of evolution. During this time, natural climate cycles have caused forests to adapt by changing the types of vegetation that live within them and by migrating to new habitats as conditions gradually change. As this has occurred in the past, it has happened at a much slower pace, allowing forests to successfully adapt. Today, changes in climate (temperature, precipitation, humidity, and air flow) are happening at a much more rapid pace, and forests have not had the luxury of time to adapt.

Forests are extremely valuable ecosystems—they help to regulate rainfall and are also key sources of food and medicine. They provide abundant wildlife habitat, carbon storage, clean air and water, and recreational opportunities. They also provide a bounty of natural resources, such as wood, plants, berries, water, wildlife, and aesthetic values. The health and diversity of forests is largely influenced by climate. Native forests adapt to the local climate features. For example, in the far-northern boreal forests, cold-tolerant species, such as white spruce, have adapted there; in drier areas, conifer and hardwood forests thrive because they need less water. Because air temperature affects the physiological processes of individual plants and productivity, the effects of climate change have a strong influence on the health of tomorrow's forests. Not all forests will have the same outcomes under the influence of climate change. Some forests will die back, whereas others may extend their ranges. Amounts of CO_2 will vary as well. Whatever the outcome of a particular forest, if the population cannot adapt or migrate with the changes, they will face extinction.

According to the IPCC, at least one-third of the world's remaining forests may be negatively impacted by climate change during this century. It may force plant and animal species to migrate or adapt faster than they physically can, disrupting entire ecosystems. The IPCC also predicts that forests will have changes in fire intensity and frequency and increased susceptibility to insect damage and diseases (IPCC, 2016).

Climate scientists are currently employing a variety of methods to predict the impacts of climate change on forests. On a global, or a large regional, scale, they are predicting shifts in ecosystems by combining biogeography models with atmospheric general circulation models (GCMs) that project changes in an environment where the CO_2 content has doubled. They also use biogeochemistry models to simulate the carbon cycle; flow of nutrients; and changes in precipitation, soil moisture, and temperature to study ecosystem productivity. They have developed global models that simulate worldwide changes in vegetation composition and distribution.

The IPCC states that climate change both directly and indirectly affects the health of forests. The direct impacts of warmer temperatures, changing rainfall patterns, and severe weather events can already be seen in certain tree and animal species (IPCC, 2016). Even small changes can affect forest growth and survival—especially the portions that lie along the outer edges of the ecosystem where the conditions are marginal. When it gets hotter, more water is lost through evapotranspiration, which causes drier conditions and decreases the vegetation's use of water. Water temperatures can also throw off the timing of flowering and fruiting for plants and adversely affect their growth rate. Forests will also be threatened when the seasonal precipitation patterns they have been used to change; water is not supplied when it is needed, causing drought conditions and stress, or supplied too much when it cannot be assimilated, causing flooding and mudslides.

Natural Resources Canada (2016) has found that the age and structure of the forest also plays an important role in determining how quickly a forest responds to changes in moisture conditions. Mature forests have well-established root systems, enabling them to tolerate drought better than younger forests or forests that have been disturbed in some other way, such as through disease infestation. Species type also plays a part—some species are more resistant than others.

Canadian researchers have identified 141 potential climate change indicators. They then selected seven of those to focus on for their value in raising awareness of ongoing changes and their ability to track and report on, and provide valuable information for dissemination. Those indicators fall within three general categories: climate, forest, and human. Although all indicators are affected in some way by other factors, they are useful for studying the effects of climate change and its mitigation in a practical way. The seven factors they are focusing on in their model are as follows, as an example of a practical approach in dealing with the real effects of climate change in a forest system:

1. The climate system
   a. *Drought*: Why is drought important, what has changed, what are adaptation tools and resources available?
   b. *Fire weather*: How has the length of the fire season increased/changed?
2. The forest system
   a. *Distribution of tree species*: Why is the distribution of tree species important with its interaction to the environment (water and nutrient cycling, competition, etc.).
   b. *Fire regime*: Wildland fire regimes influence the ecosystem and affect forest resource availability and human safety, health, and property.
   c. *Tree mortality*: It provides a measure of forest health and also affects the carbon balance.
3. The human system
   a. *Cost of fire protection*: Significant resources are invested to protect both forest and public resources.
   b. *Wildland fire evacuations*: Emergency evacuations prevent injury and death when wildfires threaten communities.

In the short term (50–100 years), changes due to climate change will be focused on ecosystem function. In the long term, shift of forest types will be more significant. Boreal forests will be impacted the most severely, as their ecosystem will be greatly reduced because significant warming is expected in the polar regions. According to Dmitry Schepaschenko of Austria's Institute for Applied Systems Analysis, "The changes could be very dramatic and very fast." In a study focused on climate change effects on boreal forests, he has discovered that climate change is forcing the boreal forests that cover the majority of Northern Canada to the tipping point. The study was done in conjunction with three authors, including Dmitry Schepaschenko of Austria's Institute for Applied Systems Analysis; another scientist from Natural Resources Canada who was unable to speak on record due to federal restrictions; and a third scientist who remained unidentified. They report that, although the region remains largely intact, it faces the most severe expected temperature increases anywhere on the earth. They predict that parts of Siberia will likely become 11°C warmer. The expected results include greater precipitation, but not enough to remediate the dryness caused by the hotter weather. This means that a drier boreal forest will suffer from new diseases, insect infestations, and uncontrollable wildfires (Weber, 2015).

Some of the most vulnerable temperate forests will be the "island" or isolated forest communities, such as the fragmented forests encroached upon by urban and agricultural areas. They will also be at risk because there will be no migration options. Forests located in the high elevations on mountain systems face a similar threat—as they migrate upward, eventually there will be nowhere left to migrate. Individual species that are indigenous to small geographic areas or have limited seed dispersal will also be threatened and endangered. Under climate change conditions, surviving forests of the future are expected to look very different from those of today. Although changes will vary in degree from area to area, all forests will be affected in some manner. Impacts of climate change are already manifesting, such as forests in the United States and Canada. Over the past century, a 1°C–2°C increase in air temperature and changes in precipitation have already been documented. Experts believe that higher levels of CO_2 have already caused forest dieback of forested areas along the Pacific and Atlantic coasts (Weber, 2015).

It is expected that temperate forests in the United States will migrate northward from 100 to 530 kilometers over the course of the next century. If the air temperature warms 2°C over the same time period, models have predicted that tree species will have to migrate from 1.6 to 4.8 kilometers per year, which is too fast for most temperate species, except for those whose seeds are carried by birds over greater distances. As a result, it is expected that grasslands will dominate many of these areas.

The effects of climate change are felt most at the poles, where the predicted rise in temperature could climb 5°C–10°C or higher over the next century. It is estimated that warming will negatively affect the species that live in the ecosystems and that approximately 24–40 percent of the species living in the boreal forests right now will be lost. Species that live in the north will be crowded out by species from the temperate regions that will be migrating northward in search of cooler climates. The species that will invade the present boreal forests will be today's temperate forest species and grasslands. As the boreal forest vegetation is forced out, it will migrate poleward 300–500 kilometers during the next century. Proof of this can already be seen in Western Canada, as their plant zones have already begun shifting poleward (IPCC, 2014b). Migrating vegetation will encounter severe challenges, however. For example, the soils in the tundra region are not fertile and conducive to high-density vegetation or tree growth. They lack the biota necessary for colonization. Specific seed dispersal rate and migration tolerance range are also the important factors that could play a role in keeping trees from being able to survive the poleward migration rate dictated by climate change.

There are other factors that will hurt species migration as well, such as habitat fragmentation (small isolated population clusters rather than one large cohesive unit) and competition from more hardy species. As temperatures change, it may also affect the timing and rate of seed production, which will affect the growth and strength of the trees. Trees that have limited seed dispersal mechanisms will also suffer. For instance, trees whose seeds are carried long distances by the wind will have a better chance of survival than those whose seeds fall closer to the tree.

The ability to adjust to a larger temperature range will also play an important role. Vegetation with narrow temperature tolerances will be vulnerable to extinction. The IPCC (2014b) has stated that a drastic change in species composition and loss of habitat with even a 2°C warming near the poles will damage the ability of an ecosystem to function as species richness begins to be killed off.

In a study conducted by Natural Resources Canada, an average rise in temperature of 1°C over Canada in the last century has had a negative impact on vegetation. At mid to high latitudes (45°N–70°N), plant growth and the length of the growing season have increased. In portions of Western Canada, there has been a decrease in rainfall as temperatures have risen and this has hurt the growth of some tree species, such as aspen poplar. In Alberta, aspen are now blooming 26 days earlier than they were a hundred years ago (Kerr, 2009).

Another major concern for boreal forests in a warmer climate is insect infestations. Insects commonly found in temperate forests, such as mountain pine beetle, will migrate north along with the forests and continue to infest and infect as they move northward, devastating industries such as logging and tourism.

As temperatures climb, drought-like conditions may develop. If this happens, there will also be greater incidences of wildfire. In the past 40 years, the trend has already been established that as climate warms, wildfires have become more frequent and are burning larger areas. This overall trend has been seen in places such as southern California, and also in boreal locations in Canada and Russia. As climate change continues, longer fire seasons, drier conditions, and more frequent severe electrical storms are projected to increase, causing the fire season to become more problematic and devastating as the climate continues to change.

Although it is true that some forest species' seeds are actually dispersed by fire, which will aid their migration, and burnt litter will add nutrients to the soil, over time reoccurrences of wildfire will fragment established vegetation colonies and make it more difficult for them to migrate. In addition, as older trees burn, they will add carbon to the atmosphere, and as younger trees replace these burnt areas, there will be less initial carbon storage capability.

The world's tropical forest ecosystems are very sensitive to disturbances such as overgrazing, logging, plowing, and burning. Converting natural ecosystems to agricultural and logging uses combined with climate change is the largest threat to the rain forests today. According to the Food and Agriculture Organization (FAO) of the United Nations, a land area the size of Ireland or South Carolina (13 million hectares) is lost to these uses every few years in the rain forests. From the mid-1990s to the present, more than 100 million hectares of tropical rainforest has been deforested (FAO, 2010). Developing countries are hit harder than developed countries. Unfortunately, it is often seen as necessary to farm the land in pursuit of food or sell the land to logging companies in exchange for needed income.

According to Jose Antonio Marengo of Brazil's National Space Research Institute, if the issue of climate change is not addressed today, the rain forests will be negatively impacted. They will receive less rainfall in addition to experience higher temperatures, enough so that it could transform the Amazon—the world's largest remaining tropical rain forest—into a savanna (grassland) by the end of the century. The Amazon covers almost 60 percent of Brazil and hosts one-fifth of the world's freshwater and nearly 30 percent of the world's plant and animal species. Marengo is involved in research focused on the following two different scenarios:

1. If no action is taken to slow or halt climate change or deforestation, temperatures will rise 5°C–8°C by 2100, and rainfall will decrease by 15–20 percent. This is the scenario that would transform the Amazon into a savanna.
2. If action is taken now to slow climate change, temperatures would most likely rise 3°C–5°C by 2100, and rainfall will decrease by 5–15 percent.

Marengo stresses that "if pollution is controlled and deforestation reduced, the temperature would rise by about 5°C by 2100. Within this scenario, the rain forest will not come to the point of total collapse." He also stated that he was optimistic that the worst-case scenario could be averted, but that it would require a major effort by industrialized nations to reduce emissions of greenhouse gases, such as carbon dioxide (Environmental and Energy Study Institute, 2007).

Unfortunately, many of the world's rain forests are being cut down at an accelerated rate for agriculture, pasture, mining, and logging. When these forests are cut down or burned, huge amounts of carbon are reintroduced back into the atmosphere. Tropical deforestation accounts for about 20 percent of the human-caused carbon dioxide emissions each year to the earth's atmosphere. This makes deforestation of the rain forests an important issue when dealing with the challenges of climate change.

THE FACTS OF DEFORESTATION

- Forests cover 30 percent of the earth's land.
- It is estimated that within 100 years, there will be no rainforests.
- Agriculture is the leading cause of deforestation.
- One and a half acres of forest is cut down every second.
- Loss of forests contributes 12–17 percent of annual greenhouse gas emissions.
- The rate of deforestation is equivalent to losing 20 football fields every minute.
- There are more than 121 natural remedies that we currently know of from the rainforest which can be used as medicines.

- The United States has less than 5 percent of the world's population, yet consumes nearly one-third of the world's paper.
- The Amazon rainforest produces 20 percent of the world's oxygen.
- Up to 28,000 species are expected to become extinct by the next quarter of the century because of deforestation.
- Half of the world's tropical forests have already been cleared.
- Approximately 4500 acres of forests are cleared every hour by forest fires, bull dozers, and machetes.
- Poverty, overpopulation, and unequal land access are the main causes of man-made deforestation.
- Industrialized countries consume 12 times more wood and its products per person than nonindustrialized countries.
- Soil erosion, floods, wildlife extinction, increase in climate change effects, and climate imbalance are some of the effects of deforestation.
- Worldwide, more than 1.6 billion people rely on forest products for all or part of their livelihoods.
- Tropical forests, where deforestation is most prevalent, hold more than 210 gigatons of carbon.
- Tropical rainforests contain over half of all the plant and animal species in the world.
- Deforestation affects the water cycle. Trees absorb groundwater and release the same into the atmosphere during transpiration. When deforestation occurs, the climate automatically changes to a drier one and also affects the water table.
- The world's forests store 283 billion tons of carbon present in the biomass.
- Forty-four percent of junk mail goes unopened.
- On average, each person in the United States uses more than 700 pounds of paper every year.

(*Source*: Conserve Energy Future, Deforestation facts, http://www.conserve-energy-future.com/various-deforestation-facts.php [accessed August 20, 2016], 2016.)

In view of the impacts that may be felt by the world's forests—temperate, boreal, and rain forest—it is important to begin looking at adaptation strategies today. The World Wildlife Fund has suggested the following as viable adaptation options in order to reduce the threat on forest ecosystems from climate change:

- Reduce present threats that could harm the ecosystem, such as prevent degradation and the introduction of invasive species.

- Manage large areas of land in a comprehensive way tied to the landscape. Focus on all the components that compose a large-scale area and plan for the adaptation migration of different species.
- Provide buffer zones and flexibility of land uses. The land must be managed to accommodate the movements of species migration.
- Protect mature forest stands. Mature trees are better able to withstand large-scale environmental changes and also provide a safe habitat for other species to adapt within.
- Maintain the natural fire regime. Different ecosystems have different fire ecologies, necessitating the development of fire management plans.
- Pests must be actively managed. Because climate change is associated with invasions of insects, disease, and exotic species, healthy management practices must be set in place to protect the forests. This could include measures such as prescribed burning and nonchemical pesticides.

Different forested areas will require different monitoring, planning, and protection strategies; it is imperative that humans understand the unique forest ecosystems and their value (World Wildlife Fund, 2003).

IMPACTS TO RANGELANDS, GRASSLANDS, AND PRAIRIES

Native grasslands cover about one-quarter of the world's land surface, making them a significant component of the world's vegetation. Therefore, any impact on them could seriously impact a major global ecosystem. The extent and health of natural grasslands are typically controlled by rainfall and fire. These two factors limit the extent of their range; with an increase in climate change, it is already known that there will be a decrease in rainfall in some areas and an increase in wildfire.

The world's grasslands stand to face serious consequences if climate change continues. These regions of the world provide grains and crops for the world's population, rangeland for cattle and sheep, and habitat for wildlife, and play a role in carbon sequestration and maintaining an overall healthy ecological balance in nature. Grasslands are also important for carbon storage. Whereas carbon is stored in trees, plants, and organic matter on the surface of the soil in forests, in grasslands it is stored in the soil. Because of human development, most of the earth's grasslands suitable for agriculture have already been developed. According to the FAO, as much as 70 percent of the earth's 1.3 billion hectares of grasslands have become degraded, mostly due to overgrazing (Conant, 2010).

Temperate grasslands are located north of the Tropic of Cancer (23.5°N latitude) and south of the Tropic of Capricorn (23.5°S latitude). The four major temperate grasslands of the world are the plains of North America, the veldts of Africa, the steppes of Eurasia, and the pampas of South America. Dominant vegetation types in these areas are grasses. The principal places that trees, such as cotton-wood, oak, and willow, are found are along river valleys, growing in areas near a reliable water source. The soil in temperate grasslands is nutrient rich due to the growth and decay of deep, dense grass roots. The roots serve to hold the soil together as a cohesive unit. In fact, some of the world's most fertile soils are found in the grasslands of the eastern prairies in the United States, the steppes of Russia and Ukraine, and the pampas of South America.

Tropical grasslands are located near the equator, between the Tropic of Cancer and the Tropic of Capricorn. These are the grasslands that occupy Africa, Australia, India, and South America. Tall grasses dominate these landscapes—and they are found in tropical wet and dry climates, where temperatures never fall below 18°C. Usually dry, there is a season of heavy rain, bringing an annual rainfall of 51–127 centimeters per year. One of the biggest threats to the tropical grasslands with increased climate change is desertification.

In the IPCC's Fourth Assessment Report, they state that the structure and function of the world's grasslands makes them one of the most vulnerable to global climate change of any of the terrestrial ecosystems. Grasslands are vulnerable to vegetation change caused by changing temperatures and precipitation due to climate change (IPCC, 2007b).

GCMs referenced by the IPCC predict grassland ecosystems will experience climatic changes such as higher maximum (daytime) and minimum (nighttime) temperatures, and more intensive precipitation events. Recent studies of climate change processes reveal a surprising situation—daytime high temperatures are not causing problems, whereas nighttime high temperatures are causing problems. The situation is most pronounced during late winter and early spring. Nighttime high temperatures have risen, changing the temperature regime enough that the last frost date occurs on average two weeks earlier now than it did 20 years ago. In addition, native grasses are now germinating later. Invasive and noxious plants that have invaded grasslands over the years are germinating early, taking the moisture out of the soil and using the nutrients that would have been available for the grasses. In addition, cattle grazing occurs on much of the grassland areas, and the animals do not ingest the weeds, which further promotes their invasion.

Based on the results of the IPCC studies, in the world's grasslands that are expected to receive an increase in precipitation, such as the western United States, these areas may not be faced with the drying out resulting from lack of water and rising temperatures, but these areas may suffer from accelerated nutrient cycling, which in turn could encourage the spread of more invasive species. In the hot desert grasslands (such as Australia, the Sonoran and Chihuahua deserts of Mexico, and the United States), GCMs predict an increase in the frequency of intense rainfall and flash floods. These areas are expected to have increased erosion and nutrient loss. In ecosystems that have already been disturbed in other ways, such as by wildfire or with desertification, they will be impacted even more and have a more difficult time surviving under climate change conditions (IPCC, 2007b).

There are several threats to grasslands in the face of climate change. Marginal grasslands could be converted into deserts as rainfall decreases. The threat of desertification is all too real in many parts of the world. As areas struggle to find water to support populations, cultivate crops, and support livestock, grassland ecosystems stand to face further degradation as temperatures rise during climate change. Therefore, as climate changes, it is important for land managers to adapt to the changes as they happen and manage the land in a responsible way. One of the biggest obstacles they face with climate change is the threat of invasive species encroaching on native grassland species. Invasion of species can happen via seed deliverance by livestock, off-road vehicles, road maintenance crews, and outdoor recreationalists. Invasive species need to be dealt with immediately before they become established. Land managers will also have to put management plans in place in order to be prepared for the changes that climate change will bring.

IMPACTS ON POLAR ECOSYSTEMS

The effects of climate change will be felt the strongest in the earth's polar regions. Already temperatures have climbed by 2.4°C in some areas in polar regions, whereas the global average is 0.6°C over the past century. Weather conditions are so harsh that polar ecosystems must maintain a delicate balance to survive. In addition, because of their sensitive nature, the polar regions are sometimes the first ecosystems to show warning signs of climate change, such as acceleration in the melting of glaciers. In the Arctic, climate change is expected to be rapid and extensive. As temperatures warm up and ice melts, Arctic ecosystems will be impacted heavily over the next century. There is extensive sea ice at the periphery of the Arctic Ocean that forms and melts each year. These waters are important to the fishing industry, accounting for almost half of the total global production. Glacial decline, melting sea ice, rising sea levels, impacts on wildlife habitat, melting permafrost, impacts on native tribal inhabitants, and changes to plant life are all being affected in the Arctic today because of climate change.

Arctic ice is getting thinner and more prone to breaking. The Ward Hunt Ice Shelf, located on the north coast of Ellesmere Island, Nunavut, Canada, is one example—it was the largest single block of ice in the Arctic. About 3000 years old, it began cracking in the year 2000. By 2002, it had broken all the way through. Today it is breaking into pieces. The breakup of the Ward Hunt Ice Shelf has already caused several impacts: polar bears, whales, and walrus have changed their migration and feeding patterns; native people are having to change their hunting territories; and coastal villages are being flooded (Figure 4.3).

The Northwest Passage is another example of the extreme effects of climate change and the results of warming in the fragile polar environment. There was a time when the Northwest Passage was considered impassable. It was a highly desired route sought after by maritime merchants, a route that would save time and mileage if passable, but elusive. For most

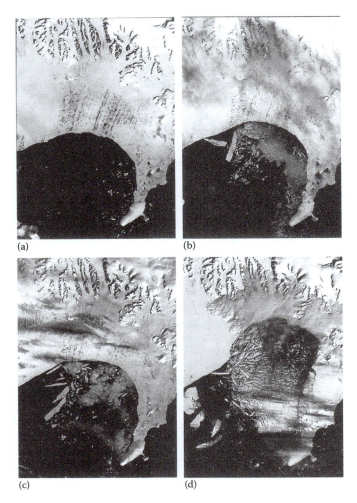

(a)

(b)

(c)

(d)

FIGURE 4.4 Map of the Northwest Passage. (Courtesy of NASA, Washington, DC.)

FIGURE 4.3 Large portion of the Larsen B Ice Shelf disintegrated in 2002, as evidenced in these MODIS satellite images from January 31, 2002 (a), February 17, 2002 (b), February 23, 2002 (c), and March 5, 2002 (d). Earth scientists predict that if global warming continues, incidences like this will become more common. (Courtesy of National Snow and Ice Data Center [NSIDC], Boulder, Colorado.)

of recorded history in North America, the Passage has been nearly impassable and deadly for those who have attempted its navigation. With the modernization of ships and warming of the earth with climate change, however, navigation through the Canadian Archipelago from Baffin Bay to the Beaufort Sea has become more common, although still a considerable challenge. Figures 4.4 through 4.7 illustrate the transition over time of the area in a warming environment.

Scientists for the United States Geological Survey (USGS) estimate that Alaska contains more than 100,000 glaciers (including about 60 active and former tidewater glaciers), which cover roughly 75,000 square kilometers, or about 5 percent of Alaska. The USGS has determined that most glaciers in every mountain range and island group in the state are currently undergoing significant retreat, thinning, or stagnation—most notable the glaciers at lower elevations. Some of these glaciers began this process as early as the mid-1700s. Although a few are currently advancing, more than 99 percent of them are retreating. In fact, during the past decade, Alaska's coastal glaciers have

FIGURE 4.5 The Northwest Passage, Arctic partially open: The Northwest Passage as it appeared on August 31, 2015, to the Visible Infrared Imaging Radiometer Suite (VIIRS) on the Suomi-NPP satellite. Note that much of the white covering the Northwest Passage in the VIIRS image is cloud cover, not sea ice. The Northwest Passage is a complex, winding maze of sounds, channels, bays, and straits that pass through often ice-choked Arctic waters. Mariners refer to two main routes: a southern passage and a northern passage. (Courtesy of NASA, http://earthobservatory.nasa.gov/IOTD/view.php?id=86589.)

FIGURE 4.6 Northwest Passage Open: On September 15, 2007, the MODIS on NASA's Terra satellite captured a largely cloud-free image of the Northwest Passage. Although the sea route had been characterized as nearly open weeks earlier, persistent cloud cover prevented a MODIS true-color image of the open route. Clouds do obscure parts of this image, but they remain confined to areas north and south of McClure Strait. As white as the clouds is the snow cover on some of the islands, especially around Parry Channel. Clear ocean water appears dark, but some sea ice still appears. In the east, along Ellesmere Island, faint white swirls indicate new sea ice forming at the end of summer melt season. McClure Strait also sports sea ice, and those blocky, broken up shapes suggest old sea ice undergoing melt. The generally dark color of McClure Strait could be concealing a thin layer of sea ice throughout the strait, possibly centimeters thick. (Courtesy of NASA, http://earthobservatory.nasa.gov/NaturalHazards/view.php?id=18964.)

FIGURE 4.7 In late July 2012, sea ice retreated from Parry Channel in the Northwest Passage. The Canadian Ice Service reported that ice cover in the channel began to fall below the 1981–2010 median after July 16, and the ice loss accelerated over the following two weeks. The MODIS on NASA's Terra satellite captured this natural-color image of Parry Channel and McClure Strait (also known as M'Clure Strait) on August 2, 2012. Although Parry Channel was almost entirely ice free, some ice lingered between Melville and Victoria islands. In this image, that ice varies in color from nearly white to dark gray. For centuries, explorers and traders have sought an Arctic shipping shortcut between Europe and Asia. In the early twentieth century, Norwegian explorer Roald Amundsen managed to navigate the southern route of the Northwest Passage, although the journey took him nearly three years. Parry Channel occupies the northern or "preferred" route, which also opened in 2007. The southern route through the Northwest Passage opened in 2008. (Courtesy of NASA, Washington, DC.)

added as much (or more) meltwater to the global ocean as the ice sheets of Greenland or Antarctica, making these glaciers a significant factor in global sea-level rise (Puckett and Molina, 2008). The photo sequence in Figures 4.8 through 4.10 illustrates tidewater glacial calving.

In a report in the *Washington Post*, scientists at the National Oceanic and Atmospheric Administration (NOAA) say that the Arctic ice cap is melting faster than that scientists had originally expected and will most likely shrink 40 percent by 2050 in most regions, which will devastate wildlife populations, such as polar bear, walruses, and other marine animals. They also predict that Arctic sea ice will retreat hundreds of kilometers farther from the coast of Alaska in the summer. Although this could be good news for fisherman and open shipping routes and new areas for oil and gas exploration, it will spell disaster for the wildlife species that inhabit the region. In 2001, the IPCC predicted that there will be "major ice loss by 2100." Then, in 2007, when they issued their next report, they remarked that "without drastic changes in greenhouse emissions, Arctic sea ice will 'almost entirely' disappear by the end of the century." In 2014, they added that the Arctic has warmed at about twice the global rate in the past three decades. For this reason, the loss of Arctic sea ice over the past three decades—a rate of about 3.5–4.1 percent—is unprecedented. It is expected that the Northern Sea Route may have up to 125 days per year that are suitable for navigation by 2050 (CarbonBrief, 2014).

FIGURE 4.9 The mass of ice cascades to the bay below, more than 30 meters (100 feet). (Courtesy of Kerr.)

FIGURE 4.8 The Aialik Glacier, located in Kenai Fjords National Park, part of the network of tidewater glaciers at Resurrection Bay, is about 24 km (15 miles) from Seward, Alaska, a major fishing port. The massive wall of ice is constantly groaning and shifting. Pieces calve off the glacier, some quite spectacular, as can be seen in this sequence of three photos. This shows a large chunk of ice break free. (Courtesy of Kerr.)

FIGURE 4.10 When the ice hits the water, the sound of the impact carries across the bay and the ice breaks apart into smaller chunks. Pieces calve off the glacier every few minutes, marking this as a glacier in retreat. (Courtesy of Kerr.)

The New York Times published a report on May 1, 2007, showing that climate scientists may have significantly underestimated the power of climate change from human-generated heat-trapping gases to shrink the sea ice in the Arctic Ocean. Dr. Julienne Stroeve, a researcher at the National Snow and Ice Data Center (NSIDC) in Boulder, Colorado, discovered that since 1953, the area of sea ice in September has declined at an average rate of 7.8 percent per decade.

"There are huge changes going on," Dr. Stroeve says. "Just with warm waters entering the Arctic, combined with warming air temperatures, this is wreaking havoc on the sea ice." (Revkin, 2007). Since that time, the NSIDC has continued to monitor Arctic sea ice, along with scientists from NASA, recording its decline. In March, 2016, they confirmed it was at a record low maximum extent for the second straight year.

"I've never seen such a warm, crazy winter in the Arctic," said NSIDC director Mark Serreze. "The heat was relentless." Air temperatures over the Arctic Ocean for the months of December, January, and February were 2°C–6°C above average in nearly every region. Sea ice extent over the Arctic Ocean in 2016 averaged 14.52 million square kilometers on March 24, beating last year's record low of 14.54 million square kilometers on February 25. Unlike 2015, the peak in 2016 was later than average in the 37-year satellite record, setting up a shorter than average ice melt season for the coming spring and summer.

The September Arctic minimum began drawing attention in 2005 when it first shrank to a record low extent over the period of satellite observations. It broke the record again in 2007, and then again in 2012. The March Arctic maximum has typically received less attention. This changed in 2015 when the maximum extent was the lowest in the satellite record.

"The Arctic is in crisis. Year by year, it's slipping into a new state, and it's hard to see how that won't have an effect on weather throughout the Northern Hemisphere," said Ted Scambos, NSIDC lead scientist (NSIDC, 2016).

The warmer it gets, the more the glaciers melt, which adds freshwater to the ocean and upsets the earth's energy balance, ocean circulation patterns, and ecosystems. Some estimates of global sea-level rise from the Arctic melt have been placed at 1 meter by 2100. According to the U.S. Environmental Protection Agency, this increase would drown roughly 58,016 square kilometers of land along the Atlantic and Gulf coasts of the United States—specifically Texas, Florida, North Carolina, and Louisiana (Figure 4.11).

Warming in the Arctic will also affect weather patterns and food production worldwide. An example of this is given by NASA: in one of their computer models, they calculated that Kansas would be 2.4°C warmer in the winter without Arctic ice, which usually sends cold air masses into the United States. Without these Arctic-induced cold air masses, winter wheat cannot be grown and soil would be 10 percent drier in the summer, wreaking havoc on summer wheat crops (AMEG, 2016).

Since its record low in 2005, when the ice cap covering the Arctic Ocean melted to a size smaller than it had been since records began to be kept a century ago, the situation has become chaotic. In Greenland, glaciers are melting and traveling now faster to the ocean, adding freshwater. These changes in the

FIGURE 4.11 This photo, taken in Denali National Park on July 21, 2016, shows the extreme glacial melting that has taken place in the Arctic environment. Visible are the remnant cirques, prominent lateral moraines, diminished central glacier at the bottom, remaining kettle lakes, and other features signaling the rapid melting of the glaciers in the area.

Arctic have the potential to impact the entire earth. One of the most serious consequences could be the disruption of the thermohaline circulation, with the ramifications discussed in Chapter 2.

The Arctic Council, an intergovernmental forum composed of eight Arctic nations (Canada, Denmark/Greenland/Faroe Islands, Finland, Iceland, Norway, Russia, Sweden, and the United States) and six Indigenous People's organizations, had an assessment prepared by the Arctic Monitoring and Assessment Programme, the Conservation of Arctic Flora and Fauna, and the International Arctic Science Committee of impacts on the Arctic as a result of climate change. They support the notion that the Arctic is extremely vulnerable to observed and projected climate impact, and that it is currently facing some of the most rapid and severe climate change on earth. Over the next 100 years, they report that climate is expected to change and these changes in the Arctic will be felt worldwide. Although some of these changes are those caused by nature, the overwhelming trends and patterns are a result of human influence, resulting from the emission of greenhouse gases.

One of the biggest key climate change indicators in the Arctic is the declining sea ice. Researchers actually use the quantitative amounts of Arctic sea ice as an early warning system of climate change. According to the Arctic Research Center, over the past 30 years, the average sea ice extent has decreased by 80 percent, which is the equivalent area of roughly one million square kilometers. By comparison, this represents an area roughly the size of Egypt. Of even more concern, the melting rate is accelerating. The most pronounced season of melting is in the late summer. It is projected that by 2100, late-summer sea ice will have decreased in a range somewhere from 50 percent to complete disappearance (Graph 4.1) (NSIDC, 2016).

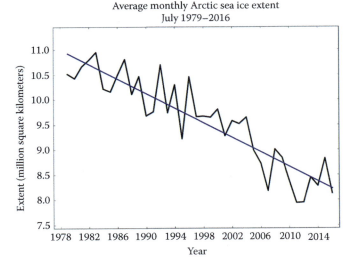

Average monthly Arctic sea ice extent
July 1979–2016

GRAPH 4.1 Monthly July ice extent for 1979–2016 shows a decline of 7.3 percent per decade. (Courtesy of National Snow and Ice Data Center [NSIDC], Boulder, Colorado.)

Another major global impact that will occur as a result of ice sheet melting in the Arctic is sea-level rise. The Greenland ice sheet is the largest ice sheet in the Arctic. In April 2016, polar researchers were stunned when nearly 12 percent of Greenland's ice sheet began melting, as observed by scientists at the Danish Meteorological Institute, beating the previous record set in May 2010 for 10 percent melt. Temperatures on the ice were in excess of 10°C in some locations. Even a weather station located at 1840 meters above sea level recorded a maximum of 3.1°C, which data analysts said would be warm for July, let alone April (The Guardian, 2016).

Sea-level rose roughly 8 centimeters over the past 20 years, and as temperatures rise, the rate is increasing. This is due to two factors: melting ice and thermal expansion of seawater. The global average sea level is projected to rise anywhere from 10 to 90 centimeters by 2100. Some climate models have predicted that the Greenland ice sheet will completely melt, causing sea levels to rise 7 meters. By comparison, a 50 centimeters rise in sea level will typically cause a shoreward retreat of coastline of 50 meters if the land is relatively flat. This would flood areas, causing significant environmental and economic impacts (Ritter and Hanley, 2011).

IMPACTS OF CLIMATE CHANGE ON DESERT ECOSYSTEMS

The world's principal deserts are found in two major zones: at 25°–35° latitude north and south of the equator. Desert ecosystems are defined as regions that are very arid and dry—receiving less than 25 centimeters of rain a year. By definition, deserts are not confined to just hot, dry areas—they can also exist in the cold, dry areas within polar regions. Because these regions are so physically harsh, the plants and animals that live within them have learned to adapt and survive in an extremely hostile environment. Covering about one-fourth of

the earth's land surface (54 million square kilometers), they are dominated by bare soil and scarce vegetative cover.

As climate change continues, current deserts will be impacted as temperatures rise, water becomes scarcer, and food sources become problematic. As desertification spreads due to climate change, ecosystems worldwide will be impacted in some way. Understanding the fragile balances in desert ecosystems and how climate change will relate to them is important as land managers look at future management options.

Climate change is expected to cause intensification in the hydrologic cycle, with a marked increase in evaporation over land and water. The higher the evaporation rate, the greater the drying will be of soils and vegetation. In areas where there is a decrease in rainfall and an increase in evaporation, droughts will occur, be more severe, and last longer.

Drought already threatens the lives of millions of people on earth. As the atmosphere continues to warm under the effects of climate change, the negative effects and hardships brought on by drought will spread across the globe as areas heat up and dry out. Regions that are already dry are expected to become drier as climate change affects the process of evapotranspiration. This will lead to increased drought in dry areas where rivers, lakes, and groundwater levels are lower than normal, and soils in agricultural areas lack sufficient moisture, negatively impacting food production. Precipitation amounts have steadily declined in the tropical and subtropical regions since the 1970s. Areas becoming continually drier include southern Africa, the Sahel region of Africa, southern Asia, the Mediterranean, and the Southwestern United States.

With this continuing trend, scientists expect that the amount of land affected by drought to increase by mid-century, as well as water resources to decline up to 30 percent. The reason for this is due to the shifting of the Hadley Cell circulation pattern. What they foresee happening is that warm air normally rises at the tropics, loses its moisture to tropical thunderstorms, and descends in the subtropics as dry air (the Hadley Cell); as the climate warms, the atmospheric circulation will shift the circulation poleward toward higher latitudes, causing storm patterns to shift with them and expanding the region that the descending drier air will affect. Because it is a larger region, more areas will be susceptible to drought (UCS, 2011).

A NOAA measure of drought from a climate model, called the Palmer Drought Severity Index (PDSI), developed in the 1960s by Wayne Palmer, predicts there will be a notable increase during this century with predicted changes in rainfall and heat around the world because of climate change. The PDSI figure for "moderate drought" is currently at 25 percent of the earth's surface. By the year 2100, it predicts it will rise to 50 percent. The figure for "severe drought" is currently at 8 percent; by 2100, it is predicted to rise to 40 percent. The figure for "extreme drought" is currently at 3 percent; by 2100, it is predicted to rise to 30 percent, based on the study conducted by the Hadley Centre (Worldwatch Institute, 2013). What this means is that one-third of the earth's land surface will no longer be able to produce food in just a matter of a few decades. To complicate matters, the areas most prone to drought are those that are currently most densely inhabited, which means that hundreds of millions of people will no longer be able to feed themselves.

For a realistic glimpse into the future, picture this: predictions are that valleys once fertile will turn dry and brown. For inhabitants of desert valleys that rely on a short rainy season each year in order to be able to grow crops and graze their animals, they will wait in anticipation for rains that will never come. Year after year, the situation will repeat itself as millions of people near starvation. Nomadic herders' animals will die; their cattle are emaciated to skin and bones now. Bleached skeletons of cows, goats, and sheep will litter the barren, dusty landscape. Nomadic herders will set out and walk for weeks without finding a watering hole or a riverbed. As people begin dying of starvation, inhabitants of different geographic areas will start fighting each other for what slim, meager resources are still in existence.

As brutal as this may sound, if climate change continues, this scenario will become commonplace. The number of food emergencies in Africa each year has almost tripled since the 1980s. Across sub-Saharan Africa, one in three people today is undernourished (Borgen Project, 2014).

In early 2007, the world witnessed public riots and protests due to food security issues. In sub-Saharan Africa, for example, chronic food security issues hit the undernourished hard, which had harsh effects on world peace and security. A year later, this was followed by rising commodity prices. It is expected that problems will continue to escalate as climate change continues and food insecurity issues increase. Governments there have not prioritized food security as a critical issue; there are low agricultural budgets, weak institutional capacity, difficulty in coordination, and declining donations, which exacerbate the problem. In 2008, the United Nations Development Program (UNDP) warned that any positive socioeconomic developments that had been achieved during the decade with the implementation of the Millennium Development Goals program (a United Nations project designed to address extreme poverty) may be seriously impacted or reversed by climate change due to new threats to water and food security, agriculture production and access, and nutrition and public health concerns. They also warned that the impacts of climate change, such as rising sea levels, droughts, heat waves, floods, and unpredictable rainfall, could cause another 600 million people to become malnourished and increase the number of people facing water scarcity to 1.8 billion by 2080 (Mburia, 2011).

DESERTIFICATION

One of the impacts that climate change may have on the surface of the Earth is to exacerbate the worldwide progression of desertification. If there is a significant decrease in the amount of rainfall reaching an arid or semiarid area, it could increase the extent of the dryland areas globally, destroying both vegetation and soils. Desertification is a degradation process inflicted on arid and semiarid landscapes due to human activities, climate change, or a combination of both. Although desertification has always existed, it has become more prominent and a much bigger concern in recent years as populations have rapidly expanded across previously untouched landscapes and arable land has been cultivated and grazed.

The first most well-known incidence of desertification in the United States was the Dust Bowl era of the 1930s, caused by a combination of drought and improper farming and land management practices. During this period of time, millions of people were forced from their homes in the biggest migration in U.S. history and forced to abandon their farms, The Great Plains were eventually restored over time through the practice of good land management plans, improved agricultural methods, and responsible conservation efforts, initiating the birth of the Soil Conservation Service. In other areas of the world, however, population explosion and increasing livestock pressure on marginally healthy land have worsened the problem of desertification, speeding it up. As the problem intensifies, the productive capability of the land decreases, destroying the original biodiversity. Different plant species produce physically and chemically different litter compositions in the ground. The litter, along with the natural biologic decomposers in the ground, helps form the soil complex and plays an active role in nutrient cycling. All the vegetation supports the primary production that provides food and wood that works together to sequester carbon, which plays an ultimate role in global climate. When these connections are broken, desertification is triggered and habitats can be jeopardized and lost. Biodiversity loss affects the health of the habitat. Climate change increases evapotranspiration and adversely affects biodiversity.

The loss of vegetation in the food chain impacts life along it. In addition, when native vegetation dies off, invasive species, such as cheat grass (*Bromus tectorum*) in the Southwestern United States, moves in and takes over. Invasive species have a better probability of survival because they lack the native species' predators. It does not take long for invasive species to overrun a landscape, pushing native species out. Once this happens, it may be that only a few invasive species are supported in an area, where at one time dozens of species once existed. In areas where this occurs, the biodiversity is significantly lowered, and unless rehabilitation efforts are put into effect, it may never be able to reestablish itself to a natural, ecologically balanced state. Even if native vegetation is reintroduced, it may not be able to survive if desertification degrades the land to the point where nutrients and water supplies are no longer available.

Biodiversity, which contributes to many of the benefits provided to humans by dryland ecosystems, is greatly diminished by desertification. For example, diversity of vegetation is critical in soil conservation and in the regulation of surface water and local climate. If these delicate balances are disrupted, it can threaten the health and existence of habitats. Desertification affects global climate change through both soil and vegetation losses. Another problem that exists involves the use of land for livestock grazing. When desertification becomes an issue and invasive weeds move into the area, the livestock may consider the new vegetation species unpalatable and refuse to graze.

It is not just one human activity that is responsible for desertification—there are many. Overgrazing of livestock and cultivation are two of the most common reasons, but deforestation is also a major contributor. If groundwater is overused or irrigation of farming areas is not done properly, they can introduce the desertification process. Desertification is also encouraged if soils collect more salt, raising their salinity levels. Climate change is being largely blamed for worldwide desertification in the earth's arid, semiarid, and subhumid areas. This means that desertification is the result of a combination of social, economic, political, physical, and natural factors, which vary from region to region.

Currently, lands that seem to be most prone currently to desertification include the areas at the fringes and outskirts of deserts. These transition areas have fragile, delicately balanced ecosystems, usually operating under various microclimates. Already delicate, once these ecosystems are stressed to their limit of tolerance, they cannot recover on their own. Grazing by livestock is especially harmful because they pack down the soil. The more the subsoil gets packed into an impermeable layer, the less water is able to percolate down into the ground. Because the scarce water is not able to penetrate the ground's subsurface, it flows off the surface, eroding the land with rills and gullies. South Africa, for example, loses 262 million metric tons of topsoil each year. In addition, as the surface dries out, the soil is carried away by the wind.

According to the USGS, there are many things scientists still do not know about desertification of productive lands and the expansion of deserts. To date, there is no consensus among researchers as to the specific causes, extent, or degree of desertification. Desertification is a subtle and complex process.

From a global context, during the past 25 years, satellites have helped scientists study climate change by providing global imagery with which to study the effects of desertification. The existence of satellite imagery, such as LANDSAT, SPOT, Quickbird, and Digital Globe, has made it possible to monitor areas over time and determine the susceptibility of the land to desertification. The problem is a global one. It is predicted that by 2100, that climate change will increase the area of desert climates by 17 percent, meaning more areas at risk of desertification. Worldwide, about 12 million hectares becomes useless for cultivation each year—an area equal to about 87 percent of the area of agricultural land in the United States.

Desertification needs to be monitored and managed as a worldwide effort as climate change intensifies. Like global uplifting, it does not stop at international borders. If it is not controlled, biodiversity will be negatively impacted. Another way desertification contributes to climate change is by allowing the carbon that has been stored in dryland vegetation and soils to be released to the atmosphere as it dries out and dies. This could have significant consequences for global climate.

The relationship between climate change and desertification is not straightforward; its variabilities are extremely complex. The best way to deal with desertification is to prevent it from happening in the first place. The most appropriate way to do this is through proper management at local, regional, and global levels. Besides being environmentally better, prevention is also more cost effective than rehabilitation. The best form of prevention requires a change in the attitudes of those living on, and working with, the land, such as employing sustainable, environmentally friendly agricultural and grazing practices. For areas that have already been degraded, rehabilitation and restoration measures can help to restore lost ecosystems, habitat, and services the drylands originally supplied. One of the most important reasons to avoid desertification is to avoid extreme poverty and hunger—when drylands become too degraded in developing countries, it leaves populations with little access to food and clean water.

According to the Ecosystems and Human Well-being: Desertification Syntheses, which is part of the report published in the 2005 Millennium Assessment, the following actions can be employed to prevent desertification (Corvalan et al., 2005):

- Implementation of a land and water management plan to protect soils from erosion, salinization, and degradation.
- Creation of economic opportunities outside of dryland areas, taking the stress off of drylands.
- Protection of vegetative cover so that it will stabilize the soil underneath and keep it from being eroded by wind and water.
- Becoming involved in alternative livelihoods that do not depend on intensive land use. These types of activities include greenhouse agriculture, tourism, and aquaculture.
- Combining areas of farming and grazing in order to centrally manage natural resources more effectively.
- Empowering local communities to effectively manage their own resources and combining traditional practices with local ones.

In areas where desertification has already become established, it is important to rehabilitate and restore the lands in order for them to return to their previous conditions. Successful restoration must be done at the local level. Several methods are commonly used, such as the following:

- Reintroduction of the original, native species that used to live there.
- Combating erosion through the systematic terracing steep areas so that water does not run down slopes eroding the land's surface.
- Establishing seed banks to ensure that species do not become endangered or extinct. Then when climate conditions exist for the plant to survive, seeds can be planted.
- Enriching the soils with nutrients, making them more fertile and conducive to vegetative growth.
- Planting additional reserves of trees.

THE STRAIGHT FACTS ABOUT DESERTIFICATION

For land managers, it is critical that appropriate, effective, and well-thought-out decisions be made about the land and its resources. The following facts reflect the shocking realities of desertification facing today's world:

- Roughly 3.6 billion hectares of the world's 5.2 billion hectares of dryland used for agriculture has been degraded by erosional processes.
- One out of every six people is directly affected by desertification; this equates to nearly one billion people.
- Desertification has forced many farmers to abandon farming and look for urban employment.
- Each year 51,800 square kilometers of land is destroyed by desertification.
- Desertification affects almost 75 percent of the land in North America.
- Dust from deserts can be blown great distances into cities. Dust has been blown from Africa to Europe and the United States. During the Dust Bowl, it was blown from Oklahoma and out across the Atlantic Ocean.
- Desertification destroys the topsoil of an area, making it unable to grow crops, support livestock, or provide suitable habitat for humans.
- Each year, the planet loses 24 billion tons of topsoil. Over the past two decades, enough has been lost to cover the entire cropland of the United States.
- Climate change can trigger desertification as well as poor land management practices.
- The destruction caused by desertification carries a hefty price tag. More than $40 billion per year in agricultural goods is lost, causing an increase in agricultural prices, negatively impacting the consumer.
- According to a United Nations study, roughly 30 percent of the earth's land is affected by drought. Every day, approximately 33,000 people starve to death.
- Desertification makes the environment more likely to experience wildfires.

Desertification is a condition that can be stopped, but it usually is not brought to the public eye until it is well under way, making rehabilitation of the landscape much more difficult and expensive. If it were brought to the public's attention sooner and not allowed to reach a critical point, it would make recovery more manageable. One way to ensure the success of this is through outreach and public education. By keeping people informed and educated about the results of their behavior on the environment, land degradation can be avoided all together. (Rinkesh, 2016).

According to the IPCC, many deserts will face a decrease in rainfall from 5 to 15 percent. Approximately half of the subhumid and semiarid parts of the southern African region are at moderate-to-high risk of desertification. Since the 1970s in West Africa, the long-term decline in rainfall has caused a 25–35-kilometer southward shift of the Sahelian, Sudanese, and Guinean ecological zones in the second half of the twentieth century (IPCC, 2014b).

The results of drought in the rain forest were that the trees managed fairly well after the first year. In the second year, they sunk their roots deeper in the ground to find moisture and still survived. However, by the third year, the trees began dying. The tallest trees were impacted first. As they toppled to the forest floor, the lower canopies were exposed to the direct sunlight. The forest floor was also exposed to direct sunlight and dried out quickly.

After the end of the third year, the biomass had released more than two-thirds of the stored CO_2 to the atmosphere. Where once the forests operated as a carbon sink, they were now a major carbon source. What has scientists concerned is that the Amazon currently holds about 81.6 billion metric tons of carbon, enough in itself to increase the rate of climate change by 50 percent. On top of that, if a wildfire were to start in these remote locations and destroy the vegetation, the rain forests would be transformed into a desert. Dr. Deborah Clark from the University of Missouri, a renowned forest ecologist, says that the research shows that "the lock has broken" on the Amazon system. "The Amazon is headed in a terrible direction." (Real Climate, 2006)

HEAT WAVES

Temperatures are obtained from monitoring stations worldwide in order to calculate global mean temperature rise. When temperatures are taken near cities, they must be corrected to eliminate the specific effect that the presence of the urban area has on the temperature reading. Because urban areas have so many dark surfaces—such as asphalt-covered roads and dark roofs on buildings, they absorb more heat than natural surfaces such as grasses, prairies, and woodlands. This absorbed heat is reradiated by the buildings and roads, and the resultant increase in temperature from these sources, as well as heat released from industry, cars, and other sources of burning fossil fuels, adds to the increased temperatures.

In order to use a reliable temperature value instead of a skewed value due to artificial inputs, collectively referred to as the urban heat island effect, this contribution to the temperature must be accounted for and removed. In addition, the various instruments and methods used worldwide must also be calibrated so that the temperatures collected are directly comparable.

In a global temperature analysis conducted by NOAA, they released a report in 2016 stating that 2015 was the warmest year in the 136-year period of record. Specifically, December 2015 was the warmest month in the period of record, at 1.11°C higher than the monthly average, breaking the previous all-time record set two months earlier in October 2015

TABLE 4.4

Sixteen Warmest Years (1880–2015)

Rank (1 = warmest)	Year	Anomaly (°C)	Anomaly (°F)
1	2015	0.90	1.62
2	2014	0.74	1.33
3	2010	0.70	1.22
4	2013	0.66	1.26
5	2005	0.65	1.19
6 (tie)	1998	0.63	1.17
6 (tie)	2009	0.63	1.13
8	2012	0.62	1.12
9 (tie)	2003	0.61	1.12
9 (tie)	2006	0.61	1.10
9 (tie)	2007	0.62	1.10
12 (tie)	2002	0.60	1.10
13 (tie)	2004	0.57	1.01
13 (tie)	2011	0.57	1.08
15	2001	0.54	0.97
15	2008	0.54	0.97

Source: National Oceanic and Atmospheric Administration, 2015.

by 0.12°C. This is the first time in the NOAA record that a monthly temperature departure from average exceeded 1°C (or reached 2°F), and the second widest margin by which an all-time monthly global temperature record had been broken. February 1998 broke the previous of record in March 1990 by 0.13°C. Table 4.4 depicts the 16 warmest years on record from 1880 to 2015 (NOAA, 2015). Figure 4.12 illustrates significant climate anomalies and events from 2015.

Figure 4.12 depicts specific climate anomalies and events from 2015.

The only reliable explanation scientists have been able to come up with this far in their models is human interference. The temperature rises in the models accurately reflect the actual temperature rises the earth has experienced so far when the effects of greenhouse gas levels from the burning of fossil fuels and deforestation are entered into the mathematical model equations. If the human interference factor is not added, the models do not work—they underestimate actual temperatures. The polar latitudes have been identified in models as those areas on earth that are being impacted the fastest and most significantly. In addition, nighttime temperatures have increased much more than daytime temperatures, keeping the earth's atmosphere warmer overall. This is significant because when the earth stays warmer at night, it retains the heat that was generated during the day in the atmosphere, starting the next day off warmer than normal.

The Union of Concerned Scientists believes that by the year 2100 there will be an increase in average surface air temperature of around 2.5°C. The result of this will mean an increase in temperature at many scales, such as days, seasons, and years. When looking at significant local levels of temperature increases, this means that some areas will succumb to more "extremely hot

summer days" during the summer and "killer heat waves" will occur (Union of Concerned Scientists, 2007).

As heat wave incidents increase, more people would be negatively affected and many of these could die. The sick, very young, the elderly, and those that cannot afford indoor air conditioning are at the most risk of dying. As an illustration, in 1995 in Chicago, nearly 500 people died during a significant heat wave.

In July and August 1999, another heat wave hit Chicago. Because emergency response network specialists learned some valuable lessons in 1995, they were better prepared for this one. Even though the response was better, 103 heat-related deaths still occurred.

Chicago is not alone. In 1980, a heat wave in the United States killed 1700 people in the East and Midwest, and again in 1988, one killed 454 people. In 1998, more than 120 people died in a Texas heat wave. Europe experienced a deadly heat wave in 2003, so hot in fact; it was considered the hottest summer in 500 years. During this period, 27,000 people died from heat-related problems. Some of those who did not die suffered irreversible brain damage from advanced fevers as a result of the intense temperatures (Environmental Defense Fund, 2011). The heat wave in the United States in 2006 was one of the worst it had ever experienced. It held the entire country in its grip and lasted for almost a month. The effects and costs of these were enormous—hundreds of people died, massive power outages were triggered, and unmanageable wildfires burned large areas of ground. Tens of thousands of people in New York went without electricity for over a week.

One thing that scientists cannot do with the overall issue of climate change is to blame a single weather event—like a heat wave, a hurricane, a blizzard, or a tornado—on climate change, because weather fluctuates naturally. However, what they can relate to are climate change are trends. Based on the fact that climate models predict more wild and unpredictable weather as a future trend, expecting an increase in heat waves does fit into the future climate scenario. One thing that both the developed and developing countries need to be aware of is that continued urbanization will increase the number of people vulnerable to these urban heat islands and heat waves.

WILDFIRE

The likelihood of a disastrous wildfire occurring increases significantly during periods of drought. The world saw the proof of this in 1997 through 1998 when an El Niño episode caused extremely dry conditions in many areas across the globe. Large forest fires occurred in Brazil, Central America, Florida, eastern Russia, and Indonesia. Wildfires can easily occur during drought periods from both natural and human-caused factors. Lightning strikes, as well as campers or hikers, commonly cause them. Regardless of the cause, during drought conditions, the vegetation is so dry that it does not take much to start a wildfire, and when they start, they burn fast and intense.

Based on a report from the Natural Resources Defense Council, the 2006 wildland fire season set new records in

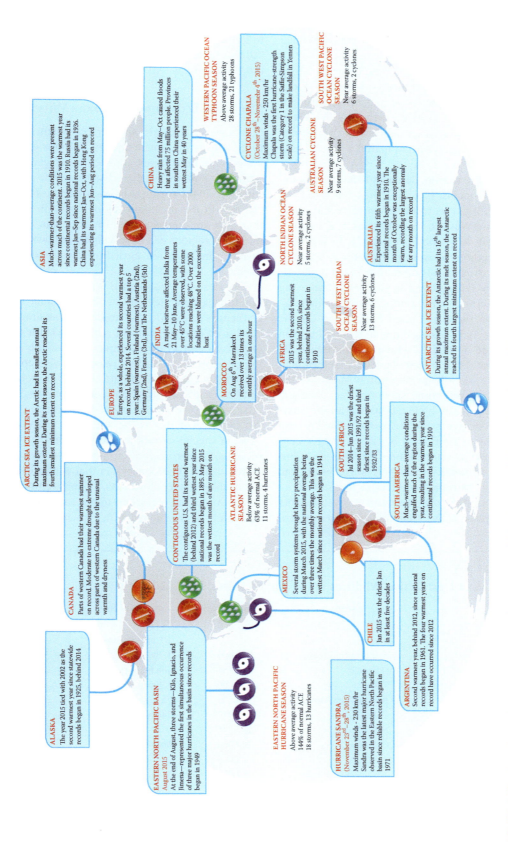

FIGURE 4.12 Significant climate anomalies and events from 2015. (Courtesy of NOAA, Silver Spring, MD.)

both the number of reported fires and the acres burned. Nearly 100,000 fires were reported and nearly 4 million hectares burned—125 percent above the 10-year average. If warming continues to spur wildfire seasons, it could be economically devastating. Fire-fighting expenditures recently have totaled $1 billion per year.

In work done by the Union of Concerned Scientists, they state that "The number of large wildfires—defined as those areas covering more than 1,000 acres—is increasing throughout the Western United States. Over the past 12 years, every state in the West has experienced an increase in the average number of large wildfires per year compared to the annual average from 1980 to 2000".

In addition, the wildfire season, defined as the time period between the year's first and last large wildfires, has also been lengthening. Although it varies from region to region, there is a trend over the entire Western United States of the season extending over the past 40 years. For example, in the early 1970s, the average length of the wildfire season was approximately five months. Currently, it is more than seven months.

Two factors that are contributing to this trend are rising temperatures and earlier snowmelt. Temperatures are increasing much faster in the Western United States than for the earth as a whole. Since 1970, average annual temperatures in the West have increased by 1.05°C, which is about twice the global average increase. Hydrologists have also determined through measurements taken from streamflow gauges in the West that the timing of springtime snowmelt has changed. Although it differs specifically by location, overall the onset of spring snowmelt is occurring 1–4 weeks earlier today than it did in the 1940s. This is significant because spring green-up happens earlier, then as temperatures climb higher during the hot, dry summer months, the vegetation dries out and dies, making conditions extremely vulnerable to wildfire.

Future wildfire projections vary between ecosystems, but higher temperatures are expected to continue to play a critical role in wildfires. The semidesert and desert regions will most likely be hit the hardest, but every type of ecosystem will feel the effects of a warming environment and increased average annual burn area. The only way to keep the vulnerable areas to a minimum will be to lower our emissions of CO_2 (UCS, 2016b). Table 4.5 lists the worst wildfires in the last decade in the United States. All were in the West.

MOUNTAIN ECOSYSTEMS IN DANGER WORLDWIDE

Mountain systems are another ecosystem that is vulnerable to the effects of climate change. Adverse effects can be seen via rising temperatures, the melting of glaciers at the higher latitudes, and the earlier disappearance of annual winter snowfalls in the spring. Also significant is the effect that these changes in the mountain regions have in the lowland areas that they are connected to. In particular, climate change in mountain regions can have a significant effect on the wildlife that lives in the region, agriculture, water supplies, and human habitat.

Climate change, coupled with the fact that many mountain ranges generate their own microclimates, presents many challenges to the life that resides on and around them. These conditions often provide for rich habitats for a large variety of organisms and wildlife, adapted specifically to the surrounding environment. For this reason, it also comes with drawbacks in a changing climate. Upsets in ecosystem balance can have a drastic effect on species that have adapted to specific niches, even to the point of threatening their existence. An example of this is the American Pika (*Ochotona princeps*), a species of rabbit that exists in the Rocky Mountain regions. Preferring cool climates, it is very sensitive to climate change. The concern is that as the climate warms, it relocates to higher elevations. At some point, however, it will run out of higher elevations to relocate and adapt to, and will ultimately be threatened with extinction (Figure 4.13).

TABLE 4.5
The Most Destructive Wildfires in the United States, 2006–2016

Date	Location	Area Burned (acres)	Property Lost	Lives Lost
March 2006	Texas	191,000	15 homes 10,000 horses/cows	0
October 2007	Southern California	500,000	1,300 homes	3
September 2010	Colorado	6,181	168 homes	0
September/October 2011	Texas	32,000	1,700 homes	2
May/July 2011	Arizona/New Mexico	538,049	75 structures	0
June/July 2012	Colorado	18,947	346 homes	0
May–July 2012	New Mexico	297,845	0	0
June/July 2012	Colorado	244,000	600+	6
June 2013	Arizona	8,000	0	19
August–October 2013	Yosemite, California	257,314	0	0
July/August 2014	Washington	250,000	300	0
September 2015	California	283,000	1,050	1

Source: (USFS, 2016).

FIGURE 4.13 American pika. (Courtesy of Sevenstar, Wikimedia Commons.)

Globally, the human-contributed greenhouse gas emissions are expected to lead to average worldwide warming of between 1.1°C and 6.4°C from 2000 to 2100. The specific amount of warming will depend on the regions, although it is expected to be greater over land and in high northern latitudes. Areas of snow cover are expected to lessen overall, whereas snowfall may increase in regions with very cold temperatures, such as high mountains. Most glaciers and ice caps will lose mass or disappear over time.

Because mountain environments are so responsive to changing temperatures, mountains are one of the best areas to see evidence of climate change. Many scientists believe that the changes happening in mountain ecosystems right now provide an early glimpse of what could come to pass in lowland environments, and that mountain areas act as an early warning system of what is to come.

Mountains exist in many regions of the world. They also occupy different positions on the globe, as well as differ in shape, extension, altitude, vegetation cover, and climate regime. They will continue to occupy different positions worldwide—altitude, vegetative cover, and climate regime—which means they will be affected differently by climate change. They do share some common features relating to climate change, however. Examples are as follows:

1. Have a unique and complex topography, making their climates variable over short distances. This makes climate modeling of mountainous areas difficult.
2. Because temperature changes with altitude, the impacts of a warmer climate are different for different elevations. The snow line is one of the most variable areas.
3. Mountains act as barriers to wind flow, which can change atmospheric wind flow patterns. These can have an effect on both precipitation amounts and patterns.

Table 4.6 shows regional climate projections for some of the earth's principal mountain regions.

A changing climate's effect on the local agriculture is also of concern in mountain ecosystems, as it plays a significant role on farming and agriculture. In mountainous regions, the major water supply originates from the annual snowmelt from

TABLE 4.6

Regional Climate Projections for Some of the Earth's Principal Mountain Regions

Mountain Range	Projection
The Andes	• Decreased annual precipitation in southern Andes
	• Other areas' precipitation will depend upon El Niño
	• Several Andes glaciers may disappear
The Rocky Mountains	• Higher elevations have already experienced a threefold increasing in temperature
	• Earlier snowmelt in the spring
	• Shift from snowfall to rainfall
	• Increased incidence of forest fires
The Hindu Kush-Himalaya	• Warming will be well above the global average
	• This range plays an important role in climate patterns
	• Warming will be well above the global average
	• Models show an increase in precipitation during the winter months
The European Alps	• The Alps have shown a higher than normal warming trend since 1979
	• Regional warming is predicted as 1.5 times higher than the average
	• Precipitation will decrease in summer
	• Substantial glacier retreat
	• Duration of snow cover will decrease

Source: From Kohler T. and D. Maselli (eds.), *Mountains and Climate Change—From Understanding to Action*, 3rd ed, Geographica Bernensia, Bern, Switzerland, 2012.

the mountains. In areas that are now experiencing early snow-melt and disappearing glaciers, this significantly affects the soil moisture and agricultural health. As a result of reduced and/or early runoff in the spring the growing seasons often result in low yields because there is no water available when it is needed most during the growing period . In some areas, such as Cotacachi, Ecuador in the Andes, the Cotacahian glacier has been the principal water source that farmers have used to water their crops. Today, it has nearly vanished due to rising temperatures, which means that the rivers in the area are nearly dry. Farmers now have to rely on the sparse rainfall in the area during the growing season, which is unpredictable and has significantly lowered their yields.

Mountain snow and glacial melt also provides drinking water for more than one billion people on earth. As the climate has warmed, this has diminished and these populations have had to change their water consumption habits and survive on lower supplies of water. On average, a home with a constant water supply will use more than 100 liters per person per day. A key goal of mountain environments and future community growth is that these developments will have to adapt to lower water availability. This means that the public will have to be educated to use their water resources responsibly.

Another area to consider is recreation. Businesses, such as ski resorts, depend on winters with above average amounts of snowfall in order to provide the best recreational opportunities for their clients. During years of less snowfall, they feel the negative impacts, not only in the length of the ski season but in its quality. In a warming world, they face the possibility of having to adapt their businesses to something that offers other activities just as important as skiing that can be done with either low amounts or no snow.

Mountain environments are also seeing economic impacts as their water supplies become more strained. As crops fail and livestock herds diminish because the land can no longer support the number of animals it previously could, this translates into serious economic losses. In areas where businesses have shrunk to a certain threshold, people have been forced to relocate to other, more desirable locations. Because each mountain environment is unique, the extent of impact depends on the current climatic and ecological situation, and on the topography of the area (Kohler and Maselli, 2012).

CHALLENGES IN MARINE ENVIRONMENTS

Climate change has a significant effect on the marine environment. It is affecting ocean temperatures, the supply of nutrients, ocean chemistry, food chains, wind systems, ocean currents, and extreme events such as cyclones. These, then affect the distribution, abundance, breeding cycles, and migrations of marine plants and animals that millions of people globally rely on for food and income. Even of more concern is that marine organisms may be responding faster to climate change than land-based plants and animals. As temperatures rise, marine plants and animals are shifting toward the poles. This, in turn, changes the dynamics of food webs and impacts the animals (including people) that depend on them.

TEMPERATE MARINE ENVIRONMENTS

There are several components of the temperate marine environment that climate change could impact, such as changes in temperature, shoreline ecology, and major storm tracks. Two elements that are the least complicated to predict are temperature and sea-level rise; others are much more complex depending on the scale and degree of interaction between other components in the environment.

It is difficult to use existing climate models to address these issues because they produce outputs that generally depict broader geographical regions than the scales that local storms occur at. Models are not refined enough at this point for their spatial resolution to be able to model areas as small as a specific bay or coast.

According to the IPCC, they predict that by 2100, the earth's near-surface temperature averaged worldwide will increase by 1.4°C–5.8°C from the 1990 levels. This means that sea-surface temperatures will also rise. The IPCC believes that this will be the highest temperature rise seen in the past 10,000 years. The aspect of the temperature rise that will cause the most significant ecological change in estuary and marine ecosystems is the rapid speed with which it is expected to happen, leaving species little time to adapt (IPCC, 2007a).

Based on data from the IPCC, globally averaged sea level rose between 10 and 20 centimeters during the twentieth century. They predict the oceans will rise another 9–88 cm between now and 2100. The rise will be the result of both (1) thermal expansion of the current ocean water (the warmer the temperature, the more water expands) and (2) melting from land-based glaciers and ice sheets.

The Pew Center on Global Climate Change is a nonprofit organization that brings together business leaders, policy makers, scientists, and other experts worldwide to create a new approach to managing the problem of climate change—what they refer to as an extremely "controversial issue." Not slanted in any particular way politically or economically, they "approach the issue objectively and base their research and conclusions on sound science, straight talk, and a belief that experts worldwide can work together to protect the climate while sustaining economic growth (EEA, 2017)."

In January 2007, the Center was one of the inaugural members of the U.S. Climate Action Partnership—an alliance of major businesses and environmental groups that calls on the federal government to enact legislation requiring significant reductions of greenhouse gas emissions. According to the Pew Center, in terms of climate change, the biggest impact on estuarine and marine systems will be temperature change, sea-level rise, the availability of water from precipitation and runoff, wind patterns, and storminess. In these often-fragile systems, temperature has a direct and serious effect. For the sea life living within the ocean, temperature directly affects an organism's biology, such as birth, reproduction, growth, behavior, and death.

Temperature differences can also influence the interaction between species, such as predator–prey, parasite–host, and other competitions that may develop over the struggle for limited resources. If temperatures change the distribution patterns of organisms, it could also change the balance of predators, prey, parasites, and competitors in an ecosystem, completely readjusting balances, food chains, behaviors, and the equilibrium of the ecosystem.

Climate change can also change the way that species interact by changing the timing of physiological events. One of the key changes that it could alter is the timing of reproduction for many species. Rising temperature could interfere with the timing of the birth being correlated with the availability of food for that species. This can be a problem, for example, for bird species that migrate and depend on a specific food source to be available when they reach their breeding grounds. If warmer temperatures have changed the timing a few weeks of when the food will be available, and it is not synchronized anymore with the migrating birds, it could leave the birds without available food, threatening their survival.

SEA-LEVEL RISE

Melting glacial and polar land ice will add to sea-level rise. The effects of sea-level rise will not be constant but will vary with location, how fast the sea level rises, and the biogeochemical responses of the individual ecosystems involved. The South Atlantic and the Gulf of Mexico have been identified as susceptible areas.

As sea levels rise, ocean water will submerge and erode the shorelines. In natural areas covered with marshes and mangroves, as sea level rises, it will flood the wetlands and waterlog the soils. Because the plants that live in the wetlands, which do not contain salty water, are not accustomed to salt, the salty ocean water would kill them. Because wetlands provide habitat for wildlife, including several migrating birds, this would also destroy their habitat.

The areas that would be hard hit are those that cannot migrate inland because urbanization has been built right up to the shoreline, effectively removing any possible potential wetland habitat. This presents a serious impact to the environment because wetlands are an important part of the biological productivity of coastal systems. Marshes provide many critical services; they function not only as habitats for wildlife but as nurseries for breeding and raising young, and as refuges from predators. Wetlands function as part of an integrated system. If they are jeopardized, their loss will affect the availability and transfer of nutrients, the flow of energy, and the availability of natural habitat needed by the multitudes of organisms already living there. One of the most unfortunate losses will be those areas where rare, threatened, or endangered plant and animal species live, such as the American alligator (*Alligator mississippiensis*), Florida black bear (*Ursus americanus floridanus*), West Indian manatee (*Trichechus manatus*), Florida panther (*Puma concolor coryi*), Southern bald eagle (*Haliaeetus leucocephalus*), snowy egret (*Egretta thula*), and roseate spoonbill (*Platalea ajaja*). If invading salt water destroys these habitats, they could become extinct.

Because the oceans are so vast, most of the predictions made about the earth's open oceans in light of climate change have been generated by computer models. The two most important functions that govern the behavior of the ocean are temperature and circulation, and these calculations are fairly straightforward in computer models and readily calculable. Another thing that makes it possible to model the oceans using computers is that human impact has not had as large an impact on the oceans as it has on the land.

TROPICAL MARINE ENVIRONMENTS

The world's tropical and subtropical marine environments represent some of the most diverse habitats on earth. Often characterized by reef-building corals, the complex arrays of marine inhabitants that occupy these waters have developed many strategies for survival. Their ecosystems are so complex that a delicate balance is needed to protect these numerous marine resources, while also accommodating an economy centered on commercial fisheries and recreation.

The tropical marine environment is also subjected to many of the environmental concerns as the temperate marine environment, such as temperature, sea-level rise, wind circulation, and algal blooms. An additional risk the tropical habitat encounters is the negative impact on reefs and corals. These ecosystems represent some of the most fragile on the earth and some of the hardest hit with the negative effects of climate change.

FRAGILE ECOSYSTEMS—REEFS AND CORALS

As climate change causes the earth's tropical oceans to heat up, they become more acidic, and strong storms become more common, the world's coral reefs are taking a beating. Rod Fujita of the Environmental Defense says, "Coral reefs may prove to be the first ecological victims of unchecked global warming." (EDF, 2007b).

Loss of coral reefs would translate into huge economic losses in coastal regions dependent on reefs. They currently provide about $30 billion each year in goods and services (Harvey, 2016).

Destruction of reef systems also represents an ecological disaster. Coral reefs are sometimes referred to as the rainforests of the ocean because they provide habitat to a rich diversity of marine life, including reef fish, turtles, sharks, lobsters, anemones, sponges, shrimp, sea stars, sea horse, and eels. Reefs attract scuba divers from around the world each year to swim among the beautiful, otherworldly shapes and color combinations. Corals actually obtain their food and color from tiny algae called zooxanthellae that live in them. They have a very narrow temperature tolerance range. In fact, if the water increases only 1.1°C above the typical maximum summer temperature, it can cause corals to expel their algae and turn white, through a process called "bleaching." If bleaching continues for a prolonged period of time, the corals will die.

Currently, 93 percent of the reefs of the Great Barrier Reef of Australia have been affected by coral bleaching due to the abnormally warm ocean temperatures accompanying climate change. Since 1998, there have been three mass bleaching events documented, and the effects have been devastating. A major contributor to the situation is the warm water flowing around the Pacific Ocean caused by El Niño events along with the warming caused by the presence of greenhouse gas emissions. Because corals bleach when they are subjected to temperatures above their normal summer maximum for at least a month (usually combined with few clouds and high levels of ultraviolet radiation), they become subjected to bleaching. Over this time interval, of a total of 911 individual reefs that make up the Great Barrier Reef complex, only 68 have not been entirely bleached. More than half have been severely bleached and 81 percent in the northernmost reaches have experienced severe bleaching. Professor Terry Hughes of James Cook University and head of the National Coral Bleaching taskforce stated, "We've never seen anything like this scale of bleaching before. In the northern Great Barrier Reef, it's like ten cyclones have come ashore all at once." Unfortunately, if coral remains bleached for a long time, it dies. Presently, the northern portion of the reef is seeing over a 50 percent mortality rate and it is expected that some reefs will exceed a 90 percent mortality (Slezak, 2016).

World Wildlife Fund has identified both the Seychelles Islands and American Samoa as locations under high stress for coral bleaching. The United Nations Educational, Scientific, and Cultural Organization recognize the Seychelles Islands as a natural World Heritage Site. They have a high coral diversity and support rare land species, such as the giant tortoise. In addition to increase ocean temperatures, these areas are also threatened by climate change because of more frequent tropical storms (which could break up the coral) and more frequent rains, flooding, and river runoff (which deposits sediments in the ocean) (WWF, 2011) (Figure 4.14).

FIGURE 4.14 Coral reefs are often referred to as the rain forests of the ocean because they provide habitat for an extremely diverse collection of life forms. Climate change is threatening their existence. (Courtesy of NOAA, Silver Spring, MD.)

OUR FRAGILE REEF SYSTEMS

The Union of Concerned Scientists has determined the following:

- Australia's Great Barrier Reef—the world's largest coral reef—is a unique marine ecosystem threatened by climate change. Damage to the reef could harm the region's biodiversity, tourism, and fisheries.
- The reef has suffered eight mass coral bleaching events since 1979, triggered by unusually high water temperatures. If there is enough time between bleaching events, the coral can often recover. However, annual bleaching is expected by the mid-century if our heat-trapping emissions continue at their current pace, thus leaving the reef vulnerable to diseases from which it may not recover.
- Ocean acidification expected to occur if atmospheric carbon dioxide surpasses 500 parts per million is likely to limit the capacity of the reef to recover from bleaching events and to cope with other stresses.
- Scientists project a significant loss of biodiversity within a decade and at worst a 95 percent decrease in the distribution of Great Barrier Reef species by late this century.

Source: UCS, 2016b.

FRESHWATER ENVIRONMENTS

Climate change is also affecting the planet's freshwater ecosystems, such as rivers, lakes, and wetlands. Observable changes include the following:

- An increase in the surface water temperature of lakes and streams, most notably those located in the high altitudes and latitudes
- An increase in the temperature of the hypolimnion layer of large deep lakes
- A reduction in lake ice cover
- Changes in mountain streams due to the melting of glaciers and permafrost, adding pollutants and solutes to surface waters

Given present trends, it is expect that not only will they continue, but the following is also likely:

- The flow in rivers and streams will change depending on the amount, intensity, seasonality, and distribution of precipitation. As temperatures climb, additional melt will cause increased flow, which will increase the transport of sediments and nutrients downstream into lakes and coastal zones.

- Water levels, habitat composition, and water residence times in wetlands will be affected with changes in precipitation, evaporation, and flooding as temperatures warm.
- Intermittent streams and small lakes in warm, dry areas may dry up and disappear. Flow in permanent streams may become intermittent, and permanent lakes may shrink and become more saline.
- Systems existing on the borderline between two conditions may become so unstable; they will abruptly degrade to the less healthy system.

There are several ecological consequences of climate change for freshwater ecosystems. The impacts that a particular ecosystem will experience will be a combination of the local stressors the region is already experiencing, combined with the effects of the changing climate. For example, a river, a lake, or a wetland ecosystem may already be subject to conditions such as acidification, input of toxic substances, hydromorphological changes, catchment land-use change, water resource management decisions, acidification, eutrophication, and invasion of exotic species. In addition to these various inputs, a specific ecosystem may also have to account for colder, temperate/warm-humid, or warm-arid climate changes to the ecosystem (EEA, 2012).

Depending on what conditional climate input existed, it would produce a different ecosystem. For instance, for cold regions, it is predicted that

- There will be an increase in primary productivity in response to the longer length of the growing season and an increase in nutrient release from catchment soils.
- Population decline or loss of colder species will occur as temperatures get warmer.

In temperate, warm-humid freshwater regions, it is expected that

- Eutrophication will be a problem.
- Temperature and nutrient balance will be harder to maintain.
- Areas with higher temperatures will have more intense algal blooms.
- Longer periods of stratification and greater oxygen depletion and increased release of phosphorus from sediments will modify the distribution of species across ecoregions.
- Species will become more susceptible to invasive species invasion.
- Overall biodiversity reduction may lead to impaired ecosystem services.

In warm-arid regions, changes in moisture balance are expected to have negative consequences for freshwater environments. Increased temperatures along with reduced precipitation are expected to cause a loss of habitat as well as changes in community composition due to a decrease in lake levels, less river flow, and increased eutrophication. All of these ecosystem changes will require both economic and ecological adaptive measures at the local scale.

SUMMING IT ALL UP

In an article published in *The Washington Post* on August 2, 2016, it stated that 2015 was the warmest year on record for earth. It was all documented in NOAA's 300-page report along with scores of other facts and information denoting the severity of 2015's climate. Titled, "State of the Climate in 2015," this notable report was authored by more than 450 scientists from 62 countries worldwide. *Every* single indicator of temperature described in the report leaves *no doubt* as to the severity of 2015's surface temperature: it far exceeded any previous years' value. It was accompanied by a record-notable El Niño—where warmer than normal tropical Pacific Ocean water also contributed to the atmospheric heat (Samenow, 2016). The article summed up the 10 most important findings of the report. They are as follows:

1. The global temperature was the highest on record: the previous record was set in 2014.
2. The average ocean surface temperature was the warmest on record: the average temperature was $0.33°C–0.39°C$ above average and was responsible for fueling global tropical cyclone activity.
3. Upper ocean heat content was highest on record: This refers to the upper ocean layer and was supported by five different datasets.
4. Global sea level was highest on record: oceans expand as they warm, and melting ice adds to the level. Sea level was about 6.99 centimeters higher than the 1993 average.
5. The El Niño event was one of the strongest on record: it was involved in elevating global temperatures, raising sea levels, and intensifying tropical cyclone activity, and led to droughts, wildfires, and the release of CO_2.
6. Greenhouse gases were the highest on record: for the first time, the level surpassed 400 ppm in the modern record. Methane and nitrous oxide also set record highs.
7. There were a record number of major tropical cyclones in the Northern Hemisphere: a total of 31 major tropical cyclones developed in the Northern Hemisphere (the previous record in 2004 was 23). Of these, 26 reached category 4 or 5—another record.
8. Arctic sea ice had its lowest maximum extent: in February 2015, the maximum sea ice extent was 7 percent below the 1981–2010 average and smallest on record.
9. Glaciers continued shrinking: 2015 was the 36th straight year of global alpine glacier retreat.
10. Extreme temperatures were the most extreme on record: 2015 had the most warm days and warm nights in western North America, Central Europe, and Central Asia. It also had the fewest number of unusually cool days on record (Samenow, 2016).

CONCLUSIONS

As evidenced from the impacts currently happening to ecosystems worldwide, the effects are not minor, nor are they localized. On the contrary, they are serious, large-scale impacts that have endangered species, degraded living conditions, destroyed habitats, impacted the health and livelihood of societies, caused starvation and extensive misery, and threatened the future of both humans and wildlife.

So far.

Yet, step back and reflect a bit.

Scientists have been warning policy makers for the past two decades of the changes humans are making to the environment—changes largely due to lifestyle preferences. And only recently has the political community been listening—Europe for the past decade has been on board, but countries such as the United States have stubbornly turned their heads the other way in almost defiant response—instead demanding that the rest of the world cut back their greenhouse gas emissions while they do nothing.

And the temperatures and parts per million carbon content in the atmosphere have just begun to rise toward what climate scientists predicted a decade ago; with a delayed response time in the Earth's atmosphere, just as they warned.

It is coming down just like clockwork.

What do you suppose it will be like when the temperature climbs another degree? Or two? What about the people in distant, undeveloped countries that never even contributed significantly to the problem that may be suffering the most, while people in developed countries are still trying to live the "business as usual" lifestyle?

Sometimes, it takes the voices of the people, who can listen to the truth, read the facts, and take a good look around their environment—and I am referring to our global environment here—to take the real lead in order to make a real difference. We all have personal choices to make—and that includes the way we treat the environment. The damage being done right now to the earth's ecosystem is oftentimes irreversible and what our direct or indirect actions are causing to be lost will not only impact those of us that inhabit the earth right now but will also impact our children and grandchildren. And for all those future generations who did not have a say and have not seen the beauty yet that this planet has to offer, the time to act is now.

REFERENCES

AMEG (Arctic Methane Emergency Group). 2016. Impacts of the rapidly changing Arctic. http://ameg.me/index.php/impact (accessed August 20, 2016).

Borgen Project. 2014. Top ten poverty in Africa facts. http://borgenproject.org/10-quick-facts-about-poverty-in-africa/ (accessed August 20, 2016).

CarbonBrief. March 28, 2014. Arctic sea ice melt: A story of winners and losers, IPCC scientist says. https://www.carbonbrief.org/arctic-sea-ice-melt-a-story-of-winners-and-losers-ipcc-scientist-says (accessed August 3, 2016).

Conant, R. T. 2010. Challenges and opportunities for carbon sequestration in grassland systems: A technical report on grassland management. *Integrated Crop Management* 9:67.

Conserve Energy Future. 2016. Deforestation facts. http://www.conserve-energy-future.com/various-deforestation-facts.php (accessed August 20, 2016).

Corvalan, C., S. Hales, and S. Hales. 2005. Ecosystems and human well-being (accessed March 27, 2011).

EDF (Environmental Defense Fund). 2011. Deadly heat waves more likely. Environmental Defense Fund. http://www.fightglobalwarming.com/page.cfm?tagID=251 (accessed March 21, 2011).

EDF. August 16, 2007a. Mother nature signals climate warnings. http://www.edf.org/article.cfm?contentID=4823 (accessed March 11, 2011).

EDF. August 16, 2007b. Threats: Coral reef bleaching. http://www.edf.org/article.cfm?contentID=4709 (accessed February 22, 2011).

EEA (European Environment Agency). November 21, 2012. Climate change, impacts, and vulnerability in Europe 2012. EEA Report No. 12/2012. http://www.eea.europa.eu/publications/climate-impacts-and-vulnerability-2012 (accessed August 21, 2016).

Environmental and Energy Study Institute. January 5, 2007. Amazon could become grassland under climate change. *Climate Change News*. http://www.eesi.org/ccn?page=199 (accessed September 21, 2010).

European Environment Agency. 2017. Pew Center on global climate change. https://www.eea.europa.eu/themes/climate/links/physical-science-on-climate/pew-center (accessed May 16, 2017).

FAO. 2010 Global Forest Resources Assessment 2010. Main Report. FAO Forestry Paper 163. Rome. http://www.fao.org/docrep/013/i1757e/i1757e00.htm (accessed May 17, 2017).

Food, fibre and forest products. Impacts, Adaptation, and Vulnerability: Working Group II, 275-303. https://www.ipcc.ch/pdf/assessment-report/ar4/wg2/ar4-wg2-chapter5.pdf (accessed May 17, 2017).

Harvey, M. 2016. Coral reefs: Importance. http://wwf.panda.org/about_our_earth/blue_planet/coasts/coral_reefs/coral_importance/ (accessed August 21, 2016).

IPCC. 2007a. *WG1: The Physical Science Basis*. Cambridge: Cambridge University Press.

IPCC, Food, fibre and forest products. *Impacts, Adaptation, and Vulnerability: Working Group II*, 275-303. https://www.ipcc.ch/pdf/assessment-report/ar4/wg2/ar4-wg2-chapter5.pdf (accessed May 17, 2017).

IPCC. 2014a. Climate change 2014 synthesis report summary for policymakers. https://www.ipcc.ch/pdf/assessment-report/ar5/syr/AR5_SYR_FINAL_SPM.pdf (accessed May 15, 2017).

IPCC. 2014b. IPCC climate change 2013—The physical change basis. http://www.cambridge.org/us/academic/subjects/earth-and-environmental-science/climatology-and-climate-change/climate-change-2013-physical-science-basis-working-group-i-contribution-fifth-assessment-report-intergovernmental-panel-climate-change?format=PB&isbn=9781107661820 (accessed August 2, 2016).

IPCC. 2016. IPCC fifth assessment report. Special report on climate change and oceans and the cryosphere (SROCC). http://www.ipcc.ch/index.htm (accessed August 2, 2016).

Kerr, P. 2009. Dare we celebrate? It's an early spring again. *ISA Ontario*. http://www.isaontario.com/content/dare-we-celebrate-its-early-spring-again (accessed August 3, 2016).

Kohler, T. and D. Maselli (eds.). 2012. *Mountains and Climate Change—From Understanding to Action*, 3rd ed. Bern, Switzerland: Geographica Bernensia.

Mburia, R. December 20, 2011. Climate change and food security issues in sub-Saharan Africa. *Climate Emergency Institute Climate Science Library.* http://www.climateemergencyinstitute. com/food_sec_subsaharan_mburia.html (accessed August 4, 2016).

National Interagency Fire Center, 2016. Statistics: Wildland fire information. https://www.nifc.gov/fireInfo/fireInfo_statistics. html (accessed May 15, 2017).

Natural Resources Canada. 2016. Forest change indicators: Trends and outlooks—Changes in Canada's forests. http://www.nrcan. gc.ca/forests/climate-change/forest-change/17768 (accessed August 3, 2016).

National Oceanic and Atmospheric Administration. 2015. Global climate report—Annual 2015. https://www.ncdc.noaa.gov/sotc/ global/201513 (accessed May 15, 2017).

NSIDC (National Snow and Ice Data Center). March 28, 2016. The Arctic sets yet another record low maximum extent. National Snow and Ice Data Center. https://nsidc.org/news/newsroom/ archive/all (accessed August 3, 2016).

Puckett, C. and B. Molnia. December 2008. Most Alaskan glaciers retreating, thinning, and stagnating, says major USGS report. United States Geological Survey. http://soundwaves.usgs. gov/2008/12/research.html (accessed August 3, 2016).

Revkin, A. C. May 1, 2007. Arctic sea ice melting faster, a study finds. *New York Times.* http://www.nytimes.com/2007/05/01/ us/01climate.html (accessed September 30, 2010).

Rinkesh. 2016. 51 facts about deforestation. Conserve Energy Future. http://www.conserve-energy-future.com/various-deforestation-facts.php (accessed August 2, 2016).

Ritter, K. and C. J. Hanley. May 3, 2011. Seas could rise up to 1.6 meters by 2100: New report. *Coastal Care News.* http:// coastalcare.org/2011/05/seas-could-rise-up-to-1-6-meters-by-2100-new-report/ (accessed May 3, 2011).

Samenow, J. August 2, 2016. The 10 most startling facts about climate in 2015—The warmest year on record. *The Washington Post.*

Slezak, M. April 19, 2016. Great Barrier Reef: 93% of reefs hit by coral bleaching. https://www.theguardian.com/environment/2016/ apr/19/great-barrier-reef-93-of-reefs-hit-by-coral-bleaching (accessed August 10, 2016).

The Guardian. April 13, 2016. Greenland sees record-smashing early ice sheet melt. https://www.theguardian.com/ environment/2016/apr/13/greenland-sees-record-smashing-early-ice-sheet-melt-climate-change (accessed May 15, 2017).

Union of Concerned Scientists. June 2007. Climate change mitigation—IPCC fourth assessment report: Climate change mitigation—IPCC highlights series, findings of the IPCC fourth assessment report.

Union of Concerned Scientists (UCS). 2011. Climate hot map: Global warming effects around the world. http://www.climatehotmap. org/global-warming-effects/drought.html (accessed May 15, 2017).

Union of Concerned Scientists. 2016a. http://www.ucsusa.org/ global_warming/science_and_impacts/impacts/infographic-wildfires-climate-change.html#.V6h6brgrKUk (accessed August 8, 2016).

Union of Concerned Scientists. 2016b. Great Barrier Reef, Australia. Climate hot map: Global warming effects around the world. http://www.climatehotmap.org/global-warming-locations/ great-barrier-reef-australia.html (accessed August 10, 2016).

Weber, B. August 21, 2015. Boreal forest being driver to tipping point by climate change, study finds. *CBC News: Technology & Science.* http://www.cbc.ca/news/technology/boreal-forest-being-driven-to-tipping-point-by-climate-change-study-finds-1.3198892 (accessed August 2, 2016).

World Wildlife Fund. 2003. *Buying Time: A User's Manual for Building Resistance and Resilience to Climate Change in Natural Systems.* WWF Climate Change Program, 246p.

World Wildlife Fund. 2011. Coral reefs bleach to death. http:// www.worldwildlife.org/what/wherewework/coraltriangle/ coral-reefs-bleaching-to-death.html. Presents critical information on the fate of the worlds' coral reefs (accessed May 17, 2017).

World Wildlife Fund. 2015, Animals Affected by Climate Change. https://www.worldwildlife.org/magazine/issues/fall-2015/ articles/animals-affected-by-climate-change (accessed May 15, 2017).

Worldwatch Institute. 2013. Scientists issue new warnings of climate change and severe impacts. http://www.worldwatch. org/node/4641 (accessed August 4, 2016).

SUGGESTED READING

Bridges, A. February 16, 2006. Greenland dumps ice into sea at faster pace. *Live Science.* https://archive.li/VJ7nj (accessed May 17, 2017).

Britt, R. R. June 30, 2005. Scientists put melting mystery on ice. *LiveScience.* http://www.msnbc.msn.com/id/8421342/40041707 (accessed December 2, 2010).

Christianson, G. 1999. *Greenhouse: The 200-Year Story of Global Warming.* New York: Walker.

Houghton, J. 2004. *Global Warming: The Complete Briefing.* New York: Cambridge University Press.

Environmental News Network Staff. November 2, 2000. Vicious cycle: Global warming feeds fire potential. *CNN.* http://archives. cnn.com/2000/nature/11/02/global.warming.enn/ index.html (accessed October 5, 2008).

FAO (Food and Agriculture Organization of the United Nations). 2001. Pan-tropical survey of forest cover changes 1980–2000. *Forest Resources Assessment.* Rome, Italy: Food and Agriculture Organization of the United Nations. http://www.fao.org/docrep/ 004/y1997e/y1997e1f.htm (accessed September 17, 2010).

Fountain, H. January 8, 2008. More acidic ocean hurts reef algae as well as corals. *The New York Times.*

Gore, Al. 2006. *An Inconvenient Truth.* Emmaus, PA: Rodale.

Greenfact Millennium Ecosystem Assessment. 2005. *Ecosystems and Human Well-being: Desertification Syntheses.* Washington, DC: Island Press.

IGES (Institute for Global Environmental Strategies). December 20, 2010. 2010 top news on the environment in Asia (Provisional Version).

Intergovernmental Panel on Climate Change (IPCC). 2014. Climate change 2014 synthesis report summary for policymakers. https:// www.ipcc.ch/pdf/assessment-report/ar5/syr/AR5_SYR_FINAL_ SPM.pdf (accessed May 15, 2017).

IPCC (Intergovernmental Panel on Climate Change). 2001. Climate change 2001: Synthesis report: Summary for policymakers. http://www.ipcc.ch/pub/un/syreng/spm.pdf (accessed December 9, 2010).

IPCC. 2007b. Food, fibre and forest products. *Impacts, Adaptation, and Vulnerability: Working Group II, 275–303.* http://www.ipcc-wg2. gov/AR4/website/05.pdf.

Ives, J. November 15 and 16, 2002. Mountain disasters. *International Year of the Mountain Conference,* University of Colorado at Boulder.

Jardine, K. 1994. *The Carbon Bomb: Climate Change and the Fate of the Northern Boreal Forests.* Amsterdam, the Netherlands: Greenpeace International.

Karoly, D., J. Risbey, and A. Reynolds. 2002. *Global Warming Contributes to Australia's Worst Drought*. World Wildlife Fund. http://www.wwf.org.au/publications/drought_report/ (accessed November 23, 2010).

Karl, T. and K. Trenberth. 1999. The human impact on climate. *Scientific American* 281(6):100–105.

Kerr, P. April 9, 2006. Dare despair: Special report, global warming. *Time*, pp. 30–36.

Kluger, J. March 26, 2006. Global warming heats up. *Time*. http://www.time.com/time/printout/0,8816,1176980,00.html.

Ludi, E. September 18, 2008. Hunger on the rise. FAO: FAO home. http://www.fao.org/newsroom/EN/news/2008/1000923/index.html (accessed November 27, 2011).

Mburia, Robert. 2011. Climate Change and Food Security Issues in Sub-Saharan Africa. https://www.climateemergencyinstitute.com/food_sec_subsaharan_mburia.html (accessed May 17, 2017).

McCarthy, M. October 4, 2006. The century of drought. *The Independent*. http://www.commondreams.org/headlines06/1004-02.htm (accessed November 22, 2008).

MSNBC News. October 12, 2006. Africa's tallest mountains nearly bare of ice: UN warns of human disaster if Kilimanjaro, Kenya Run dry in 50 years. http://www.msnbc.msn.com/id/15238801/ns/world_news-world_environment/ (accessed March 28, 2011).

NASA Earth Observatory. Mapping the decline of coral reefs. http://earthobservatory.nasa.gov/Features/Coral/ (accessed February 28, 2011).

National Geographic News. February 1, 2002. Mountain ecosystems in danger worldwide, UN says. http://news.nationalgeographic.com/news/2002/02/0201_020201_wiremountain.html (accessed January 4, 2009).

Natural Resources Canada. 2002. Climate change impacts and adaptation: A Canadian perspective. Prepared by the Climate Change Impacts and Adaptation Directorate. http://adaptation.nrcan.gc.ca (accessed January 11, 2009).

PNNL (Pacific Northwest National Laboratory). Model casts cloud over mountain snowfall. Pacific Northwest National Laboratory, Department of Energy. http://www.pnl.gov/main/news/global.html (accessed March 28, 2011).

PNNL. 2004. Global warming to squeeze western mountains dry by 2050. DOE/Pacific Northwest National Laboratory. http://www.scienceblog.com/community/older/2004/3/20042513.shtml (accessed February 16, 2011).

Rohter, L. December 11, 2005. A record Amazon drought, and fear of wider ills. *The New York Times. Science Daily*. http://www.sciencedaily.com/releases/2001/01/010111073831.htm (accessed November 9, 2010).

Shah, M., G. Fischer, and H. Van Velthuizen. 2008. *Food Security and Sustainable Agriculture. The Challenges of Climate Change in Sub-Saharan Africa*. Laxenburg, Austria: International Institute for Applied Systems Analysis.

Space Daily. February 18, 2004. Global warming to squeeze western mountains dry by 2050. http://www.spacemart.com/reports/Global_Warming_To_Squeeze_Western_Mountains_Dry_By_2050.html (accessed January 11, 2009).

SwissInfo. September 9, 2003. Global warming threatens alpine plants. http://www.swissinfo.ch/eng/Home/Archive/Global_warming_threatens_alpine_plants.html?cid=3501194 (accessed March 28, 2011).

UNDP (United Nations Development Program). 2008. *Fighting Climate Change—Human Solidarity in a Divided World*. New York: UNDP.

UNEP/AMAP. 2011. *Climate Change and POPS: Predicting the Impacts*. http://arctic-council.org/climate_change_and_pops_predicting_the_impacts (accessed April 1, 2011).

UniSci. 2001. Many Alaskan glaciers are thinning, USGS study says. *Daily University Science News*. http://www.unisci.com/stories/20014/1211011.htm (accessed October 6, 2010).

5 Sociological and Psychological Considerations to Climate Change

OVERVIEW

This chapter first takes a look at the sociological issues of climate change and explores several of the current key aspects, such as the publics' reactions to environmental issues in general and the reasons for those reactions—why some people seem to care so deeply about the environment around them and its future, yet others seem to give it no thought at all. It also looks at the public's slow shift in recent years toward understanding the basic concepts of climate change. In fact, part of the reason for changing the name from global warming to climate change was partly for sociological reasons: the label global warming implied to the public that rising temperatures were the *only characteristic* of the phenomenon. For this reason, for areas that were experiencing cooler temperatures or flooding, the public was not making the connection that it was part of the same phenomenon. Therefore, scientists began to transition to the more accurate term *climate change* to make it more understandable—that there was more involved to the issue than rising temperatures and that those changes were dangerous to the environment and future life on the planet.

This chapter also discusses how social systems and cultural values are tied to a region's biodiversity. It also examines the organization of environmental movements and the unusual chain of events that occurred in order for the first Earth Day to come about more than 45 years ago. Next it takes a close look at the effects that climate change has on society and which societies feel the brunt of change more than others. Following that, it delves into immigration issues that are expected to arise as the climate changes and how that could present more than just a casual problem to some countries. It next examines some interesting theories about what may get people to listen to messages about the environment and be willing to change their lifestyle.

We then shift gears and move into the human psychological component to examine the future of climate change from a psychologist's perspective. We will look at perception—the way people individually view the important issues will ultimately go into the collective mix and determine how we change collectively as a society. Many people's perception today are shaped by what they obtain from the media—how the programs are shown and the stories are reported—accurate or not and the effect they can have on the development of scientific issues such as climate change.

We will explore the power of the media, and with it the responsibility they have of reporting accurately and what the ramifications can be when stories are not. We will define its importance, as well as how it can become a roadblock in controversial scientific issues such as climate change, and yet can recognize the difference. This chapter then discusses the advancement of scientific theories and the subsequent evolution of thought and how the media often uses that to discredit controversial scientific issues. Finally, it covers the occasional data flaw and the interesting results that can happen with media interaction in relation to human psychology.

INTRODUCTION

An important component of the whole climate change picture is the sociological aspect; yet it is one of the least looked at and understood. Even though societies around the Earth are currently being affected and scientists have projected future scenarios for several locations worldwide and issued warnings to begin preparing by investing in mitigation efforts, planning defense strategies, becoming proactive in changing lifestyle choices, and planning for various future scenarios, the general public still displays tendencies that are reluctant and lukewarm.

What seems puzzling in this case is the general feeling of having to push people to act and respond to climate change when it is a situation that affects their very lives, futures, and the future of their children. This raises questions as to where the communication process is failing—is it an issue of education, politics, culture, economics, lifestyle, mind-sets, or some other unidentified factor? Or a combination of part, or all, of them? There has even been concern expressed in the sociological field that sociologists are not even paying enough attention to the issue. Yet, the sociological aspect of the big picture is one of the most crucial components. With such a diversity of cultures worldwide, there are several sociological issues to consider.

Because climate change is such a controversial issue, when dealing with the media, it is important for the viewers to take upon themselves the personal responsibility of becoming thoroughly educated about the topic and being aware of what climatologists and other specialists know, what they suspect, and what the controversies are about. Fortunately, today there are several organizations whose purpose is to educate others about the most critical environmental subjects—such as climate change—and teach environmental responsibility through conservation. From both the governmental and private sectors, there are many organizations that offer opportunities to get involved in fighting climate change, becoming educated, and educating others about the latest discoveries and developments, such as the Pew Environmental Group, National Aeronautics and Space Administration (NASA), NASA Goddard Institute for Space Studies (GISS), Environmental Defense Fund, National Geographic, and the U.S. Environmental Protection Agency (EPA), to name just a few. These organizations are beneficial because they are generally well connected with political and administrative information. They also keep up with the latest research techniques and present a good source of unbiased scientific information. Many also attend the formal political negotiations and conferences, and offer reports that are generally more neutral and objective, often giving the reader better, more direct information.

Again, this is another example of the value of education. Becoming knowledgeable about the issues is critical. Although this may sound obvious, on many occasions misinformation has been released to the public and has muddled and damaged the progression of solving critical issues—and climate change is an excellent case in point. Through the release of wrong or antagonistic information, the damage caused often slows the progress of research and mitigation efforts—sometimes even significant political headway—making the acquisition of a sound understanding of the psychological aspects of media attention important to have.

DEVELOPMENT

One of the most overlooked aspects of both the study and the management of climate change is the sociological aspect. Although some sociologists have repeatedly brought this issue to public and scientific attention, it still remains one area significantly lacking in the attention it deserves, which is ironic because both the short- and long-term effects of climate change affect the environment in multiple ways that impact human lives every day. For this reason, this chapter takes a head-on approach of pointing out the close tie between humans and the environment. Through seeing the close cause-and-effect interactions, it then becomes possible to see the way the society not only needs a healthy environment but cannot live productive, healthy lives without one. Then, by pointing out the reasons why people, both as individuals and as a society, disregard obvious warnings or choose to delay action, it identifies areas where we could improve. Communication is the first step. We have to identify, understand, and talk about it first to come up with plans of action in order to deal with and solve it. So let us begin.

THE ENVIRONMENT AND SOCIOLOGY

Let us start with a little background. Climate change and its effect on the society fall under the sociological study of societal–environmental interactions, or environmental sociology. The principal focus of the field is the relationship between the society and the environment—specifically social factors that lead to environmental problems, the resultant societal impacts of those problems, and the subsequent efforts to adequately solve those problems. A fairly new specialized emphasis in sociology, it did not gain real distinction until the environmental movement of the 1960s and early 1970s, particularly in the United States and Europe.

This time period in the United States became characterized as the "Environmental Decade" and is marked by the time interval when the EPA was created, and such acts as the Endangered Species Act, the Clean Air Act, and the Clean Water Act were implemented. This was also the time frame when Earth Day was instituted, which represented a significant paradigm shift in public thought and is discussed in greater detail in the section: Environmental Movements—The Classic Case of Earth Day.

Over the past few decades, adding additional public awareness toward the environment, there have also been several notable anthropogenic environmental disasters that have played a significant role in making the public more aware of damage that can be caused to the environment—a damage that is sometimes permanent. As the society becomes more aware of the environment and how altering it can have a negative impact, it has spurred in recent years the development of many environmental awareness groups committed to helping preserve our surroundings (no, we did not always have them). Fortunately, these groups have had an effect in educating political representatives, policymakers, educators, and the general public. Nevertheless, we still have a long way to go and have had too many significant environmental disasters.

One of the worst human-caused disasters occurred was the Russian nuclear power plant explosion in Chernobyl in 1986, when a reactor exploded during a failed cooling system test and ignited a massive fire that burned steadily for 10 days. The accident released radioactivity 400 times more intense than that of the Hiroshima bomb during World War II. The accident affected a huge area—parts of the Ukraine, Belarus, and Russia (the former Soviet Union), covering an area of approximately 233,000 square kilometers. The ground was contaminated with unhealthy levels of radioactive elements. The town of Pripyat, which was built specially to house the employees of Chernobyl, was evacuated after the accident, displacing 50,000 people. Even today, more than 30 years later, the town has never been "officially" reoccupied. The people who lived in the area at the time of the accident that were not killed outright in the explosion suffered many serious health problems; including radiation sickness, thyroid cancer, leukemia, and birth defects causing cancer and heart disease. Approximately seven million people were affected by this accident.

It is unarguable that the world's worst nuclear accident occurred on April 26, 1986, when the Chernobyl nuclear power plant's reactor no. 4 blew up after a cooling capability test. The resulting nuclear fire lasted 10 days and released 400 times as much radiation as the bomb dropped on Hiroshima. Even 30 years later, reactor no. 4 stews in its container of concrete and metal—now rusted, cracked, and leaking radiation. In February 2012, part of the container's roof collapsed, releasing 200 tons of lava-like radioactive material, and the new "safe" container that is being built to contain it is not finished.

Unfortunately, this has not deterred a group of 130 very determined, elderly Russian women who were part of a larger group that came back to occupy their old homeland—which just happens to be within the danger zone. This is an area where the soil, water, and air are considered some of the most highly contaminated on earth. In fact, the reactor sits at the center of a 2590-square-kilometer area called the "exclusion zone," which is a quarantined area guarded, monitored, and controlled. Yet, these 130 women, called "self-settlers," live today in the Chernobyl exclusion zone. Mostly in their 1970s and 1980s, they are the last survivors of an original group of 1200 that defied authorities and illegally returned to their homes after the accident. What some find interesting is that these remaining women have outlived many of those that were relocated and have since died from the exposure to radiation (Morris, 2013).

Another major human-caused environmental disaster was the nuclear accident on the Susquehanna River in Pennsylvania at Three Mile Island in March 1979 (Figure 5.1). This was the disaster that began the controversy over the safety of using nuclear energy in the United States. In this incident, the water pumps in the cooling system failed, causing cooling water to drain away from the reactor, which partially melted the reactor core. The accident released about one-thousandth the amount of radiation as the Chernobyl disaster did. The reactor core barely escaped meltdown due to the implementation of safety measures. Scientists do not know for certain how much radiation was released during the accident. There was an evacuation of an 8-kilometer radius as a safety precaution. Experts do believe that several elderly people died from being exposed to the radiation. Dairy farmers reported the deaths of many of their livestock. Some local residents developed cancer. Some studies also indicate that premature death and birth defects resulted as well. The cleanup for the accident began in August 1979, and finally ended in 1993, with a price tag of $975 million. Nearly 91 metric tons of radioactive fuel was removed from the area.

Times Beach, Missouri, was the site of another well-known environmental scare that got the public's attention when high dioxin levels were found in the soil. Dioxin is a hazardous chemical used in Agent Orange, a highly toxic chemical warfare agent.

FIGURE 5.1 Three Mile Island was the site of one of the nation's worst anthropogenic environmental disasters. (Courtesy of Environmental Protection Agency [EPA], Washington, DC.)

Levels in the soil were determined to be 100 times higher than the threshold considered to be toxic to humans. The dioxin had been mistakenly added to an oil mixture that was used to spray the roads in the 1970s to keep the dust problem under control. Many illnesses, miscarriages, and animal deaths at the time were blamed on the levels of dioxin in the area. This episode was one of the environmental disasters in the United States that spurred the enactment of Federal action and the implementation of the Comprehensive Environmental Response, Compensation, and Liability Act—a piece of legislation commonly referred to as Superfund because of the fund established within the act to help the cleanup of locations such as Times Beach.

In the Baia Mare gold mine in Romania in January 2000, cyanide, which is used to purify gold from rocks, overflowed into the Tisza River. The wastewater that the cyanide was in also contained lead and other hazardous materials. By February, it had impacted the Danube River, a major river that also flowed through Serbia and Hungary. This incident poisoned many of the fish in the river, and many people along the river had to be treated for eating contaminated fish.

According to the EPA, the Love Canal is one of the most horrifying environmental tragedies in U.S. history. Love Canal was originally planned to be the "perfect" community on the eastern edge of Niagara Falls in New York. In order to generate power for the community, a canal was to be dug between the upper and lower Niagara rivers so that power could be generated inexpensively for the residents. The canal was never finished for generating energy, however. In the

1920s, all that was left was a remnant of the original digging, and it was turned into a municipal and industrial chemical dumpsite. Then, in 1953, the Hooker Chemical Company, of Niagara Falls, NY, who then owned the property and the canal, covered all the waste with dirt and sold it to the city for the sum of $1. Unfortunately, the city used the ground for a new development and constructed about 100 homes and a school on it. During an extremely wet period, where the area received ample rainfall, the leaching process began. The waste disposal drums began to corrode and started breaking up through the resident's backyards. The vegetation in the area—trees, shrubs, gardens, and grass—began to die and turn black. Chemicals began to pool in people's backyards and basements. Children got burns on their hands and faces when they played outside. An increase in birth defects began to occur as well.

At the time, according to a report issued by the EPA, one father whose child was born with birth defects as a result of the tragic situation remarked: "I heard someone from the press saying that there were *only* five cases of birth defects here. When you go back to your people at EPA, please don't use the phrase '*only* five cases.' People must realize that this is a tiny community. Five birth defect cases here are terrifying" (Beck, 1979).

On August 7, 1978, New York Governor Hugh Carey told the residents at Love Canal that the New York State Government would purchase the homes that were affected by the chemicals. The same day, President Jimmy Carter approved emergency financial aid for the area. These were the first emergency funds ever to be approved for a human-caused disaster rather than a natural disaster. In addition, the United States Senate approved a "sense of Congress" amendment saying that Federal aid should be forthcoming to relieve the serious environmental disaster that had occurred.

President Carter remarked of the situation: "The presence of various types of toxic substances in our environment has become increasingly widespread—one of the grimmest discoveries of the modern era" (Beck, 1979).

A total of 221 families had to be relocated as a result of this disaster. Today, agencies such as the EPA, working under governing laws such as the Clean Air and Water Acts, the Pesticide Act, the Resource Conservation and Recovery Act, and the Toxic Substances Control Act, strive to protect the American public and the environment.

The Valley of the Drums in Kentucky gained national attention in 1979 and quickly became known as one of the United State's worst abandoned hazardous waste sites. Over a period of 10 years in Bullitt County, people had disposed of thousands of barrels of hazardous wastes. They had been haphazardly thrown in pits or trenches, or just strewn about (Figure 5.2). The drums sat so long exposed to the outdoor elements that they began to deteriorate and leak. When it rained, the barrels would fill with water, overflow, and wash the chemicals inside into nearby Wilson Creek, which led to the Ohio River. Chemicals such as toluene (associated with liver and kidney damage, respiratory illness, harmful to developing fetuses, and can cause death) and benzene (which causes leukemia, neurological problems, and weaken immune systems) were found in them. This incident gained so much national attention; it also spurred the creation of the Superfund.

In March 1989, the Exxon Valdez struck Bligh Reef in Prince William Sound, Alaska, creating the largest oil spill in U.S. history. The oil slick spread more than 7,770 square kilometers and onto more than 563 kilometers of

FIGURE 5.2 Valley of the drums. This site was used as a disposal area for hazardous chemical waste, causing serious environmental pollution. Today it is still remembered as an example of environmental irresponsibility. (Courtesy of Environmental Protection Agency [EPA], Washington, DC.)

FIGURE 5.3 The tanker Exxon Valdez accident in Prince William Sound, Alaska, in March 1989. NOAA responders survey the oil-soaked beaches of Prince William Sound. (Courtesy of NOAA.)

fragile ecosystems. It killed approximately 250,000 sea birds, 2800 sea otters (*Enhydra lutris*), 250 bald eagles (*Haliaeetus leucocephalus*), and roughly two dozen killer whales (*Orcinus orca*).

The spill released more than 11 million gallons and cost more than $3.5 billion to clean up. This was one of the environmental disasters that received a great deal of media attention and brought a significant environmental awareness to the public. Many people were outraged at the damage done to the wildlife habitat and the fragile ecosystems, and this single incident in particular served as a strong reminder that human behavior can have far-reaching impacts on the environment.

beaches in Prince William Sound, at that time known as one of the most pristine and beautiful natural areas in the world (Figure 5.3). The spill polluted about 1,900 kilometers of shoreline and was devastating to the wildlife in the

DISASTROUS HUMAN IMPACT ON THE ENVIRONMENT

The sad truth is that some of the most devastating impacts on the environment have been caused by human interaction. Ironically, many of these impacts are also related to the very activities that are leading to the negative effects of climate change. The 25 worst human-caused environmental disasters in history are discussed in the sections that follow (Dimdam, 2013) (Figure 5.4).

Ash slurry spill	Exxon valdez	Pacific gyre	Jilin chemical	Castle bravo
Three mile island	Kuwait oil fires	"Door to hell"	Palomares incident	Sidoarjo mud flow
Libby, Montana Asbestos	Deep water horizon (BP) spill	Amoco cadiz	Vietnam ecocide	Al-Mishraq fire
Love canal	Gulf of Mexico dead zone	Minamata disease	Seveso	E-waste
Baia mare cyanide spill	Aral sea	Bhopal	Chernobyl	Great smog of '52

FIGURE 5.4 Summary of the world's 25 worst human-caused environmental disasters. (From Dimdam, E. List 25. History: 25 biggest man-made environmental disasters in history. http://list25.com/25-biggest-man-made-environmental-disasters-in-history/5/ [accessed August 14, 2016], 2013.)

TVA Kingston Fossil Plant Coal Fly Ash Slurry Spill

In an 84-acre solid waste containment area, an ash dike ruptured on December 22, 2008. This fossil plant in Roane County, Tennessee, held 1.1 billion gallons of coal fly ash slurry, which was released causing a mudflow wave. Although there were no reported fatalities or injuries, it damaged several properties and government facilities.

The Exxon Valdez Oil Spill

On March 24, 1989, 260,000–750,000 barrels of crude oil was spilled in Prince William Sound, Alaska, by the oil tanker Exxon Valdez after it ran into Bligh Reef. This was one of the most devastating human-caused environmental disasters of all time, in both the short and long terms. Immediate effects included the deaths of 100,000 to as many as 250,000 seabirds, at least 2800 sea otters (*E. lutris*), 300 harbor seals (*Phoca vitulina*), 250 bald eagles (*H. leucocephalus*), 22 orcas (*O. orca*), and an unknown number of salmon (*Salmo salar*) and herring (*Clupea harengus*).

Pacific Gyre Garbage Patch

Another example of the irresponsible effects of human waste is the Pacific Gyre Garbage Patch—an enormous collection of swirling marine debris in the central North Pacific Ocean. This patch, which is characterized by high concentrations of plastics, chemical sludge, and other debris, formed gradually as a result of the marine pollution gathered by oceanic currents. Researchers are still trying to determine how much debris makes up the Great Pacific Garbage Patch. It is not just a large collection of debris floating on the surface of the ocean, but rather debris (mostly small bits of matter) floating throughout the entire column of ocean water depth. It is too large for scientists to trawl; denser debris can sink several meters beneath the surface, making the vortex's area nearly impossible to measure. About 80 percent of the debris comes from land-based activities in North America and Asia. Trash from the coast of North America takes about 6 years to reach the Garbage Patch, whereas trash from Japan and other Asian countries takes about a year. The remaining 20 percent of debris comes from boaters, offshore oil rigs, and large cargo ships that dump or lose debris directly into the water. The majority of this debris—about 705,000 tons—is fishing nets. More unusual items, such as computer monitors and LEGOs (a word derived from the Danish phrase leg godt, which means "play well."), come from dropped shipping containers.

Although many different types of trash enter the ocean, plastics make up the majority of debris for two reasons: (1) plastic is commonly used because it is inexpensive and durable and (2) plastic goods do not biodegrade but instead break down into smaller pieces. In the ocean, the sun breaks down these plastics into tinier and tinier pieces, a process known as photodegradation. There are up to 750,000 bits of microplastics in a single square kilometer of the Garbage Patch, most of it coming from plastic bags, bottle caps, plastic water bottles, and Styrofoam cups. Unfortunately, this marine debris can be very harmful to marine life in the gyre. For example, loggerhead sea turtles often mistake plastic bags for jellyfish, their favorite food. Albatrosses mistake plastic resin pellets for fish eggs and feed them to their chicks, which then die of starvation or ruptured organs. Seals and other marine mammals are especially at risk. They can get entangled in abandoned plastic fishing nets, which are being discarded more often because of their low cost. Seals and other mammals often drown in these forgotten nets—a phenomenon known as "ghost fishing."

Also hazardous, plastics both leach out and absorb harmful pollutants. As they break down through photodegradation, they leach out colorants and chemicals, such as bisphenol A, that have been linked to environmental and health problems. Plastics can also absorb pollutants, such as PCBs (Polychlorinated biphenyls), from the seawater. These chemicals can then enter the food chain when consumed by marine life.

When it comes to cleaning up the Garbage Patch, because it is so far from any country's coastline, the sad reality is that no nation will take responsibility or provide the funding to clean it up. Scientific researcher and sea captain, Charles Moore, who discovered the vortex in 1997, says cleaning up the Garbage Patch would "bankrupt any country" that tried it. Many individuals and international organizations, however, are dedicated to preventing the patch from growing.

Cleaning up marine debris is not as easy as it sounds. Many microplastics are the same size as small sea animals, so nets designed to scoop up trash would also capture these animals. Even if nets could be designed that would just catch garbage, the sheer size of the project makes it impractical. The National Ocean and Atmospheric Administration's Marine Debris Program has estimated that it would take 67 ships for one year to clean up less than 1 percent of the North Pacific Ocean (National Geographic, 2016).

Charles Moore continues to raise awareness through his own environmental organization, the Algalita Marine Research Foundation. During a 2014 expedition, Moore and his team used aerial drones to assess from above the extent of the trash below. He was able to determine that there are 100 times more plastic by weight than previously measured. His team also discovered more permanent plastic features, or islands, some over 15 meters in length.

Scientists and explorers have stressed the importance of limiting or eliminating our use of disposable plastics and increasing our use of biodegradable resources. They believe that this will be the best way to clean up the Garbage Patch. Organizations such as the Plastic Pollution Coalition and the Plastic Oceans Foundation are using social media and direct action campaigns to support individuals, manufacturers, and businesses in their transition from toxic, disposable plastics to biodegradable or reusable materials.

Jilin Chemical Plant Explosions

The Jilin chemical plant explosions were a series of explosions that occurred on November 13, 2005, in the No. 101 Petrochemical plant in Jilin City, Jilin Province, People's Republic of China. These explosions were responsible for the deaths of six workers and the injury of dozens of others, causing the evacuation of tens of thousands of residents. In addition, these explosions severely polluted the Songhua River with an estimated 100 tons of pollutants containing benzene and nitrobenzene whose exposure reduces white blood cell count and is linked to leukemia.

Castle Bravo

The code name Castle Bravo was given to the first United States test of a dry fuel thermonuclear hydrogen bomb. The bomb was detonated on Bikini Atoll, Marshall Isalnds, on March 1, 1954, as the first test of Operation Castle and was the most powerful nuclear device ever detonated by the United States at that time. This test led to the most significant accidental radiological contamination ever caused by the United States.

Fallout from the detonation fell on residents of Rongelap and Utirik atolls and spread around the world. The islanders were not evacuated until three days after the detonation and they suffered radiation sickness. They were returned to the islands 3 years later but were removed again when their island was determined to be "unsafe." In addition, the crew of the Japanese fishing vessel *Daigo Fukuryū Maru* ("Lucky Dragon No. 5") was also contaminated by fallout, killing one crew member. The blast ended up creating an international reaction about atmospheric thermonuclear testing.

Three Mile Island Nuclear Explosion

The Three Mile Island accident was a partial nuclear meltdown that occurred in one of the two United States nuclear reactors on March 28, 1979. Located on the Three Mile Island in Dauphin County, Pennsylvania, it was the worst accident in U.S. commercial nuclear power plant history with the partial meltdown resulting in the release of small amounts of radioactive gases and radioactive iodine into the environment.

The Kuwait Oil Fires

Approximately six million barrels of oil was lost from January to November 1991. Six hundred to 732 oil wells, along with an unspecified number of oil-filled low-lying areas, such as oil lakes and fire trenches, were set on fire as part of the "scorched earth policy" by the retreating Iraqi military forces. Scorched earth policy is a military strategy, of burning or destroying buildings, crops, or other resources that might be of use to an invading enemy force. In total $1.5 billion was spent by Kuwait to extinguish the fires that caused heavy pollution to the soil and air.

Door to Hell

Derweze, Turkmenistan, a drilling rig made by Soviet geologists in 1971, gave way to a large hole measuring 70 meters in diameter, exposing a large methane gas reservoir. The area, composed of sedimentary rock, triggered a large-scale collapse in other places as well, resulting in several craters opening up. The largest of the craters measured about 70 meters across and

20 meters deep. The escaping methane gas began killing the animal life in the surrounding desert. The high levels of the gas also posed the potential for an explosion, so the geologists decided to burn it off—a process called *flaring*. Anticipating that it would only take a few weeks to completely burn off, they eventually discovered that there was more gas there than they realized. Today, they still do not know how much natural gas is feeding the burning crater, as it is still burning today—nearly a half of a century later. Today, tourists come to visit it, and on the more bizarre side, for some reason, it attracts desert spiders that plunge into the pit by the thousands, lured to their deaths by the glowing flames (Geiling, 2014).

The Palomares Incident

The crash of the B-52G bomber of the United States Air Force (USAF) Strategic Air Command on January 17, 1966, led to the plutonium contamination of Palomares, a small village in the municipality of Cuevas del Almanzora, Almería, Spain. The jet-powered strategic bomber carried non-nuclear explosives that detonated causing a political conflict between the United States and Spain. Forty years later, traces of the blasts are still evident.

Sidoarjo Mud Flow

Sidoarjo (the largest mud volcano in the world), also known as the Lapindo mud, exists today due to gas blowout wells drilled by PT Lapindo Branta. Branta denies this, however, and claims that the mud flows were created by an earthquake. Each day 180,000 cubic meters of mud is spewed at its peak and has been in eruption since May 2006.

Libby, Montana Asbestos Contamination

Vermiculite mines in Libby, Montana, gave the local residents jobs and helped the local economy. However, due to the mine's high use of asbestos, the residents suffered related disorders such as mesothelioma. Because of mine activities that started back in 1919, residents are still affected today.

Deepwater Horizon (BP) Oil Spill

The Deepwater Horizon oil spill (also referred to as the BP oil spill) in the Gulf of Mexico, which began on April 20, 2010, is considered the largest accidental marine oil spill in the history of the petroleum industry. The oil spill was a direct result of the explosion and sinking of the Deepwater Horizon oil rig that claimed 11 lives. Following the explosion and sinking of the Deepwater Horizon oil rig, a seafloor oil gusher flowed for 87 days, until it was capped on July 15, 2010. Total oil wasted was estimated at 4.9 million barrels (780,000 m³). After several failed efforts to contain the flow, the well was declared sealed on September 19, 2010. Reports in early 2012, however, indicated the well site was still leaking.

There was a massive effort to protect beaches, wetlands, and estuaries from the spreading oil, yet extensive damage to marine and wildlife habitats, and fishing and tourism industries were reported. In Louisiana, two million kilograms of oily residue was removed off the beaches in 2013, more than twice the amount collected in 2012. Oil cleanup crews worked four days a week on 89 kilometers of Louisiana shoreline throughout 2013. Oil still continued to be found as far from the spill site as the waters off the Florida Panhandle and Tampa Bay, where scientists said the oil–dispersant mixture was embedded in the sand (Juhasz, 2012, Elliott, 2013, Pittman, 2013, Viegas, 2013). In 2013, it was reported that dolphins and other marine life continued to die in record numbers with infant dolphins dying at 6 times the normal rate. One study released in 2014 reported that tuna and amberjack that were exposed to oil from the spill developed deformities of the heart and other organs that would be expected to be fatal or at least life-shortening, and another study found that cardiotoxicity might have been widespread in animal life exposed to the spill (Sahagun, 2014, Wines, 2014).

Amoco Cadiz

A Liberian crude carrier ran aground on Portsall Rocks, 5 kilometers from the coast of Brittany, France, on March 16, 1978, and split into three parts and sank, releasing 1,604,500 barrels (219,797 tons) of light crude oil and 4000 tons of fuel oil. Severe weather caused the complete breakup of the ship before any oil could be pumped out of the wreck, resulting in the entire cargo of crude oil and fuel oil being spilled into the sea. This made it the largest oil spill of its kind at that time and resulted in the largest loss of marine life ever recorded from an oil spill.

Ecocide in Vietnam

The term ecocide is the destruction of our natural environment. Defined as the extensive damage to, destruction of, or loss of ecosystems of a given territory, it covers all major environmental disasters. Ecocide in reference to Vietnam refers to the use of nearly 76 million liters of herbicide that was sprayed on the jungles of Vietnam between 1959 and 1975—much of it deadly Agent Orange (Johnson, 2012). During the Vietnam War, destruction of the farmland and rice paddies that fed the enemy was promulgated by the American military strategists. Other than these areas which were the source of food and livelihood of the Vietnamese folk, the jungle along with its flora and fauna was also devastated.

The Al-Mishraq Fire

Al-Mishraq is a state-run sulfur plant near Mosul, Iraq, which in June 2003 was the site of the largest human-made release of sulfur dioxide ever recorded. The smoke plume was visible on satellite imagery for kilometers. A fire thought to have been deliberately started burned for almost a month spewing 21,000 tons of sulfur dioxide a day into the atmosphere. The plume contained a mixture of contaminants, such as particulate matter and varying concentrations of sulfur dioxide and hydrogen sulfide. A total of approximately 600,000 tons of sulfur dioxide was released into the atmosphere.

The Love Canal

In the 1940s, 21,000 tons of toxic industrial waste, containing highly toxic dioxin, was buried by Hooker Chemical (now Occidental Petroleum Corporation), which led to an adverse affect on nearby residents of Love Canal. The Love Canal neighborhood started gaining international attention as the health impact of such pollution became evident with miscarriages, cancers, and birth defects, and has been described as a "national symbol of a failure to exercise a sense of concern for future generations" (see the section "The Environment and Sociology").

Gulf of Mexico Dead Zone

The most infamous hypoxic zone in the United States, the Gulf of Mexico's "dead zone," is the dumping area for nitrogen and phosphorus, just two of the many high nutrient runoffs. These substances come from the Mississippi River, which is the drainage area for almost half of the continental America. An area of low to no oxygen that can kill fish and marine life in 2015 measured 16,768 square kilometers, which is above the average in size and larger than was forecast by the National Oceanic and Atmospheric Administration (NOAA). An area roughly the size of Connecticut and Rhode Island combined adversely affects the nation's coastal resources and habitats in the Gulf.

These "deal zones," also called hypoxia areas, are caused by nutrient runoff from agricultural and other human activities in the watershed and are highly affected by river discharge and nitrogen amounts. The nutrients stimulate an overgrowth of algae that sinks decompose, and consume the oxygen needed to support life in the Gulf. These dead zones present a critical water-quality issue. There are more than 550 of them worldwide. The dead zone in the Gulf of Mexico is the second largest one in the world (NOAA, 2015).

Minamata Disease

Considered one of the four major pollution diseases in the history of Japan, Minamata is caused by severe mercury poisoning that attacks the nervous system. Other effects include sensory disturbances, constriction of the visual field, auditory disturbances, tremors, and lesions to the brain. Fetuses are also affected. In 1956, Chisso Corporation's industrial wastewater containing methylmercury was released into Minamata Bay and the Shiranui Sea. Since the spill, 2252 people have been diagnosed with the disease, and almost half have since died from the contamination.

The Seveso Disaster

In July 1976, an explosion at a chemical manufacturing plant north of Milan, Italy, released tetrachlorodibenzo-p-dioxin into the atmosphere adversely affecting the nearby town of Seveso. Shortly thereafter 3300 animals died and many more were put down in order to prevent the spread of contamination into the food chain. Children were hospitalized with skin inflammation and nearly 500 people were found to have skin lesions.

In studies completed in 1991, 15 years after the accident and rechecked in 2009 (33 years after), the most prevalent long-term effect was the presence of chloracne—skin lesions caused by exposure to chlorine (similar to acid burns). Investigations also raised possible issues with liver function, immune function, neurologic impairment, and reproductive effects. Clinical tests did find excess mortality from cardiovascular and respiratory diseases, as well as an increase in diabetes cases. An increased occurrence of cancer of the gastrointestinal sites and of the lymphatic and hematopoietic tissue was also higher than expected. There was also a positive connection to the occurrence of breast cancer.

E-waste in Guiyu, China

Guiyu, China, is the location of what may be the largest electronic waste (e-waste) site on earth. Guiyu's soil, water, air and people are paying a high price, however. In small workshops and in the open countryside, thousands of men, women, and children are dissembling the toys and equipment of the developed world—such as old computers, monitors, printers, DVD players, photocopying machines, telephones and phone chargers, music speakers, car batteries, and microwave ovens. They use primitive methods that leave them exposed to environmental hazards. For example, circuit boards and other computer parts are burned, individually, over open fires to extract metals. This smelting process releases large amounts of toxic gases into the air.

Plastics are graded by quality and other parts are burned to separate plastics from scrap metal. After this thorough dismembering, any remaining combustibles are left to burn in open fires, filling the air with the acrid stench of plastic, rubber, and paint.

Manually, the workers carefully reduce every piece of equipment to its smallest components. These are then farmed off to "specialists," workers dedicated to stripping wires for the copper they contain or melting the lead solder from circuit boards. The streams in the area are highly polluted and contaminated with heavy metals, toxic pollution, and industrial waste. The groundwater in Guiyu is not drinkable. As a result of this exposure to all these components, 82 percent of the children in the area suffer from lead poisoning. The area also has a high rate of miscarriages. Another study discovered a high incidence of skin damage, headaches, vertigo, nausea, chronic gastritis, and gastric and duodenal ulcers. Sadly, this province is referred to as the "electronic graveyard." One solution to this problem is to force manufacturers to begin producing greener electronics.

Baia Mare Cyanide Spill

After the Chernobyl incident in Russia, this cyanide spill in Baia Mare, Romania, is aptly called the worst environmental disaster in Europe. On January 30, 2000, 100,000 cubic meters of cyanide-contaminated water leaked out from a dam, spewing out 100 tons of cyanide. An incredible amount of fish and aquatic plants was killed and up to 100 people were hospitalized after eating contaminated fish.

The Shrinking of the Aral Sea

Dubbed as "one of the planet's worst environmental disasters," 10 percent of Aral Sea's 68,000 square kilometers has disappeared due to the diversion of rivers for irrigation. This percentage that was once a part of the fourth largest inland body of water is now a plain of highly saline soil with depleting marine life. The two rivers that feed the Aral Sea are the Amu Darya and Syr Darya rivers, respectively, reaching the Sea through the south and the north. The Soviet government decided in the 1960s to divert those rivers so that they could irrigate the desert region surrounding the Sea in order to grow agriculture rather than supplying the Aral Sea basin. Because this natural environment has been so drastically disrupted by digging irrigation canals and diverting major water sources, the consequences of those invasive actions have become irreparably destructive. These consequences range from unexpected climate feedbacks to public health issues, affecting the lives of millions of people in and out of the region. It also affected soil salinity. The level of salinity rose from approximately 10 grams per liter to often more than 100 grams per liter in the remaining southern Aral (Columbia University, 2016).

The Bhopal Disaster

Known as the world's worst industrial disaster, more than half a million people were exposed to methyl isocyanate gas and other toxic chemicals in Bhopal, Madhya Pradesh, India, on the night of December 2–3, 1984. The Union Carbide India Limited pesticide plant's leak of the poisonous gas claimed 2259 casualties.

The Chernobyl Nuclear Explosion

Being one out of two accidents classified as level 7 on the International Nuclear Event Scale, Chernobyl is known as the worst nuclear power plant incident in history. It was the result of a flawed reactor design that was operated with inadequately trained personnel, and the resulting steam explosion and fires released at least 5 percent of the radioactive reactor core into the atmosphere and downwind. Two Chernobyl plant workers died on the night of the accident, and an additional 28 people died within a few weeks as a result of acute radiation poisoning. Cancers, deformities, and other long-term illnesses were the scars of not only human inhabitants but animals as well (World Nuclear Association, 2016).

The Great Smog of 1952

Thousands died and a hundred thousand fell ill because of a blanket of smog that covered London for five days in 1952. Cold weather combined with windless conditions collected airborne pollutants from the use of coal to form a thick layer of smog over the city. Recent research showed that 12,000 premature deaths can be attributed to this smog (Dimdam, 2013).

One topic that has been in the news quite a bit the past few years and has subsequently made the public more environmentally aware is the melting of the polar ice as climate change becomes more of a problem. *Time* magazine has run several special editions covering the melting of glaciers, rising seas, and diminishing icepack (such as in the issues of April 9, 2001; April 3, 2006; April 9, 2007; and October 1, 2007). Likewise, *National Geographic* featured an article on melting ice caps and rising sea levels in their June 2007 issue. NASA prints its own briefs in an attempt to keep the public apprised of what is happening in the climate change arena. For example, in late February 2008, the Wilkins Ice Shelf on the Antarctic Peninsula disintegrated, an indication of warming temperatures in the region. The Moderate Resolution Imaging Spectroradiometer sensors on NASA's Terra and Aqua satellites provided some of the earliest evidence of the Wilkins Ice Shelf disintegration (Figure 5.5). Then in January 2010, they provided coverage of the Ronne-Filchner Ice Shelf shattering in Antarctica, as further evidence of climate change, in an attempt to convey the seriousness of the situation to the public (Figure 5.6). The piece of ice that broke free in this incident was larger than the state of Rhode Island.

Environmental scientists have said that because the Arctic/Antarctic ecosystems are so sensitive and fragile, that when they are stressed, they respond quickly. For this reason, they provide early warning to the effects of climate change. Scientists at NOAA and NASA believe the Arctic will be one

of the first areas to react to the effects of global warming and show subsequent signs of impact. They are warning the public today that the average temperatures in the Arctic region are rising twice as fast as they are anywhere else in the world. NASA has taken satellite images of the Polar ice cap and has determined that it is shrinking at a rate of 9 percent each decade. They predict that if this trend continues, there may not be any ice left in the Arctic during the summer season by the end of this century. This would upset the food chain and negatively impact the balance of the ecosystem, affecting the native people, wildlife, and vegetation.

Scientists are already seeing changes in the feeding and migration patterns of walrus, seals, polar bears, and whales. This makes it difficult for native inhabitants of the area (such as the Eskimo and Inuit) to hunt and obtain the food supply that they rely on for survival. Melting ice also contributes to rising sea levels. Today, many of the native villages that have existed for centuries along the coastlines are being flooded and swamped as the sea levels rise and put the coasts under water. This is a very serious example of how climate change is threatening the identity, culture, way of life, and very existence of cultures.

Climate change scientists have recently announced that rising temperatures are already affecting Alaska. Deborah Williams, executive director of the Alaska Conservation Foundation, states that annual temperatures have increased (6.7°C–8.3°C) and winter temperatures have warmed 13°C–17°C—more than any other place on Earth; more than four times the global average. Permafrost is melting, causing more than 600 families so far to lose their homes.

Steven Amstrup, a polar bear specialist with the U.S. Geological Survey, says, "As the sea ice goes, that will direct to a very great extent what happens to polar bears" (Carlton, 2005). Polar bears could become extinct within the next century because they have adapted to hunting on the ice. If they try to swim, they are more likely to tire and drown.

There is no lack of scientists warning the public of the ill consequences to the society if climate change is allowed to continue. In fact, they have delivered ample warnings over the past two decades, which have been well publicized (Figure 5.7). Why then do certain populations display a general trend toward ignoring the significance of the warnings? In particular, why does the United States not heed the warnings? Although studies are being published, newscasts being aired of the negative effects currently in progress, and projections being announced that will impact every human life in the near future, the business as usual attitude seems to prevail in the United States. Other countries, such as those of the European Union, Canada, and Australia, for example, seem to be much more aware of the situation—to the point where they have already incorporated meaningful changes into their lives to help mitigate, prepare, and accommodate those changes as they occur. For example, Europe is much more oriented toward public transportation rather than driving personal vehicles.

In a study conducted by Bassett et al. (2008), participants from Europe, North America, and Australia were studied to determine if there was a relationship between active

February 28, 2008

March 17, 2008

FIGURE 5.5 Wilkins ice shelf break up, Antarctica: In late February 2008, the Wilkins ice shelf on the Antarctic Peninsula disintegrated. (Courtesy of NASA, http://earthobservatory.nasa.gov/Features/WilkinsIceSheet/.)

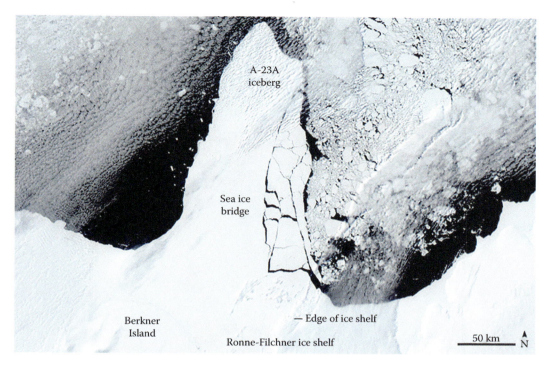

FIGURE 5.6 Ronne-Filchner ice shelf, Antarctica: Within a 24-hour space, an area of sea ice larger than the state of Rhode island broke away from the Ronne-Filchner ice shelf and shattered into many smaller pieces. (Courtesy of NASA, http://earthobservatory.nasa.gov/ Natural Hazards/view.php?id=42302.)

transportation (defined as the percentage of trips taken by walking, bicycling, and public transit) and obesity rates. What was interesting about this study from a climate change aspect were the percentages of active transportation participants between the three geographic areas. Europe dominated the active transportation—they walked more than residents of the United States: 382 versus 140 kilometers per person per year. They also bicycled more: 188 versus 40 kilometers per person per year. The conclusion of the study was that Europeans walked and bicycled more per year that did those from the United States, Australia, and Canada, and had the lowest obesity rates (Bassett et al., 2008). From a climate change aspect, this is a more desirable behavior because they are using less fossil fuel. The study cited the following reasons as to why Europeans did not drive as much:

- They have densely packed cities that generate shorter trips.
- Europe has more restrictions on car use (car-free zones, no "through" zones, etc.).
- They have extensive, safe, convenient facilities for cycling and walking.
- They provide bikeways, pedestrian walkways, and bicycle parking lots in conjunction with ample public transportation areas.
- Traffic regulations and enforcement policies favor pedestrians and cyclists over motorists.
- Owning and operating a personal car is extremely expensive.

There are probably many reasons why people choose to utilize public transportation, walk, bicycle, carpool, or drive a personal vehicle; and these reasons are diverse, depending on cultural values, economic opportunities, educational opportunities, geographic factors, and other key reasons. Each of these diverse reasons, however, has an impact on climate change, because the choices people make—especially those that involve the burning of fossil fuels somewhere along the production line—are going to come into play as temperatures rise.

In a study conducted by Constance Lever-Tracy in 2008 and 2010, she stated that, although the public had become aware over the years of human-caused climate change, most sociologists, as a general rule (excluding environmental sociologists), had surprisingly little to say about the possible future social avenues that may occur as a result of it. She also stated that in general because sociologists were usually not in a position to either validate or not validate the claims of natural scientists, there was an established tendency to "look the other way." In the case of climate change, however, not paying attention and becoming directly involved could spell disaster for the future of sociology because the consequences of climate change could spell disaster for the society. Therefore, it would be in the best interests of sociologists to pay close attention, because as she said, "these developments can affect the very core of our discipline's concerns. We need a cooperative multidisciplinarity of social and natural scientists working together."

Her observation is very astute, because the consequences do directly affect society. Perhaps if more sociologists

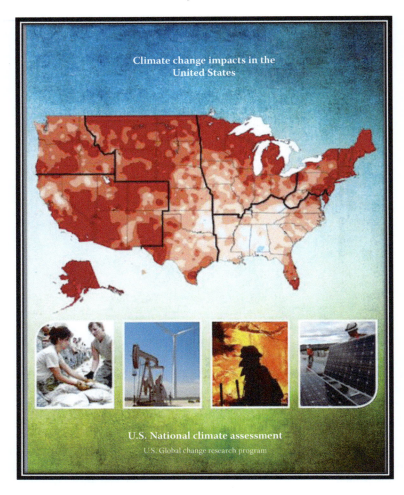

FIGURE 5.7 There is no shortage of magazines currently reporting on climate change in an effort to keep the public educated on the topic. (Courtesy of Wasatch Images.)

analyzed the potential effects and what that could mean to the future of society as a whole, then perhaps more people would be prone to listen. She points to the multitude of events that occurred in 2005 as the crucial "tipping point" that should have rallied the professional attention of sociologists. In particular, she discusses evidence, such as the 26 tropical storms and three category 5 hurricanes that occurred that year (including Hurricane Katrina), and that social responses to these disasters were mixed. The announcement by NASA that the oceans were getting warmer, the escalation of greenhouse effect, the drastic melting of the Siberian permafrost releasing trapped methane, and the unexpected shrinking glaciers on the Antarctic ice sheet shocked the scientific community, yet received no real reaction from sociologists.

Yet, her explanation for the non-alarmist response does make sense. She explains it as a flaw in the science of sociology and attributes it to two factors. The first factor is the mirroring of an indifference sociologists find in contemporary society toward the future. The second factor is sociologists' continuing foundational suspicion of naturalistic explanations for social facts, which generally leads sociologists to question or ignore the authority of natural scientists.

In the first reason, she explains that the rate of change of natural processes has been shrinking toward the timescales of human society. For human society, there is a strong emphasis on immediate gratification and a decline in long-term direction or plans. This "desensitizing" has made it so that even threats just decades away now barely register. It is as if people today live in a "permanent present" reality. In other words, people today care much less about future scenarios: they do not think about the future as much because they live in the now. This is the here and now, the disposable society, the time when people focus on immediate consumption. And it extends further that politicians are more interested in the next election than the climate 50 years from now. Business leaders are more focused on the next annual report than what sea level will be in 20 years and what islands will no longer exist and which coastal areas will be long submerged. Lever-Tracy's suggestion to solve this disparity is for both social and natural scientists to work together toward a common end, and that it is just as necessary to focus into the future. As she sums up, "The alternative is irrelevance or worse—an effective complicity with the vested interests of fossil fuel corporations" (Lever-Tracy, 2008, 2010).

In response to Lever-Tracy's thoughts on climate change, Steven R. Brechin (2008) responded. He believes that climate change presents the most significant challenge facing the world today, and that in itself it should be more than sufficient to mobilize the world to act and respond rapidly in a meaningful way. He also believes that the human response will be made, in part, to conditions not yet fully realized, which presents the society with a unique challenge: it is not a matter of simply reacting, but in thoroughly thinking things through and being prepared. He also believes that it is most likely that only after global societies are restructured by human altered natural processes will sociology have a new focal point. One of the biggest obstacles he points out is that it will be more difficult for the more affluent Northern countries to willingly change because they will not be as directly affected as other countries. For this reason, they will not feel like they are "living on the edge." Therefore, they will be harder to motivate toward a new mind-set or paradigm. To them, he states, "environmental concerns have become and continue to be seen as no more than background noise." He also states, "social change is difficult, especially when response must be made to conditions not yet fully realized."

As frustrating as it might be, the human factor is a significant part of the big picture, and motivating people to get prepared for disaster or be proactive about an environmental concern that may be a generation away in their eyes can sometimes feel like an impossible task.

The Human Need for Biodiversity

There is a strong link between biodiversity and human well-being (Figure 5.8). Biodiversity contributes both directly and indirectly to many factors of human well-being, such as security, basic materials for a good life, good social relations, good health, and freedom of choice and action. Over the past century, many societies have benefited from converting natural ecosystems to human-dominated ecosystems and exploiting biodiversity. However, not everyone has benefited. Others have experienced a decreased well-being, and some social groups have experienced an increase in poverty. For example, with the issue of deforestation, the societies doing the harvesting have benefited, but the societies living in the forest have experienced a lowering of biodiversity and resulting poverty.

The way humans treat the biodiversity of an area is critically important, because if the biodiversity is lowered, then the environment is not as resilient to abrupt, or unexpected, change. Then, if a significant change does occur, the area may not be able to support enough diversity to remain strong enough, which can cause the ecosystem to be compromised and die. Following that logic, if the ecosystem becomes unhealthy, it cannot support the human and animal populations that depend on it, which then compromises the existence of the life within it.

Over the past few hundred years, human activity has significantly changed the surface of the Earth. As a result, the Earth's climate is also being changed. Species becoming extinct is not new news; in fact, it is happening at rates

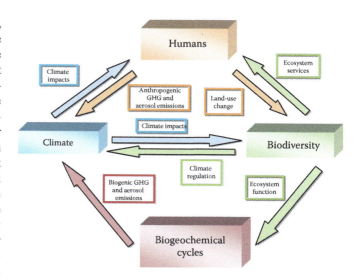

FIGURE 5.8 The relationships between climate change, biodiversity, and human well-being. (Courtesy of Wasatch Images.)

faster than it ever has before. The disregard being shown toward the Earth's ecosystems today is taking its toll on their health. In particular, according to the Millennium Ecosystem Assessment in 2005 (MEA, 2005), there are four categories of services provided by ecosystems to society, as shown in Figure 5.9: supporting, regulating, provisioning, and cultural services. Supporting services are what all other ecosystem services depend upon and include carbon cycling and water and nutrient cycling. Regulating services provide the mechanisms that moderate the impact of stresses and shocks on ecosystems and include climate and disease regulation. This also

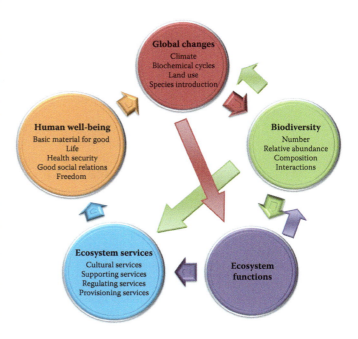

FIGURE 5.9 The relationships between ecosystem services, biodiversity, and global change. (Courtesy of Wasatch Images.)

determines the distribution of provisioning services, such as food, fuel, and fiber, and cultural services, such as spiritual and aesthetic clues (Kinzig et al., 2007).

An assessment of ecosystem transformation is largely determined by who is doing the assessment and exploitation. If it is the individual who has benefited, then the transformation will be valued; if it is an individual who did not benefit, then he was short-changed. The study was readdressed in 2013 and a quantitative index system to measure human well-being was developed, providing a better understanding of the relationships between ecosystem services and well-being (Yang et al., 2013).

ENVIRONMENTAL MOVEMENTS—THE CLASSIC CASE OF EARTH DAY

One of the most notable days in support of the environmental cause via a socialistic voice was the inception of Earth Day in 1970—a truly landmark event. Founded by U.S. Senator Gaylord Nelson, the idea for it actually began in 1962. An environmentalist himself, he was openly concerned that the environment was not an issue that the politicians in the United States seemed to care anything about. Therefore, he met with President Kennedy and persuaded him to give the visibility to the issue by going on a national conservation tour. The outcome was a 5-day, 11-state conservation tour in September 1963. To Nelson's surprise, the tour was not a success: it did not succeed in putting the issue onto the national political agenda.

Following his first failure, Nelson did not give up. He continued to speak on environmental issues to a variety of audiences in approximately 25 states. Everywhere he went, he saw evidence of environmental degradation. He also found something else: public interest and support. For the next six years, however, he was unable to get the concept of environmentalism onto the political agenda. Then, in the summer of 1969, Nelson was out West on a speaking tour, and he took note of the Vietnam War demonstrations, called "teach-ins." They had spread to college campuses all across the nation and had the ability to rally large groups together.

They gave Nelson the idea he needed. He decided to organize a huge grassroots protest over what was happening to the environment. He was convinced that if he could tap into the environmental concerns of the general public and generate that same student antiwar energy he had seen on campuses into the environmental cause, he could generate a demonstration that would force this issue onto the political agenda. A gamble it was, but he was game for a try.

At a conference in Seattle in September 1969, he announced that in the spring of 1970, there would be a nationwide grassroots demonstration on behalf of the environment and invited everyone to attend and participate. The news services carried the story from coast to coast. The response was electric. It took off with enthusiasm. Crowds of people chipped in the effort and sent letters and telegrams, and made phone calls. Finally, the American people felt that they had

a forum to express their concern about what was happening to the land, rivers, lakes, and air—and they all joined in. The four months following that Nelson ran an Earth Day affairs office out of his office in Washington, DC, coordinating all the details. Then, five months before the scheduled day for the rally, the *New York Times* ran an article, and in it the following was quoted:

> Rising concern about the environmental crisis is sweeping the nation's campuses with an intensity that may be on its way to eclipsing student discontent over the war in Vietnam … a national day of observance of environmental problems … is being planned for next spring … when a nationwide environmental "teach-in" … coordinated from the office of Senator Gaylord Nelson is planned…

When the day arrived on April 22, 1970, it became immediately obvious that it was a spectacular success—a credit to the spontaneous response at the grassroots level. There were more than 20 million demonstrators and thousands of schools and local communities that participated. According to Gaylord Nelson, "That was the remarkable thing about Earth Day. It organized itself" (Envirolink, 2011).

RAMIFICATIONS OF CLIMATE CHANGE ON SOCIETY

When discussing the sociology and the effects of climate change, it is also important to look at environmental sustainability. Along with the concept of the here and now mentality, it would seem in order to be thinking and planning ahead for methods to sustain the society. There are several major concerns with the society: the inequitable use of resources and hence the inequitable impact on the environment; the extreme variance of mind-sets ranging from those who practice strictly green lifestyles and pull more than their weight being environmentally responsible and those who live each day like there were four Earth's resources at their disposal; the uneven natural distribution of resources; the uneven distribution of wealth; the variable control of trade and commerce; the disparate distribution of power; and the list goes on and on. All of these social factors play a role in how climate change will affect an individual's outcome. For example, the cartogram is a graphic statistical representation of carbon emissions. As expected, the United States is bloated, as is much of Europe and China (Figure 5.10). Then if you look at Africa, it is hardly visible, along with South America. That being graphic enough, the even harder piece of information to swallow is what the map would look like if we were illustrating those negatively impacted by the effects of climate change. That map would appear much different. In that version, the United States and Europe would be very small, whereas Africa would be blown up like a balloon, as would Asia. Unfortunately, there is nothing fair about this picture: with climate change, the wealthiest nations are the biggest offenders (for the most part) but will be impacted the least; and the world's poorest nations, who emit relatively few greenhouse gases (GHGs), are destined to receive the brunt of the detrimental impacts from climate change.

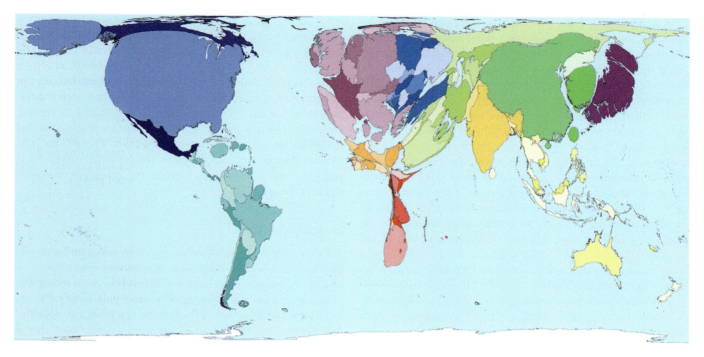

FIGURE 5.10 A cartogram depicting world climate change. This illustrates the world in terms of carbon emissions. The United States, Europe, and China are disproportionately large, whereas Africa is barely visible. (Courtesy of SASI Group, University of Sheffield, Sheffield, UK.)

CLIMATE JUSTICE

In an article by Randy Poplock (2007), the issue of "climate justice" is introduced. He discusses the way in which low-income and minority populations generally bear the brunt of human health consequences during emergency situations, such as Hurricane Katrina in 2005, because they tend to have limited access to health insurance. In general, the thought behind climate justice is that it is an effective tool to work at the intersection of environmental degradation and the racial, social, and economic inequities it perpetuates. When an activity occurs that threatens the environment, climate justice is the means to protect the victims.

Every person depends on the environment. Sadly, the current system consolidates wealth in the hands of a corporate minority, while threatening the health and security of all people. The unfortunate effect is that it is the least privileged global citizens and the most vulnerable individuals who are the first to feel the effects of the climate crisis and who end up suffering the most damage. Every person on this earth deserves a livable climate, and under climate justice, it is unreasonable for wealthier nations to exploit less fortunate nations for their resources and through practices that harm the world's climate, such as deforestation, oil exploration, and certain agricultural practices. Civilization requires a livable climate, and climate change causes unlivable conditions, such as drought, floods, and resource scarcity. This leads to famine, civil unrest, armed conflict, human suffering, and government oppression. The sad reality is that as the climate degrades, so does the quality of human life (Peaceful Uprising, 2016).

In May 2015, Shell Oil brought its enormous Polar Pioneer drilling rig into Seattle's port, planning to use the city as a staging ground for its drilling operations off the northwest coast of Alaska (Hymas, 2015). Seattle residents, working with Poplock and other climate activists from the Northwest, boarded kayaks and greeted the rig on its arrival in the port, blocking its entry, making a loud statement of protest against climate injustice. Shell had environmentally questionable Arctic drilling attempts in 2012, igniting anger among those protesting their current activities. The protest was both land and ocean based and lasted three days.

There are many components of climate change that affect the less fortunate and innocent. Heat waves—one of the effects common to climate change—are particularly devastating to low-income and minority residents because they generally lack air conditioning, live in higher crime areas (where doors and windows remain closed for security reasons, cutting air circulation), and congregate in highly urbanized neighborhoods (where temperatures are usually higher due to the urban heat effect).

Heat waves are not the only danger putting minorities at higher risk, either. Floods and wildfires are just as dangerous and devastating—and they are also associated with climate change. In this case, low-income populations are less mobile, often lack access to warning systems (such as the Internet), or do not understand English warnings, which make them more susceptible to catastrophe. It is also this population that tends to lack adequate property or homeowner's insurance, making wildfire and flood damage even more devastating.

Another serious issue is raising prices for commodities. As items become more in demand, they will become more expensive, such as water. Electricity rates could start climbing, making it difficult for some people to have access to power. As farms begin suffering from projected water shortages, food may start becoming scarce, causing grocery prices to escalate, furthering impacting low-income families. He also points out, with some irony, low-income families are the ones who get penalized, but are generally not the ones who contribute the most to the emission of GHGs. He also points out that, although the news commonly portrays these types of scenarios about other countries, it also happens in the United States as well. For example, when Hurricane Katrina hit the South in 2005, those that were hit the hardest were the low-income minorities. Poplock says, "Climate justice must be considered in policy decisions and public dialogue. Every environment issue poses a social justice concern, and climate change is no exception" (Poplock, 2007).

In a paper by Cameron et al. (2013) at the World Resources Institute, they discuss what needs to be included in a new climate agreement in order to have climate justice and equity. They believe that climate justice demands an effective and equitable agreement that places people at the center, not government or business interests. In this manner, it protects the most vulnerable and equitable shares the burdens and benefits of our responses to climate change. This approach is already consistent with the priorities of many countries, least developed to highly developed alike. For all countries, the following concepts apply: (1) Despite their current level of development, they have vulnerable citizens and ecosystems to protect from the devastating impacts of climate change, (2) all countries want to develop and prosper, and (3) all countries want to create opportunities for job creation and growth. A key concept of climate justice is that it is possible to take action on climate change while pursuing sustainable development, while protecting people and treating them equitable.

The World Resources Institute recommends the following aspects of climate justice that could provide equity in a new climate agreement (Cameron et al., 2013):

1. The voices of the most vulnerable to climate change must be heard and acted upon when designing a new climate agreement.
2. A basic element of good international practice is the requirement for transparency in decision-making and accountability for decisions that are made.
3. Human rights can also play a role in informing how equity is applied in a new climate regime with a view to protecting the rights of all people.
4. The principle of intergenerational equity, which is central to sustainable development and climate justice, should be concretely stated as a priority, equally important to intragenerational equity, in the new climate agreement.
5. Different countries will take different actions, in different forms, and in different time frames—but all will need to act, all have responsibility to protect human rights, and all can benefit from the transition to a new type of economic growth.

With this plan, countries will be judged in terms of both the quality of their commitment and the extent to which it qualifies as a fair share of the global effort required to prevent dangerous climate change. Commitments are not just related to GHG emissions, they also relate to actions taken to adapt to climate change, and how well they protect citizens from risk, and how well governments take action through investments and technology.

IMMIGRATION ISSUES

Another very real issue that needs to be considered today is the ramification of immigration. It is commonplace today to pick up a newspaper and read headlines about mass extinction of species, global warming, deforestation, melting of polar ice caps, the rapid depletion of fossil fuels, sea-level rise, drought, failed crops, dust storms, and polar bears stranded on icebergs to get a taste of the consequences of climate change. We also hear about environmental nonsustainability, political instability, economic hardships, and poverty. These are all relevant social issues that bear extreme weight and need to be dealt with in some way.

Another issue not heard as much is that of climate models and their analytical results. Most climate models agree that by the end of this century, the polar and subpolar areas will receive more precipitation, and the subtropics—the area between the tropical and temperate zones—will receive substantially less. These correspond with the areas that currently depend heavily on rainfall for crop irrigation—areas that have been suffering from drought. This will have negative ramifications on food production, prohibiting those regions from being able to produce enough food to adequately support their populations, increasing the incidence of starvation. For example, in Africa, where conservative estimates show that only 6 percent of its cropland receives irrigation, it may experience greater starvation and political unrest than it currently does. Elizabeth Kolbert's article, "Changing Rains" in the April 2009 issue of *National Geographic*, states that changes in rainfall patterns could be contributing to the conflict in Darfur, for nomadic herders and farmers constantly fight over croplands. She also points out that there are already several tensions in the Middle East over allocation of water in places with severe shortages. In addition, several ecologists have warned that future wars will be waged all over the world not over oil, but over water (Kolbert, 2009).

In addition to wars over water, this will also trigger another social side effect of climate change with international refugees. Water shortages will result in food shortages. When this occurs, it will leave people with only one feasible option: to immigrate to countries with modest water supplies and available food. Climate scientists have predicted that rising sea levels could displace billions of people—mostly low-income families—who live along the coastlines of Africa, Asia, and Latin America. Every country that could be a potential destination will suddenly be faced with how to handle a mass international immigration. This, in turn, could lead to extreme political instability. These are considerations that climate scientists are warning about now, and have been for years, yet the general public seems complacent to respond. It is a problem, however, that remains critical to address.

ONE EXPLANATION TO ENCOURAGE GOING GREEN

So how do you get someone's attention and convince them to go green? According to the author Steve Stillwater, there are several motivations that could be effective: the green economy is expected to produce new jobs, alternative energy technologies is being developed to replace fossil fuel-based energy sources, you do less harm to the environment, you can live more efficient and be less wasteful, or you can save money. But according to the *Wall Street Journal*, none of those is the correct answer. The correct answer is as follows: the strongest motivating factor to cause someone to go green is good, old-fashioned peer pressure (Simon, 2010).

Based on a recent experiment, this illustrates the *Wall Street Journal*'s point superbly: two different placards were placed in hotel bathrooms to encourage guests to reuse their towels. Some of the rooms' placards stated, "Show your respect for nature." Other rooms' placards stated, "Join fellow guests in helping to save the environment." After they had run the experiment and tallied the results, only 25% of the guests reused their towels that had the first placard; yet 75% of the guests reused their towels that had the second placard in their rooms. Based on these results, the hotel tried a second round of experiments and made the placard specific to the guest in the particular room. The placards then read similar to this: "75% of the guests who stayed in Room 295 reused their towels." This sign achieved an even higher compliance percentage, illustrating that simple, old-fashioned peer pressure works and it is more effective than using some rationale about the benefits to the environment.

Based on these results, Stillwater made a reference to companies marketing green products and the obvious implications for "using peer pressure and creating a guilt complex." He noted that most people would rather not be singled out as not being willing to go along with most other people in protecting the environment, and that it was a much more powerful motivator than just quietly going green because you think it is the right step to take for the environment. So the logical next question is as follows: how do sociologists deal with that? And if that is a legitimate mind-set along with only living in the here and now and wanting to plan for the future, it sounds like sociologists have their work cut out for them, at the very least, on this one. What is also interesting is that maybe it is not really necessary to spend millions on studies to figure out how the human mind works and how to get the public to respond and adapt in a positive manner to climate change—perhaps the solution is much simpler than sociologists realize.

OTHER SOCIAL FACTORS TO CONSIDER

As climate change progresses, whether countries are large or small, wealthy or impoverished, there will be some social issues all will have to deal with. Some of those concerns include drought and desertification, soil condition, geographic location, ability to cope, cultural restrictions issues, the plight of urban populations, and the necessities of relocation.

Drought and Desertification

Drought and desertification has become an issue that affects the world. No longer confined to isolated areas, its effects will reach across the globe as temperatures climb. Today, it also seems that even the wealthier nations are not any better prepared to cope with the devastation it can wreak (Figure 5.11a and b). According to Raymond Anselmo (2009) of the Model United Nations of the Far West, Santa Clara, California, even highly developed nations are experiencing the same difficulties, as evidenced by the significant droughts in portions of the United States recently. He points out that, although there have been moves toward solving the problems associated with drought and desertification and addressing the accompanying socioeconomic complications, the problems are as great—if not greater—than they have ever been. He stated that even back in 1990, there was an estimated 15 million acres worldwide that were lost to desertification each year, along with 26 billion tons of topsoil due to erosion, even though experts back then were warning that decisive, comprehensive action was needed in order to rectify the situation. Today, the problem still continues, and each acre lost is an acre lost to the future of crop production. With growing populations and decreasing viable agricultural areas, it is as if the society is willingly heading to its own destruction.

Anselmo also points out that, although there are several root causes of drought and desertification, with a few exceptions, most are human caused; most causes are because humans overtax ecosystems. For this reason, he believes their social impact is fourfold in nature:

1. A lack of food means starvation and ill health, causing many to either die or become incapacitated.
2. As people begin to starve, they flee to other locations, causing internal and external refugee problems, which often lead to cultural clashes.
3. The lack of supplies, massive floods, and not being able to find work in the new location where refugees move causes a breakdown of traditional family structure. This can also cause people to rebel against their own government if they do not move elsewhere.
4. It can harm political advancement, such as when a government has to limit freedoms or tighten its economic belt in order to deal with the new conditions.

Anselmo also points out that solutions have been proposed from all areas and viewpoints, and a few nations have already put plans in place to solve their problems. Several nations currently have proposals and plans for drought and

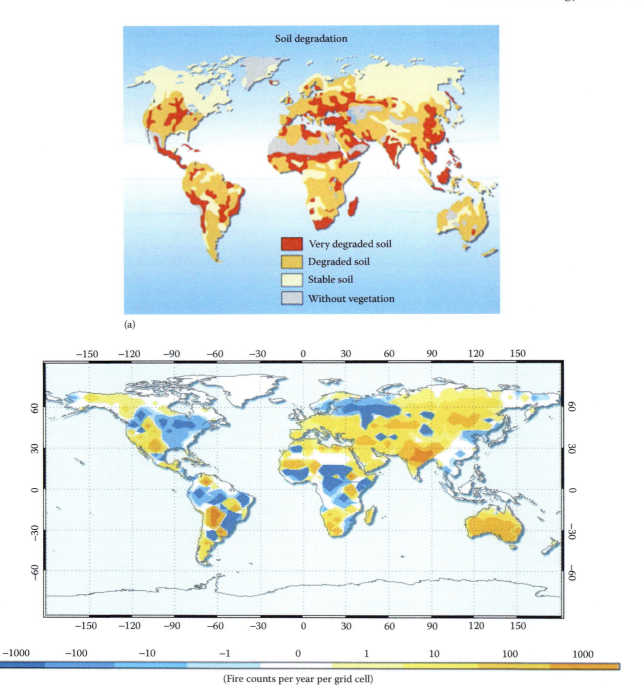

FIGURE 5.11 (a) This map of soil degradation and desertification worldwide shows some of the areas most prone to the negative effects of climate change. These areas will struggle with shortages of water and lack of food. (Courtesy of United Nations Environmental Programme [UNEP].) (b) This image shows a new ecologically-based global fire map derived from fire activity data, Net Primary Productivity (NPP) data, and Normalized Difference Vegetation Index (NDVI) data. The color intensity in each ecoregion represents an ecoregion's fire activity index (from 0 to 1, unitless). The NPP data set was developed by Marc Imhoff, former Terra project scientist at the NASA Goddard Space Flight Center, and colleagues. (Courtesy of NASA; J. G. Pausas and E. Ribeiro, *Global Ecol. Biogeogr.*, 22, 728–736, 2013.)

desertification problems such as encouraging the private sector to develop new technology, dams and storage reservoirs in monsoon areas, and others prone to regular flooding, canals, windbreaks, better irrigation, water resource and forest management, tree and grass planting programs, higher reserves for economic protection, and reducing fossil fuel burning. The drawback, however, is that actually applying these ideas costs money, which is often in short supply.

Soil Condition

Soil condition is another factor that affects the society as a whole. Many scientists fear that the world society, at its current rate of consumption, is running out of useable topsoil. Concerning the future of soil and soil erosion, John Crawford, a professor at the University of Sydney, Sydney, Australia, says we have about 60 years of topsoil left, and 40 percent of soil currently used for agriculture worldwide is classed as either degraded or seriously degraded. Seriously degraded is defined as "70 percent of the topsoil (the layer allowing plants to grow) is gone." Undesirable farming methods are being lost at unsustainable rates—methods that strip the soil of carbon and make it weaker in nutrients, causing it to be lost between 10 and 40 times faster than it can be naturally replenished. This also applies to the well-maintained farming land in Europe. Unfortunately, soil replenishes itself very slowly: perhaps only an inch a century, if conditions remain right. Problems are also introduced with too much disturbance, such as overplowing, overfertilizing, and overgrazing.

If this problem is not addressed, there are two key issues: the first issue is the loss of soil productivity. If nothing is done, we can expect to produce 30 percent less food over the next 20–50 years. Compared with an expected population growth and demand of wealthier people in countries such as China and India to eat more meat, a requirement to be able to grow 50 percent more food is expected (World Economic Forum, 2012). The second issue is that water will reach a crisis point. The countries that will be impacted the most are China, India, Africa, and parts of South America. The resulting scarcity of food will force prices up, and the poorest countries will be hit the hardest, especially those that rely on imports for survival. Less available food will lead to violence as populations strive to acquire enough food to feed themselves. Wealthier countries and those with adequate food sources will have to contend with immigrants fleeing to their countries. This will also bring violence.

We currently have about a quarter of a hectare per person of land, and we are using half of the total land area on the globe for agriculture. This means that the land is being overstressed by both overpopulation and the modern diet. We are subsidizing unsustainable food production at the cost of our health and environment. Crop breeding is also contributing to the problem, as it has half the micronutrients of older strains. This is promoting a nutrient deficit in the population's nutrition, such as in iron. If farming practices were corrected, however, much of the carbon could be returned to the soil. Sustainable farming practices could be employed, eliminating erosion, providing environmental health, and promoting soil health (World Economic Forum, 2012).

Geographic Location

Certain groups of people will be affected more by climate change than others because of where they live. Some of the fastest growing population centers are the coastal areas. The Southern and Western regions of the United States are also rapidly expanding. These represent the most vulnerable areas to devastating coastal storms, air pollution, drought, and heat waves. It is also expected that the Mountain West (Utah, New Mexico, Colorado, and Wyoming) will face water shortages and increased wildfires.

Arctic areas will most likely experience problems with melting permafrost and reduced sea ice. Coastal areas, which already have very dense populations, will face challenges as changes in climate put demands on continued growth and existing resources (Environmental Protection Agency, 2016).

Ability to Cope

Different groups have different abilities to cope with different levels of stress. People who live in poverty may have a difficult time coping with changes. They may have limited resources to cope with the downfalls and hardships climate change brings. The elderly may also have a difficult time adjusting. They currently make up a larger share of the population in the warmer areas, and if these are the areas that become warmest and hardest hit first, they may be the first to suffer. Young children, as well, are one of the most vulnerable groups because their immune systems are still developing and rely heavily on others.

Cultural Restrictions

Cultural groups, such as Native Americans, are considered to be at some risk because they may be tied to one area, such as the land assigned to them on their reservations. They may also have tribal land that is not as well developed as others and may not be able to change their jobs or lifestyles. Because their land is tied to a specific area with a certain boundary, relocation would be a challenge, making their choices limited in what they can do once the climate begins to change (EPA, 2016). For example, tribes located in the Southwest United States are projected to experience changes in water quality and availability on their lands which are already some of the most arid in the United States (USGCRP, 2009). Additionally, climate change may significantly affect cultural traditions practiced by tribes. For example, certain Alaskan Natives have cultural ties with animals, such as seals and caribou, which will experience changes to their own habitats.

Urban Populations

City residents and urban infrastructure have distinct sensitivities to climate change impacts. Heat waves may be amplified in cities because cities absorb more heat during the day than suburban and rural areas (urban heat island effect). Cities are more densely populated than suburban or rural areas—more than 80 percent of the U.S. population lives in urbanized regions. As a result, increases in heat waves, drought, or violent storms in cities will affect a larger number of people (USGCRP, 2009). Urban residents may also be at the mercy of the vulnerabilities of aging infrastructure, including drainage and sewer systems, flood and storm protection, transportation systems, and power supplies during periods of peak demand, which typically occur during summer heat waves (EPA, 2016).

Relocation

Many communities will end up being forced to relocate as they are continually exposed to rising sea levels or extreme rainfall. Small island developing states are the most vulnerable areas and are currently feeling the effects of climate change today (Byrd and DeMates, 2014). For example, in 2014, Fiji had its first village—the village of Vunidogoloa—relocated. It was required to be moved 1 kilometer further inland as part of the country's climate change response program because seawater had already begun to flood residents' homes. Fiji expects to relocate 34 other villages as their coastlines become inundated with rising seawater and flooding increases. In the near future, the entire nation of Kiribati (a small island state in the Pacific) expects to become uninhabitable because of sea-level rise. The country has recently purchased land in Fiji so that they can relocate. What this means for the citizens of Kiribati is an acceptance of an entirely new way of life, although they abandon the one that has existed for centuries. The culture and future of these people will be changed forever as they relocate to a new geographic area and make a new life.

One thing important to note is that because the effects of climate change are not just simply environmental, but also economic, cultural, and social, new and existing policies must take a universal approach and include affected disciplines, ideas, values, and cultures. Effectively managing and dealing with climate change is a larger concern than merely scientific, and on a much more serious and inclusive scale than our society has ever had to deal with. It requires a unique approach that conveys the concept that the social factors need to be considered along with the scientific.

CLIMATE CHANGE, HUMAN PSYCHOLOGY, CULTURAL VALUES, AND THE MEDIA

The media has an enormous influence on what the public hears about. It is the media that disseminates information, through newscasts, magazines, newspapers, and the Internet, giving them an unparalleled opportunity not only to inform the public of the latest issues but also to play a role in how that information is perceived. Another component that contributes to how information is received is different for each person and is based on preferences, perceptions, and beliefs that are influenced by psychology and value systems. These are the sometimes-subtle forces at work shaping people's opinions about highly controversial subjects, such as climate change.

According to Dr. H. Steven Moffic, a professor of psychiatry and behavioral medicine at the Medical College of Wisconsin, Milwaukee, Wisconsin, "Global warming is a concept that everyone hears about, but many are slow to respond to. The problems and risks of climate change seem to be far in the future—they might be 25 or 50 years away—so why would people pay attention to those issues when there are so many day-to-day problems to deal with?"

Dr. Moffic believes the ability to ignore climate change is very human. "Our brains in many ways have not evolved much from when humans started to develop thousands of years ago. We are hardwired to respond to immediate danger—we call this the 'fight or flight response'—but there is no similar mechanism that alerts us to long-term dangers."

He believes that these reactions are just part of human nature. "People are so preoccupied with immediate problems like jobs and health and the economy that it's hard to pay attention to climate change, and to willingly take on another challenge."

"The issue of how much humans contribute to the cause of global warming may also contribute to why we tend to ignore its impact. Who wants to believe they might be guilty for contributing to a problem that could destroy the Earth?"

In order to put the issue into perspective, Dr. Moffic suggests everyone identify and do simple things that do not require big changes. He believes that each individual can have a large effect on others and, through example, influence others to take action. He also suggests that everyone "try to make global warming a more immediate issue—whether it is thinking about your kids, grandkids, the future of the whole Earth, or your health. Try to think about ways in which this issue is important to you right now" (Bloom, 2008).

Dr. Moffic has not left it at that, however. In a more recent study, he believes that the health risks from climate change go further than the medical arena was initially aware of. He presents the idea that health risks associated with climate instability also include post-traumatic stress disorder, drug abuse, autism, and what he refers to as "solastalgia." Moffic says, "It's sort of a corollary of nostalgia. But it's this kind of environmental grief where you live gets changed, against your will, obviously. You can't leave, and you feel this sadness for what you've lost right in front of you... I think we're seeing that in Appalachia now with the coal mining—mountaintop stripping, the same kind of thing" (Allegheny Front, 2014).

Moffic believes that mental health issues get unfairly ignored when it comes to climate change. He sees that as a tremendous shortcoming, as the mental health risks of continuing climate change are just as significant, if not more so than the physical health issues associated with it. He believes the mental health risks will potentially affect as many, if not more, of the population than other health concerns. To support his theories, he has developed a list of studies that he cites as "emerging climate change manifestations and their psychiatric implications." They are as follows (Allegheny Front, 2014):

- Linear increase in violence, especially in warm climates and the inner city
- Increase in alcohol and substance abuse
- Increase in attempted and completed suicides
- Increase in heat strokes from psychiatric medication side effects
- Increase of post-traumatic stress disorders and other disorders after environmental disasters, more so if those traumas are felt to be manmade
- Climate refugees with added loss and cultural bereavement
- Group conflict and competition for resources
- New syndrome of solastalgia

In work done by Elke U. Weber at the Center for Research on Environmental Decisions at Columbia University, New York, on why the subject of climate change has not scared more people yet, she attributes it to universal characteristics of human nature. According to Weber, behavioral decision research over the past 30 years has given psychologists a good understanding about the way humans respond to risk, specifically in the decisions they make to take action to reduce or manage those risks. One of the biggest motivators to respond to risk is worry. When people are not alarmed about a risk or hazard, their tendency is not to take precautions.

Weber points out that with the issue of climate change, personal experiences with notable and serious consequences are still rare in many regions of the world. In addition, when people base their decisions on statistical descriptions about a hazard provided by others, it is not a big enough motivator for action (Weber, 2006).

An example of this can be seen in a scenario such as the rapid rise in the price of gasoline in 2008. When prices skyrocketed at the pumps, it caught the public's attention and raised an immediate interest in hybrid cars, alternative fuels, and public transportation, because the consumer was hit hard financially. Then, when gasoline prices dropped again, consumers thought less about energy conservation and alternative fuels because they were no longer immediately suffering the direct consequences. Human nature dictates that if something negative happens elsewhere in the world, the mind-set of an individual is "it only happens to others."

The stark reality about climate change is the inertia it engenders. Other locations may be suffering through droughts (such as Africa) or sea-level rise (such as the Pacific or Caribbean Islands), but people think it will not happen in the United States or wherever else they may live. Sadly, when it eventually does, it will already be too late. And just like it is human nature to procrastinate when an immediate threat is not looming, eventually the public will be caught in the mind-set: "I wish I had done something about it sooner."

Weber also believes that the reason people tend to avoid taking action against long-term risks is related to two psychological factors: the finite pool of worry hypothesis and the single action bias. The finite pool of worry hypothesis suggests that people can only worry about so many issues at one time, and of the issues they worry about they are prioritized from the greatest to the least. Generally, the greatest worries are those most directly affecting their lives at the moment. As an example, Weber pointed out that the finite pool of worry was demonstrated by the fact that in the United States, there was a rapid increase in concern about terrorism after the attacks on 9/11. Because of the intense focus on terrorism, other important issues—such as environmental degradation or restrictions on civil liberties, take an immediate backseat. The single action bias is described by Weber as follows: decision makers are very likely to take one action to reduce a risk that they encounter and worry about, but are much less likely to take additional steps that would provide incremental protection or risk reduction. The single action taken is not necessarily the most effective one, nor is it the same for different

decision makers. However, regardless of which single action is taken first, decision makers have a tendency to not take any further action, presumably because the first action suffices in reducing the feeling of worry or vulnerability. He concludes that based on behavioral research over the past 30 years attention-grabbing and emotionally engaging information interventions may be required to ignite the public concern for action in response to climate change (Weber, 2006).

What Weber did notice in 2011, however, was that although getting the public's attention with climate change has proven very challenging—in part because they see the problem as too remote—new evidence has led him to believe that direct experience with one anticipated impact—flooding—has increased people's concern and willingness to step up and now make a serious effort to do something about climate change, such as saving energy. He also notes that they appear to be strongly influenced by whether they think their behavior will be effective—or in other words: they are beginning to care about the issue (Weber, 2011).

THE POWER OF THE MEDIA

Reporting about climate change by the media has run the gambit in recent years. Because there are many points of view, the question is where does the truth lie. Reports and stories concerning climate change have ridiculed scientists and environmental groups. Reports have shown big businesses and countries (such as the United States) openly challenging the facts of climate change. Industries, such as oil companies, have accused the media of misinforming the public about the ill effects of burning fossil fuels. Other news stories have accused the Bush administration of silencing critics, including leading government climate scientists who have warned the public openly of the consequences of climate change.

As further reports about climate change continue to reveal a bleaker future, some are concerned that it will encourage fear tactics from environmentalists, whitewashing by some business interests, and a show by governments to illustrate reductions in emissions.

A few media reports claim the climate change to be a fraud; still others claim it is simply a cause designed to harm the U.S. economy and make the United Nations more powerful. Others say it is driven by academia and the simple desire of climate scientists to make a lot of money by using fear as a tool to earn more research grants.

All of this misinformation presents a challenge. A trend that has emerged is that the mainstream media in recent years has turned toward reporting actions and solutions. But there does seem to be a fine line on what the public expects. Some climate change researchers have expressed concern that too much reporting will lead to climate fatigue whereby the public will become desensitized to the issue. Others feel that the media should be used as an educational tool and that there is so much potential to educate the general public in ways that are not fatiguing. As an example, consider a billboard advertising a pickup truck dependent on fossil fuels and not rated with high fuel efficiency. The advertisement, however, has emotional appeal by suggesting the luxury, comfort, and status that will be bestowed upon the buyer of the vehicle. It focuses on the importance of having fun in the vehicle and being envied by others. When the public looks at this type of advertisement, they are not reminded of climate change issues or the health of the environment and future generations. Rather, they are urged to continue to support the fossil fuel industry by purchasing a vehicle that is not fuel efficient or helpful toward the solution of current environmental issues.

Now consider a movie poster that sends an entirely different message. If the movie is a nature film and is focused on the Earth and those who live on it, it may have breathtaking photography of the planet and the animal and plant life on it. Imagine that it has an engaging storyline and communicates very well the connections between life on land and in the oceans, and the overall connection to everything on Earth. This type of media presentation would serve not only as an effective visual graphic for the entertainment it represents but also as a strong positive approach toward public education, applicable to people of all ages. Instead of causing environmental fatigue, it sparks environmental interest and curiosity through the movie's creative storyline and breathtaking photography, giving the viewer a glimpse of the diversity and fragility of life on Earth that they probably would never see otherwise.

CONFRONTING THE DISINFORMATION CAMPAIGN

Put simply, disinformation is defined as intentionally false or inaccurate information that is spread deliberately. It is the purposeful act of deception and use of false statements to convince someone of untruth. It is often cleverly and deceptively done in order to lead a group along in a thought process toward an end goal—the goal to believe a concept based on mistruths and lies. The fact is that it is not uncommon in the society. Some may refer to forms of it as misinformation and propaganda. Since the introduction of social media, it has spread like wildfire in Internet rumors, gossip mills, and so forth. Certain periodicals cater to it. It is popular in some circles for its dramatic effect. Unfortunately, it is also present in important issues—issues that affect people's lives in critical ways. One of those is climate change, and it works to promote climate change denial. In cases like this, it can be argued that the role of ethics is paramount. In an article by Donald A. Brown (2012), he discusses the ethical aspects of disinformation campaigns in relation to climate change. Brown stresses the importance of treating whatever scientific uncertainties that come up about scientific change ethically, precautionary, and with respect. By approaching it this way, it allows the acknowledgment of the important role of skepticism in science as a healthy role to be used to improve science, not to destroy the hard-won work that has been accomplished. According to Brown, climate skepticism should be encouraged rather than vilified, provided that skeptics play by the rules of science including publishing in the peer-reviewed literature, not making claims unsupported by scientific evidence. He cautions against the use of the following tactics that climate scientists have had to deal with as they have worked to understand climate change and warn the public (Brown, 2012):

- Reckless disregard for the truth
- Focusing on unknowns while ignoring knowns
- Claims of "bad" science
- Creation of "front groups"
- Manufacturing pseudoscience
- Think tank campaigns
- Misleading PR campaigns
- Creation of astroturf groups
- Cyber-bullying of scientists and journalists

Brown sees the climate change disinformation movement composed of many organizations and participants, such as conservative think tanks, front groups, astroturf groups, conservative media, and individuals. The goal of the disinformation campaign is to use scare tactics. They attempt to make people believe them by saying the following:

- There is no warning.
- It is not caused by humans.
- Reducing GHG emissions will cause more harm than good.

Unfortunately, having to waste precious time dealing with nonproductive and nonforward-thinking groups slows down progress. It also masks the truth and steers people in the wrong direction, hurting important progress toward finding a solution.

KEEPING A JOURNALISTIC BALANCE

Journalistic balance—giving both sides of an issue a voice—is an important concept. The organization Fairness and Accuracy in Reporting, New York, states that a new study found that in U.S. media coverage of climate change, superficial balance—telling both sides—can actually be a form of informational bias. As the Intergovernmental Panel on Climate Change (IPCC), for example, has reiterated that human activities have had a discernible influence on the global climate and that climate change is a serious problem that must be addressed immediately, the media, in the name of balance, have given disproportionate air play to the small group of climate change skeptics and allowed them to have their views greatly amplified.

When reading reports from the media, it is important to clearly note who is being interviewed and whether or not the source is reputable and noteworthy. For example, the IPCC is a reputable organization. It consists of top scientists from around the globe, and they employ a decision-by-consensus approach. To back their reputability, D. James Baker, administrator of the U.S. NOAA and former undersecretary for oceans and atmosphere at the Department of Commerce under the Clinton administration, has said, "There's no better scientific consensus on this or any issue I know—except maybe Newton's second law of dynamics" (Powell, 2011).

In 1996, the Society of Professional Journalists removed the term *objectivity* from its ethics code. Today, the trend seems to lean more toward fairness, balance, and accuracy. Journalists currently are taught to identify the most dominant, widespread position and then tell both sides of the story. Robert Entman, a media scholar, says, "Balance aims for neutrality. It requires that reporters present the views of legitimate spokespersons of the conflicting sides in any significant dispute and provide both sides with roughly equal attention" (Boykoff, 2013).

In an attempt to make a coverage balanced, it is important to understand that it does not mean the coverage is accurate. In terms of global warming, "balance" may allow skeptics—of which many are funded by the various carbon-based industries—to be frequently and inappropriately consulted and quoted in news reports on climate change. It is important here to make a distinction: when the issue is of a political or social nature, fairness—presenting arguments on both sides with equal weight—helps ensure that the reporting is not biased. This method can cause problems when it is applied to scientific issues, however, because it results in journalists presenting competing points of view on a scientific question as though they had equal scientific weight, when they most certainly do not. This is what has happened with climate change. There have been incidences where the media has let skeptics have too much voice and it has enabled them to confuse the public and distort the seriousness of the problem.

Unfortunately, in giving equal time to opposing views, the major mainstream media has occasionally seriously trivialized scientific understanding of the important role that humans play in the climate change process. Although there is a value to presenting multiple points of view, it does the reader an extreme disservice when scientific findings that the world's top scientists and experts have come to a well-represented consensus over (representing enormous investments of research and time) are presented next to, and with equal weight, the opinions of a few skeptical scientists. This results in confusing the reader, leaving them with the frustration that they no longer know what to believe because there are so many conflicting reports. Situations like this slow down the constructive progression of climate change research and it is a travesty to true science.

An example of this can be seen in a *New York Times* article from April 10, 2009. Marc Morano sponsors a website (ClimateDepot.com) dedicated solely to the downplay of climate change. His chief goal is to debunk climate change as a serious issue. Kert Davies, the research director for Greenpeace, commented that he would "like to dismiss Mr. Morano as irrelevant, but could not. He is relentless in pushing out misinformation. In denying the urgency of the problem, he definitely slows things down on the regulatory front. Eventually, he will be held accountable, but it may be too late."

As scientists who are actively involved in climate change research look into Mr. Morano's claims, they say "he may be best known for compiling a report listing hundreds of scientists whose work he says undermines the consensus on climate change. Environmental advocates, however, say that many of the experts listed as scientists on Morano's website have no scientific credentials and that their work persuaded no one not already ideologically committed" (Kaufman, 2009).

One of Morano's recent reports entitled "More than 700 International Scientists Dissent over Man-Made Global Warming Claims" was far from balanced. Kevin Grandia, who manages Desmogblog.com, which describes itself as dedicated to combating misinformation on climate change, says the report is filled with so-called experts who are really weather broadcasters and others without advanced degrees (Kaufman, 2009). Grandia also said Morano's report misrepresented the work of legitimate scientists. He pointed to Steve Rayner, a professor at Oxford, who was mentioned for articles criticizing the Kyoto Protocol. Dr. Rayner, however, in no way disputed the existence of climate change or that human activity contributes to it, as Morano's report implied. In e-mail messages, he had asked to be removed from the Morano report, but his name was not; it was published with it included. When asked about it, Morano replied that he had no record of Dr. Rayner's asking to be removed from the list and that the doctor must "not be remembering this clearly" (Morano, 2009). In cases like these, it is imperative that any information obtained about climate change—or any scientific issue, for that matter—be looked at critically and its validity assessed as to its scientific soundness and quality (Figure 5.12).

Scientists' Mind-Sets and Data Change

One way the media has negatively impacted the advancement of climate change research is to attack scientists when they have changed their theories or their positions from a scientific viewpoint. For example, the media brought up a theory postulated back in the 1970s that did not pan out and allowed outspoken critics to use it in an attempt to diminish the reputations of scientists today. Several mainstream media sources republished the stories from the 1970s about a coming age of global cooling and the climate disaster it would trigger. Because this nearly 40-year-old theory never panned out, some skeptics have said climate change will not pan out either. But scientists say that is an unfair comparison.

Dr. William Connolley, a climate modeler for the British Antarctic Survey, says that "Although the theory got hype from the news media in the 1970s, it never got much traction within the scientific community; but that new data and research over the decades have convinced the vast majority of scientists that global warming is real and under way" (Kirtley, 2013).

The issue in the 1970s centered around the possibility that nearly three decades of cooling experienced in the Northern Hemisphere since World War II might be the beginning of a new ice age. Data suggested that perhaps the huge increase in dust and aerosols from pollution and development might be stepping up the cooling process. The investigation did not last long, however, because temperatures began to rise again and the issue was abandoned. Today, improved climate methodologies have revealed that, although aerosols did have a cooling effect, CO_2 and other GHGs were more potent in bringing

FIGURE 5.12 It is imperative to be selective about which literature is reviewed and believed. Any item in print from a source that is not reputable should not be assumed to carry the same weight and impact as these sources shown here, which are reputable and only publish quality, researched, scientifically-sound material. (Courtesy of Wasatch Images.)

about atmospheric change on a global scale. Improvements in technology over recent years have greatly aided the advancement and accuracy of scientific research, which continues to evolve and improve.

To go back to the issue of climatologists changing their minds, however, R. Stephen Schneider, a professor in the department of biological sciences at Stanford University, Stanford, California, and a senior fellow at the Center for Environment Science and Policy of the Institute for International Studies, at Berkeley, University of California, says, "Scientists are criticized by global-warming skeptics for making new claims and revising theories, as if we are required to stay politically consistent. But that goes against science. We must allow for new evidence to influence us. For some, the original speculation was that dust and aerosols would increase at a rate far beyond CO_2 and lead to global cooling. We didn't know yet that such effects were

so regionally located. By the mid-1970s, it was realized that greenhouse gases were perhaps more likely to be shifting climate on a global scale" (Azios, 2007). Connolley stated, "Climate science was far less advanced in the 1970s, only beginning in a way, and ideas were explored in a tentative way that has later been abandoned" (Azios, 2007).

This represents an inherent issue of science in general. As additional knowledge is gained about a subject, processes and outcomes of phenomena may change. Scientists need to remain open-minded and objective. If they do not remain open-minded, they will miss critical pieces of scientific information and possibly risk the outcome of scientific breakthroughs (Azios, 2007).

One thing remains clear, however. The media, if used correctly, have the enormous potential to guide the public and can play a significant role in helping people understand the science, the relevant issues, and the options for a better future.

The Occasional Data Flaw

Another thing the public seems to forget and the media seems to be completely unforgiving of, if not antagonistic about, is the occasional data flaw. Unfortunately, there seems to be the public perception that scientists are up on some imaginary pedestal and everything they release is cast in stone, as we have just discussed, with no room for change. This also applies to the occasional data flaw. Unfortunately, just like everyone else experiences, the occasional error does occur. Although this may seem obvious and not warrant mentioning here, it is important, because when it happens, both the media first, then the public often seems quick to jump on it, openly criticizing the source, often in an attempt to discredit not only the scientist and agency it originated from but also the issue in general.

An example of this can be illustrated by an incidence that happened to NASA's GISS. GISS is one of the foremost research units on climate change, led by Dr. James E. Hansen, one of the most world-renowned climate change experts. Dr. Hansen has been involved for more than 30 years in global temperature data analysis, and the GISS, under his leadership, has provided the scientific community with exceptional data analysis, which has led to the remarkable advancements seen in the climate change field in recent years. Yet, even an organization as distinguished as this one is not exempt from public criticism, perhaps even more vulnerable to it because of their status.

As an illustration, in 2001 the GISS published an updated analysis of a temperature record which had additional data added to it, with the objective of improving the long-term temperature record. The change consisted of using the newest United States Historical Climatology Network (USHCN) analysis for several United States stations that were part of the USHCN network. The improvement, developed by the NOAA researchers, adjusted station records that included station moves or other discontinuities. Unfortunately, GISS failed to recognize initially that the station records obtained electronically from NOAA did not contain the necessary adjustments. Therefore, there was a discontinuity in 2000 in the records of these stations (1999 and prior already contained the adjustments; 2000 had not been adjusted yet). Once GISS caught the error, it was immediately corrected. The resultant error that was originally reported (the data without the correction applied to the year 2000 data) averaged 0.15°C over the contiguous 48 states. This area covers approximately 1.5 percent of the globe, making the global error negligible, however. Therefore, one may think that catching the negligible error, correcting it, and subsequently rerunning and re-releasing the analysis would be sufficient. But that is not how the scenario played out once the media got hold of what had happened. The printed story was embellished and distributed to news outlets throughout the country with a headline along the genre of "NASA Cooks the Data." The following headlines and excerpts are just a few of what was actually released:

1. NASA, NOAA cooking the data
   Excerpt: The cooks—er, "scientists"—at NASA's Goddard Institute for Space Studies (GISS) have released their latest sky-is-falling temperature findings….
   (*Source*: http://www.iceagenow.com/NASA_NOAA_cooking_the_data.htm)
2. Global warming hysteria: Did NASA cook the books on warming statistics?
   Excerpt: I have no idea if this is true, but if it is—there should be a Congressional inquiry…
   (*Source*: http://www.firstthings.com/blogs/secondhandsmoke/2011/01/30/global-warming-hysteria-did-nasa-cook-the-books-on-warming-statistics/)
3. James Hansen: Cooking the NASA books for climate change
   Excerpt: …much of the world uses NASA's continually-revised data and graphs to determine weather history and policy. Which is unfortunate. Particularly, because NASA simply cannot be trusted to provide scientifically unbiased information on this subject. When not demonstrating incomprehensible incompetence, NASA cooks the books…
   (*Source*: http://deathby1000papercuts.com/2008/11/james-hansen-cooking-the-nasa-books-for-climate-change/)
4. NASA, NOAA create global warming trend with cooked data
   Excerpt: There is a major problem with the NASA and NOAA numbers, according to skeptical researchers who have dissected the data: They are inaccurate, the result of cherry-picking, computer manipulation and "best guess" interpretation…
   (*Source*: http://www.examiner.com/seminole-county-environmental-news-in-orlando/nasa-noaa-create-global-warming-trend-with-cooked-data)

Is this responsible journalism? Hardly. Yet, one has to wonder how many people read this and believed it. And the bigger question is as follows: how many people are led astray and kept from looking further into the truth and doing something constructive about fighting climate change because of media stories like these? This is a noteworthy illustration as to why a reader must be very careful of the sources of information obtained and being educated in one's interpretation of that source's story.

The next logical question that must be asked is as follows: Why? What is the media's purpose of printing these types of stories? One only has to look at the tabloids obtainable at the grocery store, perhaps. There seems to be an insatiable need for drama, gossip, and sensationalism in certain cultures, and perhaps this is simply an extension of it. But when it is applied to a critical issue—such as climate change—and can do more harm than good, should not that be where the proverbial line is drawn? With regard to the incident discussed previously, media stories also went on to say that "NASA has now silently released corrected figures, and the changes are truly astounding. The warmest year on record is now 1934" (Malkin, 2007). But this report was also incorrect, however, because when both the before and after graphs prepared by NASA were compared, both 1998 and 1934 were tied for the warmest year (prior to 2000) in *both* graphs, meaning the media had publically misinterpreted the graphs. Therefore, the obvious misinformation of the media reports, along with the subsequent absence of any effort to correct the stories after NASA pointed out the misinformation, may appear as if the media's aim may instead have been to create distrust or confusion in the minds of the public rather than to transmit accurate information.

This brings to mind a serious question: how are scientists supposed to handle a situation where a data error—no matter how quickly it is corrected—provides ammunition for people who may be more interested in launching a public relations campaign, rather than reporting science? Is it possible to eliminate these occasional data flaws or an unintentional occurrence of disinformation? Of course not. Scientists are human beings, too, and occasionally mistakes will occur. But when they are recognized, they are also fixed and data are rerun. It is when intentional attacks on integrity somehow enter the scientific picture where the true damage can be done. As in this example, it can serve to delay, or even weaken, the forward movement of advancements that need to happen for the benefit of the public.

CONCLUSIONS

When dealing with the issues of climate change, it is always easy to get caught up in the cause/effect relationships of the physical environment because many of them are so visually obvious: polar bears drowning in the seas, barren landscapes cracked and dried that have not seen water for months, wilted crops that have failed during their growing season, angry floods ravaging everything in their path, and rising sea levels consuming coastlines. But what is critical—that unfortunately gets overlooked—is the human factor, the impact climate change is having on social systems and cultures around the world. Even more horrifying is the damage done to the societies in undeveloped countries that never contributed significantly to the problem in the first place, feeling the most severe of the impacts. And to add insult to injury is the sad realization that many that live in the industrialized countries that emitted the GHGs in the first place causing the problem to escalate to the point it is at today will never know of their suffering and discontent. What a shame, indeed. But it does not need to be that way and it is not too late to change it.

One of the critical factors that is absolutely necessary for the solution of the problem is education. These are the news stories that should be reaching prime time. Education is important in the entire scheme of things with climate change. Knowledge is power, and through knowledge this problem can be researched, understood, and mitigated. Yet even in educated, developed countries, it is surprising the number

of the general population that still does not truly understand the causes and ramifications of climate change. This is a topic ready for sociologists to study, explore, and embrace. There are people right now feeling the harsh effects of climate change, and there will be more to come. The time to begin making a difference is as soon as it is understood there is a problem.

As evidenced from the examples presented throughout this chapter, the media has enormous power and control over the general public. But along with that power needs to come the responsibility and responsible journalism, which is especially critical when it comes to controversial and critical scientific issues such as climate change. The general public can be very susceptible to what is said in the news and, unfortunately through a misrepresented story, much damage can be done to a sound scientific concept, causing setbacks in progress being made toward mitigation and public understanding. Each of us has a responsibility as well. It is also important that as readers and viewers, we research and process what we are exposed to, thereby being able to define the difference between responsible accounting and hearsay. It is also imperative that we realize that our perceptions of issues are naturally processed through our own personal biases and cultural backgrounds, and it is up to us not to cloud issues by letting them cross personal boundaries, thereby reducing the inherent value of discoveries being made or slowing the direction that the society needs to take to ensure its protection and well-being.

REFERENCES

Allegheny Front. August 1, 2014. Doctor: Climate instability shakes mental health. *The Allegheny Front.* http://archive.alleghenyfront.org/story/doctor-climate-instability-shakes-mental-health.html (accessed August 17, 2016).

Anselmo, R. 2009. Social and economic impact of drought and desertification. *Economic and Social Council, Model United Nations Far West.* http://www.munfw.org/archive/41st/eco-soc1.htm (accessed January 13, 2011).

Azios, T. October 11, 2007. Global-warming skeptics: Is it only the news media who need to chill? *The Christian Science Monitor.* http://www.csmonitor.com/2007/1011/p13s03-sten.html (accessed August 23, 2016).

Bassett, D. R. Jr., J. Pucher, R. Buehler, D. L. Thompson, and S. E. Crouter. 2008. Walking, cycling, and obesity rates in Europe, North America, and Australia. *Journal of Physical Activity and Health* 5:795–814.

Beck, E. C. January 1979. The love canal tragedy. Environmental Protection Agency. https://www.epa.gov/aboutepa/love-canal-tragedy (accessed August 23, 2016).

Bloom, D. December 3, 2008. American psychiatrist explores the psychology of global warming. *RushprNews.* http://www.rush-prnews.com/2008/12/03/american-psychiatrist-explores-the-psychology-of-global-warming (accessed August 23, 2016).

Boykoff, J. 2013. The suppression of dissent: How the state and mass media squelch US American social movements. https://books.google.com/books?id=BYzdAAAAQBAJ&pg=PA249&lpg=PA249&dq=Balance+aims+for+neutrality.+It+requires+that+reporters+present+the+views+of+legitimate+spokespersons+of+the+conflicting+sides+in+any+significant+dispute+and+provide+both+sides-+with+roughly+equal+attention&source=bl&ots=dwkYP

UKGRQ&sig=pXB6VUzPq8Zjpm-z7XsWycmOEGA&hl= en&sa=X&ved=0ahUKEwjwo9GS-svOAhXM7SYKHYl- CA0kQ6AEIJTAC#v=onepage&q=Balance%20aims%20f or%20neutrality.%20It%20requires%20that%20reporters% 20present%20the%20views%20of%20legitimate%20spokes persons%20of%20the%20conflicting%20sides%20in%20an y%20significant%20dispute%20and%20provide%20both%2 0sides%20with%20roughly%20equal%20attention&f=false (accessed August 18, 2016).

Brechin, S. R. 2008. Ostriches and change: A response to 'global warming and sociology.' *Current Sociology* 56(3):467– 474. http://csi.sagepub.com/cgi/content/abstract/56/3/467 (accessed August 23, 2016).

Brown, D. A. February 19, 2012. Ethical analysis of disinformation campaign's tactics: Reckless disregard for the truth, specious claims of 'bad' science. *ThinkProgress.* https://thinkprogress. org/ethical-analysis-of-disinformation-campaigns-tactics- reckless-disregard-for-the-truth-specious-75fe5bfaba46#. e4en1zfso (accessed August 15, 2016).

Byrd, R. and L. DeMates. December 6, 2014. 5 reasons why cli- mate change is a social issue, not just an environmental one. *The Huffington Post.* http://www.huffingtonpost.com/ rosaly-byrd/climate-change-is-a-socia_b_5939186.html (accessed August 22, 2016).

Cameron, E., T. Shine, and W. Bevins. September 2013. Climate jus- tice: Equity and justice informing a new climate agreement. World Resources Institute, Mary Robinson Foundation Climate Justice. http://www.wri.org/sites/default/files/climate_ justice_ equity_and_justice_informing_a_new_climate_agreement. pdf (accessed August 15, 2016).

Carlton, J. December 15, 2005. Is global warming killing the polar bears? Published in *Wall Street Journal.* https://www. commondreams.org/scriptfiles/headlines05/1215-02.htm (accessed May 17, 2017).

Columbia University. 2016. The Aral sea crisis. http://www. columbia.edu/~tmt2120/introduction.htm (accessed August 14, 2016).

Dimdam, E. June 7, 2013. List 25. History: 25 biggest man-made environmental disasters in history. http://list25.com/25- biggest-man-made-environmental-disasters-in-history/5/ (accessed August 14, 2016).

Elliott, D. 2013. For BP cleanup, 2013 meant 4.6 million pounds of oily gunk. http://www.npr.org/2013/12/21/255843362/for- bp-cleanup-2013-meant-4-6-million-pounds-of-gulf-coast-oil (accessed May 17, 2017).

Envirolink. 2011. How the first earth day came about. http://www. earthday.org/about/the-history-of-earth-day/?gclid=CIKfo eXW2M4CFYRnfgodRgcJaQ (accessed August 23, 2016).

Environmental Protection Agency. August 8, 2016. Climate impacts on society. https://www3.epa.gov/climatechange/ impacts/society.html (accessed August 16, 2016).

Geiling, N. May 20, 2014. This Hellish desert pit has been on fire for more than 40 years. http://www.smithsonianmag.com/ travel/giant-hole-ground-has-been-fire-more-40-years- 180951247/?no-ist (accessed August 14, 2016).

Hymas, L. May 15, 2015. Shell says eff you to seattle. *Grist.* http://grist.org/climate-energy/shell-says-fu-to-seattle- brings-rig-to-city-despite-objections/ (accessed August 15, 2016).

Johnson, M. December 17, 2012. Ecocide in Vietnam. *Prezi.* https://prezi.com/frn1pajpiitd/ecocide-in-vietnam/ (accessed August 14, 2016).

Juhasz, A. April 18, 2012. Investigation: Two years after the BP spill, a hidden health crisis festers. *The Nation.*

Kaufman, L. April 9, 2009. Dissenter on warming expands his campaign. *New York Times.* http://www.nytimes. com/2009/04/10/us/politics/10morano.html?_r=0 (accessed August 23, 2016).

Kinzig, A. P. and C. Perrings. 2007. Biodiversity and human wellbe- ing. Center for Climate Change and Environmental Studies, Nigeria. http://www.center4climatechange.com/biodiversity. php (accessed May 17, 2017).

Kirtley, D. June 2013. The 1970s ice age myth and time magazine covers. http://scienceblogs.com/gregladen/2013/06/04/the- 1970s-ice-age-myth-and-time-magazine-covers-by-david- kirtley/ (accessed August 18, 2016).

Kolbert, E. April 2009. Changing rains. *National Geographic.* http://ngm.nationalgeographic.com/2009/04/changing-rains/ kolbert-text (accessed August 23, 2016).

Lever-Tracy, C. 2008. Global warming and sociology. *Current Sociology* 56(3):445–466. http://csi.sagepub.com/cgi/content/ abstsract/56/3/445 (accessed August 23, 2016).

Lever-Tracy, C. August 31, 2010. Sociology still lagging on cli- mate change. http://www.socresonline.org.uk/15/4/15.html (accessed August 14, 2016).

Malkin, M. 2007. Hot news: NASA quietly fixes flawed tempera- ture data: 1998 was NOT the warmest year in the millen- nium. MichelleMalkin.com, August 9. http://michellemalkin. com/2007/08/09/hot-news-nasa-fixes-flawed-temperature-data- 1998-was-not-the-warmest-year-in-the-millenium/ (accessed May 17, 2017).

Millennium Ecosystem Assessment. 2005. *Ecosystems and Human Well-being: Biodiversity Synthesis.* Washington, DC: World Resources Institute. Discusses the effects of climate change on the world's ecosystems.

Morano, M. March 16, 2009. More than 700 international scientists dissent over man-made global warming claims. *Free Republic.* http://www.freerepublic.com/focus/f-news/2207835/posts (accessed August 23, 2016).

Morris, H. November 7, 2013. After chernobyl, they refused to leave. http://www.cnn.com/2013/11/07/opinion/morris-ted- chernobyl/ (accessed August 13, 2016).

National Geographic Encyclopedia. 2016. Great Pacific Garbage Patch. http://nationalgeographic.org/encyclopedia/great- pacific-garbage-patch/ (accessed August 14, 2016).

NOAA. August 4, 2015. 2015 Gulf of Mexico dead zone 'above aver- age'. http://www.noaanews.noaa.gov/stories2015/080415- gulf-of-mexico-dead-zone-above-average.html (accessed August 14, 2016).

Pausas, J. G. and E. Ribeiro. 2013. The global fire–productivity rela- tionship. *Global Ecology and Biogeography*, 22, 728–736.

Peaceful Uprising. 2016. What is climate justice? http://www.peace- fuluprising.org/what-is-climate-justice (accessed August 15, 2016).

Pittman, C. August 20, 2013. Oil from BP spill pushed onto shelf off Tampa Bay by underwater currents, study finds. *Tampa Bay Times.*

Poplock, R. 2007. http://www.seattlepi.com/local/opinion/article/ The-poor-are-hit-hardest-by-climate-change-but-1246942. php (accessed August 23, 2016).

Powell, J. L. 2011. *The Inquisition of Climate Change Science.* New York: Columbia University Press.

Sahagun, L. February 13, 2014. Toxins released by oil spills send fish hearts into cardiac arrest. *Los Angeles Times.*

Simon, S. 2010. The secret to turning consumers green. *The Wall Street Journal*, October 18. http://online.wsj.com/article/ SB10001424052748704575304575296243891721972.html (accessed May 17, 2017).

USGCRP (United States Global Change Research Program). 2009. *Global Climate Change Impacts in the United States*. T. R. Karl, J. M. Melillo, and T. C. Peterson (eds.). United States Global Change Research Program. New York: Cambridge University Press.

Weber, E. U. 2006. Experience-based and description-based perceptions of long-term risk: Why global warming does not scare us (yet). *Climate Change* 77(1–2):103–120. http://www.springerlink.com/content/mj4213574nkj60v2/ (accessed August 23, 2016).

Wines, M. March 24, 2014. Fish embryos exposed to oil from BP spill develop deformities, a study finds. *The New York Times*.

World Economic Forum. December 14, 2012. What if the world's soil runs out? *Time*. http://world.time.com/2012/12/14/what-if-the-worlds-soil-runs-out/ (accessed August 16, 2016).

World Nuclear Association. June 2016. Chernobyl accident, 1986. World Nuclear Association. http://www.world-nuclear.org/information-library/safety-and-security/safety-of-plants/chernobyl-accident.aspx (accessed August 14, 2016).

Yang, W., T. Dietz, D. B. Kramer, X. Chen, and J. Liu. May 22, 2013. Going beyond the millennium ecosystem assessment: An index system of human well-being. *PLoS One Journal*. http://journals.plos.org/plosone/article?id=10.1371/journal.pone.0064582 (accessed August 15, 2016).

SUGGESTED READING

Boyce, T. and J. Lewis. 2009. *Climate Change and the Media*. New York: Peter Lang Publishing, 261p.

Boykoff, J. and M. Boykoff. November/December 2004. Journalistic balance as global warming bias. *FAIR*. http://www.fair.org/index.php?page=1978 (accessed August 23, 2016).

Boykoff, M. T. 2011. *Who Speaks for the Climate? Making Sense of Media Reporting on Climate Change*. Cambridge: Cambridge University Press, 238p.

Calthorpe, P. 2010. *Urbanism in the Age of Climate Change*. Washington, DC: Island Press. Discusses how modern cities can be planned and built with climate change mitigation in mind.

CCC. 2011. Climate targets. Committee on Climate Change. https://www.theccc.org.uk/tackling-climate-change/reducing-carbonemissions/carbon-budgets-and-targets/ (accessed August 23, 2016).

CCSP. 2008. Analyses of the effects of global change on human health and welfare and human systems. A Report by the U.S. Climate Change Science Program and the Subcommittee on Global Change Research. Gamble, J. L. (ed.), Ebi, K. L., F. G. Sussman, T. J. Wilbanks (Authors). Washington, DC: U.S. Environmental Protection Agency.

Crate, S. A. and M. Nuttall (eds.). 2009. *Anthropology and Climate Change: From Encounters to Actions*. Walnut Creek, CA: Left Coast Press. Provides a well-rounded understanding of how societies around the globe perceive and adapt to climate change from the perspective of their own unique socio-cultural framework.

Hansen, J. E. 2009. *Storms of My Grandchildren: The Truth about the Coming Climate Catastrophe and Our Last Chance to Save Humanity*. New York: Bloomsbury.

Norgaard, K. M. 2011. *Living in Denial: Climate Change, Emotions, and Everyday Life*. Cambridge, MA: The MIT Press.

Pooley, E. 2010. *The Climate War: True Believers, Power Brokers, and the Fight to Save the Earth*. New York: Hyperion.

Stillwater, S. Green living motivations: What makes someone go green? http://EzineArticles.com/?expert=Steve_Stillwater (accessed August 23, 2016).

The Ismaili. March 27, 2009. A changing climate: Exploring the social impact of global warming. http://www.theismaili.org/cms/684/A-changing-climate-Exploring-the-social-impactof-global-warming (accessed August 23, 2016).

The Royal Society. June 2007. Biodiversity-climate interactions: Adaptation, mitigation, and human livelihoods. www.royal-soc.org.

United Nations Human Settlement Program. 2011. *Cities and Climate Change: Global Report on Human Settlements 2011*. London: Earthscan Publications. Discusses the effects of climate change on the world's cities.

Viegas, J. April 2, 2013. Record dolphin, sea turtle deaths since Gulf spill. *Discovery News*.

Washington, H. and J. Cook. 2011. *Climate Change Denial: Heads in the Sand*. London: Earthscan Publications.

Weber, E. U. 2006. Experience-based and description-based perceptions of long-term risk: Why global warming does not scare us (yet). *Climate Change* 77(1–2):103–120. http://www.springerlink.com/content/mj4213574nkj60v2/ (accessed October 5, 2010). Discussed why people are not more concerned about climate change.

Weber, E. U. March 20, 2011. Climate change hits home. *Psychology* 1. www.nature.com/natureclimatechange (accessed August 16, 2016).

6 Geopolitical Considerations

OVERVIEW

This chapter discusses the current political attitude both in the United States and internationally on the climate change issue. It treats the United States and its policy development separately because of its unique nature—over the past two decades its policy has developed independently from the rest of the participating countries; the majority of the other developed countries of the world have gone through a process of cohesion, while the United States has chosen to stand on the sidelines watching, but doing very little politically toward curbing the emission of GHGs. For example, it did not join with other countries and adopt the Kyoto Protocol back in 1997. Within the past few years, however, it has taken significant action toward progressive change; which this chapter looks at, specifically what the EPA is doing, the Climate Action Plan, and The Clean Power Plan. It then shifts to an international perspective and examines the progress of key countries, as well as provides an overview of the Paris Climate Agreement; a historic advance forward for the management of climate change.

INTRODUCTION

In order to get climate change effectively under control, it will take the efforts of every country worldwide. Because of the immensity of the issue, the backing of national governments is critical—both legislatively and economically. And even more important is whether or not individual countries will be able to work together in accomplishing both their short- and long-term goals. For many of the world's countries, new negotiations post-Kyoto Protocol are just a continuation and evolution in a long line of negotiations and agreements already forged. But for some nations—who are also major GHG emitters, such as the United States and China—these new negotiations resulting from the Paris Agreement are breaking new territory and only time will tell how well these new players adjust to the rigor of GHG emission commitments and control and whether or not they will be successful. For the world as a community, the present will open up a new era of opportunity and potential achievement. Only time will tell as each country determines their true commitment to the issue at hand. But one thing is certain, the political decisions and actions of each country weigh heavily on the future—a future that we all hold stock in.

DEVELOPMENT

Historically, the United States has not been a leader in stressing the importance of the climate change issue. In fact, they openly chose not to join with the rest of the countries in even adopting the Kyoto Protocol in 1997, let alone ratifying it, stating that it was unfair China and India were not held accountable at that time as undeveloped nations for their GHGs, when they were rapidly becoming developed and still would not be held to the same standards as developed countries. Although some politicians have tried to get involved in environmental issues, the overall trend has been one of inaction. However, times do seem to be changing. Severe weather events are occurring, species are becoming endangered, glaciers are melting, and areas are suffering from drought. The media seems to have finally taken on the role of making the public aware of the effects of a warming world. The big question still remains to be answered, however: To what extent are the Americans willing to accept responsibility for the threat, take action, and make the personal sacrifices necessary to control the problem? To be specific—are Americans finally willing to pay slightly more for alternate, renewable energy and significantly change their lifestyles in order to reduce their use of fossil fuels?

One thing that has frustrated many Americans is that the U.S. government—typically a leader in global issues—has seemed to move so slowly to take action to halt the emissions of GHGs. Fortunately, state governments are not holding back and waiting any longer. Governors from half of the states have put into effect agreements to lower GHGs. Even federal courts have ordered the executive branch to start regulating GHGs.

In a political presentation given on January 26, 2009, President Obama delivered a speech concerning jobs, energy, and climate change, during which he made the following points about his policy on climate change:

- "Year after year, decade after decade, we have chosen delay over decisive action. Rigid ideology has overruled sound science. Special interests have overshadowed common sense. Rhetoric has not led to the hard work needed to achieve results and our leader's raise their voices each time there is a spike on gas prices, only to grow quiet when the price falls at the pump."
- "Now America has arrived at a crossroads. Embedded in American soil, in the wind and the sun, we have the resources to change. Our scientists, businesses, and workers have the capacity to move us forward."
- "It falls on us to choose whether to risk the peril that comes with our current course or to seize the promise of energy independence. And for the sake of our security, our economy and our planet, we must have the courage and commitment to change."

- "It will be the policy of my administration to reverse our dependence on foreign oil while building a new energy economy that will create millions of jobs."
- "Today I am announcing the first steps on our journey toward energy independence, as we develop new energy, set new fuel efficiency standards, and address greenhouse gas emissions."
- "We will make it clear to the world that America is ready to lead. To protect our climate and our collective security, we must call together a truly global coalition. I have made it clear that we will act, but so too must the world. That is how we will deny leverage to dictators and dollars to terrorists, and that is how we will ensure that nations like China and India are doing their part, just as we are now willing to do ours."
- "We have made our choice: America will not be held hostage to dwindling resources, hostile regimes, and a warming planet. We will not be put off from action because action is hard. Now is the time to make the tough choices. Now is the time to meet the challenge at this crossroad of history by choosing a future that is safer for our country, prosperous for our planet, and sustainable."

Obama stressed that the federal government must work with, not against, the individual states to control climate change. His plan also outlined the goal of requiring cars to meet a 35 MPG fuel efficiency standard by 2020 and vowed to "help the American automakers prepare for the future, build the cars of tomorrow, and no longer ignore facts or science." Climate change is real, and his energy policy will be dictated to deal with climate change and will free U.S. dependence on foreign oil for security purposes.

Those pledges were made seven years ago. Let us take a look at where the United States stands today in terms of those goals.

THE REALITIES OF THE U.S. CLIMATE CHANGE BELIEF SYSTEM

This chapter focuses first on the United States because it presents one of the most difficult and frustrating pathways leading to the identification, acceptance, and solution of the climate change problem of any country on the planet, and the reason is not singular, or direct. It ranges from economic to political to institutional to lifestyle preferences to existing infrastructure to personal reasons, and the gambit in between. It has been debated tirelessly. Green movements have rallied for change. Yet, even amid wide-ranging research, progress, public education (with enthusiasm and support), the United States *is still* sorely behind the rest of the world. We have already looked at the sociology and psychology behind it and the spin of the media—and that does provide some insight. On a positive note, we also know the environmental movement

seems to be healthy and growing in the United States—lots of homeowners are going solar, many drivers have invested in hybrid transportation. People are conserving, buying local, and doing their part. Somewhere, however, the United States is still missing the mark.

In a Yale study conducted by Anthony Leiserowitz in 2015, a national poll was taken in order to build a unique type of statistical model: one that is able to *accurately* predict the publics' climate change beliefs, risk perceptions, and policy preferences. Not only that, its creators believe it has the predictive power to define these preferences down to the levels of: (1) state, (2) county, and (3) congressional district. What this means, in a nutshell, is no matter where an individual lives in the United States, this model is able to successfully predict what that individual believes about climate change. So, bottom line: local perceptions about climate science and policy are predictable at a level of detail that even ordinary public polling methods cannot replicate.

Sounds impossible? The researchers say it is more than possible. Using a technique called "multilevel regression and poststratification," they claim the method uses information from a series of nationally representative online panels that examined participants' climate beliefs. Those are then blended with other data so that the models predictive capabilities are strengthened (Figure 6.1).

Researchers claim this stratification tool is critical in understanding peoples' perceptions about climate change (and other important issues). They argue that professionals such as social scientists, advocates, educators, planners, and policy-makers need this level of understanding in order to make critical decisions (Cushman, 2015). For instance, this type of information could be used to understand where there would already be support for renewable energy initiatives, in making decisions about public transportation changes, and to understand how much support policy-makers would have for climate change policy initiatives. It could be a tool for politicians to use when trying to decide whether or not to tax or cap emissions of carbon dioxide, spend more on researching alternatives to fossil fuels, or join an international climate treaty.

The study also confirmed that geographic patterns in beliefs are generally consistent with political patterns. For instance, California and New York shows a relatively high concern about climate change; yet Wyoming and Oklahoma show a lower concern. These "red flag" relationship warnings can be useful when discussing potential policy matters and needing to come to more "creative" solutions. When looking at other patterns that emerged from the data, the researchers discovered that of all the racial and ethnic groups in the United States, the Latinos are the most concerned about climate change and urban dwellers are more likely to believe in climate change than rural regions. In college towns, the scientific aspect on climate change is especially likely to be understood.

Estimated % of adults who think global warming is mostly caused by human activities, 2014

Display model output: Global warming is caused mostly by human activities

FIGURE 6.1 Map from Yale Project on Climate Change Communication. New tool that shows/predicts by geographic location whether people think that global warming is caused mostly by human activities. (From Howe, P. et al., *Nat. Clim. Change*, 5, 596–603, 2015.)

In another study at Yale, researchers found that the majority of the American public accepts the scientific consensus and believes in the reality of climate change and favors action in combating it. A major problem, however, is that many lawmakers still stand on the side of denial in the matter. When researchers compared votes cast by U.S. senators on a 2015 January climate change measure with their constituents' opinions, they found a noteworthy difference in opinion. The vote was on an amendment to recognize that: (a) climate change is real, and (b) human activity is a significant contributor to climate change. Specific results included Sen. Marco Rubio (Florida) who voted no, while 56 percent of Floridians support the fact that climate change is real and human-caused. Cory Gardner (Colorado) voted no, yet 58 percent of Coloradans accepted the evidence and agreed we were to blame. Dean Heller (Nevada) also voted no, and 57 percent of Nevadans disagreed. (Whitehouse, Climate Change. 2016).

These trends are very telling. If politicians' views are not up to par with the citizens of the United States, the ramifications imply that very little, if anything, is going to move forward to solve the problem. And is not that exactly what we have been seeing? Anthony Leiserowitz, director of the Yale Project on Climate Change Communication, remarked "This gap in climate belief is a major problem. These senators are some of the key decision makers in American society, and their decisions affect all of us."

Leiserowitz is correct. These political leaders are key decision makers. Not only do some of them balk at acknowledging climate change, but it also has a partisan component to it, a gridlock between the parties, which further degrades or stifles any talk or agreement. Furthermore, their decisions do affect the entire country. So, whose fault is that? The politicians, for not educating themselves, or for being bought out by fossil fuel companies or other special interest groups? Yes. But only partly. The greater responsibility falls on the American public. The same American public who elected them.

Yale has additionally discovered that even though partisan gridlock is currently in full swing in Congress, businesses have taken a lead and launched their own energy efficiency, and sustainability programs. They have stepped up and willingly invested millions of dollars toward fighting climate change. Americans are getting on board, as well. In Yale's study, 63 percent of the public supported climate action. Solar power in homes are an illustration of the forward movement underway. In 2006, approximately 30,000 homes were solar powered; in 2013 that number had climbed to 400,000. By the beginning of 2017, more than one million homes will be solar powered, and by 2020, it is projected that upward of 3.8 million homes will have converted to solar power (Figure 6.2). In addition, 61 percent of Americans think the federal government should force corporations to pay a price for carbon pollution. To date, however, none of these concepts have carried any political ramifications for legislators, but that may change. As Leiserowitz noted, "Climate change is not currently at the top of most voters' minds when choosing who to vote for, but that's starting to change. There are numerous organizations now working to build public and political will for climate action, including making it a more important electoral issue" (Bagley, 2015a).

The bottom line is this: anyone in office as president can have the noblest goals toward fixing the problem of climate change. But if the American public does not elect members

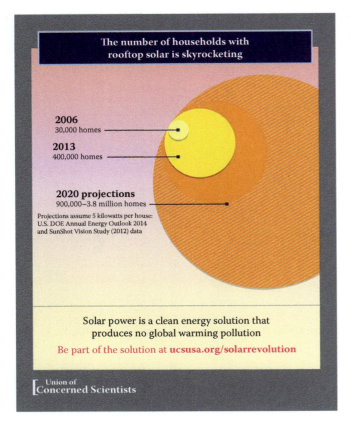

FIGURE 6.2 The number of houses with rooftop solar in the United States. (Courtesy of Union of Concerned Scientists [UCS].)

of Congress that have the same goals (to combat climate change), we are getting nowhere. We will never progress. Congress will stew in a bipartisan gridlock, doing nothing while we continue to pump CO_2 into the atmosphere at an accelerated rate. So, until American citizens put people into office that genuinely care more about fixing the climate than arguing or taking bribes, it is never going to change. The American public must demand it and settle for nothing less. It is up to each American citizen to demand climate policy and it begins by who you vote for. Your vote does count. If we want the change we claim we do, we need to collectively make sure it happens.

In an analysis completed by Stuary Capstick a scientist at Cardiff University in Wales, has determined that the United States still lags behind much of the rest of the developed world in the understanding of climate change. Conversely, the Middle East, Africa, Russia, Commonwealth States, and developing nations of Asia still trail the United States in understanding the threat of climate change. Capstick also found an interesting trend over several decades: during the 1980s and 1990s, there was an increasing global awareness and public concern about climate change. Following that, in many countries, a trend of skepticism seemed to take hold around 2005–2010. Close on the heels of skepticism, the issue quickly became a partisan one, largely due to global recession and concerns that taking any kind of action could hurt various economies. Many countries then drifted away from the partisan aspect of

the issue. In countries, like the United States, Australia, and the United Kingdom, however, the divide has only widened. Capstick believes this reflects fossil fuel-funded denial campaigns and the widening ideological divide between conservatives and progressives. According to Capstick, "There is a tendency to focus on how many people don't accept climate change, but that's not the reality" (Bagley, 2015c). Again, it is back to the media's portrayal of the issue. When climate deniers fund public denial campaigns and this is what the public hears, they tend to believe that rather than researching the facts for themselves.

Also presenting fodder for climate change doubters are those saying that there cannot be a warming climate as long as ice has been extending at the ice caps at the South Pole. The fact that the ice has been advancing in spite of warming oceans has puzzled scientists for a few years now as well as providing fodder for the doubters. Finally in May 2016, NASA researchers were able to answer that question. The answer was discovered in a new study involving the wind and ocean currents which circle the frozen continent. What they discovered is that the icy winds blowing off Antarctica, as well as a powerful ocean current that circles Antarctica, are much larger factors in the formation and persistence of Antarctic sea ice than changes in temperature, as previously thought. What is happening is the Southern Ocean Circumpolar Current is preventing warmer ocean water from reaching the Antarctic sea ice zone, which is helping to isolate the continent. The winds within that ice zone are keeping the water extremely cold. This has enabled the sea ice cover to grow in recent years—even as global temperatures have risen markedly.

This information was obtained from satellite readings of Antarctic sea ice movement and thickness, along with new, detailed interpretation of charts which showed the shape of the ocean bottom around Antarctica. During this time period, scientists have been warning that the sea ice and glaciers in the Arctic have been shrinking rapidly. They have also been observing that the ice shelves in West Antarctica are being undermined by warm currents where they are connected to the ocean floor. This is where they are warning that the melting is going to contribute to rapidly rising sea levels. The difference with sea ice, however, is that it does not have a big effect on sea level: it grows and melts every season. The ice shelves are different. They are the floating extensions of the huge, land-based ice sheets and glaciers; and as they disintegrate, the flow of land ice toward the sea speeds up, causing sea-level rise.

Antarctic sea ice has increased slightly over the past decade. In fact, between 2012 and 2014, it reached record high during the winter months. In September 2014 it exceeded 20.1 million square kilometers (Berwyn, 2016). The phenomenon causing this could not be adequately determined by models, however, so while scientists could see this was happening, they were at a loss for the correct explanation as to why. It was during this time that Oklahoma Republican Senator James Inhofe called global warming a hoax, along with other prominent deniers of climate change, and declared scientists should focus on "what is not melting" instead (Congressional Record Volume, 2014). During the past four decades, the Arctic has lost roughly 53,872 square kilometers of sea ice per year, while the Antarctic has gained an annual average of 18,907 square kilometers. The climate models could not explain the mechanism causing the freezing in the Antarctic. It was not until Son Nghiem, a researcher at NASA's Jet Propulsion Laboratory began studying the daily satellite imagery instead of relying solely on the climate models that he began to see what was going on. This shows the importance of direct human observation in unusual cases like this one. The satellite readings showed that as sea ice forms early in the season, wind blowing off the cold Antarctic ice cap pushes it offshore and northward. As the ice moves away from the shore, it breaks and is pushed around by wind, eventually forming thicker ridges of ice (similar to a reef) that protects the constantly-forming younger ice from being eroded by wind and waves. Nghiem states: "The ice at the front of the ice pack is the older ice. It's thicker and rougher; it forms a great wall that protects the ice inside, so the internal ice opens up, stretches out. In the open water, ice can grow ten times as fast as at the front."

He also discovered that the Southern Ocean Circumpolar Current—which helps determine sea-ice extent—is steered by submerged ridges and canyons along the edge of the Antarctic continental shelf, rather than by climate change or other climatic conditions. Simultaneously, the central portion of Antarctica is still so cold that even if the air masses flowing off the continent warm slightly, it stays cold enough to form new ice, a phenomena observed by Marcel Nicolaus, an ice physics scientist at the Alfred Wegener Polar Research Institute in Germany.

The significance of this is that the results of direct research have shown that some theories have changed about the way ice forms and behaves in Antarctica. While politicians and the media have used it to deny the existence of climate change, it has instead not only reinforced its existence, but also has given the scientific community better insights into how climate change is affecting both the Arctic and Antarctic climates, and while more than half the U.S. population believes climate change is human caused, and the percentage of politicians is much less; polls have found approximately 98 percent of working climate scientists accept the evidence for climate change being human-caused.

Perhaps it is appropriate to mention at this point while we are discussing why the United States is lagging behind the curve in climate change understanding, acceptance, and action, the toll that "climate denial" takes on scientists. When scientists spend time refuting denialist theories, one study shows that they add credence to their antagonists' campaigns. The realism of it is, climate denial campaigns have helped slow the public's acceptance of man-made climate change and delay political action for years, but a trend has emerged lately that when scientists spend time and resources addressing denialists' claims, researchers also now often downplay future climate risks to avoid being labeled an "alarmist" by climate contrarians. Stephan Lewandowsky, a cognitive scientist at

the University of Bristol in England and lead of the study, said that "Scientists are reticent to begin with, and if they know that everyone wants to ream at them for being alarmists, that adds to their reticence. They are human, just like you and I... and some of the actions by deniers can be quite nasty." He commented that hate mail and complaints directed to their research institutions are very common (Bagley, 2015b).

Kevin Trenberth, a climate scientist at the National Center for Atmospheric Research (NCAR) in Boulder, Colorado supported this. He commented, "Climate denial campaigns can absolutely influence what you do and what you write about. Part of the reason they do it is to distract you and get you to waste your time. Instead of publishing the good science needed to advance our understanding of climate change, scientists are left defending their work and debunking false claims." Perhaps first understanding the reality and then taking mindful action against the denial movement is a good place to focus some efforts.

The Beginnings of Change in Legislation

The ultimate goal of political action on climate change is to limit and/or reduce the concentration of GHGs in the atmosphere. Political action is a critical component necessary to make any significant global change because without the implementation of the necessary laws and regulations—such as GHG emissions limits, regulatory frameworks within which carbon trading markets can operate, reportable and trackable systems of accountability, and tax incentives or funding assistance—productive and long-term change is not feasible.

Although the United States had a slow start toward addressing the climate change issue, current legislation is now percolating, and progress is slowly being made. One of the most significant advancements for the climate change issue was when it made its way to the Supreme Court. On April 2, 2007, in one of its most important environmental decisions in years, the U.S. Supreme Court ruled that the EPA had the authority to regulate heat-trapping gases in automobile emissions. The Court further stipulated that the EPA could in no manner "sidestep its authority to regulate the greenhouse gases that contribute to global climate change unless it could provide a scientific basis for its refusal." This gave the EPA the right to regulate CO_2 and other heat-trapping gases under the Clean Air Act.

According to Justice John Paul Stevens, "The only way the agency could avoid taking further action was if it determined that greenhouse gases do not contribute to climate change or provides a good explanation why it cannot or will not find out whether they do" (Greenhouse, 2007).

The Supreme Court also heard another case concerning the Clean Air Act, giving the EPA a broader authority over factories and power plants that want to expand or increase their emissions of air pollutants. Under this broader reading, they made a ruling of 9 to 0 against the Duke Energy Corporation of North Carolina in favor of the EPA, which made environmentalists ecstatic, marking a historic occurrence in the U.S. Supreme Court as a positive step toward the mitigation of climate change. Interestingly, since the ruling on the first case, there has been a growing interest among various industrial groups in working with environmental organizations on proposals for emissions limits.

On April 17, 2009, the EPA took it even further and formally declared CO_2 and five other GHGs to be pollutants that endanger public health and welfare. This landmark decision put in motion a process that leads to the regulation of GHGs for the first time in U.S. history. According to the EPA, "The science supporting the proposed endangerment finding was compelling and overwhelming." The decision received diverse reactions. Many Republicans in Congress and industry spokesmen warned that regulation of CO_2 emissions would raise energy costs and kill jobs. Democrats and environmental advocates, however, said the decision was long overdue and would bring long-term social and economic benefits.

Lisa P. Jackson, the EPA administrator, said, "This finding confirms that greenhouse gas pollution is a serious problem now and for future generations. Fortunately, it follows President Obama's call for a low-carbon economy and strong leadership in Congress on clean energy and climate legislation" (Greenhouse, 2007).

This opens the door for the EPA to determine specific targets for reductions of heat-trapping gases and set new requirements for energy efficiency in vehicles, power plants, and industry. It also allows them to regulate climate-altering substances under the Clean Air Act. Historically, Washington's efforts in finding new solutions to energy demand and efficiency were to develop more fuel-efficient cars, not alternative-fuel cars, making this new approach by Congress significant. Other portions of the bill are equally groundbreaking. The bill calls for a significant increase in the amount of ethanol used in the nation's fuel supply. Congress is proposing to double the nation's current level of production to 57 billion liters. It also foresees that by 2022, an additional 79 billion liters a year of ethanol or other biofuels will be produced by developing technology that can obtain useful energy from biomass such as straw, tree trimmings, corn stubble, and even common garbage.

Another reason why political involvement is crucial is that in order to accomplish these goals, the nation's key scientists and business leaders will need political and financial support to successfully deal with the technical, environmental, and logistical obstacles they will encounter. Some environmentalists remain uneasy because ethanol produced from corn still requires energy and fertilizer involving the use of natural gas, oil, and coal. Some food producers argue that the plan would require growing eight million hectares of corn—leaving fewer farming acres for fruits, vegetables, soybeans, alfalfa, and other crops and leading to higher food prices. As with all important issues, there are always pros and cons that must be taken into account when making decisions.

New Data Released by the National Oceanic and Atmospheric Administration

The long-debated hiatus in "global warming" as it is been called by climate denialists for years, who have tried to claim it proved scientists' projections on climate change are inaccurate and exaggerated is now being countered by new evidence discovered by the National Oceanic and Atmospheric Administration (NOAA). In fact, there was not even a slowdown in the warming.

The "stall" that deniers have pointed to during the past 15 years can now be explained: some of it was masked by incomplete data records that have been improved and expanded in recent years (Bagley, 2015d).

Tom Karl, director of NOAA's National Centers for Environmental Information and lead of the study said, "The rate of temperature increase during the last half of the twentieth century is virtually identical to that of the twenty-first century." This is just one example in a rapidly growing number of similar studies refuting the idea of a slowdown or stop in global warming. This is backed by other climate and earth scientists, also agreeing that there is no true "pause" or "hiatus" in warming (Figure 6.3).

Previous calculations had estimated the earth had warmed 0.113°C per decade from 1950 to 1999, and 0.039°C per decade from 1998 to 2012, according to the IPCC. This translates to a smaller increasing linear trend over the past 15 years than over the past 30 to 60 years. This is what the IPCC listed in their Fifth Assessment report. Within the last decade, however, thousands of new weather stations have been built in previously under-reported areas on land and a vast network of buoys have been deployed that more accurately measure sea

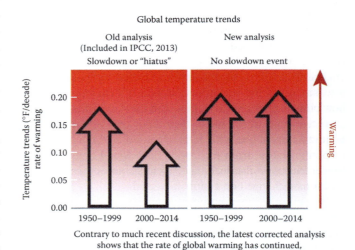

Contrary to much recent discussion, the latest corrected analysis shows that the rate of global warming has continued, and there has been no slow down.

FIGURE 6.3 In the latest corrected analysis by NOAA in 2015, it shows that global warming is still continuing and not slowing down. (Courtesy of NOAA; From Karl, T. R. et al., The recent global surface warming hiatus: Fact or artifact of data biases, *Science*, 2015.)

surface temperatures. This data was used to update the record and correct any discrepancies. NOAA found that the earth had warmed 0.086°C per decade between 1998 and 2012—more than double the previous estimates. Then, when they included 2013 and 2014 readings (the years of record-breaking heat worldwide), the warming per decade was 0.116°C. This corrected value not only obliterates any "hiatus" critics saw in the record, but also the scientists now believe the findings may even underestimate the world's warming because they do not consider what has happened in the Arctic, where temperatures have increased rapidly in recent decades, but where there are still a limited number of weather recording stations. The only slowdown in warming scientists can find occurred from 2000 to 2012 in the Pacific, but the event was local and barely impacted the global mean temperature at the time. Karl does stress that "There certainly is variability from year to year, and one can find periods in the record where there are small changes, but over the long term, the world is still warming at an alarming rate." Gavin Schmidt, a climate scientist and director of the NASA Goddard Institute for Space Studies reminded us all: "There will be a very predictable chorus of 'data manipulation' and 'fraud' by the deniers as they see a talking point disappear, and so it will just continue as before. Just remember, their objections have little or nothing to do with science" (Bagley, 2015d).

What the Environmental Protection Agency Is Doing about Climate Change

The Environmental Protection Agency (EPA) is playing a highly significant role in the management of climate change today. To their credit, they have taken a common-sense approach to developing carbon pollution standards for both new and existing power plants. In September 2013, they announced proposed standards for new power plants and initiated outreach efforts to help inform the development of emission guidelines for existing plants. In June 2014, they released the Clean Power Plan, which are the first-ever carbon pollution standards for existing power plants that will protect the health of future generations and finally put the United States in line with achieving a 30 percent reduction in carbon pollution from the power sector by 2030. The Clean Power Plan will also lead to climate and health benefits with an estimated worth of $55 billion to $93 billion per year in 2030 and will cut pollution that leads to soot and smog by over 25 percent in 2030 (whitehouse.gov). The Clean Power Plan is discussed in detail later in this chapter.

Because the EPA is responsible for protecting human health and safeguarding the natural environment, including air, water, and land, they are responsible for, and have jurisdiction over, the following initiatives: BACT Guidance, Tailoring Rule, Endangerment Findings, Mandatory Greenhouse Gas Reporting Rules, Federal Vehicle Standards, and Renewable Fuel Standards (RFSs). A summary of each follows:

BACT guidance: On November 10, 2010, the EPA released its guidance for the Best Available Control Technology (BACT) requirements for GHG emissions from major new or modified stationary sources of air pollution. This falls under the requirement in the Clean Air Act that stipulates that any major new source or any major modifications to existing sources that emit GHGs must employ technologies that now limit GHG emissions. For each facility, the BACT requirements must address the specific conditions of that particular facility and direct them to complete the maximum degree of emission reduction that has been demonstrated through available methods, systems, and the particular environmental considerations of that facility. What it does not provide for, however, is carbon capture and sequestration technology unless the facility happens to be located next to an operating oil field whose operator agrees to purchase carbon dioxide for enhanced oil recovery.

The Tailoring Rule: Finalized on May 13, 2010, the Tailoring Rule is designed to focus new source review and permitting requirements on the largest stationary sources. Under the Clean Air Act, any pollutant subject to regulation under any provision of the Act triggers additional requirements. For example, if an industry begins making a new product or makes modifications to an old process, they are subject to new regulation reviews by the EPA for compliancy. On April 1, 2010, the EPA issued its first standards for GHG emissions when it finalized the rules it was using. Beginning January 2, 2011, sources currently subjected to new source review permitting requirements for other pollutants would also have to meet the requirements for GHG emissions if they exceeded 75,000 tons per year. Beginning July 1, 2011, these requirements also apply to new sources with GHG emission greater than 100,000 tons per year no matter what the amount is of emissions of other pollutants they emit.

Endangerment finding: When the Supreme Court's decision in Massachusetts v. EPA 549 U.S. 497 (2007) found that GHG emissions met the definition of "air pollutants" under the Clean Air Act, it required that EPA determine whether or not emissions of GHGs (from new motor vehicles) cause or contribute to air pollution which may reasonably be anticipated to endanger public health or welfare, or whether the science was too uncertain to make a reasoned decision. On April 17, 2009, EPA issued its proposed endangerment finding. This proposal laid out the basis for the proposed determination that six key GHGs—carbon dioxide (CO_2), methane (CH_4), nitrous oxide (N_2O), hydrofluorocarbons (HFCs), perfluorocarbons

(PFCs), and sulfur hexafluoride (SF_6)—were responsible for contributing to climate change which results in a threat to the public health and welfare of current and future generations. In addition, under the Clean Air Act, EPA also had to determine that emissions from particular sources "cause or contribute" to emissions which threaten public health and welfare. As part of its proposal, EPA also proposed to find that the combined emissions of these well-mixed GHGs from new motor vehicles and their engines contribute to the GHG pollution which threatens public health and welfare.

On December 7, 2009, EPA issued its final endangerment finding and its final cause or contribute finding for light-duty vehicles. Based on an overwhelming body of scientific evidence, they officially recognized that GHG emissions are a danger to public health and welfare. This action did not, by itself, impose any restrictions on any entities, but it was a required step in the process leading to specific regulations of GHGs. EPA proposed to regulate GHG emissions from light-duty vehicles and engines in a joint proposal with the National Highway Traffic and Safety Administration. These regulations were slated to become final in March 2010, marking them as the first actions to limit GHGs consistent with the Supreme Court's decision and under authority of the Clean Air Act.

Following that, Senator Murkowski (R-AK) sponsored a "disapproval resolution" which was intended to prevent the EPA from regulating GHG emissions under the existing Clean Air Act in order to stop its progression. In June 2010, the Senate defeated the "disapproval resolution," which sent a clear message that the Senate needs to act on the issue in a positive and responsible manner.

Even though the endangerment finding, by itself, does not regulate any sources, it laid the necessary foundation for future EPA regulations, such as the light duty vehicle and engine rule—a standard designed to cut CO_2 through the improvement of automobile fuel economy. This measure is expected to cut 6 billion metric tons of GHG over the lifetimes of the vehicles sold in model years 2012–2025, save families more than $1.7 trillion in fuel costs, and reduce the United States' dependence on oil by more than 2 million barrels per day in 2025 (EPA, 2016a).

EPA is also working with state and regional partnerships. Currently, there are 23 states that either have or are developing programs to regulate GHG emissions. In addition, several court decisions (e.g., Conn. vs. AEP) have opened the door for common law nuisance claims against firms emitting GHG emissions.

Mandatory Greenhouse Gas Reporting Rule

On September 22, 2009, the EPA announced that it now requires large emitters of GHGs to begin collecting data under a new reporting system. Fossil fuel and GHG suppliers, manufacturers of both motor vehicles and engines, and facilities that emit 25,000 metric tons or more of CO_2 equivalent per year are now required to report their GHG emissions each year directly to the EPA. Data collection began in January 2010, and were compiled into the first annual report in January 2011. This represented a significant development in climate change and its control in the United States. According to Lisa P. Jackson (EPA's current administrator):

> "For the first time, we begin collecting data from the largest facilities in the country, ones that account for approximately 85 percent of the total U.S. emissions. The American public and industry itself, will finally gain critically important knowledge and with this information we can determine how best to reduce those emissions."

Federal Vehicle Standards

On May 19, 2009, President Obama made a historic announcement to set new national GHG and corporate average fuel economy standards for light-duty vehicles. Then, on September 15, 2009, the EPA and the Department of Transportation (DOT) issued a joint proposal to enhance these standards, which were finalized on April 1, 2010.

These new standards account for over 60 percent of the GHG emissions from the transportation sector. New Federal standards accelerate fuel economy improvements required under the Corporate Average Fuel Economy (CAFE) program, which is administered by the National Highway Traffic Safety Administration (NHTSA), under the DOT. This is a change to the Energy Independence and Security Act of 2007, which aimed for an average fuel economy of 35.7 miles per gallon by 2020. It sped it up by requiring an average fuel economy of 25.5 mpg by 2016. It also included a GHG emission limit per vehicle, which would be set by EPA using its authority under the Clean Air Act. The tables illustrate the Projected Emissions Target Under the CO_2 Standards (g CO_2/mi) and Projected Fuel Economy Standard (mpg), respectively (Tables 6.1 and 6.2).

TABLE 6.1

Projected Fleet-Wide Emissions Compliance Targets under the Footprint-Based CO_2 Standards (g/mi) and Corresponding Fuel Economy (mpg)

	2016 (base)	2017	2018	2019	2020	2021	2022	2023	2024	2025
Passenger cars (g/mi)	225	212	202	191	182	172	164	157	150	143
Light trucks (g/mi)	298	295	285	277	269	249	237	225	214	203
Combined cars and trucks (g/mi)	250	243	232	222	213	199	190	180	171	163
Combined cars and trucks (mpg)	35.5	36.6	38.3	40.0	41.7	44.7	46.8	49.4	52.0	54.5

TABLE 6.2
Projected Fuel Economy Standard (mpg)

	2012	2013	2014	2015	2016	2017	2018
Passenger cars	33.6	34.4	35.2	36.4	38.2	39.6	41.1
Light trucks	25	25.6	26.2	27.1	28.9	29.1	29.6
Combined cars and trucks	29.8	30.6	31.4	32.6	34.3	35.1	36.1

	2019	2020	2021	2022	2023	2024	2025
Passenger cars	42.5	44.2	46.1	48.2	50.5	52.9	55.3
Light trucks	30.0	30.6	32.6	34.2	35.8	37.5	39.3
Combined cars and trucks	37.1	38.3	40.3	42.3	44.3	46.5	48.7

The estimated benefits from this are significant—just over the lifetime of the vehicles sold from 2012 to present (2016), the reduction of CO_2 emissions is equal to 950 million metric tons and 1.8 billion barrels of oil have been saved. This also reflects a reduction on our reliance of foreign oil—another cost savings (EPA, 2009). The finalized standards will also reduce CO_2 emission from the light-duty vehicles by 21 percent in 2030.

Renewable Fuel Standard

The RFS is a requirement that a certain percentage of petroleum transportation fuels be displaced by renewable fuels (EPA, 2016b). RFS1 began with the Energy Policy Act of 2005. Congress then updated the standard in 2007 and it became the Energy Security and Independence Act of 2007. This new RFS is known as RFS2 and is an RFS for biofuels that require *only* obligated parties to sell a certain amount of biofuels per year through 2022. This standard contains a four-part mandate for lifecycle GHG emissions levels relative to a 2005 baseline of petroleum for: (1) renewable fuel, (2) advanced biofuel, (3) biomass-based diesel, and (4) cellulosic biofuel. In order to be classified under one of these categories, a fuel must meet the percentage reduction in life-cycle GHG emissions. As shown in Figure 6.4, the RFS2 slowly ramps up advanced biofuels (cellulosic, biomass-based diesel, and noncellulosic advanced) until they overtake conventional biofuels in consumption levels by 2022. RFS2 was published by the EPA on March 26, 2010. The following are the identified impacts:

- RFS2 will displace about 13.6 billion gallons of petroleum-based gasoline and diesel fuel in 2022, which represents approximately 7 percent of expected annual gasoline and diesel consumption in 2022.
- RFS2 will decrease oil imports by $41.5 billion, and will result in additional energy security benefits of $2.6 billion by 2022.
- By 2022, gasoline costs should decrease by 2.4 cents per gallon and diesel costs should decrease by 12.1 cents per gallon because of the increased use of renewable fuels.
- RFS2 will reduce GHG emissions by 138 million metric tons in 2022, which is equivalent to taking about 27 million vehicles off the road.
- RFS2 will increase net farm income by $13 billion dollars (equivalent to 36 percent) in 2022.
- RSF2 will increase the cost of food $10 per person in 2022.

The U.S. Department of Energy's Recovery Act Spending is also a part of the Executive Branch Action for Climate Change. The American Recovery and Reinvestment Act of 2009 (Recovery Act or ARRA) is the economic stimulus package that was passed by Congress on February 13, 2009 and signed by President Obama on February 17, 2009. About $63 billion from the Recovery Act is targeted for energy, transportation, and climate

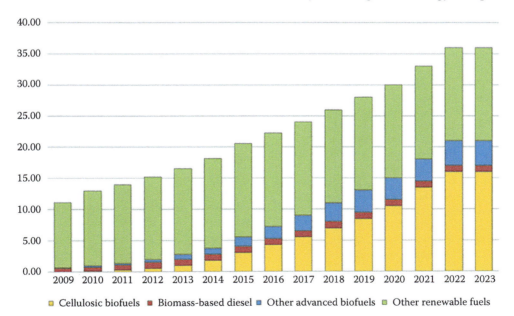

FIGURE 6.4 Renewable fuel volume requirement (in Billion Gallons). (Courtesy of Environmental Protection Agency [EPA].)

research spending, with an additional $21 billion in climate-energy tax incentives. The Department of Energy received $36.7 billion of the climate and energy-related funds, with nearly $33 billion for direct grants and the remainder for loan guarantees.

As of July 31, 2016, qualifying technologies included biomass, combined heat and power, fuel cells, geothermal, incremental hydropower, landfill gas, marine hydrokinetic, microturbine, municipal solid waste, solar, and wind. Projects statistics included:

- Total number of projects funded: 105,178
- Amount of funding committed: $24.9 billion
- Total installed capacity of funded projects: 33.3 GW
- Total estimated annual electricity generation from funded projects: 88.8 TWh[*]

The following table lists some of the other notable activities the EPA is involved in to combat climate change.

What the EPA Is Doing about Climate Change

Collecting Emission Data

EPA collects different types of GHG emissions data. This data assists policy-makers, businesses, and the EPA track GHG emissions trends and identify opportunities for reducing emissions and increasing efficiency. Examples include *The Inventory of U.S. Greenhouse Gas Emissions and Sinks* and the *Greenhouse Gas Reporting Program*.

Obtaining Reductions

The EPA reduces GHG emissions and promotes a clean energy economy through successful partnerships and regulatory initiatives. Examples include their *Developing Common-sense Regulatory Initiatives, Partnering with the Private Sector,* and *Reducing EPA's Carbon Footprint*.

Evaluating Policy Options, Costs, and Benefits

EPA conducts economy-wide analyses to understand the economic impacts and effectiveness of proposed climate polities.

Advancing Scientific Knowledge

EPA contributes to world-class climate research through the U.S. Global Change Research Program and the IPCC. EPA's Office of Research and Development conducts research to understand the environmental and health impacts of climate change and to provide sustainable solutions for adapting to and reducing the impact from a changing climate.

Participating in International Partnerships

EPA participates in a variety of international activities to advance climate change science, monitor the environment, and promote activities that reduce greenhouse gas emissions. They establish partnerships, provide leadership, and share technical expertise to support these activities.

Partnering with States, Localities, and Tribes

EPA's State and Local Climate and Energy Program provides technical assistance, analytical tools, and outreach support on climate change issues to state, local, and tribal governments.

Assist Communities in Adaptation Measures

EPA's Climate Ready Estuaries and Climate Ready Water Utilities programs help coastal resource managers and water utility managers plan and prepare for climate change.

Source: https://www3.eps.gov/climatechange/EPAactivities.html.

---

[*] 88.8 TWh is the equivalent of 8.1 million homes.

THE U.S. CLIMATE ACTION PLAN

President Obama announced the U.S. Climate Action Plan on June 25, 2013. A comprehensive plan, it is the first of its kind designed to achieve three goals:

1. To cut domestic carbon pollution
2. To prepare the United States for climate change impacts
3. To lead international efforts to address global climate change

This is the most comprehensive climate plan presented by a U.S. president to date. The first objective, to cut domestic carbon pollution, is centered on reducing GHG emissions by 26–28 percent below 2005 levels by 2025. Plans to do this involve reducing GHG emissions from the power sector and promoting energy efficiency and clean energy projects nationwide. The plan will accomplish this by cutting emissions from light-duty vehicles through 2025 (EESI, 2015). The next section on The Clean Power Plan gives more information on this objective.

The World Resources Institute believes that this goal can be met with serious dedication and action. They have identified four areas with the greatest opportunity for reductions: power plants, energy efficiency, hydrofluorocarbons (HFCs), and methane. Fortunately, all these are specifically included in the plan. The World Resources Institute believes the plan is a welcome step to putting the United States on a pathway to a safer future (Morgan and Kennedy, 2013).

Energy efficiency goals include reducing CO_2 pollution by a total of 3 billion metric tons through 2030 through new and existing efficiency standards for appliances and federal buildings. This equals a significant reduction—the equivalent of removing nearly two years' worth of emissions from coal power plants. Energy efficiency is one of the most cost-effective ways to reduce emissions, as more efficient equipment uses less energy and saves consumer's money.

This part of the plan also calls for doubling renewable energy in the United States by 2020 and opening public lands for renewable energy development, which would be equal to an additional 10 gigawatts of installed renewable capacity on those lands by 2020. This could generate enough energy to power 2.6 million American homes. The federal government is also opening up some public lands for clean energy projects.

The second goal—preparing for climate change—will provide federal agencies and communities throughout the country with the proper resources they will need to improve their resiliency against rising sea levels, extreme weather events, drought, and other destructive impacts caused by climate change.

The third goal—leading international efforts to address global climate change—will help establish the United States as a capable world leader in climate change action. This can help the United States make up for lost ground on climate change action in the past. The United States is currently involved in encouraging and forming agreements between single, or groups of, countries in order to accelerate the transition from energy generated by fossil fuels to the rapid investment and development of clean energy technologies.

The following list outlines some of the plan's notable highlights (whitehouse.gov):

Goal 1: Cutting carbon in the United States:

- Directs the EPA to work closely with states, industry, and other interested parties to establish carbon pollution standards for both new and existing power plants.
- Provides $8 billion in loans for the development of innovative energy efficiency technologies.
- Enables DOI to issue enough permits on public lands for renewables projects (such as wind and solar) so that by 2020 more than 6 million homes can be powered; designates the first-ever hydropower project for priority permitting; and sets a new goal to install 100 megawatts of renewables in federally assisted housing by 2020; and maintains the commitment to use renewables on military installations.
- Under the President's Better Building Challenge, focuses on helping commercial, industrial, and multi-family buildings cut waste and become at least 20 percent more energy efficient by 2020.
- Sets goals to reduce carbon pollution by at least 3 billion metric tons by 2030. This represents more than half of the annual carbon pollution from the U.S. energy sector and will be accomplished through efficiency standards for appliances and federal buildings.
- Commits to partnering with industry and others to develop fuel economy standards for heavy-duty vehicles to save families money at the pump and further reduce reliance on foreign oil and fuel consumption after 2018.
- Finds new opportunities to reduce pollution of highly-potent GHGs known as hydrofluorocarbons; directs agencies to develop a comprehensive methane strategy. Also commits to protect our forests and critical landscapes.

Goal 2: Preparing the United States for the impacts of climate change:

- Directs agencies to support local climate-resilient investments by removing barriers or counterproductive policies and modernizing programs. It also establishes a short-term task force of state, local, and tribal officials to advise on key actions that the Federal government can take to help strengthen communities on the ground.
- Attempts to create sustainable and resilient hospitals in the face of climate change through a public-private partnership with the healthcare industry.

- Maintains agricultural productivity by delivering tailored, science-based knowledge to farmers, ranchers, and landowners. It helps communities prepare for drought and wildfire by launching a National Drought Resilience Partnership and by expanding and prioritizing forest- and rangeland-restoration efforts to make areas less vulnerable to catastrophic fire.
- Provides climate preparedness tools and information needed by state, local, and private-sector leaders through a centralized "toolkit" and a new Climate Data initiative.

Goal 3: Lead international efforts to address global climate change:

- Commits to expand major new and existing international initiatives, such as bilateral initiatives with China, India, and other major emitting countries.
- Leads the global sector in public financing toward cleaner energy by calling for an end to U.S. government support of public financing of new coal-fired power plants overseas (except for the most efficient coal technology available in the world's poorest countries, or those facilities that utilize carbon capture and sequestration technologies).
- Strengthen global resilience to climate change by expanding government and local community planning and response abilities.

THE CLEAN POWER PLAN

Power plants account for nearly 40 percent of U.S. CO_2 emissions. That equals more than every car, truck, and airplane in the United States combined. On August 3, 2015, the EPA finalized a new set of standards designed to reduce carbon emissions from power plants for the first time in U.S. history. Previously, power plants had no restrictions on how much carbon pollution they could release into the atmosphere. The Clean Power Plan, represents the first control imposed on power plants. It was developed under the Clean Air Act, which was a Congressional act requiring the EPA to take steps to reduce air pollution deemed harmful to public health. The significance of The Clean Power Plan is that it represents the most significant opportunity to help curb the growing consequences of climate change.

The Plan establishes state-by-state targets for carbon emissions reductions while offering a flexible framework under which states may meet those targets. The goal is to reduce the national electricity sector emissions by an estimated 32 percent below 2005 levels by 2030. The Plan is also flexible; it provides a number of options to cut carbon emissions and determines state emissions reduction targets by estimating the extent to which states can take advantage of each of them. States have the following options for cutting emissions:

- Investing in renewable energy
- Using natural gas

- Becoming more energy efficient
- Using nuclear power
- Shifting away from coal-fired power

The options allow states to vary with their target plans because of each state's unique mix of electricity-generation resources; and also because of technological feasibilities, costs, potentials, and other factors. Each state can choose to combine any of the available options in a flexible approach in order to meet their targets. States can also team up in multi-state or regional compacts to find the lowest cost options to reduce their CO_2 emissions. States must submit a final plan, or an initial plan with an extension request, by September 6, 2016.

Critics are concerned that if the regulation shuts down a significant amount of generation capacity with existing power plants, the supply of electricity may not remain reliable, causing severe brown- or blackouts across the country. While EPA claims there will be the potential to retire approximately 33 gigawatts (GW) of coal capacity alone, they do not agree that power outages will be a problem and that the lights will remain in service to back this claim, they have taken several steps to insert a "safety valve," which will require several procedural hoops for states, but EPA believes it will eliminate any possibility of reliability disruptions. They did acquiesce, however, in extending the initial compliance period from 2020 to 2022. The EPA has also agreed to work with the Department of Energy and the Federal Energy Regulatory Commission to ensure continued electric reliability (AAF, 2015). The combined climate and health benefits of the Clean Power Plan are expected to outweigh the costs. It will also deliver billions of dollars in net benefits each year, such as $26–45 billion in 2030 (UCS, 2016).

On February 9, 2016, the Supreme Court stayed implementation of the Clean Power Plan pending judicial review. The Court's decision was not on the merits of the rule. EPA firmly believes the Clean Power Plan will be upheld when the merits are considered because the rule rests on strong scientific and legal foundations. Individual states have the choice to continue to work to cut carbon pollution from power plants and can still seek the agency's guidance and assistance. If they do, until all this is resolved, the EPA will continue to provide tools and support.

FIVE STEPS THE UNITED STATES MUST TAKE NOW

In a forward-thinking study by Dr. Michael Shank, Director of Foreign Policy at Friends Committee on National Legislation and Professor at George Mason University School for Conflict Analysis and Resolution; and Matt Lichtash, coauthor of *The Plan: How the U.S. Can Help Stabilize the Climate and Create a Clean Energy Future*, they have identified five steps that the United States should immediately take in order to successfully combat the negative effects of climate change. Part of the drive behind their suggestions lies in the fact that

the United States must cut emissions more than twice as much as currently provided for in the U.S. Climate Action Plan in order to avoid the worst impacts of climate change. Their suggestions are as follows:

Step 1: Implement a greenhouse gas fee: A GHG Fee that would put a price on all energy sources that cause GHG emissions and contribute to climate change should be implemented. This is seen as the single most effect tool that could be enacted in order to control the current emissions by making everyone responsible for the contribution they are making to the problem. When it comes to the partisan conflicts on contrasting views of how to deal with climate change, the originators of the study note that both parties, who currently claim to be trying to deal with the national debt problem, should be pleased with the unique revenue-raising opportunity presented in this option. With the revenue raised from the GHG Gas Fee, it could also be used to fund critical research and development; long-overdue electrical grid enhancements nationwide; offer incentives for smart land-use practices; and protect the most vulnerable communities from climate change, such as those most vulnerable to sea-level rise, drought, wildfire, and so forth. Shank and Lichtash believe that, depending on how the fee is structured, it could reduce overall GHG emissions from 2005 levels by 18–32 percent in 2020 and 50–95 percent in 2050 (Shank et al., 2013).

Step 2: Establishment of a National Green Bank: The establishment of an independent, government-backed National Green Bank will provide greater certainty for renewable energy investors at no net cost to the American taxpayer. This would mean lower borrowing rates, which ultimately would decrease the price of clean energy. In fact, the Coalition for Green Bank estimates that a National Green Bank could cut clean energy borrowing costs in half, create more than a million new jobs, and save more oil than we currently import from Saudi Arabia. The advantage of utilizing a Green Bank would guarantee that the United States' energy infrastructure remains innovative, sustainable, and resilient from petro-politics.

Step 3: Increase investments in public transportation: According to the DOT, the nation should increase investments in public transportation by $1.5 million per year (Department of Transportation Report to Congress, 2010). They believe this would serve to double ridership growth. DOT also points out that 97 percent of all U.S. transportation energy use is petroleum-based and this is unsustainable. They believe that further research and development of alternative technologies geared to provide a clean alternative for the transportation sector is imperative. Such new energy sources could include advanced batteries, hydrogen fuels, and cellulosic biofuels. DOT estimates that "if significant advances were to occur in battery technology, total transportation-related emissions could be reduced by 26–30 percent by 2050."

Step 4: Repeal all exemptions listed in various acts that favor hydraulic fracturing: It is necessary to repeal the exemptions listed in the Clean Air Act, the Clean Water Act, the Safe Drinking Water Act, and other environmental regulations that put hydraulic fracturing at an unfair advantage relative to other energy sources that play by the rules. By eliminating loopholes that exist in these various Acts, it will ensure that natural gas production becomes fully responsible for the impacts of its pollution and is forced to compete fairly with all other energy sources. Currently there are several negative impacts associated with the production process that the industry is not held responsible for, such as methane leakage—a particularly potent GHG.

Step 5: Discontinue all fossil fuel subsidies: It is suggested that the most wasteful fossil fuel subsidies should be abandoned. It is not suggested that the Home Energy Assistance Program be discontinued, however. This benefits many low-income American families that rely on the program to help pay for heating bills. The wasteful ones, however, could be discontinued and would have no impact on gas prices. It would also raise approximately $40 billion in revenue over the next decade. Also significant, it would send a strong message to the international community for other nations to follow our lead. Even though this would not cause U.S. emissions to decline substantially, there does exist a large reduction potential worldwide if other governments join in the same effort and participate in responsible subsidy reform.

Shank and Lichtash have noted that their plan could be countered by the cost to implement. They believe that while initial increases in investments are needed, the technologies required to transition to a clean energy economy are already becoming cost-competitive and energy efficiency improvements and new innovations will further reduce costs. They also point out that detailed in a study conducted in 2012, inaction on climate change will cost the United States 2.1 percent of GDP by 2030, which is greater than our current annual spending on Medicaid (DARA, 2012). What this means is that the cost of inaction now is far greater than spending the necessary funding to put these aggressive plans in place (Shank and Lichtash, 2013).

THE INTERNATIONAL POLITICAL ARENA

In July 2015, the United States and Brazil made a joint commitment to use more renewable resources to produce

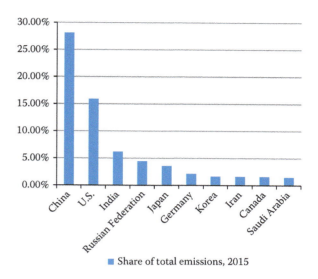

FIGURE 6.5 The world's top 10 emitters of CO_2 in 2015.

electricity. Separately, China formalized details of its goal to reduce carbon emissions. These three countries are in the top ten carbon emitting countries in the world, with China and the United States the top two emitters, respectively. These two emitters accounted for 36 percent of global carbon emissions from 1990 through 2011, making their latest announcement welcome news throughout the climate change community (Murphy, 2015) (Figure 6.5).

Brazil and the United States both pledged to produce 20 percent of their electric power using renewable resources by 2030. In addition, Brazil pledged to restore 120,000 square kilometers of rainforests over the same period of time. They issued a joint statement detailing how they will cooperate to achieve the goals through a partnership and utilize other international resources. One of the primary goals is the halt of illegal deforestation. China also submitted its climate change action plan to the United Nations. China's President Xi Jinping presented a plan to reduce carbon emissions based on a reduction of the economy's 2005 carbon intensity by 60–65 percent by 2030. While not as aggressive as what some activists were hoping for, their plan has been met with cautious optimism and the hope that they will intensify it once they begin moving in the right direction. Their current plan would not keep them below the 2 degree limit, but would result in an improvement in carbon intensity of 70 percent.

PROGRESS IN KEY COUNTRIES

Let us take a look at the world's largest emitters and what they plan on doing as of 2015 (Sturmer et al., 2015). Percent reductions are based on 2005 levels, unless otherwise noted.

Summary of International Climate Plans

China
Rank: #1
Goals:
- It says its emissions will peak by 2030 at the latest.
- It has vowed to reduce the carbon intensity of its economy 60–65 percent by 2030.

Status:
- While their target is insufficient, current policies would actually result in an improvement in carbon intensity of 70 percent and an earlier peaking of China's emissions.
- This, however, will still not be enough to limit warming to below 2 degrees.
- Stricter targets need to be implemented.

United States
Rank: #2
Goals:
- Committed to a 26–28 percent reduction by 2025.
- Power plant owners have been ordered to cut CO_2 emissions 32 percent by 2030.
- Long-term, it has committed to an 83 percent reduction by 2050.

Status:
- While it is currently in line to meet its 2020 target, greater cuts will be needed to reach the 2025 pledge.
- It says the U.S. decision to cut emissions by tightening regulations is reasonable, and that could put it on track to reach the 2025 pledge.

India
Rank #3
Goals:
- Has not submitted a post-2020 target yet
- Has pledge to reduce its GDP emissions intensity 20–25 percent by 2020.

Status:
- Their current pledge is not strong enough to limit warming to less than 2 degrees unless other countries make deeper reductions.

Russia
Rank #4
Goals:
- Committed to a 25–30 percent reduction by 2030 (based on 1990 levels)

Status:
- Their current pledge is inadequate as it is heavily based on forestry emissions and does not provide information on the accounting rules it used to come to its target.

Indonesia
Rank #5
Goals:
- Pledged to reduce emissions so that by 2020 they are 26 percent below 2020 forecasts.
- If sufficient international support is in place, they have pledged to cut by 41 percent.

Status:
- Current policies will not meet this pledge, but they are in line to decrease emissions about 13 percent by 2020.

Brazil
Rank #6
Goals:
- Pledge that by 2020 its emissions will be 36.1 percent to 38.9 percent lower than 2020 emissions forecasts.

Status:
- Brazil was one of the first major developing countries to set an emissions target and is on track to meet the target with current policies.

Japan
Rank #7
Goals:
- It has committed to a 26 percent reduction by 2030 (based on 2013 levels).

Status:
- Their 2030 target is currently inadequate to keep climate change below 3–4 degrees in the twenty-first century. In fact, it can be achieved almost without Japan taking any further action.

Canada
Rank #8
Goals:
- Committed to cut emissions 30 percent by 2030.
- Will tighten regulations around electricity generation and transportation, land use change, and purchase carbon credits internationally.

Status:
- They will miss their target by a wide margin.
- They are relying too much on land-use change, which is too hard to measure.

Germany
Rank #9
Goals:
- Committed to a 55 percent reduction by 2030 (based on 1990 levels).
- Germany is part of the European Union (EU) and the EU is not on track to meet this goal.

Status:
- The EU is currently on track to reduce emissions by 25–35 percent of 1990 levels.
- More needs to be done to meet the 2030 goals.

Mexico
Rank #10
Goals:
- It has set a goal to reduce GHGs and short-lived climate pollutants emissions 25 percent by 2030, based on forecasts for the nation's 2030 emissions.
- They expect emissions intensity per unit of GDP to reduce by around 40 percent from 2013 to 2030.

Status:
- Mexico's efforts are not consistent with keeping warming below 2 degrees, but has committed to making deeper cuts if other countries follow suit.

Iran
Rank #11
Goals:
- Iran has some of the world's largest gas reserves and is a major crude oil exporter.
- Rising domestic demand has created a gas supply and vehicle pollution crisis in some of their cities.

Status:
- Iran plans to set up a carbon trading market to reduce industrial emissions of climate-warming gas.
- Due to heavy subsidization of fossil fuels, there is little incentive for private investments in wind or solar power projects.

South Korea
Rank #12
Goals:
- It has committed to a 37 percent reduction by 2030 compared to 2030 forecasts.

Status:
- They have been criticized, accused that it is the equivalent of limiting emissions at 536 Mt but that emissions need to be at least below 500 $MtCO_2$ per year to even place it within the medium risk category.

(Continued)

Australia

Rank #13

Goals:

- The highest per capita emitter among the top 15 countries, it has pledged to cut carbon emissions by 26–28 percent by 2030.

Status:

- To keep to the 2C commitment, it needs to reduce emissions by 40–60 percent by 2030 (based on emissions from 2000). That equates to an emission reduction of 45–65 percent.

United Kingdom

Rank #14

Goals:

- It has committed to a 50 percent reduction by 2025 based on 1990 levels.
- It has committed to cut emissions at least 80 percent by 2050 based on 1990 levels.

Status:

- It is currently on track to reduce emissions by 25–35 percent of 1990 levels.
- More needs to be done to ensure the 2030 goals are met.

Saudi Arabia

Rank #15

Goals:

- It has not set emissions reduction targets for either 2020 or 2030.

Status:

- It is predicted their GHG emissions will increase 30 percent by 2020 compared with 2010 levels and 60 percent by 2030 if current policies are not changed.

UNITED NATIONS CLIMATE TALKS, PARIS, 2015

In December 2015, representatives of 195 countries came together in Paris and approved a plan to drive down global emissions and avoid the most disruptive and catastrophic consequences of climate change. Called the Paris Agreement—it provided a framework to gear up ambition over time: creating a transparent system for reporting and review; establishing regular assessments of progress; and strengthening commitments every five years, beginning no later than 2020. And a major milestone: nearly every country on earth, covering nearly 99 percent of global emissions, has submitted commitments to finally take action on the climate.

Signed by 196 nations, the Paris Agreement is the first comprehensive global treaty to combat climate change. It will carry on from the Kyoto Protocol when it ends in 2020, entering into force once it is ratified by at least 55 countries, covering at least 55 percent of global GHG emissions. The key points include the following:

- *The warming level*: The agreement commits nations to keep temperatures "**well below 2°C** above preindustrial levels and to **pursue efforts to limit the temperature increase to 1.5°C.**"
- *The carbon budget*: In order to keep warming below 2°C, we can emit a total of approximately 3.6 trillion tons of GHGs. But this only gives us a 66 percent change. For a better chance, or for a lower warming

limit, we will have to emit much less. The Paris Agreement calls for **global emissions to peak "as soon as possible"** and for a balance to be achieved between the rate of GHG emissions and the removal of these gases from the atmosphere by sometime between 2050 and 2100 (Figure 6.6).

- *The coverage of country pledges*: At the Conference, 185 countries submitted pledges ahead of Paris (Figure 6.7). They cover:
  - 94 percent of global emissions
  - 97 percent of global population
- *The pledges and the 2°C target*:
  - The targets are not yet enough to limit warming to below 2°C (Figure 6.8).

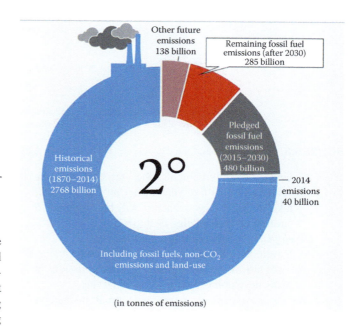

FIGURE 6.6 The new carbon budget. (From Climate Action Tracker, marcia.rocha@climateanalytics.org; t.kuramochi@newclimate.org; earth-syst-sci-data.net.)

FIGURE 6.7 Country pledges, Paris talks. (From Climate Action Tracker, climateactiontracker.org/indcs.html.)

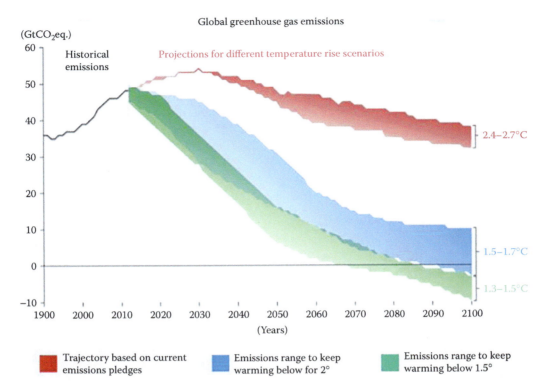

FIGURE 6.8 The pledges and the 2°C target. (From Climate Action Tracker, climateactiontracker.org/global.html.)

- Under the agreement, the targets will be reviewed every five years, after an initial stocktake in 2023. The countries will be expected to strengthen their pledge each time.
- *Developed versus developing countries*:
  - Developed countries are responsible for the majority of historical GHG emissions, while emissions in many developing nations are still growing.
- The Paris Agreement makes no formal distinction between developed and developing countries' responsibility to cut emissions.
- Wealthy countries are required to "continue taking the lead by undertaking economy-wide absolute emission reduction targets."
  - Developing countries are "encouraged" to do this "over time.'
- *Finance issues*:
  - Developing countries will need financial assistance to help reduce their emissions.
  - Under the agreement, developed countries will provide at least US$100 billion a year in climate finance from 2020.
- *Legality*:
  - The agreement is legally binding, but some specific details, such as the amount of climate finance is involved, are not.
- *Transparency and accountability*:
  - Developed countries will have to disclose their GHG emissions, progress on targets, climate adaptation, and finance at least every two years. Other countries can do so voluntarily.

The Paris Agreement, signed on Earth Day 2016 in New York, is historic for many reasons. It signifies that the world as a whole agreed on a solid path forward in a unified fight against climate change. After years of lukewarm international commitments, this is indeed a significant turning point. It signals a proactive turning point in the path to a genuine low-carbon economy, a future open to continued innovation welcome to technology, energy, creative finance, and true conservation. As of August 23, 2016, there were 180 signatories to the Paris Agreement.

It is also important to note that this did not just happen overnight. This was years in the making. When negotiations to address climate change failed in 2009, many countries walked away with a better sense of how to make an agreement work. Rather than settle for failure, they used past failures as a guide to create a "from-the-bottom-up approach," where each country set its own goals. This is considered one of the major factors that enabled the Paris agreement to finally work for everyone involved; it ensured change because everyone was able to be fully invested in what they felt would work for them (Nature Conservancy, 2016).

These agreements are also historic because they ask any nation signing to reduce GHG emissions and to regularly increase their ambitions. The agreement requires that ratifying nations "peak" their GHG emissions as soon as possible and then pursue the highest possible ambition that each country can achieve. This forces countries to constantly strive to improve and become more energy efficient as technology advances. Also significant with the Agreement is that countries have agreed to strive to keep warming *well below* the critically-identified tipping-point marker of 2°C. This is the first time in the history of climate change negotiations that any effort has been pursued to limit temperature increases to a newly-set goal of 1.5°C.

The Paris Agreement is uniquely written to have some aspects that are binding and others that are not. The binding aspects include the requirements to report on progress toward lowering emissions. Nonbinding elements include the setting of emission-reduction targets. This is a critical aspect that has not been introduced until now. Previous attempts at an agreement required similar measures to be adopted by all signing parties. This proved nonworkable, however, because economies, cultures, and nations all differ so greatly that a common denominator was hard to construct and make feasibly work. It set the entire end goal up for failure because of it. By allowing ratifying countries to determine the best way forward for them, individually, is what garnered immense support for this Agreement.

These Agreements also looked to the natural environment for assistance in mediating climate change—in particular, forests. Duncan Marsh of the Nature Conservancy sums up: "The Agreement affirms the important role that ecosystems, biodiversity, and land use can play in reducing greenhouse gas emissions and helping communities and countries reduce risks and adapt to climate change impacts. It also promotes sustainable management of land, which can range from conserving and restoring forests to improving agriculture." (Nature Conservancy, 2016)

This sentiment reflects the importance in investing in nature. It is well known among the scientific community that healthy ecosystems can protect people from the effects of climate change. This will become increasingly important as the effects of a changing climate increase over the coming years.

These agreements seem to be better suited to involve countries in a more personally-tailored way, giving them a say in how to meet their personal goals. It gives them a vested interest, bringing them on board, enabling all participating countries to share globally in a win-win situation.

CONCLUSION

Addressing the challenge of climate change will ultimately require a comprehensive set of approaches, including support for the development of noncarbon energy sources, the use of market mechanisms to put a price on GHG emissions, tax incentives, efficiency standards to promote the use of efficient products and technologies, and limits on GHG emissions.

The success of the effort—national and international—largely hinges on domestic action by the country itself—both its politicians and their constituency. This means that it is the public which is first responsible to elect those willing to take the forefront and act as true leaders, willing to make the right choices for the current population and that of the future. The perpetual reign of inadequacy and nonaction will no longer suffice. It is also important to note that strong leadership can be a good example and an impetus for other countries to follow. Like a domino effect, this can create a ripple

effect, moving the world in the right direction for future generations. It is also time for the United States to take a role of leadership—in technology, commitment, and management. It has taken a backseat with no significant advancement toward a meaningful solution for too long. Today, however, it appears to be in a better position to take on the types of binding commitments needed to ensure a sustained and effective global effort. It is now time to step up to the plate and take part in perhaps one of the most serious global efforts of all time—an effort that reaches toward future generations.

It is also time for rapidly developing nations, such as China and India, to realize they are walking into an industrialized world much different than the one the original industrialized world began to create two centuries ago. This is an era of much greater understanding of the environment and the detrimental effects humans can have on it, and because of it, there has been a collective enlightening—an understanding of what nonrenewability means. With that newfound knowledge of human impact on the environment and the detrimental consequences comes a greater responsibility to protect it, and hence, a new requirement. There can no longer be a "business as usual" approach. Each nation must carry the weight of their due responsibility or not become or remain developed. It is a changing world and the key is to change it for the better, for future generations. There is no going back when the future is at stake.

REFERENCES

AAF (American Action Forum). August 18, 2015. What will keep the lights on? Inside the clean plan's safety valve. https://www.americanactionforum.org/insight/what-will-keep-the-lights-on-inside-the-clean-power-plans-safety-valve/ (accessed August 19, 2016).

Bagley, K. April 17, 2015a. Republican senators lag behind voters on climate change. *InsideClimate News.* https://insideclimatenews.org (accessed August 25, 2016).

Bagley, K. May 11, 2015b. Climate denial takes a toll on scientists—and science. *Inside Climate News.* https://insideclimatenews.org/print/39558 (accessed August 25, 2016).

Bagley, K. May 23, 2015c. U.S. is laggard among developed nations in understanding climate change. *Inside Climate News.* https://insideclimatenews.org (accessed August 25, 2016).

Bagley, K. June 4, 2015d. Global warming's Great Hiatus gets another debunking. *Inside Climate News.* https://insideclimatenews.org/news/04062015/global-warming-great-hiatus-gets-debunked-NOAA-study (accessed August 25, 2016).

Berwyn, B. May 31, 2016. Why is Antarctica's sea ice growing while the Arctic melts? Scientists have an answer. *Inside Climate News.* https://insideclimatenews.org/print/43198 (accessed August 26, 2016).

Congressional Record Volume. March 11, 2014. Climate change: Congressional record online through the Government Publishing Office. https://www.gpo.gov/fdsys/pkg/CREC-2014-03-11/html/CREC-2014-03-11-pt1-PgS1507.htm (accessed August 25, 2016).

Cushman, J. H. April 6, 2015. Climate change beliefs, parsed on a local level, paint a valuable portrait. *Inside Climate News.* https://insideclimatenews.org (accessed August 25, 2016).

DARA. 2012. Climate vulnerability monitor. http://daraint.org/climate-vulnerability-monitor/climate-vulnerability-monitor-2012/report/ (accessed August 28, 2016).

Department of Transportation, U.S. April 2010. Transportation's role in reducing U.S. greenhouse gas emissions: Volume 1 synthesis report. Report to congress. http://ntl.bts.gov/lib/32000/32700/32779/DOT_Climate_Change_Report_-_April_2010_-_Volume_1_and_2.pdf (accessed August 28, 2016).

EESI. August 5, 2015. Fact sheet: Timeline of progress made in President Obama's climate action plan. http://www.eesi.org/papers/view/fact-sheet-timeline-progress-of-president-obama-climate-action-plan (accessed August 20, 2016).

Environmental Protection Agency. 2009. Proposed Rulemaking to Establish 2012-2016 Light-Duty Vehicle CAFÉ and GHG Standards. https://www.epa.gov/sites/production/files/2015-01/documents/10062009mstrs_charmley.pdf (accessed May 17, 2017).

EPA. February 11, 2016a. Supreme court stays clean power plan. https://www.epa.gov/cleanpowerplan/regulatory-actions (accessed August 19, 2016).

EPA. February 23, 2016b. Regulations and standards: Light-duty. https://www.usnews.com/opinion/blogs/world-report/2013/10/15/five-steps-america-must-take-now-to-combat-global-warming (accessed May 19, 2017).

Greenhouse, L. April 3, 2007. Justices say EPA has power to act on harmful gases. *The New York Times*. http://www.nytimes.com/2007/04/03/washington/03scotus.html (accessed May 17, 2017).

Howe, P., Mildenberger, M., Marlon, J. and Leiserowitz, A., 2015. Geographic variation in climate change public opinion in the U.S. *Nature Climate Change*. doi:10.1038/NCLIMATE2583.

Karl, T. R., A. Arguez, B. Huang, J. H. Lawrimore, J. R. McMahon, M. J. Menne, T. C. Peterson, R. S. Vose, and H.-M. Zhang. 2015. The recent global surface warming hiatus: Fact or artifact of data biases. *Science*.

Morgan, J. and K. Kennedy. June 25, 2013. First take: Looking at President Obama's climate action plan. World Resources Institute. http://www.wri.org/blog/2013/06/first-take-looking-president-obama%E2%80%99s-climate-action-plan (accessed August 21, 2016).

Murphy, T. July 1, 2015. World's biggest carbon emitters make major climate commitments. *Humanosphere*. http://www.humanosphere.org/environment/2015/07/worlds-biggest-carbon-emitters-make-major-climate-commitments/ (accessed August 28, 2016).

Nature Conservancy. 2016. Climate change: The Paris agreement. http://www.nature.org/ourinitiatives/urgentissues/global-warming-climate-change/the-paris-agreement-what-does-it-mean.xml?src=sea.AWG.prpari.crv1&gclid=CjwKEAj wuo--BRDDws3×65LL7h8SJABEDuFRNUiv8T6AThSL C3KYheW6sX-AK_jLiTmzmoJahM4N1RoCpNXw_wcB (accessed August 28, 2016).

Shank, M. and M. Lichtash. October 15, 2013. Five steps America must take now to combat climate change. *U.S. News*. https://www.usnews.com/opinion/blogs/world-report/2013/10/15/five-steps-america-must-take-now-to-combat-global-warming (accessed May 19, 2017).

Sturmer, J., T. Leslie, M. Liddy, and C. Gourlay. August 10, 2015. What the world's 15 biggest emitters are promising on climate change. http://www.abc.net.au/news/2015-08-11/climate-change-what-top-15-emitters-are-promising/6686548 (accessed August 28, 2016).

Whitehouse, Climate Change. June 21, 2016. Climate change: The white house. https://www.whitehouse.gov/energy/climate-change (accessed August 19, 2016).

UCS (Union of Concerned Scientists). 2016. The clean power plan: A climate game changer. http://www.ucsusa.org/our-work/global-warming/reduce-emissions/what-is-the-clean-power-plan#.V7bwbZgrKUk (accessed August 26, 2016).

SUGGESTED READING

BBC. March 10, 2016. Scotland's climate change progress 'exemplary'. http://www.bbc.com/news/uk-scotland-scotland-politics-35772636 (accessed August 30, 2016).

Editorial Board. January 2, 2016. 2015: A year of progress and buffoonery on climate change. *The Washington Post*. https://www.washingtonpost.com/opinions/2015-a-year-of-progress-and-buffoonery-on-climate-change/2016/01/02/9ad69 55c-af33-11e5-9ab0-884d1cc4b33e_story.html?utm_term=.eb7c65198350 (accessed August 31, 2016).

Kamarck, E. December 3, 2015. The real enemy to progress on climate change is public indifference. *Brookings*. https://www.brookings.edu/blog/fixgov/2015/12/03/the-real-enemy-to-progress-on-climate-change-is-public-indifference/ (accessed August 30, 2016).

Murphy, T. July 1, 2015. World's biggest carbon emitters make major climate commitments. *Humanosphere*. http://www.humanosphere.org/environment/2015/07/worlds-biggest-carbon-emitters-make-major-climate-commitments/ (accessed August 28, 2016).

Steyer, T. August 28, 2016. Can progress on climate change keep UP? *The Washington Post*. http://www.realclearpolitics.com/2016/08/28/can_progress_on_climate_change_keep_up_390168.html (accessed August 28, 2016).

7 Public Perception and Economic Issues

OVERVIEW

In order to understand the issue of climate change and be able to work with others in finding a solution to the problem, it is important to know where others stand on the issue. Understanding others' perceptions and opinions is important in order to have clear, concise, meaningful communication. Without that productive negotiations are not possible. This chapter begins by looking at worldwide public perception of the issue and what people think in general, from an ideological standpoint, demographically, how they view responsibility and mitigation, and where they place climate change amid other world threats in terms of importance.

Furthermore, one of the most significant things that we must not fail to address is the economic issues associated with a changing climate and the decisions we make concerning it. Because of this, we will look at how climate change relates to economic considerations overall, and then delve into specific sectors of the economy to see how they will be affected as temperatures begin to change, rainfall and wind patterns shift, and seasonal timing becomes altered. We will explore the impacts of drought and desertification on the economy, what can be expected in the future in the fishing and forestry industries, what it means for agriculture and food production, and how significant the impacts to the recreation and tourism industry are expected to be. We will also look at how climate change is expected to affect the healthcare system. We will conclude by addressing how we, as a world community, can meet the challenges of these climate change impacts.

INTRODUCTION

Public perception is powerful. As we saw in the last chapter, the media can shape public opinion in one direction or another, which is why when the issue is something as important as climate change, it is imperative that it be represented factually. Public opinion is also critical when laws are being made. If the public has been educated with the facts and has been able to draw accurate conclusions, they can be proactive toward their government representatives and lawmakers and have an influence on them to put into law those acts which can help lower emissions and help mediate the negative effects of climate change in other ways. Rather than hear that a society is either in favor of something, or not, it is often more helpful to look closer at different breakdowns of the population—age, income, gender, political group, and so forth—to get a better idea of group ideology and why they feel as they do. This leads to a better understanding between groups and helps build public support.

Next, we will look at the economic issues associated with climate change. The direct costs of not taking on the challenges imposed by climate change are often neglected or ignored—and unfortunately, in the past, have not been calculated into a long-term solution. This makes it much more expensive in the long run when these costs finally have to be addressed. Impacts will vary globally and across different sectors of the economy and are often linked with each other, making the chain reactions and affects difficult to predict. The indirect effects of climate change are considered less frequently in negotiations or solutions, yet they are generally substantial in the long run. In reality, the true economic impact of climate change is full of "hidden" costs. It is not just the replacement value of infrastructure, as it often appears at first glance—there are also costs of less obvious components, such as workdays and productivity lost, provisions of temporary shelter and supplies, potential relocation and retraining costs, transportation costs, and other issues specific to the situation. Because of all these interrelationships, it is imperative to gain a working knowledge of the economic interactions of the various components involved in global climate change. Through an understanding of these complicated interactions, it is then possible to plan for the future in the most efficient way possible.

Because of this, we will take a look at some of the economic considerations that are often overlooked, yet affected by climate change, such as the economic impacts of drought, desertification, the fishing and forestry industries, impacts to the recreation and tourism industry, and healthcare and labor-related impacts. We will conclude by addressing how to meet the diverse challenges of climate impacts.

DEVELOPMENT

In the last chapter we got a feel for the geopolitical aspect of climate change and the relative difficulties and challenges various countries are facing not only in getting the general public on board with support of the issue, but also the politicians, especially—who can be unusually difficult for many reasons. Obtaining their support is critical because without their endorsement, the country is not able to pass laws to move in a productive direction. We have seen that public opinion—on average more than 50 percent—is in agreement with the climate change issue. They identify and accept that it is one of the current problems that needs to be proactively dealt with. Yet, the percentage of lawmakers that directly influence our forward progress also identify the issue as a real problem are much fewer in number than the general population. This chapter will take a closer look at how different populations worldwide view climate change.

WHAT THE PUBLIC THINKS ABOUT CLIMATE CHANGE

To begin, we will lay the foundation of who in which sectors of the population worldwide are on board with what aspects of climate change. We will look at climate change in general, how responsibility and mitigation should be handled, what different areas' ideological views are, and who should take responsibility. We will also look at how the world in general looks at the issue of world threats and where climate change is ranked. This is important because the larger the perceived threat to a country an issue is, the more attention and funding that issue generally receives.

Climate Change in General

In 2015, the Pew Research Center conducted some global surveys prior to the Paris Talks in December to get an idea of how different populations worldwide felt about the climate crisis. They were curious about a broad range of related topics, such as:

- How various country's populations view the concept of climate change?
- Whether they could already see the effects they had been hearing scientists warn about for so long.
- Which countries had a division of opinion based on political views?
- What other social or political issues influenced how they looked at climate change?
- Whether or not someone's concern over climate change promoted them to act or not.
- Who should have to clean up the environment—those who made the mess or everyone?
- Whether or not we should drastically change our lifestyles now or wait for technology to fix the problem.
- Are opinions consistent within each country or is the country divided?
- Is there an age-group stratification within countries concerning alternative energy?

Although a lot to consider, all these issues are important, because if we cannot come to a working agreement within each country, then finding a way to work internationally will be nearly impossible. So, let us take a look at what people from around the world think about the world's climate.

Many people see the effects of climate change right now, and there is a well-defined group of those who believe they will see its effects within the next few years. On a scale of present concern over the issue, many people are extremely worried. While often the focus is on the future—what is coming in the future, what we can expect to happen, and so forth—a good share of the world's population not only are identifying with what is happening right now as climate change (drought, unusual and violent weather occurrences, melting glaciers, unusual patterns of precipitation, wildfires, and so forth), but also 51 percent of people across the earth's nations believe that climate change is harming people around

the world *right now*. Approximately eight in ten report that global climate change currently impacts people or will do so in the next few years, compared with a median of only 13 percent who say that climate change will not impact people for many years or will never affect people. The following two charts sum up those results: Figure 7.1, Global Climate Change Concern Scale, and Figure 7.2, Immediacy of Climate Change Worries.

In Latin America, about three-quarters of the population reports that climate change is currently hurting them and the detrimental effects are affecting their daily lives. It is such a part of their lives already that Brazilian President Dilma Rousseff's recent pledge aimed at cutting the country's carbon emissions 37 percent by 2025. Deforestation is a constant reminder of the devastative effects of environmental degradation and climate change. In Europe, about 60 percent of the population believe that climate change is already negatively affecting people now. Surprisingly, Americans are among the most likely to believe that the effects of climate change are still a long way off. While 41 percent report that climate change is already harming people around the world, 29 percent still believe that it will not harm people for many years or may never harm people, which is the greatest mindset of this kind in any nation polled. Finally, the study stated with a median of 26 percent, Middle Easterners are the least likely to believe that climate change is currently harming people.

Of all the aspects concerning climate change, the one that seemed to concern most people was that of drought (or water shortage). When compared with severe weather occurrences, extreme heat, and rising sea levels, drought was the most pressing. A median of 44 percent said they were most concerned about this possibility. Of the remaining, 25 percent said severe weather, such as floods and intense storms, 14 percent cited unusually hot weather, and only 6 percent stated rising sea levels as their top concern. Latin America weighed in heavily with drought (59 percent overall), with Brazil at 78 percent. Water shortages seemed to be near the top of everyone's lists. Sub-Saharan Africans were in a similar boat—for them 59 percent stated drought as the scariest effect.

The largest portion of Americans who are most concerned with drought are those who reside within the Western and Midwestern portions of the United States. Much of the landscape in the Western United States is considered desert. These areas for the past few years have experienced several water shortages where water has had to be rationed in years following meager snowfall. Not only have residents seen these consequences, but also others that are closely related, such as poor range conditions for cattle, poor crop seasons for agriculture, and low reservoirs and rivers for drinking water. Lack of appropriate water resources for recreation, such as skiing, fishing, boating, and so forth have become common place. The severe lack of water also impacts wildlife habitat and dries out the natural vegetation, causing extremely high wildfire potential. When a wildfire does occur, it can be extremely devastating to personal life, wildlife, resources, and property.

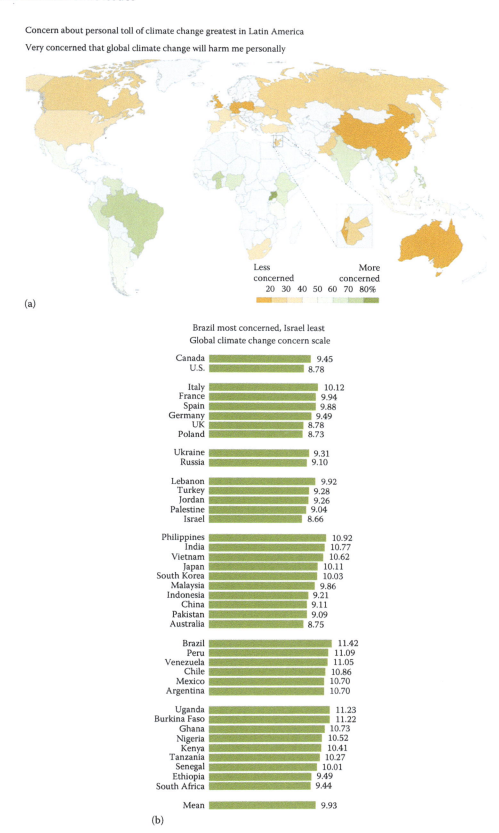

FIGURE 7.1 (a) A map showing globally which countries are most concerned about the issue of climate change. (Courtesy of Pew Research Center, Washington, DC.) (b) This shows the relative level of concern (between 0 and 10), the level of concern a country has of climate change having adverse affects. The score is a composite score based on three questions worth four points each. (Courtesy of Pew Research Center, Washington, DC.)

Immediacy of climate change worries Latin Americans, Europeans most
Global climate change is harming/will harm people around the world...

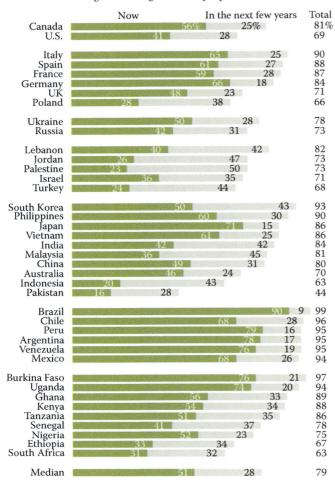

	Now	In the next few years	Total
Canada	56%	25%	81%
U.S.	41	28	69
Italy	65	25	90
Spain	61	27	88
France	59	28	87
Germany	66	18	84
UK	48	23	71
Poland	28	38	66
Ukraine	50	28	78
Russia	42	31	73
Lebanon	40	42	82
Jordan	26	47	73
Palestine	23	50	73
Israel	36	35	71
Turkey	24	44	68
South Korea	50	43	93
Philippines	60	30	90
Japan	71	15	86
Vietnam	61	25	86
India	42	42	84
Malaysia	36	45	81
China	49	31	80
Australia	46	24	70
Indonesia	20	43	63
Pakistan	16	28	44
Brazil	90	9	99
Chile	68	28	96
Peru	79	16	95
Argentina	78	17	95
Venezuela	76	19	95
Mexico	68	26	94
Burkina Faso	76	21	97
Uganda	74	20	94
Ghana	56	33	89
Kenya	54	34	88
Tanzania	51	35	86
Senegal	41	37	78
Nigeria	52	23	75
Ethiopia	33	34	67
South Africa	31	32	63
Median	51	28	79

FIGURE 7.2 Immediacy of climate change worries: This illustrates the percentage of populations that believe (a) global climate change is harming and (b) will harm people around the world, for several countries. Latin Americans and Europeans come in at the top of the scale. Data for "Not for many years," "Never" and volunteered category "climate change does not exist" not shown. (Courtesy of Pew Research Center, Washington, DC.)

Figures 7.3 and 7.4 detail information about climate change and public perception and drought.

Ideological Views

The political spectrum in several countries has caused a significant divide in the view of climate change. In the United States, there is a distinct divide along partisan lines. Democrats are much more likely than Republicans to see climate change as a very serious problem, believe its effects are currently being felt, agree that those effects will definitely harm them personally, and be very supportive of the United States participating in international agreements designed to limit the emission of greenhouse gases (GHGs). The Republicans, on the other hand, do not see climate change in this manner; many even deny its existence and do not see the need to be entering into any type of agreements to deal with the issue. Many are also

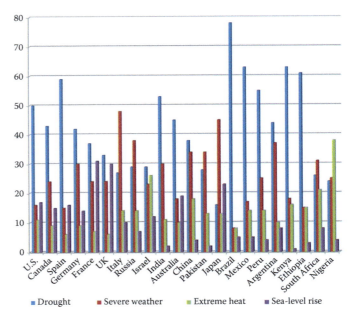

FIGURE 7.3 Drought is the top global climate concern. (Created by author.)

West and midwest most concerned
about drought

Americans' top concern about the effects of global
climate change is droughts or water shortages

FIGURE 7.4 The Western and Midwest sections of the United States are most concern about drought. Nationally, 50% say drought is their biggest concern. (Courtesy of Pew Research Center, Washington, DC.)

very supportive of the fossil fuel industry. They feel strongly that the effects of climate change are not harming people now and they are not concerned that any negative effects will personally harm them (Figure 7.5).

The United States is not alone in this type of stifling dichotomy. Several other countries are experiencing the same phenomena when it comes to the issue of having to make major changes in the way they live. For example, in Canada, there is a three-way partisan split with 86 percent of the New Democratic Party (NDP) agreeing that lifestyles will have to adjust to fit the effects of global warming, 75 percent of the liberals agree, yet only 57 percent of the conservatives

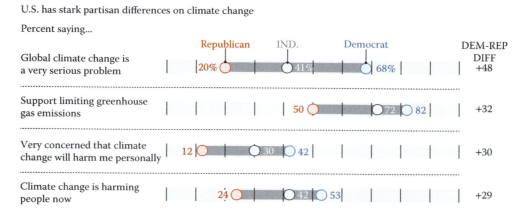

FIGURE 7.5   The partisan differences on climate change in the U.S. (Courtesy of Pew Research Center, Washington, DC.)

see it that way. In Germany, 89 percent of the Greens see a lifestyle change in order, yet only 68 percent of the Christian Democratic Union of Germany (CDU) and Christian Social Union in Bavaria (CSU) [referred to as the CDU/CSU]. The United Kingdom and Australia are similarly divided on the issue. Figure 7.6 shows the range of opinions in the United States, Germany, Canada, the United Kingdom, and Australia. Similar spreads also exist in Italy, France, Spain, and Poland.

Climate Change Demographics

Also interesting is the demographics within countries of which groups of people agree and disagree with the existence and effects of climate change. Several countries were polled about (1) whether they believed that climate change was a serious problem and (2) whether they supported their country limiting GHG emissions as a part of an international agreement. Some very telling results emerged, as seen in the illustration (Figure 7.6). Getting the public to support the active limiting of GHG emissions is a huge step forward because if signifies a possible change in lifestyle, such as driving less or becoming more conservative. In the case with this survey, climate action was consistently more significant than climate concern. Globally, a median of 54 percent in the survey considered climate change to be a *very* serious problem (a median of 85 percent said it is at least somewhat serious). But a much higher median (78 percent) support their country signing an international agreement limiting GHG emissions from the burning of coal, natural gas, and petroleum.

One of the most interesting responses was China—the nation responsible for the greatest annual release of CO_2 into the atmosphere. About 71 percent of the Chinese people support an international treaty to curtail emissions, but only 18 percent of the public surveyed expressed intense concern about climate conditions. This is a wide spread—a 53-percentage point differential. What this suggests is that the Chinese government has general public support for its recent initiatives to deal with global warming even though the Chinese people are not overly concerned about global warming.

As illustrated in Figure 7.7, the public's willingness to support limitations on emissions far exceeds the intensity of people's climate concern in a number of other major carbon-emitting countries. What the Pew Research Institute has noticed are the countries that have a low perception of climate change, but are extremely supportive of limiting GHG emissions (Stokes et al., 2015) (Figure 7.7).

People who reside in countries with high per-capita levels of carbon emissions are less concerned about climate change, as shown in the previous world map. Some of the most concerned countries are those which have very low emissions per capita, such as Africa, Latin America, and Asia. To these people, climate change does not seem to be a distant threat; they are seeing its effects today, in such ways as extreme drought in some locations. Latin America tops the list. In Brazil alone, 90 percent say that climate change is harming people right now, as compared to America with only 41 percent sharing the same belief. The same beliefs hold true for the belief of personal toll. Those in Latin America, as well as Sub-Saharan Africa report that they feel they are particularly vulnerable to the effects of climate change personally in their lives right now. In addition, three-quarters of Filipinos say they are very concerned that global climate change will harm them personally over the course of their lives. When you look at drought, water shortages, spread of disease, and sea-level rise, it makes sense. These events are occurring right now in those parts of the world. People in nations likely to be less vulnerable to the effects of climate change, according to the UN's World Risk Index, generally reflect less concern.

What is unfortunate is that there has also been a trend of significant declines of concern from several key countries. In China, for example, which is now the largest emitter of GHGs, the number of people saying climate change is a very serious problem has decreased by 23 percentage points since 2010. South Korea has decreased by 20 percent, Japan by 13 percent, and Russia by 10 percent. The study by Pew Research discovered that there is a notable trend that the high CO_2 emitters are less intensely concerned about the climate change issue (Stokes et al., 2015) (Figures 7.8 and 7.9).

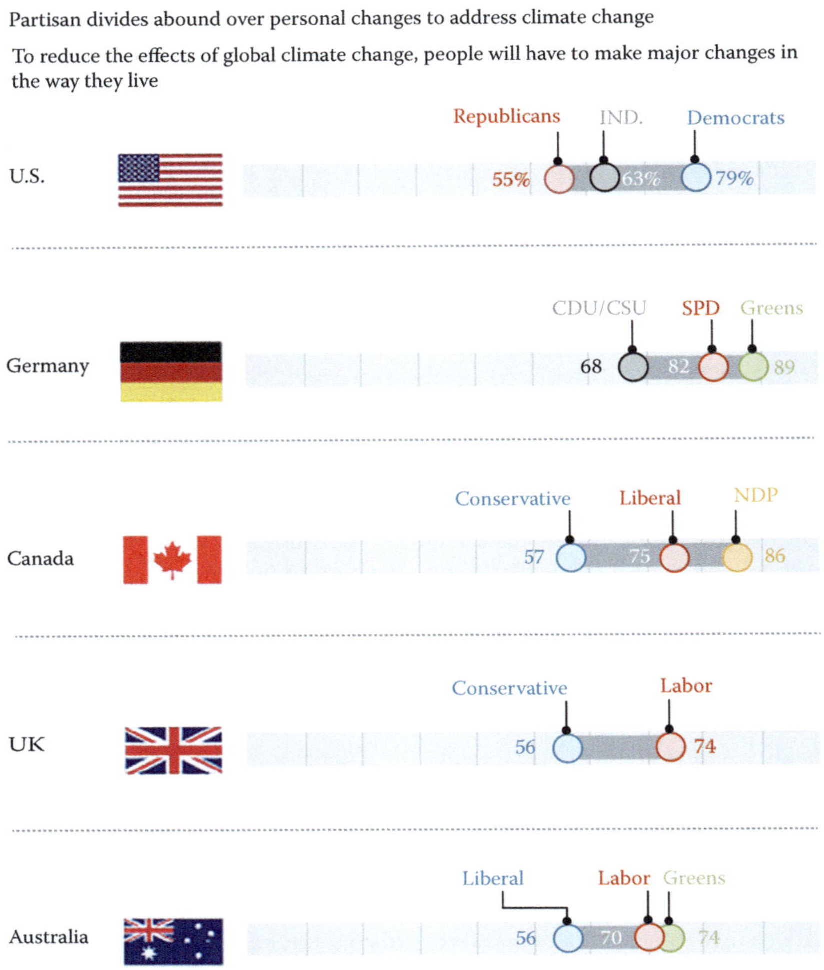

Partisan divides abound over personal changes to address climate change

To reduce the effects of global climate change, people will have to make major changes in the way they live

FIGURE 7.6   Several countries worldwide have partisan differences over climate change issues. Dot color is color used by each party; it does not necessarily indicate ideological similarity. (Courtesy of Pew Research Center, Washington, DC.)

In the United States, the demographic breakdown for those groups that voice greater concern for climate change and its effects include those with a lower income, women, and younger people. About half of American women say global climate change is a very serious problem, compared with 39 percent of their male counterparts. Women are also more likely than men to believe that climate change is harming people now and that its effects will personally harm them at some point in their lives.

Climate Change Responsibility and Mitigation

Many people are of the opinion that the wealthy countries should take more of the responsibility in addressing climate change. According to the study looked at here, 54 percent of the countries worldwide that were surveyed felt that the wealthier countries, such as the United States, Japan, and Germany should do more than the developing countries to address climate change because they have produced most of the world's GHGs so far. They share the mindset that because they have contributed most to the existing problem, they have a larger responsibility to mitigate it. In contrast, only 38 percent say that developing countries should do just as much, because

they will produce most of the world's emissions in the future. In addition, dealing with climate change may require a combination of technological innovation and changes in people's lifestyles. The majority of those surveyed worldwide believe that such lifestyle changes will be more important than technological change in combating climate change. There is also a generational trend in opinions over who should bear the greatest burden in cutting back GHGs. Of all the participants in the survey, it was the younger (ages 18–29) Americans, Japanese, Indonesians, and Australians that were more likely than their elders (ages 50 and older) to say that rich countries should do more than developing nations to address climate change. This is attributed to the fact that the younger generation may see itself as most likely to have to live with the consequences of climate change.

One of the most surprising findings was that overwhelmingly countries were in favor of governments limiting GHG emissions, as shown in the graph (Figure 7.10). This public support for taking action is noteworthy; it illustrates the fact that populations do want government involvement and support in managing and mitigating the effects of climate change. Significant for the United States, nearly 70 percent favor Washington agreeing to a

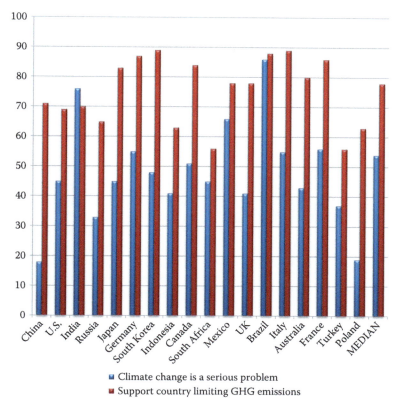

FIGURE 7.7 Climate change belief versus climate change support. (Created by author.)

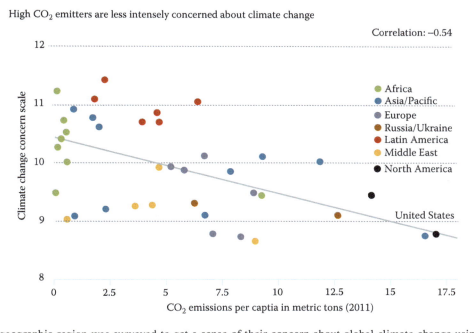

FIGURE 7.8 Each geographic region was surveyed to get a sense of their concern about global climate change using an index ranging from 3 to 12, with 12 representing being the most concerned about the impacts of climate change. Those surveyed were coded as a 4 if they felt (1) climate change was a very serious problem, (2) it was harming people now, and (3) they felt that climate change would harm them personally. The mean score for each country was used in the analysis. What the study illustrated was that the highest CO_2 emitters were traditionally the least concerned about the harmful effects caused by the emissions. (Courtesy of Pew Research Center, Washington, DC.)

FIGURE 7.9 Pedestrians outside in pollution-filled air in China. Some industrialized areas in China are so bad that people must wear masks to protect themselves from the unhealthy levels of pollutants in the air. (From http://www.purelivingchina.com/residential/blog/.)

multilateral commitment to limit the burning of pollutants such as coal, natural gas, and petroleum. This represents significantly more than the 45 percent of the American public who see climate change as a very serious problem. But this general support for taking action masks significant differences in the United States over signing an international climate accord. Eighty-five percent of Americans ages 18-29 back the idea, while only 60 percent of the age group of 50 and older do. Approximately 82 percent of Democrats favor limiting GHG emissions; but only half of the Republicans do.

CURRENT THREAT PERCEPTION

Various populations worldwide name climate change as the top worldwide threat, according to the Pew Research Center

(Carle, 2015). Other threats that ranked fairly high were ISIS (Europeans and Middle Easterners frequently ranked them first), economic instability (a large concern for many countries), Iran's nuclear program (Israelis and Americans are highly concerned), cyberattacks on governments, banks or corporations limited to a few nations, and territorial disputes between China and surrounding countries remain generally regional concerns. The perception of threat around the world are shown in the map (Figure 7.11). Of the 40 countries surveyed, 19 of them cited climate change as their biggest worry, making it the most widespread concern of any issue included in the survey. The secondary concern was the global economy, such as in Russia, Brazil, Venezuela, Ghana, Uganda, and Ukraine. What is significant to see in the work done here is that climate change consistently ranks high as a global concern today, placing it on the forefront for action and mitigation. Because of its concern, that assures it will be put in a priority position for attention and a workable solution. In the remainder of this chapter, we will discuss the various economic considerations brought about because of climate change.

CLIMATE CHANGE AND ECONOMIC CONSIDERATIONS

The economic effects of climate change touch virtually every sector of the economy and have far-reaching effects. It affects agriculture and food production, which affects domestic and global trade. It affects the fishing and forestry industries. The issues of sea level rise and intense weather events affect tourism, and the recreation, construction, and insurance industries. It also impacts transportation networks, has energy considerations, and relates to healthcare and labor issues. Due to its far-reaching impacts, it cannot be ignored. Instead, it must be dealt with now in order to plan for, mitigate, and

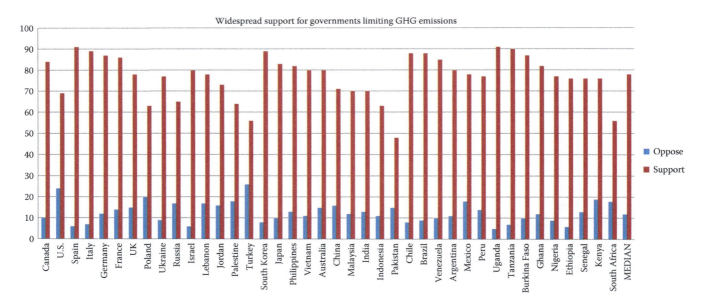

FIGURE 7.10 Of all the countries polled in the Pew Research study, there was an overwhelming consensus supporting governments taking an active role in limiting greenhouse gas emissions. (Created by author.)

Greatest threats around the world

Top concern

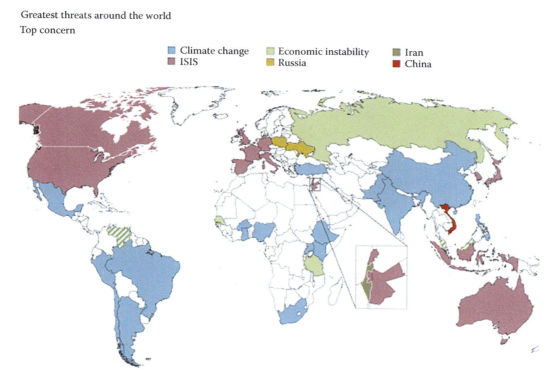

FIGURE 7.11 Major threats as seen by countries around the world. Of the 40 countries surveyed, 19 cited climate change as their biggest worry, making it the most widespread concern of any issue. In Malaysia and Venezuela, both climate change and economic instability are top concern. (Courtesy of Pew Research Center, Washington, DC.)

adapt as efficiently and wisely as possible in order to reduce negative impacts, conserve resources, and provide for future generations.

New research shows that, from an economic perspective, addressing climate change now is far more important than was previously thought. Even though scientists have done a fair amount of research on the impacts of rising temperatures on certain components of the economy, such as agricultural output and worker productivity, for example, the effects of temperature on overall economic productivity are still not very well understood. In a research study done by Burke (2015), at the University of California, Berkeley, determined that the global economy is linked to global climate change at a macro scale. He found that, unmitigated, climate change could reduce global gross domestic product (GDP) by more than 20 percent by 2100—an amount roughly 5–10 times larger than the current estimates.

The authors came to this conclusion by assembling historical data on economic output and climate from over 150 countries. They then used the data to examine how changes in economic growth were related to changes in temperature. They noticed a couple of patterns. First, the economies of countries that were currently colder than average (such as Canada or Norway) tend to grow faster when their own national temperatures warm up. The opposite appears true for countries that are currently hot (think of countries in Africa or South Asia)—they tend to grow much slower when temperatures rise. At a global scale, growth appears

to peak at an annual average temperature of 13°C (the same as the average annual temperature of New York City). They observed the same relationships in the historic data, independent of the country's financial status. They felt this had more to do with the concept of adaptation rather than with time or financial status. They also believe that we cannot assume that rich countries are less sensitive to changes in climate than poor countries, not that—despite a lot of development over the past half-century—we have gotten better at dealing with climate over time.

What the scientists were able to discover was that under a "business-as-usual" warming scenario, the world economy could be more than 20 percent smaller by 2100 than it would have been had temperatures remained fixed at today's levels. This does not mean that the world will be poorer in 2100; it means that the world will be substantially *less rich* than *it would have been* had temperatures not warmed (Figure 7.12).

What Burke discovered is that about 20 percent of countries around the world that are currently cooler than optimal could benefit from future temperature increase. But the 80 percent of countries that are either at the current global optimum or already past will be harmed as temperatures warm—and this includes both poor countries in the tropics, as well as many wealthy countries, such as Japan and the United States. What they determined from their findings is that the global economic benefit of emissions reductions could be much larger than previously assumed. They acknowledge that benefits must be weighed against the costs of undertaking emissions

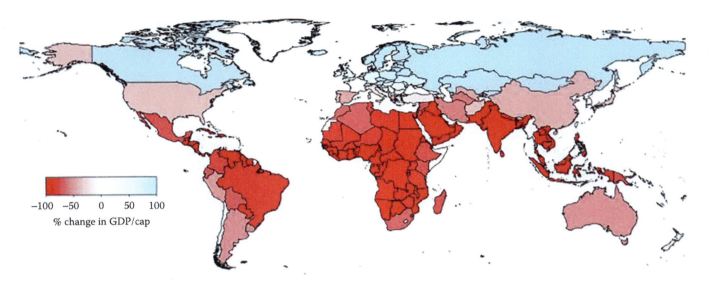

FIGURE 7.12 Potential impact to the world's future GDP under a "business-as-usual" approach to climate change. (From World Economic Forum, Cologny, Switzerland.)

reductions—such as the costs of switching to alternative energy sources, for example—but suggest that various mitigation options that were previously considered too costly now be put back into consideration.

They warn that if societies continue to function as they have recently, climate change will most likely reshape the global economy by substantially reducing global economic output and possibly even exaggerate the existing global economic inequalities, relative to a world without climate change.

ECONOMIC IMPACTS OF DROUGHT AND DESERTIFICATION

Desertification is a form of land degradation by which land becomes more arid. It is generally defined as "the process of fertile land transforming into desert typically as a result of deforestation, drought, or improper/inappropriate agriculture" (Princeton University Dictionary). This process usually results in the desertified land losing its vegetation, water bodies (lakes and streams), and wildlife. This, coupled with changing weather patterns due to climate change can spell disaster for certain geographical areas.

The principal cause of desertification is the removal of vegetation. This causes removal of nutrients from the soil, which makes the land infertile and unusable for arable farming. The two principal causes for this are:

- Human activity—cutting down trees to allow more grazing, or over-grazing of land by farmed animals
- Climate change—warming of air temperatures and decreases in precipitation can cause drought conditions and prevent the sustained growth of vegetation

With these processes combined, around 12 million hectares of productive land become barren every year due to desertification and drought alone (UNCCD, 2014). Desertification is a major problem worldwide, most commonly in dryland areas (including large areas in Africa, such as the Sahel). Approximately 40 percent of the earth's land surface is covered by drylands, and these areas are currently inhabited by more than 2 billion people. Drylands are highly vulnerable to natural and human destruction, due to the low water content of the soil (UNCCD, 2014). It is important to note, however, that land degradation can affect all regions—drought and desertification is not always synonymous with dryland areas. Roughly one-third of all global agricultural land is either highly or moderately degraded due to poor farming practices and is also vulnerable to degradation and desertification. The map shows which areas worldwide are *most vulnerable* to desertification (Figure 7.13).

The impacts of drought, desertification, and land degradation through climate, often severely experienced in land-locked developing countries (LLDCs), are felt by the overall economy mainly through key areas such as agriculture, water, health, and environment. The figure illustrates that the impacts of climate change, drought, and desertification—such as extreme weather events, severe droughts, reduced water availability, increased risk of diseases, and other effects on these key areas—ultimately lead to slow economic growth. They also lead to unsustainable development (Figure 7.14). Today, the combined effects of drought and desertification are taking a heavy toll on the economies of developing countries—especially the landlocked ones. The United Nations Convention to Combat Desertification (UNCCD) estimates that about 6 million hectares of productive land worldwide has been lost every year since 1990 due to land degradation. A larger proportion of the land loss occurs predominantly in areas classified as drylands—home to most LLDCs. It is estimated that the land loss translates to income losses of $42 billion per year (Makaudze, 2013).

Two-thirds of the arable land in Africa is expected to be lost by 2025, which represents an annual loss of more than 3 percent of agricultural GDP for the entire Sub-Saharan

Desertification vulnerability

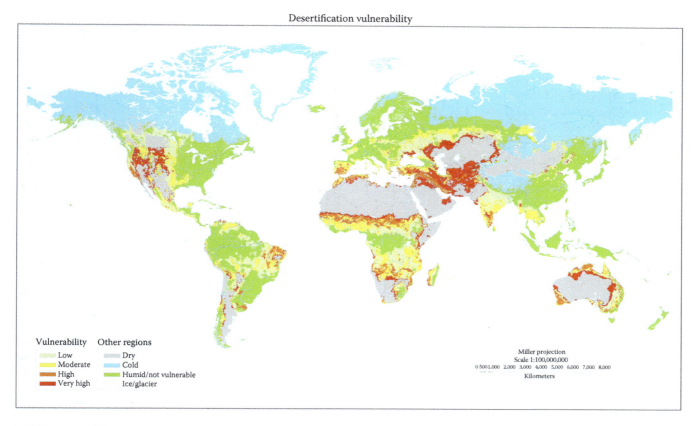

FIGURE 7.13 This map shows global desertification vulnerability. It is based on a reclassification of the global soil climate map and global soil map by the Natural Resource Conservation Service (soil type is closely linked to climate, and so varies across the globe in response to the local environmental conditions). (From U.S. Department of Agriculture [USDA], Naches, WA.)

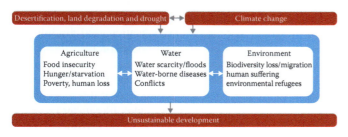

FIGURE 7.14 Impacts of drought and desertification to landlocked developing countries. (Created by author.)

Africa region. Countries that have already experienced heavy losses due to drought and desertification include Ethiopia, Uganda, and Zimbabwe. GDP loss from reduced agricultural productivity due to land degradation is estimated at $130 million per year. Uganda's land degradation in their drylands is causing a heavy toll on their economy; and in Zimbabwe, deforestation is rapidly transforming large areas into a desert (Makaudze, 2013).

The United States is also at risk. In a study published by NASA in February 2015, droughts in the Southwest and Central Plains during the last half of this century could be drier and longer than drought conditions seen in the same

regions in the last 1000 years. Based on several climate models, research pointed to continued increases in human-produced GHG emissions driving up the risk of severe droughts in these regions.

Ben Cook, climate scientist at NASA's Goddard Institute for Space Studies and the Lamont-Doherty Earth Observatory at Columbia University in New York City, and lead author of the study, commented, "Natural droughts like the 1930s Dust Bowl and the current drought in the Southwest have historically lasted maybe a decade or a little less. What these results are saying is we're going to get a drought similar to those events, but it is probably going to last at least 30 to 35 years." He added that the current likelihood of a megadrought—a drought lasting more than three decades—is 12 percent. If GHG emissions do not stop increasing in the mid-twenty-first century, Cook and his colleagues project the likelihood of megadrought to reach more than 60 percent (Cole and McCarthy, 2015).

Projections worldwide for the next few decades suggest that with continued climate change and the projected increase in global temperatures, coupled with reduced rainfall in many dryland regions, there may be considerable impacts for land cover and desertification. Over the coming decades, the average river run-off and water availability in some dry regions is projected to decrease by 10–30 percent. According to the United Nations Convention to Combat Desertification:

- By 2020, an estimated 60 million people could move from the desertified areas of Sub-Saharan Africa toward North Africa and Europe.
- By 2025, up to 2.4 billion people across the world may be living in areas subject to periods of intense water scarcity. This may displace as many as 700 million people.
- By 2050, 200 million people worldwide may be permanently displaced environmental migrants.

(*Source*: UNCCD, 2014)

Climate change and its impacts on desertification are large-scale issues which require regional and global-scale measures. It is important to keep in mind, however, that the climatic effects on land occur at ecosystem and landscape levels. Because of this, individual and community efforts to rehabilitate the land are crucial and must be part of the larger-scale plan to deal with the effects of climate change (Figure 7.13).

THE FISHING AND FORESTRY INDUSTRIES

Climate change is already impacting the world's oceans and will have serious consequences for the hundreds of millions of people who depend on fishing for their livelihoods. Impacts on fisheries that have already been observed include an increase in the frequency and intensity of extreme weather events, such as the El Niño phenomenon in the South Pacific; the warming of the world's oceans, with the Atlantic in particular showing signs of warming deep below the surface, and warmer-water species increasing toward the South and North Poles. There has also been an increase in salinity in near-surface waters in hotter regions, whereas the opposite is happening in colder areas due to more precipitation, melting ice, and other processes. The oceans are also becoming more acidic, which is harmful to coral reefs. The fishing communities in the high latitudes, as well as those that rely on the coral reef systems will be most exposed to the impact of climate change. Fisheries located in deltas, coral atolls, and ice-dominated coasts will be vulnerable to flooding and coastal erosion due to rises in sea level. There are approximately 260 million people employed in marine fisheries jobs worldwide (Teh and Sumaila, 2013). In addition, fish is also the world's most widely traded foodstuff and a key source of export earnings for many poorer countries. The sector has particular significance for small island states.

The distribution of marine fish and plankton are predominantly determined by climate, and true to expectations, marine species have been documented moving northward in U.S. waters and the timing of plankton blooms is also shifting. Extensive shifts in the ranges and distributions of both warmwater and coldwater species of fish have also been documented (Janetos et al., 2008). The polar areas are key

areas where climate change has already made a significant impact, and impacts to the fishing industry are no exception (ThinkProgress, 2014). The waters around Alaska are already undergoing significant alterations in marine ecosystems with important implications for fisheries and the residents there who depend on them. Alaska is one of the leading areas for salmon, crab, halibut, and herring catch. In addition, many native communities depend on local harvests of fish, walrus, seals, whales, seabirds, and other marine species for their food supply. Climate change causes significant alterations in marine ecosystems with important impacts on the fisheries economy. Ocean acidification associated with increasing amounts of CO_2 concentrations is a serious threat to coldwater marine ecosystems.

One case in point is the northern Bering Sea off Alaska's west coast. An extremely productive fishing area off of Alaska's west coast, the Bering Sea Pollock fishery is the world's largest single fishery—producing more than 40 percent of the nation's total annual fish catch. It has experienced major declines over the past decade as air and water temperatures have risen and sea ice has rapidly declined. Resident populations of fish, seabirds, seals, walruses, and other species there depend on plankton blooms that are regulated by the extent and location of the ice edge in the spring. As sea ice retreats, however, the location, timing, and species composition of the plankton blooms changes, reducing the amount of food reaching the living creatures on the ocean floor, which radically changes the species composition and populations of fish and other marine life forms, which in turn, is causing devastating results for the fishing industry.

In a new study designed to understand the changes caused by climate change and the adaptation that will be necessary by the species living there, the National Oceanic and Atmospheric Administration (NOAA) in conjunction with the University of Washington are taking a multi-disciplinary approach and combining physical oceanography and fisheries science to provide abundance estimates for key fish stocks and potential management options for the future. The scientists involved in the study will produce several projections of what the Bering Sea ecosystem will look like under different climate and fishing scenarios. For example, rising sea temperatures will cause some species to thrive, others to decline. With reliable predictions, this will allow fishermen and coastal communities to be able to plan for the future. It will help resource managers ensure that fisheries and seafood supplies will remain stable over the long term. It will also benefit other areas by developing new methods of interdisciplinary research. They will be sharing their results with the IPCC in the future so that other areas around the world can use their models as a basis for their own future planning to enhance and preserve ecosystem health (NOAA Fisheries, 2015).

It is estimated that the global fishing industry will lose $17 to $41 billion by 2050 due to the effect of climate change on the marine environment, according to a 2014 report published by Sustainable Fisheries Partnership and the University of Cambridge. It presents a range of challenges that increasing ocean temperatures and acidification will bring to the seafood industry. Their conclusions are based on the findings from the IPCC's Fifth Assessment Report. The study determined that climate change will put the 400 million people who depend heavily on fish for food at risk, especially small-scale fishermen in the Tropics. The reasons is because yields are anticipated to fall by 40–60 percent in the Tropics and Antarctica. In the high latitudes, however, the report said yields are expected to increase 30–70 percent (University of Cambridge, 2014).

Some fish stocks will be able to migrate to cooler or more food- or oxygen-rich waters, which is a welcome news for those fish populations. It can lead to conflicts, however, among countries as to which nations are entitled to the displaced stocks, which could also lead to more illegal fishing. The report also emphasized the danger ocean acidification and warming waters pose to coral reefs, emphasizing that those in Southeast Asia and parts of the Pacific are some of the highest at risk. Reefs serve as nurseries or habitats for up to 12 percent of fish caught in tropical countries, and coral reef fisheries worldwide are already being fished unsustainably. The findings in the study estimate coral reef fish production in the Pacific could decrease by up to 20 percent by 2050, but partially to the habitat damage that climate change will cause to the reefs.

The aquaculture industry has also been targeted to be at risk. Oyster farmers off the coasts of the United States have already been impacted by ocean acidification. One case in point was Goose Point Oyster Company in Willapa Bay, Washington. They were forced to move their oyster larvae operations to Hawaii after high acidity levels off the coast of Washington caused larvae to start dying a decade ago.

Climate change also worsens other human-caused stressors on marine ecosystems—such as overfishing and habitat destruction. Repairing the human-caused destruction, such as restoring reefs that have been damaged by human activities, can restore their health and can build back their effectiveness. Restoring coastal and marine environments can also help mitigate climate change. For example, in areas where rehabilitation has used mangrove forests, salt marshes, and sea grass beds have acted as major carbon sinks as well, which serve to bury organic carbon in ocean sediment. These habitats can also serve a dual purpose by acting as fish nurseries to help restore fish stocks.

Forests provide many services that are critical to the well-being of humans. They provide air and water quality maintenance, watershed protection, and water flow regulation; wildlife habitat and biodiversity conservation; recreational opportunities and esthetic fulfillment; raw materials for wood and paper products; and climate regulation and carbon storage. As climate change continues, it will change the balance of the forest ecosystems, and the changes are not projected to be for the better.

In the United States, forests cover about 304 million hectares, which is roughly one-third of the country's total land mass. These forests provide many of the mentioned benefits and services to its population: clean water, recreational benefits, wildlife habitat, carbon storage, and a variety of forest products. Climate plays a significant role in the function of forest ecosystems and an essential part in forest health. As the climate changes it may worsen many of the threats to forests, such as pest outbreaks, wildfires, human development, and drought. Both direct and indirect changes have an effect on the growth and productivity of forests. Direct effects are due to changes in atmospheric CO_2 and climate. Indirect changes occur through complex interactions in forest ecosystems. An unstable climate also affects the frequency and severity of many forest disturbances.

Along with the projected impacts of climate change, forests also face impacts from land development, suppression of natural periodic forest fires, and air pollution. Although it is difficult to identify how each component is impacting forests separately, the combined impact is obvious and changes are already being seen. As these changes continue, we will be able to see a compromise in the quality of goods and services. Impacts are expected on forest growth and productivity. If trees have sufficient water and nutrients, then increases in CO_2 may enable trees to be more productive. Higher future CO_2 levels could benefit forests with fertile soils in the Northeast. However, increased CO_2 may not be as effective in promoting growth in the West and Southeast, where water is limited.

Warming temperatures could increase the length of the growing season. However, warming could also shift the geographic ranges of some tree species. Habitats of some types of trees are likely to move northward or to higher altitudes. Other species may be at risk locally or regionally if conditions in their current geographic range are no longer suitable. For example, species that currently exist only on mountaintops in some regions may die out as the climate warms since they cannot shift to a higher altitude.

Climate change may increase the risk of drought in some areas and the risk of extreme precipitation and flooding in others. Increased temperatures would alter the timing of snowmelt, affecting the seasonal availability of water. Although many trees are resilient to some degree of drought, increases in temperature could make future droughts more damaging than those experienced in the past. Drought increases wildfire risk, since dry trees and shrubs provide fuel for fires. Drought also reduces trees' ability to produce sap, which protects them from destructive insects, such as pine beetles. So what do all these predictions really mean? Disturbances can interact with one another—or with changes in temperature and precipitation—to increase risks to forests (EPA, 2016a).

The question still remains as to whether or not forests will migrate to cooler ground. This would involve all species, including plants. Some species will seek higher altitudes, others will move further poleward. In temperate regions, plant and tree species can migrate naturally by 25–40 kilometers a century. However, if, for instance, there was a 3°C increase in temperature over a hundred year period in a particular region, the conditions in that area would undergo dramatic change, equivalent in ecological terms to a shift of several hundred kilometers.

In the past few decades scientists have observed the first signs of this process taking place in the northern hemisphere caused by the rising temperatures associated with climate change. There have been a number of bird, tree, scrub, and herb species migrated by an average of six kilometers every 10 years, or species have sought higher altitudes of between one and four meters. Several trees and plants in the northern hemisphere have begun flowering earlier—on average advancing by two days every 10 years—which increase the risk of buds being killed by late frost. Slightly higher temperatures and a larger accumulation of CO_2 in the atmosphere have accelerated growth rates of species in forest ecosystems. Forests in temperate regions have seen about a 15 percent productivity gain since the beginning of the twentieth century. In addition, CO_2 fertilization along with increased nitrogen levels and additional soil moisture, have all contributed to greater forest productivity over the last century (EPA, 2016a).

As climate change continues, the effects from wildfires will increase. The 2015 wildfire season in the United States broke records; it was unlike anything most career firefighters had ever seen. Across the United States, fires erupted not only in dry woodlands, but also in grasslands, rainforests, and tundra; ignited by lightning strikes and careless campers. Fires burned lichen-covered trees in the rainforests of the Pacific Northwest. In Alaska, they ate into the permafrost. Two hundred U.S. military personnel were called in to battle the blazes in emergency situations all across the West; as were Canadians, Australians, and New Zealanders. The 2015 wildfire season saw the burning of a record-breaking 4 million hectares (10.1 million acres). The fire season saw the destruction of 4,500 homes and the tragic loss of 13 wildland firefighters' lives (Figure 7.15).

And the contributing factor: climate change. Tom Vilsack, Secretary of Agriculture, commented: "While the news that more than 10 million acres burned is terrible, it's not shocking, and it is probable that records will continue to be broken."

The burned area surpassed the previous record set in 2006 of 9.9 million acres. The price tag is not cheap, either. During just a one-week period, wildfire suppression across the nation cost a record $243 million (Dickie, 2016). When the funds run dry, monies have to be transferred from other programs to fight the fires, such as forest restoration, trail work, and watershed management. Congress had failed to put the fire funding item in the budget as the proposed Wildfire Disaster Funding Act, as was proposed, or the fire year would have been treated as a national disaster. Because of the Congressional failure, it put the Forest Service $1 billion beyond their operating budget. Pulling money away from forest restoration halts operations such as forest thinning and prescribed burns that can reduce the spread and intensity of fires.

Climate change is projected to increase the extent, intensity, and frequency of wildfires in certain areas of the country, making it a critical issue today. Fires can also contribute to climate change, since they can cause rapid large releases of CO_2 to the atmosphere.

AGRICULTURE AND FOOD PRODUCTION

The agricultural and food production industries will experience some of the greatest disruptions and changes of all from climate change. While climate change affects agriculture, climate is also affected by agriculture, which currently contributes 13 percent of all human-induced GHG emissions globally. In the United States, agriculture represents 9 percent of the country's total GHG emissions, including 80 percent of its nitrous oxide emissions and 31 percent of its methane emissions (Russell, 2014, EPA, 2016b).

Increased agricultural production will be necessary in the future in order to feed growing global populations, and agricultural productivity is directly tied to the climate and land resources. Climate change can have both beneficial and detrimental impacts on plants. As GHG levels rise in the atmosphere, projected climate changes are likely to increasingly

FIGURE 7.15   The Aggie Creek Fire burns 30 miles northwest of Fairbanks, Alaska. (Courtesy of U.S. Forest Service [USFS], Washington, DC.)

challenge farmer's capacities to efficiently produce food, feed, fuel, and livestock products. Crops in the future, as climate change progresses, are expected to respond to three factors: rising temperatures, changing water resources, and increasing CO_2 concentrations. Generally, warming causes plants that are below their optimum temperature to grow and produce faster. While this can be a good thing, it is not always. For instance, when cereal crops grow faster there is less time for the grain itself to grow and mature, which reduces the yields. For some annual crops, late season stress can be avoided by adjusting their planting date.

The time period during which the wheat grows and matures (called the grain-filling period) and other small grain shorten dramatically with rising temperatures. Moderate increases in temperature will decrease yields of corn, wheat sorghum, beans, rice, cotton, and peanut crops. In addition, some crops are sensitive to high nighttime temperatures, which have been rising even faster than daytime temperatures. Nighttime temperatures are expected to continue to rise in the future. These changes in temperature are critical to the reproductive phase of growth because warm nights increase the respiratory rate and reduce the amount of carbon that is captured during the day by photosynthesis to be released in the fruit or grain. As temperatures continue to rise and drought periods increase, crops will be more frequently exposed to temperature thresholds at which pollination begins to fail and the quality of the vegetable crops decrease. Grain, soybean, beans, rice, cotton, peanut crops, and canola crops have relatively low optimal temperatures, and will have reduced yields and will begin to experience failure as warming increases.

Temperature also affects the length of the growing season and water availability. For crops that do better in higher temperatures, such as sweet potatoes, the growing season will be longer; but for those that are temperature sensitive, such as lettuce and broccoli, their growing season will be truncated. As temperatures rise, plants will also be required to use more water to keep cool. But that only works to a certain point and when that level is reached and temperatures exceed the threshold for that species, it will not produce seed anymore, which means it will no longer reproduce.

Similar to forests, temperature change will also cause a migrational shift in species. If temperatures in a region warm, species will shift poleward; if they cool, they will migrate toward the equator. Therefore, temperature will be one of the major factors determining where they will be able to be efficiently grown.

Higher CO_2 levels typically cause plants to grow larger. This can also pose a problem, because for some crops, this makes them less nutritious. On the other hand, CO_2 also makes some plants more water-use efficient, so that they produce more plant materials with less water. This could be a benefit in dry or drought areas, depending on the type of plant and specific location.

In some cases, there may be adaptation measures to accommodate for these changed conditions. In the most simple of cases, changing the planting dates and planting a crop later in the year could overcome the problems of excessive heat with the least economic impact. The economic feasibility of this,

for the farmer, would depend on several factors, such as the optimum planting date for maximum profits and timing delivery of produce with market demand and competition.

In some cases, the typical fruits grown in a region may have to be discontinued if they require longer winter chilling periods than what the area currently receives. Another issue that needs to be considered is any increase in heavy downpours; or less frequent, but more intense, precipitation episodes. Changes of precipitation patterns are predicted to increase in areas across the United States. One of the early consequences of excessive rainfall is delayed spring planting, which in turn lowers profits for farmers that typically produce high-value early season crops such as melon, sweet corn, and tomatoes. When cultivated fields flood during the growing season, it causes crop losses due to low oxygen levels in the soil, increased susceptibility to root diseases, and increased soil compaction because of the heavy farm equipment used on wet soils. As an example of the magnitude of losses felt by farmers, in the spring of 2008, when heavy rains caused the Mississippi River to rise two meters above flood stage, it inundated hundreds of thousands of hectares of cropland, which coincided with the same time the farmers were harvesting wheat and planting corn, soybeans, and cotton. The farmers suffered economic losses of about $8 billion, with many farmers going bankrupt and others taking many years to recover (NOAA, 2008). Heavy downpours can also reduce the quality of the crops, causing significant economic losses. When vegetable and fruit crops are sensitive to short-term, minor stresses, extreme weather events can cause significant economic damage.

Another major concern with rising temperatures is the northward expansion of invasive weeds. Today, southern farmers lose more of their crops to weeds than do northern farmers. Based on the 2009 U.S. Global Change Research Program Report, southern farmers in the United States lose 64 percent of their soybean crop to invasive weeds, while northern farmers lose 22 percent. One especially problematic, very aggressive invasive weed that plagues the southern regions of the United States—kudzu—has traditionally been confined to areas where winter temperatures do not drop below certain thresholds. But as temperatures continue to rise with the changing climate, it will expand its range (along with other similar invasive weeds) northward into key agricultural areas, causing economic losses in expanded areas. Currently, kudzu has invaded approximately 1 million hectares of the Southeast and is a carrier of the fungal disease soybean rust, which seriously threatens the soybean production business, playing havoc with that part of the economy (Karl et al., 2009).

Controlling weeds represents a huge economic expense. In the United States alone, it currently costs more than $11 billion annually, with the majority of that expenditure on herbicides, which suggests that both herbicide use and costs will increase as temperatures and CO_2 levels rise with climate change. In addition, the most commonly used herbicide in the United States—glyphosate (commonly known as RoundUp®), actually loses its efficacy on weeds grown at CO_2 levels that are projected to occur in the coming decades. This means that higher concentrations of the chemical and more frequent usage will be necessary, which in turn increases both the economic and environmental costs associated with the use of chemicals.

Insects and diseases also thrive in warming temperatures, necessitating an increased use of pesticides. As temperatures climb, both insects and pathogens will expand their ranges northward. In addition, as temperatures warm, it will allow more insects to survive over the winter, whereas cold winter temperatures in the past served to control their populations. Insects present two negative aspects toward crop production—they not only damage crops but also carry disease, which kills crops and vegetation. Crop diseases are expected to increase as earlier springs and warmer winters promote proliferation and higher survival rates of disease pathogens and parasites. Consequently, the longer growing season will also allow some insects to produce more generations during a single season, which will greatly increase the insect population. In addition, plants that grow in higher CO_2 concentrations may grow larger, but they tend to be less nutritious, so insects need to consume more in order to meet their protein requirements, which cause greater crop destruction (Hatfield et al., 2008).

Rangeland grazing for cattle and other livestock will also take a toll under rising temperatures. In the United States, eastern pasturelands are planted and managed, and western rangelands are native pastures, not seeded and much more arid. These rangelands are already experiencing the effects of climate change due to increasing levels of CO_2. Specific grasses are beginning to dominate that do better under higher CO_2 levels and the quality of the forage is beginning to drop, meaning that more acreage is needed to provide the animals with the same nutritional value, resulting in an overall decline in livestock productivity. Woody shrubs and invasive cheat-grass are encroaching into once-fertile grasslands, which further reduce the forage value of the rangeland. The combination of these factors leads to a decline in livestock productivity and economic decline for the rancher.

It is also expected that livestock production will fall as temperatures rise. Temperature and humidity interact to cause stress in animals, just as in humans, and the higher the heat and humidity, the greater the stress and discomfort, and the larger the reduction in the animals' ability to produce milk, gain weight, and reproduce. Milk production will decline in dairy operations, the number of days it takes for cows to reach their target weight will grow longer in meat operations, conception rate in cattle is projected to fall, and swine growth rates is predicted to decline due to increased heat. As a result, swine, beef, and milk production are all projected to decline as temperatures rise (Hatfield, et al., 2008).

IMPACTS TO THE RECREATION AND TOURISM INDUSTRY

Recreation and tourism play important roles in the economy and are very important components in the lives of many people worldwide. Tourism is one of the largest economic sectors in the world, and it is also one of the fastest growing. The employment opportunities created by recreational tourism provide economic benefits not only to individuals but also to communities. Nearly 212 million Americans participated in an outdoor activity at least once in 2015 (Physical Activity Council, 2016). In some regions tourism and recreation are major employment centers, creating jobs for many in primary and secondary industries. For example, in the United States, areas where popular national parks are located, such as Yellowstone National Park in Wyoming or Zion National Park in Utah, there are not only employment opportunities at the parks, but also in the regions around the parks in a variety of sectors, including lodging facilities, restaurants, other tourist attractions, retail stores, and so forth. These businesses bring billions of dollars to the regional economies in which they are located. The same applies globally, making tourism and recreation one of the world's key industries. Recreational activities, such as fishing, hunting, skiing, snowmobiling, diving, beach-going, and other outdoor activities make important economic contributions and are a part of family traditions that have value that goes beyond financial returns.

According to the Outdoor Recreation Economy in their report in 2015, the outdoor recreation industry has an average annual $646 billion impact on the U.S. national economy. They reported that, "This amount factors in the amount Americans spend on outdoor trips and gear, the companies that provide that gear and related services, and the companies that support them (Outdoor Recreation Economy, 2016)." The outdoor industry also supports 6.5 million jobs. This equates to one in 20 U.S. jobs, and generates about $88 billion in federal and state tax revenue while stimulating 8 percent of all consumer spending (Outdoor Foundation, 2014).

Since most recreation and tourism activities occur outside, increased temperature and precipitation have a direct effect on the enjoyment of these activities and on the desired number of visitor days and associated level of visitor spending as well as tourism employment. Weather conditions are a critical factor influencing tourism. The availability and quality of natural resources is another important factor that determines the quality of outdoor recreation and tourism, such as the condition of beaches, wetlands, forests, wildlife, and snow—all of which are affected by climate change. Climate change, however, is currently changing the outcome of that for many areas and will continue to in the future.

It will equate to reduced opportunities for some activities and locations and expanded opportunities for others. According to the IPCC, the effects of climate change on tourism in a particular area will depend in part on whether the tourist activity is summer- or winter-oriented and, for the latter, the elevation of the area and the impact of climate on alternative activities (IPCC, 2007). Therefore, the range of effects on each area will differ depending on the climate change effects at that particular location. In the short term, the length of the season for, and desirability of, several of the most popular outdoor activities, such as hiking, beach-going, backpacking, horseback riding, and camping, will most likely be enhanced due to warmer temperatures and a longer summer season. Other activities, however, are likely to be negatively impacted by even slight increases in warming, such as snow- and ice-dependent activities, such as skiing, snowboarding, ice climbing, snowmobiling, and ice fishing. Therefore, areas today that serve as a ski resort may not have adequate snow in the future, so may have to become a mountain resort used for other activities, such as hiking, mountain biking, or horseback riding. This is already occurring in some of the alpine areas in Europe, where some ski resorts have now become "mountain beach" resorts (Fox, 2014). According to an article in the United Kingdom *Telegraph*, many of Europe's picturesque alpine towns are transforming their venues into golfing, cycling, and cross-country trekking to offset economic losses from lack of adequate snow cover to support their dwindling skiing industry.

The results of a Swiss study reveal snow cover suitable for winter tourism has fallen by 60 percent in the past 60 years. As a result, instead of their traditional snow-themed venue, they are moving into golfing, cycling, and cross-country trekking in order to offset losses caused by declining snowfall (de Quetteville, 2008).

According to Stephan Lerendu, the director of the Tourist Office at Avoriaz, France, "People think snow is a resource that will never run out but it's difficult right now. Tennis courts, saunas, and gyms are must-haves. Golf courses are indispensable for ski stations. We are adapting ski schools for the summer, with downhill biking and Nordic walking instead. We are creating green, blue, red, and black pistes for cycling. It's the future." A piste is a trail with an artificially prepared surface of packed snow.

Currently in Crans-Montana, Switzerland, golf courses already host European PGA tournaments. It also offers canyoning and paragliding. The statistics for snow days in the area are shocking: at low altitudes—up to 800 meters—the number of snow days (days with enough snow to make a snowman), has fallen from 28 days to 13. At mid altitudes—up to 1,300 meters—days suitable for cross-country skiing have gone from 55 to 38. And at high altitudes, for downhill skiing, days have declined from 93 to 74. Dr. Christoph Marty, who led the research, commented, "There is nothing similar in 1,000 years of records. Since this impact takes time to filter through the environment, what we are seeing now is only the fruit of what we did 20 years ago" (de Quetteville, 2008).

It has also become a critical problem for the ski industry in the western portion of the United States. In Squaw Valley, California, which averaged 1,143 centimeters of snowfall per year between 2008 and 2014, they received less than one-third of that in 2015. In fact, the snowpack accumulation was so poor that the International Ski Federation canceled the ski-cross and snowboardcross World Cup two weeks before it was scheduled to begin in early March. Of all the recreation industries, the ski industry has felt the impact of climate change the greatest. Ski resort towns economies rely almost entirely on snowfall. When snowfall accumulation is meager, tourists do not show up; and when they do not show up, it is not just the ski resorts that suffer, but the entire economy for the region— restaurants, grocery stores, gas stations, hotels, and so forth. California, currently suffering through a 100-year "mega-drought" is not the only state feeling these effects, either. The entire western United States is feeling the negative effects of a warming world. In a 2008 study conducted by the University of Maryland, they estimated that Colorado (which claims the largest ski-based economy in the country) would lose $375 million in revenue and 4,500 jobs by 2017 due to skier attrition from lack of snow (Williamson et al., 2008).

Executive director, Diana Madson, of the nonprofit Mountain Pact, a new advocacy group focused on stemming the impacts of climate change on ski towns remarked, "It's a bigger issue than, 'Oh, we can't ski.' It's loss of jobs and major environmental degradation." In Tahoe, Madson is coming up with ways to help mountain towns survive the trend in increasingly warm winters. She studied the effect of climate change on mountain towns at the Yale School of Forestry, and discovered that drought, lack of snow, and more frequent fires have been eroding the economic base of ski towns across the country. She also discovered along with that, that towns were spending their resources addressing the symptoms but not doing anything about the real cause—climate change. This realization led her to some innovative thinking and ideas. She decided that if all the ski towns were to ban together and pool their resources and knowledge, they could become a formidable political bloc. With a common framework they could be heard. The result of her efforts was the formation of the Mountain Pact group—the formal joining of ski communities to lobby for the kind of environmental policies that will help them adapt to climate change. As Madson says, "Climate conversation has been focused on urban areas and coastal areas. Mountain towns are not being heard" (Hansman, 2015).

Supporting Madson's claims, Dr. Elizabeth Burakowski, a researcher at the National Center for Atmospheric Research says that there is already strong evidence that mountain towns are suffering from winter warmth. The country's $12.2 billion winter tourism industry cannot survive on snow-gun slopes, and ski resorts account for 36 percent of winter tourism-related employment. Dr. Burakowski stresses, "In order to protect winter—and the hundreds of thousands whose livelihoods depend upon a snow-filled season—we must act now to support policies that protect our climate, and in turn, our slopes."

Because of the hardship that ski resorts have been experiencing for a while now, many have already made sustainable practices a way of business. For example, Aspen Resort, in Colorado, is capturing methane from a nearby coal plant to power its snow guns; Deer Valley, in Utah, runs its snowcats on biodiesel. Ski resorts across the West have banded together and joined the Mountain Pact in order to be heard. Before that, only Park City, in Utah, had a climate adaptation plan in place since 2009. The partners of Mountain Pact have currently drawn up a list of common problems, identifying drought mitigation, water quality, forest health, wildfire, and new business plans as the most pressing concerns. Currently, they are focusing on three objects: adaptation, mitigation, and federal prioritization. They also support the Wildfire Disaster Funding Act and want to pressure the federal government to block expanded coal leasing on Department of Interior lands, and to keep carbon emissions in check in favor of climate change mitigation (Hansman, 2015).

If climate change negatively impacts the recreational uses of mountain areas, these changes will have a ripple effect of economic consequences throughout the local, national, and international economy, particularly in business sectors such as the travel industry, the transportation industry, sports equipment industry, food industry, and others related to leisure activities. For example, according to the U.S. Global Change Research Program, some western ski resorts, such as those in Utah, Colorado, and California could face a 90 percent decrease in snow pack, making the country's most iconic ski locations facing obliteration.

Some of the possible identified effects that will occur in the recreation and tourism industries include:

- Declines in cold water and cool-water fish habitat may affect recreational fishing.
- Shifts in migratory bird populations may affect recreational opportunities for bird watchers and wildlife enthusiasts.
- Coastal regions may lose pristine beaches due to sea-level rise.
- Winter recreation (skiing, snowmobiling, ice fishing) is likely to be affected by reduced snow pack and fewer cold days and costs to maintain these industries will rise.
- Tourism and recreation in locations such as Alaska will most likely undergo climate-driven transformation through loss of wetlands and a reduction in habitat for migratory birds.
- Arctic breeding and nesting areas for migration birds will be lost, affecting bird watching activities.
- Melting permafrost in the Arctic will pose economic damage to structures and other infrastructures to residents there.
- Hunting and fishing will change as animals' habitats shift and as relationships among species in natural communities are disrupted by their different responses to rapid climate change.
- Water-dependent recreation in areas projected to get drier, and beach recreation in areas that are expected to see rising sea levels, will suffer.
- Some regions will see an expansion of the season for warm weather recreation such as hiking and bicycle riding.

In summary, the net economic effect of near-term climate change on recreational activities is probably going to be positive. In the longer term, however, as the effect on ecosystems and seasonality become more pronounced, the net economic effect on tourism and recreation is not known with clear certainty and will vary from location to location.

HEALTHCARE AND LABOR RELATED IMPACTS

Climate change presents a unique challenge to human health. Unlike specific health threats caused by one particular disease vector, there are several different ways that climate change can lead to potentially serious health effects. For instance, there are the directly measurable health effects that occur as a result from incidents such as heat waves and severe storms. There are also health issues that result because, or are made worse because of, pollution and airborne allergens. The third class of health issues is the climate-sensitive infectious diseases.

When trying to assess the economic impacts of the health-related issues of climate change, it is necessary to also factor in the need to manage new and changing climate conditions. Potential health risks must be recognized so that they can be avoided, minimized, or mitigated effectively. Diseases being introduced from elsewhere globally must also be considered, because society is global in nature. It is currently expected that developing nations will suffer even greater health consequences from climate change. With global travel so common, however, there are no real barriers or dividing lines anymore, requiring all areas to plan accordingly.

In addition, climate extremes—such as severe storms and drought—can undermine public health infrastructure, further stress environmental resources, destabilize economies, and potentially create security risks anywhere in the world. An increase in the risk of illness and death related to extreme heat and heat waves are very likely in the future. Some reduction in the risk of death related to extreme cold is also expected. Also of note, the population—for example in the United States—is aging and older people are more vulnerable to hot weather and heat waves. Today, the U.S. population over age 65 is 12 percent. By 2050, however, it is projected there will be 21 percent over 65, or more than 86 million people. Diabetics are also at greater risk of heat-related death.

According to a report issued by the 13-agency U.S. Global Change Research Program released in 2009, which discusses extreme weather and how climate change will affect life in the United States, both the current and projected effects through 2099 under various GHG emissions scenarios, that portion of the earth will see some drastic changes (Figure 7.16). For example, by the end of the century, summers will be much more severe across the country. In the New England area, summers will resemble those of present-day North Carolina. The Great Lakes region can expect to feel more like Oklahoma. Texas, typically hot now with 10–20 days a year over 38°C will jump to approximately 100 days of that temperature range—over three months instead of three weeks. Chicago, which has experienced severe heat waves and resulting high death tolls in the past can expect that kind of relentless heat up to three times a year. The Southwest portion of the country, which has experienced multiple droughts and resultant wildfires in the recent past can expect to face even more frequent droughts, as spring rains decline by as much as half. Cities such as Los Angeles, Salt Lake City, and Phoenix, which have large populations will face difficult times as snow packs shrink and melt earlier and water evaporates more rapidly, resulting in reservoirs remaining at only partial capacity.

Another major health concern as temperatures rise is the urban heat island effect. Large amounts of concrete and asphalt in cities absorb and hold heat. Tall buildings prevent heat from dissipating and reduce airflow. Simultaneously, there is generally little vegetation to provide shade and

Recent past 1961–1979

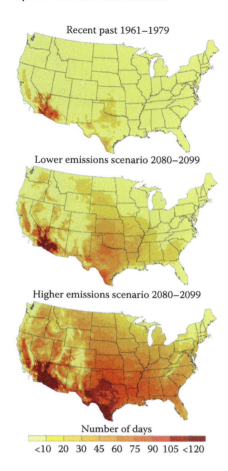

Lower emissions scenario 2080–2099

Higher emissions scenario 2080–2099

Number of days

<10 20 30 45 60 75 90 105 <120

FIGURE 7.16 The number of days in which the temperature exceeds 100°F by late this century, compared to the 1960s and 1970s, is projected to increase strongly across the United States. For example, parts of Texas that recently experienced about 10–20 days per year over 100°F are expected to experience more than 100 days per year in which the temperature exceeds 100°F by the end of the century under the higher emissions scenario. The lower emissions scenario assumes a CO_2 concentration of 550 ppm, and the higher emissions scenario assumes a concentration of 850 ppm. (From U.S. Global Change Research Program, Washington, DC.)

evaporative cooling. As a result, parts of cities can be up to 6°C warmer than the surrounding rural areas, compounding the temperature increases that people experience as a result of human-induced warming, such as heat given off from automobile traffic, machinery, and other means. As human-induced warming is projected to raise average temperatures by approximately 3.5°C–6.5°C by the end of this century under the higher emissions scenario, heat waves are expected to continue to increase in frequency, severity, and duration. As an illustration, by the end of the century, the number of heat-wave days in Los Angeles is projected to double and the number in Chicago to quadruple if emissions are not reduced. What this means, is that cities like Chicago can expect that the average number of deaths due to heat waves could more than double by 2050 under a lower emissions scenario and quadruple under a higher one (Figure 7.17).

The U.S. Global Change Research Program recognizes that the full effect of climate change on heat-related illness and

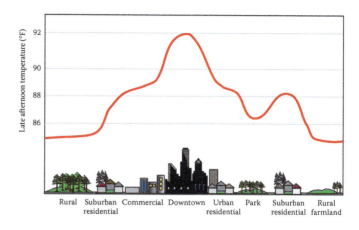

FIGURE 7.17 The urban heat island effect is projected to be responsible for many heat-related deaths as temperatures climb in urban areas throughout the remainder of the century. (Courtesy of U.S. Global Change Research Program, Human Health Sector, https://nca2009.globalchange.gov/urban-heat-island-effect/index.html.)

death involves a number of factors including actual changes in temperature; and human population characteristics, such as age, wealth, and fitness. In addition, adaptation at the scale of a city includes options such as heat wave early warning systems, urban design to reduce heat loads, and enhanced services during heat waves, which all have an economic impact on the supporting society.

Climate change and warming temperatures will also make it more challenging to meet air quality standards necessary to protect public health. Higher temperatures and associated stagnant air masses are expected to make it more challenging to meet air quality standards, especially for ground-level ozone (a component of smog). According to the U.S. Global Change Research Program it has been established that breathing ozone results in short-term decreases in lung function and damages the cells lining the lungs. It also increases the incidence of asthma-related hospital visits and premature deaths. Vulnerability to ozone effects is greater for those who spend time outdoors, especially with physical exertion, because this results in higher amounts reaching the lungs. As a result, those most at risk include children, outdoor workers, and athletes. Ground-level ozone concentrations are affected by several factors, including weather conditions, emissions of gases from vehicles and industry that lead to ozone formation (especially nitrogen oxides and volatile organic compounds [VOCs]), natural emissions of VOCs from plants, and pollution blown in from other locations. Warmer temperatures are projected to increase the natural emissions of VOCs, accelerate ozone formation, and increase the frequency and duration of stagnant air masses that allow pollution to accumulate, which will exacerbate health symptoms. The formation of ground-level ozone occurs under hot and stagnant conditions, which are the same weather conditions that accompany heat waves are expected to increase as climate change continues. Oftentimes, heat waves and air pollution occur together because they both occur under the same weather conditions. These interactions will more than likely increase as climate change continues, as well.

Extreme weather events are also responsible for emotional trauma injury, illness, and even death is known to result from extreme weather occurrences. The health impacts of extreme storms can also contribute to physical and mental health problems, which are projected to increase.

When looking at severe weather events—such as Hurricane Katrina—injury, illness, emotional trauma, and death do result from extreme weather events. The number and intensity of severe events are already increasing and are projected to increase in the future. Human health impacts are expected to be more severe in poorer countries where the emergency preparedness and public health infrastructure are less developed. Early warning and evacuation systems and effective sanitation lessen the health impacts of extreme events, and developing countries often lack these.

Indirect health impacts of extreme storms go beyond direct injury and death to indirect effects, such as carbon monoxide poisoning from portable electric generators in use following hurricanes and mental health impacts such as depression and post-traumatic stress disorder.

Several disease-causing agents that are commonly transmitted by water, food, or animals are susceptible to changes in replication, survival, persistence, habitat range, and transmission as a result of climate change factors such as increasing temperature, precipitation, and extreme weather events. Issues that will become problems as temperatures rise include:

- Cases of waterborne *Cryptosporidium* and *Giardia* increase following heavy downpours. These parasites can be transmitted in drinking water and through recreational water use.
- Heavy rain and flooding can contaminate certain food crops with feces from nearly livestock or wild animals, increasing the likelihood of food-borne disease associated with fresh produce.
- Climate change affects the life cycle and distribution of the mosquitoes, ticks, and rodents that carry West Nile virus, equine encephalitis, Lyme disease, and Hantavirus. Moderating factors, such as housing quality, land use patterns, pest control programs, and a robust public health infrastructure are likely to prevent the large-scale spread of these diseases in the United States.
- As temperatures rise, tick populations that carry Rocky Mountain spotted fever are projected to shift from south to north.

In addition, rising temperatures and CO_2 concentration increase pollen production and prolong the pollen season in several plants with highly allergenic pollen, presenting health risks for those with allergies. Climate change has caused an earlier onset of the spring pollen season in some areas such as the United States, as these areas presently warm up earlier in the year. Based on data from Obi et al. laboratory studies suggest that increasing CO_2 concentrations and temperatures increase ragweed pollen production and prolong the ragweed pollen season.

MEETING THE CHALLENGES OF CLIMATE IMPACTS

Now that the public is becoming more educated and seeing first hand some of the effects of climate change, businesses are beginning to take a stronger role, attempting to project ahead and take steps to adequately prepare for what lies ahead. Some of the action taken so far has been reactive—businesses responding and adapting to changes that have already taken place. An example of this—reacting, such as getting out of harm's way—would be a business relocating somewhere further inland (in a safer zone) after losing their place of business to Hurricane Katrina. Another example is farmers who are currently interested in obtaining drought- and flood-resistant seeds in response to increased weather extremes.

The PEW Center identifies the fact that successful adaptation over the long term requires recognizing and acting on threats from an early stage—ideally before they occur—and identifying appropriate responses (Sussman, 2008).

One thing Sussman points out is that with the changes currently happening with the climate, it may be necessary to alter traditional business practices, requiring companies to put themselves through a paradigm shift, no longer relying on historical trends and decisions or keeping a "business-as-usual" mindset. Long-term mission assessments must now be made with climate change a central theme in order to plan ahead. Sussman cites a major reason to complete a risk assessment to determine the likelihood of changes, with the best approach being proactive adaptation where future climate change is consciously anticipated and appropriate options incorporated into decision making. For example, in locating and building a new facility, it will be important to take into account the location of rivers if a water draw for electric power generation is necessary or cooling water is needed. Sussman believes that screening to identify the potential risks of near-term and long-term climate change is the first step in determining whether or not a risk assessment is necessary to identify further actions.

The purpose of the screening is to determine whether the business might be at risk; what aspects are at risk and from what; and whether a more complete risk assessment is needed to determine exactly what, if any, further actions are needed. The goal of the screening is to classify/screen risks into one of three categories, as illustrated in Figure 7.18:

- Class A: What to assess now.
- Class B: What can wait and be studied at a later date.
- Class C: What requires no action.

It is important to look at the business from a long-term perspective and determine the role climate change may have on it. This means identifying the projected climate and physical effects of climate change that could possibly threaten the success of the business operation, such as flooding, sea level rise, or drought. Looking at the immediacy of an issue is also

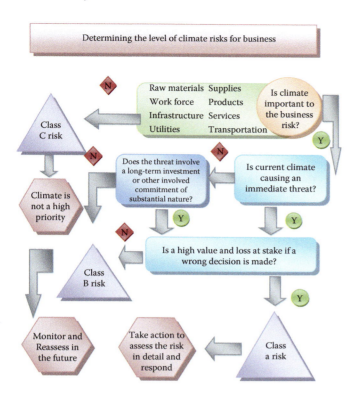

FIGURE 7.18 It is an imperative part of any successful business implementation plan today to include the influence of any climate change impacts that may be significant to the operation of a business. (From Gines, J. K., *Climate Management Issues: Economics, Sociology, and Politics*, CRC Press, New York, 2012.)

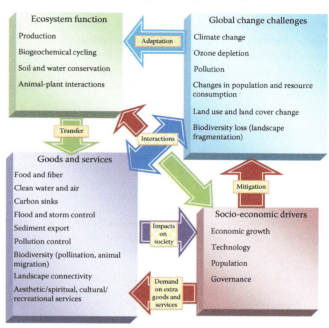

FIGURE 7.19 Economics are influenced interactively by ecosystems and their resultant goods and services, as well as socioeconomic drivers, and global change. These components can act independently, dependently, or interdependently to form the complete economic picture, and a working knowledge of all components is necessary in order to make sound, long-term business decisions. (From Gines, J. K., *Climate Management Issues: Economics, Sociology, and Politics*, CRC Press, New York, 2012.)

important, as it can prevent costly business decisions and potential mistakes and poor investment choices. For example, choosing a location of a business that may later need to be relocated could prove costly enough to threaten the survival of the business. Any sizable cost of a poor decision, whether it be a significant change in cash flow, a downturn in an investment, sizable restriction of growth goals, negative impacts on the firm's reputation, or any other significant measure of success necessitates comprehensive risk assessment and management (Sussman, 2008).

Making long-term decisions based on developing carefully thought-out future plans is a wise economic move. Decisions that are made today as to where to locate a business and what design standards to build to, will have implications decades down the road. What Sussman emphasizes is that decisions made now will help plan for and avoid future vulnerabilities. It is important to gain a firm understanding in a long-term business solution of the systematic integration of global change in general and the relationships between and among ecosystem functions, socio-economic drivers, goods and services, and global change challenges (Figure 7.19). Everything human populations do in their production of goods and services, which are made available through ecosystem functions, affect global systems. Those effects are not only be far-reaching, but also cascade through various ecosystem functions. Although we may not always directly see the effects of our actions on

the environment—or see them manifested immediately—the effects are still felt, not only by the human component but also by the natural world.

CONCLUSIONS

Scientific evidence continues to mount that climate change will directly or indirectly affect all sectors and regions of the global economy, although not equally. While there may be temporary benefits from a changing climate, the costs of climate change will rapidly exceed any benefits briefly gained in the few areas of the world that may realize any short-term benefits, and the costs of climate change will rapidly exceed any benefits, placing major stress on public sector budgets, personal income, and job security. Because of the extreme total economic costs of climate change, any delayed action or inaction will, in the long-term, be the most costly economic solution. Because of this, workable, well-thought-out economic plans calling for immediate action to mitigate GHG emissions combined with solid plans to adapt to the unavoidable impacts from damage already set in motion will significantly reduce both the overall short- and long-term costs of a changing climate and environment.

In addition, not just one country or region can solve this global issue. A wide range of resources must be brought into the solution. It must be a multinational effort, and in this regard, governments need to team up with the latest researchers at universities and private research institutions for the latest information in order to make the most informed decisions. There needs to be a seamless mesh in place between the scientific community and the decision-making sector, so that the most prudent policies and investment decisions are made.

REFERENCES

Burke, M. December 9, 2015. The global economic costs from climate change may be worse than expected. *Brookings.* https://www.brookings.edu/blog/planetpolicy/2015/12/09/the-global-economic-costs-from-climate-change-may-be-worse-than-expected/ (accessed September 2, 2016).

Carle, J. July 14, 2015. Climate change seen as top global threat. Pew Research Center. http://www.pewglobal.org/2015/07/14/climate-change-seen-as-top-global-threat/ (accessed May 20, 2017).

Cole, S. and L. McCarthy. February 13, 2015. Carbon emissions could dramatically increase risk of U.S. megadroughts. National Aeronautics and Space Administration (NASA). http://climate.nasa.gov/news/2238/carbon-emissions-could-dramatically-increase-risk-of-us-megadroughts/ (accessed September 2, 2016).

de Quetteville, H. May 26, 2008. Global warming forces European ski resorts to offer summer sports. *The Telegraph.* http://www.telegraph.co.uk/news/worldnews/2032879/Global-warming-forces-European-ski-resorts-to-offer-summer-sports.html (accessed September 2, 2016).

Dickie, G. January 11, 2016. 2015 wildfires burned a record-breaking 10.1 million acres. *High Country News.* https://www.hcn.org/articles/wildfires-burned-a-record-breaking-10-1-million-acres-in-2015 (accessed September 2, 2016).

EPA. 2016a. Forests: Climate impacts on forest. https://www3.epa.gov/climatechange/impacts/forests.html (accessed September 2, 2016).

EPA. 2016b. Agriculture and food supply impacts. https://www3.epa.gov/climatechange/impacts/agriculture.html (accessed September 2, 2016).

Fox, P. February 7, 2014. The end of snow? *The New York Times.* http://www.nytimes.com/2014/02/08/opinion/sunday/the-end-of-snow.html?_r=0 (accessed September 2, 2016).

Gines, J. K. 2012. *Climate Management Issues: Economics, Sociology, and Politics.* New York: CRC Press.

Hansman, H. February 26, 2015. How ski resorts are fighting climate change. *Outside.* http://www.outsideonline.com/1930841/how-ski-resorts-are-fighting-climate-change (accessed September 2, 2016).

Hatfield, J., K. Boote, P. Fay, L. Hahn, C. Izaurralde, B. A. Kimball, T. Mader et al. 2008. Agriculture. In: *The Effects of Climate Change on Agriculture, Land Resources, Water Resources and Biodiversity in the United States*, Backlund, P., O. Ort, W. Polley, A. Thomson, D. Wolfe, M. G. Ryan, S. R. Archer et al. (eds.). Synthesis and Assessment Product 4.3. Washington, DC: U.S. Department of Agriculture, pp. 21–74.

IPCC. 2007. Climate change 2007: Synthesis report. *Contribution of Working Groups I, II and III to the Fourth Assessment Report of the Intergovernmental Panel on Climate Change*, Core Writing Team, Pachauri, R. K. and Reisinger. A. (eds.).

Geneva, Switzerland: IPCC. https://www.ipcc.ch/publications_and_data/publications_ipcc_fourth_assessment_report_synthesis_report.htm (accessed September 13, 2016).

Janetos, A., L. Hansen, D. Inouye, B. P. Kelly, L. Meyerson, B. Peterson, and R. Shaw. Biodiversity. 2008. In: *The Effects of Climate Change on Agriculture, Land Resources, Water Resources, and Biodiversity in the United States.* Synthesis and Assessment Product 4.3. Washington, DC: U.S. Department of Agriculture, pp. 151–181.

Karl, T. R., J. M. Melillo, and T. C. Peterson (Eds.). 2009. *Global Climate Change Impacts in the United States.* London: Cambridge University Press. Presents a comprehensive report on what types of changes to expect in the United States between now and the end of the century.

Makaudze, E. 2013. The impact of climate change, desertification and land degradation on the development prospects of landlocked developing countries. United Nations Office of the High Representative for the Lease Developed Countries, Landlocked Developing Countries and Small Island Developing States (UN-OHRLLS). http://unohrlls.org/custom-content/uploads/2015/11/Impact_Climate_Change_2015.pdf (accessed September 2, 2016).

NOAA. July 9, 2008. 2008 midwestern U.S. floods. National Climatic Data Center. Discusses the flooding that occurred during the spring of 2008 in the Midwestern U.S.

NOAA Fisheries. August 26, 2015. New comprehensive bering sea climate change study to focus on fish and fishing and provide insights for management in a changing marine environment. http://www.afsc.noaa.gov/News/BS_climate-change-study.htm (accessed September 2, 2016).

Outdoor Foundation. 2014. Outdoor research participation report. *Outdoor Industry Association.* http://www.outdoorfoundation.org/pdf/ResearchParticipation2014.pdf (accessed September 2, 2016).

Outdoor Recreation Economy. 2016. Outdoor recreation economy report. https://outdoorindustry.org/research-tools/outdoor-recreation-economy/ (accessed September 2, 2016).

Physical Activity Council. 2016. 2016 participation report. http://www.physicalactivitycouncil.com/pdfs/current.pdf (accessed September 2, 2016).

Russell, S. May 29, 2014. Everything you need to know about agricultural emissions. World Resources Institute. http://www.wri.org/blog/2014/05/everything-you-need-know-about-agricultural-emissions (accessed September 2, 2016).

Stokes, B., R. Wike, and J. Carle. November 5, 2015. Public support for action on climate change. Pew Research Center: Global Attitudes & Trends. http://www.pewglobal.org/2015/11/05/2-public-support-for-action-on-climate-change/ (accessed September 2, 2016).

Sussman, F. G. 2008. *Adapting to Climate Change: A Business Approach.* Arlington, VA: Pew Center on Global Climate Change.

Teh, L. C. L. and U. R. Sumaila. March 2013. Fish and fisheries. *Wiley Online Library.* http://onlinelibrary.wiley.com/doi/10.1111/j.1467-2979.2011.00450.x/abstract (accessed September 2, 2016).

ThinkProgress. May 29, 2014. The fishing industry is poised to lose billions due to climate change, report finds. https://thinkprogress.org/the-fishing-industry-is-poised-to-lose-billions-due-to-climate-change-report-finds-3e041c753fa#.v3bnssvm9 (accessed September 2, 2016).

UNCCD. 2014. Desertification: The invisible frontline. http://www.unccd.int/Lists/SiteDocumentLibrary/Publications/Desertification_The%20invisible_frontline.pdf (accessed September 2, 2016).

University of Cambridge. May 2014. Climate change: Implications for fisheries and aquaculture. http://cmsdevelopment.sustainablefish.org.s3.amazonaws.com/2014/05/27/IPCC_AR5_Fisheries_Summary_FINAL_Web-1001a188.pdf (accessed September 3, 2016).

Williamson, S., M. Ruth, K. Ross, and D. Irani. July 2008. Economic impacts of climate change on Colorado. University of Maryland, The Center for Integrative Environmental Research. http://cier.umd.edu/climateadaptation/Colorado%20Economic%20Impacts%20of%20Climate%20Change.pdf (accessed September 3, 2016).

SUGGESTED READING

Bagley, K. December 24, 2015. As climate change imperils winter, the ski industry frets. *Inside Climate News.* https://insideclimatenews.org/news/23122015/climate-change-global-warming-imperils-winter-ski-industry-frets-el-nino (accessed September 2, 2016).

Bull, S. R., D. E. Bilello, J. Edmann, M. J. Sale, and D. K. Schmalzer. 2007. Effects of climate change on energy production and distribution in the United States. In: *Effects of Climate Change on Energy Production and Use in the United States,* Wilbanks, T. J., V. Bhatt, D. E. Bilello, S. R. Bull, J. Edmann, W. C. Horak, Y. J. Huang et al. (eds.). Synthesis and Assessment Product 4.5. U.S. Climate Change Science Program, Washington, DC, pp. 45–80.

Ebi, K. L., J. Balbus, P. L. Kinney, E. Lipp, D. Mills, M. S. O'Neill, and M. Wilson. 2008. Effects of global change on human health. In: *Analyses of the Effects of Global Change on Human Health and Welfare and Human Systems,* Gamble, J. L. (ed.), Ebi, K. L., F. G. Sussman, and T. J. Wilbanks (authors). Synthesis and Assessment Product 4.6. Washington, DC: U.S. Environmental Protection Agency, pp. 39–87.

Fox, P. February 7, 2014. The end of snow? *The New York Times.* http://www.nytimes.com/2014/02/08/opinion/sunday/the-end-of-snow.html?_r=0 (accessed September 2, 2016).

Hansman, H. February 26, 2015. How ski resorts are fighting climate change. *Outside.* http://www.outsideonline.com/1930841/how-ski-resorts-are-fighting-climate-change (accessed September 2, 2016).

Helm, D. and C. Hepburn. 2010. *The Economics and Politics of Climate Change.* Cary, NC: Oxford University Press.

IPCC. 2001. Summary for policymakers, climate change 2001: Synthesis report. *A Contribution of Working Groups I, II, and III to the Third Assessment Report of the Intergovernmental Panel on Climate Change.* New York: Cambridge University Press. http://www.ipcc.ch/publications_and_data/publications_and_data_reports.htm (accessed September 9, 2016).

Saad, L. and J. M. Jones. March 16, 2016. U.S. concern about global warming at eight-year high. *Gallup.* http://www.gallup.com/poll/190010/concern-global-warming-eight-year-high.aspx (accessed September 3, 2016).

Scott, M. J. and Y. J. Huang. 2007. Effects of climate change on energy use in the United States. In: *Effects of Climate Change on Energy Production and Use in the United States,* Wilbanks, T. J. V. Bhatt. D. E. Bilello, S. R. Bull, J. Edmann, W. C. Horak, Y. J. Huang et al. (eds.). Synthesis and Assessment Production 4.5. U.S. Climate Change Science Program. Washington, DC, pp. 8–44.

Mills, E. 2006. Synergisms between climate change mitigation and adaptation: An insurance perspective. *Mitigation and Adaptation Strategies for Global Change* 12(5):809–842.

Mills, E. 2009. A global review of insurance industry responses to climate change. *The Geneva Papers on Risk and Insurance—Issues and Practice.*

Orr, J. C., V. J. Fabry, O. Aumont, L. Bopp, S. C. Doney, R. A. Feely, A. Gnanadesikan et al. 2005. Anthropogenic ocean acidification over the twenty-first century and its impact on calcifying organisms. *Nature* 437(7059):681–686.

ThinkProgress. October 16, 2015. Poll finds fewer Americans than ever doubt climate change is happening. https://thinkprogress.org/poll-finds-fewer-americans-than-ever-doubt-climate-change-is-happening-16257790947d#.ovk3dkvy2 (accessed September 3, 2016).

8 Climate Change Mitigation Options

OVERVIEW

Throughout the world, countries are adopting policies in an attempt to make progress against climate change. Common approaches include increasing renewable energy generation and encouraging energy efficiency. The positive effects of these are reducing dependence on petroleum, reducing our vulnerability to energy price spikes, promoting development of local economies, and improving air quality. Another major approach is addressing the CO_2 we have already generated and doing something about managing that. One answer currently available is carbon sequestration—the actual storing of CO_2 in designated repositories so that it can no longer add to the climate change problem. This chapter examines several other mitigation and management options being utilized today including cap and trade as a policy tool, how the carbon trading market works in an international arena, the need for global action, and possible economic implications. It also examines the carbon pricing rationale and compares it with the cap and trade approach. Next it looks at both direct and indirect carbon sequestration and storage and then examines both geologic formation and deep ocean sequestration. Following that, this chapter reviews the economic benefits of carbon storage and concludes by exploring the value and necessity of various global adaptation strategies.

INTRODUCTION

As the earth's atmosphere continues to warm and more people become aware and educated about climate change—including its effects and ramifications for the future—efforts worldwide are being made to reduce its impacts, find ways to mitigate the situation, and adapt to the present environment, as well as prepare for the future. Every person on this earth will have to, in some way, adapt to the effects of climate change, focusing on ways to help solve the problem, reduce what impacts are possible, find ways to be environmentally responsible, and learn to cope in a positive way to permanent change. On a community, national, and international level, there are also adaptation and mitigation options available—some already successfully in operation, others just beginning, and still others on the horizon. The important thing is that we keep progressing and moving ahead; taking the action possible to minimize what impacts we can, prevent what we can, and adapt to the rest. In order to make that possible, our forward actions cannot pause.

DEVELOPMENT

Climate change mitigation refers to efforts taken to reduce or prevent the emission of GHGs. Mitigation can mean several things. It can mean using new technologies and renewable energies, making older equipment more energy efficient, or changing the management practices or consumer behavior. It can be as complex as a plan for a new city, or as simple as an improvement to an air conditioner design—it is whatever enables less GHGs to be emitted. Worldwide, there are efforts being made to achieve these goals; from high-tech transportation systems, construction of urban and inner-city bicycle paths, to convenient recycling centers that serve communities and everything in between. Mitigation also includes protecting existing natural carbon sinks, such as oceans and forests, or creating new sinks through the development of green agriculture or silviculture. Any method of mitigation that is a good fit for a community and helps lower harmful GHG emissions into the atmosphere is a success.

CLIMATE CHANGE MITIGATION OPTIONS

Mitigation can apply to several general categories: agriculture, forest, energy, manufacturing, transportation, tourism, buildings, cities, and waste. We will discuss each one of these sectors in this section.

The Agricultural Sector

One of the biggest challenges to be faced with climate change is the adequate production of food. As the climate changes—temperatures warm, droughts occur, and land becomes degraded—areas that we have typically been able to rely on for food production are going to become stressed to the point that it will become difficult, if not impossible, to continue food production at our present level.

Current farming methods are depleting the earth's resources and producing huge quantities of GHGs. Agricultural operations currently produce 13 percent of human-based global GHG emissions (UNEP, 2016). This means that while current agricultural practices are engaging in unsustainable practices such as deforestation and promoting biodiversity loss in much of the world, the environment is paying a huge price there, as well as in contributing to the GHG problem. Conventional agriculture also has other harmful side effects, such as erosion, leeching chemicals into water sources, and lowering soil fertility, while the global economy pumps billions of U.S.

dollars into the effort. Turning toward sustainable agricultural methods makes sense from several angles. It will ease pressure on the environment, help cope with climate change, create opportunities to diversify economies, increase yields, reduce costs, generate jobs, reduce poverty, and increase food security. In fact, increasing farm yields and improving ecosystems services will greatly help the 2.6 billion people who make their livelihood from the agricultural industry—especially in developing nations where most farmers live on small parcels in rural areas.

One form of mitigation that would make a notable difference would be to simply reduce agricultural waste and inefficiency. Almost 50 percent of food produced is lost through crop loss or waste during storage, distribution, marketing, and household use. Some of these inefficiencies—most notable crop and storage losses—could be solved with an investment in simple storage technology, and buying locally.

The Forestry Sector

Forestry is an enormous industry—it supports the livelihoods of over 1 billion people. Most of these are in the developing countries of the world. Forestry is also species rich. More than half of all the earth's species live in forests. They play a valuable part in our lives for many reasons: they help regulate our climate through the carbon cycle (they are a CO_2 sink) and protect some of our purest watersheds. Unfortunately, approximately 13 million hectares is destroyed, deforested, or degraded every year. It is estimated that if $40 billion per year from 2010 to 2050 was invested in reforestation and payment to landholders for conservation, it could raise the value of the forest industry by 20 percent, and at the same time increase forest carbon storage by 28 percent. Creating a stable global market that would attract investment in forest-derived goods and assure their equitable and sustainable production could possibly offer one of the best possibilities for establishing a workable plan. In order for this to work, it would be necessary to create financial value for forest carbon storage. Adding value for conservation measures, forest management, and enhancing forest stocks would make it even more attractive (UNEP, 2016).

The Energy Sector

One of the most rapidly expanding sectors is energy. As populations continue to grow and societies continue to develop, there is an ever-increasing demand for energy. Unfortunately, this sector is the most difficult to curtail because it is such a fundamental part of our lives. Energy is needed in everything we do, as shown in Figure 8.1. As we continue to indulge in the use of fossil fuels, we not only have to take responsibility for the GHGs we are adding to the atmosphere at a furious rate but we also have to deal with other significant issues as well, such as energy security, air pollution, and environmental degradation. The current fossil fuel-heavy energy system we use is not only environmentally unsustainable but also highly inequitable, leaving approximately 1.4 billion people without access to electricity (UNEP, 2016). Added to this, much of the growing energy demand is occurring in developing countries,

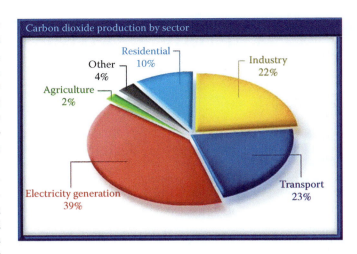

FIGURE 8.1 CO_2 production by sector. (Courtesy of U.S. Energy Information Administration [EIA], Washington, DC.)

where rising fossil fuel prices and resource constraints are adding further pressure on both the environment and the economy.

Because of these downfalls, mitigation is becoming critical. Investing in renewable energy sources is where the future of energy must go. The good news is that technology is rapidly advancing and it is becoming affordable for both businesses and private consumers to invest in it as a primary energy source. In 2015, renewable energy sources accounted for about 10 percent of total U.S. energy consumption and about 13 percent of electricity generation (U.S. Energy Information Administration, 2016). The changes and growth of renewable energy sources and use from 2001 to 2013 are shown in Figure 8.2.

With the government rebates being offered for solar panels, many homeowners are investing in solar power. It is projected that by the end of 2016, more than 1 million American homes will have solar panels. In 2006, only about 30,000 homes had solar panels. At that time, the cost was $9 per watt of power generated by solar panels. Today, the cost has been reduced to $3.79 per watt (Harrington, 2015). Solar panels have also increased in efficiency. American homes are now able to meet more than 85 percent of their electricity demand (EnergySage Solar Research, 2015). Chief Executive and Founder of EnergySage Solar Marketplace says "People love solar; there's very little not to like about it. No noise, no emissions, out of sight, produces electricity, it's beautiful, and it makes financial sense. I think as more people find out about it and more people become comfortable shopping for solar, and they don't feel like they're being sold but they have control, I think the sky's the limit."

As more people become aware of the climate change issue and realize the negative effects happening right now, as we saw in the previous chapter, the incentive to invest in renewable energy will continue to grow, at all levels: government, business, and in the private sectors. This is the best mitigation option available.

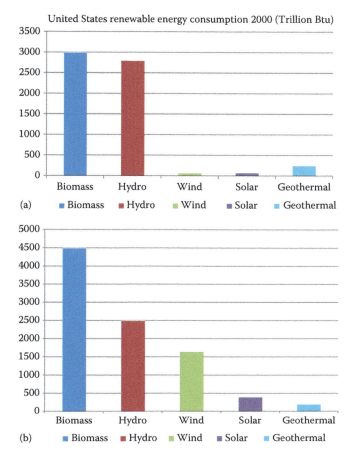

FIGURE 8.2 (a) U.S. renewable energy use by sector in 2001. (b) U.S. renewable energy use by sector in 2013. (Created by author.)

The Manufacturing Sector

Manufacturing is responsible for 35 percent of global electricity use, 20 percent of CO_2 emissions, and one-fourth of primary resource extraction. It has a major impact on the environment and plays a significant role in climate change. Simultaneously, manufacturing accounts for 23 percent of worldwide employment makes it a crucial sector of the worldwide economy. Therefore, its importance cannot be ignored; instead, a healthy and workable solution must be found to mitigate the climate change problem. The best way to approach this is to change the manufacturing process. In some cases, it may be possible to re-design a product, which may improve not only the product's life span but also lead to a more efficient use of resources, easier recycling, and the generation of less pollution during the manufacturing process and life of the product. As an example, two innovations that can save both resources and money are recycling heat waste and closed-cycle manufacturing. Also viable approaches are remanufacturing and reconditioning. Both are labor-intensive approaches, which can serve to create jobs and also require relatively little capital investment.

The Transportation Sector

The transportation sector is one of the largest challenges being faced today concerning CO_2 emissions, especially in developed countries. Not only is it a problem for emissions but also it is hazardous for human health and well-being and the environment in general. Transportation uses more than half of the earth's liquid fossil fuels and is responsible for nearly one-fourth of energy-related GHG emissions. It is responsible for widespread air pollution, over a million fatal traffic accidents per year, and needless traffic congestion—impacts that costs countries more than 10 percent of the gross domestic product.

Already a problem in developed countries—such as the United States, which has its transportation system geared toward motor vehicle traffic rather than mass transit due to the vast size of the country—but its use in developing countries is increasing at an alarming rate, as well. This includes countries such as China, India, and Russia. It is increasing so rapidly, that current projections estimate that the number of motor vehicles will triple by 2050. Studies conducted by the United Nations Environmental Programme (UNEP) show that investments in public transportation and vehicle efficiency could yield exceptional economic returns. They have determined that a green, low-carbon transportation sector could reduce GHG emissions by as much as 70 percent, with minimal additional investment. Furthermore, if sustainable regulatory policies are added to that, it could be improved further (UNEP, 2016). UNEP outlines that for this transformation to happen, there needs to be a major shift in the way the public thinks about investing in transportation. They proposed what they call a "three-pronged strategy: Avoid—Shift—Clean." It is defined as follows: Help users *avoid* or reduce trips—without restricting mobility—through smarter city planning and land use options. *Shift* passengers away from private vehicles to public and nonmotorized transportation, and freight users from trucks to rail or water transport. Finally, make vehicles cleaner, through both efficiency improvements and cleaner fuels. UNEP is currently working toward this paradigm shift through several initiatives and programs.

The Tourism Sector

This may not be a sector that people often think of, but tourism can be fossil-fuel intensive. Tourism is one of the top five export earners in 150 countries, and the number one export in 60. This may seem like welcome news for national economies, but if not properly managed it can be unwelcome news for the environment and local populations. Tourists are traveling more often and to more distant destinations, using more energy-intensive, fossil fuel-based transport, and the sector's GHG contribution has increased to 5 percent of global emissions. Other unsustainable practices, such as excessive water use, waste generation, and habitat encroachment, are threatening ecosystems, biodiversity, and local culture.

If done with sustainability in mind, however, tourism can be a positive endeavor for both the local economy and the environment. Green tourism aims to reduce poverty by creating local jobs and stimulating local business, while establishing ecologically sustainable practices that preserve resources and reduce pollution. One drawback today is that far too little tourism profits reach the people living in and near tourist destinations. By increasing local involvement, it can not only generate income but also encourage communities to protect their environment. Investing in energy efficiency and waste management can reduce GHG emissions and pollution and also save hotel owners and service providers' money. If it is done right, natural areas, biodiversity, and cultural heritage can all receive the direct benefits of sustainable tourism.

The Building Sector

Roughly one-third of the world's energy use takes place inside buildings. This makes the building sector one of the largest contributors to GHG emissions. In addition, the construction industry consumes more than one-third of the earth's natural resources and generates huge quantities of solid waste. This puts buildings and all that which is associated with it in line for mitigation, especially the acquisition of natural resources that are used in the industry.

If energy efficiency is improved in buildings through greener construction methods and retrofitting existing structures, it can make a huge difference in reducing GHG emissions. In addition, many of these improvements can be completed at a low cost, utilizing existing technology. Green construction can also have a positive effect on productivity, public health, and employment opportunities. According to estimates, every $1 million invested could result in 10 to 14 jobs (UNEP, 2016).

Cities present many opportunities for mitigation, and it is critical that they do. They are developing rapidly, especially in developing countries. Urban areas are now home to nearly half of the earth's population, which use approximately 60 percent of available energy and account for more than half of the carbon emissions. Other side effects include a heavy impact on water supplies, stresses on public health, environmental impacts, and pressures on the quality of life—especially for the poor. With continued growth, fundamental changes in urban development will have to occur in order to create a sustainable future.

Because of the density of urban areas, they are in a prime position to enable a strong collaboration between local governments, private partnerships between businesses, and academia, who together can work toward building a more sustainable society. With the right policies, practices, and infrastructures in place, cities can become green models for efficient transportation, water treatment, construction, and responsible resource utilization.

The Waste Management Sector

Another key area that must become more sustainable is waste management. As countries' economies grow, so does the volume of their garbage. According to recent estimates, approximately 11.2 billion metric tons of solid waste are currently being collected around the world every year, and the decay of the organic portion is contributing around 5 percent of global GHG emissions (UNEP, 2016). The most rapidly growing type of waste in both developing and developed countries is electrical and electronic products, which contain hazardous substances that make disposal a challenge. Human health and the environment are becoming increasingly at risk, especially when dumpsites are uncontrolled or volume becomes unmanageable. Some of the major impacts include illnesses and infections, ground water pollution, GHG emission, and ecosystem destruction.

In order to mitigate this problem, it is possible to turn it into an economic opportunity instead. Managing waste, from collection to recycling, is currently a growing market estimated at $410 billion per year, not including the substantial informal segment in developing countries. Recycling, in fact, actually creates more jobs than it replaces. Investing in this greener style of waste management could produce many environmental and economic benefits for those who choose to get involved. Benefits include resource savings, nature protection, along with employment and business opportunities. It is also important to remember that the best way to manage waste is to produce less of it. Minimizing waste is the first step toward green living. The ultimate goal to keep in mind is to produce as little waste as possible, recycle, or remanufacture as much as possible, and treat any unavoidable waste in a manner that is the least harmful to the environment and humans.

TEN EASY SOLUTIONS TO KEEP CLIMATE CHANGE AT BAY

Scientific American recommends the following easy steps that everyone can follow to keep the negative effects of a changing climate from wreaking havoc on our environment:

1. *Forego fossil fuels*: Eliminate the burning of coal, oil and, eventually, natural gas. Definitely an enormous challenge, there are alternatives that can be used when possible, such as commodities like plant-derived plastics, biodiesel, and wind and solar power.
2. *Infrastructure upgrade*: Investing in new infrastructure or drastically upgrading existing highways and transmission lines would help cut greenhouse gas emissions and help promote economic growth in developing countries. Using energy-efficient buildings and improved cement-making processes is important to include in this process to keep GHG at a lower level.
3. *Move closer to work*: In the United States, transportation is the second leading source of GHG emissions. For example, burning one gallon of gasoline produces 20 pounds of CO_2. This input of GHGs into the atmosphere could be heavily curtailed, however, if people moved closer to their place of employment, used mass transit, or switched to walking, cycling, or using some other type of transportation that did not require anything other than human energy. Other

considerations include teleworking (working from home) several days a week.

4. *Consume less*: The most obvious way to cut back on GHG emissions is to buy less stuff. This can be not using a car to relying on reusable grocery bags to buying less to buying local. One idea is to think purchasing green products—buying products that will have the least impact on the environment. For instance, when purchasing a car, look at a hybrid. Or when grocery shopping, buy in bulk so that you buy the product where less packaging material was used.

5. *Be efficient*: This is the idea of "doing more with less." When driving, make sure your car is well-maintained, your tires are properly inflated, and you do not speed (this can limit the amount of GHG emissions from a vehicle). Buy energy efficient appliances, such as Energy Star certified items. When you are home, turn the lights off when you are not in the room. Most of this is just common sense—you just need to be aware of it.

6. *Eat smart*: Much of the agriculture in the United States requires barrels of oil for the fertilizer to grow it and diesel fuel to harvest and transport it. Some grocery stores that stock organic produce that do not require fertilizer is still shipped from far away, so it still contributed GHGs due to the transportation part of the process. Buying local at farmers markets not only helps the local economy but also eliminates the GHGs that would have been added to the atmosphere if it had been transported a longer distance. Meat requires pounds of feed to produce just a single pound of protein. The problem with food from the grocery store is that there is no way to know for sure how far it has had to travel to get to its endpoint. University of Chicago researchers have estimated that each meat-eating American produces 1.5 tons more GHGs through their food choice than do vegetarians. In addition, it takes less land to grow crops that it does to raise animals.

7. *Stop cutting down trees*: Each year, 13 million hectares of forests are cut down. In the tropics alone, 1.5 billion metric tons of carbon are added to the atmosphere. Roughly 20 percent of human-caused GHG emissions come from this source, which could be avoided. If agricultural practices were improved, more people recycled paper, and better forest management practices were implemented (balancing the amount of wood harvested with trees being planted), it would offset this emission. To also help in this area, try to purchase and refurbish used goods or if buying new, make sure the wood is certified to have been sustainably harvested.

8. *Unplug*: It is a fact that U.S. citizens spend more money on electricity to power devices when off than when on. Televisions, computers, and so forth actually consume more energy when in sleeping mode or "switched off," so it is just better to completely unplug them. Just as a comparison, 1 billion kilowatt-hours of electricity ($100 million at current electricity prices) is the equivalent of the release of more than one million metric tons of greenhouse gases.

9. *One child*: There are at least 7.4 billion people living today. The United Nations predicts it will be at least 9 billion by 2050. UNEP currently estimates that it takes 22 hectares to sustain each person today. This includes food, clothing, and other resources. Projecting that to future populations, this is unsustainable.

10. *Future fuels*: Probably the greatest challenge of this century will be replacing fossil fuels with something else. Current ideas include ethanol derived from crops or hydrogen electrolyzed out of water, but they all have their own drawbacks and none is immediately available on a global scale.

These are just a starting point of ideas. The good thing about science is that it is innovative, creative, and new ideas come up often from creative minds. With everyone working together, it is possible to lower all of our collective GHG emissions.

(Source: Biello, 2007)

ECONOMICS OF MITIGATION

Several models have been developed that attempt to reasonably calculate the expected costs of climate mitigation. Unfortunately, there is simply not an easy, one-size-fits-all answer for that. The latest analysis provided by the fifth IPCC Assessment on Mitigation of Climate Change puts it at 0.06 percent a year of global GDP growth. Others have put it slightly higher, others slightly lower. The problem is, there are so many variables involved and circumstances vary for place to place, making it a very hard number to hone in on. Besides that, models are predicting all the way through the year 2100, and it is nearly impossible to predict anything that far out. So, let us look at what we do know, perhaps a bit more realistically.

The models that climate economists use are called integrated assessment models (IAMs). They integrate models of energy markets and land use with GHG projections. They attempt to look at all the relevant factors but do come under some criticism because they can have inherent weaknesses, such as overstating costs of climate mitigation or understating the need for action, as discussed in a study of economic models by economists Richard Rosen and Edeltraud Guenther (Rosen and Guenther, 2015). What the climate models are good for, however, are for determining the effects of marginal changes in one or a small set of economic variables. These models can be used when isolating specific variables and looking at just them for shorter periods of time. The concept of being able to predict costs for time periods of 50–100 years into the future is not considered realistic, however, as too much can happen during that time period, changing the scenario. For example, if an oil crisis or a financial crisis were to occur that was not initially figured into the model it would disrupt the projections and the model would fail if it was not recalibrated (assuming the crisis had not been predicted). If a significant weather event were to occur, that would require that the model also be recalibrated to a new baseline, and so forth. Otherwise, the model is left to operate in a state of instability, eventually failing or giving false results.

Based on logic and the chance of random events occurring, Rosen and Guenther, when asked about whether or not it mattered if 100-year models could be reliably produced, candidly remarked "No, because humanity would be wise to mitigate climate change as quickly as possible without being constrained by existing economic systems and institutions, or risk making the world uninhabitable. This conclusion is clear from a strictly physical and ecological perspective, independent of previously projected economic trade-offs over the long run, and it is well-documented in the climate change literature. As climate scientists constantly remind us, even if the world successfully implemented a substantial mitigation program today, a much warmer world is already built into the physical climate system. And since we can never know what the cost of a hypothetical reference case would be, and since we must proceed with a robust mitigation scenario, we will never be able to determine the net economic benefits of mitigating climate change, *even in hindsight*."

Perhaps that is the most sound, straight-forward advice of all: taking action now. Projecting so far into the future may fall under the guise of "planning" ahead, but better served is the near future and what we can do now to make a difference. That said, looking more near term, the following is what the IPCC did have to say in their 2014 report on Mitigation of Climate Change as far as projections on where they foresee the direction countries need to be headed in mitigation efforts right now to get us headed down that 100-year road (Kille, 2014):

- Reducing carbon emissions from electrical generation is one of the most cost-effective ways to slow climate change. Plans include moving to low-carbon electricity supplies of up to 80 percent by 2050.
- The EPA proposed its "power plant rules" in June 2014, which are designed to accelerate the shift to natural gas and provide incentives to encourage the deployment of industrial-scale carbon capture and storage for coal-fired plants.
- Wind and other renewables have continued to grow rapidly and per-kilowatt prices are falling to near-parity with fossil fuels. States currently play a central role in the growth of renewables, especially wind and solar, and there are currently a wide range of incentives in place.
- Over the next 20 years, annual investments in renewables, nuclear, and electricity generation with carbon capture and storage are projected to rise by $147 billion, while those for fossil-fuel electrical generation capacity will decline by about $30 billion. (While the relative changes are significant, the average annual investment in energy systems is $1.2 trillion.)

Table 8.1 lists mitigation options suggested by the IPCC in their Fourth Assessment Report (IPCC, 2007).

CAP AND TRADE

Cap and trade is the most environmentally and economically realistic approach to controlling GHG emissions. The "cap" sets a limit on emissions, which is lowered over time to reduce the amount of pollutants released into the atmosphere. The "trade" creates a market for carbon allowances, helping companies innovate in order to meet, or come in under, their allocated limit. The less they emit, the less they pay, so it is in their best interest by being an economic incentive to pollute less.

The cap is seen as the only real way to limit pollution. It sets a maximum allowable level of pollution and then penalizes companies that exceed their emission allowance. No other system can guarantee to lower emissions. There are various advantages to having a cap. They are as follows:

- *The cap acts as a limit*: It limits the amount of pollution that can be released. It is measured in billions of tons of carbon dioxide (or equivalent) per year. It is set based on science.
- *The cap covers all major sources of pollution*: The cap limits emissions economy-wide, covering electric power generation, natural gas, transportation, and large manufacturers.
- *Emitters can release only limited pollution*: Permits or "allowances" are distributed or auctioned to polluting entities—one allowance per ton of carbon dioxide, or CO_2 equivalent heat-trapping gases. The total amount of allowances will be equal to the cap. A company or utility may only emit as much carbon as it has allowances for.
- *Industry can plan ahead*: Each year, the cap is ratcheted down on a gradual and predictable schedule. Companies can plan well in advance to be allowed fewer and fewer permits—less global warming pollution—each year.

The trading portion of the plan is what leads to investment and innovation. Some companies find it fairly simple to reduce their pollution to match their number of permits, while others find it more difficult. Trading lets companies buy and sell allowances, leading to more cost-effective pollution cuts, as well

TABLE 8.1
Mitigation Options

Sector	Key Mitigation Technologies and Practices Currently Commercially Available. *Key mitigation technologies and practices projected to be commercialized before 2030 shown in italics*	Policies, Measures, and Instruments Shown to be Environmentally Effective	Key Constraints and Opportunities to Implementation (Normal font = constraints; *italics = opportunities*)
Energy supply	Improved supply and distribution efficiency; fuel switching from coal to gas; nuclear power; renewable heat and power (hydropower, solar, wind, geothermal and bioenergy); combined heat and power; early applications of carbon dioxide capture and storage (CCS); *CCS for gas, biomass and coal-fired electricity generating facilities; advanced nuclear power; advanced renewable energy, including tidal and wave energy, concentrating solar, and solar photovoltaics*	Reduction of fossil fuel subsidies; taxes or carbon charges on fossil fuels Feed-in tariffs for renewable energy technologies; renewable energy obligations; producer subsidies	Resistance by vested interests may make them difficult to implement *May be appropriate to create markets for low-emissions technologies*
Transportation	More fuel-efficient vehicles; hybrid vehicles; cleaner diesel vehicles; biofuels; modal shifts from road transport to rail and public transport systems; nonmotorized transport (cycling, walking); land-use and transport planning; *second generation biofuels; higher efficiency aircraft; advanced electric and hybrid vehicles with more powerful and reliable batteries*	Mandatory fuel economy; biofuel blending and CO_2 standards for road transport Taxes on vehicle purchase, registration, use and motor fuels; road and parking pricing Influence mobility needs through land-use regulations and infrastructure planning; investment in attractive public transport facilities and nonmotorized forms of transport	Partial coverage of vehicle fleet may limit effectiveness Effectiveness may drop with higher incomes *Particularly appropriate for countries that are building up their transportation systems*
Buildings	Efficient lighting and daylighting; more efficient electrical appliances and heating and cooling devices; improved cook stoves, improved insulation; passive and active solar design for heating and cooling; alternative refrigeration fluids, recovery and recycling of fluorinated gases; *integrated design of commercial buildings including technologies, such as intelligent meters that provide feedback and control; solar photovoltaics integrated in buildings*	Appliance standards and labeling Building codes and certification Demand-site management programs Public sector leadership programs, including procurement Incentives for energy service companies	Periodic revision of standards needed Attractive for new buildings. Enforcement can be difficult Need for regulations so that utilities may profit *Government purchasing can expand demand for energy-efficient products* *Success factor: Access to third party financing*
Industry	More efficient end-use electrical equipment, heat and power recovery; material recycling and substitution; control of non-CO_2 gas emissions; and a wide array of process-specific technologies; *advanced energy efficiency; CCS for cement, ammonia, and iron manufacture; inert electrodes for aluminum manufacture*	Provision of benchmark information; performance standards; subsidies; tax credits Tradable permits Voluntary agreements	May be appropriate to stimulate technology uptake. Stability of national policy important in view of international competitiveness Predictable allocation mechanisms and stable price signals important for investments Success factors include: clear targets, a baseline scenario, third-party involvement in design and review and formal provisions of monitoring, close cooperation between government and industry
Agriculture	Improved crop and grazing land management to increase soil carbon storage; restoration of cultivated peaty soils and degraded lands; improved rice cultivation techniques and livestock and manure management to reduce CH_4 emissions; improved nitrogen fertilizer application techniques to reduce N_2O emissions; dedicated energy crops to replace fossil fuel use; improved energy efficiency, *improvements of crop yields*	Financial incentives and regulations for improved land management; maintaining soil carbon content; efficient use of fertilizers and irrigation	*May encourage synergy with sustainable development and with reducing vulnerability to climate change, thereby overcoming barriers to implementation*

(Continued)

TABLE 8.1 (*Continued*)

Mitigation Options

Sector	Key Mitigation Technologies and Practices Currently Commercially Available. *Key mitigation technologies and practices projected to be commercialized before 2030 shown in italics*	Policies, Measures, and Instruments Shown to be Environmentally Effective	Key Constraints and Opportunities to Implementation (Normal font = constraints; *italics = opportunities*)
Forestry/ forests	Afforestation; reforestation; forest management; reduced deforestation; harvested wood product management; use of forestry products for bioenergy to replace fossil fuel use; *tree species improvement to increase biomass productivity and carbon sequestration; improved remote sensing technologies for analysis of vegetation/soil carbon sequestration potential and mapping land-use change*	Financial incentives (national and international) to increase forest area, to reduce deforestation and to maintain and manage forests; land-use regulation and enforcement	Constraints include lack of investment capital and land tenure issues. *Can help poverty alleviation*
Waste	Landfill CH_4 recovery; waste incineration with energy recovery; composting of organic waste; controlled wastewater treatment; recycling and waste minimization; *biocovers and biofilters to optimize CH_4 oxidation*	Financial incentives for improved waste and wastewater management Renewable energy incentives or obligations Waste management regulations	*May stimulate technology diffusion* Local availability of low-cost fuel Most effectively applied at national level with enforcement strategies

Source: IPCC, Fourth Assessment Report.

as incentives to invest in cleaner technology. Companies also have the flexibility to trade with companies anywhere because all carbon dioxide goes into the upper atmosphere and has a global effect. Therefore, it does not matter whether the factory making the emission cuts is in Los Angeles, New York, Paris, or Singapore—it reduces global emissions. Companies that participate in the cap and trade program see several advantages to it. Some of them include:

- Companies can turn pollution cuts into revenue. If a company is able to cut its pollution easily and cheaply, it can end up with extra allowances. It can then sell its extra allowances to other companies. This provides a powerful incentive for creativity, energy conservation, and investment.
- The option to buy allowances gives companies more flexibility. Conversely, some companies might have trouble reducing their emissions, or want to make longer-term investments instead of quick changes. Trading allowances gives these companies another option for how to meet each year's cap.
- The same amount of pollution cuts is achieved. While companies may exchange allowances with each other, the total number of allowances remains the same and the hard limit on pollution is still met every year. The goal is achieved through teamwork and cooperation.

The international trade in carbon credits is intended to promote investment in energy efficiency, renewable energy, and other ways of reducing emissions. In the majority of developed, industrialized countries, GHG-emitting companies have taken on the responsibility of running,

regulating, and facilitating the trade of carbon credits in the carbon market.

THE ECONOMICS OF CAP AND TRADE

Nat Keohane, PhD, newest member of the U.S. National Economic Council as of January 2011, serving as advisor on environmental and energy policy to President Obama, states that aggressive cap and trade is not only affordable but also critical to both the earth and humanity's future. The cost to the economy will be minimal—for example, it is estimated to be less than 1 percent of the U.S. GDP in 2030. Keohane also stresses that the longer action is delayed the more expensive it will be to make emission cuts. In addition, the more time that passes without addressing the issues, the more irreversible damage will be done by climate change. Through the use of economic models, Dr. Keohane determined that by continuing with a business as usual approach, the U.S. economy would reach $26 trillion by January 2030. With a cap on GHG emissions, however, the economy will reach the same level only 2–7 months later. Therefore, the impact on the economy would not be significant—"just pennies a day," according to Dr. Keohane (EDF, 2009).

He also stresses that total job loss would be minimal (the manufacturing sector would experience some impact), and the new carbon market would create a multitude of new jobs. Households will be most affected by energy costs, but even there the increase would be modest. Overall costs would be small enough to allow programs to be developed that would take any burden off low-income households.

Dr. Keohane believes that cap and trade is the best means to fight climate change because it not only gives each company the ability to choose how to cut their emissions, it gives the economy the most flexibility to reduce pollution in the most cost-effective way. He also says it turns market failure into market success: "Global warming is a classic example of what economists term 'market failure.'" GHG emissions have sky-rocketed because their hidden costs are not factored into business decisions—factories and power plants pay for fuel but not for the pollution they cause. Putting a dollar value on the pollution fixes that failure and gives industry incentive to pollute less" (Keohane and Goldmark, 2008).

He also says it taps American ingenuity and that history clearly shows that Americans can overcome steep challenges. In two short years during World War II, for example, Americans redirected much of the U.S. economy. Manufacturers produced different goods against tight deadlines. Detroit converted car factories to munitions production. Fireworks factories made military explosives. A. C. Gilbert, a maker of model train engines, produced airborne navigational instruments. Therefore, based on past performance, given the right incentives, the United States can transform the way energy is made as well.

He also cautioned that we must act immediately, or costs and risks will rise. The longer we wait to curb pollution, the steeper the cuts must be to avoid catastrophic climate change. Time is required to develop new technologies and build infrastructure. Plus, developing countries like China and India are waiting for the United States to act before they take action. Because of these reasons, there is very little time remaining to cap GHG emissions before a large risk of climate catastrophe and heavy economic costs are incurred. But if action is taken now, it can be successfully done—affordably.

CARBON PRICING

Carbon pricing is a market-based strategy for lowering global warming emissions. The aim is to put a price on carbon emissions—an actual monetary value—so that the costs of climate impacts and the opportunities for low-carbon energy options are better reflected in our production and consumption choices. Carbon pricing programs can be implemented through legislative or regulatory action at the local, state, or national level.

What is important to note is that in most cases, the real costs of climate change impacts today—such as the growing costs of wildfires (the loss of lives, property, habitat, and resources); public health and the damaging costs of heat waves (heat stroke, death, damage to power systems, and crop failure); the costs of catastrophic weather events (flooding, heavy downpours, hurricanes, and droughts)—are all currently borne by taxpayers, insurance companies, and the individuals who are directly affected. What is not happening, is that those who play a key part and hold an important stake in the environmental responsibility are not being held accountable, and that is the producers and consumers of the carbon-intensive goods that are causing the GHG emissions. While many do not stop to think about it like that, it is the producers and consumers of the products that are contributing

to the climate change situation, yet the victims of the impacts are bearing the costs.

Putting a price on carbon helps to incorporate climate risks into the cost of doing business. Emitting carbon becomes more expensive, and consumers and producers seek ways to use technologies and products that generate less of it. The market then operates on an efficient means to cut emissions, which is geared to encourage a shift to a clean energy economy and move innovation toward low-carbon technologies. Along with this, complementary renewable energy and energy efficiency policies are also critical to cost-effectively drive down emissions.

Carbon pricing is considered to be a powerful, efficient, and flexible tool for helping address climate change. It is supported by experts, businesses, policymakers, various civil groups, investors, states, and several countries. In fact, carbon pricing programs are already in use in many states and countries, such as California, the Regional Greenhouse Gas Initiative States (a cooperative effort among Connecticut, Delaware, Maine, Maryland, Massachusetts, New Hampshire, New York, Rhode Island, and Vermont) to reduce CO_2 emissions from the power sector, and Europe (EU emissions trading system).

There are two basic approaches to putting a price on carbon: one is the *cap-and-trade program*, the other is with a *carbon tax*. The cap-and-trade has already been discussed. With a carbon tax, laws or regulations are enacted that establish a fee per ton of carbon emissions from a sector or the whole economy. Owners of emission sources that are subject to the tax would have an incentive to lower their emissions, by transitioning to cleaner energy and using energy more efficiently. A rising carbon tax helps ensure a decline in emissions over time. Figure 8.3 shows the areas worldwide that participate in carbon pricing.

There are also *hybrid approach* options. These include programs that limit carbon emissions but set bounds on how much the price can vary in order to keep them within a specified range. Other approaches could include tailoring the tax to meet specific emission reduction goals, or creating a hybrid model for an area where some participants might pay a carbon tax and others might operate under a cap-and-trade arrangement. Still, others might have a cross between a cap-and-trade and renewable electricity. These arrangements can be as flexible and innovative as needed to meet the end goal. On another level, gasoline taxes, severance taxes for coal mining and natural gas or oil drilling are yet other ways of indirectly factoring a price on carbon into consumer or business decisions, with the end goal of ultimately cutting GHG emissions.

CARBON SEQUESTRATION AND STORAGE

In a computer model developed by Scott Doney of the Woods Hole Oceanographic Institution (WHOI), it indicates that the land and oceans will absorb less carbon in the future if current trends of emissions continue, which could mean significant shifts in the climate system. According to Doney, "Time is of

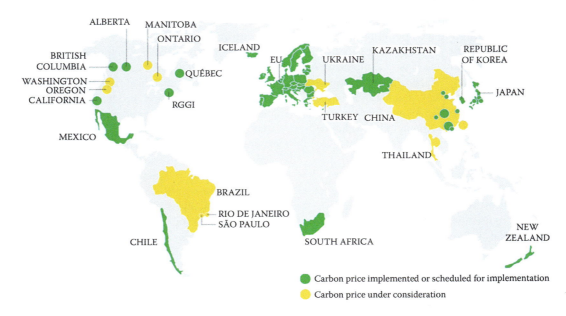

FIGURE 8.3 Carbon pricing worldwide. Thirty-nine countries and 23 subnational jurisdictions have some form of carbon pricing in place, covering 12 percent of all greenhouse gas emissions. (Courtesy of World Bank, Washington, DC.)

the essence in dealing with greenhouse gas emissions. We can start now or we can wait 50 years, but in 50 years we will be committed to significant rapid climate change, having missed our best opportunity for remediation." He also stressed that the Earth's ability to store carbon in its natural reservoirs is inversely related to the rate at which carbon is added to the atmosphere. In other words, as soon as humans cut GHG emissions, the easier it will be for the Earth to naturally store carbon. He stresses that the study suggests that land and oceans can absorb carbon at a certain rate, but at some point they may not be able to keep up (ScienceDaily, 2010). Therefore, because of all the excess GHG that we are pumping into the atmosphere at a furious rate, we can no longer rely on natural mechanisms to solve the problem. One possible solution is to store the excess CO_2 somewhere—a process known as carbon sequestration (or capture).

Carbon sequestration is a geoengineering technique focused on mitigating climate change by finding methods to store CO_2 or other forms of carbon released from fossil fuel combustion. It is one method designed to slow the atmospheric and marine accumulation of GHGs. It is not a way to cut back the production of GHGs, it is a way to store those already emitted. There are two basic categories: indirect and direct sequestration. Indirect sequestration does not require human-controlled manipulation of CO_2. Instead, it is accomplished through natural processes, such as the uptake of CO_2 by living organisms—such as photosynthesis. Direct sequestration is an active process, which is the deliberate human-controlled separation and capture of CO_2 from other by-products of combustion. It is then transferred to some nonatmospheric reservoir for permanent (or semi-permanent) storage.

Indirect Sequestration

Currently, the primary means of indirect sequestration is through the growth of forests. The least expensive method, it is limited of course by the amount of the Earth's land held in forests. In this setting, the plants accumulate CO_2 naturally. This type of sequestration is actually lowering every year due to accelerating rates of deforestation. As more land is destroyed, less land remains forested, resulting in less land holding CO_2 in natural reservoirs. Not only does this deforested land no longer store CO_2, but as forests are burned during the deforestation process, the CO_2 that has been stored long-term is also released to the atmosphere.

Photosynthesis is the core process of indirect sequestration. This is what converts airborne CO_2 along with water into energy that is stored in plant tissue as glucose ($C_6H_{12}O_6$) or other compounds for later use. Oxygen is released from the plant as a by-product. We are producing roughly 9 Gt of CO_2 annually. Annual rates of sequestering in a growing forest could be as high as 1–3 kg CO_2/m^2. Therefore, more than 400 million hectares of growing forest would need to be added annually in order to accomplish this. Unfortunately, deforestation continues. On the brighter side, however, world deforestation has slowed down as more forests are becoming better managed and lands are being put under protection (FAO, 2015).

There are several advantages of indirect sequestration. They include:

- They are a relative low-cost alternative.
- Forests can prevent flooding and reduce topsoil runoff in some regions, thereby preventing soil erosion and loss of nutrients.

- Forests are capable of moderating extremes in local climate in the vicinity of the forest.
- Forests provide, with careful management practices, sustainable harvesting of forest products, such as specialized timbers, edible products, resins, extracts, and so forth.
- They provide a source of employment opportunities for local populations.

There are also some concerns about indirect sequestration, such as:

- It is only a short-term solution.
- The future costs are unknown.
- Projects must be certified.
- There is an element of uncertainty in carbon retention properties during climate change. For example, if a drought occurs and the trees perish, they can no longer store CO_2.
- There is a need to maintain forests in equilibrium long-term.
- There may be unexpected costs in the future.

There have been proposals for man-made projects, similar in nature to the natural indirect tree model. Global Research Technologies (GRT) has devised a new technology prototype above-ground collector system to perform the same basic function as what forests do: collect atmospheric CO_2 over the next few years to evaluate it as a secondary energy collections system (ScienceDaily, 2010).

Direct Sequestration

There are several mechanisms in effect for direct sequestration. Presently, about one-third of all human-generated carbon emissions have dissolved in the ocean, but how fast the ocean can remove CO_2 from the atmosphere depends on atmospheric CO_2 levels, ocean circulation, and mixing rates. The more CO_2 in the atmosphere, the more the ocean needs to absorb. Also, the more rapid the circulation the greater the volume of water that is exposed to the higher CO_2 levels, which increases uptake by the ocean. Climate change, however, will cause ocean temperatures to rise, and warmer water holds less dissolved gas, which means the oceans will not be able to store as much anthropogenic CO_2 as climate change progresses. This means that the rate will slow down and lower the effectiveness of ocean sequestration as an efficient mechanism for carbon sequestration. Another negative side effect on the oceans from climate change is that increasing amounts of CO_2 in the water will increase its acid content. When CO_2 gas dissolves in ocean water, it combines with water molecules (H_2O) and forms carbonic acid (H_2CO_3). The acid releases hydrogen ions into the water. The more hydrogen ions in a solution, the more acidic it becomes. According to the WHOI, hydrogen ions in ocean surface waters are now 25 percent higher than in the pre-industrial era, with an additional 75-percent increase projected by 2100.

Therefore, it is also necessary to look at direct carbon storage options, as well. Carbon capture and storage (CCS) is a process that involves capturing the CO_2 arising from the combustion of fossil fuels (such as in power generation or refining fossil fuels), transporting it to a storage location, and isolating it long-term from the atmosphere. Before CO_2 gas can be sequestered from power plants and other point sources, however, it must be captured as a relatively pure gas. The U.S. Department of Energy reports that, on a mass basis, CO_2 is the 19th largest commodity chemical in the United States, and it is routinely separated and captured as a by-product from industrial processes such as synthetic ammonia production and limestone calcinations (DOE, 2011) (Figure 8.4). CCS has the potential to reduce overall mitigation costs, but its widespread application would depend on overall costs, the ability to successfully transfer the technology to developing countries, regulatory issues, environmental issues, and public perception. The capture of CO_2 would need to be applied to large point sources, such as energy facilities or major CO_2-emitting industries to make it cost effective. Potential storage areas for the CO_2 would be in geological formations (such as oil and gas fields, nonminable coal seams, and deep saline formations), in the ocean (direct release into the ocean water column or onto the deep sea floor), and industrial fixation of CO_2 into inorganic carbonates.

Current technology captures roughly 85–95 percent of the CO_2 processed in a capture plant. A power plant that has a CCS system (with an access geological or ocean storage) uses approximately 10–40 percent more energy than a plant of equivalent output without CCS (the extra energy is for the capture and compression of CO_2). The final result with a CCS is that there is a reduction of CO_2 emissions to the atmosphere by 80–90 percent compared to a plant without CCS.

When CO_2 is captured, it must be separated from a gas stream. Techniques to do this have existed for 60 years. Used in the production of town gas by scrubbing the gas stream with a chemical solvent, CO_2 removal is already used in the production of hydrogen from fossil fuels. This practice helps remove CO_2 from contributing to climate change. When the CO_2 is transported to its storage site, it is compressed in order to reduce its volume; when it is compressed it only occupies 0.2 percent of its normal volume. Each year, several million tons of CO_2 are transported by pipeline, ship, and road tanker.

Today, there are several options for storing CO_2. Initially, it was proposed to inject CO_2 into the ocean where it would be carried down into deep water where it would stay for hundreds of years. In order for any CCS scheme to be effective, however, it needs to sequester huge amounts of CO_2—comparable to what is currently being submitted into the atmosphere—in the range of gigatons per year. Due to the size requirement, the most feasible storage sites are the Earth's natural reservoirs, such as certain geological formations or deep ocean areas.

The technology of injecting CO_2 underground is very similar to what the oil and gas industry uses for the exploration and production of hydrocarbons. It is also similar to the underground injection of waste practiced in the United States.

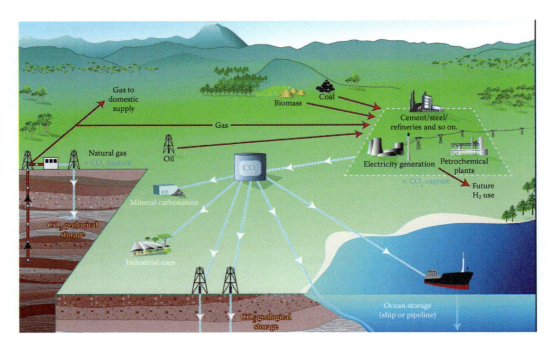

FIGURE 8.4 The PICC's schematic diagram of possible CCS systems, showing the sources for which CCS might be relevant, transport of CO_2, and storage options. (Courtesy of Rubin, E. et al., IPCC special report: Carbon dioxide capture and storage technical summary, http://www.ipcc.ch/pdf/special-reports/srccs/srccs_technicalsummary.pdf, 2011.)

In the same manner, wells would be drilled into geological formations and the CO_2 would be injected. This is also the same method used today for enhanced oil recovery. In some areas, it has been proposed to pump CO_2 into the ground for sequestration while simultaneously recovering oil deposits. There are arguments both for and against this strategy; on one hand, recovering oil would offset the cost of sequestration. On the other hand, burning the recovered oil as a fossil fuel adds additional CO_2 to the atmosphere, which offsets some of the positive effects of the sequestration.

Two other strategies involve injecting CO_2 into saline formations or into nonminable coal seams. The world's first CO_2 storage facility, located in a saline formation deep beneath the North Sea, began operation in 1996. Other alternatives have been proposed as well, such as using CO_2 to make chemicals or other products, fixing it in mineral carbonates for storage in a solid form, such as solid CO_2 (dry ice), CO_2 hydrate, or solid carbon. Another option is to capture CO_2 from flue gas using micro-algae to make a product that can be turned into a biofuel. In order to decide where to find feasible sites for carbon sequestration, it is important to know where large carbon sources are geographically distributed in order to assess their potential. This enables managers to estimate the costs of transporting CO_2 to storage sites.

The IPCC believes that more than 60 percent of global CO_2 emissions originate from the power and industry sectors. Geographically, 66 percent of these areas occur in three principal regions worldwide: Asia (30 percent), North America (24 percent), and Western Europe (12 percent). In the future, however, the geographical distribution of emission sources is expected to change. Based on data from the IPCC, by 2050 the bulk of emission sources will be from the developing regions, such as China, South Asia, and Latin America. The power generation, transport, and industry sectors are still expected to be the leading contributors of CO_2.

Global storage options are focusing primarily on geological or deep ocean sequestration. It is expected that CO_2 will be injected and trapped within geological formations at subsurface depths greater than 800 meters where the CO_2 will be supercritical and in a dense liquid-like form in a geological reservoir, or injected into deep ocean water with the goal of dispersing it quickly or depositing it at great depths on the ocean floor with the goal of forming CO_2 lakes. Current estimates place both types of sequestration as having ample potential storage space—estimates range from hundreds to tens of thousands of gigatons (Gt) of CO_2.

GEOLOGICAL FORMATION SEQUESTRATION

Many of the technologies required for large-scale geological storage of CO_2 already exist. Because of extensive oil industry experience, the technologies for drilling, injection, stimulations, and completions for CO_2 injection wells exist and are being patterned after current CO_2 projects. In fact, the design of a CO_2 injection well is very similar to that of a gas injection well in an oil field or natural gas storage project. Capture and storage of CO_2 in geological formations provides a way to eliminate the emission of CO_2 into the atmosphere by capturing it from large stationary sources, transporting it (usually by pipeline) and injecting it into suitable deep rock formations. Geologic storage of CO_2 has been a natural process within the Earth's upper crust for millions of years; there are vast reservoirs of carbon held today in coal, oil,

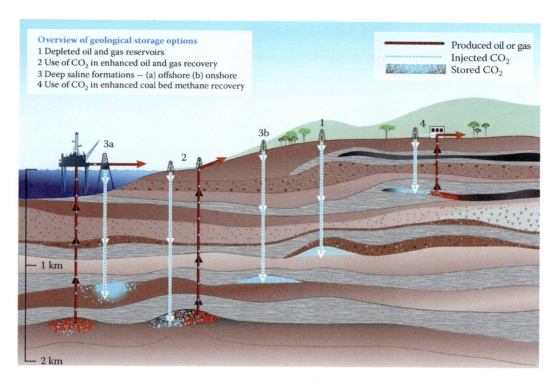

FIGURE 8.5 CO_2 can be sequestered in deep underground geological formations. (Courtesy of Rubin, E. et al., IPCC special report: Carbon dioxide capture and storage technical summary, http://www.ipcc.ch/pdf/special-reports/srccs/srccs_technicalsummary.pdf, 2011.)

gas, organic-rich shale, and carbonate rocks. In fact, over eons CO_2 has been derived from biological activity, igneous activity, and chemical reactions that have occurred between rocks, and fluids and gases have naturally accumulated in the subsurface layers (Figure 8.5).

The first time CO_2 was purposely injected into a subsurface geological formation was in Texas in the early 1970s as part of an "enhanced oil recovery" effort. Based on the success of this effort, applying the same technology to store anthropogenic CO_2 as a GHG mitigation option was also proposed around the same time, but not much was done to pursue any actual sequestration. It was not until nearly 20 years later, in the early 1990s that research groups began to take the idea more seriously. In 1996, Statoil and its partners at the Sleipner Gas Field in the North Sea began the world's first large-scale storage project. Following their lead, by the end of the 1990s, several research programs had been launched in the United States, Europe, Canada, Japan, and Australia. Oil, coal mining, and electricity-generating companies spurred much of the interest in this technology as a mitigation option for waste by-products in their respective industries.

Since this initial push, environmental scientists—many connected with the IPCC—have become involved in geologic sequestration as a viable option to combat climate change. The significant issues now are whether the technique is: (1) safe, (2) environmentally sustainable, (3) cost effective, and (4) capable of being broadly applied. Geologic storage is feasible in several types of sedimentary basins, such as oil fields, depleted gas fields, deep coal seams, and saline formations. Formations can also be located both on and offshore.

Offshore sites are accessed through pipelines from the shore or from offshore platforms. The continental shelf and some adjacent deep-marine sedimentary basins are also potential sites, but the abyssal deep ocean floor areas are not feasible because they are often too thin or impermeable. Caverns and basalt are other possible geological storage areas.

Not all sedimentary basins make good candidates, however. Some are too shallow, some not permeable enough, and others do not have the ability to keep the CO_2 properly contained. Suitable geologic formations require a thick accumulation of sediments, permeable rock formations saturated with saline water, extensive covers of low porosity rocks to act as a seal, and structural simplicity. In addition, a feasible storage location must also be economically feasible, have enough storage capacity, and be technically feasible, safe, environmentally and socially sustainable, and acceptable to the community. Most of the world's populations are concentrated in regions that are underlain by sedimentary basins. The table lists some of the current geological storage projects around the world (Table 8.2).

The most effective geologic storage sites are those where the CO_2 is immobile because it is trapped permanently under a thick, low-permeability seal, is converted to solid minerals, or is absorbed on the surface of coal micropores or through a combination of physical or chemical trapping mechanisms. If done properly, CO_2 can remain trapped for millions of years. When converting possible local and regional environmental hazards, the biggest danger is if CO_2 were to seep from storage, human exposure to elevated amounts of CO_2 could cause respiratory problems. This is why these storage facilities are closely monitored.

TABLE 8.2

Current Carbon Sequestration Projects Within Geological Formations

Country	Name	Location	CO$_2$ Capacity (Mtpa[a])	Project Start Date	Storage Formation (Depth and Type)
United States	Shute Creek	Wyoming	7 Mtpa	1986	3400 m sandstone/limestone
United States	Lost Cabin	Wyoming	0.9 Mtpa	2013	1400 m sandstone
United States	Illinois Industrial	Illinois	1 Mtpa	2016	2130 m sandstone
United States	Coffeyville	Kansas	1 Mtpa	2013	914 m sandstone
United States	Enid Fertilizer	Oklahoma	0.7 Mtpa	1982	2865 m carboniferous deposit
United States	Val Verde	Texas	1.3 Mtpa	1972	2135 m limestone
United States	Texas Clean Energy	Texas	2.4 Mtpa	2019	Not specified
United States	Century Plant	Texas	8.4 Mtpa	2010	Not specified
United States	Petra Nova	Texas	1.4 Mtpa	2016	2066 m sandstone
United States	Air Products	Texas	1 Mtpa	2013	1700 m sandstone
United States	Kemper County	Mississippi	3 Mtpa	2016	Not specified
United States	Riley Ridge	Wyoming	2.5 Mtpa	2020	Not specified
Canada	Quest	Alberta	1 Mtpa	2015	2 km Cambrian Basal Sands
Canada	Boundary Dam	Saskatchewan	1 Mtpa	2014	1.5 km Weyburn Oil Unit
Canada	Great Plains Synfuel Plant	Saskatchewan	3 Mtpa	2000	1500 m carbonate
South America	Petrobras Lula	Brazil	0.7 Mtpa	2013	7000 m Pre-salt carbonate
Europe	Snøhvit	Norway	0.7 Mtpa	2008	2670 sandstone
Europe	Sleipner	Norway	0.85 Mtpa	1996	1100 m sandstone
Europe	Caledonia	United Kingdom	3.8 Mtpa	2022	2200 m sandstone
Europe	Don Valley	United Kingdom	1.5 Mtpa	2020	Offshore deep saline formations
Europe	Rotterdam Opslag	Netherlands	1.1 Mtpa	2019	3500 m sandstone
Africa	In Salah	Algeria	Under revision	2004	1900 m sandstone
Middle East	Uthmaniyah	Saudi Arabia	0.8 Mtpa	2015	2100 m mudstone
UAE	Abu Dhabi	Abu Dhabi	0.8 Mtpa	2016	Complex carbonate
Europe	Teesside Collective	United Kingdom	2.8 Mtpa	2020s	Under evaluation
South Pacific	Gorgon	Australia	4 Mtpa	Not specified	2.3 km sandstone
South Pacific	Southwest Hub	Australia	2.5 Mtpa	2025	3000 m sandstone
South Pacific	Carbon Net	Australia	5 Mtpa	2020s	1500 m sub-sea
Asia	China Resources	China	1 Mtpa	2019	Not specified
Asia	Korea-CCS 1	Korea	1 Mtpa	2020	Under evaluation
Asia	PetroChina Jilin	China	0.5 Mtpa	2017	2.4 km oil-bearing formation
Asia	Korea-CCS 2	Korea	1 Mtpa	2023	Under evaluation
Asia	Sinopec Qilu	China	0.5 Mtpa	Retrofit	3000 m
Asia	Huaneng GreenGen IGCC	China	2 Mtpa	2020	3000 m
Asia	Shanxi Intl Energy	China	2 Mtpa	2020	Not specified
Asia	Yanchang Integrated CCS	China	0.44 Mtpa	2017	2200 m Yanchang Formation
Asia	Shenhua Ningxia	China	2 Mtpa	2020	Not specified

Source: Global CCS Institute. The global status of CCS: 2015 summary report. https://www.globalccsinstitute.com/projects/large-scale-ccs-projects#map.

[a] Mtpa: million tons per annum.

OCEAN SEQUESTRATION

Various technologies have been identified to enable and increase ocean CO$_2$ storage. One suggested option would be to store a relatively pure stream of CO$_2$ that has been captured and compressed. The CO$_2$ could be loaded onto a ship and injected directly into the ocean or deposited on the sea floor. CO$_2$ loaded on ships could be either dispersed from a towed pipe or transported to fixed platforms feeding a CO$_2$ lake on the sea floor. The CO$_2$ must be deeper than 3 kilometers, because at this depth, CO$_2$ is denser than seawater. Relative to CO$_2$ accumulation in the atmosphere, direct injection of CO$_2$ into the ocean could reduce maximum amounts and rates of atmospheric CO$_2$ increase over the next several centuries. Once released, it would be expected that the CO$_2$ would dissolve into the surrounding seawater, disperse, and become part of the ocean carbon cycle.

C. Marchetti was the first scientist to propose injecting liquefied CO$_2$ into waters flowing over the Mediterranean sill into the mid-depth North Atlantic, where the CO$_2$ would be isolated from the atmosphere for centuries—a concept that relies on the slow exchange of deep-ocean waters with the surface to isolate CO_2 from the atmosphere. Marchetti's objective was to transfer CO_2 to deep waters because the degree of isolation from the atmosphere increases with depth in the ocean. Injecting the CO_2 below the thermocline would enable the most efficient storage. In the short-term, fixed or towed pipes would be the most viable method for oceanic CO_2 release because the technology is already available and proven.

One proposed option would be to send the CO_2 down as "dry ice torpedoes." In this option, CO_2 could be released from a ship as dry ice at the ocean's surface. If CO_2 has been formed into solid blocks with a density of 1.5 tm^{-3}, they would sink quickly to the seafloor and could potentially penetrate into the seafloor sediment. Another method, called "direct flue-gas injection," would involve taking a power plant fire gas and pumping it directly into the deep ocean without any separation of CO_2 from the flue gas. Costs for this are still prohibitive, however.

It would be possible to monitor distributions of injected CO_2 using a combination of shipboard measurement and modeling approaches. Current analytical monitoring techniques for measuring total CO_2 in the ocean are accurate to about ± 0.05 percent. According to the IPCC, measurable changes could be seen with the addition of 90 metric tons of CO_2 per 1 km^3. This would mean that 1 metric gigaton of CO_2 could be detected even if it were dispersed over an area 10^7 km^3 (or 5,000 km × 2,000 km × 1 km), if the dissolved inorganic carbon concentrations in the region were mapped out with high-density surveys before the injection began (Figure 8.6).

In the case of monitoring the injection of CO_2 into the deep ocean via a pipeline, several monitoring techniques could be employed. At the point of entry from the pipeline into the ocean, an inflow plume would be created of high CO_2/low pH water extending from the end of the pipeline. The first monitoring array would consist of sets of chemical, biological, current sensors, and underwater cameras in order to view the end

of the pipeline. An array of moored sensors would monitor the direction and magnitude of the resulting plume around the pipe. Monitors would also be set along the pipeline to monitor leaks. A shore-based facility would provide power to the sensors and could receive real-time data. In addition, a forward system would monitor the area and could provide data over broad areas very quickly. Moored systems could monitor the CO_2 influx, send the information to surface buoys, and make daily transmissions back to the monitoring facility via satellite.

Kurt Zenz House, a Harvard researcher who was one of the first to propose undersea carbon storage, says "Under immense pressure and cold temperature below the seafloor, CO_2 forms a very dense liquid that is much heavier than sea water. In addition, gravity would prevent the liquefied gas from seeping upward, just as it prevents water in a well from flying into the air" (Doughton, 2008).

ECONOMIC BENEFITS OF CARBON STORAGE

There are several economic benefits of carbon storage. Both the carbon tax and cap-and-trade, and their hybrid options, all with any auctioned allowances are able to generate significant revenues. The use of the revenues is currently a hot topic. They need to be used to further mitigate climate change, improve air quality, create jobs in the sustainable energy sector, develop low-carbon or zero emission passenger transportation, and so forth. Some potential uses of carbon revenues could include, for example,

- Offsetting the disproportionate impacts of higher energy prices for low-income households (e.g., rebates on electricity bills for low income households).
- Investing in communities that are exposed to a high amount of pollution from fossil fuels.

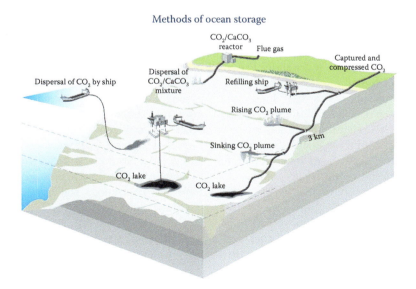

Methods of ocean storage

FIGURE 8.6   There are several proposed methods of CO_2 sequestration in the world's oceans. (Courtesy of Rubin, E. et al., IPCC special report: Carbon dioxide capture and storage technical summary, http://www.ipcc.ch/pdf/special-reports/srccs/srccs_technicalsummary.pdf, 2011.)

- Investing in climate-resilient infrastructure or relocation costs for communities at high risk.
- Contributing to efforts to cut carbon and prepare for climate change in developing countries.
- Providing transition assistance to workers and communities that depend on fossil fuels for their livelihoods (such as funding for job training and investments in economic diversification).
- Investing in renewable energy technology, such as clean vehicles, fuels, and transit options.

There are limitless opportunities to invest in technology that can make a difference. There are also opportunities to educate society about climate change so that everyone understands what the real issues are, enabling them to make better choices and become informed in policymaking and key political decisions that must be decided in the future. The more you know and understand, the wiser and better informed choices you can make.

ADAPTATION STRATEGIES

In a report issued by NASA in 2009, due to the increasing challenges caused by climate change, several scientists and policymakers in the United States came together to take part in the newly established United States National Assessment on the Potential Consequences of Climate Variability and Change—called the National Assessment for short (USGCRP, 2009). The National Assessment currently consists of 16 separate regional projects. Project leaders are charged with assessing their region's most vulnerable aspects—the resources that would be impacted most by climate change. These include resources such as water supply and quality, agricultural productivity, and human health issues. Once potential impacts are identified, strategies are proposed and developed to cope and adapt to climate change impacts should they occur.

Michael MacCracken, head of the coordinating national office, says "The goal of the assessment is to provide the information for communities, as well as activities to prepare and adapt to the changes in climate that are starting to emerge."

The more successful mitigation strategies apply toward the effects of climate change today, the less human populations will have to adapt in the short- and long-term. According to the PEW Center on Global Climate Change, however, recognition that the climate system has a great deal of inertia and is increasing, mitigation efforts alone are now insufficient to protect the Earth from some degree of climate change. Even if extreme measures to combat climate change were taken immediately to slow or even stop emissions, the momentum of the Earth's climate is such that additional warming is inevitable. Some of the warming that is unstoppable now is due to emissions of GHG that were released into the atmosphere decades ago. Because of this, humans have no choice but to adapt to the damage that has already been done.

Adaptation is not a simple, straight-forward issue for humans or ecosystems. Each system has its own "adaptive capacity."

In systems that are well managed (such as in developed countries and regions like the United States, Canada, Western Europe, and Australia), wealth, the availability of technology, responsible decision-making capabilities, human resources, and advanced communication technology help tremendously in successful adaptation to climate change. Societies that are able to anticipate environmental changes and plan accordingly ahead of time are also more likely to succeed.

The ability of natural ecosystems to successfully adapt is another issue, however. While biological systems are usually able to adapt to environmental changes and inherent genetic changes, the time scales are usually much longer than a few decades or centuries (such as the case with climate change). With changes in climate, even minor changes can be detrimental to natural ecosystems. An example of this is the polar bear in the Arctic. Today, sea ice is melting at a rapid rate leaving the polar bear with limited areas to breed and hunt. The situation has already become so grave in a short period of time that the polar bear's survival is now in jeopardy. The polar bear is now listed as threatened under the Threatened and Endangered Species Act based on evidence that the animal's sea ice habitat is shrinking and is likely to continue to do so over the next several decades. The polar bear was listed as threatened on May 14, 2008. Listing the polar bear as "threatened" means the animal is at risk of becoming an "endangered" species—in danger of extinction—in the foreseeable future if its habitat continues to be destroyed or adversely changed.

Like the polar bear, many of the world's ecosystems are stressed by several types of disturbances, such as pollution, fragmentation (isolation of habitat), and invasion of exotic species (this includes weed invasions). These factors coupled with climate change are likely to impact ecosystems' natural resiliency and prevent them from being able to adapt over the long term.

As far as human adaptability, some adaptation will involve the gradual evolution of present trends; other adaptations may come as unexpected surprises. Changes will involve sociopolitical, technological, economic, and cultural aspects. Because of the reality that populations are increasing, more people live in coastal areas, and more people live in flood plains and in drought-prone areas, adaptation measures will be required as climate changes. Fortunately, however, technology has developed to a point that there are better means today to successfully respond to climate change than there were in the past. For example, agricultural practices have evolved to the point where most crop species have been able to be translocated thousands of miles from their regions of origin by resourceful farmers.

A critical key to success is reactive adaptation; how willing will populations be to permanently change behaviors in order to adapt to changing climates and environmental conditions? Situations populations will have to adapt to will encompass issues such as:

- Rationed water
- Changes in water use habits
- Changes in crop type
- Resource conservation plans
- Mandatory use of renewable energy
- Restricted transportation types

Waiting to act until change has occurred can be more costly than making forward-looking responses that anticipate climate change, especially with coastal and floodplain development. A "wait-and-see" approach would be unwise with regard to ecosystem impacts. According to the PEW Center, "Proactive adaptation, unlike reactive adaptation, is forward-looking and takes into account the inherent uncertainties associated with anticipating change. Successful proactive adaptation strategies are flexible; they are designed to be flexible under a wide variety of climate conditions" (Kerr Gines, 2012).

An extremely important part in adaptation that cannot be overlooked is government influence and public policy. Governments have a strong influence over the magnitude and distribution of climate change impacts and public preparedness.

When climate and environmental disasters occur, it is usually government institutions that provide the necessary funding, develop the technologies, management systems, and support programs to minimize the occurrence of a repeat situation. A well-known example of this is the Dust Bowl that occurred in the Midwestern United States in the 1930s. It was through the efforts of the U.S. government that conservation efforts were started in order to properly manage the nation's soil and agriculture in order to prevent a repeat disaster of that nature. Table 8.3 lists some adaptation strategies suggested by the IPCC.

In view of climate change today and the already unstoppable effects into the future, adaptation and mitigation are necessary (and complementary concepts). Adaptation is

TABLE 8.3
Adaptation Strategies

Sector	Adaptation Option/Strategy	Underlying Policy Framework	Key Constraints and Opportunities to Implementation (Normal font = constraints; *italics = opportunities*)
Water	Expanded rainwater harvesting; water storage and conservation techniques; water reuse; desalination; water-use and irrigation efficiency	National water policies and integrated water resources management; water-related hazards management	Financial, human resources and physical barriers; *integrated water resources management; synergies with other sectors*
Agriculture	Adjustment of planting dates and crop variety; crop relocation; improved land management, e.g., erosion control and soil protection through tree planting	R&D policies; institutional reform; land tenure and land reform; training; capacity building; crop insurance; financial incentives, e.g., subsidies and tax credits	Technological and financial constraints; access to new varieties; markets; *longer growing season in higher latitudes; revenues from "new" products*
Infrastructure/settlement (including coastal zones)	Relocation; seawalls and storm surge barriers; dune reinforcement; land acquisition and creation of marshlands/wetlands as buffer against sea level rise and flooding; protection of existing natural barriers	Standards and regulations that integrate climate change considerations into design; land-use policies; building codes; insurance	Financial and technological barriers; availability of relocation space; *integrated policies and management; synergies with sustainable development goals*
Human health	Heat-health action plans; emergency medical services; improved climate-sensitive disease surveillance and control; safe water and improved sanitation	Public health policies that recognize climate risk; strengthened health services; regional and international cooperation	Limits to human tolerance (vulnerable groups); knowledge limitations; financial capacity; *upgraded health services; improved quality of life*
Tourism	Diversification of tourism attractions and revenues; shifting ski slopes to higher altitudes and glaciers; artificial snow-making	Integrated planning (e.g., carrying capacity; linkages with other sectors); financial incentives, e.g., subsidies and tax credits	Appeal/marketing of new attractions; financial and logistical challenges; potential adverse impact on other sectors (e.g., artificial snow-making may increase energy use); *revenues from "new" attractions; involvement of wider group of stakeholders*
Transport	Realignment/relocation; design standards and planning for roads, rail and other infrastructure to cope with warming and drainage	Integrating climate change considerations into national transport policy; investment in R&D for special situations, e.g., permafrost areas	Financial and technological barriers; availability of less vulnerable routes; *improved technologies and integration with key sectors (e.g., energy)*
Energy	Strengthening of overhead transmission and distribution infrastructure; underground cabling for utilities; energy efficiency; use of renewable sources; reduced dependence on single sources of energy	National energy policies, regulations, and fiscal and financial incentives to encourage use of alternative sources; incorporating climate change in design standards	Access to viable alternatives; financial and technological barriers; acceptance of new technologies; *stimulation of new technologies; use of local resources*

Source: IPCC.

a key requirement in order to lessen future damage. It is important to understand, however, that even though society will have to adapt, losses suffered will be inevitable and certain geographical areas will experience more extreme losses than others, particularly the developing countries.

While it is understood today that both mitigation and adaptation must occur simultaneously, each country will be faced with different issues to resolve and overcome Developing nations may face different issues of climate change than developed nations. Varying geographic locations will also experience differing ranges of change. The only way to manage this overwhelming issue is for nations to work together in a global effort to control GHG emissions, working to understand universal cause-and-effect relationships. Without international cooperation, there is little hope of stopping the problem before it is too late.

CONCLUSIONS

While it is true that many underdeveloped countries are now beginning to cause a climate change problem because they are industrializing and making the same mistakes that currently developed countries once made (which can be corrected with assistance and guidance from developed countries), there is also the issue of the undeveloped countries that are facing the worst effects of climate change—such as sea level rise—who have not ever significantly contributed to the problem. This suggests an uneven balance of responsibility. If a global climate policy is going to work, many argue that all countries must participate in the solution to some degree because emerging and developing economies are expected to produce 70 percent of global emissions during the next 50 years. Other experts add that any framework that does not include large and fast-growing economies (China, India, Brazil, and Russia) would be very costly and politically unwise. Others believe that undeveloped countries should not be held accountable. Like most issues, reality—and morality—most likely lies somewhere in between.

This chapter has presented several financial and technological strategies to handle the mitigation of climate change. Whichever methods are used will ultimately depend on the region, available technology, available finances, political policy, and prevailing social paradigm. What is critical is that action be taken immediately to fight climate change in order to lower the negative consequences of sea-level rise, flooding, drought, disease, and other disasters. Perhaps facing the issue realistically is through a combination of some workable means of adaptation, mitigation, and prevention. All three approaches have been discussed in this book and all are viable, workable components to the ultimate solution. While changing our personal behavior, mindset, and attitudes is perhaps the most critical element in the mix, we cannot overlook the assistance and good that can come from mitigation efforts, either, and where they are technologically, environmentally, and economically feasible, they are also worth looking at. And as we have learned, climate change is already in progress and its effects are not completely eradicable at this point, so adaptation is also a necessity. In this instance, developed countries need to help those that need assistance in a struggling world. The battle needs to be fought—and won—by all.

REFERENCES

Biello, D. November 26, 2007. 10 solutions for climate change. *Scientific American.* http://www.scientificamerican.com/article/10-solutions-for-climate-change/ (accessed September 6, 2016).

DOE. 2011. Carbon capture research. U.S. Department of Energy. http://fossil.energy.gov/programs/sequestration/capture/ (accessed September 9, 2016).

Doughton, S. July 14, 2008. Storing carbon dioxide under NW seafloor proposed. *The Seattle Times.* http://www.seattletimes.com/seattle-news/storing-carbon-dioxide-under-nw-seafloor-proposed/ (accessed September 8, 2016).

EDF. October 2, 2009. Cost of cutting carbon: Pennies a day. Environmental Defense Fund. http://www.edf.org/article.cfm?contentID=5405 (accessed September 9, 2016).

EnergySage Solar Research. 2015. Solar marketplace activity. https://www.energysage.com/data/#reports (accessed September 5, 2016).

Food and Agriculture Organization of the United Nations (FAO). September 7, 2015. World deforestation slows down as more forests are better managed. *FAOUN.* http://www.fao.org/news/story/en/item/326911/icode/ (accessed September 6, 2016).

Global CCS Institute. 2015. The global status of CCS: 2015 summary report. https://www.globalccsinstitute.com/projects/large-scale-ccs-projects#map (accessed May 20, 2017).

Harrington, R. October 13, 2015. The US is about to hit a big solar energy milestone. *Tech Insider.* http://www.techinsider.io/solar-panels-one-million-houses-2015-10 (accessed September 5, 2016).

Intergovernmental Panel on Climate Change. 2007. WG1—The physical science basis. http://www.ipcc.ch/ (accessed May 20, 2017).

Keohane, N. and P. Goldmark. 2008. What will it cost to protect ourselves from global warming? Environmental Defense Fund. http://www.edf.org/documents/7815_climate_economy.pdf (accessed September 9, 2016).

Kerr Gines, J. 2012. *Climate Management Issues: Economics, Sociology, and Politics.* Boca Raton, FL: CRC Press.

Kille, Leighton Walter. April 24, 2014. How to mitigate climate change: Key facts from the U.N.'s 2014 report. *Journalist's Resource.* http://journalistsresource.org/studies/environment/climate-change/united-nations-ipcc-working-group-iii-report-climate-change-mitigation (accessed September 5, 2016).

Rosen, R. A. and E. Guenther. February 2015. The economics of mitigating climate change: What can we know? *Technological Forecasting and Social Change* (accessed September 5, 2016).

Intergovernmental Panel on Climate Change (IPCC). 2005. IPCC special report on carbon dioxide capture and storage. Prepared by Working Group III of the IPCC Cambridge University Press. https://www.ipcc.ch/pdf/special-reports/srccs/srccs_wholereport.pdf (accessed May 21, 2017).

ScienceDaily. Comprehensive look at human impacts on ocean chemistry. *ScienceDaily.* June 21, 2010. http://www.sciencedaily.com/releases/2010/06/100617185131.htm (accessed September 9, 2016).

UNEP (United Nations Environment Programme). 2016. Climate change mitigation. http://www.unep.org/climatechange/mitigation/Home/tabid/104335/Default.aspx (accessed September 5, 2016).

U.S. Energy Information Administration (EIA). April 4, 2016. FAQs about renewables. http://www.eia.gov/tools/faqs/faq.cfm?id=92&t=4 (accessed September 5, 2016).

USGCRP. 2009. The national climate assessment. *United States Global Change Research Program*. http://www.globalchange.gov/what-we-do/assessment (accessed September 9, 2016).

SUGGESTED READING

Carbon Capture and Storage Association. 2016. What is CCS? *Carbon Capture and Storage Association*. http://www.ccsassociation.org/what-is-ccs/ (accessed September 9, 2016).

Environmental Protection Agency. August 9, 2016. Adapting to climate change. https://www3.epa.gov/climatechange/adaptation/ (accessed September 9, 2016).

Hower, M. July 14, 2016. 7 companies to watch in carbon capture and storage. *GreenBiz*. https://www.greenbiz.com/article/companies-watch-carbon-capture-and-storage (accessed September 9, 2016).

Kato, T. May 29, 2008. Implications of climate change for Africa. Fourth Tokyo International Conference on African Development (TICAD IV). http://www.imf.org/external/np/speeches/2008/052908a.htm (accessed September 9, 2016).

UNEP (United Nations Environment Programme). 2016. Climate change adaptation. http://www.unep.org/climatechange/adaptation/ (accessed September 9, 2016).

Section II

Energy Sources, Their Future, and Sustainable Living

9 Energy
Basics, Concepts, and Quantification

OVERVIEW

Before we can begin discussing energy sources and their future, we must first gain a clear understanding of exactly what energy is—and what it is not. This chapter provides an introduction to the concepts of energy: what it is, how different aspects of it were discovered over centuries of time, and what the different forms and sources of energy are. It discusses the various ways that energy can be transported and how it can be converted from one form to another. It explains how different types of energy are typically measured and compared and discusses the many ways energy is utilized in society and how the trends of use have changed over time, and introduces what the implications of those changes mean to society today. Finally, it offers insight into the conservation of energy, why energy efficiency has become such an important issue today, and why switching to renewable energy after relying for so long on traditional resources makes so much sense in the rapidly-changing world that we live in.

INTRODUCTION

It seems these days that everyone is concerned about energy. Concerns about whether there is enough of it, concerns about the costs of different types. Many worry that energy resources may be running out or fear that energy sources are bad for our health. Some are concerned that some energy resources might ruin the environment. Many talk about the latest heat waves and the fact that blackouts and brown-outs are getting more common in large urban areas. There is interest in new technology and what the scientific world might come up with next to power this gadget and that equipment. Homeowners that are interested in getting off the electric grid are suddenly investing in solar panels for their home's roof, and now proudly producing their own electricity. Do you know any of these people? Are you one of them?

The truth is, we *should* be concerned about energy today. In a warming world with a stressed environment, the resources we depend on for our energy resources is an important thing to be thinking about today because it is an investment in our future. Traditionally, we have depended on fossil fuels to provide energy for our transportation, heating, industry, manufacturing, and so forth. But now we know those choices have harmed the environment and our future.

Now it is time to think of energy in a different way. It is time for a paradigm shift from nonrenewable resources to renewable ones. We need to shift from fossil fuels (petroleum products, coal, and nuclear energy) to renewable resources (solar, wind, and water); for our sakes, our children and the environment. This chapter will review energy and the reasons why a shift of thought in how we look at the energy sources we utilize are a major key to future growth, healthy living, and peace of mind.

DEVELOPMENT

If you ask 10 people what energy is, you just may get 10 different answers. So, to begin, let us define what we mean when we talk about energy, as it is used in multiple ways, in many different contexts. For instance, you may say:

> Energy is the stuff we need to be able to accomplish physical actions such as hiking, running, lifting a chair, cooking a pizza, or charging a smart phone.

With this description, you would be partially right, but not completely. This description describes some of the things energy is used for; but not what it is, how it behaves, or what happens to it after it is used. It also does not define it as whether it is a thing, has properties, a condition, a state, whether it is permanent, temporary, exists forever, or can be destroyed. Let us look a little deeper into what energy is.

What Is Energy?

To understand the concept of energy, it is important to understand the concept of work from the standpoint of physics. As an illustration, think of pushing a stalled car across an intersection. You would probably agree that is a form of work. In physics, the "work" of pushing that car is equal to the force you pushed with, multiplied by the distance over which you did the pushing. Or:

$$Work = Force \times Distance$$

Therefore, the scientific definition of energy is:

The energy of an object (or system), is how much work the object (or system) can do on some other object (or system).

In this example, you were able to push the car because your body has a certain amount of *chemical energy* in your body from the food you eat. This chemical energy is released to generate *force* via your muscles, which you then directed to push the car across the intersection. The change in your body's stored chemical energy is exactly equal to the work you did on the car, plus any *heat energy* generated in your body while you did the work.

There are several ways in which an object or a system can possess energy, and each way corresponds to having a different form of energy. The bottom line, however, is that energy always means the *ability to do work*.

THE DISCOVERY OF ENERGY

The discovery of mechanical energy was not straight forward and simple. It actually came about as a series of discoveries. The understanding of both *kinetic* and *potential* energy did not just come to some lucky or intuitive scientist on a whim. Both types of energy are quantifiable by exact mathematical formulas, whose properties had to be proved. It took many people a lot of hard iterative work over more than a 1000 years to overcome misconceptions about what energy was—and was not—before they finally derived the correct formulas. Early on, scientists such as Galileo Galilei (1564–1642) and Isaac Newton (1643–1727), discovered how forces are related to acceleration. The information they discovered is summed up by Newton's three famous Laws of Motion, which are:

1. An object at rest will remain at rest and an object in motion will remain in motion at constant velocity unless acted upon by a net force.
2. The net force on an object is equal to its mass times its acceleration (F = ma).
3. For every action there is an equal and opposite reaction.

Although these laws today may seem fairly simple and obvious, it really was not so to scientists centuries ago. Aristotle (384–322 BCE), for instance, believed that all objects eventually come to a natural state of rest. His conclusion was drawn from his practical experience—most objects in our experience do exactly that: they eventually stop because they are subject to forces, such as friction with the air, and these forces do bring objects in motion to rest with respect to the ground. The problem with his theory was that it did not account for objects that were *not* subject to interactions with other objects, such as the ground. Those objects will keep moving unchanged. Then Galileo came along 700 years later and realized that fact. The reason why it took several centuries before it was discovered was because it was an abstract notion—in the real world it is not possible to turn off all interactions. After that was discovered, it took the contributions of Isaac Newton and Gottfried

Liebniz (1646–1716), who both developed (independently) the body of mathematical techniques known as calculus, and applied it to analyze these laws. Part of their contributions include formalizing certain combinations of variables with special names—some of them are the different forms of energy that we are familiar with today, such as mechanical energy. Therefore, through this long series of discoveries, we eventually were given the mathematical definition of energy:

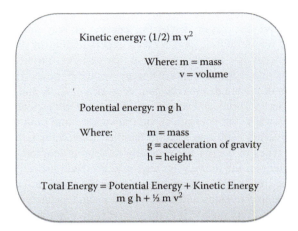

Kinetic energy: $(1/2) \, m \, v^2$

Where: m = mass
v = volume

Potential energy: m g h

Where: m = mass
g = acceleration of gravity
h = height

Total Energy = Potential Energy + Kinetic Energy
$m \, g \, h + \frac{1}{2} \, m \, v^2$

In the total energy equation, it is inferred that the (initial) potential energy must have been completely converted into the (final) kinetic energy. Because these two energies change exactly together in a way that keeps the energy expression constant, they stay constant in time, which is what makes them so useful.

Once these relationships were established, new physics equations were developed to show that when objects interact (exert forces on each other), then the work exerted by one object on another, defined as:

$$Work = Force \times Distance$$

is equal to the loss in energy that the object experiences while doing that work. It also implies that the work is equal to the energy that the object being acted on gains. This relationship is referred to as the *work-energy theorem* and is the reason that energy is conserved.

The discovery of heat, which is the microscopic motion of molecules and also a form of energy, was still not known a century after Newton's time. Scientists thought it might be some kind of substance unrelated to energy that was contained in objects and could flow between things. They believed it was released when objects were burned or worn away by friction. It was referred to as *caloric fluid*. It was not until English physicist James Prescott Joule (1818–1889), proved that heat is a form of energy by showing how it could come from conversion of other forms of energy—such as mechanical or chemical energy—and that when heat is considered in the calculation of the total energy, it is conserved.

Two other forms of energy—electromagnetic radiation (light) and nuclear energy ("rest-mass" energy of matter)—represent two physical extremes of energy. "Rest-mass energy" refers to the *intrinsic* energy that an object has by virtue of its simply having mass; whereas light is a "pure energy state," and has zero "rest-mass." Ordinary objects that have both rest-mass and kinetic energy can be thought of as being in a state somewhere between these two extremes. There were several debates and inconsistencies at the time Einstein was working on his Special Theory of Relativity about "rest-mass" energy, as well as the true properties of electromagnetic waves. In the process of clearing up all the confusion, Einstein deduced that matter itself is actually a form of energy, and the exact amount of the rest mass energy is given by his famous formula:

$$E = mc^2$$

where:
  E is the energy
  m is an object's mass
  c is the speed of light

What Einstein deduced from this formula is that there is an enormous amount of energy bundled up inside an ordinary matter. So large, in that, that only a small fraction of that energy is released in nuclear reactions in nuclear reactors and nuclear weapons. More importantly, the energy given off by the sun comes mostly from rest-mass *converted* into energy when hydrogen nuclei in the sun fuse to form helium nuclei (fusion). What Einstein actually did was extend Newton's theory to apply to speeds faster that the speed of light.

Electromagnetic radiation is considered a pure form of energy. Not all of it is visible to our eyes; in fact only a very small portion of the electromagnetic spectrum is visible. The entire spectrum from shortest wavelength to longest includes:

Region	Wavelength (meters)
Gamma Ray	10^{-13}
X-Ray	10^{-10}
Ultraviolet	10^{-8}
Visible	10^{-6}
Infrared	10^{-4}
Microwave	10^{-2}
Radio	10

Initially, light was believed to be tiny "units of energy, which Newton suggested. In the nineteenth century, it was shown that light was a wave with both electric and magnetic properties. This discovery also led to the realization that accelerating charged particles can generate light, which led to the invention or radio, microwaves, and other equipment. In the early twentieth century, Einstein and others discovered that light can also travel as discrete packets of energy, called photons. This gave light a unique property; it traveled both as particles and as waves, which is what the theory of quantum mechanics in modern physics is based on. It is also the property at the core of particle physics.

FORMS AND SOURCES OF ENERGY

Energy exists in many forms. Some forms are easy to identify as energy. Other forms cannot be seen until the energy is released. Some forms of energy, such as light and sound, carry energy from one place to another. Vibrating objects, which have kinetic energy, make sound. The vibrations spread through the air in waves, which carry the energy with them. Particles in the air vibrate as the sound waves go by but do not actually move along. The louder the sound, the more energy there is in the vibrations of the particles. In water—such as ocean waves—when the wave moves past, the water does not move along; it just shifts up and down. Energy moves through the wave, away from the point of origin. The higher the wave, the more energy there is.

Light is also a form of energy. It travels in straight lines called rays. A light ray carries energy from where the light is made. There is more energy in bright light than dim light, which is why the farther away a light source is, the dimmer it looks. Light is part of the family of rays called the *electromagnetic spectrum*. The electromagnetic spectrum includes radio waves, microwaves, infrared rays, ultraviolet rays, and X-rays. This energy is often called radiation.

Chemical energy is the energy stored in substances. Signs of this energy are only visible when the substance takes part in a chemical reaction. For example, when a fuel burns, the chemical energy in it is released as heat and light energy. Therefore, the energy stored in a battery is actually chemical energy. When the battery is connected to an electric circuit, the energy is released as electrical energy. Electrical energy is a useful form of energy because it can be sent along wires used to work in many different types of machines.

Nuclear energy is the energy stored in the nucleus of an atom. When a nuclear reaction occurs, energy can be released as heat and light and other radiation, such as X-rays.

When somebody moves an object, energy is used. Energy is transferred to the object. This is called mechanical energy. An object can have three types of mechanical energy—kinetic energy, elastic energy, and gravitational energy. Any object that is moving has kinetic energy. The faster an object moves and the faster it is, the more kinetic energy it has. Springs have elastic energy. This energy is stored when an object is bent, twisted, or stretched. When the object is released, it springs back into its original shape, releasing elastic energy.

When an object falls downward, it has gravitational energy because it is being pulled downward by the earth's gravity. The higher up the object—or the heavier it is—the more gravitational energy it has. When an object is moved upward against the force of gravity, it gains more gravitational energy. When the energy is stored within an object and can be released later, this is called potential energy. Physics and chemistry are the two principle scientific disciplines that study the movement, motion, actions, interactions, and reactions of materials.

TABLE 9.1
Major Types of Energy

Kinetic (Working) Energy

Electrical Energy

Electrical energy is delivered by tiny charged particles called electrons, typically moving through a wire. Lightning is an example of electrical energy in nature.

Radiant (Light) Energy

Radiant energy is electromagnetic energy that travels in transverse waves. Radiant energy includes visible light, X-rays, gamma rays, and radio waves. Light is one type of radiant energy. Sunshine is radiant energy, which provides the fuel and warmth that make life on earth possible.

Thermal (Heat) Energy

Thermal energy, or heat, is the vibration and movement of the atoms and molecules within substances. As an object is heated up, its atoms and molecules move and collide faster. Geothermal energy is the thermal energy in the earth.

Motion Energy

Motion energy is energy stored in the movement of objects. The faster they move, the more energy is stored. It takes energy to get an object moving, and energy is released when an object slows down. Wind is an example of motion energy. A dramatic example of motion is a car crash, when the car comes to a total stop and releases all its motion energy at once in an uncontrolled instant.

Sound Energy

Sound is the movement of energy through substances in longitudinal (compression/rarefaction) waves. Sound is produced when a force causes an object or substance to vibrate. The energy is transferred through the substance in a wave. Typically, the energy in sound is smaller than other forms of energy.

Potential (Stored) Energy

Chemical Energy

Chemical energy is energy stored in the bonds of atoms and molecules. Batteries, biomass, petroleum, natural gas, and coal are examples of stored chemical energy. Chemical energy is converted to thermal energy when people burn wood in a fireplace or burn gasoline in a car's engine.

Stored Mechanical Energy

Mechanical energy is energy stored in objects by tension. Compressed springs and stretched rubber bands are examples of stored mechanical energy.

Nuclear Energy

Nuclear energy is energy stored in the nucleus of an atom—the energy that holds the nucleus together. Large amounts of energy can be released when the nuclei are combined or split apart.

Gravitational Energy

Gravitational energy is energy stored in an object's height. The higher and heavier the object, the more gravitational energy is stored. When a person rides a bicycle down a steep hill and picks up speed, the gravitational energy is being converted to motion energy. Hydropower is another example of gravitational energy, where gravity forces water down through a hydroelectric turbine to produce electricity.

Most energy is initially derived from the energy of the sun. Once it leaves the sun, it is converted into other forms of energy, such as the energy that contributes to the growth of plants—which in turn gives many life-forms energy in the form of food. Humans get their energy from food. It is the energy from the sun that allowed plants to grow and animals to live back in prehistoric times and that has since provided humans with energy sources in the form of oil, gas, and coal—commonly called fossil fuels (Table 9.1).

The sun also provides the mechanisms that drive the water cycle, allow humans to tap into the power of moving water to generate electricity. It also provides the power to create wind energy—another form of energy used to generate electricity.

When energy is expended, it is converted into other forms. For example, the chemical energy of gasoline is converted into movement energy as it powers cars, trucks, airplanes, lawnmowers, weed eaters, leaf blowers, motorcycles, and other objects that are powered by an engine. Lighting fixtures convert electrical energy into light energy. Brakes on a car convert movement energy into heat energy. Stereos convert electrical energy into sound energy. When humans move, work, and participate in sports, they are converting the chemical energy derived from food to movement energy.

When energy is used in its various forms and then converted to other forms, some of the energy is converted into heat energy. When heat energy is used—such as in a furnace—it is usable energy. When it is not used, however, it is wasted energy. An example of wasted energy is the heat given off from a truck's engine during and after it has been running. The higher the efficiency of a system, the less waste (usually heat) it will produce.

In order to change the form of energy, an energy converter must be used. Energy converters range from simple to complex. The blades of a waterwheel are a simple form of energy converter that uses the energy of falling water to turn the huge wheel. The turning wheel then does the work. Waterwheels can operate many types of machinery—and have for hundreds of years worldwide—such as a mill, which grinds grain into flour. Steam, gasoline, and electrical engines are examples of more complex converters. These can make machines work or make vehicles move. Through the use of converters, energy can be changed into mechanical or electrical energy. Electrical energy supplies the heat, light, and power humans need every day. Mechanical energy moves machines and other objects.

ENERGY FROM THE SUN

The heat energy from the sun that reaches the earth is only one-thousandth of one-millionth of the heat produced by the sun. The sun is about 150 million kilometers away from the earth. If it was closer, the weather would be too hot for life, and everything would dry up and be burned. If the sun was farther away, the climate would be too cold for life to exist, and the earth would be covered with a layer of ice.

Many commonly used systems lose up to 40 percent of their energy as wasted heat. Engineers are continually trying to solve this problem through technology. Today, many converters can change energy from different sources directly into electrical energy with no resulting heat loss. The solar battery is an example of this type of converter.

Sources of energy are not always close to where they are needed. This means that energy has to be transported from one place to another. Because energy is central to our lives, enormous amounts are transported every day—much of it in the form of stored energy, or fuel.

The three principal fuels used for energy—oil, gas, and coal—are bulky and have to be transported as inexpensively as possible in order to keep the energy source affordable. Oil and gas can be moved by several methods through pipelines and by supertankers. A well-known example of oil being transported through a pipeline is the oil fields in northern Alaska (Figure 9.1). The fields are remote, and the subsequent lack of roadways and the harsh climate make it difficult to gain access by any other way than by pipeline. The pipe is raised up on stilts above ground for much of its journey. It cannot be buried for most of its length because the ground is frozen underneath with permafrost and the surface only thaws during the summer months, making it dangerous to bury the pipeline. The oil is kept moving at all times by pumps to keep it from freezing in the pipe (Figure 9.2a and b).

Because coal is a solid, transporting it requires ship, train, or truck. Coal is often transported from the coal mine to a

(a)

(b)

FIGURE 9.2 The Alaska pipeline (a) an example of the scraper pigs that keep the pipeline clear of wax build-up that forms when the oil cools as it passes through the pipeline. The oil enters the pipeline at a temperature of 49°C (111°F) and (b) the foils spaced at intervals on the tops of the support stilts draw off heat to the atmosphere so that it does not enter the ground and melt the permafrost. (Courtesy of Kerr.)

coal-using power station. Electricity, on the other hand, can be transported for long distances cheaply by using thick power cables suspended from tall pylons. Natural gas is very bulky, and it is only convenient to transport it as a gas through pipelines. Natural gas can also be cooled under pressure, converting it into a liquid. This type of gas can be used in camping stoves and barbecues.

Because there are many different kinds of energy—such as solar, chemical, electrical, mechanical, and nuclear energy—it is important to understand that one form of energy can be changed into another form. No matter what form energy takes, all are equal. An example of this is turning on a flashlight. As soon as the switch is flipped, the battery's chemical energy has instantly been changed into light energy. When a television is turned on, it converts electrical energy into light and sound energy. When a tree is cut with a wood saw, mechanical energy is changed into thermal energy.

Another important aspect about energy is that it never disappears. Although it can change forms, it is never created,

FIGURE 9.1 The Alaskan pipeline, running south of Fairbanks, Alaska. (Courtesy of Kerr.)

nor destroyed. It is only converted from one form to another. Science is based on this fact—no energy can be made without using an equal amount of another energy form.

HOW MUCH ENERGY DOES THE UNITED STATES USE OF THE TOTAL WORLD'S SHARE?

In 2013, world total primary energy consumption was about 543 quadrillion British thermal units (Btu), and U.S. primary consumption was about 97 quadrillion Btu, which is equal to 18 percent of world total primary energy consumption.

Energy can also be divided into renewable and nonrenewable groups. A renewable energy source is one that can be easily replenished. A nonrenewable one is one that cannot be realistically replenished. Renewable and nonrenewable energy sources can be used as primary energy sources to produce useful energy such as heat or used to produce secondary energy sources such as electricity. As an illustration, the electricity used in a home was probably generated from burning coal or natural gas, a nuclear reaction, or from a hydroelectric plant on a river. The gasoline that is used to fuel cars is made from crude oil (a nonrenewable energy) and may even contain a biofuel (a renewable energy) such as ethanol, which is made from processed corn. The graphic shows the energy sources used in the United States in 2015. Nonrenewable energy sources accounted for roughly 90 percent of all energy used. Biomass, which includes wood, biofuels, and biomass waste, is the largest renewable energy source, and it accounts for about half of all renewable energy and about 5 percent of the total U.S. energy consumption (Figure 9.3).

The nonrenewable energy sector includes five major categories. Most of the energy consumed in the United States is from these sources. They will be discussed in greater detail in Chapter 10. They are:

- Petroleum products
- Hydrocarbon gas liquids
- Natural gas
- Coal
- Nuclear energy

Crude oil, natural gas, and coal are commonly referred to as fossil fuels because they were formed over millions of years by the action of heat from the earth's core and pressure from rock and soil on the remains (or fossils) of dead plants and creatures like microscopic *diatoms*. Most of the petroleum products consumed in the United States are made from crude oil, but petroleum liquids can also be made from natural gas and coal. Nuclear energy is produced from uranium, a nonrenewable energy source whose atoms are split (through a process called nuclear fission) to create heat and, eventually, electricity.

The renewable energy sector includes five main sources (this will be discussed in detail in Chapter 11):

- Solar energy: This is energy derived from the sun to solar panels.
- Geothermal energy: This is the utilization of natural heat from inside the earth.
- Wind energy: This is the capture of air motion through the use of a large wind turbine.
- Biomass: This is the acquisition of plant material for the transformation of native vegetation into energy.
- Hydropower: The creation of energy derived from flowing, or dropping, water.

Most renewable energy comes either directly or indirectly from the sun. Sunlight, or solar energy, can be used directly for heating and lighting homes and other buildings, for generating electricity, and for hot-water heating, solar cooling, and a variety of commercial and industrial uses. The sun's heat also drives the wind, whose energy is captured with wind turbines. Then, the wind and the sun's heat cause water to evaporate. When this water vapor turns into rain or snow and flows downhill into rivers or streams, its energy can be captured using hydroelectric power.

Along with the rain and snow, sunlight causes plants to grow. The organic matter that makes up those plants is known as biomass. Biomass can be used to produce electricity, make transportation fuels, or chemicals in an energy form called biomass energy.

Hydrogen can also be found in many organic compounds as well as water. It is the most abundant element on earth, but it does not occur naturally as a gas. It is always combined with other elements, such as with oxygen, to produce water. Once separated from another element, hydrogen can be burned as a fuel or converted into electricity. Wood is a renewable resource because more trees can be grown to make more wood. Renewable energy resources also include geothermal energy from inside the earth.

Renewable energy, once jut a concept for the future, is becoming more mainstream every day. The public is realizing the benefits of switching to renewable energy. From installing

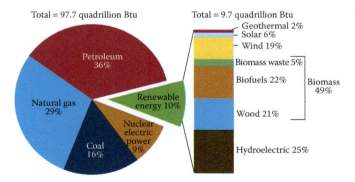

FIGURE 9.3 U.S. energy consumption by energy source, 2015. Note: Sum of components may not equal 100% because of independent rounding. (From U.S. Energy Information Administration, *Monthly Energy Review*, Tables 1.3 and 10.1 [April 2016], Preliminary Data.)

solar panels at home to using waste products to provide heat and power for communities to utility companies offering power from renewable energy sources (wind, solar, etc.) the future of renewable energy sources is now a reality for several areas of the world.

Both nonrenewable and renewable energy can be converted into secondary energy sources, such as electricity and hydrogen. A secondary energy source is simply an energy resource that is made using a primary resource. Electricity is a secondary resource because it can be generated by a number of different primary resources, such as coal or hydropower, which are both primary energy sources.

Most of the energy we use today is nonrenewable. Traditionally, this is what we have always had to rely on, but it is unsustainable. Because these resources are unrenewable, when they are used up, they will be gone. The geologic processes involved to produce them took place over such a long time that compared to the human time span they will not be produced again in our lifetime. In addition, a repeat of the exact same geologic processes and conditions it took to produce them initially is not likely. These energy resources—produced hundreds of millions of years ago, such as coal and fossil fuels—are examples of nonrenewable energy sources—when they are gone, they are gone. Today, they are the most common energy sources used to generate electricity, heat buildings, power cars, and manufacture products.

Fortunately, the use of renewable energy is growing. Renewable energy sources include biomass, geothermal energy, hydropower, solar energy, and wind energy. They are referred to as renewable energy sources because they are naturally replenished regularly. Day after day, the sun shines, the wind blows, and rivers flow. Renewable energy sources are used for electricity generation, for heat generation, and for transportation fuels. The illustration shows U.S. energy consumption by source in 2015 (Figure 9.4).

Secondary sources of energy are different than other sources because they are energy carriers—they are used to store, move, and deliver energy in an easily useable form. Another energy source must be used to make secondary sources of energy like electricity and hydrogen.

LAWS OF ENERGY

As previously stated, energy is neither created nor destroyed. The term that is used to describe this is *conservation of energy*. When people use energy, it does not simply disappear after it is used. Instead, energy is changed from one form into another. For example, a car engine burns gasoline; which converts the chemical energy contained in gasoline into mechanical energy. Solar cells change radiant energy into electrical energy. The energy changes form, but the total amount of energy in the universe remains the same.

The term energy efficiency is the amount of useful energy obtained from any type of system. If a machine is perfectly energy-efficient, it would convert all the energy put into the machine to useful work. In reality, converting one form of energy into another form of energy always involves a loss of useable energy. Most energy transformations are not very efficient. The human body is one very good example. Our bodies are similar to a machine, with food being the fuel that they require to keep them running. It is food—the energy provider—that gives a person energy to move, breathe, and think. But the human body is not very efficient at converting food into useful work. In reality, it is less than 5 percent efficient most of the time. The remaining 95 percent of the energy is lost as heat.

In any discussion of energy, it is important to understand that it revolves around the fact that there are three laws that pertain to energy, as follows:

- *First Law of Energy*: Energy can neither be created nor destroyed. This means that energy cannot be made out of nothing. The total amount of energy in the universe is constant. (Note here that this applies to a *closed system*. The earth is not a closed system, it receives constant energy from the sun).
- *Second Law of Energy*: The second law refers to the state of energy and is reflected in a measurement of the degree of disorder (a measurement called entropy). When a lump of coal is burned (which represents a material in a very ordered state), a change occurs which results in a more disordered state and you can never combine the resultant products (heat and gas) back to form the exact original lump of coal (First Law). The Universe, according to scientific evidence, is winding down, and the sun will eventually die. When an energy source is used, it is not destroyed; but enters a more disordered state. This makes the energy less available to use and in converting the energy to power means some loss of the original available energy.
- *Third Law of Energy*: The third law is that everything comes to a halt only when the temperature reaches

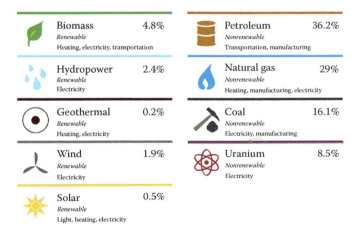

Biomass	4.8%	Petroleum	36.2%
Renewable		Nonrenewable	
Heating, electricity, transportation		Transportation, manufacturing	
Hydropower	2.4%	Natural gas	29%
Renewable		Nonrenewable	
Electricity		Heating, manufacturing, electricity	
Geothermal	0.2%	Coal	16.1%
Renewable		Nonrenewable	
Heating, electricity		Electricity, manufacturing	
Wind	1.9%	Uranium	8.5%
Renewable		Nonrenewable	
Electricity		Electricity	
Solar	0.5%		
Renewable			
Light, heating, electricity			

FIGURE 9.4 U.S. Energy Consumption by Source, 2015. Sum of individual percentages may not equal 100 because net imports of coal coke and of electricity are not included. (From EIA.)

−273.15°C (absolute zero). At this point, the entropy measurement is zero. Entropy is a measure of the unavailable energy in a closed thermodynamic system that is also considered to be a measure disorder.

Together, these laws help form the foundation of modern science. These are absolute physical laws—meaning that everything in the observable universe is subject to them.

TRANSPORTING ENERGY

Energy is constantly being transported from place to place and transferred between objects. When an object that possesses energy moves from one place to another, it *transports* its energy with it. Kinetic energy can be *transferred* from one object to another when objects collide. When this happens, the total energy generated during the collision is conserved in the process. The most important modes of transfer for renewable energy technology are the following:

- Light propagation in space
- Light propagation in materials:
  - *Transmission*: Transparent or translucent
  - *Reflection*: Coherent or diffuse
  - *Absorption*
- Heat propagation in materials and in space:
  - *Conduction*
  - *Convection*
  - *Radiation*
- Electrical

Light

Light (both photons and electromagnetic waves) propagates by itself in a vacuum at the speed of light, which is 299,792,458 m/s. When light interacts with materials it behaves in different ways depending on the nature of the material it is interacting with. It will either be transmitted, reflected, or absorbed. When it is transmitted, it passes through an object. This happens when an object is either transparent, enabling the light to pass straight through, or translucent, allowing the light to pass through, but causing its direction to be scattered by the material. When light is reflected, the light bounces off the material. Reflection can either be coherent, where the angle of incidence equals the angle of reflection; or diffuse, where the reflected direction is randomly scattered (Figure 9.5). When light enters a material but does not pass through, it is absorbed. When this happens, its energy is converted into heat, microscopic vibrations of the material, or is absorbed by chemical reactions triggered by the light, which is a photochemical effect.

Heat

There are three methods that heat energy can be transferred or transported: conduction, convection, and radiation. Conduction and convection refer to transfer of the thermal energy. Radiation is a conversion of energy to a different form (photons of light) and the subsequent travel (transport) of those

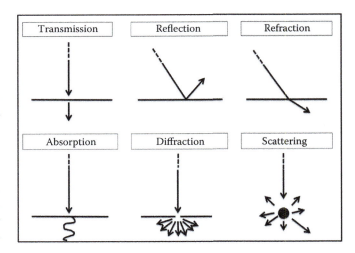

FIGURE 9.5 The simplest interaction with light is transmission, which occurs when light passes through the object without interacting. Light coming through window is a simple example of transmission. Reflection occurs when the incoming light hits a very smooth surface like a mirror and bounces off, like a mirror. Refraction occurs when the incoming light travels through another medium, from air to glass for example. When this happens the light slows down and changes direction. This change in direction is dependent on the light's wavelength so its spectrum of wavelengths are separated and spread out into a rainbow. Diffraction occurs when light hits an object that is similar in size to its wavelength. When light passes through a sufficiently-thin slit it will diffract and spread. If it is visible light, this will also create a rainbow. Absorption occurs when the incoming light hits an object and causes its atoms to vibrate, converting the energy into heat which is radiated. Anyone with a dark-colored car on a hot day will experience the effects of adsorption. Scattering occurs when the incoming light bounces off an object in many different directions. A good example of this is known as Rayleigh scattering, where sunlight is scattered by the gasses in our atmosphere. This is what gives the sky its blue color. (Courtesy of Kerr.)

photons. Conduction is the diffusion of thermal energy through a substance, which happens because hotter molecules—those that are vibrating, rotating, or traveling faster—interact with colder molecules, and in the process transfer some of their energy. For example, conduction of thermal energy is what makes the handle of a metal frying pan on the stove get hot, even though the handle is not exposed to the heating element. Metals are some of the best conductors of heat energy. Materials such as plastic or wood are poor conductors, which is why the handles of pans are often made of wood. Materials that are not good conductors are referred to as insulators.

This relationship can be expressed mathematically. The rate H, at which heat conducts through a material across a fixed temperature difference ΔT, for example, from the inside of a warm house to the outside through a wall, is given by the area A of the surface, times the temperature difference ΔT, divided by the thermal resistance R:

$$H = \frac{(A\,\Delta T)}{R}$$

R is also called the "R-factor" of the material. When considering the insulating power of the walls of a house, R is likely to have units of square feet divided by Btus per hour (ft²/(Btu-hr)). Btu stands for British Thermal Unit, and is the amount of energy needed to raise the temperature of one pound of water one degree Fahrenheit. Discussions involving conduction may also refer to the thermal conductivity K of a material, which is related to the R-factor:

$$R = \frac{L}{K}$$

where L is the thickness of the material. Therefore, if you look up the thermal conductivity of a material, you can calculate the R value for a piece of that material with thickness L.

Convection is the transfer of heat energy by the movement of a substance, such as a heated gas or liquid from one place to another. For example, hot air rising to the ceiling is an example of convection. This would be an example of a convection current.

Radiation applies to both the light waves (photons) and rays consisting of other subatomic particles, such as electrons (beta rays) and helium nuclei (alpha rays) that are emitted by radioactive materials.

With heat transfer, however, the term "radiation" refers just to light (electromagnetic waves), and in particular, to the fact that all objects, even those that are in equilibrium (at equal temperature) with their surroundings, continuously emit, or radiate electromagnetic waves (light waves) into their surroundings. The source of this radiation is the thermal energy of the materials—the movement of the object's molecules.

The amount of lightwave radiation radiated by ordinary objects is surprisingly large, even though we usually do not notice it. For example, when an object is at 21°C (room temperature), it radiates about 460 watts per square meter of its surface. It does not grow cold right away, however, because it is completely surrounded by other objects that are the same temperature—and even by the air itself—and these objects are also radiating energy. This means that the energy loss from radiation leaving the objects is balanced by the incoming radiation coming from the other objects.

As an example, if the sky is cloudy, then the heat radiating from the ground will be mostly absorbed and reradiated back to earth by the clouds, keeping the air near the surface warm. On a clear night, however, the ground and nearby air can cool very dramatically by radiating out into space, an effect called "radiation cooling."

The Black-Body Spectrum of Radiation from Objects

The spectrum of light radiated by objects—how much energy is radiated at each frequency—has approximately the same mathematical form for all objects. It depends only on the *temperature of the object and not on the specific kind of material*. The spectrum is called the "black-body spectrum," because it is most perfectly illustrated by objects which absorb all the light falling upon them (which means they are perfectly *black* in color). For relatively low temperatures, such as room temperature, most of the black-body spectrum is at long wavelengths of light, that is, in the infrared or longer wavelengths, which are invisible to the human eye, while at high temperatures the spectrum lies at shorter wavelengths, and can become visible if the temperature is very high.

As an example, an ordinary object sitting on a desk appears not to radiate anything (although it does), because most of its radiation is at wavelengths longer than light waves in the visible range. On the other hand, when an electric stove burner starts to glow red, it does so because it is reaching a temperature at which the black body spectrum is starting to strongly overlap the region of visible light.

Another example of a good black body radiator is the sun. Its peak lies in the visible range of the spectrum because it is so hot. For any temperature, the wavelength at which the black-body spectrum has its peak is given by *Wien's Displacement Law*:

$$\text{Peak wavelength in micrometers} = \frac{2900}{T}$$

where T is the temperature in degrees Kelvin. For the sun, which has a surface temperature of about 6000 Kelvin, we determine that the sun's peak wavelength is about 0.5 micrometers, which corresponds roughly to the color yellow, approximately in the middle of our visible range. This means that our eyes are well adapted to the peak wavelengths given off by the sun.

Electrical Energy

It is also possible to transfer energy via electrical transmission. Within a wire this is accomplished through electric fields associated with electrons in the metal wire. The electrons literally push on each other and convey forces through the wire, which thereby transfer energy. The electrochemical processes in a battery create positive and negative electric charges at the battery contacts which push on and force the moveable electrons in the wires to move. Electrical energy is converted to heat when some of the electrons encounter resistance. This means the electrons are pushed through materials causing the atoms of the material to start vibrating. In other configurations, the vibrations of electrons may create electric and magnetic fields—such as in the coils of a motor, which do work.

CONVERTING ENERGY AND UNITS OF MEASUREMENT

Energy can be converted from one form to another. This occurs through three basic methods:

1. Through the action of forces
2. When atoms absorb or emit photons of light
3. When nuclear reactions occur

When energy is converted through the action of forces, it can involve gravitational, electrical and magnetical, and frictional forces. Gravitational forces are involved when gravity accelerates a falling object. This converts its potential energy to kinetic energy. Conversely, when an object is lifted, the gravitational field stores the energy exerted by the lifter as potential energy in the earth-object system. When the object is dropped, the potential energy is converted to kinetic energy.

In electric and magnetic force fields, charged particles—upon which electrical fields exert forces—possess potential energy in the presence of an electric field similar to that of an object in a gravitational field. These force fields are able to accelerate particles, converting a particle's potential energy into kinetic energy. Charged particles can interact via the electric and magnetic fields they create, transferring energy between them, and in the case of an electrical current in a conductor, it causes molecules to vibrate. This converts electrical potential energy into heat.

With frictional forces, the macroscopic (large-scale) energy of an object—the potential and kinetic energy associated with the position, orientation, or motion of the entire object not counting the thermal or heat energy of the system, can be converted into thermal energy (heat), whenever the object slides against another object. The sliding causes the molecules on the surfaces of contact to interact via electromagnetic fields with one another and start vibrating.

The second method of energy conversion—atoms absorbing or emitting photons of light—occurs when light falls on an object. An incident photon may either pass through the object, be reflected by the object, or be absorbed by the atoms making up the object. If most of the photons pass through, the object is considered transparent. Depending on the smoothness of the surface on the scale of the photon's wavelength, the reflection may be either diffuse (rough surface) or coherent (smooth surface). If the photon is absorbed, the photon's energy may also be split up and converted in the following ways:

1. Photothermal effect: The absorbed energy may simply produce thermal energy, or heat in the object. In this case the photon's energy is converted into vibrations of the molecules called *phonons*. This is heat energy.
2. Photoelectric effect: the absorbed energy may be converted into the kinetic energy of conduction electrons, which becomes electrical energy.
3. Photochemical effect: The energy may bring about chemical changes which effectively store the energy.

The third method of energy conversion is in nuclear reactions. This is when rearrangements of the subatomic particles that make up the nuclei of atoms are involved. There are two basic types of nuclear reactions:

- Fission: when nuclei combine
- Fusion: when nuclei split apart

Measuring Energy: Units and Calculators

It is often useful to know how much energy is being changed from one form to another. For example, natural gas companies need to have a way to measure the fuel demands of their customers so they can provide enough when it is most needed. In most scientific experiments there is some form of energy measurement. A meter is an instrument that measures energy—such as the power and gas meters located on the sides of houses—and the standard unit of energy in the metric system is called a joule (J), named after James Joule.

Energy can be measured in many different units. A Calorie is defined as the amount of heat energy needed to raise the temperature of 1 kilogram of water by 1°C. Power is simply the rate at which the energy is changed. Energy can also be measured in units called foot-pounds. One foot-pound is the amount of work done in moving an object one foot against a one-pound force. Therefore, if a three-pound weight is lifted five feet off the ground, it is using 15 foot-pounds of energy.

Horsepower is another measurement describing energy. But it is a measure of power, not of energy itself. Power is not the same as energy. Energy is the ability to do work. Power is a measure of how quickly it is done. Mechanical power is measured in units called horsepower. James Watt, a Scottish engineer, first suggested the term when working with the newly invented steam engine. When doubtful farmers asked how many horses a steam engine could replace, this term was invented in order to make a logical comparison. Watt measured the amount of work a horse did in eight hours. Many types of power are now measured in units called watts.

MEASURING ENERGY

Energy is measured in joules, which are very small amounts of energy. A mug of hot chocolate cooling down at room temperature will release about 100,000 joules. The calorie is an old-fashioned unit often used to measure the energy contained in food. A slice of bread contains about 70 calories. One calorie equals about 4000 joules. Power is the rate at which energy is given off or used, and it is measured in watts. The use of one joule of energy every second is one watt. A 60-watt light bulb uses 60 joules of energy every second to give off heat and light.

Physical units reflect measures of distance, area, volume, height, weight, mass, force, impulse, and energy. Different types of energy are measured by different physical units:

- Barrels or gallons for liquid petroleum fuels (such as gasoline, diesel fuel, and jet fuel) and biofuels (ethanol and biodiesel)
- Cubic feet for natural gas
- Tons for coal
- Kilowatt-hours for electricity

To compare different fuels, it is necessary to convert the measurements to the same units. In the United States, the unit of measure most commonly used for comparing fuels is the British thermal unit (Btu), which is the amount of energy required to raise the temperature of one pound of water one degree Fahrenheit. One Btu is approximately equal to the energy released in the burning of a wood match. Because energy used in different countries comes from different places, Btu content of fuels varies slightly from country to country. Other common

TABLE 9.2

Btu Content of Common Energy Units, 2015

1 barrel of crude oil (42 gallons)	5,729,000 Btu (for U.S. produced crude oil)
1 gallon of gasoline	120,405 Btu
1 gallon of heating oil	138,500 Btu
1 gallon of diesel fuel	137,381 Btu (distillate fuel with less than 15 parts per million sulfur content)
1 barrel of residual fuel oil	6,287,000 Btu
1 cubic foot of natural gas	1,032 Btu
1 gallon of propane	91,333 Btu
1 short ton of coal	19,882,000 Btu
1 kilowatt hour of electricity	3,412 Btu

Source: U.S. Energy Information Administration, 2016a.

units for comparing energy include barrels of oil equivalent, metric tons of oil equivalent, metric tons of coal equivalent, and terajoules. Btu content of each fuel provided below is the average heat content for fuels consumed in the United States in 2015.

Table 9.2 illustrates the Btu content of common energy units.

It is often helpful to convert energy units in order to compare energy sources to make appropriate decisions. For example, converting different energy sources to Btu to decide whether or not to invest in a natural gas furnace or one that uses heating oil. Suppose you have a natural gas furnace in your home that used 81,300 cubic feet of natural gas for heating last winter. Your neighbor has a furnace that burns heating oil that used 584 gallons of heating oil over the same time period. You can convert the natural gas and heating oil consumption data into Btu to determine which home used more energy for heating, as follows:

Natural gas:
81,300 cubic feet (your home) × 1,032 Btu per cubic foot = 83,901,600 Btu

Heating oil:
584 gallons (neighbor's home) × 138,500 Btu per gallon = 80,884,000 Btu

Final result: You used more energy to heat your home.

Let us look at another example:

Let us suppose you are now looking for a new furnace for your home and you are comparing heating systems that use natural gas with systems that use heating oil. One factor you need to consider is the cost of the fuels. You can compare the price of the fuels on an equal basis by dividing the price per unit of the fuels by the Btu content of the fuels in million Btu per unit to get the price in dollars per million Btu, as follows:

Natural gas:
$10.40 per thousand cubic feet ÷ 1.032 million Btu per thousand cubic feet = $10.08 per million Btu

Heating oil:
$2.70 per gallon ÷ 0.1385 million Btu per gallon = $19.49 per million Btu

Final result: The cost per million Btu for natural gas is about half the cost of heating oil.

HOW BIG' IS A BARREL? BE MINDFUL OF WHICH 'BARREL' YOU ARE TALKING ABOUT!

A barrel is a unit of volume or weight that is different depending on who uses the term and what it contains. Here are some illustrations:

1 barrel (b) of petroleum or related products = 42 gallons
1 barrel of Portland cement = 376 pounds
1 barrel of flour = 196 pounds
1 barrel of pork or fish = 200 pounds
1 barrel of (U.S.) dry measure = 3.29122 bushels or 4.2104 cubic feet

A barrel may be called a "drum," but a drum usually holds 55 gallons.

MEASURING ELECTRICAL ENERGY

In the International System of Units, which are based on metric units, and which form the basis for the electrical units we use, both work and energy have the same unit—the Joule (pronounced: jewel). Named after English physicist James Prescott Joule (1818–1889) he showed that heat is a type of energy. In order to understand what a Joule is, it is important to understand what the unit of force (used in the International System of Units) called the Newton is. Named after Isaac Newton, a Newton of force is defined as a force that can accelerate a mass of 1 kilogram so that it picks up 1 meter per second of velocity during each second that the force is exerted. Therefore, a Joule is the amount of energy we expend as work if we exert a force of 1 Newton of force over a distance of one meter. As an illustration, 1 Joule is about how much energy it takes to lift one pound 9 inches. Generally, what we want to know when talking about powering appliances in our home with electricity, is the *rate of energy use*, or how much energy per unit time the appliance draws. This quantity is called the *power*, and is written as

$$\text{Power} = \frac{\text{Energy}}{\text{Time}}$$

For electrical power, the *Watt* is used:

$$1 \text{ Watt} = \frac{1 \text{ Joule}}{\text{Second}}$$

Power and energy are not the same thing, so do not confuse them. Power is the *rate* at which energy is delivered, not an amount of energy itself:

$$\text{Energy} = \text{Power} \times \text{Time}$$

As an example, a 100-watt light bulb is a device that converts 100 joules of electrical energy into 100 joules of electromagnetic radiation (light) every second. If you leave a 100-watt light on for one hour (3600 seconds), then the total energy you used was:

$$\text{Energy} = \text{Power} \times \text{Time}$$

$$(100 \text{ Joules/Second}) \times (3600 \text{ Seconds}) = 360,000 \text{ Joules}$$

Watts are a very convenient unit when working with appliances, such as specifying the power of light bulbs. If you want to calculate total energy use, however, it is better to work with the "kilowatt hour."

To give you a feeling for how much power the sun provides, consider that on a sunny day, at solar noon, the sunlight at the surface of the earth delivers about 1000 watts (1 kilowatt) per square meter. A typical photovoltaic solar cell can convert about 15 percent of this to electricity, that is, about 150 watts (the best cells in the laboratory can go somewhat higher, up to about 34 percent, or 340 watts).

Let us look next at how much power you would need to power your home. Assuming 15 percent efficient solar cells (so that we can capture 150 watts per square meter when the sun is shining), the total power will be given by

$$\text{Power} = (\text{Area of solar panels}) \times 150 \text{ watts/m}^2$$

Plugging this into the formula above for energy, and the hours of sunlight for the time, we find:

$$\text{Energy generated per day} = (\text{Area of solar panels})$$

$$\times 150 \text{ watts/m}^2 \times (\text{hours of sunlight})$$

Assuming that the energy generated per day is equal to the energy used per day, and solving for the Area, we find:

$$\text{Area of panels required} = \frac{(\text{Energy used per day})}{(150 \text{ watts/m}^2 \times \text{hours of sunlight})}$$

U.S. residences presently use about 14 kilowatt-hours of electrical energy a day on average (which is probably unnecessarily high and could be easily lowered by switching to more efficient appliances). Suppose you have five good hours of sunlight during the day. Then, using the formula above, the area in solar panels you would need to obtain the average household draw of 14 kilowatt-hours per day would be

$$\text{Area needed} = \frac{14,000 \text{ watt-hours}}{(150 \text{ watts/m}^2 \times 5 \text{ hours})}$$

$$= 18.6 \text{ square meters} = 200 \text{ square feet}$$

It can be seen that this figure is an area of 10 feet by 20 feet, much less than the roof area of a typical house. Therefore, the sun provides ample power for household electricity.

MEASURING THERMAL ENERGY

The basic unit for thermal energy in home heating applications is the *therm*, which is defined to be 100,000 Btu's:

$$1 \text{ therm} = 100,000 \text{ Btu's}$$

Remember: Btu stands for British Thermal Unit and is the amount of energy required to raise the temperature of 1 lb of water one degree Fahrenheit. Intuitively, it is helpful to think of a Btu as approximately equivalent to the heat given off by burning one match head.

A Btu is equivalent to 1055 Joules, and from this you can calculate that a therm is about 105,500,000, or 105.5 million Joules. To get a feeling for how much energy a therm is, a home furnace is typically rated at somewhere around one therm per hour.

For comparison, typical heating loads for a 1,800 feet^2 home would typically be (due to climatic and geographic differences):

- Phoenix, Arizona: 389 therms
- Santa Fe, New Mexico: 1,444 therms
- Great Falls, Montana: 1,728 therms

For these same areas, the annual energy from the sun (called insolation) is listed as follows. This is the annual energy falling on the same surface area (due to latitude):

- Phoenix, Arizona: 12,600 therms
- Santa Fe, New Mexico: 12,110 therms
- Great Falls, Montana: 8,870 therms (less because it is further north)

What is interesting to note here is that there exists abundant solar energy for heating homes, even in the locations further north, such as Montana. When a home is built to capture just the right amount of solar energy without overheating, it is referred to as a *passive solar design*.

DEGREE DAYS

Cold winter weather or sweltering summer heat can increase the cost of utility bills. Individuals can determine how much of the rise in utility bills is a result of the weather by using a unit of measure called the degree day. A degree day compares the outdoor temperature to a standard temperature of 65°F. The more extreme the temperature, the higher the number of degree days. A higher number of degree days will require more energy for space heating or cooling.

Hot days are measured in cooling degree days. On a day with a mean temperature of 80°F, for example, 15 cooling degree days would be recorded. Cold days are measured in heating degree days. For a day with a mean temperature of 40°F, 25 heating degree days would be recorded. Two such cold days would result in a total of 50 heating degree days for the two-day period. By analyzing the degree day patterns in an area, it is possible to assess the climate and the heating and cooling needs based on time and season.

Degree day data can be weighted according to the population of a region to estimate energy consumption. The U.S. Energy Information Administration (EIA) uses *population-weighted degree days* to model and project energy consumption for the United States and for U.S. Census regions and divisions. Figure 9.5 shows the average heating and cooling degree days for the United States (Figure 9.6a, b).

CHANGING TRENDS IN ENERGY USE IN THE UNITED STATES

The three major fossil fuels—petroleum, natural gas, and coal—have dominated the U.S. energy mix for more than the past 100 years. In the last few years, however, several recent changes in U.S. energy production have occurred:

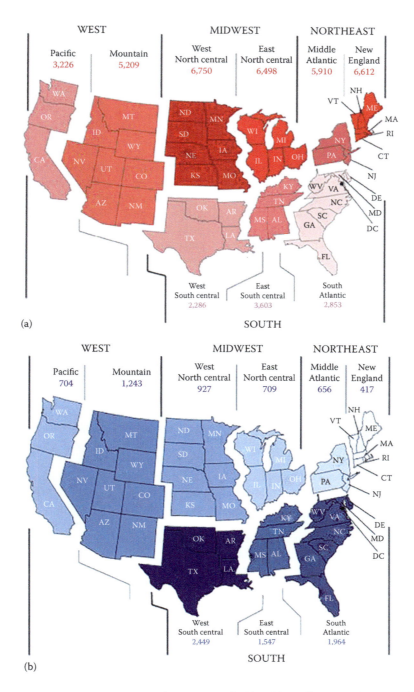

FIGURE 9.6 (a) Heating degree days by census region: This shows that the West North Central region normally requires ore heating than other regions. (b) The cooling degree days by census region: This shows that the West South Central region normally requires more cooling than other regions. (From EIA.)

- The production of coal peaked in 2008 and has been trending downward ever since. By 2015, coal production was at the same level it was in 1981. The prime reason for this decline in production is the decline in the use of coal for electricity generation.
- Natural gas production was higher in 2015 than in any previous year. This is because more efficient and cost-effective drilling and production techniques have led to increased production of natural gas from shale formations over the past decade. Natural gas is viewed as more environmentally friendly.
- Crude oil production has decreased each year between 1970 and 2008. In 2009, the trend reversed and production began to rise. More cost-effective drilling and production technologies helped to increase production, most notable in Texas and North Dakota. In 2015, crude oil production was at nearly the same level as in 1972.
- Natural gas plant liquids (NGPL) are hydrocarbon gas liquids that are extracted from natural gas before it is put into pipelines for transmission to consumers. NGPL production has increased along with increases in natural gas production. In 2015, NGPL production was about twice as much as it was in 2005.
- Total renewable energy production and consumption both reached record highs of approximately 9.7 quadrillion Btu in 2015. Hydroelectric power production in 2015 was roughly 18 percent below the 50-year average, but increases in energy production from wind and solar helped to increase the overall energy production from renewable sources. Energy production derived from wind and solar sources were at record highs in 2015.

Uses of Energy

In the United States, there are five principal energy consuming sectors: industrial, transportation, residential, commercial, and electric power. The five sectors utilize energy as follows:

- Industrial sector: this sector includes all the facilities and equipment used for manufacturing businesses, agriculture, mining, and the construction industry.
- The transportation sector includes vehicles that transport people or goods, such as cars, trucks, buses, motorcycles, trains, aircraft, boats, barges, and ships.
- The residential sector takes into account all homes and apartments.
- The commercial sector takes into account all offices, malls, stores, schools, hospitals, hotels, warehouses, restaurants, and places of worship and public assembly.
- The electric power sector is what consumes the primary energy to generate most of the electricity consumed by the other four sectors.

Figure 9.7 illustrates total U.S. energy consumption by end-use sector, from 1949 to 2015. (Figure 9.7).

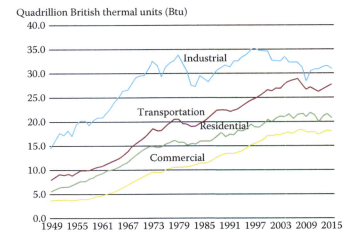

FIGURE 9.7 U.S. energy consumption by end-use sector, 1949–2015. (From EIA.)

Each sector consumes primary energy. Primary energy is energy in the form that it is first accounted for in an energy balance, before any transformation to secondary or tertiary forms of energy. For example, let us look at coal. Coal can be converted to synthetic gas, which can then be converted to electricity. Therefore, in this example, coal is *primary energy*, synthetic gas is *secondary energy*, and electricity is *tertiary energy*.

The industrial, transportation, residential, and commercial sectors also use most of the electricity (a secondary energy source) that the electric power sector produces. They are called *end-use* sectors because they purchase or produce energy for their own consumption and not for resale.

Energy consumption in the United States was three times greater in 2015 that it was in 1949. In every year during this time span, except for 19 of the years, energy consumption increased over the previous year. A big upset in the trend was seen, however, in 2009. This marked the year of the economic recession. In 2009, the real gross domestic product (GDP) fell 2.8 percent compared to the previous year, and total energy consumption decreased by almost a full 5 percent. This was the largest single-year decrease in both real GDP and in total energy consumption from 1949 through 2015. Decreases in energy consumption occurred in all four major end-use sectors in 2009; ending with residential dropping 3 percent, commercial 3 percent, industrial 9 percent, and transportation 3 percent. The years after that proved to be unsteady in all four sectors, affected by economic growth, weather, and fuel price fluctuations. Figure 9.8 illustrates the percentage of energy consumed by each sector in 2015 (Figure 9.8).

Energy Efficiency and Conservation

Energy is used every day for transportation, cooking, heating and cooling rooms, manufacturing, lighting, entertainment, and many other uses. The choices people make about how they use energy—such as turning machines off when they

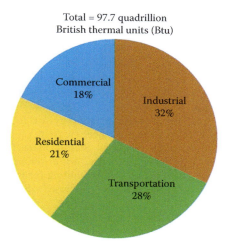

Total = 97.7 quadrillion
British thermal units (Btu)

FIGURE 9.8 Share of total U.S. energy consumed by end-use sector in the United States, 2015. (From EIA, 2016a.)

are not using them or choosing to buy fuel-efficient vehicles and energy-efficient appliances—affects the environment and everyone's lives. When it comes to energy use, we can strive to use energy efficiently and conservatively. Both are notable goals. But both are also different in concept. Being conservative with energy means to strive to use less energy. Being efficient in energy use means to use energy more wisely. The difference between the two concepts is that energy conservation is any behavior that results in the use of less energy. Unplugging appliances when not in use, turning off the lights when not in the room, and recycling aluminum cans are all ways of conserving energy. Energy efficiency is the use of technology that requires less energy to perform the same function. Using a compact fluorescent light bulb that requires less energy rather than using an incandescent bulb to produce the same amount of light is an example of energy efficiency, as is driving a hybrid car that gets better mileage.

The Necessity of Renewable Energy

There are many reasons today in support of incorporating, and then switching, to renewable energy for our energy needs now and permanently into our future. In order to do this, however, it will require not only some changes in infrastructure, but even more importantly, changes in lifestyle and mindset, also. Developed nations, such as the United States, have been almost completely reliant on fossil fuels in almost every aspect of their lives. To be this restricted to a single energy option leaves us vulnerable to many issues that we cannot control and others that are not in our best interests. Issues such as changes in supply and demand or having to contend with unrest in countries that fossil fuels are acquired from is not in our best interest and can have long-reaching negative effects.

One of the major arguments for switching to renewable energy is that developed societies have been almost completely reliant on fossil fuels for nearly every aspect of energy since their introduction, which presents one obvious problem: they are nonrenewable. Whether it is in a decade or a century,

they will eventually run out, and putting complete dependence on a limited resource is not wise. At some point they are going to become too expensive to realistically use. While we have been warned about this for years now, many still believe that this is an issue that developers should simply resolve. But it is much more complicated than just redesigning cars to run on something else, for example. We are really referring to every aspect of our lives: transportation, heating, cooling, building materials, consumer materials, food, supplies, housing, manufacturing and industry, and so forth. It affects all of us at all levels, which means that we must all contribute to the solution. Because of this, you cannot just decide to stop using fossil fuels and turn them off overnight. Humans rely too much on the fossil fuel infrastructure they have developed, which has now put us in a compromised position.

Fossil fuels also contribute greatly to climate change. As one of the leading contributors to GHGs in the atmosphere, we must shift away from their use and make the shift to renewable energy. Several utility companies today are adding renewables—such as solar and wind—to their mix of energy that they offer. Keep in mind, however, that manufacturing the solar panels and wind turbines does use fossil fuels in the manufacturing process, which is another reason why we need to get switched over to renewable energy and manufacturing processes in the near future. Using fossil fuels in the manufacturing process applies to many different types of items, which puts an additional urgency in taking our focus off of fossil fuels and focusing on getting the infrastructure of the United States and other similarly positioned countries off of one single finite input. This also means we do not have to rely on fluctuating prices or shortages from outside energy producers. Renewable energy is now affordable for private homes. As of 2016, the federal government was still offering significant tax rebates for homeowners that were choosing solar and wind options for their own homes.

Renewables are also an economically sound decision. Greenpeace (2007) has estimated that the world could save about $180 billion a year by switching 70 percent of the earth's electricity production to renewable options. The World Resources Institute stated in their 2014 Renewable Energy Report that studies conducted by several different energy research laboratories suggests that increased renewable energy generation has the potential to save the American public tens of billions of dollars a year over the current mix of electric power options. Although they acknowledge that upgrading the infrastructure will require an initial investment, they show that ultimately renewable energy cuts costs system-wide by replacing power plants that are expensive to operate, mostly due to fossil fuel expenses. In their study, New York State would save $1.3 billion in power plant operation costs per year by 2018—which is approximately $65 per person. In the Midwest and mid-Atlantic areas, savings by 2020 (Midwest) and 2026 (mid-Atlantic) are estimated at $12.2 billion and $14.5 billion per year, respectively (Ryor and Tawney, 2014). Ryor and Tawney also point out the economic benefits of renewable energy creating new jobs. In Germany, for example, nearly one-fourth of their national energy demand is met by renewable resources. This has led to an increase in employment within their energy sector by nearly 380,000 new jobs (Ryor and Tawney, 2014).

Renewable energy is also better for the environment. Each year, power plants in the United States put more than 2.5 million tons of CO_2 into the atmosphere. Fossil fuels are also responsible for significant land, water, and air pollution beyond their contribution to GHGs. For instance, coal mining brings solid wastes to the surface that would normally remain underground and the areas around a mine can remain barren for generations. Mines that are not properly reclaimed also present a safety hazard. As people participate in recreational activities on public lands, un-reclaimed mines pose a safety hazard with toxic chemicals and open adits and mine entrances. The burning of coal for energy also produces several different types of particulate matter that pollutes the atmosphere. The finest particles, when inhaled, can cause various respiratory health problems. When these pollutants make their way into the water cycle and fall to the ground as acid rain, it can destroy the landscape and pollute large bodies of water.

Renewable energy resources do have their challenges, too. Wind turbines can impact migrating bird species when birds fly into them. Dams can disrupt the ecology of the surrounding areas as well as disrupt the habitat of some species.

CONCLUSIONS

If there is one thing most people would probably agree on with regard to energy, it is that we all depend on it. We depend on it for just about everything we do, from work to our lives at home, getting from one place to another, and just about every activity we are involved in. The challenges we have today are that with a growing population and dwindling resources, we must utilize our resources in a smarter way. That begins by changing our mindset and the way we look at using energy. If we can transition now to a lifestyle utilizing renewable energy, conserving, and being efficient, we will build a better, cleaner world for tomorrow.

REFERENCES

Greenpeace. July 6, 2007. Energy revolution = Money saved. *Greenpeace International.* http://www.greenpeace.org/international/en/news/features/energy-revolution-money-save-060707/ (accessed September 19, 2016).

Ryor, J. and L. Tawney. June 18, 2014. Shifting to renewable energy can save U.S. consumers money. *World Resources Institute.* http://www.wri.org/blog/2014/06/shifting-renewable-energy-can-save-us-consumers-money (accessed September 19, 2016).

United States Energy Information Administration. September, 2012. Annual energy review 2011. http://www.eia.gov/energyexplained/index.cfm?page=about_degree_days (accessed September 15, 2016).

United States Energy Information Administration (EIA). 2016a. Energy explained. http://www.eia.gov/energyexplained/index.cfm?page=about_laws_of_energy (accessed September 22, 2016).

United States Energy Information Administration (EIA). 2016b. April, 2016. Monthly energy review. http://www.eia.gov/energyexplained/index.cfm?page=us_energy_use (accessed September 15, 2016).

SUGGESTED READING

Porter, E. August 30, 2016. The challenge of cutting coal dependence. *The New York Times.* http://www.nytimes.com/2016/08/31/business/economy/the-challenge-of-cutting-coal-dependence.html?rref=collection%2Ftimestopic%2FSolar%20Energy&action=click&contentCollection=energy-environment&region=stream&module=stream_unit&version=latest&contentPlacement=2&pgtype=collection&_r=0 (accessed September 19, 2016).

University of Houston. September 19, 2016. More efficient way to split water, produce hydrogen. *ScienceDaily.* https://www.sciencedaily.com/releases/2016/09/160919154228.htm (accessed September 21, 2016).

10 Nonrenewable Energy Resources

OVERVIEW

This chapter illustrates on nonrenewable energy and how it has been a key part of our lives—and lifestyles—for a long time now. We will begin by explaining about crude oil and petroleum products and the critical roles of gasoline, diesel fuel, heating oil, hydrocarbon gas, and natural gas. We will then read about coal and the issues associated with it. Subsequently, we will take a look at the nuclear industry, what it takes to make energy in those environments, and our utilization of them. We will also look at other less-developed forms of fossil energy— tar sands and oil shale—and what lies ahead for them.

INTRODUCTION

At some point in the future, nonrenewable energy sources will be a thing of the past. They are considered nonrenewable because they cannot be replenished within a human lifetime— or even within several human lifetimes. In fact, some of these energy sources, such as coal and petroleum products, took millions of years to form. Because these resources have been abundant and technology was developed early in our society to utilize them in every practical aspect—transportation, industry, manufacturing, commercial and residential building, business, and so forth—they've become deeply engrained into our society and into our perceptions and lifestyles. Moreover, they were inexpensive and convenient. We had incorporated them into every facet of our lives—and that wasn't necessarily a problem, until recently. As we became more aware of our environment, we began to make some important observations and come to a few key realizations. With that, we had to finally admit that fossil fuels were not good for our health, our environment, or our future; and that we needed to make some changes before it was too late. We'd finally realized that these energy sources were the root cause of polluting our atmosphere and were a key factor for contributing to a changing global climate and degraded environment.

This chapter examines those nonrenewable resources that we have become so dependent on, where they come from, how we access them, and the ways that we use them. In this chapter, you will be able to see how dependent we have grown over the decades on their reliable presence and how entrenched our infrastructure is in support of their continued existence. What it will take now is the combined efforts and dedication of everyone to move forward together in a common, cleaner direction.

DEVELOPMENT

Most nonrenewable energy sources are fossil fuels: coal, petroleum, and natural gas. Because carbon is the main element in fossil fuels, the time period when fossil fuels are formed—about 360 to 300 million years ago—is called the Carboniferous Period. These are sources that will run out or will not be replenished in our lifetimes—or even in many lifetimes from now. All fossil fuels are formed in a similar way. Hundreds of millions of years ago, even before the time of the dinosaurs, earth had a very different landscape. It was covered with wide, shallow seas and swampy forest environments. There were plants, algae, and plankton that grew in ancient wetlands that absorbed sunlight and created energy through photosynthesis. When they died, the organisms settled at the bottom of lakes and seas. There was still energy stored in all the biomass—both plants and animals—when they died. Over eons, the dead plants were compacted under the seabed. Rocks and other sediments were piled on top of them, creating high heat and pressure underground in this closed environment. The plant and animal remains eventually turned into fossil fuels (coal, natural gas, and petroleum). Today, there are huge underground pockets, called reservoirs, of these nonrenewable sources of energy all around the world.

There are several advantages and disadvantages of nonrenewable resources. Fossil fuels have been a valuable source of energy for a long time. They are relatively inexpensive to extract and they can be stored, piped, and shipped virtually anywhere in the world. Because of these conveniences, they have become a crucial part of our lifestyle.

The problem with these energy resources is that they are harmful to the environment and are causing the earth's climate to change at an alarming rate. Let's look at each of these resources in detail.

CRUDE OIL AND PETROLEUM PRODUCTS

Petroleum has been used by civilization since at least 3000 BCE. Mesopotamians used "rock oil" in architectural adhesives, ship caulks, medicines, and roads. Two thousand years ago, the Chinese refined crude oil for use in lamps and for heating homes. In the past, petroleum was collected in small containers from where it oozed from the ground. In America, the Indians, doctors, and pharmacists used it as a form of medicine. For example, the Indians had used it for hundreds of years to treat skin sicknesses and breathing difficulties.

Civilizations then began to find other uses for petroleum. Eventually, it replaced whale oil for lighting, and the oil lamp was invented. Most experts credit Edwin L. Drake with starting the oil industry on a large scale. In 1858, the Seneca Oil Company, which was interested to use oil as a fuel, hired Drake. He was hired to drill a well near Titusville, Pennsylvania. Drake worked with Billy Smith, a well digger, and dug one pit after another. The men used a wooden rig and a steam-run drill. Because each pit they dug was threatened with water and cave-ins, Drake ran an iron pipe deep into the ground and drilled inside it. The pipe acted as a casing and kept Drake's path clear for drilling. About a year later, Drake and Smith had dug a well 21.2 meters deep. On August 27, 1859, the oil suddenly swept up the shaft.

Gas was first manufactured in the late eighteenth century. Scientists discovered that gas could be produced from heating coal without air, which is the same process that is used to make coke from coal. Hence, gas was called coke oven gas, or coal gas.

William Murdock, a British engineer, is known as the father of the gas industry. In 1792, he was using gas to light his own home. Coal gas was first used for public lighting with gaslit street lamps and public buildings. From there, its use spread to private homes.

In order to transport the gas, it was stored in elastic water skins that were stowed in horse-drawn wagons. A long tube fitted with a tap was inserted into the water skins. This controlled the dispensing of the gas. The wagon delivered the gas to individual homes by plugging the tube into the customer's tank. The tap was then opened and straps were tightened around the water skins to force the gas from the skin into the tank. Eventually, gas was stored in cylinders, making it less cumbersome to deliver.

Gas began to be used as a fuel in other modes of transportation, as well. Trains, steamboats, and ships carried gas in bottles. As the demand for gas increased, more efficient delivery methods were experimented with, and eventually pipes were placed underground in which to directly deliver the gas from factory to home. By this time, gas was not only used for lighting, but for cooking and heating, as well. That marked the beginning of using oil and gas as we know it.

Crude oil is a mixture of hydrocarbons that formed from the ancient plants and animals that lived millions of years ago. Crude oil is a fossil fuel, and it exists in liquid form on underground pools or reservoirs, in miniscule spaces within sedimentary rocks, and near the surface in tar (or oil) sands. Petroleum products are fuels made from crude oil and other hydrocarbons contained in natural gas. Petroleum products can also be made from coal, natural gas, and biomass.

After crude oil is removed from the ground, it is sent to a refinery where different parts of the crude oil are separated into useable petroleum products. These petroleum products include gasoline, distillates such as diesel fuel and heating oil, jet fuel, petrochemical feedstocks, waxes, lubricating oils, and asphalt. A 42-gallon barrel of crude oil yields about 45 gallons of petroleum products in the United States due to the increase in volume during the processing. Figure 10.1 shows the yield from a barrel of crude oil.

Crude oil was formed from the remains of animals and plants, called diatoms, which lived millions of years ago in a marine environment. Over millions of years, the remains of these animals and plants were covered by layers of sand, silt, and rock. Heat and pressure from these layers transformed them into crude oil. The word petroleum means *rock oil* or *oil from the earth*. Figure 10.2 illustrates this process.

Petroleum refineries convert crude oil and other liquids into many petroleum products that people use daily, although the key product is transportation fuels. Other uses include heating,

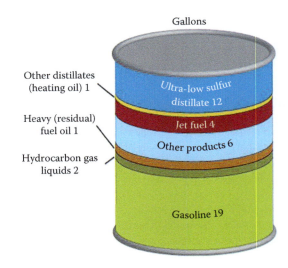

FIGURE 10.1 Products made from a barrel of crude oil. (From EIA, Washington, DC.)

paving roads, and generating electricity. Petroleum products are also used as feedstocks for making chemicals. On average, U.S. refineries produce about 19 gallons of motor gasoline, 12 gallons of ultra-low sulfur distillate fuel, most of which is sold as diesel fuel, and 4 gallons of jet fuel from a 42-gallon barrel of crude oil. More than a dozen other petroleum products are also produced in refineries. In addition, refineries produce liquids that are used by the petrochemical industry to make a variety of chemicals and plastics. Refining breaks down crude oil into its various components, which are then selectively reconfigured into new products. All refineries have three basic steps: separation, conversion, and treatment. The illustration shows many of a refinery's most important processes (Figure 10.3).

Separation

Modern separation involves piping crude oil through hot furnaces. The resulting liquids and vapors are discharged into distillation units. Inside the distillation units, the liquids and vapors separate into petroleum components called fractions according to their weight and boiling point. Heavy fractions are on the bottom and light fractions are on the top. The lightest fractions, including gasoline and liquefied petroleum gas (LPG), vaporize and rise to the top of the distillation tower, where they condense back to liquids. Medium weight liquids, including kerosene and diesel oil distillates, stay in the middle of the distillation tower. Heavier liquids, called gas oils, separate lower down in the distillation tower, while the heaviest fractions with the highest boiling points settle at the bottom of the tower.

Conversion

After distillation, the heavy, lower-value distillation fractions can be processed further into lighter, higher-value products such as gasoline. This is where fractions from the distillation units are transformed into streams (intermediate components) that eventually become finished products. The most widely used conversion method is called cracking because it uses heat and pressure to crack heavy hydrocarbon molecules into lighter ones. A cracking unit consists of one or more tall,

FIGURE 10.2 (a) Diatoms magnified under a microscope. (From Jmpost, University of Tasmania Scanning Electron Microscope.) and (b) the process of oil and gas formation. (From Department of Energy [DOE].)

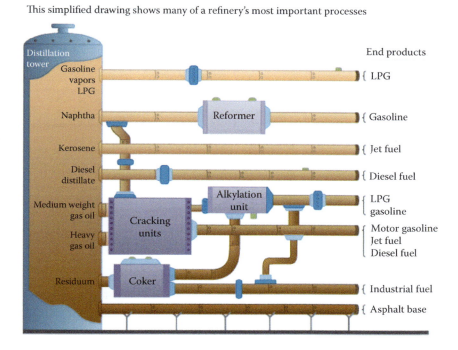

FIGURE 10.3 This drawing illustrates a refinery's most important processes. (From EIA, Washington, DC.)

thick-walled, rocket-shaped reactors and a network of furnaces, heat exchangers, and other vessels. Cracking is not the only form of crude oil conversion. Other refinery processes rearrange molecules instead of splitting them. These include processes such as alkylation, which makes gasoline components by combining gaseous byproducts of cracking; and the reforming process, which uses heat, pressure, and catalysts to obtain high-octane gasoline components.

Treatment

The finishing touches occur during the final treatment. To make gasoline, refinery technicians carefully combine a variety of streams from the processing units. Octane level, vapor pressure ratings, and other special considerations are what determine the gasoline blend.

Storage

Both incoming crude oil and the outgoing final products need to be stored. These liquids are stored in large tanks on a tank farm near the refinery. Pipelines then carry the final products from the tank farm to other tanks across the country.

Petroleum Sources

In the United States, crude oil is produced in 31 of the states and in U.S. coastal waters. In 2015, about 65 percent of U.S. crude oil production came from these five states only:

- Texas (37 percent)
- North Dakota (12 percent)
- California (6 percent)
- Alaska (5 percent)
- Oklahoma (5 percent)

In 2015, about 16 percent of U.S. crude oil was produced from wells located offshore in the federally administered waters of the Gulf of Mexico. Although total U.S. crude oil production generally declined between 1985 and 2008, annual production increased from 2009 to 2015. More cost-effective drilling technology helped to boost production, especially in Texas, North Dakota, Oklahoma, New Mexico, and Colorado (Figure 10.4a and b).

Worldwide, about 100 countries produce crude oil. In 2014, 47 percent of the world's total crude oil production came from these five countries:

- Russia (13 percent)
- Saudi Arabia (13 percent)
- United States (11 percent)
- China (5 percent)
- Canada (4 percent)

The United States is the world's largest petroleum consumer, and it consumed about 19 million barrels (MMbbl/d) of petroleum products per day in 2014 (about 20 percent of world total). The United States is the world's third-largest crude oil producer, but only part of the nation's petroleum needs are met by crude oil and other liquids produced in the United States. U.S. dependence on imported petroleum has declined since peaking in 2005. This trend is the result of a variety of factors including a decline in consumption and shifts in supply patterns. The economic downturn following the financial

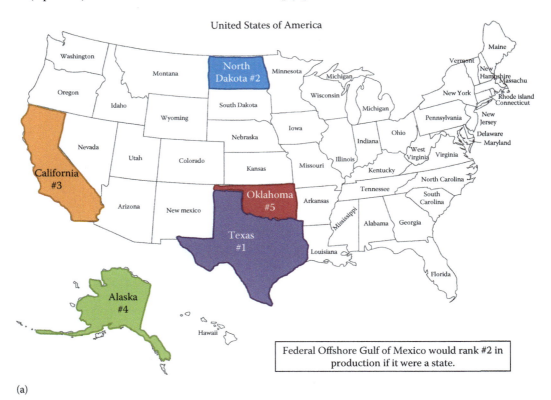

United States of America

Federal Offshore Gulf of Mexico would rank #2 in production if it were a state.

(a)

FIGURE 10.4   (a) Top five crude oil producing states in the U.S., 2015.   (Continued)

Selected producers, 1973–2015

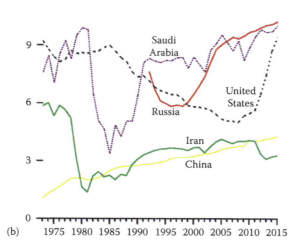

(b)

FIGURE 10.4 (Continued)   (b) Top five crude oil producing countries, 1980–2014.

crisis of 2008, improvements in efficiency, changes in consumer behavior, and patterns of economic growth all contributed to the decline in petroleum consumption. Additionally, increased use of domestic biofuels (ethanol and biodiesel) and strong gains in domestic production of crude oil and natural gas plant liquids expanded domestic supplies and reduced the need for imports. In addition to crude oil, the United States also imports refined petroleum products such as gasoline. Although the United States produces most of the petroleum products it consumes (using imported and domestically produced crude oil and other liquids), it imported about 1.9 MMbbl/d of finished petroleum products in 2014.

In addition to imports of finished petroleum products such as gasoline, diesel fuel, and jet fuel, the United States also imports unfinished products used as refinery inputs and blending components. Unfinished oils are refined from crude oil at refineries outside the United States. Imported unfinished oils are used as inputs to U.S. refineries for processing into finished petroleum products. Imported gasoline blending components and fuel ethanol are blended at U.S. refineries and terminals to produce finished gasoline. Figure 10.5 illustrates U.S. import and domestic petroleum use in 2014.

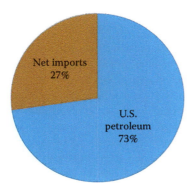

FIGURE 10.5   U.S. use of petroleum, both domestic and import. (From EIA, Washington, DC.)

DID YOU KNOW?

OPEC and Persian Gulf countries are not the same.

The Organization of the Petroleum Exporting Countries (OPEC) was found in 1960 for the purpose of negotiating with oil companies on matters of oil production, prices, and future concession rights. Of the 12 countries currently in OPEC, only 6 of them are in the Persian Gulf.

OPEC	Persian Gulf
• Iran	• Iran
• Iraq	• Iraq
• Kuwait	• Kuwait
• Saudi Arabia	• Saudi Arabia
• Qatar	• Qatar
• United Arab Emirates	• United Arab Emirates
• Algeria	• Bahrain
• Angola	
• Ecuador	
• Libya	
• Nigeria	
• Venezuela	

Offshore Resources

When someone stands at an ocean beach, they are not at the very edge of the United States, because the shoreline is not the nation's border. The border is actually 322 kilometers (200 miles) away from the land. This area around the country is called the Exclusive Economic Zone (EEZ). In 1983, U.S. President Ronald Reagan claimed the EEZ in the name of the United States. In 1994, all countries were granted an EEZ of 322 kilometers miles from their coastline according to the United Nations Convention on the Law of the Sea.

The ocean floor extends from the beach into the ocean on a continental shelf that gradually descends to a sharp drop, called the continental slope. The width of the U.S. continental shelf varies from 16 to 400 kilometers. The water on the continental shelf is relatively shallow, rarely more than 150 to 200 meters deep. The continental shelf drops off at the continental slope, ending in abyssal plains that are three to five kilometers below sea level. Many of the plains are flat, while others have jagged mountain ridges, deep canyons, and valleys. The tops of some of these mountain ridges form islands where they extend above the water.

Several federal government agencies manage the natural resources in the EEZ. The U.S. Department of Interior's Bureau of Ocean Energy Management and the Bureau of Safety and Environmental Enforcement (BSEE) manage the development of offshore energy resources by private companies. These companies lease areas for energy development and pay the federal government royalties on the energy resources they produce from the ocean. Individual states control the waters off their coasts out to 4.8 kilometers for most states and between 14.5 and 19 kilometers for Florida, Texas, and some other states.

Most of the energy the United States gets from the ocean is oil and natural gas from wells drilled on the ocean floor. Other energy sources are also being developed offshore. Wind turbines are located offshore in several countries, and wind energy projects are being considered in several areas off the Atlantic coast of the United States. Wave energy, tidal energy, ocean thermal energy conversion, and methane hydrates are other energy sources currently being explored.

World Petroleum Consumption and Prices

Worldwide petroleum consumption in 2013 was 91.2 million barrels a day. The three largest petroleum consuming countries in 2013 and their share of total world petroleum consumption are:

- United States (21 percent)
- China (11 percent)
- Japan (5 percent)

Crude oil prices are determined by global supply and demand, with economic growth being the biggest factor that affects demand. Growing economies require energy, and the petroleum products made from crude oil and other hydrocarbon liquids account for about 34 percent of the energy consumed worldwide. The Organization of the Petroleum Exporting Countries (OPEC) can have a major influence on prices by setting production targets for its members. OPEC members produced about 42 percent of the world's crude oil in 2014. If OPEC members limit production, they can influence global crude oil supply and prices. OPEC countries have close to three-quarters of the estimated world crude oil reserves and most of the spare crude oil production capacity.

Crude oil and petroleum product prices can be affected by events that may disrupt their supply to market, including geopolitical and weather-related events. These events can lead to actual crude oil supply or demand disruptions and can create uncertainty about future supply or demand. This can lead to higher volatility in prices. The volatility of oil prices is tied to the low responsiveness of supply and demand to price changes in the short term. Crude oil production capacity and equipment that use petroleum products as the main source of energy are relatively fixed in the near term. It takes years to develop new supply sources or to vary production, and it is hard for consumers to switch to other fuels or to increase fuel efficiency in the near term when prices rise. Under these conditions, a large price change may be necessary to rebalance physical supply and demand.

Most of the crude oil reserves in the world are located in regions that have been prone to political upheaval, or in regions that have had oil production disrupted because of political events. Several major oil price shocks have occurred at the same time as supply disruptions triggered by political events, most notably the Arab Oil Embargo in 1973–1974, the Iranian revolution, the Iran–Iraq war in the 1980s, and the Persian Gulf War in 1990–1991. More recently, disruptions to supply from political events have occurred in Nigeria, Venezuela, Iraq, Iran, and Libya.

Given the history of oil supply disruptions caused by political events, market participants constantly assess the possibility of future disruptions. In addition to the size and duration of a potential disruption, market participants also consider the availability of crude oil stocks and the ability of other producers to offset a potential supply loss. When spare capacity and inventories are low, a potential supply disruption may have a greater impact on prices than might be expected if only current demand and supply were considered.

Weather also plays a significant role in the supply of crude oil. In 2005, hurricanes in the Gulf of Mexico region shut down oil production operations and many refineries that process crude oil into petroleum products. As a result, petroleum product prices increased sharply as supplies to the market dropped. Severe cold weather can strain product markets as producers attempt to supply enough of the product, such as heating oil, to consumers in a short amount of time to meet demand. This can result in higher prices for the commodity.

In addition to the previous examples, other events such as refinery outages or pipeline problems can restrict the flow of crude oil and petroleum products to market. These types of events can also lead to a temporary supply disruption that could increase prices for these commodities.

The influence of all of these factors on crude oil prices tends to be relatively short term. Once the supply disruption subsides, oil and product flows return to normal, and prices usually return to their previous levels. Even so, this can represent a stressful market situation with supplies not always a guarantee due to political or physical situations on top of the fact that the resource is dwindling in supply.

GASOLINE

Gasoline is a fuel made from crude oil and other petroleum liquids. It is mainly used as an engine fuel in vehicles. Refineries in the United States produce about 19 gallons of gasoline from every 42-gallon barrel of crude oil that is refined. Refineries and companies that produce the *finished* motor gasoline sold in retail gasoline fueling stations may add various liquids so that the gasoline burns cleaner and meets air pollution control standards and requirements. Most of the gasoline now sold in the United States also contains about 10 percent fuel ethanol by volume. This is required by a federal law intended to reduce the amount of oil that the United States imports from other countries. There are three main grades of gasoline sold at retail gasoline refueling stations, based on their octane levels:

- Regular
- Midgrade
- Premium

The octane level of a fuel refers to its resistance to combustion. A fuel with a higher octane level is less prone to pre-ignition and detonation, also known as engine knocking. Refiners charge more for higher octane, making premium grade gasoline the most expensive. Some companies have different names for these grades of gasoline, like *unleaded, super,* or *super premium,* but they all indicate the octane rating, which reflects the antiknock properties of gasoline. Higher ratings result in higher prices. Manufacturers recommend the grade of gasoline for use in each model of a vehicle. However, most gasoline-fueled vehicles will operate on regular gasoline, which is usually the least expensive grade.

The Evolution of Gasoline

Edwin Drake dug the first oil well in Pennsylvania in 1859 and distilled the oil to produce kerosene for lighting. Although other petroleum products, including gasoline, were also produced in the distillation process, Drake did not want these other products, so he discarded them. It wasn't until 1892 with the invention of the automobile that gasoline was recognized as a valuable fuel. By 1920, there were 9 million vehicles on the road powered by gasoline, and service stations selling gasoline were opening around the country. Gasoline is the fuel used by most of the light-duty vehicles in the United States.

By the 1950s, cars were becoming bigger and faster. Gasoline octane increased, and lead was added to improve engine performance. Unleaded gasoline was introduced in the 1970s when health problems from lead became apparent. In the United States, leaded gasoline for use in on-road vehicles was completely phased out as of January 1, 1996. Most other countries have also stopped the use of leaded gasoline in vehicles.

In 2005, the U.S. Congress enacted a Renewable Fuel Standard (RFS) that set minimum requirements for the use of renewable fuels, including ethanol, in motor fuels. In 2007, the RFS renewable fuel use targets were set to increase steadily to a level of 36 billion gallons by 2022. In 2014, about 13 billion gallons of ethanol were added to the gasoline consumed in the United States. In most areas of the country, the retail gasoline that is now sold is about 10 percent ethanol by volume.

Gasoline is a toxic and highly flammable liquid. The vapors given off when gasoline evaporates and the substances produced when it is burned (carbon monoxide, nitrogen oxides, particulate matter, and unburned hydrocarbons) contribute significantly to air pollution. Burning gasoline is also a key contributor to recent climate change.

The goal of the Clean Air Act (the Act) is to reduce air pollution in the United States. Specifically, the Act (first passed in 1970) and its amendments require less-polluting engines and fuels. As a result, emissions control devices on passenger vehicles were required beginning in 1976. In the 1990s, the EPA established emissions standards for other types of vehicles and for engines used in gasoline-burning nonroad equipment.

In the 1970s, lead also became a public health concern. Steps to move away from leaded gasoline began in 1976 when catalytic converters were installed in new vehicles to reduce the emissions of toxic air pollutants. Vehicles equipped with a catalytic converter cannot operate on leaded gasoline because the presence of lead in the fuel damages the catalytic converter. Following those changes, the Clean Air Act Amendments of 1990 required cleaner burning reformulated gasoline to reduce air pollution in metropolitan areas that had significant ground-level ozone pollution.

DID YOU KNOW THIS?

- Burning a gallon of gasoline (that does not contain ethanol) produces about 19.6 pounds of carbon dioxide?
- The midgrade option of gasoline was introduced in 1986 as leaded gasoline was being phased out. Most gas stations already had the pumps for the leaded, unleaded, and premium gas. The midgrade was offered as an additional choice for motorists who wanted a higher octane than the regular grade. This proved also to be a convenient solution to continue to use all three pumps—just create a new type of gas of fuel.
- Gasoline changes with the seasons.
    - The main difference between winter- and summer-grade gasoline is vapor pressure. Gasoline vapor pressure is important for an automobile engine to work properly. During winter months, vapor pressure must be high enough for the engine to start easily.
- Gasoline evaporates more easily in warm weather, releasing more volatile organic compounds that contribute to health problems and to the formation of ground-level ozone and smog. To cut down on pollution, the U.S. Environmental Protection Agency requires petroleum refiners to reduce the vapor pressure of gasoline during summer months.

Varieties by Grade and Formulation

In addition to the different grades of gasoline, the formulation of gasoline may differ depending on the location or the season. Federal and state air pollution control programs that aim to reduce carbon monoxide, smog, and air toxins require oxygenated, reformulated, and low-volatility gasoline. Some areas of the country are required to use specially formulated gasoline to reduce certain emissions, and the formulation may change during winter and summer months. These area-specific requirements mean that gasoline is not a homogenous product nationwide. Gasoline produced for sale in one area of the United States may not be authorized for sale in another area.

The characteristics of the gasoline depend on the type of crude oil that is used and the setup of the refinery where the gasoline is produced. Gasoline characteristics are also affected by other ingredients that may be included in the blend, such as ethanol. Most of the fuel ethanol added to gasoline is made from corn grown in the United States.

After crude oil is refined into gasoline and other petroleum products, the products must be distributed to consumers. Most gasoline is shipped first by pipeline to storage terminals near consuming areas and then loaded into trucks for delivery to individual gas stations. Gasoline and other products are sent through shared pipelines in *batches*. Since these batches are not physically separated in the pipeline, some mixing, or *commingling* of products occurs. This mixing is why the quality of the gasoline and other products must be tested as the products enter and leave the pipeline to make sure they meet appropriate specifications. Whenever the products fail to meet local, state, or federal product specifications, they must be removed and trucked back to a refinery for further processing.

Gasoline is sold at more than one hundred thousand retail outlets across the nation, and many are *unbranded* dealers that may sell gasoline produced by different companies. *Branded* stations may not necessarily sell gasoline produced by its refineries. This mixing of brands occurs because gasoline from different refineries is often combined for shipment by pipeline, and companies owning service stations in the same area may be purchasing gasoline at the same bulk storage and distribution terminal.

The only difference between the gasoline at station X and the gasoline at station Y may be the small amount of additives that those companies add to the gasoline before it gets to the pump. Even if it was possible to determine which company's refinery produced the gasoline, the source of the crude oil used at that refinery may vary. Most refiners use a mix of crude oils from various domestic and foreign sources. The mix of crude oils can change based on the relative cost and availability of crude oil from different sources (Figure 10.6).

The Gasoline Market

Americans used about 375 million gallons of gasoline per day in 2014. With about 319 million people in the United States in 2014, that calculates to more than a gallon of gasoline every day for each person. Gasoline is one of the major fuels consumed in the United States, and it is the main product refined from crude oil. About 10 percent of the volume of finished motor gasoline now consumed in the United States is ethanol. In 2014, gasoline (including fuel ethanol) accounted for about 65 percent of all the energy used for transportation, 47 percent of all petroleum consumption, and 17 percent of total U.S. energy consumption. About 45 barrels of gasoline

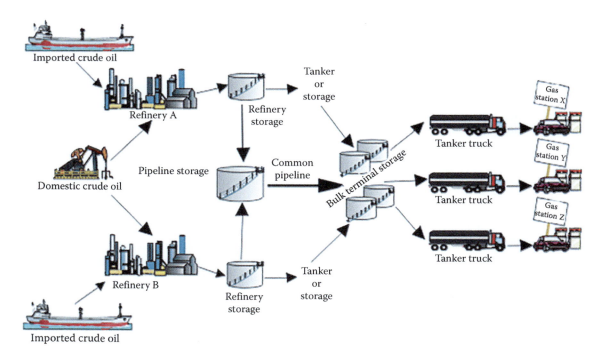

FIGURE 10.6 Flow of crude oil and gasoline to the local gas station. (From EIA, Washington, DC.)

FIGURE 10.7 Annual price fluctuation to produce gasoline so that costs can be compared. (From EIA, Washington, DC.)

are produced in U.S. refineries from every 100 barrels of oil refined to make petroleum products.

Gasoline is the main petroleum product consumed in the United States, followed by distillate fuel oil, hydrocarbon gas liquids (HGL), propane, and jet fuel. Distillate fuel oil includes diesel fuel and heating oil. Diesel fuel is used in the diesel engines of heavy construction equipment, trucks, buses, tractors, boats, trains, some automobiles, and electricity generators. Heating oil, also called fuel oil, is used in boilers and furnaces to heat homes and buildings, for industrial heating, and for producing electricity. Total distillate fuel oil consumption in 2014 was more than 4 million barrels per day, or 21 percent of total petroleum consumption.

Hydrocarbon gas liquids (HGL), the third most-used category of petroleum, include propane, ethane, butane, and other hydrocarbon gas liquids that are produced at natural gas processing plants and oil refineries. HGL consumption in 2014 was about 2.4 million barrels per day. The petrochemical industry uses HGL as a feedstock for making several products.

Propane, a heavily consumed HGL, is also used in homes for space heating and water heating, clothes drying, cooking, heating greenhouses and livestock housing, drying crops, and as a transportation fuel. Jet fuel, the fourth-most used petroleum product in the United States, had almost 1.5 million barrels per day used in 2014. The following table shows the five top gasoline consuming states in 2014:

Top Five Gasoline Consuming States, 2014

State	Million Barrels/Day	Million Gallons/Day	Share of Total U.S. Consumption
California	0.89	37.55	11 percent
Texas	0.86	36.26	10 percent
Florida	0.47	19.94	6 percent
New York	0.34	14.36	4 percent
Ohio	0.30	12.63	4 percent

Source: U.S. Energy Information Administration, Petroleum and Other Liquids—Prime Supplier Sales Volumes.

There are two ways to compare recent gasoline prices with historical prices. One way to compare gasoline prices is to examine the price actually paid at the pump, or the *nominal* price. The other way to compare gasoline prices is to examine the *real* price, which shows past prices in today's dollar value.

The figure shows the average annual nominal and real prices of retail regular gasoline from 1976 to 2015. The real price is based on the value of the dollar in December 2015 (Figure 10.7).

The retail price of gasoline includes four main components:

- *The cost of crude oil*: This is determined by supply and demand, disruptions in international and domestic supplies, lack of supply, unrest and conflict, among other things. This is the most crucial factor.
- *Refining costs and profits*: These vary seasonally and by region.
- *Distribution and marketing costs and profits*: These involve distributing, marketing, and retail dealer costs and profits are also included in the retail price of gasoline.
- *Taxes*: Federal, state, and local taxes are added to the cost, partly because of the different gasoline formulations required to reduce air pollution in different parts of the country.

Prices can fluctuate further if any event slows or stops the production of gasoline, whether it is human-caused (like a refinery shutdown) or nature-caused (like Hurricane Katrina) events. Prices also change seasonally, gradually rising in the spring and peaking in late summer when people drive more as they participate in summer vacations and other activities. Prices generally decline during winter months. Summer month prices typically rise about 7 percent higher than other months of the year. The gasoline formulations also change seasonally. Environmental regulations require that gasoline sold in the summer be less prone to evaporate during warmer weather. This means that refiners must replace cheaper but more evaporative gasoline components with less evaporative but more expensive components. From 2004 to 2014, the

FIGURE 10.8 Long lines at a gas station in Maryland in 1979 when gasoline demand was high but supply was low. (Courtesy of Warren K. Leffler.)

average monthly price of U.S. retail regular gasoline in June, July, and August was about 47 cents per gallon higher than the average price in January. Prices can also vary in different regions of the United States depending on how far the gasoline is sold from the source of supply. Supply sources include refineries, ports, and pipeline and blending terminals.

Prices are also affected by worldwide supply and demand. Rapid gasoline price increases occurred in response to crude oil shortages caused by the Arab oil embargo in 1973, the Iranian revolution in 1978, the Iran/Iraq war in 1980, and the Persian Gulf War in 1990. World crude oil prices reached record levels in 2008 as a result of high worldwide oil demand relative to supply. Other factors that contribute to higher crude oil prices include political events and conflicts in some major oil producing regions, as well as other factors such as the declining value of the U.S. dollar (the currency at which crude oil is traded globally). The Organization of the Petroleum Exporting Countries (OPEC) has significant influence on world oil prices, because its members produce about 40 percent of the world's crude oil. OPEC members are also the only countries that have spare production capacity and the ability to bring more oil into production relatively quickly. Since it was organized in 1960, OPEC has tried to keep world oil prices at a target level by setting production levels for its members.

Gasoline prices tend to increase as the available supply of gasoline grows smaller relative to real or expected demand or consumption. The supply of gasoline is a function of crude oil supply and refining, imports of refined gasoline, and gasoline inventories (stocks). Stocks are the cushion between major short-term supply and demand imbalances, and stock levels can have a significant impact on gasoline prices. If refinery or pipeline problems and/or reductions in imports cause supplies to decline unexpectedly, gasoline inventories can drop rapidly. This drop in inventories can cause wholesalers to bid higher for available supply because they're worried that there may not be enough in the immediate future for their needs. Like most businesses, it is largely controlled by supply and demand (Figure 10.8). The fluctuating prices that have occurred in the market send a clear signal about the balance of supply and demand. Rising prices indicate that additional supply is needed, and falling prices indicate there is too much supply for current demand. The large changes in world oil prices in the past 8 years demonstrate how factors such as political instability and market control can influence oil prices, and they demonstrate the difficulty in relying so heavily on petroleum as a principal energy resource—consumers are constantly vulnerable to the issues discussed earlier.

DIESEL FUEL

Diesel fuel, is used in motor vehicles that use the compression ignition engine named for its inventor, German engineer Rudolf Diesel. He patented his original design in 1892. Compression ignition means that in a diesel engine, air is compressed inside the cylinders of the engine, which causes the air in the cylinders to heat up. When fuel is injected into the cylinders of hot, compressed air, the fuel ignites. Although diesel engines are capable of burning a wide variety of fuels, most of the diesel fuel consumed is refined from crude oil.

Nearly all semi-trailer trucks, delivery vehicles, buses, trains, ships, boats, and barges have diesel engines. Farm vehicles, construction vehicles, and military vehicles and equipment also have diesel engines. Some cars and small trucks are equipped with them. It is also used in diesel engine generators to generate electricity. Most remote villages in Alaska, and in other locations, use diesel generators to supply their electricity. Many industrial facilities, large buildings, institutional facilities, hospitals, and electric utilities also have diesel generators for backup and emergency power supply.

Diesel fuel is a type of distillate fuel. On average, about 12 to 13 gallons of distillate are produced from each

42-gallon barrel of crude oil in U.S. refineries. Before 2015, diesel fuel sold in the United States contained high quantities of sulfur. Sulfur in diesel fuel produces air pollution emissions that are harmful to human health. In 2006, the U.S. EPA issued requirements for the reduction of the sulfur content of diesel fuel. The requirements phased in over time, beginning with diesel fuel sold for vehicles used on roadways and eventually covering all diesel fuel. All diesel fuel now sold in the United States must have a sulfur content of no more than 15 parts per million. In 2015, distillate fuel oil accounted for about 21 percent of total petroleum consumption by the transportation sector.

Source of Diesel

U.S. refineries produce diesel fuel from domestically produced crude oil and imported crude oil. In 2014, the United States produced 53 percent of the crude oil processed in U.S. refineries. A small amount of diesel fuel is imported. In 2014, the volume of imported diesel fuel was equivalent to only 3 percent of the amount that was consumed.

Most diesel fuel is transported by pipeline from refineries and ports to terminals near major consuming areas, where it is then loaded into tanker trucks for delivery to retail service stations. A small amount of diesel fuel is transported by barge and rail.

Diesel fuel and other products are also sent through shared pipelines in *batches*. Because these batches are not physically separated in pipelines, this causes some mixing, or *commingling*, of products to occur. This possible mixing is why the quality of the diesel fuel and other products has to be tested to make sure they meet regulations as they enter and leave pipelines. Whenever any product fails to meet a local, state, or federal product specification, it must either be removed and trucked back to a refinery for further processing or be sold as a different product.

Importance of Diesel

Blends of up to 20 percent biodiesel with 80 percent petroleum diesel (B20) can be used in unmodified diesel engines. Inventor Rudolf Diesel originally designed his engine to use coal dust as fuel. However, he experimented with vegetable oil (biodiesel) before the petroleum industry began making diesel fuel.

The first diesel engine automobile trip was completed on January 6, 1930. Nearly an 800-mile trip, it stretched from Indianapolis, Indiana to New York City, and demonstrated the potential value of the diesel engine design, which has been used in millions of vehicles since its inaugural trip, now well-known for its dependability.

Diesel fuel has been, and continues to be, important to the U.S. economy. As a transportation fuel, it offers a wide range of performance, efficiency, and safety features. Diesel fuel contains between 18 and 30 percent more energy per gallon than regular gasoline. Diesel fuel also offers a greater power density than other fuels, so it provides more power per volume.

In 2015, diesel fuel accounted for about 21 percent of the petroleum fuels consumed by the U.S. transportation sector.

Diesel is the fuel of choice in the transportation of goods and commodities, transporting nearly all the products the public consumes. It is also commonly used in public transportation, such as in public and school buses. Both the agricultural sector and construction industry rely heavily on it. Farm equipment is run on it nearly exclusively; construction equipment depends on the power that diesel is capable of providing. Diesel engines can handle demanding construction work, such as lifting steel beams, digging foundations and trenches, drilling wells, paving roads, and moving soil safely and efficiently.

Military forces also rely heavily on diesel fuel. It is a mainstay in tanks and trucks because diesel fuel is less flammable and less explosive than other fuels. Diesel engines are also less likely to stall than gasoline-fueled engines. It has proven its worth in diesel engine generators in order to generate electricity. Many industrial facilities, large buildings, institutional facilities, hospitals, and electric utilities, have diesel generators for backup and emergency power supply. Most villages in Alaska use diesel generators for their electricity.

The cost of producing and delivering diesel fuel to customers includes the costs of crude oil, refinery processing, marketing and distribution, and retail station operation. The retail pump price reflects the costs and the profits (and sometimes losses) of the refiners, marketers, distributors, and retail station owners. The relative share of these cost components to the retail price of diesel fuel varies over time and varies among regions of the country.

Before 2004, the average price of diesel fuel was often lower than the average price of regular gasoline. In some winters when the demand for distillate heating oil was high, the price of diesel fuel rose above the gasoline price. Since September 2004, the price of diesel fuel has been generally higher than the price of regular gasoline throughout the year for several reasons. Worldwide demand for diesel fuel and other distillate fuel oils steadily increased, with strong demand in China, Europe, and the United States. In the United States, the transition to ultra-low sulfur diesel (ULSD) fuel affected diesel fuel production and distribution costs. Also, the federal excise tax on diesel fuel is 6 cents higher per gallon than the federal excise tax on regular gasoline.

HEATING OIL

Heating oil is a petroleum product used by many Americans, especially in the Northeast, to heat their homes. Heating oil and diesel fuel are closely related products because they are both distillates. The main difference between the two fuels is that heating oil may contain a greater amount of sulfur than diesel fuel. This fact was more so the case in the past, but it is less so today. All diesel fuel now sold in the United States must have a sulfur content of no more than 15 parts per million. Some of this *low-sulfur distillate/diesel fuel oil* may actually be sold as heating oil. Several states in the Northeast have begun to require a reduction in the level of sulfur allowed in heating oil (Today in Energy, 2012).

DID YOU KNOW?

- Heating oil is dyed red.
- The U.S. Internal Revenue Service (IRS) requires heating oil and distillate fuel oils that are not for highway use to be colored with a red dye. The red color indicates that the fuel is exempt from federal, state, and local taxes applied to fuels sold for use on public roadways. The red color also indicates that the fuel cannot be used legally in vehicles that normally operate on roadways.

DID YOU ALSO KNOW?

- Refiners may delay producing heating oil for the winter if consumer demand is high for a seasonal product like gasoline. This may result in lower heating oil inventories at the beginning of the heating season. This was the case in September and October 2005, after Hurricane Katrina and Hurricane Rita resulted in the shutdown of some Gulf Coast refineries. As gasoline prices increased to more than $3.00 per gallon, refiners had an incentive to produce more gasoline at a time when they would normally concentrate on heating oil production.

Origins of Heating Oil

Heating oil is refined from crude oil. The United States has two sources of heating oil:

- Domestic oil refineries
- Imports from foreign countries

Refineries produce heating oil as a part of the distillate fuel category of petroleum products. Distillates include heating oil and diesel fuel. Heating oil generally has higher sulfur content than diesel fuel. Distillate products are moved throughout the United States by pipelines, tankers (ships), barges, trains, and trucks.

Refiners have limitations on the amount of heating oil they can make to meet consumer demand during the winter heating season. Refiners can increase heating oil production in the winter, but that means they also have to produce greater amounts of other petroleum products. If there is not a market for the larger volumes of other petroleum products, this may limit the amount of extra heating oil that they produce. Therefore, some of the heating oil that refiners produce in the summer and fall months is stored for sale in the winter. During the coldest winter months, these inventories are used to help meet demand.

Heating oil is imported from other countries to supplement U.S. refinery production and inventories. Most U.S. imports of distillate come from Canada and Russia and are imported into the East Coast. Heating oil is brought to storage terminals

by refiners and other suppliers. For example, heating oil may be delivered to a central distribution area, such as New York Harbor, where it is then redistributed by barge to other consuming areas like New England. Once heating oil is in the consuming area, it is redistributed by truck to smaller storage tanks closer to a retail dealer's customers, or it is redistributed directly to residential customers.

USES OF HEATING OIL

About six million households in the United States use heating oil as their main space heating fuel. Some households also use heating oil for water heating, but in much smaller amounts than the amounts used for space heating. Some commercial and institutional buildings also use heating oil for space and water heating. Because space heating is the primary use for heating oil, the demand is seasonal, and it is affected directly by temperatures. Most heating oil use occurs during the heating season (October through March).

Most of the U.S. households that use heating oil are located in the Northeast. In 2014, about 13.6 billion liters of heating oil were sold to residential consumers in the Northeast, equal to 88 percent of the total U.S. residential fuel oil sales. In 2014, about 37 percent of total commercial sector consumption of heating oil was in the Northeast. The number of houses with heating systems that use heating oil has declined over time because heating systems in homes that previously used heating oil are now being converted to other types of systems as upgrades are being made to these homes. Most new homes use natural gas or electricity for heating (Figure 10.9a and b).

THE HEATING OIL MARKET

Heating oil prices paid by consumers are determined by several things: the cost of crude oil, the cost to produce the product, the cost to market and distribute/deliver the product, and the profits (and sometimes losses) of refiners, wholesalers, and local distributors. From the winter of 2005–2006 through the winter of 2014–2015, crude oil accounted for 57 percent of the average cost of a liter of heating oil during winter months. The next largest component, distribution and marketing costs and profits, accounted for approximately 27 percent of the cost of a liter of heating oil. Refinery processing costs and profits accounted for another 16 percent.

In order to keep consumer prices as low as possible for the consumer, a common practice is to fill up heating oil tanks filled in the late summer/early autumn, just before the cold weather sets in. This is generally when the prices for the oil are the lowest. Some heating oil dealers may offer their customers a budget plan to help even out monthly bills. Some dealers may also offer fixed-price protection programs that may help keep costs down. To help keep expensive heating costs down, consumers must do their part, as well. Consumers can reduce heating oil consumption by caulking and weather stripping windows and doors to seal out cold air, by installing proper insulation in the attic and walls, and by reducing temperature settings on thermostats.

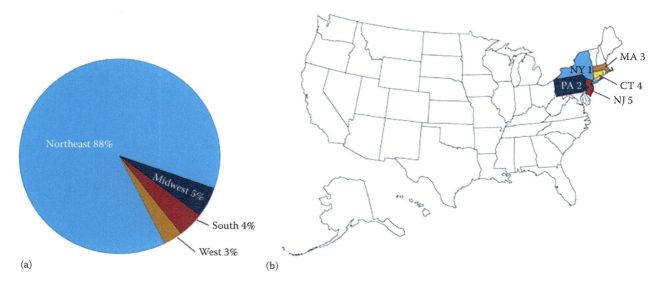

(a)

(b)

FIGURE 10.9 (a) Sales of residential heating oil by region in the United States, 2014. (From EIA, Washington, DC.) and (b) top five heating oil consuming states, 2014. (From EIA, Washington, DC.)

Like the other fossil fuels we have already discussed, heating oil prices fluctuate, as well. Changes in the cost of crude oil are a major component of the price of heating oil. Crude oil prices are determined by worldwide supply and demand. Demand can vary worldwide based on the economy and weather in addition to other factors. Supply can be influenced by weather events in the United States and by political events in other countries. The supply of oil produced by members of the OPEC can also affect world crude oil prices.

When crude oil prices are stable, home heating oil prices tend to rise in the winter months when demand is highest. A homeowner in the Northeast might use 3,217 to 4,542 liters of heating oil during a typical winter, and then they may consume very little during the rest of the year.

Crude oil prices are also determined by worldwide supply and demand. Demand can vary worldwide based on the economy and weather in addition to other factors. Supply can be influenced by weather events in the United States and by political events in other countries.

The number of heating oil suppliers in a particular region can affect the level of price competition in that area. Heating oil prices and service offerings can vary substantially in locations with few suppliers compared to areas with a large number of competing suppliers. Consumers in rural locations may pay higher prices for heating oil because there are fewer competitors.

Home heating oil prices can sometimes change dramatically for a variety of reasons. When refiners, wholesalers, dealers, and consumers have enough heating oil in storage, and if temperatures do not drop or if crude oil prices do not increase rapidly, retail prices may hold fairly steady. A large cold weather system can impact supply, demand, and prices. People typically use more fuel at the same time delivery systems are interrupted, such as during a winter storm.

During cold weather, the amount of heating oil in storage may be used much faster than it can be replenished, and refineries may not be able to keep up with demand. Wholesale buyers become concerned that supplies are not adequate to cover short-term customer demand, and they bid up prices for available product.

In the Northeast, for example, additional supplies of heating oil may have to be delivered from other parts of the world like the Gulf Coast or Europe. Transporting heating oil from these sources to the Northeast is expensive, and delivery can take several weeks. During that time, storage inventories drop further, buyers' anxiety about available short-term supply rises, and then prices rise—sometimes sharply—until new supply arrives.

HYDROCARBON GAS LIQUIDS

Natural gas and crude oil are mixtures of different hydrocarbons. Hydrocarbons are molecules of carbon and hydrogen in various combinations. *Hydrocarbon gas liquids* (HGL) are hydrocarbons that occur as gases at atmospheric pressure and as liquids under higher pressures. HGL can also be liquefied by cooling. The specific pressures and temperatures at which the gases liquefy vary by the type of HGL. HGL may be described as being *light* or *heavy* according to the number of carbon atoms and hydrogen atoms in an HGL molecule. HGL are categorized chemically as:

- Alkanes, or paraffins
- Ethane—C_2H_6
- Propane—C_3H_8
- *Butanes*: Normal butane and isobutane—C_4H_{10}
- Natural gasoline or pentanes plus—C_5H_{12} and heavier
- Alkenes, or olefins
- Ethylene—C_2H_4
- Propylene—C_3H_6
- Butylene and isobutylene—C_4H_8

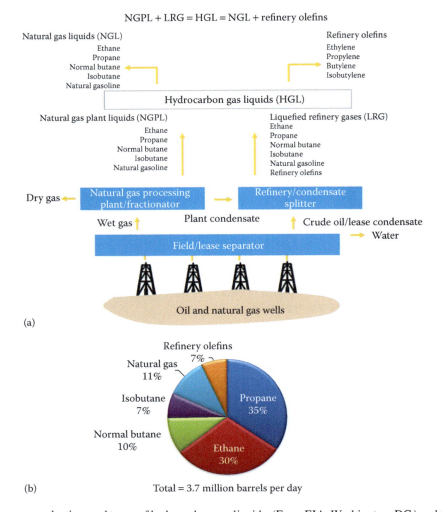

FIGURE 10.10 (a) Sources, production, and types of hydrocarbon gas liquids. (From EIA, Washington, DC.) and (b) U.S. hydrocarbon gas liquids production by type, 2014. (Courtesy of Kerr.)

HGL are found in raw natural gas and crude oil, and they are extracted when natural gas is processed at natural gas processing plants and when crude oil is refined into petroleum products. Natural gas plant liquids (NGPL), which account for most of HGL production in the United States, fall solely into the alkanes category. Refinery production accounts for the remainder of U.S. alkanes production. Greater volumes of olefins are produced at petrochemical plants from HGL and heavier feedstock. Figure 10.10 illustrates the sources, production, and types of hydrocarbon gas liquids.

Because HGL straddle the gas/liquid boundary, their versatility and high energy density in liquid form make them useful for many purposes:

- Feedstock in petrochemical plants to make chemicals, plastics, and synthetic rubber
- Fuels for heating, cooking, and drying
- Fuels for transportation
- Additives for motor gasoline production
- Diluent (a diluting or thinning agent) for transportation of heavy crude oil

In 2014, total HGL use accounted for about 13 percent of total U.S. petroleum consumption.

DID YOU KNOW?

Propane and butane were discovered in 1912 by Dr. Walter Snelling, a U.S. scientist. He identified these gases in gasoline, and he found that cooling and pressuring these gases changed them to liquid. He also learned that the liquefied gases could be stored and transported in pressurized containers.

Propane is produced from crude oil refining and natural gas processing. Its production is generally consistent throughout the year because it is part of the natural gas processing procedure. In natural gas processing plants, propane enters the plant as a part of the wet natural gas stream received from natural gas wells. When the raw, wet natural gas is cooled and pressurized in the plant, the heavier hydrocarbons, including ethane, propane, normal butane, isobutane, and natural gasoline, turn into liquids and separate from the natural gas stream.

The mixed liquids, also referred to as Y-grade, are then refined and separated in a fractionator into purity products (products having a minimum of 90 percent of one type of hydrocarbon gas liquid).

Propane can be produced in oil refineries at two stages of the crude oil refining process. The first refining process that yields propane is the atmospheric distillation column, where crude oil undergoes initial distillation. In modern, complex refineries, propane (and propylene) is also produced in the fluid catalytic cracker (FCC) when long-chain hydrocarbon molecules are cracked under high temperatures and pressures to break these molecules into lighter hydrocarbons that are more suitable for motor gasoline production. In addition to those gasoline blending products, the refinery crackers produce lighter molecules including propane, normal butane, isobutane, and their olefins.

The United States has historically imported propane to supplement U.S. propane production, especially during the autumn and winter months mainly to meet demand from farmers and for space heating in buildings. Most propane imports are received by rail from Canada into the Midwest, East Coast, and Rocky Mountain regions of the United States (Figure 10.11).

Transporting and Storing

Hydrocarbon gas liquids (HGL) that are extracted from natural gas or that are produced at petroleum refineries may be transported as liquids in mixtures of HGL or as separate HGL products in pipelines, rail cars, trucks, ships, and barges.

There are five main forms in which HGL are transported:

- Y-grade (raw, unseparated HGL)
- E-P mix (most frequently 80 percent ethane and 20 percent propane)
- P/P mix (refinery-grade propane-propylene mixture)
- LPG (mixture of liquefied propane, normal butane, and isobutane)
- Purity products (separate, distinct products; mostly ethane, propane, and normal butane)

Most of the HGL produced in the United States are transported in pipelines from where they are produced to places where they are used or to places where they are stored for distribution. There are many regions in the United States (like the West, New England, and Florida) that are not served by HGL pipelines. In these areas, railroads often transport large volumes of HGL to wholesale and bulk purchasers in pressurized railroad tank cars. Railroads and trucks are also used to transport HGL to consumers. The primary HGL product delivered to consumers is consumer-grade propane, which is transported by truck in pressurized tanks to homes, farms, and businesses where it can be used as engine fuel, for crop drying, space and water heating, and for cooking.

Special ships are used to transport HGL (usually LPG) to and from shipping ports in the United States. The ships, called *gas tankers*, vary in size and by the method used to keep the HGL in liquid form. The HGL may be pressurized, refrigerated, or both. Over short distances, propane and normal butane are also moved by barge along intercoastal waterways and navigable rivers.

Storage of HGL is required because HGL may be produced in volumes that exceed the capacity of different modes of transportation used to carry the HGL to consumers. Production may also not match HGL seasonal requirements. For example, production of propane is relatively consistent throughout the year, but demand for propane is usually lower in the summer and higher in the fall and winter. Propane is stored when demand is low, and propane is withdrawn from storage when demand is high.

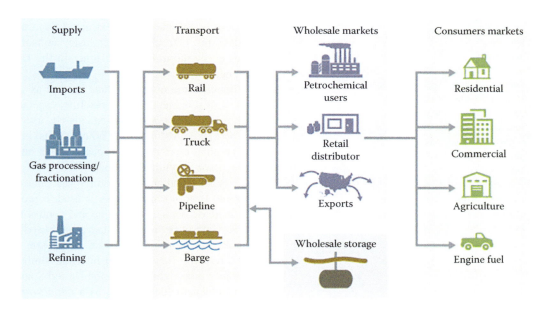

FIGURE 10.11 Propane supply and disposition. (From EIA, Washington, DC.)

Large volumes of HGL are primarily stored as a pressurized liquid in underground caverns. Most of the caverns are in salt formations, but there are also propane storage caverns mined out of shale, granite, and limestone rock. In regions where geology is not well suited for underground caverns, large above-ground tanks may be used. The above-ground tanks are the primary storage method for propane and butanes in New England.

Once HGL are transported close to consumers, they are stored in pressurized (and sometimes cooled) tanks located above or below the ground. LPG, which can refer to propane, butanes, or a mixture of the two, is stored and distributed in many different sizes of tanks, from the small canisters used for torches and camping stoves, to 90,000-gallon bullet-shaped tanks used at industrial facilities.

Uses of Hydrocarbon Gas Liquids

Hydrocarbon gas liquids (HGL) are versatile products used in every end-use sector, such as the residential, commercial, industrial (manufacturing and agriculture), electric power, and transportation. Their uses, however, vary; as shown in the following table:

Hydrocarbon Gas Liquids, Uses, Products, and Consumers

HGL	Uses	End-Use Products	End-Use Sectors
Ethane	Petrochemical feedstock for ethylene production; power generation	Plastics; antifreeze; detergents	Industrial
Propane	Fuel for space heating, water heating, cooking, drying, and transportation; petrochemical feedstock	Fuel for heating, cooking, and drying; plastics	Industrial (includes manufacturing and agriculture), residential, commercial, and transportation
Butanes: normal butane and isobutane	Petrochemical and petroleum refinery feedstock; motor gasoline blending	Motor gasoline; plastics; synthetic rubber; lighter fuel	Industrial and transportation
Natural gasoline (pentanes plus)	Petrochemical feedstock; additive to motor gasoline; diluent for heavy crude oil	Motor gasoline; ethanol denaturant; solvents	Industrial and transportation
Refinery olefins (ethylene, propylene, normal butylene, and isobutylene)	Intermediate feedstocks in the petrochemical industry	Plastics; artificial rubber; paints and solvents; resins	Industrial

Source: EIA

Most of the propane consumed in the United States is used as a fuel, generally in areas where the supply of natural gas is limited or not available. This use is highly seasonal, with the largest consumption occurring in the fall and winter months. There are four major consumer uses of propane:

- In homes, for space heating and water heating; for cooking; for drying clothes; and for fueling gas fireplaces, barbecue grills, and backup electrical generators
- On farms, for heating livestock housing and greenhouses, for drying crops, for pest and weed control, and for powering farm equipment and irrigation pumps
- In businesses and industry, to power fork lifts, electric welders, and other equipment
- As a fuel for on-road internal combustion engine vehicles such as cars, school busses, or delivery vans, and nonroad vehicles such as tractors and lawn mowers

The nonconsumer market for propane is the petrochemical industry. The primary use of propane in the petrochemical industry is as a feedstock, along with ethane and naphtha, in petrochemical crackers to produce ethylene, propylene, and other olefins. Propane can also be used as a dedicated feedstock in the petrochemical industry for on-purpose propylene production. Propylene and the other olefins may be converted into a variety of products, mostly plastics and resins, but also glues, solvents, and coatings.

Ethane is mainly used to produce ethylene, which is then used by the petrochemical industry to produce plastics. Ethane consumption in the United States has increased over the past several years. Ethane can also be used directly as a fuel for power generation, either on its own or blended with natural gas.

The presence of ethane in dry natural gas gives it a higher heat value—so when it's not extracted from pipeline-delivered natural gas, then that gas has a higher heat value. Because it is not completely extracted from natural gas, not only does the petrochemical industry consume ethane, but so does every natural gas consumer in the United States to some degree.

Although some normal butane is used as a fuel for lighters, most of it is blended into gasoline, especially during the cooler months. Normal butane can also be used as a feedstock in the petrochemical industry. When normal butane is used in petrochemical cracking, the process yields butadiene, which is a precursor to synthetic rubber.

Isobutane, whether from natural gas plants, refineries, or isomerized from normal butane, is used to produce alkylates, which increase octane in gasoline and control the volatility of gasoline. High-purity isobutane can also be used as a refrigerant.

Natural gasoline (also known as pentanes plus) can be blended into the fuels used in internal combustion engines, particularly motor gasoline. In the United States, natural gasoline is added to fuel ethanol as a *denaturant* to make the ethanol undrinkable, which is required by law. Some ethanol producers use natural gasoline to make E85.

About half of U.S. natural gasoline production is exported to Canada where it is used as a diluent (to reduce viscosity) for heavy crude oil, so that the crude oil can be more easily moved in pipelines and railcars.

Imports and Exports of Hydrogen Gas Liquids

The United States typically produces more hydrocarbon gas liquids (HGL) than it uses each year. Occasionally, the U.S. must import some amount of HGL to supply high, seasonal demand and to supply some regions of the country that are not supplied with enough domestic sources. Certain HGL, such as propylene, are also imported because U.S. falls short of satisfying the total petrochemical demand. In 2014, HGL imports of 142,600 barrels per day (b/d) accounted for about 7.5 percent of total U.S. imports of petroleum products (this does not include crude oil). More than 62 percent of the HGL imports were propane. Normal butane and isobutane together accounted for slightly more than 10 percent of the HGL imports, followed by natural gasoline at slightly less than 10 percent Most U.S. imports of propane and butanes are received by rail from Canada into the Midwest and Northeast regions of the United States. These imports are highly seasonal, with two-thirds of all imports occurring October through March. This reflects propane's use as a heating fuel, and butanes' use in gasoline blending during colder months when gasoline vapor pressure requirements allow its use in higher quantities. In 2014, 93 percent of U.S. HGL imports were received from Canada, and the rest was imported on ships from overseas. In 2014, U.S. HGL imports were only one-fifth as large as U.S. HGL exports.

Exports of HGL increased from about 70,000 b/d in 2007 to 703,000 b/d in 2014, which was equal to 18 percent of total U.S. exports of petroleum products in 2014. These exports were largely driven by annual U.S. production exceeding annual U.S. demand. The increases in HGL production were mainly the result of increases in production of wet natural gas from shale gas and tight oil resources. Propane accounted for about 60 percent of the total HGL exports in 2014, and natural gasoline (pentanes plus) accounted for about 24 percent. Between 2011 and 2014, exports of HGL accounted for more than half of all growth in petroleum product exports out of the United States.

Propane was exported from the United States to 52 countries in 2014. These are the top five destinations and their share of total U.S. propane exports:

- *Mexico*: 15 percent
- *Japan*: 12 percent
- *Brazil*: 12 percent
- *Netherlands*: 11 percent
- *Panama*: 7 percent

In 2014, 99 percent of natural gasoline exports were sent to Canada, with the remainder exported to 29 countries in minor quantities as a denaturant in fuel ethanol. Most of the natural gasoline exported to Canada is mixed into the oil produced from oil sands in Alberta, Canada, so that the oil can be transported by pipelines and rail, primarily back to U.S. destinations.

NATURAL GAS

Natural gas is a gas that occurs deep beneath the earth's surface. It consists mainly of methane, a gas (or compound) with one carbon atom and four hydrogen atoms. Natural gas also contains small amounts of hydrocarbon liquids and nonhydrocarbon gases. Natural gas can be used as a fuel or to make materials and chemicals.

Origins of Natural Gas

Millions of years ago, the remains of plants and animals (diatoms) decayed and built up in thick layers, sometimes mixed with sand and silt. Over time, these layers were buried under more sand, silt, and rock. Pressure and heat changed some of this organic material into coal, some into oil, and some into natural gas. In some places, the natural gas moved into large cracks and spaces between layers of overlying rock. Natural gas is also contained in tiny pores within some formations of shale, sandstone, other types of sedimentary rock, and in coal. The search for natural gas begins with geologists, by locating the types of rock that are likely to contain natural gas deposits. Some of these areas are on land, and some are offshore and deep under the ocean floor.

Today, geologists use seismic surveys to find the right places to drill wells. Seismic surveys use echoes from a vibration source at the earth's surface (usually a vibrating pad under a truck built for this purpose) to collect information about the rocks beneath. Sometimes it is necessary to use small amounts of dynamite to provide the vibration needed.

If a site seems promising, an exploratory well may be drilled to collect data on the formation to find out if it contains enough natural gas to be economically feasible. Once a formation is proven for economical production, one or more production wells are drilled down into the formation, and natural gas flows up through the wells to the surface. In the United States and in a few other countries, natural gas is produced directly from shale and other types of rock formations that contain natural gas in pores within the rock. The rock formation is fractured by forcing water, chemicals, and sand down a well. This releases the natural gas from the rock, and the natural gas flows up the well to the surface. Wells drilled to produce oil may also produce associated natural gas.

The natural gas withdrawn from a well is called wet natural gas because it usually contains liquid hydrocarbons and nonhydrocarbon gases. Methane and other useful gases are separated from the wet natural gas near the site of the well or at a natural gas processing plant. The processed gas is called *dry* or *consumer-grade* natural gas. This natural gas is sent through pipelines to underground storage fields and/or to distribution companies, and then to consumers.

Coal may contain *coalbed methane*, which can be captured when coal is mined. Coalbed methane can be added to natural gas pipelines without any special treatment. Another source of methane is biogas that is produced in landfills and in machines called *digesters*.

Most of the natural gas consumed in the United States is produced in the United States. Some natural gas is imported from Canada and Mexico in pipelines. A small amount of natural gas is also imported as liquefied natural gas.

DID YOU KNOW?

Because natural gas is colorless, odorless, and tasteless, mercaptan (a chemical that smells like sulfur) is added before distribution, to give natural gas a distinct unpleasant odor (it smells like rotten eggs). This added smell serves as a safety device by allowing it to be detected in the atmosphere in cases where leaks occur.

Delivery and Storage

Transporting natural gas from the wellhead to consumers requires many processing steps, and it includes several physical transfers.

The natural gas delivery infrastructure can be grouped into three categories (Figure 10.12):

- Processing
- Transportation
- Storage

Natural gas fed into the mainline natural gas transportation system in the United States must meet specific quality measures so that the pipeline network (or grid) can provide uniform quality gas. Natural gas produced at the wellhead may contain contaminants and natural gas liquids, which must be removed before the natural gas can be safely delivered to the high-pressure, long-distance pipelines that transport natural gas to consumers.

Processing the wellhead natural gas into pipeline-quality dry natural gas can be complex and usually involves several processes to remove oil, water, natural gas liquids, and other impurities such as sulfur, helium, nitrogen, hydrogen sulfide, and carbon dioxide.

A natural gas processing plant typically receives natural gas from a gathering system of pipelines from natural gas wells and sends out processed gas to one or more major pipeline networks. Liquids removed at the processing plant may be sent to petrochemical plants, refineries, and other gas liquids consumers.

The number of stages and the type of techniques used to create pipeline-quality natural gas depends on the composition of the natural gas produced at the well.

A natural gas transmission line is a wide-diameter, long-distance portion of a natural gas pipeline system, located between the gathering system (production area), the natural gas processing plant the other receipt points, and the principal consumer service areas. There are three types of transmission pipelines:

- *Interstate natural gas pipelines:* These operate and transport natural gas across state borders.
- *Intrastate natural gas pipelines:* These operate and transport natural gas within a state border.
- *Hinshaw natural gas pipelines:* These receive natural gas from interstate pipelines and deliver it to consumers for consumption within a state border.

When natural gas gets to the communities where it will be used (usually through large pipelines), it flows into smaller pipelines called *mains.* Small lines, called *services,* connect to the mains and go directly to homes or buildings where the natural gas will be used.

Underground natural gas storage provides pipelines, local distribution companies, producers, and pipeline shippers with an inventory management tool, seasonal supply backup, and access to natural gas when necessary, to avoid imbalances between receipts and deliveries on a pipeline network.

There are three main types of natural gas underground storage facilities used in the United States today:

- *Depleted natural gas or oil fields:* Most of the existing natural gas storage in the United States is in depleted natural gas or oil fields that are close to consumption centers.
- *Salt caverns:* Salt caverns provide high withdrawal and injection rates relative to their working gas capacity. Base gas requirements are relatively

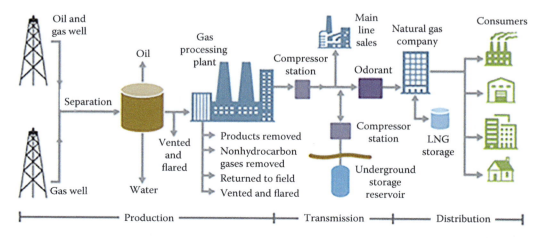

FIGURE 10.12 Natural gas production and delivery. (From EIA, Washington, DC.)

low. Most salt cavern storage facilities have been developed in salt dome formations located in the Gulf Coast states. Salt caverns have also been leached from bedded salt formations in states in the Midwest, Northeast, and Southwest.

- *Aquifers:* In some areas, most notably in the Midwest, natural aquifers have been converted to natural gas storage reservoirs. An aquifer is suitable for natural gas storage if the water bearing sedimentary rock formation is overlaid with an impermeable cap rock (Figure 10.13).

Natural Gas Pipelines

The U.S. natural gas pipeline network is a highly integrated network that moves natural gas throughout the continental United States. The network is an intricate transportation system made up of about 4.8 million kilometers of mainline and other pipelines that links natural gas production areas, and storage facilities with consumers. This natural gas transportation network delivered more than 680 million cubic meters (Tcf) of natural gas during 2014 to about 73 million customers.

About 486,000 kilometers of wide-diameter, high-pressure interstate transmission pipelines (pipelines that cross state boundaries) and intrastate transmission pipelines (pipelines that operate within state boundaries) transport natural gas from the producing and processing areas to storage facilities and distribution centers. Compressor stations (or pumping stations) located along the length of the pipeline network keep the natural gas flowing forward through the pipeline system. More than 300 companies operate mainline transmission pipelines. More than 1,100 local distribution companies deliver natural gas to end users through hundreds of thousands of kilometers of small-diameter service lines. Local distribution companies reduce the pressure of the natural gas received from the high-pressure mainline transmission system to a level that is acceptable for use in residences and commercial establishments. A map of the natural gas lines in the

A Salt caverns
B Mines
C Aquifers
D Depleted reservoirs
E Hard-rock caverns

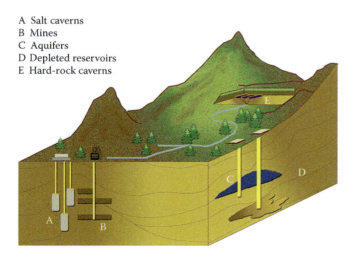

FIGURE 10.13 Possible underground storage facilities. (From EIA, Washington, DC.)

U.S. is shown in Figure 10.14. About 228,530 kilometers of the current 486,000 kilometers of the mainline natural gas transmission network were installed in the 1950s and 1960s as consumer demand for natural gas more than doubled following World War II. The period of greatest local distribution pipeline growth happened more recently. In the 1990s, pipelines were installed to provide service to many new commercial facilities and housing developments that wanted access to natural gas supplies.

Natural gas prices, along with oil prices, increased substantially between 2003 and 2008. Higher prices provided natural gas producers an incentive to expand development of existing fields and to begin exploration of previously undeveloped natural gas fields. Consequently, new pipelines have been constructed, and others are being built to link these expanded and new production sources to the existing mainline transmission network. From 2000 to 2014, about 34,260 more miles of distribution pipelines were added to the national transmission network. As you can see, this is an extensive, complex system.

Liquefied Natural Gas

Liquefied natural gas (LNG) is natural gas that has been cooled to a liquid state, at about −162°C, for shipping and storage. The volume of natural gas in its liquid state is about 600 times smaller than its volume in its gaseous state. This process, which was developed in the nineteenth century, makes it possible to transport natural gas on specially designed ships to places pipelines do not reach. Liquefying natural gas is a way to move natural gas long distances when pipeline transport is not feasible. Markets that are too far away from producing regions to be connected directly to pipelines have access to natural gas because of LNG. In its compact liquid form, natural gas can be shipped in special tankers to terminals in the United States and in other countries. At these terminals, the LNG is returned to its gaseous state and transported by pipeline to distribution companies, industrial consumers, and power plants.

The United States imports and exports LNG. In 2014, the United States imported about 1.2 billion cubic meters of LNG. Nearly all LNG imports were from Trinidad and Tobago, Norway, and Yemen. Pipeline imports of natural gas are mostly from Canada. Most U.S. natural gas exports are delivered to Canada and Mexico by pipeline. Because LNG is more energy dense than gaseous natural gas, there is increasing interest in using LNG as a fuel for heavy-duty vehicles and other transportation applications.

Remaining Natural Gas Resources

As we learned with petroleum products, underground reservoirs hold oil and natural gas. A *reservoir* is a location where large volumes of methane, the major component of natural gas, can be trapped in the subsurface of the earth at places where suitable geological conditions occurred at the right times. Reservoirs are made up of porous and permeable rocks that can hold significant amounts of oil and natural gas in pore spaces in the rock. *Proved reserves* of natural gas are estimated quantities

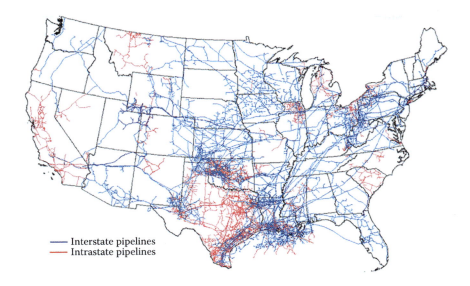

FIGURE 10.14   Map of U.S. interstate and intrastate natural gas pipelines. (From EIA, Washington, DC.)

that analyses of geological and engineering data have demonstrated to be economically recoverable from known reservoirs in the future. As more areas are inventoried, proved reserves are added each year with successful exploratory wells and as more is learned about fields where current wells are producing natural gas. The application of new technologies can convert previously uneconomic natural gas resources into proved reserves. U.S. proved reserves of natural gas have increased every year since 1999. The U.S. total natural gas proved reserves, estimated as *wet* gas—which includes natural gas plant liquids—set a record of nearly 11 trillion cubic meters (Tcm) in 2014. The dry gas portion of these reserves (after liquids are removed) equaled 9.5 Tcm. Major advances in natural gas exploration and production technologies have resulted in increased U.S. natural gas proved reserves.

In addition to the proved natural gas reserves, there are large volumes of natural gas classified as *undiscovered technically recoverable resources*. Undiscovered technically recoverable resources are expected to exist because the geologic settings are favorable even though their existence is uncertain at this point. Undiscovered technically recoverable resources are also assumed to be producible over a time period using existing recovery technology. The U.S. Energy Information Administration has estimated that as of January 1, 2013, the United States had 56 Tcm of undiscovered technically recoverable resources of dry natural gas (Figure 10.15).

Uses of Natural Gas

Currently, natural gas is a major energy source in the United States, which used about 759 billion cubic meters (Bcm) of natural gas in 2014, the equivalent of 27.5 quadrillion British thermal units (Btu) and 28 percent of total U.S. energy use.

Natural gas is presently used as a fuel to produce steel, glass, paper, clothing, brick, and electricity. It is also used as

a raw material for many products, including paints, fertilizer, plastics, antifreeze, dyes, photographic film, medicines, and explosives. It is a major fuel used to heat buildings. About half of the homes in the United States use natural gas as their main heating fuel. It is also used in homes and businesses for cooking, heating water, drying clothes, and for outdoor lighting.

The top five consumers of natural gas in the United States in 2014 were:

- Electric power sector—229 million cubic meters
- Industrial sector—210 million cubic meters
- Residential sector—144 million cubic meters
- Commercial sector—99 million cubic meters
- Oil and natural gas industry—45 billion cubic meters

While natural gas is used throughout the United States, there were five states that accounted for approximately 36 percent of the total U.S. natural gas consumption in 2014:

- Texas—12.7 percent
- California—8.6 percent
- Louisiana—4.9 percent
- New York—4.9 percent
- Florida—4.9 percent (Figure 10.16)

The pricing for natural gas is divided into two components: the commodity cost and the transmission/distribution cost. The commodity cost includes the cost of the natural gas itself, either as produced natural gas or as natural gas purchased at a market trading hub or purchased under a contract by marketers. The transmission and distribution costs include the cost to transport the gas via pipeline from the location it is produced to the local natural gas companies and the cost to transport it to the consumer.

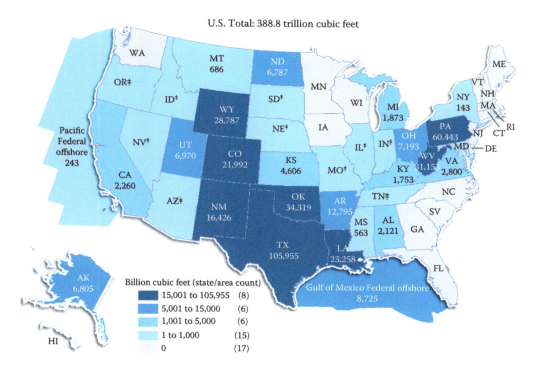

U.S. Total: 388.8 trillion cubic feet

Billion cubic feet (state/area count)
- 15,001 to 105,955 (8)
- 5,001 to 15,000 (6)
- 1,001 to 5,000 (6)
- 1 to 1,000 (15)
- 0 (17)

FIGURE 10.15   Natural gas proved reserves by state/area, 2014. (From EIA, Washington, DC.)

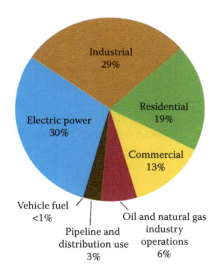

FIGURE 10.16   Natural gas use in the United States. (From EIA, Washington, DC.)

Prices for natural gas have declined recently but they do vary across the nation. There are nine major factors that ultimately determine the price at any particular location:

- Distance away from the area that produces the natural gas
- Availability and capacity of the transmission pipelines that deliver the natural gas
- Volume and characteristics of the customer demand, such as the timing of when the demand occurs

- Costs of distribution, taxes, and other charges for a particular area
- Individual state regulations
- Availability of competing suppliers
- Variations in the amount of natural gas production
- The volume of natural gas being imported and exported
- The amount of gas being kept in storage facilities (called *current storage levels*)

With that said, consumers are always interested in lowering their bills—as they should be. The good news is that it is possible. Here are some easy ways to do that (we will discuss other ways in greater detail in a later chapter):

- Shop around for lower-priced natural gas if a customer choice program is available.
- Participate in a local gas company's yearly budget plan to spread natural gas bills evenly throughout the year (sometimes referred to as an *equal payment plan*).
- Turn down your thermostat setting, especially when your home is not occupied.
- Have an energy audit on your home to identify ways your home is losing energy or is inefficient. Once you've identified them, get them corrected.
- Check natural gas space-heating equipment to make sure it is operating efficiently.

Like all other commodities, natural gas prices are also a function of market supply and demand. Because of limited alternatives for natural gas consumption or production in the near term, even small changes in supply or demand over a short period can result in large price movements that bring supply and demand back into balance. Decreases in supply cause prices to rise, increases in supply tend to result in lower prices. Other demand factors can also affect prices, such as:

- Levels of current economic growth
- Variations in winter and summer weather
- Prices of competing fuels
- Increases in demand, which lead to price increases; while decreases in demand most often lead to lower prices

In the United States, most of the natural gas prices are driven by supply. With more efficient, cost-effective drilling and completing techniques—such as in shale and other similar geologic formations, more natural gas has been able to be recovered. The result is the supply of natural gas greatly increases, causing supplies to rise drastically. With more supply than demand, this drives the price of natural gas down (Today in Energy, 2016). Conversely, when the supply is low, but the demand is high, it becomes a scarce commodity, and this drives the prices up.

One thing that can disrupt normal production and cause a supply shortage, and hence a price hike, is severe weather. Severe weather events, such as hurricanes can drastically disrupt production. For instance, with all the destructive hurricanes during the 2005 hurricane season along the U.S. Gulf Coast, it shut down about 4 percent of the total U.S. natural gas production between August 2005 and June 2006. The shortage in natural gas this triggered caused the prices of gas to climb drastically. The effect proved debilitating to a significant portion of the population.

COAL

Coal is a combustible black or brownish-black sedimentary rock with a high amount of carbon and hydrocarbons. Coal is classified as a nonrenewable energy source because it takes millions of years to form. It contains the energy stored by plants that lived hundreds of millions of years ago in swampy forests and were covered by layers of rock and debris. Buried under extreme pressure and heat, the plants were slowly converted into what we now know as coal (Figure 10.17).

The Chinese were using coal as early as 1000 BCE to bake porcelain. The ancient Greeks also wrote about it in their history. In the Western countries, many of the forests had been destroyed by the 1100s to build houses and ships. Wood to heat houses became expensive, so people began looking for alternate ways to provide heat to their homes.

By the 1600s, coal had grown in popularity as a heating source. It was also used in breweries, glass-making, brick making, and many other businesses. Coal continued to grow in popularity. Originally, it was mined from shallow near-surface deposits. As this coal was used, miners began building mine shafts in order to go deeper into the ground where many coal deposits existed.

Coal mining was very hazardous. Shafts and tunnels had to be supported by beams. There was also the constant danger of cave-ins, explosions, or fires. By the beginning of the 1700s, coal was used in tall furnaces. This marked the beginning of the Industrial Revolution. England had many coal deposits, and during that time, it became a very wealthy nation because of this. With the invention of the steam engine, the demand for coal further increased. Coal is the energy source that significantly changed civilization and triggered the development of the modern world.

Coal is classified into four main types, or ranks: anthracite, bituminous, subbituminous, and lignite. The ranking depends on the types and amounts of carbon the coal contains and on the amount of heat energy the coal can produce. The rank of a coal deposit is determined by the amount of pressure and heat that acted on the plants over time.

Anthracite contains 86–97 percent carbon, and generally has the highest heating value of all ranks of coal. Anthracite only accounted for less than 1 percent of all the coal mined in the United States in 2014. All of the anthracite mines in the United States are located in northeastern Pennsylvania. Anthracite is mainly used by the metals industry.

Bituminous coal contains 45–86 percent carbon. Bituminous coal in the United States is between 100 and 300 million years old. It is the most abundant rank of coal found in the United States, and it accounted for 48 percent of total U.S. coal production in 2014. Bituminous coal is used to generate electricity, and it

How coal was formed

Before the dinosaurs, many giant plants died in swamps

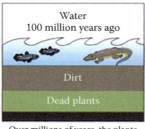
Over millions of years, the plants were buried under water and dirt

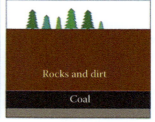
Heat and pressure turned the dead plants into coal

FIGURE 10.17   The process of coal formation. (From EIA, Washington, DC.)

is an important fuel and raw material for making iron and steel. West Virginia, Kentucky, Pennsylvania, Illinois, and Indiana were the five main bituminous coal-producing states in 2014, and were responsible for 70 percent of total bituminous production.

Subbituminous coal generally contains 35–45 percent carbon, and it has a lower heating value than bituminous coal. Most subbituminous coal in the United States is at least 100 million years old. About 44 percent of total U.S. coal production in 2014 was subbituminous, and nearly 90% percent was produced in Wyoming.

Lignite contains 25–35 percent carbon and has the lowest energy content of all coal ranks. Lignite coal deposits tend to be relatively young and were not subjected to extreme heat or pressure. Lignite is crumbly and has high moisture content, which contributes to its low heating value. Lignite accounted for 8 percent of total U.S. coal production in 2014. About 92 percent of total lignite production is mined in Texas and North Dakota, where it is burned at power plants to generate electricity. A facility in North Dakota also converts lignite to *synthetic natural gas* and pipes it to natural gas consumers in the eastern United States.

Mining and Transportation of Coal

Coal miners use huge machines designed to remove coal from the earth. Many coal deposits in the U.S., called *coal beds* or *seams*, are near the earth's surface, while others lie deep underground. About two-thirds of U.S. coal production comes from surface mines. Modern mining methods allow coal miners to easily reach most of the nation's coal reserves and to produce about three times more coal in one hour than was possible back in 1978.

There are two principal mining methods: surface mining and underground mining. *Surface mining* is often used when coal is less than 61 meters underground. In surface mining, large machines remove the topsoil and layers of rock known as overburden to expose coal seams. *Mountaintop removal* is a form of surface mining where the tops of mountains are dynamited and removed to access coal seams. Once the coal is removed, the disturbed area may be covered with topsoil for planting grass and trees. Surface mines produce most of the coal in the United States because surface mining is less expensive than underground mining.

Underground mining—also called deep mining—is necessary when the coal is several hundred meters below the surface. Some underground mines are more than 300 meters deep and extend for kilometers. Miners ride elevators down deep mine shafts and travel on small trains in long tunnels to get to the coal. The miners use large machines to dig out the coal.

Once the coal is removed from the ground, the miners often send it to a preparation plant near the mining site. The plant cleans and processes the coal to remove rocks, dirt, ash, sulfur, and other unwanted materials. This process increases the coal's heating value.

Coal can be transported from mines and processing plants to consumers in several different ways:

- Conveyors, trams, and trucks move coal around mines and short distances from the mines for loading onto other modes of long-distance transportation.
- Trains transport nearly 70 percent of coal deliveries in the United States for at least part of the way from mines to consumers.
- Barges transport coal on rivers and lakes.
- Ships transport coal on the Great Lakes and the oceans to consumers in the United States and other countries.
- Slurry pipelines move mixtures of crushed coal and water. This method is not currently in use in the United States.

Transporting coal can be more expensive than the cost of mining coal. Some coal consumers, such as coal-fired electric power plants, are often situated near coal mines in order to lower the transportation costs (Figure 10.18a–c).

Sources of Coal

In 2014, about 1 billion short tons of coal were produced in 25 U.S. states. Surface mines were the source of two-thirds of total U.S. coal production and accounted for 62 percent of the total number of mines. About 1.6 million tons, or about 0.2 percent of total coal production, was considered refuse recovery coal. Refuse recovery is the recapture of coal from a mine. The resulting product has been cleaned to reduce the concentration of noncombustible materials found on it. Five individual states accounted for approximately 70 percent of total U.S. coal production in 2014:

- Wyoming (40 percent)
- West Virginia (11 percent)
- Kentucky (8 percent)
- Pennsylvania (6 percent)
- Illinois (6 percent)

Coal is found in three principal regions of the United States: the Appalachian coal region, the Interior coal region, and the Western coal region (includes the Powder River Basin), as illustrated in the map (Figure 10.19).

The Appalachian coal region includes Alabama, Eastern Kentucky, Maryland, Ohio, Pennsylvania, Tennessee, Virginia, and West Virginia. Nearly one-third of the U.S. coal production occurs in this region, with West Virginia the largest coal-producing state in the region, and the second-largest coal-producing state in the country. Nearly three-quarters of the mines in this region are underground. The Interior coal region includes Arkansas, Illinois, Indiana, Kansas, Louisiana, Mississippi, Missouri, Oklahoma, Texas, and Western Kentucky. Roughly 20 percent of the total U.S. coal is mined in this coal region, with Illinois the largest coal producer in this area. Just over half of the mines here are underground. The Western coal region encompasses Arizona, Colorado, New Mexico, North Dakota, Utah, Wyoming, Washington,

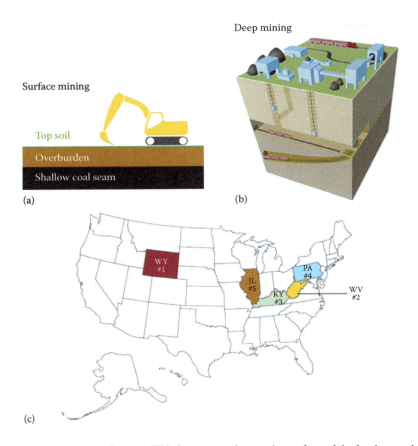

FIGURE 10.18 (a) Surface mining procedure for coal. This leaves scarring on the surface of the landscape that must be reclaimed when the mining extraction is complete, (b) an example of an underground mining operation, consisting of shafts and tunnels, and (c) a map of the top five coal producing states in the U.S., 2014. (From EIA, Washington, DC.)

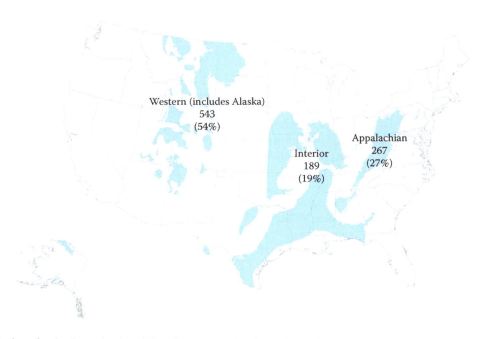

FIGURE 10.19 Coal production by region in million short tons and regional share of total production, 2014. (From EIA, Washington, DC.)

Montana, and Alaska. Just over half of the total U.S. coal production comes from this region, with Wyoming producing nearly three-fourths of the region's total. Surface mines account for 90 percent of the producing mines here.

U.S. Imports and Exports

Although the United States produces a large amount of coal—about 997 million short tons in 2014—some power plants along the Gulf Coast and the Atlantic Ocean sometimes find it less costly to import coal from other countries than it is to obtain coal from U.S. coal-producing regions. In 2014, the United States consumed nearly 917 million short tons of coal. Of this amount of coal, about 1 percent (11 million short tons) was imported, mostly from South America.

The United States is a net exporter of coal, meaning that it exports more coal to other countries than it imports. Between 2000 and 2010, about 5 percent of the coal produced in the United States, on average, was exported to other countries. In 2012, 12 percent of U.S. coal production was exported, and coal exports rose to a record high of 126 million short tons. Exports of coal declined to about 118 million short tons in 2013 and to 97 million short tons in 2014.

Metallurgical coal and steam coal are two general types of coal that are exported. Metallurgical coal can be used for steel production, and steam coal can be used for electricity generation. Metallurgical coal dominates U.S. coal exports (Figure 10.20).

How Much Coal Is Left?

Several measures are used to determine how much coal is left in the United States, which are based on two major factors: (1) various degrees of geologic certainty and (2) on the economic feasibility of mining the coal.

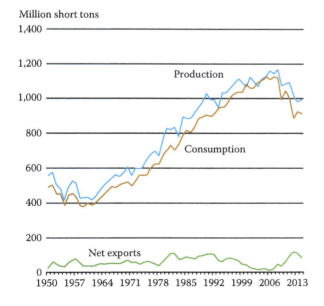

FIGURE 10.20 U.S. coal consumption and net exports 1950–2014. (From EIA, Washington, DC.)

Coal mining companies report to the U.S. Energy Information Administration (EIA) the amount of recoverable reserves at U.S. coal mines that produced at least 25,000 short tons of coal (or 10,000 short tons of anthracite coal) in a year. As of January 1, 2015, about 19.4 billion short tons of recoverable reserves were identified at producing mines. The amount of coal reserves at producing mines is a small portion of the total amount of coal that exists in the United States.

The total amount of coal that exists in the United States is difficult to estimate because it is buried underground. The following are examples of these estimates (Figure 10.21).

- *Total resources* are EIA's best estimate of the total amount of coal (including undiscovered coal) in the United States. Total resources are estimated to be about 3.9 trillion short tons.
- The *Demonstrated Reserve Base* (*DRB*) is the sum of coal in both measured and indicated resource categories of reliability. This represents 100 percent of the in-place coal that could be mined commercially at a given time. EIA estimates that the DRB in 2014 was 478.4 billion short tons.
- *Estimated recoverable reserves* include only the coal that can be mined with today's mining technology, after considering accessibility constraints and recovery factors. EIA estimates there are 255.8 billion short tons of U.S. recoverable coal reserves, which is about 53 percent of the DRB.

Based on U.S. coal production in 2014 of about one billion short tons, the U.S. estimated recoverable coal reserves would last about 256 years. The actual number of years that those reserves will last depends on changes in production, reserves estimates, and new discoveries.

WHAT ARE INTERNATIONAL COAL RESERVES?

As of December 31, 2011, total world proved recoverable reserves of coal were estimated at 979.8 billion short tons. Just five countries have nearly three-fourths of the world's coal reserves (Figure 10.22).

- United States—27 percent
- Russia—17 percent
- China—13 percent
- Australia—9 percent
- India—7 percent

Uses of Coal

Coal was the source of about 33 percent of the electricity generated in the United States in 2015. Power plants can make steam by burning coal. The steam then turns turbines (machines for generating rotary mechanical power) to generate electricity. Many industries and businesses have their

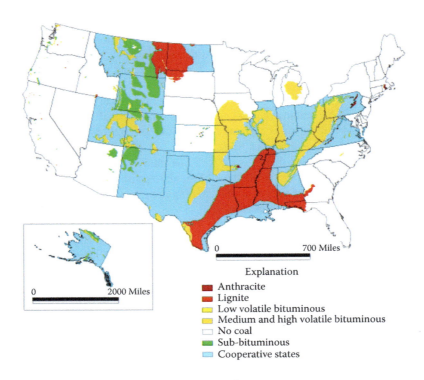

FIGURE 10.21   U.S. coal resource regions. (From EIA, Washington, DC.)

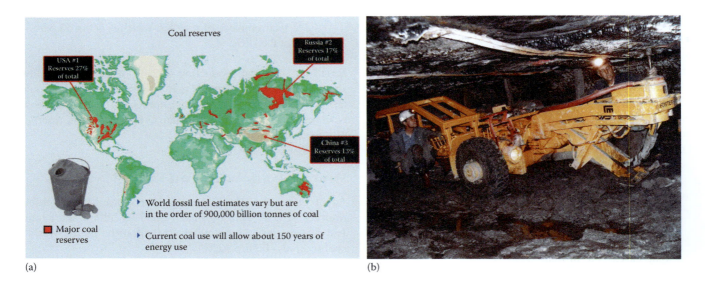

(a)   (b)

FIGURE 10.22   (a) World coal reserves. (From EIA, Washington, DC.) and (b) the predominant underground mining method in the U.S. is the "room and pillar" method, a term derived from the mining pattern of a series of excavated areas (rooms) and unexcavated areas (pillars) which are left to support the roof. (Courtesy of U.S. Department of Energy, Washington, DC.)

own power plants, and some use coal to generate electricity, mostly in combined heat and power plants. A combined heat and power (CHP) plant is a plant designed to produce both heat and electricity from a single heat source.

Many industries use coal and coal byproducts. The concrete and paper industries burn large amounts of coal to produce heat. The steel industry uses coal indirectly to make steel. Steel plants use coal that is baked in furnaces to make

coal coke, and the coke is used to smelt iron ore into iron to make steel. Coal coke is a solid carbonaceous residue derived from low-ash, low-sulfur bituminous coal from which the volatile constituents are removed by baking in an oven at temperatures as high as 1,093°C so that the fixed carbon and residual ash are fused together. Coke is used as a fuel and as a reducing agent in smelting iron ore in a blast furnace. The high temperatures created by burning coke give steel the strength and flexibility needed for bridges, buildings, and automobiles.

Coal can be turned into gases and liquids that can be used as fuels or processed into chemicals to make other products. These gases or liquids are sometimes called *synthetic fuels* or *synfuels*. Synthetic fuels are made by heating coal in large vessels. They produce fewer air pollutants when burned than burning coal directly.

In North Dakota, the Great Plains Synfuels Plant converts coal into synthetic natural gas—referred to as *syngas*. Syngas produced from coal can also be used to produce electricity and hydrogen. Currently, no commercial operating facilities in the United States produce liquids from coal, but coal has been converted to liquids in South Africa for decades.

Prices and Future of Coal

The price of coal varies by coal rank and grade, mining method, and geographic region. Remember, it is classified into four main ranks (subbituminous, lignite, bituminous, and anthracite), depending on the amounts and types of carbon it contains and the amount of heat energy it can produce. Coal with high heat content is generally priced higher.

For comparison, the average sale prices of coal at mines producing each of the four major ranks of coal in 2014 were as follows:

- *Subbituminous*: $14.72 per ton
- *Lignite*: $19.44 per ton
- *Bituminous*: $55.99 per ton
- *Anthracite*: $90.98 per ton

Surface-mined coal is usually priced lower than underground-mined coal. In locations where coal beds are thick and near the surface, such as in Wyoming, mining costs and coal prices tend to be lower than in locations where the beds are thinner and deeper, such as in Appalachia. The higher cost of coal from underground mines reflects the more difficult mining conditions and the need for more miners (Figure 10.22b).

Once coal is mined, it must be transported to consumers. Transportation costs add to the delivered price of coal. In some cases, like in long-distance shipments of Wyoming coal to power plants in the East, transportation costs can be more than the price of coal at the mine. Most coal is transported by train, barge, truck, or a combination of these modes. All of these transportation modes use diesel fuel. Increases in oil prices can significantly affect the cost of transportation, which affects the final delivered price of coal.

In the *Annual Energy Outlook 2016* (AEO), the U.S. Energy Information Administration (EIA) projects that coal prices will generally increase through the year 2040. The amount that coal prices increase depends on projections for coal demand and coal mining productivity. The implementation of the U.S. Environmental Protection Agency's Clean Power Plan is a major factor in the projections for coal demand in the AEO (U.S. Energy Information Administration, 2016).

Nuclear

Atoms are the tiny particles in the molecules that make up gases, liquids, and solids. Atoms themselves are made up of three particles—protons, neutrons, and electrons. An atom has a nucleus (or core) containing protons and neutrons, which is surrounded by electrons. Protons carry a positive electrical charge and electrons carry a negative electrical charge. Neutrons do not have an electrical charge. There is enormous energy present in the bonds that hold the nucleus together. This nuclear energy can be released when those bonds are broken. The bonds can be broken through nuclear fission, and this energy can be used to produce electricity.

In nuclear fission, atoms are split apart, which releases energy. All nuclear power plants use nuclear fission, and most nuclear power plants use uranium atoms. During nuclear fission, a neutron hits a uranium atom and splits it, releasing a large amount of energy in the form of heat and radiation. More neutrons are also released when a uranium atom splits. These neutrons go on to hit other uranium atoms, and the process repeats itself over and over again. This is called a nuclear chain reaction. This reaction is controlled in nuclear power plant reactors to produce a desired amount of heat.

Nuclear energy can also be released in nuclear fusion, in which atoms are combined or fused together to form a larger atom. This is the source of energy in the sun and stars. Nuclear fusion is the subject of ongoing research as a source of energy for heat and electricity generation, but it is not yet clear whether or not it will be a commercially viable technology because of the difficulty of controlling a fusion reaction.

Uranium is the fuel most widely used by nuclear plants for nuclear fission. It is considered to be a nonrenewable energy source, even though it is a common metal found in rocks worldwide. Nuclear power plants use a certain kind of uranium, referred to as U-235, for fuel because its atoms are easily split apart. Although uranium is about 100 times more common than silver, U-235 is relatively rare.

Most U.S. uranium ore is mined in the western United States. Once uranium is mined, the U-235 must be extracted and processed before it can be used as a fuel (Figure 10.23).

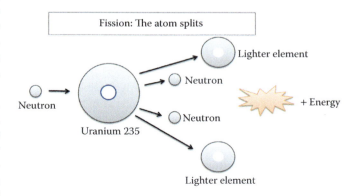

FIGURE 10.23 The nuclear fission process. (Courtesy of Kerr.)

Nuclear Power Plants

All nuclear power in the United States is used to generate electricity. Many power plants, including nuclear power plants, heat water to produce electricity. These power plants use steam from heated water to spin large turbines that generate electricity. Nuclear power plants use heat produced during nuclear fission to heat water. Fission takes place inside the reactor of a nuclear power plant. At the center of the reactor is the core, which contains uranium fuel. The uranium fuel is formed into ceramic pellets. Each ceramic pellet produces roughly the same amount of energy as 150 gallons of oil. These energy-rich pellets are stacked end-to-end in 12-foot metal fuel rods. A bundle of fuel rods, sometimes hundreds, is called a fuel assembly. A reactor core contains many fuel assemblies.

The heat produced during nuclear fission in the reactor core is used to boil water into steam, which turns the turbine blades. As the turbine blades turn, they drive generators that make electricity. Afterward, the steam is cooled back into water in a separate structure at the power plant called a cooling tower. The water can then be reused.

The Nuclear Fuel Cycle

Most nuclear reactors use uranium dioxide (UO_2) as a fuel to create electricity. In 2015, about 21.3 million kilograms of uranium were loaded into commercial U.S. nuclear power reactors. These reactors generated 797 billion kilowatt hours of electricity, or about 20 percent of total U.S. electricity in 2015.

The nuclear fuel cycle consists of *front end* steps that prepare uranium for use in nuclear reactors and *back end* steps to safely manage, prepare, and dispose of highly radioactive spent nuclear fuel. Chemical processing of spent fuel material to recover any remaining product that could undergo fission again in a new fuel assembly is technically feasible, but is not permitted in the United States.

Exploration

The nuclear fuel cycle starts with exploration for uranium and the development of mines to extract the uranium ore. Several techniques are used to locate uranium, such as airborne radiometric surveys, chemical sampling of groundwater and soils, and exploratory drilling to analyze the underlying geology. Once uranium ore deposits are located, the mine developer then samples the area with more closely spaced *in fill*, or development drilling, to determine how much uranium is available and what it might cost to recover it.

Uranium Mining

When ore deposits that are economically feasible to recover are located, the next step in the fuel cycle is to mine the ore using one of the following techniques depending on the nature of the uranium deposit:

- Underground mining
- Open pit mining
- In-situ (in-place) solution mining
- Heap leaching

Before 1980, most U.S. uranium was produced using open pit and underground mining techniques. Today, most U.S. uranium is produced using a solution mining technique commonly called in-situ-leach (ISL) or in-situ-recovery (ISR). This process extracts uranium that coats the sand and gravel particles of groundwater reservoirs. The sand and gravel particles are exposed to a solution with a pH that has been elevated a small amount by using oxygen, carbon dioxide, or caustic soda. The uranium dissolves into the groundwater, which is pumped out of the reservoir and processed at a uranium mill. Heap leaching involves spraying an acidic liquid solution onto piles of crushed uranium ore. The solution then drains down through the crushed ore and leaches uranium out of the rock, which is recovered from underneath the pile. Heap leaching is no longer used in the United States.

Uranium Milling

After the uranium ore is extracted from an open pit or underground mine, it is refined into uranium concentrate at a uranium mill. The ore is crushed, pulverized, and ground into a fine powder. Chemicals are added to the fine powder, which causes a reaction that separates the uranium from the other minerals. Groundwater from solution mining operations is circulated through a resin bed to extract and concentrate the uranium.

The concentrated uranium product is a bright yellow or orange powder called *yellowcake* (U_3O_8). The solid waste material from pit and underground mining operations is called *mill tailings*. The processed water from solution mining is returned to the groundwater reservoir where the mining process is repeated.

Uranium Conversion

The next step in the nuclear fuel cycle is to convert yellowcake into uranium hexafluoride (UF_6) gas at a converter facility. Three forms (isotopes) of uranium occur in nature: U-234, U-235, and U-238. Current U.S. nuclear reactor designs require a stronger concentration (enrichment) of the U-235 isotope to operate efficiently. To separate the three types of uranium isotopes, the UF_6 gas is sent to an enrichment plant where the individual uranium isotopes are separated.

Uranium Enrichment

The uranium hexafluoride gas produced in the converter facility is called *natural UF_6* because the original concentrations of uranium isotopes are unchanged. The United States currently has two operating enrichment plants (where isotope separation takes place). One plant uses a process called gaseous diffusion to separate uranium isotopes, and the other plant uses a gas centrifuge process. Because the smaller U-235 isotopes travel slightly faster than U-238 isotopes, they tend to leak (diffuse) faster through the porous membrane walls of a diffuser, where they are collected and concentrated. The final product has about 4–5 percent concentration of U-235 and is called *enriched UF_6*. Enriched UF_6 is sealed in canisters and allowed to cool and solidify before it is transported to a nuclear reactor fuel assembly plant by train, truck, or barge.

Another enrichment technique is the gas centrifuge process, where UF_6 gas is spun at high speed in a series of cylinders to separate $235UF_6$ and $238UF_6$ isotopes based on their different atomic masses. Atomic vapor laser isotope separation (AVLIS) and molecular laser isotope separation (MLIS) are new enrichment technologies currently under development. These laser-based enrichment processes can achieve higher initial enrichment (isotope separation) factors than the diffusion or centrifuge processes and can produce enriched uranium more quickly than other techniques.

Uranium Reconversion and Nuclear Fuel Fabrication

Once the uranium is enriched, it is ready to be converted into nuclear fuel. The United States has five nuclear reactor fuel fabrication facilities where the enriched UF_6 gas is reacted to form a black uranium dioxide powder. The powder is then compressed and formed into small ceramic fuel pellets. The pellets are stacked and sealed into long metal tubes that are about one centimeter in diameter to form fuel rods. The fuel rods are then bundled together to make up a fuel assembly. Depending on the reactor type, each fuel assembly has about 179–264 fuel rods. A typical reactor core holds 121–193 fuel assemblies.

At the Reactor

Once they are fabricated, trucks transport the fuel assemblies to the reactor sites. The fuel assemblies are stored onsite in *fresh fuel* storage bins until the reactor operators need them. At this stage, the uranium is only mildly radioactive, and all radiation is contained within the metal tubes. Typically, reactor operators change out about one-third of the reactor core (40–90 fuel assemblies) every 12–24 months.

The reactor core is a cylindrical arrangement of the fuel bundles, about 3.7 meters in diameter and 4.3 meters high that are encased in a steel pressure vessel with walls that are several inches thick. The reactor core has no moving parts except for a small number of control rods that are inserted to regulate the nuclear fission reaction. Placing the fuel assemblies next to each other and adding water initiates the nuclear reaction.

The Back End of the Nuclear Fuel Cycle

After use in the reactor, fuel assemblies become highly radioactive and must be removed and stored at the reactor under water in a spent fuel pool for several years. Even though the fission reaction has stopped, the spent fuel continues to give off heat from the decay of radioactive elements that were created when the uranium atoms were split apart. The water in the pool serves to both cool the fuel and block the release of radiation. From 1968 through June 2013, 241,468 fuel assemblies had been discharged and stored at 118 commercial nuclear reactors operating in the United States.

Within a few years, the spent fuel cools in the pool and may be moved to a dry cask storage container for storage at the power plant site. An increasing number of reactor operators now store their older spent fuel in these special outdoor concrete or steel containers with air cooling.

The final step in the nuclear fuel cycle is the collection of spent fuel assemblies from the interim storage sites for final disposition in a permanent underground repository. The United States currently has no permanent underground repository for high-level nuclear waste (Figure 10.24).

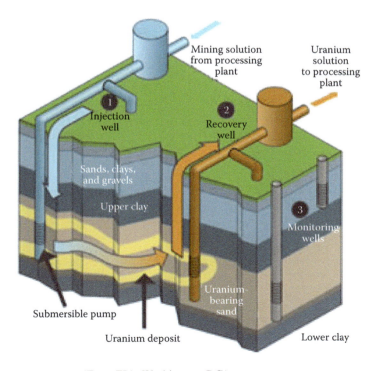

FIGURE 10.24 The in situ recovery process. (From EIA, Washington, DC.)

URANIUM SOURCES

Uranium is a common metal found in rocks all over the world. Uranium occurs in combination with small amounts of other elements. Economically recoverable uranium deposits have been discovered primarily in the western United States, Australia, Canada, Central Asia, Africa, and South America.

After uranium is mined, the U-235 must be extracted and processed before it can be used as a fuel. Mined uranium ore typically yields one to four pounds of uranium concentrate (U_3O_8 or *yellowcake*) per ton, or 0.05–0.20 percent yellowcake.

The United States imports most of the uranium it uses as fuel. Owners and operators of U.S. nuclear power reactors purchased the equivalent of 25.7 million kilograms of uranium in 2015. Only 6 percent of the uranium delivered to U.S. reactors in 2015 was produced in the United States and 94 percent came from other countries:

- 37 percent from Kazakhstan, Russia, and Uzbekistan
- 30 percent from Canada
- 17 percent from Australia
- 10 percent from Malawi, Namibia, Niger, and South Africa

THE NUCLEAR INDUSTRY

The United States has 100 nuclear reactors at 60 operating nuclear power plants located in 30 states. Thirty-six of the plants have two or more reactors, and 46 plants are located east of the Mississippi River. Nuclear power has supplied about 20 percent of annual U.S. electricity since 1990. The average age of U.S. nuclear reactors is about 35 years old. The oldest operating reactors, Nine Mile Point Unit 1 and Oyster Creek, began commercial operation in December 1969. Thirty-four reactors began commercial operation between 1985 and 1996. Four reactors were permanently shut down in 2013, and one reactor was taken out of service in 2014. The newest reactor to enter service, Watts Bar Unit 2, came online in June 2016 and is expected to begin commercial operation by the end of 2016.

Since 1990, the share of total annual U.S. electricity generation provided by nuclear power has averaged about 20 percent. Nuclear generation has generally increased through power plant modifications to increase capacity (known as uprates) and by shortening the length of time reactors are offline for refueling.

Most of the commercial reactors in the United States are located east of the Mississippi River. Illinois has the largest number of commercial reactors (11) and, as of December 31, 2015, had the largest nuclear net summer electricity generation capacity at about 11,590 megawatts (MW). Net summer capacity is the maximum output, commonly expressed in megawatts (MW), that generating equipment can supply to system load, as demonstrated by a multi-hour test, at the time of summer peak demand (period of June 1 through September 30.) This output reflects a reduction in capacity due to electricity use for station service or auxiliaries. The largest reactor in the United States, with a capacity of more than 1,350 megawatts, is the Grand Gulf Nuclear Station, located in Port Gibson, Mississippi. The smallest reactor, with a capacity of 478 megawatts, is at Fort Calhoun, Nebraska.

Of the 31 countries in the world that have commercial nuclear power plants, the United States has the most nuclear capacity and generation. France has the second-highest nuclear electricity generation and obtains about 75 percent of its total electricity from nuclear energy. Fifteen other countries generate more than 20 percent of their electricity from nuclear power (Figures 10.25 and 10.26).

Nuclear power plants are generally used more intensively than other power plants

- For cost and technical reasons, nuclear power plants are generally used more intensively than coal units or natural gas units. In 2015, the nuclear share of total U.S. electricity generating capacity was 9 percent, while the nuclear share of total electricity generation was about 20 percent.

Recent U.S. nuclear construction activity

- In June 2016, the Tennessee Valley Authority's (TVA) Watts Bar Unit 2 in Tennessee became the first new U.S. reactor to come online since 1996. Watts Bar Unit 2 is expected to begin commercial operation by the end of 2016.

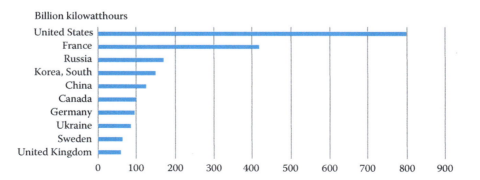

FIGURE 10.25   Nuclear electricity generation by top-10 countries, 2014. (From EIA, Washington, DC.)

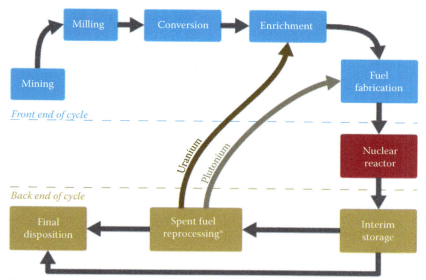

Spent fuel reprocessing is omitted from the cycle in most countries, including the United States

FIGURE 10.26 The nuclear fuel cycle. (From EIA, Washington, DC.)

- In February 2012, the U.S. Nuclear Regulatory Commission (NRC) voted to approve Southern Company's application to build and operate two new reactors, Units 3 and 4, at its Vogtle plant in Georgia. The Vogtle reactors are the first new reactors to receive construction approval in more than 30 years. In March 2012, the NRC voted to approve South Carolina Electric and Gas Company's application to build and operate two new reactors, Units 2 and 3, at its Virgil C. Summer plant in South Carolina.

When will new reactors in the United States come online?

- Although four nuclear reactors were retired in 2013 and one was retired in 2014, nuclear generation capacity at the end of 2015 was about the same as nuclear capacity in 2002, when there were 104 operating reactors. Power plant modifications to increase capacity (uprates) at existing power plants have made it possible to maintain the same nuclear capacity. These uprates, combined with high capacity utilization rates, have allowed nuclear power to consistently maintain a share of about 20 percent of total annual U.S. electricity output since 1990. With many nuclear plants operating at or near capacity, maintaining the current share will depend on new reactors being built, as electricity demand increases.
- Four new reactors that are now under construction (Vogtle Units 3 and 4, and Summer Units 2 and 3) are expected to come online between 2019 and 2020.
- As of May 2016, the NRC had about 20 applications for new reactors in various stages of review. The NRC application review process can take up to

5 years to complete. Under current licensing regulations, a company that seeks to build a new reactor can use off-the-shelf reactor designs that the NRC has previously approved. When the applicant uses an NRC-certified reactor design, that means that all safety issues related to the design have been resolved, and the focus of the NRC's review is the quality of construction. Construction of a nuclear power plant may take 5 years or more. The U.S. Energy Information Administration projects that new nuclear electricity generation capacity will be added through 2040, but that capacity retirements and derates will result in only a small net increase in generation capacity from nuclear power by 2040.

OTHER FOSSIL ENERGY

Two other sources of fossil energy include tar sands and oil shale. Not a mainstream source of energy, they are on the list as potential sources, but have their limitations. This section discusses their role as an energy source.

Tar Sands

Tar sands (also referred to as oil sands) are a combination of clay, sand, water, and bitumen—a heavy black viscous oil. Tar sands can be mined and processed to extract the oil-rich bitumen, which is then refined into oil. Bitumen is made of hydrocarbons—the same molecules in liquid oil—and is used to produce gasoline and other petroleum products. The bitumen in tar sands cannot be pumped from the ground in its natural state; instead tar sand deposits are mined, usually using strip mining or open pit techniques, or the oil is extracted by underground heating with additional upgrading. Extracting bitumen from tar sands—and refining it into products like gasoline—is significantly costlier and more difficult than extracting and refining liquid oil.

Tar sands are mined and processed to generate oil similar to oil pumped from conventional oil wells, but extracting oil from tar sands is more complex than conventional oil recovery. Oil sands recovery processes include extraction and separation systems to separate the bitumen from the clay, sand, and water that make up the tar sands. Bitumen also requires additional upgrading before it can be refined. Because it is so viscous (thick), it also requires dilution with lighter hydrocarbons to make it transportable by pipelines (UCS, 2016).

Much of the world's oil (more than two trillion barrels) is in the form of tar sands, although it is not all recoverable. While tar sands are found in many places worldwide, the largest deposits in the world are found in Canada (Alberta) and Venezuela, and much of the rest is found in various countries in the Middle East. In the United States, tar sands resources are primarily concentrated in Eastern Utah, mostly on public lands. The in-place tar sands oil resources in Utah are estimated at 12 to 19 billion barrels (ostseis.anl.gov).

The Tar Sands Industry

Currently, oil is not produced from tar sands on a significant commercial level in the United States; in fact, only Canada has a large-scale commercial tar sands industry, though a small amount of oil from tar sands is produced commercially in Venezuela. The Canadian tar sands industry is centered in Alberta. Currently, more than one million barrels of synthetic oil are produced from these resources per day. Currently, tar sands represent about 40 percent of Canada's oil production, and output is expanding rapidly. Approximately 20 percent of U.S. crude oil and products come from Canada, and a substantial portion of this amount comes from tar sands. Tar sands are extracted both by mining and in situ recovery methods. Canadian tar sands are different than U.S. tar sands in that Canadian tar sands are water wetted, while U.S. tar sands are hydrocarbon wetted. As a result of this difference, extraction techniques for the tar sands in Utah will be different, if extracted, than for those in Alberta.

Rising crude oil prices and political unrest make tar-sand-based oil production in the United States commercially attractive, and both government and industry are currently interested in pursuing the development of tar sands oil resources as an alternative to conventional oil.

Tar Sands Extraction and Processing

Tar sands deposits near the surface can be recovered by open pit mining techniques. New methods introduced in the 1990s, considerably improved the efficiency of tar sands mining, which reduces the cost. These systems use large hydraulic and electrically powered shovels to dig up tar sands and load them into enormous trucks that can carry up to 320 tons of tar sands per load (Figure 10.27).

After mining, the tar sands are transported to an extraction plant, where a hot water process separates the bitumen from sand, water, and minerals. The separation takes place in separation cells. Hot water is added to the sand, and the resulting slurry is piped to the extraction plant where it is agitated. The combination of hot water and agitation releases bitumen from the oil sand, and causes tiny air bubbles to attach to the bitumen droplets, that float to the top of the separation vessel, where the bitumen can be skimmed off. Further processing removes residual water and solids. The bitumen is then transported and eventually upgraded into synthetic crude oil.

About two tons of tar sands are required to produce one barrel of oil. Roughly 75 percent of the bitumen can be recovered from sand. After oil extraction, the spent sand and other materials are then returned to the mine, which is eventually reclaimed.

In-situ production methods are used on bitumen deposits buried too deep for mining to be economically recovered. These techniques include steam injection, solvent injection, and firefloods, in which oxygen is injected and part of the resource burned to provide heat. So far steam injection has been the preferred method. Some of these extraction methods require large amounts of both water and energy (for heating and pumping).

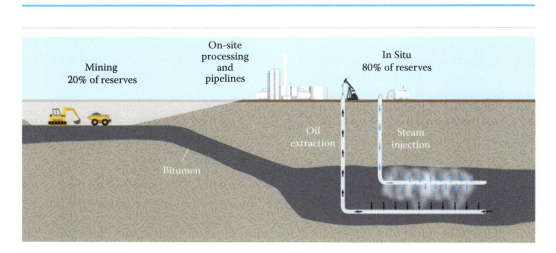

FIGURE 10.27 Extracting oil from tar sands. (From Union of Concerned Scientists [UCS], Cambridge, MA.)

Oil Shale

The term oil shale generally refers to any sedimentary rock that contains solid bituminous materials (called kerogen) that are released as petroleum-like liquids when the rock is heated in the chemical process of pyrolysis. Oil shale was formed millions of years ago by deposition of silt and organic debris on lake beds and sea bottoms. Over long periods of time, heat and pressure transformed the materials into oil shale in a process similar to the process that forms oil; however, the heat and pressure were not as great. Oil shale generally contains enough oil that it will burn without any additional processing, and it is known as "the rock that burns." Geologists believe there is more oil shale out there in the rocks of the world—three trillion barrels worth of fuel—than there is oil in existing reserves globally.

Oil shale can be mined and processed to generate oil similar to oil pumped from conventional oil wells; however, extracting oil from oil shale is more complex than conventional oil recovery and currently is more expensive. The oil substances in oil shale are solid and cannot be pumped directly out of the ground. The oil shale must first be mined and then heated to a high temperature (a process called retorting); the resultant liquid must then be separated and collected. An alternative but currently experimental process referred to as *in situ* retorting involves heating the oil shale while it is still underground, and then pumping the resulting liquid to the surface.

Oil Shale Resources

Oil shale has been mined extensively in Brazil, China, Estonia, Germany, Israel, and Russia, but up to two-thirds of the world's supply lies in the Green River basin of the western United States, including parts of Wyoming, Utah and Colorado. To date, these American oil shale resources remain virtually untapped (Figure 10.28a and b). Estimates of the oil resource in place within the Green River Formation range from 1.2 to 1.8 trillion barrels. Not all resources in place are recoverable; however, even a conservative estimate of 800 billion barrels of recoverable oil from oil shale in the Green River Formation is three times greater than the proven

oil reserves of Saudi Arabia. Present U.S. demand for petroleum products is about 20 million barrels per day. If oil shale could be used to meet a quarter of that demand, the estimated 800 billion barrels of recoverable oil from the Green River Formation would last for more than 400 years (Oil Shale and Tar Sands Programmatic EIS Information Center, 2012).

More than 70 percent of the total oil shale acreage in the Green River Formation, including the richest and thickest oil shale deposits, is under federally owned and managed lands. Thus, the federal government directly controls access to the most commercially attractive portions of the oil shale resource base. The U.S. was pro-development for oil shale during the 1970s oil shortages, but when gas prices fell again and availability returned, the enthusiasm for oil shale dropped (Figure 10.29).

American companies didn't look into mining domestic oil shale again until 2003—again, due to spiking oil prices. President George W. Bush's Energy Policy Act of 2005 officially opened federal lands to oil shale extraction. But then once again lowered oil prices, along with environmental concerns and growing enthusiasm for renewable energy sources left oil shale's future in the U.S. again uncertain.

Oil shale extraction was also met with enormous opposition from environmental groups. They quickly brought to the forefront that extracting operations destroy affected landscapes, forcing plants and animals out, with regeneration unlikely for decades. They also reiterated that oil shale extraction is heavy on water usage and the deposits in the U.S. are located in a traditionally water-starved area. The process requires as much as five barrels of water—for dust control, cooling and other purposes—for every barrel of shale oil produced. This, again, dropped the interest in developing the resource (LeDoux, 2016).

The Oil Shale Industry

While oil shale has been used as fuel and as a source of oil in small quantities for many years, few countries currently produce oil from oil shale on a significant commercial level. Many countries do not have significant oil shale

FIGURE 10.28 A sample of oil shale. (Courtesy of Fafner.)

(a)

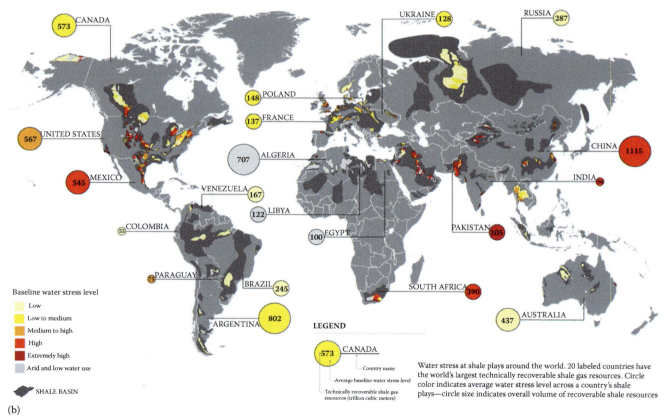

(b)

FIGURE 10.29 (a) Location of the green river formation oil shale in the United States. (From http://ostseis.anl.gov.) and (b) worldwide location of oil shale deposits. (From World Resources Institute, Washington, DC.)

resources, but in those countries that do have significant oil shale resources, the oil shale industry has not developed because historically, the cost of oil derived from oil shale has been significantly higher than conventionally pumped oil. The lack of commercial viability of oil shale-derived oil has in turn inhibited the development of better technologies that might reduce its cost (Westenhaus, 2010).

Relatively high prices for conventional oil in the 1970s and 1980s stimulated interest and some development of better oil shale technology, but oil prices eventually fell, and major research and development activities largely ceased. More recently, prices for crude oil have again risen to levels that may make oil shale-based oil production commercially viable, and both governments and industry are interested in pursuing the development of oil shale as an alternative to conventional oil.

Oil Shale Mining and Processing

Oil shale can be mined using one of two methods: underground mining using the room-and-pillar method or surface mining. Room and pillar mining is a system in which the mined material is extracted across a horizontal plane, creating horizontal arrays of rooms and pillars. The pillars support the ceiling and the material is extracted from the "room." After mining, the oil shale is transported to a facility for retorting, a heating process that separates the oil fractions of oil shale from the mineral fraction. The vessel in which retorting takes place is known as a retort. After retorting, the oil must be upgraded by further processing before it can be sent to a refinery, and the spent shale must be disposed of. Spent shale may be disposed of in surface impoundments, or as fill in graded areas; it may also be disposed of in previously mined areas. Eventually, the mined land needs to be reclaimed.

Surface Retorting

While current technologies are adequate for oil shale mining, the technology for surface retorting has not been successfully applied at a commercially viable level in the United States, although technical viability has been demonstrated. Further development and testing of surface retorting technology is needed before the method is likely to succeed on a commercial scale.

In Situ Retorting

An in situ conversion process (ICP) is currently being developed by Shell Oil. The process involves heating underground oil shale, using electric heaters placed in deep vertical holes drilled through a section of oil shale. The volume of oil shale is heated over a period of 2–3 years, until it reaches 343°C–371°C, at which point oil is released from the shale. The released product is planned to be gathered in collection wells positioned within the heated zone.

To date, the financial and environmental costs—costs of blasting, transporting, crushing, heating, and adding hydrogen, as well as the safe disposal of huge quantities of waste material—are great enough that large-scale production has not occurred yet in the United States. Oil shale has been used as a low-grade fuel in Estonia and other countries by burning it directly. It also has several environmental issues associated with it. Future development remains to be seen.

CONCLUSIONS

As we've seen in this chapter, nonrenewable energy sources are those sources that drain fossil reserves deposited over eons of time. They're considered nonrenewable because they cannot be regenerated within our lifetimes or within several lifetimes. This results in the depletion of these energy reserves—once they are gone, they are gone forever as far as we are concerned. There are several countries which have achieved significant reduction of these sources and are currently suffering from the side effects of drilling these energy reserves from deep underground, such as China and India. The environmental impact is currently so great that the visual evidence is striking, such as unhealthy air pollution conditions.

There are several locations worldwide that are currently experiencing rapid degradation of nonrenewable sources in terms of fossil fuels. Eventually they will be gone, which is an issue we must address now. It also must be addressed now for an even more important reason: these sources of energy are degrading our environment at a rapid rate. With increased exploitation of fossil fuels, there are many associated environmental effects such as land and air pollution. The release of GHGs is one of the most critical issues we face. In a world with a rapidly changing climate, we must address the ramifications of continuing down the path of nonrenewable energy now and find a way to phase them out and make the transition to renewable resources. Not only is it one of the most daunting challenges we face in modern times, but it is one of the most critical. With human ingenuity and commitment, however, facing it as a surmountable challenge rather than a lost cause will yield success. Whenever humankind has been committed to a cause in the past, success has always been within reach.

REFERENCES

LeDoux, L. 2016. Can oil shale be used as a power source? Why the U.S. hasn't tapped this resource for energy. http://www.scientificamerican.com/article/can-oil-shale-provide-power/ (accessed September 25, 2016).

Oil Shale and Tar Sands Programmatic EIS Information Center. 2012. Oil shale/tar sands environmental impact statement. http://ostseis.anl.gov/index.cfm (accessed September 25, 2016).

Today in Energy. April 18, 2012. Sulfur content of heating oil to be reduced in Northeastern states. Energy Information Administration. http://www.eia.gov/todayinenergy/detail.cfm?id=5890 (accessed September 21, 2016).

Today in Energy, May 5, 2016. Hydraulically fractured wells provide two-thirds of U.S. natural gas production. http://www.eia.gov/today http://www.eia.gov/todayinenergy/detail.cfm?id=26112inenergy/detail.cfm?id=26112 (accessed September 23, 2016).

UCS. 2016. What are tar sands? http://www.ucsusa.org/clean-vehicles/all-about-oil/what-are-tar-sands#.V-fI8_ArKUk (accessed September 25, 2016).

U.S. Energy Information Administration. *Short-Term Energy Outlook*, Table WFO1. January 2016.

Westenhaus, B. July 15, 2010. Shale oil's place in our energy future. *Oil Price* http://oilprice.com/Energy/Crude-Oil/Shale-Oils-Place-In-Our-Energy-Future.html (accessed September 25, 2016).

SUGGESTED READING

Jabusch, G. July-August 2016. The structural (and possibly abrupt) decline of fossil fuels. *Green Money Journal.* http://www.greenmoneyjournal.com/july-august-2016/the-structural-and-possibly-abrupt-decline-of-fossil-fuels/ (accessed September 21, 2016).

Klare, M. T. July 14, 2016. Bad news: We're actually using more fossil fuels than ever. *The Nation.* https://www.thenation.com/article/bad-news-were-actually-using-more-fossil-fuels-than-ever/ (accessed September 25, 2016).

Nesbit, J. August 13, 2015. Fossil fuels are doomed. *U.S. News.* http://www.usnews.com/news/blogs/at-the-edge/2015/08/13/fossil-fuels-are-doomed (accessed September 15, 2016).

Randall, T. June 12, 2016. The world nears peak fossil fuels for electricity. *Bloomberg.* http://www.bloomberg.com/news/articles/2016-06-13/we-ve-almost-reached-peak-fossil-fuels-for-electricity (accessed September 23, 2016).

Rapier, R. June 8, 2016. World sets record for fossil fuel consumption. *Forbes.* http://www.forbes.com/sites/rrapier/2016/06/08/world-sets-record-for-fossil-fuel-consumption/ (accessed September 22, 2016).

11 Renewable Energy Resources

OVERVIEW

This chapter will focus on renewable energy: what it is, how it works, what forms of renewable energy are currently available for use and which ones are on the horizon slated for use in the near future. We will examine each in turn: solar, geothermal, wind, hydropower, biofuels, and even secondary sources, such as electricity and hydrogen, in an effort to determine which direction humans need to take in order to protect the landscape and keep the environment as healthy as possible.

INTRODUCTION

There are many reasons to look seriously at renewable energy resources today. The most obvious reason is that the nonrenewable energy resources we have been using will not last forever and we need to find a replacement for them. They are also a cleaner, more environmentally friendly choice. If we keep our environment cleaner and free of pollutants, it gives us many benefits, some tangible, others less so. For example, there are environmental benefits that an atmosphere with less carbon dioxide would provide, and in turn, less damage done by a changing climate and future threat of that change. Renewable energy technologies are also clean sources of energy with a much lower environmental impact overall. For example, switching from coal-fired power plants eliminates the destructive mining process. Renewable energy is also desirable because it never runs out. This ensures that there will be a continuous and plentiful energy supply for future generations. We can be guaranteed that we can leave the earth energy resource rich and know that there will still be energy sources for our future generation. Renewables also provide jobs for the economy. Most renewable energy investments are spent on materials and workmanship to build and maintain the facilities, rather than on costly energy imports. Renewable energy investments are usually spent within our own home country—often locally, which is even more attractive. This means that we can benefit personally by the jobs that are created. Renewables also represent energy security. With political uprisings and conflicts, the world has been held hostage repeatedly with nonrenewables, putting us in a state of fluctuating supply and uncontrollable prices because of our dependence on fossil fuels. Renewables eliminate those problems, giving us energy security. Because this is the direction of the future, let us look at the options we have and the choices we need to make now for a cleaner future.

DEVELOPMENT

More than 150 years ago, wood supplied nearly 90 percent of the nation's energy needs. As more consumers began using coal, petroleum, and natural gas, the United States relied less on wood as an energy source. Today, the use of renewable energy sources is increasing, especially biofuels, solar, and wind. In 2015, approximately 10 percent of total U.S. energy consumption was from renewable energy sources (or about 9.7 quadrillion British thermal units (Btu). More than half of U.S. renewable energy is used for producing electricity, and about 13 percent of U.S. electricity generation was from renewable energy sources in 2015 (Figure 11.1). Renewable energy plays an important role in reducing greenhouse gas (GHG) emissions. When renewable energy sources are used, the demand for fossil fuels is reduced. Unlike fossil fuels, nonbiomass renewable sources of energy (hydropower, geothermal, wind, and solar) do not directly emit GHGs.

The consumption of biofuels and nonhydroelectric renewable energy sources more than doubled from 2000 to 2015, mostly due to state and federal government mandates and incentives for renewable energy. The U.S. Energy Information Administration (EIA) projects that the use of renewable energy in the United States will continue to grow through 2040 (EIA, 2016).

It is still true that renewable energy is more expensive to produce and to use than fossil fuel energy. Suitable renewable resources are often located in remote areas, and it can be expensive to build power lines from the renewable energy sources to the cities that need the electricity. In addition, renewable sources are not always available, for reasons such as:

- Clouds reduce electricity from solar power plants.
- Days with low wind reduce electricity from wind farms.
- Droughts reduce the water available for hydropower.

Today, however, there are several different types of renewable energy available for use. They include biomass: municipal solid waste (MSW), wood, biofuels, hydropower, geothermal, wind, and solar.

Biomass energy is produced from nonfossilized materials derived from plants. Wood and wood waste are the largest sources of biomass energy, followed by biofuels and energy from waste. Wood biomass includes wood pellets; wood chips from forestry operations; residues from lumber, pulp/paper, and furniture mills; and fuel wood for space heating. The largest single source of wood energy is *black liquor*, a residue of pulp, paper, and paperboard production. Garbage, or MSW, contains biomass (or biogenic) materials such as paper, cardboard, food scraps, grass clippings, leaves, wood, leather products, and other nonbiomass combustible materials, mainly plastics and other synthetic materials made from petroleum. MSW can be recycled, composted, sent to landfills, or used in waste-to-energy plants. There are hundreds of landfills in the United States that recover biogas, or methane, which forms as waste decomposes in low-oxygen (anaerobic) conditions.

The methane is burned to produce electricity and heat. Biofuels include alcohol fuels such as ethanol and *biodiesel*, a fuel made from grain oils and animal fats. Biofuels used in the United States are mostly fuel ethanol, which is produced from corn.

Hydropower is electricity produced from flowing water. Hydropower output varies depending on rainfall. Most hydropower is produced at large facilities built by the federal government. There are two types of hydropower—conventional hydropower and pumped storage. Conventional hydropower

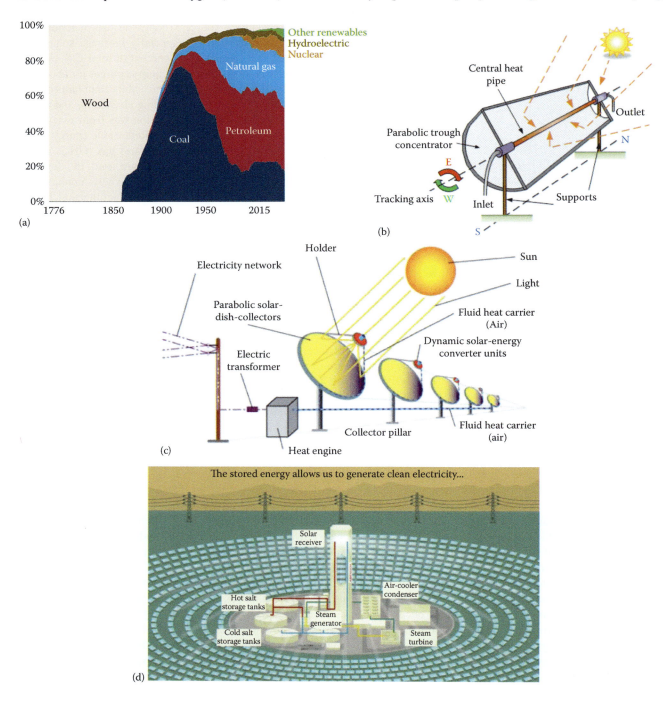

FIGURE 11.1 (a) The historical use of energy in the United States, 1776–2015. (From EIA, Washington, DC.) (b) Schematic of a parabolic-trough collector. (From Wikimedia commons.) (c) A schematic of a solar dish collector. (From Energy.gov.) (d) Drawing of a solar power tower. (From Energy.gov.) *(Continued)*

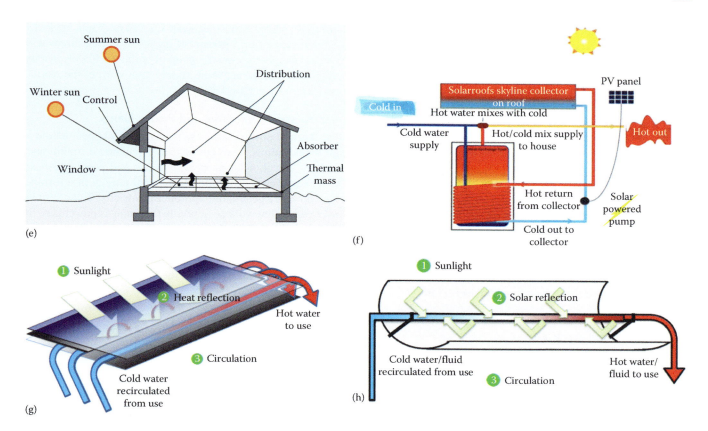

FIGURE 11.1 (Continued) (e) An example of a passive solar system. (From Energy.gov.) (f) A design of an active solar water heating system. (From Energy.gov.) (g) A flat-plate solar collector used for water or air heating. (From EPA.) (h) A solar concentrating collector. (From Energy.gov.)

uses dams to collect and release water to spin a turbine and create electricity. Pumped storage generates electricity by moving water between two reservoirs at different elevations.

Geothermal is energy from the hot interior of the earth. Fissures on the earth's crust allow water heated by geothermal energy to rise naturally to the surface at hot springs and geysers. Wells drilled into the earth allow steam or water to escape to the surface in a controlled manner to operate steam turbines and to generate electricity.

Wind turbines use blades to collect the wind's kinetic energy. Wind flows over the blades creating lift, which causes the blades to turn. The blades are connected to a drive shaft that turns an electric generator, which produces electricity.

Solar energy systems use radiation from the sun to produce heat and electricity. There are three basic categories of solar energy systems: solar thermal systems, solar thermal-electric power plants, and photovoltaic (PV) systems.

Solar thermal systems are for heating buildings and water. These systems use solar collectors to absorb solar radiation to heat water or air for space heating and water heating.

Solar thermal-electric power plants use solar collectors to focus the sun's rays to heat fluid to a high temperature. This working fluid can then be used to generate steam to operate a turbine, which is then used to produce electricity in a generator. The three types of solar thermal power systems used in the United States are parabolic trough, solar dish, and solar power towers.

PV systems use solar electric cells that convert solar radiation directly into electricity. Individual PV cells are configured into modules of varying electricity producing capacities. PV applications range from single solar cells for powering watches to large installations with hundreds of modules for electric power production.

This chapter will look at each one of these renewable energy sources in turn, focusing on their potential as sources to provide viable energy for a sustainable future.

SOLAR

When converted to thermal energy, solar energy can be used to heat water for use in homes, buildings, or swimming pools; to heat spaces inside homes, greenhouses, and other buildings; and to heat fluids to high temperatures to operate turbines that generate electricity. Solar energy can be converted into electricity in two ways:

- PV devices or *solar cells* that change sunlight directly into electricity. Individual PV cells are grouped into panels and arrays of panels that can be used in a variety of applications ranging from single small cells that charge calculator and watch batteries, to systems that power single homes, to large power plants covering many acres.

- Solar thermal/electric power plants generate electricity by concentrating solar energy to heat a fluid and produce steam that is then used to power a generator.

There are two main benefits of solar energy—solar energy systems do not produce air pollutants or carbon dioxide; and when located on buildings, solar energy systems have minimal impact on the environment. Solar energy is not without its limitations, however. The drawbacks with solar energy are as follows:

- The amount of sunlight that arrives at the earth's surface is not constant. The amount of sunlight varies depending on location, time of day, time of year, and weather conditions.
- Because the sun does not deliver that much energy to any one place at any one time, a large surface area is required to collect the energy at a useful rate.

Photovoltaics and Electricity

PV cells convert sunlight into electricity. A PV cell, commonly called a solar cell, is a nonmechanical device that converts sunlight directly into electricity. Some PV cells can convert artificial light into electricity. Sunlight is composed of photons, or particles of solar energy. These photons contain varying amounts of energy that correspond to the different wavelengths of the solar spectrum. A PV cell is made of a semiconductor material. When photons strike a PV cell, they may be reflected, pass right through, or be absorbed by the semiconductor material. Only the absorbed photons provide energy to generate electricity. When enough sunlight (solar energy) is absorbed by the material, electrons are dislodged from the material's atoms. Special treatment of the material surface during manufacturing makes the front surface of the cell more receptive to the dislodged or *free* electrons, so the electrons naturally migrate to the surface of the cell.

When the electrons leave their position, holes are formed. When many electrons, each carrying a negative charge, travel toward the front surface of the cell, the resulting imbalance of electrical charge between the cell's front and back surfaces creates a voltage potential like the negative and positive terminals of a battery. Electrical conductors are placed on the cell to absorb the electrons. When the conductors are connected in an electrical circuit to an external load, such as an appliance, electricity flows in the circuit.

There are two types of systems: concentrating collectors and PV systems. Concentrating solar energy technologies use mirrors to reflect and concentrate sunlight onto receivers that collect solar energy and convert it to heat. This thermal energy can then be used to produce heat or electricity with a steam turbine or a heat engine driving a generator. PV cells convert sunlight directly into electricity. PV systems can range from systems that provide tiny amounts of electricity for watches and calculators to systems that provide the amount of electricity used by hundreds of homes. Hundreds and thousands of houses and buildings around the world have PV systems on their roofs. Many multi-megawatt PV power plants have also been built. Covering 4 percent of the world's desert areas with PV could supply the equivalent of all of the world's electricity. The Gobi Desert alone could supply almost all of the world's total electricity demand.

Solar Thermal Collectors and Power Plants

Solar thermal power plants use the sun's rays to heat a fluid to high temperatures. The fluid is then circulated through pipes so that it can transfer its heat to water and produce steam. The steam is converted into mechanical energy in a turbine, which powers a generator to produce electricity. Solar thermal power generation works essentially the same as power generation using fossil fuels, but instead of using steam produced from the combustion of fossil fuels, the steam is produced by heat collected from sunlight. Solar thermal technologies use concentrator systems to achieve the high temperatures needed to produce steam.

There are three main types of solar thermal power systems:

- Parabolic trough
- Solar dish
- Solar power tower

Parabolic Troughs

Parabolic troughs are used in the longest operating solar thermal power facility in the world, which is located in the Mojave Desert in California. The Solar Energy Generating System (SEGS) has nine separate plants. The first plant, SEGS 1, has operated since 1984, and the last SEGS plant that was built, SEGS IX, began operation in 1990. The SEGS facility is one of the largest solar thermal electric power plants in the world.

A parabolic-trough collector has a long parabolic-shaped reflector that focuses the sun's rays on a receiver pipe located at the focus of the parabola. The collector tilts with the sun as the sun moves from east to west during the day to ensure that the sun is continuously focused on the receiver.

Because of its parabolic shape, a trough can focus the sun from 30 to 100 times its normal intensity (concentration ratio) on the receiver pipe located along the focal line of the trough, achieving operating temperatures higher than 399°C.

The solar field has many parallel rows of solar parabolic-trough collectors aligned on a north-south horizontal axis. A working (heat transfer) fluid is heated as it circulates through the receiver pipes and returns to a series of heat exchangers at a central location. Here, the fluid circulates through pipes so it can transfer its heat to water to generate high pressure, superheated steam. The steam is then fed to a conventional steam turbine and generator to produce electricity. When the hot fluid passes through the heat exchangers, it cools down, and is then recirculated through the solar field to heat up again.

The power plant is usually designed to operate at full power using solar energy alone, given sufficient solar energy. However, all parabolic trough power plants can use fossil fuel combustion to supplement the solar output during periods of low solar energy.

Solar Dishes

Solar dish/engine systems use concentrating solar collectors that track the sun, so they always point straight at the sun and concentrate the solar energy at the focal point of the dish. A solar dish's concentration ratio is much higher than a solar trough's concentration ratio, and it has a working fluid temperature higher than 749°C. The power-generating equipment used with a solar dish can be mounted at the focal point of the dish, making it work well for remote operations or, as with the solar trough, the energy may be collected from a number of installations and converted into electricity at a central point.

The engine in a solar dish/engine system converts heat to mechanical power by compressing the working fluid when it is cold, heating the compressed working fluid, and then expanding the fluid through a turbine or with a piston to produce work. The engine is coupled to an electric generator to convert the mechanical power to electric power.

Solar Power Tower

A solar power tower, or central receiver, generates electricity from sunlight by focusing concentrated solar energy on a tower-mounted heat exchanger (receiver). This system uses hundreds to thousands of flat, sun-tracking mirrors called heliostats to reflect and concentrate the sun's energy onto a central receiver tower. The energy can be concentrated as much as 1,500 times that of the energy coming in from the sun.

Energy losses from thermal-energy transport are minimized because solar energy is being directly transferred by reflection from the heliostats to a single receiver, rather than being moved through a transfer medium to one central location, as with parabolic troughs.

Power towers must be large to be economical. This is promising technology for large-scale grid-connected power plants. The U.S. Department of Energy (DOE), along with a number of electric utilities, built and operated a demonstration solar power tower near Barstow, California, during the 1980s and 1990s.

There are two operating solar power tower projects in the United States:

- A 5-megawatt, two-tower project, located in the Mojave Desert in southern California
- A 392-megawatt project located in Ivanpah Dry Lake, California (Figure 11.2)

Solar thermal energy can be used to heat water or air. It is most often used for heating water in buildings and in swimming pools. Solar thermal energy is also used to heat the insides of buildings. Solar heating systems can be classified as *passive* or *active*.

Passive solar space heating happens in a car on a hot summer day. The sun's rays pass through the windows and heat up the inside of the car. In passive solar heated buildings, air is circulated past a solar heat-absorbing surface and through the building by natural convection. No mechanical equipment is used for passive solar heating (Figure 11.1d–g).

FIGURE 11.2 Ivanpah solar electric generating system with all three towers under load, 2013. (By Sbharris - Own work, CC BY-SA 3.0, https://commons.wikimedia.org/w/index.php?curid=31089494.)

Active solar heating systems use a *collector* and a fluid to collect and absorb solar radiation. Fans or pumps circulate air or heat-absorbing liquids through collectors and then transfer the heated fluid directly to a room or to a heat storage system. Active water heating systems usually include a tank for storing water heated by the system.

Solar collectors are either nonconcentrating or concentrating. In nonconcentrating collectors, the collector area (the area that intercepts the solar radiation) is the same as the absorber area (the area absorbing the radiation). *Flat-plate collectors* are the most common type of nonconcentrating collectors and are used when temperatures lower than 93°C are sufficient. Nonconcentrating collectors are often used for heating water or air for space heating in buildings and in swimming pools.

There are many flat-plate collector designs, but they generally have four specific components:

- A flat-plate absorber that intercepts and absorbs the solar energy
- A transparent cover that allows solar energy to pass through but reduces heat loss from the absorber
- A heat-transport fluid (air or liquid) flowing through tubes to remove heat from the absorber
- A layer of insulation on the back of the absorber

Concentrating collectors have a configuration where the area intercepting the solar radiation is greater, sometimes hundreds of times greater, than the absorber area. The collector focuses or concentrates solar energy onto an absorber. The collector usually moves so that it maintains a high degree of concentration on the absorber (Figure 11.3).

Using solar energy does not produce air or water pollution and does not produce GHGs, but using solar energy may have some indirect negative impacts on the environment. For example, there are some toxic materials and chemicals that are used in the manufacturing process of PV cells, which convert sunlight into electricity. Some solar thermal systems

FIGURE 11.3 A house has solar collectors on the roof to collect solar energy. (Courtesy of Gray Watson, Creative Commons.)

use potentially hazardous fluids to transfer heat. U.S. environmental laws regulate the use and disposal of these types of materials.

As with any type of power plant, large solar power plants can affect the environment where they are located. Clearing land for construction and the placement of the power plant may have long-term impacts on plant and animal life by reducing habitat areas for native plants and animals. Power plants may require water for cleaning solar collectors or concentrators and may require water for cooling turbine generators. Using ground water or surface water in some arid locations with significant solar potential may affect the ecosystem. In addition, birds and insects can be killed if they fly into a concentrated beam of sunlight created by a *solar power tower*.

Solar Hot Water

The shallow water in a lake is generally warmer than the deeper water. That is because the sunlight can heat the lake bottom in the shallow areas, and subsequently heats the water. This is a natural form of solar water heating. The sun can also be used the same way to heat water used in buildings and swimming pools.

Most solar water heating systems for buildings have two principle components: a solar collector and a storage tank. A flat-plate collector is the most common type of collector. They are generally mounted on the roof, and consist of a thin, flat, rectangular box with a transparent cover that faces the sun. Small tubes run through the box and carry the fluid—either water or other fluid, such as an antifreeze solution—to be heated. The tubes are attached to an absorber plate, which is painted black to collect the heat. As heat builds up in the collector, it heats the fluid passing through the tubes. The system's storage tank—generally a well-insulated container—collects and holds the hot liquid. Systems that use fluids other than water usually heat the water by passing it through a coil of tubing in the tank, which contains hot fluid.

Solar water heating systems can be either active or passive. The most common type is active systems. Active systems rely on pumps to move the liquid between the collector and the storage

tank, while passive systems rely on gravity and fact that water naturally circulates as it is heated. Swimming pool systems that use this concept are much simpler. The pool's filter pump is used to pump the water through a solar collector, which is usually made of black plastic or rubber. The pool itself serves as the water storage container (Renewable Energy World, 2016).

Passive Solar Heating and Daylighting

The power of the sun is enormous. All you have to do is walk outside into a hot, sunny day and feel the sun's rays to feel its power. Taking advantage of its energy is not new. Civilizations have taken advantage of natural heating for centuries by facing their homes in the direction of the sunlight to take advantage of its natural heating power. Following in these traditions, many buildings today are designed to take advantage of this natural resource through the use of passive solar heating and daylighting.

The south side of a building always receives the most sunlight. Because of this simple fact, buildings designed for passive solar heating usually have large, south-facing windows. Materials that absorb and store the sun's heat can then be built into the sunlit floors and walls. The floors and walls will then heat up during the day and slowly release heat at night, when the heat is needed most. This type of passive solar design feature is called *direct gain*.

There are other passive solar heating designs as well. Two others that are commonly used are *sunspaces* and *Trombe walls*. A sunspace (which is similar to a greenhouse) is built on the south side of a building. As sunlight passes through glass or other glazing, it warms the sunspace. Proper ventilation then allows the heat to circulate into the building. In contrast, a Trombe wall is a very thick, south-facing wall, which is painted black and made of a material that absorbs a lot of heat. A pane of glass or plastic glazing, which is installed a few inches in front of the wall, helps to hold in the heat. The wall heats up slowly during the day. Then as it cools gradually during the night, it gives off its heat inside the building.

Many of the passive solar heating design features also provide daylighting. Daylighting is basically the use of natural sunlight to brighten up a building's interior. To lighten up north-facing rooms and upper levels, a *clerestory*, which is a row of windows near the peak of the roof, is commonly used along with an open floor plan inside that allows the light to bounce throughout the building.

One caution is that too much solar heating and daylighting can pose a challenge during the hot summer months. Fortunately, there are many design features that help to keep passive solar buildings cool in the summer. One option is that overhangs can be designed to shade windows when the sun is high in the summer, blocking some of its effects. Sunspaces can also be closed off from the rest of the building. Buildings can also be designed to use fresh-air ventilation in the summer, keeping the building cool throughout the day.

Space Heating and Cooling

While commercial and industrial buildings can use the same solar technologies that residential buildings can use, such as

PVs, passive heating, daylighting, and water heating, they also have other options available to them. They can use solar energy technologies that would be impractical for a home, such as ventilation air preheating, solar process heating, and solar cooling technologies.

Many large buildings need ventilated air to maintain indoor air quality. In cold climates, heating this air uses enormous amounts of energy. Solar ventilation systems are able to preheat the air, saving both energy and money. This type of system uses a transpired collector, which consists of a thin, black metal panel mounted on a south-facing wall to absorb the sun's heat. As air passes through several small holes in the panel, a space behind the perforated wall allows the air streams from the holes to mix together. The heated air is then sucked out from the top of the space into the ventilation system.

Solar process heating systems are designed to provide large quantities of hot water or space heating for nonresidential buildings. These systems include solar collectors that work in tandem with a pump, a heat exchanger, and/or one or more large storage tanks. The two main types of solar collectors used are either an *evacuated-tube collector* or a *parabolic-trough collector*. They can operate at high temperatures with high efficiency. An evacuated-tube collector is a shallow box full of many glass, double-walled tubes and reflectors to heat the fluid inside the tubes. A vacuum between the two walls insulates the inner tube, holding in the heat. Parabolic troughs are long, rectangular, U-shaped curved mirrors that are tilted to focus sunlight on a tube, which runs down the center of the trough. This heats the fluid within the tube.

The heat from a solar collector can also be used to cool a building. It may seem surprising to be able to use heat to cool a building; however, in reality, the solar heat is really just an energy source. Just as a home air conditioner uses an energy source—electricity—to create cool air, solar absorption coolers use a similar approach, combined with some very complex chemistry applications, to create cool air from solar energy. Solar energy can also be used with evaporative coolers (also known as "swamp coolers") to extend their application to more humid climates, using another chemistry application called *desiccant cooling*.

GEOTHERMAL

The word geothermal comes from the Greek words geo (earth) and therme (heat). Geothermal energy is heat from within the earth. This heat can be recovered as steam or as hot water, and it can be used to heat buildings or to generate electricity. Geothermal energy is a renewable energy source because the heat is continuously produced inside the earth. The energy is generated in the earth's core. Temperatures hotter than the sun's surface are continuously produced inside the earth caused by the slow decay of radioactive particles, a process that happens in all rocks (Figure 11.4). The earth has a number of different layers:

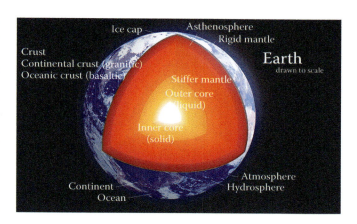

FIGURE 11.4 The layers inside the earth. (Courtesy of Kelvinsong, Wikimedia Commons.)

- The core has a solid inner core and an outer core made of hot melted rock called magma.
- The mantle surrounds the core and is about 2,897 kilometers thick. The mantle is made up of magma and rock.
- The crust is the outermost layer of the earth. The crust forms the continents and ocean floors. The crust can be 5–8 kilometers thick under the oceans and 24–56 kilometers thick on the continents.

The earth's crust is broken into pieces called plates. Magma comes close to the earth's surface near the edges of these plates. This is where volcanoes occur. The lava that erupts from volcanoes is partly magma. The rocks and water absorb the heat from this magma deep underground. The rocks and water found deeper underground have higher temperatures. People around the world use geothermal energy to heat their homes and to produce electricity by drilling deep wells and pumping the hot underground water or steam to the surface. People can also make use of the stable temperatures near the surface of the earth to heat and cool buildings. Geothermal reservoirs are naturally occurring areas of hydrothermal resources. They are deep underground and are largely undetectable above ground. Geothermal energy finds its way to the earth's surface in three ways:

- Volcanoes and fumaroles (holes where volcanic gases are released)
- Hot springs
- Geysers

The most active geothermal resources are usually found along major tectonic plate boundaries where earthquakes and volcanoes are located. One of the most active geothermal areas in the world is called the Ring of Fire (Figure 11.5). This area encircles the Pacific Ocean.

When magma comes near the earth's surface, it heats ground water trapped in porous rock or water running along fractured rock surfaces and faults. Hydrothermal features

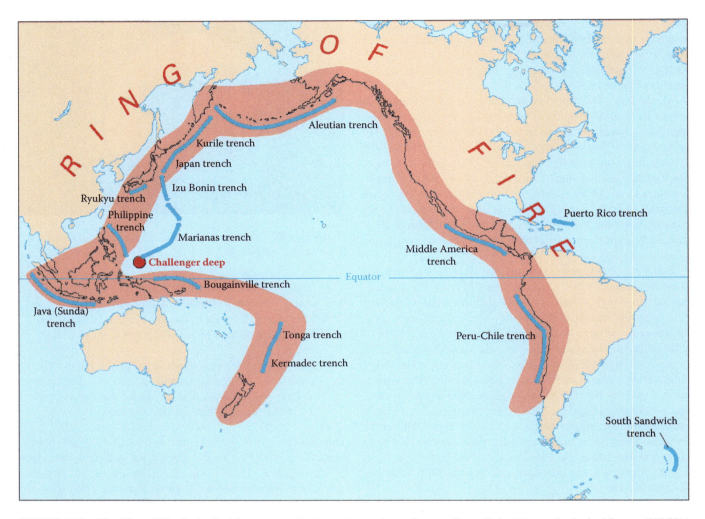

FIGURE 11.5 The Ring of Fire in the Pacific ocean marks the active geothermal areas. (From United States Geological Survey [USGS].)

have two common ingredients, such as water (hydro) and heat (thermal).

Geologists use various methods to find geothermal reservoirs. Drilling a well and testing the temperature deep underground are the most reliable method for locating a geothermal reservoir.

Geothermal Locations

Most of the geothermal power plants in the United States are located in western states and Hawaii, where geothermal energy resources are close to the earth's surface (Figure 11.6). California generates the most electricity from geothermal energy. The Geysers dry steam reservoir in northern California is the largest known dry steam field in the world and has been producing electricity since 1960.

Some applications of geothermal energy use the earth's temperatures near the surface, while others require drilling miles into the earth. There are three main types of geothermal energy systems:

- Direct use and district heating systems
- Electricity generation power plants
- Geothermal heat pumps

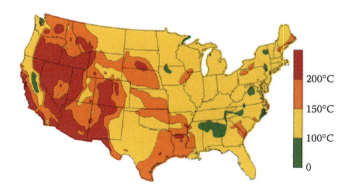

FIGURE 11.6 U.S. geothermal resource map. (From U.S. Department of Energy, Office of Energy Efficiency & Renewable Energy.)

Direct Use and District Heating Systems

Direct use and district heating systems use hot water from springs or reservoirs located near the surface of the earth. Ancient Roman, Chinese, and Native American cultures used hot mineral springs for bathing, cooking, and heating. Today, many hot springs are still used for bathing, and many people believe the hot, mineral-rich waters have natural healing powers. Geothermal energy is also used to heat buildings through district heating systems. Hot water near the earth's surface is piped directly into buildings and industries for heat. A district heating system provides heat for most of the buildings in Reykjavik, Iceland (Figure 11.7). Industrial applications of geothermal energy include food dehydration, gold mining, and milk pasteurizing. Dehydration, or the drying of vegetable and fruit products, is the most common industrial use of geothermal energy (Figure 11.8).

FIGURE 11.7 The Blue Lagoon, a geothermal spa located near Reykjavík, Iceland. (By McKay Savage from London, UK (Iceland - Blue Lagoon 09) [CC BY 2.0 (http://creativecommons.org/licenses/by/2.0)], via Wikimedia Commons.)

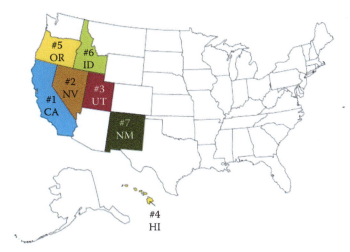

FIGURE 11.8 The United States has significant geothermal resources. The seven highest ranking states are all in the West (plus Hawaii).

Geothermal Electricity Generation

The United States leads the world in the amount of electricity generated with geothermal energy. Geothermal electricity generation requires water or steam at high temperatures (150°C–370°C). Geothermal power plants are generally built where geothermal reservoirs are located within a mile or two of the earth's surface. In 2015, U.S. geothermal power plants produced about 16.8 billion kilowatt-hours (kWh), or 0.4 percent of total U.S. electricity generation. In 2015, seven states had geothermal power plants. The following table shows the share of U.S. geothermal electricity produced by the listed states, 2015:

Electricity Produced with Geothermal Energy in the United States, 2015

State	Percentage of Total Produced (%)
California	74
Nevada	20
Utah	3
Hawaii	1.2
Oregon	1.1
Idaho	0.6
New Mexico	0.1

Uses of Geothermal Energy

Geothermal energy contributes a significant share of electricity generation in several countries. In 2013, 20 countries, including the United States, generated a total of about 70 billion kWh of electricity from geothermal energy (EIA, 2016). Indonesia was the second-largest geothermal power producer after the United States, at about 11 billion kWh of electricity, which equaled approximately 5 percent of Indonesia's total power generation. Iceland was the seventh-largest producer at about 5 billion kWh of electricity, but the largest share of its total electricity generation was from geothermal energy at 27 percent.

Geothermal power plants use hydrothermal resources that have both water (hydro) and heat (thermal). Geothermal power plants require high-temperature (300°F–700°F) hydrothermal resources that come from either dry steam wells or from hot water wells. People use these resources by drilling wells into the earth and then piping steam or hot water to the surface. The hot water or steam is used to operate a turbine that generates electricity. Some geothermal wells may be as deep as 3.2 kilometers.

Types of Geothermal Power Plants

There are three basic types of geothermal power plants:

- Dry steam plants
- Flash steam plants
- Binary cycle power plants

Dry steam plants use steam piped directly from a geothermal reservoir to turn generator turbines. They are the simplest and oldest design. The geothermal steam needs to be 150°C or greater to turn the turbines. The first geothermal power plant was built in 1904 in Tuscany, Italy, where natural steam erupted from the earth. *Flash steam plants* take high-pressure hot water from deep inside the earth, pull it into lower-pressure tanks, and convert it to steam to drive generator turbines. Fluid temperatures must be at least 180°C. The geothermal reservoirs used contain water with temperatures greater than 360°C. The hot water flows up through wells in the ground under its own pressure. As it flows upward, the pressure decreases and some of the hot water boils into steam. The steam is then separated from the water and used to power a turbine/generator. When the steam cools, it condenses to water and is then injected back

into the ground to be used again. Most geothermal power plants are flash steam plants. *Binary cycle power plants* are the most recent development. They transfer the heat from geothermal hot water to another liquid. The heat causes the second liquid to flash vaporize, which is then used to drive a generator turbine. This is the most common type of geothermal electricity station being built today. The differences between dry steam, flash steam, and binary cycle power plants are shown in the diagrams (Figures 11.9 and 11.10a–c).

FIGURE 11.9 Geothermal energy is a site-specific form of energy. Where the geologic conditions are right, the earth releases immense amounts of heat. This energy can be harnessed and converted into electricity. This plant is located in Sonoma and Lake Counties, California, and is called the Geysers Complex. (Courtesy of U.S. Department of Energy.)

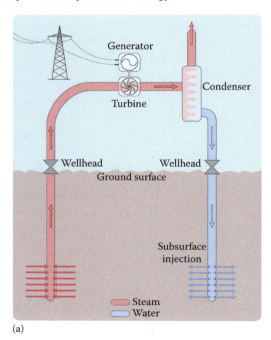

(a)

FIGURE 11.10 The differences between dry steam, flash steam, and binary cycle power plants are shown in the three diagrams. (a) Dry steam power plant. (Courtesy of Goran Tekan, Wikimedia Commons.)   *(Continued)*

Flashed steam geothermal process diagram

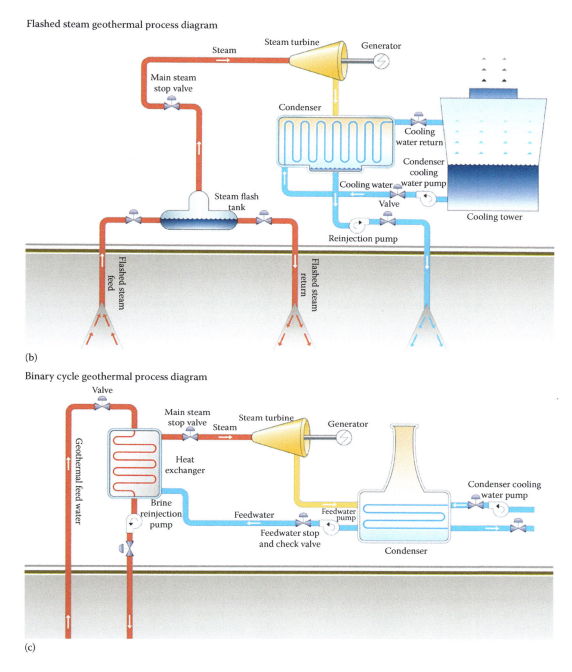

(b)

Binary cycle geothermal process diagram

(c)

FIGURE 11.10 (Continued) The differences between dry steam, flash steam, and binary cycle power plants are shown in the three diagrams. (b) flash steam power plant. (Courtesy of Goran Tekan, Wikimedia Commons.) and (c) binary cycle power plant. (From U.S. Department of Energy, Energy Efficiency & Renewable Energy.)

Geothermal Heat Pumps

The factor that makes geothermal energy possible is its consistency. Although temperatures above ground change depending on time of day and season, the temperatures 10 feet below the earth's surface are consistently between 10°C and 16°C. For most areas, this means that the soil temperatures are usually warmer than the air in winter and cooler than the air in summer. Geothermal heat pumps use the earth's constant temperatures to heat and cool buildings. They transfer heat from the ground (or water) into buildings during the winter and reverse the process in the summer. This consistency in temperature makes geothermal heat pumps energy efficient and cost effective. The EPA says that geothermal heat pumps are the most energy efficient, environmentally clean, and cost-effective systems used for temperature control; and they can be used for all types of buildings, including homes, office buildings, schools, and hospitals.

The environmental impact of geothermal energy depends on how geothermal energy is used or how it is converted to useful energy. Direct use of geothermal energy and heat pumps has virtually no negative impact on the environment. Instead, it can be considered to have a positive effect, because it reduces the use of other types of energy that do have a negative impact on the environment. In addition, geothermal power plants do not burn fuel to generate electricity, so the levels of air pollutants they emit are low. They release less than 1 percent of the carbon dioxide emissions released by a fossil fuel power plant. Geothermal power plants further limit air pollution through the use of scrubber systems that remove hydrogen sulfide. Hydrogen sulfide is naturally found in the steam and in the hot water used to generate geothermal power. Geothermal plants also emit 97 percent less acid rain-causing sulfur compounds than are emitted by fossil fuel power plants. After the steam and water from a geothermal reservoir are used, they are injected back into the earth.

WIND

Wind is another usable energy source that can be traced back to thousands of years. The first mechanical device that was built to use the wind as a source of power was the sailboat. Sailboats and larger sailing ships allowed humans to begin exploring and trading with other civilizations. For this, harbors were built, creating towns and cities along coastlines.

The first sailing ships had simple square sails that carried them in the direction the wind blew. It was Arab sailors who discovered how to sail into the wind using a triangular sail called the lateen. These were similar to the triangular sails that are still used today.

The first machine designed to use wind power to do work on land was the windmill, which was invented in the seventh century in Persia (now Iran). This new technology then spread to the Middle East and other countries such as India and China. The windmill was one of the first large-scale inventions designed to make peoples' workloads easier. Windmills were initially used to grind grain between heavy millstones to make flour. They were also used to pump water from rivers to irrigate crops. In sawmills, people used the motion of the shaft to run a saw, which slid up and down to slice rough logs into pieces of lumber.

Windmills were not built in Europe until the twelfth century. The Europeans used them to generate power. They invented a new type of mill, called the post mill. A center post grounded this type of windmill, and the whole building turned around the center post so that the sails always faced into the wind. The stationary windmill—where only the top of the structure in which the sails were mounted turned into the wind—followed this. Wind-powered grain mills and sawmills were replaced by more efficient machinery in the early 1900s. Some farmers still use windmills today, however, to pump water and drain flooded areas and to pull up underground water for irrigating crops. Today, high-powered windmills, called wind turbine generators, are used to make electricity.

Generating wind energy is possible in natural wind corridors where amble and consistent amount of wind is generated by the uneven heating of the earth's surface by the sun.

Because the earth's surface is covered by different features (water, mountains, plains, hills, valleys), they absorb the sun's heat at different rates. One example of this uneven heating can be seen in the daily wind cycle—such as in coastal areas between the land and sea. During the day, the air above the land heats up faster than the air above water. The warm air over the land expands and rises, and the heavier, cooler air rushes in to take its place, creating wind. At night, the winds are reversed because the air cools more rapidly over land than over water. This happens on various scales—locally and regionally. In the same way, the atmospheric winds that circle the earth are created because the land near the earth's equator is heated more by the sun than the land near the North Pole and the South Pole. Due to this heating and cooling relationship, there are some areas that have a consistent wind system associated with it. These areas are the most suitable candidates for wind-generated electricity and where you generally will see wind farms built for the generation of electricity (Figure 11.11b). Wind turbines use blades to collect the wind's kinetic energy. Wind flows over the blades creating lift, which causes them to turn. The blades are connected to a drive shaft that turns an electric generator, which produces electricity.

In 2014, wind turbines in the United States generated about 4 percent of total U.S. electricity generation. While this may seem like a small share of the country's total electricity production, it was equal to the electricity use of about 17 million U.S. households. The amount of electricity generated from wind has grown significantly in recent years. Electricity generation from wind in the United States increased from about 6 billion kilowatt hours (kWh) in 2000 to about 182 billion kWh in 2014. New technologies have decreased the cost of producing electricity from wind, and growth in wind power has been encouraged by both government and industry incentives.

Locations of Wind Power Projects

Wind power plants require specific planning. There is more to it than just putting up a few turbines in a windy area. Each turbine is carefully positioned, and the wind speed, direction, and consistency information has been factored into the exact specifications of how the turbine will function. Wind speed typically increases with altitude and increases over open areas without windbreaks. Good sites for wind turbines include the top of smooth, rounded hills; open plains and water; and mountain gaps that funnel and intensify wind. Wind speeds vary throughout the United States. Wind speeds also vary throughout the day and from season to season. The wind energy potential in the United States is about 10 times the amount of electricity consumption for the entire country. Wind potential is not the same everywhere, however. The strongest wind resources—which would make the best sites for productive energy-generating sites—vary by region and topography. Wind energy projects are developed by companies that seek out the areas with the strongest wind resource but also review other critical factors such as access to land, access to the transmission lines, ability to sell the electricity, public involvement, and other development-related factors. Once a site is identified, a developer will conduct wind

(a)

Valley breeze

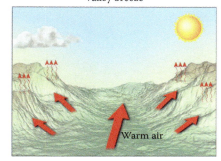

Warm air

➤ Forms during the day
➤ Warm air flows up the canyon

(b)

Mountain breeze

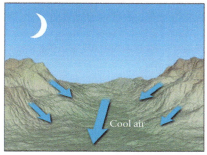

Cool air

➤ Forms during the night
➤ Cool air flows down toward the valley

FIGURE 11.11 (a) Wind turbines at Tehachapi Pass, California. Here, the wind blows more frequently from April through October than it does in the winter. This fluctuation is a result of the extreme heat of the Mojave Desert during the summer months. As the hot air over the desert rises, the cooler, denser air above the Pacific Ocean rushes through the Tehachapi mountain pass to take its place. Fortunately, the seasonal variations in wind speeds in California match the electricity demands of consumers. In California, people use more electricity during the summer for air conditioners. This wind farm, with 5,000 wind turbines, produces enough electricity to meet the needs of 350,000 people each year. (From NREL, Golden, CO.) (b) Mountain and valley breezes are an example of local winds caused by the geography of an area. During the day, the sun heats the valley floor and warms the air above it. Warm air from the valley moves upward and creates a valley breeze. At night, the mountains cool faster than the valleys. Cold air sinks from the mountain peaks, creating a mountain breeze. These winds are consistent and serve as favorable places to locate wind farms. (Courtesy of Kerr.)

resource assessment, siting and permitting, transmission studies over a period of several years. The majority of wind projects are located on private land, where the developer leases the land from the original landowner providing lease payments. After early stages of development, a developer will seek out a contract with a purchaser of electricity, work with finance markets, order wind turbines, and hire a specialized construction company to build the project. Once a project is built and delivering electricity to the power grid, a project owner or operator will maintain the project for its 20–30 year life (Figure 11.11a).

In the United States, wind power projects with one or more large wind turbines were located in 40 states in 2015. The five states with the most electricity generation from wind in 2015 were Texas, Iowa, Oklahoma, California, and Kansas. These states combined produced about 50 percent of total U.S. wind electricity generation in 2015. From an international perspective, many countries generate electricity with wind energy. Most wind power projects are located in Europe and in the United States where government programs have supported wind power development. China and India have increased wind electricity generation in recent years and were among the top five producers of electricity generation from wind in 2013. As of 2015, China is the leading country for wind-produced energy (Figure 11.12). The waters off the coasts of the United States have significant potential for electricity generation from wind energy, and several offshore wind projects in New England are in the planning stages. Europe has a number of operating offshore wind energy projects, as well.

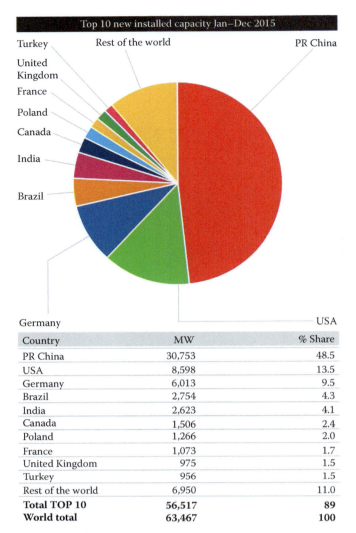

Country	MW	% Share
PR China	30,753	48.5
USA	8,598	13.5
Germany	6,013	9.5
Brazil	2,754	4.3
India	2,623	4.1
Canada	1,506	2.4
Poland	1,266	2.0
France	1,073	1.7
United Kingdom	975	1.5
Turkey	956	1.5
Rest of the world	6,950	11.0
Total TOP 10	**56,517**	**89**
World total	**63,467**	**100**

FIGURE 11.12 New wind power installed capacity by country in 2015. (Courtesy of Global Wind Energy Council (GWEC, Durham, NC.)

(a)

(b)

FIGURE 11.13 (a) Vertical axis wind turbines located in Altamont Pass in California. (From NREL, Golden, CO.) (b) The skyline of Androssan, North Ayrshire, Scotland, is dominated by an enormous wind farm. The houses in front are on Eglinton Road. (Courtesy of Vincent van Zeijst, Creative Commons.)

Types of Wind Turbines

There are two basic types of wind turbines:

- Horizontal-axis turbines
- Vertical-axis turbines

The size of wind turbines varies widely. The length of the blades helps to determine the amount of electricity a wind turbine can generate. Small wind turbines used to power a single home or business may have a capacity of less than 100 kilowatts (100,000 watts). The largest turbines have capacities of 5–8 million watts. Larger turbines are often grouped together to create wind power plants, or *wind farms* that provide power to electricity grids. Most of the wind turbines currently in use are horizontal-axis turbines. Horizontal-axis turbines have blades like airplane propellers. The horizontal-axis turbines used on wind farms can be as tall as 20-story buildings (about 60 meters) and can have blades more than 30 meters long.

Taller turbines with longer blades generate more electricity. Horizontal-axis turbines commonly have three blades.

Vertical-axis turbines have blades that are attached to the top and the bottom of a vertical rotor. The most common type of vertical-axis turbine—the Darrieus wind turbine, named after the French engineer Georges Darrieus who patented the design in 1931—looks like a giant, two-bladed egg beater. Some versions of the vertical-axis turbine are 30 meters tall and 15 meters wide. Very few vertical-axis wind turbines are in use today because they do not perform like horizontal-axis turbines (Figure 11.13).

Benefits of Wind Energy

Wind energy is a clean, renewable form of energy that uses virtually no water and pumps billions of dollars into our economy every year. Since 2008, the United States wind industry has generated more than $100 billion in private investment. In addition, wind energy presents a positive business venture in many parts of the country, providing economic investment to rural communities through lease payments to landowners. Wind energy helps avoid a variety of environmental impacts due to its low impact emitting zero GHG emissions or conventional pollutants and consuming virtually no water. One negative aspect of it, however, is when birds fly into the blades of the windmills.

One example of the success of wind energy is that of Scotland's use of wind power. In August, 2016, under the power of the gusty winds sweeping across the country's highlands, a significant milestone was reached. For the country's first time, their arsenal of signature spinning white windmills that grace the lush countrysides was able to generate enough electricity to power the entire country. That marvelous achievement put them in the ranks with the growing league of nations that have achieved the same astounding feat: Portugal, Denmark, and Costa Rica. And for the rest of the world's nations—it has given us all hope that we may someday be able to achieve the same goal.

Scotland has made several increasingly ambitious renewable energy targets over the past decade and has been able to surpass every one of them. Today, more than half of the entire country's electricity comes from zero-carbon sources—such as wind, hydro, and solar—and the latest target of 100 percent by 2020 may become the reality they are aiming for. The citizens of Scotland have stood in support of renewable energy for years—they fully understand the connection between human causes and climate change. The windmills have even become a visual icon representing the city of Glasgow, similar in the way that shipbuilding did historically, and the citizens are—and should rightfully be—proud of it.

Ironically, the hurdles they face in the future of outstanding achievements are political in nature. The British government (under whom they are ruled) has shifted their support dramatically of recent from a planned offshore wind farm and solar power plant to nuclear energy and fracking. The shift began under then Prime Minister David Cameron and has continued under his successor, Theresa May. So far, subsidies for solar and onshore wind have been removed. Scotland is considering another vote for its independence over it, while also hoping to attract independent business ventures. As Niall Stuart, the Scottish Renewables chief says, "This has to be the beginning of the story. It can't be the end." (Washington Post, 2016).

HYDROPOWER

Hydropower is one of the oldest sources of energy for producing mechanical and electrical energy. It was used thousands of years ago to turn paddle wheels to help grind grain. Before steam power and electricity were available in the United States, grain and lumber mills were powered directly with hydropower. The first industrial use of hydropower to generate electricity in the United States occurred in 1880, when 16 brush-arc lamps were powered using a water turbine at the Wolverine Chair Factory in Grand Rapids, Michigan. The first U.S. hydroelectric power plant opened on the Fox River near Appleton, Wisconsin, in 1882. Today, hydropower is the largest renewable energy source for electricity generation in the United States. In 2015, hydropower accounted for about 6 percent of total U.S. electricity generation and 46 percent of electricity generation from all renewables. Since the source of hydroelectric power is water, hydroelectric power plants are usually located on or near a water source.

Because hydropower relies heavily on the earth's natural water cycle, understanding the water cycle is important to understanding hydropower and its benefits and constraints (Figure 11.14). The water cycle has three components:

- Solar energy heats water on the surface of rivers, lakes, and oceans, which causes the water to evaporate.
- Water vapor condenses into clouds and falls as precipitation (rain, snow, etc.).
- Precipitation collects in streams and rivers, which empty into oceans and lakes, where it evaporates and begins the cycle again.

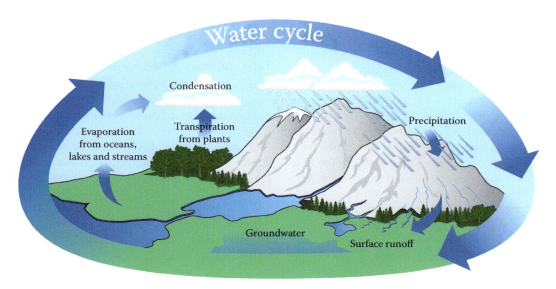

FIGURE 11.14 The earth's water cycle. (Courtesy of NASA, Washington, DC.)

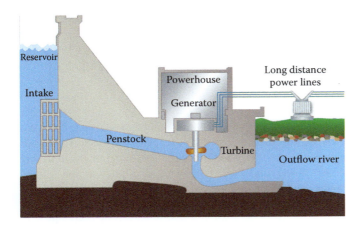

FIGURE 11.15 Schematic of a hydroelectric dam. (Courtesy of Tennessee Valley Authority.)

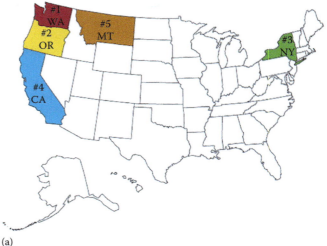

(a)

(b)

FIGURE 11.16 (a) Top hydropower producing states. (From U.S. Energy Information Administration, *Electric Power Monthly*, February, 2015.) (b) An example of a tital barrage, this is the Rance Tidal Power Station in Bretagne, France. (From User: Dani 7C3 – Own work, CC BY 2.5, https://commons.wikimedia.org/w/index.php?curid=915904.)

The amount of precipitation that drains into rivers and streams in a geographical area is what determines the amount of water available for producing hydropower. Seasonal variations in precipitation and long-term changes in precipitation patterns, such as droughts, have a significant impact on hydropower production.

The volume of the water flow and the change in elevation (or fall) from one point to another determine the amount of available energy in moving water. Swiftly flowing water in a big river, like the Columbia River that forms the border between Oregon and Washington; and water falling rapidly from a high point, such as Niagara Falls in New York, carry substantial energy in their flow. At both Niagara Falls and the Columbia River, water flows through a pipe, or *penstock*, then pushes against and turns blades in a turbine to spin a generator to produce electricity. In a *run-of-the-river system*, the force of the current applies pressure on a turbine. In a *storage system*, water accumulates in reservoirs created by dams and is released as needed to generate electricity (Figure 11.15).

Suitable Hydropower Locations

Most of the U.S. hydropower capacity is located in the West. More than half of U.S. hydroelectric capacity for electricity generation is concentrated in Washington, Oregon, and California. In 2014, approximately 30 percent of total U.S. hydropower was generated in Washington, which is the site of the Grand Coulee Dam, the largest hydroelectric facility in the United States. Although Washington has the most hydroelectric generating capacity of any state, New York has the largest hydroelectric capacity of all states east of the Mississippi River. Most hydropower is produced at large facilities built by the federal government. The West has many of the largest dams, but there are also smaller facilities operating around the country (Figure 11.16a). Only a small percentage of dams in the United States produce electricity. Most dams were constructed for irrigation and flood control.

Tidal and Wave Power Generation

Tides are caused by the gravitational pull of the moon and sun and the rotation of the earth. Near the shore, water levels can vary up to 12 meters as a result of tides. The movement of water as a result of tidal forces can be used to produce energy. Tidal power is more predictable than wind energy and solar power. A tidal range of 3 meters is needed to produce tidal energy economically.

Tidal Barrages

Some tidal energy plants use a structure similar to a dam called a barrage. The barrage can be located across an inlet of an ocean bay or lagoon that forms a tidal basin. Sluice gates (gates commonly used to control water levels and flow rates) on the barrage allow the tidal basin to fill on the incoming high tides and allows the tidal basin to empty through the turbine system on the outgoing ebb tide. There are two-way systems that generate electricity from both the incoming and outgoing tides (Figure 11.16b).

A potential disadvantage of tidal power is the negative effect a tidal station can have on plants and animals in estuaries of the tidal basin. Tidal barrages can change the tidal level in the basin and increase turbidity (the amount of matter in suspension in the water). They can also affect navigation and recreation.

There are currently six tidal power barrages operating in the world. The largest is the Sihwa Lake Tidal Power Station in South Korea with a total power output capacity of 254 megawatts (MW). The second-largest and oldest is in La Rance, France, with 240 MW capacity. The next largest is in Annapolis Royal in Nova Scotia, Canada, at 20 MW, and it is followed by the 3.7 MW Jiangxia Tidal Power Station in China, the 1.7 MW kilowatt tidal barrage in Kislaya Guba, Russia, and the 1 MW Uldolmok Tidal Power Station in South Korea. The United States does not have any tidal power plants, and it only has a few sites where tidal energy could be produced economically. France, England, Canada, and Russia have much more potential to use tidal power.

Tidal Fences

Tidal fences can also harness the energy of tides. A tidal fence has vertical axis turbines mounted in a fence. All the water that passes is forced through the turbines. Tidal fences can be used in areas between two landmasses like channels. Tidal fences are cheaper to install than tidal barrages, and they have less impact on the environment than tidal barrages. However, tidal fences can disrupt the movement of large marine animals. A tidal fence is planned for the San Bernardino Strait in the Philippines (Figure 11.17).

Tidal Turbines

Tidal turbines are basically wind turbines in the water that can be located anywhere there is strong tidal flow. Because water

is about 800 times denser than air, tidal turbines have to be much sturdier than wind turbines. Tidal turbines are heavier and more expensive to build but capture more energy. There is an operating 1.5 MW tidal turbine in Strangford Lough, Scotland, and at Uldolmok, South Korea.

Wave Power

Waves are caused by the wind blowing over the surface of the ocean. There is tremendous energy in ocean waves. It is estimated that the total energy potential of waves off the coasts of the United States is 252 billion kilowatt hours a year, the equivalent of about 6 percent of U.S. electricity generation in 2014. The west coasts of the United States and Europe and the coasts of Japan and New Zealand have good wave energy potential sites for harnessing wave energy because of wave energy potential.

One way to harness wave energy is to bend or focus the waves into a narrow channel, increasing their power and size. The waves can then be channeled into a catch basin or used directly to spin turbines that generate electricity. Several other methods of capturing wave energy are currently being developed. These methods include placing devices on or just below the surface of the water and anchoring the devices to the ocean floor. The U.S. DOE is currently in the process of developing technology to harness the power of waves in the future (Figure 11.18).

Ocean Thermal Energy Conversion

Energy from the sun heats the surface water of the ocean. In tropical regions, surface water can be much warmer than deep water. This temperature difference can be used to produce electricity. The Ocean Thermal Energy Conversion (OTEC) system

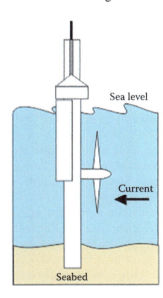

FIGURE 11.17 Tidal turbine. (Adapted from National Energy Education Development Project.)

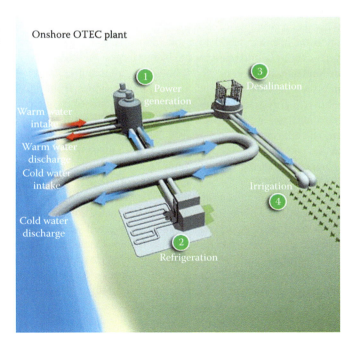

FIGURE 11.18 Wave energy site. (Adapted from National Energy Education Development Project.)

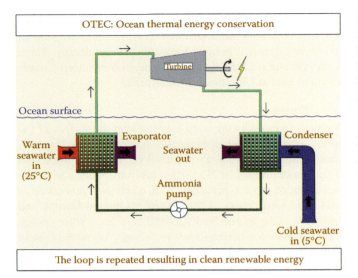

FIGURE 11.19 Ocean thermal energy conversion system. (From NOAA.)

uses a temperature difference of at least 77°F to operate a turbine to produce electricity. The United States became involved in OTEC research in 1974 with the establishment of the Natural Energy Laboratory of Hawaii Authority. The laboratory is one of the world's leading test facilities for OTEC technology (Figure 11.19). There is no large-scale operation of OTEC today, mainly because there are many challenges associated with the technology although it is currently being researched.

BIOENERGY

Biomass energy, or bioenergy, which is just energy generated from organic matter, has been utilized for thousands of years. It has been a staple of energy ever since civilizations have relied on burning wood to cook food or build fires to keep warm. Today, wood continues to be our largest biomass energy resource. In addition to wood, however, other sources of biomass can now be used, including plants, residues from agriculture or forestry, and the organic components of municipal and industrial wastes. Even the fumes from landfills can be used as a biomass energy source. The use of biomass energy has the potential to greatly reduce GHG emissions. Biomass generates about the same amount of carbon dioxide as fossil fuels, but every time a new plant grows, carbon dioxide is actually removed from the atmosphere. The net emission of carbon dioxide will be zero as long as plants continue to be replenished for biomass energy purposes. These energy crops, such as fast-growing trees and grasses, are called biomass feedstocks. The use of biomass feedstocks can also help increase profits for the agricultural industry. Bioenergy can be broken down into the subcategories: biofuels, biopower, and bioproducts (Bhattarai, et al, 2011). The following sections discuss each.

Biofuels

Biofuels are transportation fuels such as ethanol and biodiesel that are made from biomass materials. These fuels are usually blended with petroleum fuels (gasoline and diesel fuel), but they can also be used on their own. Using ethanol or biodiesel means less gasoline and diesel fuel is burned, which can reduce the amount of crude oil imported from other countries. Ethanol and biodiesel are also cleaner-burning fuels than pure gasoline and diesel fuel.

Ethanol

Ethanol is an alcohol fuel made from the sugars found in grains such as corn, sorghum, and barley. Other sources of sugars to produce ethanol include

- Sugarcane
- Sugarbeets
- Potato skins
- Rice
- Yard clippings
- Tree bark
- Switchgrass

Most of the fuel ethanol used in the United States is distilled from corn. Researchers are working on ways to make ethanol from all parts of plants and trees rather than just grain. Farmers are experimenting with fast-growing woody crops such as small poplar and willow trees and switchgrass to see if they can be used to produce ethanol.

Sugarcane and sugarbeets are the most common ingredients used to make ethanol in other parts of the world. Since alcohol is created by fermenting sugar, sugar crops are the easiest ingredients to convert into alcohol. Brazil, the world's second-largest fuel ethanol producer, makes most of its ethanol from sugarcane. Most of the cars in Brazil are capable of running on pure ethanol or on a blend of gasoline and ethanol.

Ethanol can also be produced by breaking down cellulose in woody fibers. Cellulosic ethanol is considered as an advanced biofuel and involves a more complicated production process than the process used to make conventional ethanol. Trees and grasses are potential feedstocks (the raw material needed to make a product) for cellulosic ethanol production. Trees and grasses require less energy, fertilizers, and water than grains and can also be grown on lands that are not suitable for growing food. Scientists have developed fast-growing trees that grow to full size in 10 years. Many grasses can produce two harvests a year for many years without annual replanting.

Nearly all of the gasoline now sold in the United States is about 10 percent ethanol by volume. Any gasoline-powered engine in the United States can use E10 (gasoline with 10 percent ethanol), but only specific types of vehicles can use mixtures with fuel containing more than 10 percent ethanol. A flexible-fuel vehicle can use gasoline with ethanol content greater than 10 percent. The EPA ruled in October 2010 that cars and light trucks of model year 2007 and newer can use E15 (gasoline with 15 percent ethanol). E85, a fuel that contains 51 to 83 percent ethanol, depending on location and season, is mainly sold in the Midwest and can only be used in a flexible-fuel vehicle.

Biodiesel

Biodiesel is a fuel made from vegetable oils, fats, or greases—such as recycled restaurant grease. Biodiesel fuel can be used in diesel engines without changing the engine. Pure biodiesel is nontoxic, biodegradable, and produces lower levels of most air pollutants than petroleum-based diesel fuel. Biodiesel is usually sold as a blend of biodiesel and petroleum-based diesel fuel. A common blend of diesel fuel is B20, which is 20 percent biodiesel.

As global fossil fuel supplies dwindle and GHG concentrations rise, it is becoming more important to find suitable biofuel alternatives to petroleum products. The use of biodiesel fuel has increased substantially since 2001. Because of biodiesel's environmental benefits, its use in the United States alone grew from about 10 million gallons in 2001 to 1.5 billion gallons in 2015. Part of its dramatic consumption increase was attributed to the Renewable Fuels Standard in 2011, but its use has continued to climb in the years since then.

Any diesel engine can use biodiesel at blend levels of 5 percent by volume (B5) or lower. Many vehicles, such as school and transit buses, snowplows, garbage trucks, mail trucks, and military vehicles, can use biodiesel blends of 20 percent by volume (B20). Fueling stations that sell biodiesel blends to the public are available throughout the United States. In 2012, 6.44 billion gallons of biodiesel were consumed in about 64 countries; 54 percent was consumed in just 5 countries, as shown in the chart.

World Biodiesel Consumption, 2012

Country	Billion Gallons	Share of Total (%)
World Total	6.44	
United States	0.92	14
Germany	0.75	12
Brazil	0.74	11
France	0.66	10
Spain	0.42	7
All Others	2.96	46

From a climate and environmental perspective, it is important to remember that biodiesel burns much cleaner than petroleum diesel. Biodiesel is nontoxic and biodegradable. Compared to petroleum diesel fuel, which is refined from crude oil, biodiesel produces fewer air pollutants such as particulates, carbon monoxide, sulfur dioxide, hydrocarbons, and air toxics. Biodiesel does slightly increase emissions of nitrogen oxides, however.

Biodiesel use may also reduce GHG emissions. Using a gallon of biodiesel produced in the United States does not produce the CO_2 emissions that result from burning a gallon of petroleum diesel. Biodiesel may be considered carbon-neutral because the plants used to make biodiesel, such as soybeans and palm oil trees, absorb CO_2 as they grow. The absorption of CO_2 offsets the CO_2 produced while making and using biodiesel. Most of the biodiesel produced in the United States is made from soybean oil. Some biodiesel is also produced from used oils or fats, including recycled restaurant grease.

In some parts of the world, large areas of natural vegetation and forests have been cleared and burned to grow soybeans and palm oil trees to make biodiesel. The negative environmental impacts of land clearing may be greater than any potential benefits of using biodiesel produced from soybeans and palm oil trees.

Historic Development of Biofuels

In the 1850s, ethanol was a major lighting fuel. During the Civil War, a liquor tax was placed on ethanol to raise money for the war. The tax increased the price of ethanol so much that it could no longer compete with other fuels such as kerosene. Ethanol production declined sharply because of this tax, and production levels did not begin to recover until the tax was repealed in 1906.

In 1908, Henry Ford designed his Model T, a very early automobile, to run on a mixture of gasoline and alcohol. Ford called this mixture the fuel of the future. In 1919, when prohibition began, ethanol was banned because it was considered an alcoholic beverage. It could only be sold when mixed with petroleum. Ethanol was used as a fuel again after prohibition ended in 1933.

Ethanol and biodiesel were the fuels used in the first automobile and diesel engines, but lower-cost gasoline and diesel fuel made from crude oil became the dominant vehicle fuels. Its use increased temporarily during World War II when oil and other resources were scarce. In the 1970s, interest in ethanol as a transportation fuel was revived as oil embargoes, rising oil prices, and growing dependence on imported oil increased public interest in alternative fuels. Since that time, ethanol use and production have been encouraged by tax benefits and by environmental requirements for cleaner-burning fuels. The federal government has promoted ethanol use in vehicles to help reduce oil imports since the mid-1970s. In 2005, Congress enacted a Renewable Fuel Standard (RFS) that set minimum requirements for the use of renewable fuels, including ethanol. In 2007, the RFS renewable fuel use targets were set to rise steadily to a level of 36 billion gallons by 2022. In 2014, about 13 billion gallons of ethanol were added to the gasoline consumed in the United States.

Growing plants for biofuels is controversial, however, because the land, fertilizers, and energy used to grow biofuel crops could be used to grow food crops instead. Also, in some parts of the world, large areas of natural vegetation and forests have been cut down to grow sugarcane for ethanol and soybeans and palm oil trees to make biodiesel. The United States government supports efforts to develop alternative sources of biomass that do not compete with food crops and that use less fertilizer and pesticides than corn and sugarcane. The U.S. government also supports methods to produce ethanol that require less energy than conventional fermentation. For example, ethanol can also be made from waste paper, and biodiesel can be made from waste grease, oils, and even algae. Using ethanol or biodiesel means less gasoline and diesel fuel is burned, which can reduce the amount of crude oil imported from other countries. Ethanol and biodiesel are also cleaner-burning fuels than pure gasoline and diesel fuel.

DID YOU KNOW?

E85: E85 is an alternative fuel that contains up to 85 percent ethanol. Although E85 is used mainly in the Midwest, there are about 2,600 E85 fueling stations located around the country. Only flexible-fuel vehicles can use E85.

Flexible-fuel vehicles: Flexible-fuel (flex-fuel) vehicles can run on any mixture of ethanol and gasoline up to E85. Although there are more than 100 million flex-fuel vehicles in the United States, only about 10 percent of them use E85.

Ethanol Is Nontoxic and Biodegradable

Unlike gasoline, pure ethanol is nontoxic and biodegradable, and it quickly breaks down into harmless substances if spilled. Chemical denaturants are added to fuel ethanol (about 2 percent by volume), and many of the denaturants used are toxic. Similar to gasoline, ethanol is a highly flammable liquid and must be transported carefully.

Ethanol Can Reduce Pollution

Ethanol and ethanol–gasoline mixtures burn cleaner and have higher octane levels than pure gasoline, but they also have higher evaporative emissions from fuel tanks and dispensing equipment. These evaporative emissions contribute to the formation of harmful, ground-level ozone and smog. Gasoline requires extra processing to reduce evaporative emissions before it is blended with ethanol.

Biomass/Biopower

Biomass is a renewable energy source that comes from organic material from plants and animals (Figure 11.20). It contains energy stored from the sun. Plants absorb the sun's energy through photosynthesis. Then, when biomass is burned, the chemical energy is released as heat. Biomass can be burned directly or converted to liquid biofuels and biogas that are burned as fuels. The following are examples of biomass and their uses for energy:

- Wood and wood processing wastes: These are burned to heat buildings, to produce process heat in industry, and to generate electricity.
- Animal manure and human sewage: These are converted to biogas, which can be burned as a fuel.
- Agricultural crops and waste materials: These are burned as a fuel or converted to liquid biofuels.
- Food, yard, and wood waste in garbage: These are burned to generate electricity in power plants or converted to biogas in landfills.

Burning biomass is only one of the ways to release its energy. It can also be converted to other usable forms of energy—such as methane gas or transportation fuels such as ethanol and biodiesel. Biomass can also produce methane gas, which is a

FIGURE 11.20 Examples of biomass. (From DOE.)

component of landfill gas or biogas that forms when garbage, agricultural waste, and human waste decompose in landfills or digesters (special containers for that purpose). Biomass fuels provide 5 percent of the energy used in the United States in 2015. Of that amount, about 43 percent originated from wood, 46 percent from biofuels (primarily ethanol), and 11 percent from municipal waste. This is still an area where research is currently being conducted in the United States.

A unique and remarkable country for renewable energy is Sweden. It has made noteworthy strides in achieving independence from fossil fuel-related energy. In 2009, the Swedish government approved a plan to have renewable energy reach 50 percent of the total energy consumed in the country by the year 2020. Its goal is to also be totally independent of imported fossil fuels in its transportation sector by 2030 (Renewable Energy World, 2010). The total energy consumption generated from biomass in Sweden grew from 88 terawatt hours (TWh) to 115 TWh between 2000 and 2009, while the usage of oil-based products declined from 142 TWh to 112 TWh during the same period, according to Svebio (its Swedish Bioenergy Association). Biomass surpassed oil to become the number one source for energy generation in 2009, which accounted for 32 percent of the total energy consumption in the country. Their goal is to keep increasing this usage trend.

Three types of energy generation from biomass are: (1) wood and wood waste; (2) waste-to-energy (MSW); and (3) landfill gas and biogas. The next sections will describe these three.

Wood and Wood Waste

People have used wood for cooking, for heat, and for light for centuries. In fact, wood was the main source of energy for the

FIGURE 11.21 Biomass. (Courtesy of Kerr.)

world until the mid-1800s. Wood continues to be an important fuel in many countries, especially for cooking and heating in developing countries. In 2015, approximately 2 percent of total U.S. annual energy consumption was from wood and wood waste (bark, sawdust, wood chips, wood scrap, and paper mill residues). Industry, electric power producers, and commercial businesses use most of the wood and wood waste fuel consumed in the United States. The wood and paper products industry uses wood waste to produce steam and electricity, which saves money because it reduces the amount of other fuels and electricity that must be purchased. Some coal-burning power plants burn wood chips to reduce sulfur dioxide emissions. About 20 percent of total U.S. wood energy consumption in 2015 was by the residential sector, and wood accounted for about 3 percent of total residential energy consumption.

Wood is used in homes throughout the United States for heating wood in fireplaces and wood-burning appliances and as pellets in pellet stoves. In 2012, about 2.5 million U.S. households used wood as the main heating fuel. An additional nine million households used wood as a secondary heating fuel (Figure 11.21).

Waste-to-Energy

Municipal solid waste (MSW), often called garbage, is used to produce energy at waste-to-energy plants and at landfills in the United States. MSW contains biomass (or biogenic) materials such as paper, cardboard, food waste, grass clippings, leaves, wood, leather products, and other nonbiomass combustible materials such as plastics and other synthetic materials made from petroleum.

In 1960, the average U.S. citizen threw away 1.2 kilograms of trash per day. Today, the average U.S. citizen throws away about 2 kilograms of trash every day. Of those 2 kilograms, about 34 percent is recycled or composted, and about 13 percent is burned and converted to energy. The rest, about 53 percent, is discarded, mostly into landfills. About 85 percent of household trash is material that will burn, and about 61 percent of that is biogenic—material that is made from biomass (plant or animal products).

MSW is burned at special waste-to-energy plants that use the heat to make steam to generate electricity or to heat buildings. In 2013, there were about 80 waste-to-energy plants in the United States that generated electricity or produced steam. These plants burned about 27 million metric tons of MSW in 2013, and generated nearly 14 billion kilowatt-hours of electricity—about the same amount used by 1.3 million U.S. households in 2013. Many large landfills also generate electricity by using the methane gas that is produced as biomass decomposes in the landfill. Producing electricity is only one good reason to burn MSW. Burning waste also reduces the amount of material that would probably be buried in landfills. Burning MSW reduces the volume of waste by about 87 percent.

Landfill Gas and Biogas

Landfills for MSW can also serve as a source of energy. Anaerobic bacteria that live in landfills decompose organic waste to produce a gas called biogas. Biogas contains methane. Methane is the same energy-rich gas found in natural gas, which is used for heating, cooking, and producing electricity.

Landfill biogas can be dangerous to people and the environment because methane is flammable, and it is an extremely strong GHG. In the United States, there are rules requiring landfills to collect methane gas for safety reasons and to reduce GHG emissions. Some landfills control the methane gas emissions simply by burning or flaring methane gas. Methane gas can also be used as an energy source. Many landfills collect biogas, treat it, and then sell the methane. Some landfills use the methane to generate electricity.

Farmers produce biogas in large tanks called digesters where they place manure and bedding material from their barns. They cover their manure ponds (also called lagoons) to capture biogas. Biogas digesters and manure ponds contain the same anaerobic bacteria found in landfills. The methane in the biogas can be used for heating and can also be used for generating electricity on the farm.

Bioproducts

Whatever products that can be made from fossil fuels, can also be made using biomass. These bioproducts—also called biobased products—are not only made from renewable sources, they also often require less energy to produce than petroleum-based products. The process for making biofuels—releasing the sugars that make up starch and cellulose in plants—also can be used to make antifreeze, plastics, glues, artificial sweeteners, and gel for toothpaste.

Other important building blocks for bioproducts include carbon monoxide and hydrogen. When biomass is heated with a small amount of oxygen present, these two gases are produced in abundance. This mixture is called biosynthesis gas. Biosynthesis gas can be used to make plastics and acids, which can be used in making photographic films, textiles, and synthetic fabrics. When biomass is heated in the absence of oxygen, it forms pyrolysis oil. A chemical called phenol can be extracted from pyrolysis oil. Phenol is used to make wood adhesives, molded plastic, and foam insulation.

The biggest different between a bioproduct and a conventional product is that bioproducts look at the entire product life cycle and are made from plant-derived resources. Petroleum-based products are made from nonrenewable resources and do not focus on their life cycle after use. The most commonly used plant resources utilized in the production of bioproducts are soybeans and corn, but products are certainly not limited to just those plant resources. The following are also used to produce products:

- Sunflower
- Canola
- Miscanthus
- Mycelium (the vegetative part of fungus)
- Switchgrass
- Algae
- Forest-derived materials
- Sugarcane
- Flax
- Potatoes
- Wheat

While concern is often expressed that if bioproducts are made from food plants, such as corn and wheat, it will affect the world's food supply (Steer and Hanson, 2015), the fact is that many bioproducts are made from agricultural, forestry, and biological wastes. For example, one agricultural waste that can be used in producing bioproducts is wheat straw. Once wheat has been harvested and the grain and chaff have been removed, a straw product is left over. While this wheat straw could be used for animal bedding or even making baskets, it also has the potential to be used in creating composite lumber or be utilized as a biofuel.

One thing to keep in mind, as well, is that not all bioproducts may be biodegradable. This usually depends on the product's purpose. If it is a product that is meant to be long lasting, it is manufactured to be nonbiodegradable, which makes sense. After all, if you used a biobased exterior paint, you would expect that it would not be biodegradable; and it would not be. Bioproducts will have this information listed on their product labels. It is surprising how many different products are available today that are bioproducts. The following list shows some of the more common available bioproducts:

- Household cleaners
- Paints and stains
- Personal care items
- Plastic bottles and containers
- Packaging materials
- Office supplies
- Soaps and detergents
- Lubricants
- Clothing
- Plates, napkins, and cutlery
- Building materials
- Vaccines, antibiotics, and drugs
- Cosmetics
- Paper

As more research is done in this field, new discoveries are made and new applications are developed. This continues to lessen our dependence on fossil fuels and moves us further into the renewable energy arena. It is fascinating to think that today we can wear T-shirts and sleep on pillows that are made from corn products. Even better is the idea that these new items are compostable and biodegradable. This will reduce the amount of plastic products in landfills, as well as produce less waste during their production. As scientific advancements are made, the environment continues to get a little greener.

SECONDARY ENERGY SOURCES

A secondary energy source is one that is made using a primary resource. It can be an energy source that is generated by a number of different primary sources. Secondary energy sources are also referred to as energy carriers because they move energy in a useable form from one place to another. There are two well-known energy carriers:

- Electricity
- Hydrogen

For many energy needs, it is much easier to use electricity or hydrogen than it is to use the primary energy sources themselves.

Electricity

Electricity is the flow of electrical power or charge. It is both a basic part of nature and one of the most widely used forms of energy. It is classified as a secondary energy source, and it is also referred to as an energy carrier. That means that consumers use energy in the form of electricity, which is produced from the conversion of other sources of energy, such as coal, natural gas, nuclear, solar, hydro, or wind energy. The energy sources used to make electricity can be renewable or nonrenewable, but electricity itself is neither renewable nor nonrenewable.

Before electricity became available more than a century ago, other means were used to illuminate homes, such as candles and kerosene lamps. Food was cooled in iceboxes, and rooms were warmed by wood-burning or coal-burning stoves. The discovery of electricity came about as a long process of small discoveries over a period of more than 2000 years. In 600 BCE, the Ancient Greek civilization discovered that rubbing fur on amber (fossilized tree resin) caused an attraction between the two—the discovery of static electricity. Later discoveries of pots with sheets of copper inside later led researchers and archaeologists in the 1930s to believe that the ancient Romans may have developed some crude batteries and were able to produce light during their rule. Surprisingly, similar devices have been found in archaeological digs near Baghdad, which indicates that the ancient Persians may have also used an early form of batteries to produce light.

By the seventeenth century, several electricity-related discoveries had been made, such as the invention of an early electrostatic generator, the differentiation between positive and negative currents, and the classification of materials as conductors or insulators. A conductor is a material through which an electrical current may pass; an insulator is a substance that does not readily conduct electricity. In the early 1600s English scientist, Thomas Browne, was conducting carefully planned "electricity" experiments. Then in 1752, Benjamin Franklin conducted his famous experiment with a kite, a key, and a storm. His experiment only proved that lightning and tiny electric sparks were the same thing. Following him, Italian physicist Alessandro Volta discovered that particular chemical reactions could produce electricity, and in 1800, he constructed the voltaic pile that produced a steady electric current, and so he was the first person to create a steady flow of electric charge. He was also the first to create a transmission of electricity by linking positively-charged and negatively-charged connectors and driving an electrical change or voltage through them.

Electricity became available for use in technology in 1831, when Michael Faraday created the electric dynamo (which is a crude power generator), which enabled an electric current to be generated in a practical way. This opened the way to the U.S. scientist Thomas Edison and UK scientist Joseph Swan (both invented the incandescent filament light bulb around 1878) to set up a joint company to produce the first practical filament lamp. Edison employed his direct current (DC) system to provide power to illuminate first electric street lamps in New York in 1882. Following this, in the late 1800s and early 1900s Serbian American engineer, inventor, and electrical wizard Nikola Tesla, opened the way for commercial electricity. Prior to 1879, DC electricity had been used in arc lights for outdoor lighting. In the late 1800s, Nikola Tesla pioneered the generation, transmission, and use of alternating current (AC) electricity, which reduced the cost of transmitting electricity over long distances. Tesla's inventions used electricity to bring indoor lighting to homes and used electricity to power industrial machines.

Following his contributions, American inventor and industrialist George Westinghouse purchased and developed Tesla's patented motor for generating AC, and the work of Westinghouse, Tesla, and others gradually convinced the U.S. population that the future lie in AC rather than DC.

Others who worked to bring the use of electricity to where it is today include Scottish inventor James Watt, Andre Ampere, a French mathematician, and German mathematician and physicist George Ohm. Despite the fact electricity is a secondary energy source, it is one of the most important energy-related sources we have. Without it, we could not carry on the activities we do each day in our normal routines; it is of such great importance. People do not often appreciate its importance and necessity until it is not available. We use electricity every day to do many jobs—from lighting, heating, and cooling homes to powering televisions, computers, and many other functions in between.

The Science of Electricity

In order to gain an understanding of electricity, it is important to understand the dynamics of atoms. Atoms are the building blocks of the universe. Everything in the universe is made of atoms—every star, shrub, and creature. The human body is made of atoms. Air and water are also made of atoms. Atoms are so small that millions of them would fit on the head of a pin. The center of an atom is called the *nucleus*. The nucleus is made of particles called *protons* and *neutrons*. *Electrons* spin around the nucleus in shells a great distance from the nucleus. If the nucleus was the size of a tennis ball, the atom would be the size of a sphere about 1,450 feet in diameter, or about the size of the largest event stadium in the world. Atoms are mostly empty space. Electrons are held in their shells by an electrical force (Figure 11.22).

The protons and electrons of an atom are attracted to each other. They both carry an *electrical charge*. Protons have a positive charge (+) and electrons have a negative charge (−). The positive charge of the protons is equal to the negative charge of the electrons. Opposite charges attract each other. An atom is in balance when it has an equal number of protons and electrons. The neutrons carry no charge and their number can vary.

The number of protons in an atom determines the kind of atom, or *element*, it is. An element is a substance consisting of one type of atom (the Periodic Table shows all the known elements) (*see Appendix*), all with the same number of protons. Every atom of hydrogen, for example, has one proton, and every atom of carbon has six protons. The number of protons determines which element it is.

Electrons usually remain at a constant distance from the nucleus in precise shells. The shell closest to the nucleus can hold two electrons. The next shell can hold up to eight electrons. The outer shells can hold more. Some atoms with many protons can have as many as seven shells with electrons in

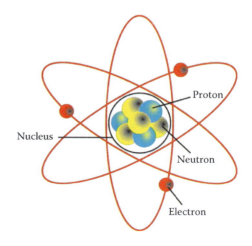

FIGURE 11.22 Schematic of an atom, illustrating the main parts: the nucleus, with its protons and neutrons; and the electrons that orbit around the nucleus in different shells. (Courtesy of Kerr.)

them. The electrons in the shells closest to the nucleus have a strong force of attraction to the protons. Sometimes, the electrons in an atom's outermost shells do not have a strong force of attraction to the protons. These electrons can be pushed out of their orbits. Applying a force can make them move from one atom to another. These moving electrons are electricity.

Lightning is a form of electricity. It is caused by electrons moving from one cloud to another or electrons jumping from a cloud to the ground. Any time you have felt a shock when you touched an object after walking across a carpet is due to static electricity. This is caused by a stream of electrons jumping from you to the object you touched. When someone rubs their head with a balloon and causes their hair to stand straight up, this is caused by rubbing some electrons off the balloon. The electrons moved into their hair from the balloon, and then tried to get far away from each other by moving to the ends of the hair. They pushed against or repelled each other and made the hair move.

Batteries, Circuits, and Transformers

The spinning of the electrons around the nucleus of an atom creates a tiny magnetic field. The electrons in most objects spin in random directions, and their magnetic forces cancel each other out. Magnets are different, because the molecules in magnets are arranged so that their electrons spin in the same direction. This arrangement and movement create a magnetic force that flows from a north-seeking pole and a south-seeking pole. The magnetic force creates a *magnetic field* around a magnet.

Magnets do not act like most objects. If you try to push the same poles together, they repel each other. But if you put different poles together, the magnets will stick together because the north and south poles are attracted to each other. Just like protons and electrons—opposites attract with magnets. These properties can be used to generate electricity. Moving magnetic fields can pull and push electrons. Metals such as copper have electrons that are loosely held, so the electrons in copper wires can easily be pushed from their shells by moving magnets. By using moving magnets and copper wire together, electric generators create electricity. Electric generators convert kinetic energy (the energy of motion) into electrical energy.

A chemical reaction between the metals and the chemicals frees more electrons in one metal than it does in the other. One end of the battery is attached to one of the metals, and the other end is attached to the other metal. The end that frees more electrons develops a positive charge, and the other end develops a negative charge. If a wire is attached from one end of the battery to the other, electrons flow through the wire to balance the electrical charge. An electrical load is a device that uses electricity to do work or to perform a job. If an electrical load—such as an incandescent light bulb—is placed along the wire, the electricity can do work as it flows through the wire. Electrons flow from the negative end of the battery through the wire to the light bulb. The electricity flows through the wire in the light bulb and back to the positive end of the battery.

Electricity must have a complete path, or electrical circuit, before the electrons can move. If a circuit is open, the electrons cannot flow. When a light switch is turned on, a circuit is closed. The electricity flows from an electric wire, through the light bulb, and back out another wire. The light bulb produces light as electricity flows through a tiny wire in the bulb, gets very hot, and glows. When the switch is turned off, the circuit is opened and no electricity flows to the light. The light bulb burns out when the tiny wire breaks and the circuit is opened. When a television is turned on, electricity flows through wires inside the television, producing pictures and sound. Sometimes electricity runs motors, such as in washers or appliances.

To solve the problem of sending electricity over long distances, U.S. physicist William Stanley developed a device called a transformer. A transformer allows electricity to be efficiently transmitted over long distances. A transformer also makes it possible to supply electricity to homes and businesses located far from electric generating plants. A transformer changes the voltage of electricity in power lines. The electricity produced by a generator travels along cables to a transformer, which then changes electricity from low voltage to high voltage. Electricity can be moved long distances more efficiently using high voltage. Transmission lines are used to carry the electricity to a substation. Substations have transformers that change the high-voltage electricity into lower voltage electricity. From the substation, distribution lines carry the electricity to homes, offices, and factories that all use low voltage electricity. Electricity is measured in units of power called Watts. A Watt is the unit of electrical power equal to one ampere under a pressure of 1 volt. One Watt is a small amount of power. Some devices require only a few watts to operate, and other devices require larger amounts. The consumption of small devices is usually measured in Watts, and the consumption of larger devices is measured in kilowatts (kW), or 1,000 Watts. Electricity generation capacity is typically measured in multiples of kW, such as megawatts (MW) and gigawatts (GW). One MW is 1,000 kW, and one GW is 1,000 MW.

A Watthour is equal to the energy of one Watt supplied to, or taken from, an electric circuit steadily for one hour. The amount of electricity a power plant generates or the amount of electricity an electric utility customer uses over a period of time is typically measured in kilowatt-hours (kWh), which is the number of kilowatts generated or consumed in one hour. For example, if you use a 60-Watt (0.06 kilowatt) light bulb for five hours, you have used 300 Watthours, or 0.3 kilowatt-hours, of electrical energy.

Electric utilities measure the electricity consumption of their customers with meters that are usually located on the outside of the customer's property where the power line enters the property. In the past, all electricity meters were mechanical devices that had to be read manually by a utility employee. Eventually automated reader devices were developed that reported electricity use from mechanical meters with an electronic signal on a periodic basis. Now, electronic smart meters can be used to measure electricity consumption on a real-time basis and can provide access to the data using wireless networks and the Internet. Some smart meters can even measure the use of individual devices and allow the utility or customer to control electricity use remotely.

The Generation of Electricity

A generator is a device that converts a form of energy into electricity. Nearly all of the electricity consumers' use is produced by generators that convert kinetic (mechanical) energy into electrical energy. Generators operate because of the relationship between magnetism and electricity. In 1831, scientist Michael Faraday discovered that when a magnet is moved inside a coil of wire, an electric current flows in the wire.

The most widely used method of producing electricity uses generators with an electromagnet—a magnet produced by electricity—not a traditional magnet. The generator has a series of insulated coils of wire that form a stationary cylinder. This cylinder surrounds a rotary electromagnetic shaft. When the electromagnetic shaft rotates, it induces a small electric current in each section of the wire coil. Each section of the wire coil then becomes a small, separate electric conductor. The small currents of the individual sections are added together to form one large current. This current is the electricity that is transmitted from generators to consumers.

An electric power plant uses a turbine or other similar machine to drive these types of generators. There are steam turbines, gas combustion turbines, water turbines, and wind turbines, depending on what the original energy source is. Steam turbines using biomass, coal, geothermal energy, natural gas, nuclear energy, and solar thermal energy produce about 70 percent of the electricity used in the United States. These power plants are about 35 percent efficient. This means that for every 100 units of primary heat energy that go into a power plant, only 35 units are converted to useable electrical energy. Other types of devices that generate or produce electricity include electrochemical batteries, fuel cells, solar PV cells, and thermoelectric generators.

Most of the electricity in the United States is produced using steam turbines. A turbine converts the kinetic energy of a moving fluid (liquid or gas) to mechanical energy. In a steam turbine, steam is forced against a series of blades mounted on a shaft. The steam rotates the shaft connected to the generator. The generator, in turn, converts its mechanical energy to electrical energy based on the relationship between magnetism and electricity. In steam turbines powered by fossil fuels (coal, petroleum, and natural gas), the fuel is burned in a furnace to heat water in a boiler to produce steam.

Most of the electricity in the United States is generated using fossil fuels. In 2015, coal was used for about one-third of the 4 trillion kilowatt-hours of electricity generated in the United States. In addition to being burned to heat water for steam, natural gas can also be burned to produce hot combustion gases that pass directly through a natural gas turbine, spinning the turbine's blades to generate electricity. Natural gas turbines are commonly used when electricity use is in high demand. In 2015, nearly one-third of U.S. electricity was fueled by natural gas.

Petroleum can be burned to produce hot combustion gases to turn a turbine or to make steam that turns a turbine. Residual fuel oil and petroleum coke, products from refining crude oil, are the main petroleum fuels used in steam turbines. Distillate (or diesel) fuel oil is used in diesel-engine generators. Petroleum was used to generate less than 1 percent of all electricity in the United States in 2015.

Nuclear power plants produce electricity with nuclear fission to create steam that spins a turbine to generate electricity. Most U.S. nuclear power plants are located in states east of the Mississippi River. Nuclear power was used to generate nearly one-fifth of all U.S. electricity in 2015.

All renewable energy sources combined provided 13 percent of electricity generated in the United States in 2015. Hydropower contributed of about 6 percent of U.S. electricity generation in 2015. Most hydropower is produced at large facilities built by the federal government, like the Hoover Dam. The West has many of the largest hydroelectric dams, but there are many hydropower facilities operating all around the country. Wind power has increased significantly in the United States since 1970 and provided almost 5 percent of U.S. electricity generation in 2015. Biomass, which includes lumber and paper mill wastes, food scraps, grass, leaves, paper, garbage, wood chips, corn cobs, and wheat, can be burned directly in steam-electric power plants, or they can be converted to a gas that can be burned in steam generators, gas turbines, or internal combustion engine generators. Biomass accounted for about 2 percent of the electricity generated in the United States in 2015. Both solar and geothermal power generated about one percent of the electricity in the United States in 2015.

Delivery and Uses of Electricity

Electricity is generated at power plants and then transmitted through a grid—which is a very complex network—of electricity substations, power lines, and distribution transformers before it reaches individual consumers. In the United States, the entire grid consists of more than 7,300 power plants, almost 257,495 kilometers of high-voltage power lines, and millions of low-voltage power lines and distribution transformers connecting about 145 million customers (Figure 11.23). The electricity that arrives at a home or other building can come from several sources. The utility, distribution company, or retail service provider selling the power may be a not-for-profit municipal entity; an electric cooperative owned by its members; a private, for-profit company owned by stockholders (often called an investor-owned utility); or a power marketer. Some federally owned authorities—such as the Tennessee Valley Authority—also buy, sell, and distribute power.

The origin of the electricity that customers consume can vary. Utilities may generate all the electricity they sell using just the power plants they own, or they can also purchase some of their supply on the wholesale market from other utilities, power marketers, independent power producers, or from a market based on membership in a regional transmission

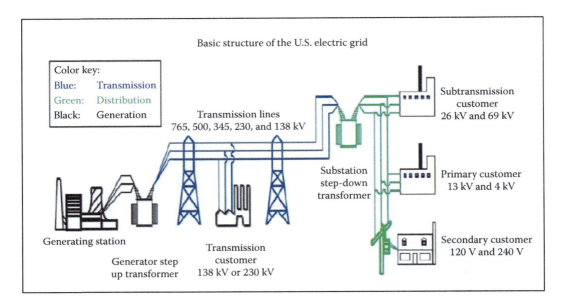

FIGURE 11.23   An electrical grid illustrating how electricity is delivered to its endpoint from its generation point. (Courtesy of Department of Energy [DOE].)

reliability organization. Most of the existing grid in the United States was built during a highly-structured, highly regulated era. The existing grid was designed to ensure that everyone in the United States had reasonable access to electricity service. Utility customers, through fees authorized and regulated by state regulatory commissions, generally pay for developing and maintaining the grid.

Many local grids are interconnected for reliability and commercial purposes. They also form larger, more dependable networks so that they can better maximize the coordination and planning of electricity supply. These networks can be big enough to extend throughout many states. The North American Electric Reliability Corporation (NERC) was established to ensure that the grid in the United States was reliable, adequate, and secure. Some NERC members have formed regional organizations with similar missions. These regional organizations are referred to as Independent System Operators (ISOs) and Regional Transmission Organizations (RTOs). They are part of a national standard design advocated by the Federal Energy Regulatory Commission (FERC). Some organizations have members who connect to lines in Canada or Mexico. Most organizations, depending on the location and the utility, are indirectly connected to dozens and often hundreds of power plants. Some electricity consumed in the United States is imported from Canada and Mexico (Figure 11.24).

Electricity is a valuable commodity—society uses it constantly. In 2014, United States electricity consumption was nearly 3,862 billion kilowatt-hours (kWh). Because it is such an essential commodity—used for lighting; heating/cooling; refrigeration; operating appliances, machinery, computers, electronics; running public transportation systems, and a lengthy list of other critical applications, electricity use in the U.S. was 13 times greater in 2014 than it was in 1950. The percentage uses of the main sectors totaled:

- Residential: 36 percent
- Commercial: 35 percent
- Industrial: 28 percent (this is the direct use of electricity only)
- Transportation: 0.2 percent (mostly by public transportation)

The single largest use for electricity in the United States is lighting. In the commercial sector, lighting is used heavily for retail outlets, office environments, educational institutions, public offices and government buildings, outdoor and street lighting. Other major commercial uses of electricity are for ventilation, refrigeration, cooling, powering computers and other office equipment, and space and water heating. There are multiple uses of electricity in the public sector that include things such as powering security, medical, scientific, and research equipment; powering elevators and escalators; running cooking, laundry, and cleaning equipment; operating maintenance equipment, and so forth. The manufacturing industry uses a large amount of electricity in order to function. About half of the electricity used by U.S. manufacturers is used to drive machinery. Process and boiler heating is the next largest use of electricity, which is about 13 percent. Electricity for lighting takes about 7 percent. Use of electricity is expected to increase in the United States in the future, but as appliances and equipment become more energy efficient, the growth rate could see a decline. In the Annual Energy Outlook 2015, total U.S. electricity use is projected to grow by an average of less than 1 percent per year from 2014 to 2040.

It is projected in the Annual Energy Outlook 2015, that member countries of the Organization for Economic Cooperation and Development (OECD) consumed 51 percent of the world's total electricity supply in 2010, but their share

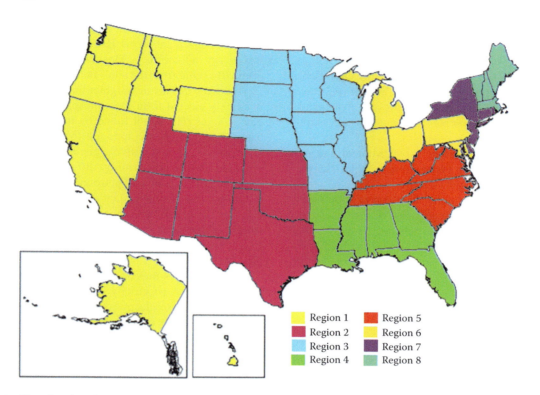

FIGURE 11.24 The electric grid zones (NERC regions) in the United States. NERC stands for North American Electric Reliability Corporation. (From U.S. Energy Information Administration.)

is projected to decline. It is projected that the non-OECD nations alone will account for 64 percent of world electricity use by 2040. The table lists the member countries of the OECD and the date they ratified the agreement.

Organization for Economic Cooperation and Development (OECD) and Ratification Date

OECD Member Country	Date of Ratification
Australia	June 7, 1971
Austria	September 29, 1961
Belgium	September 13, 1961
Canada	April 10, 1961
Chile	May 7, 2010
Czech Republic	December 21, 1995
Denmark	May 30, 1961
Estonia	December 9, 2010
Finland	January 28, 1969
France	August 7, 1961
Germany	September 27, 1961
Greece	September 27, 1961
Hungary	May 7, 1996
Iceland	June 5, 1961
Ireland	August 17, 1961
Israel	September 7, 2010
Italy	March 29, 1962
Japan	April 28, 1964
Korea	December 12, 1996
Latvia	July 1, 2016
Luxembourg	December 7, 1961
Mexico	May 18, 1994
Netherlands	November 13, 1961
New Zealand	May 29, 1973
Norway	July 4, 1961
Poland	November 22, 1996
Portugal	August 4, 1961
Slovak Republic	December 14, 2000
Slovenia	July 21, 2010
Spain	August 3, 1961
Sweden	September 28, 1961
Switzerland	September 28, 1961
Turkey	August 2, 1961
United Kingdom	May 2, 1961
United States	April 12, 1961

The cost of electricity is influenced by several factors. These include the costs to construct, finance, maintain, and operate power plants and the electricity grid (complex system of power transmission and distribution lines). There are several key factors that influence the price of electricity:

- Fuels
- Power plants
- Transmission and distribution systems
- Weather conditions
- Regulations

Fuel: Fuel costs can vary based on the per unit cost, such as dollars per ton for coal or thousand cubic feet for natural gas, and the relative cost, in dollars per million British thermal unit (Btu) equivalent. Electricity generators with relatively high fuel costs tend to be used most during periods of high demand.

Power plants: Each power plant has construction, maintenance, and operating costs.

Transmission and distribution system: Maintaining and using the transmission system to deliver electricity is an additional cost to the electricity.

Weather conditions: Rain and snow can provide water for low-cost hydropower generation. Extreme temperatures can increase the demand for electricity, especially for cooling. Severe weather can also damage power lines and add costs to maintain the electricity grid.

Regulations: In some states, prices are fully regulated by Public Service Commissions, while in other states, there is a combination of unregulated prices (for generators) and regulated prices (for transmission and distribution).

Typically, prices are higher during the summer months. The prices usually reflect the demand, availability of different generation sources, fuel costs, and power plant availability. In the summer, the total demand is high and more generation is necessary to meet the high demand. Prices also vary by customer type. Prices are usually the highest for residential and commercial consumers because it costs more to distribute electricity to those customers. Industrial consumers use more electricity and can receive it at higher voltages, so it is more efficient and less expensive to supply electricity to them. They pay close to the wholesale price. In 2014, the average retail price of electricity in the United States was 10.45 cents per kilowatt-hour (kWh). The average prices by major type of utility customers were:

Consumer Type	Price per Kilowatt Hour (kWh)
Residential	12.50 cents per kWh
Commercial	10.75 cents per kWh
Industrial	7.01 cents per kWh
Transportation	10.27 cents per kWh

Prices can also vary locally due to the availability of power plants and fuels, possible lower fuel costs in certain geographic areas, and local pricing regulations.

Hydrogen

Hydrogen is the simplest element—each atom of hydrogen has only one proton. Hydrogen is also the most plentiful gas in the universe. Stars—like the sun—are made mostly of hydrogen.

The sun, in fact, is basically a giant ball of hydrogen and helium gases. In the sun's core, hydrogen atoms combine to form helium atoms. This process—called fusion—gives off radiant energy. The radiant energy from the sun gives earth light and helps plants grow. Radiant energy is stored as chemical energy in fossil fuels. Most of the energy people use today originally came from the sun's radiant energy.

Hydrogen gas is much lighter than air, and it rises fast and is quickly ejected from the atmosphere. This is why hydrogen as a gas (H_2) is not found by itself on earth. Hydrogen gas is found only in compound form with other elements. Hydrogen combined with oxygen is water (H_2O). Hydrogen combined with carbon forms different compounds such as methane (CH_4), coal, and petroleum. It is found in all growing things and is an abundant element in the earth's crust. It also has the highest energy content of any common fuel by weight (about three times more than gasoline), but it has the lowest energy content by volume (about four times less than gasoline).

Energy carriers move energy in a useable form from one place to another. Electricity is the most well-known energy carrier. People use electricity to move the energy produced from coal, uranium, natural gas, and renewable energy sources in power plants to homes and businesses. For many energy needs, it is easier to use electricity than it is to use the energy sources themselves.

Hydrogen is an energy carrier like electricity, and it must be produced from another substance. Hydrogen can be produced from a variety of resources (water, fossil fuels, or biomass), and it is a byproduct of other chemical processes. While hydrogen is not widely used as a fuel now, it has the potential for greater use in the future.

Production of Hydrogen

Because hydrogen does not exist on earth as a gas, it must be separated from other elements. Hydrogen atoms can be separated from water, natural gas molecules, or biomass. The two most common methods used to produce hydrogen are *steam reforming* and *electrolysis* (water splitting).

Steam reforming is currently the least expensive method of producing hydrogen, and it accounts for most of the commercially produced hydrogen in the United States. This method is used in industries to separate hydrogen atoms from carbon atoms in methane (CH_4). The steam reforming process results in carbon dioxide emissions, however.

Electrolysis is another process that produces hydrogen. It splits hydrogen from water using an electric current. The process can be used on a small scale or on a large scale. Electrolysis does not produce any emissions other than hydrogen and oxygen. However, if the electricity used in the process is produced from fossil fuels, then there are pollution and carbon dioxide emissions indirectly associated with electrolysis. However, the electricity used in electrolysis can also come from renewable sources such as wind and solar.

Research is underway to develop other methods used to produce hydrogen, such as using microbes that use light to make hydrogen, converting biomass into liquids and separating the hydrogen, and using solar energy technologies to split hydrogen from water molecules.

Uses and Availability of Hydrogen

Nearly all of the hydrogen consumed in the United States is used by industry for refining petroleum, treating metals, producing fertilizer, and processing foods. The National Aeronautics and Space Administration (NASA) is the largest user of hydrogen as a fuel. NASA began using liquid hydrogen in the 1950s as a rocket fuel, and NASA was one of the first to use fuel cells to power the electrical systems on space craft.

Hydrogen fuel cells produce electricity by combining hydrogen and oxygen atoms. This combination results in an electrical current. Hydrogen fuel cells are efficient, but expensive to build. There are many different types of fuel cells that can be used for a wide range of applications. Small fuel cells have been developed to power laptop computers, cell phones, and military applications. Large fuel cells can provide electricity for emergency power in buildings and in remote areas that do not have power lines. Hydrogen use in vehicles is a major focus of fuel cell research and development.

The interest in hydrogen as an alternative transportation fuel is based on hydrogen's ability to power fuel cells in zero-emission electric vehicles, its potential for domestic production, and the fuel cell vehicle's potential for high efficiency. There are about 500 hydrogen-fueled vehicles in use in the United States. Most hydrogen-fueled vehicles are buses and automobiles with an electric motor powered by a fuel cell. A few of these vehicles burn hydrogen directly. The high cost of fuel cells and the limited availability of hydrogen have limited the number of hydrogen-fueled vehicles.

There are about 40 hydrogen refueling stations for vehicles in the United States. About 12 are available for public use, nearly all of which are located in California. Production of hydrogen cars is limited because people will not buy hydrogen cars if there are no refueling stations, and companies will not build refueling stations if there are no cars and no customers. In May 2014, the California Energy Commission started a $46.6 million program to help fund the development of 28 publicly accessible hydrogen refueling stations in California to promote a consumer market for zero-emission fuel cell vehicles. Once the infrastructure becomes more developed and both hydrogen cars and refueling stations can be supported, this industry could become highly successful in the future.

RENEWABLE PROGRAMS AND INCENTIVES

Today, it is encouraged for both the commercial and residential sectors to consider investing in renewable energy. To go along with this, there are several programs and incentives currently available to help financially assist those interested in becoming involved in investing in renewable resources. For example, the following represent a few:

- Government financial incentives
- Renewable portfolio standards (RPSs) and state mandates or goals
- Renewable Energy Certificates (RECs)
- Net metering programs
- Feed-in tariffs
- Green power programs
- Ethanol and other renewable motor fuels
- Renewables Research and Development

For government financial incentives, several federal government tax credit, grant, and loan programs are available for qualifying renewable energy technologies and projects, such as the Renewable Energy Production Tax Credit (PTC), the Business Energy Investment Tax Credit (ITC), personal income tax credit, and grant and loan programs available from several government agencies, including the U.S. Department of Agriculture, the U.S. DOE, and the U.S. Department of the Interior. Every state has some type of financial incentive to support or subsidize the installation of renewable energy equipment, including grants, loans, rebates, and tax credits.

A RPS generally requires that a percentage of electric power sales come from renewable energy sources. Some states have specific mandates for power generation from renewable energy, while others have voluntary goals. Most states have RPSs and goals, as shown in the illustration (Figure 11.25). Compliance with RPS policies can sometimes require or allow for the trading of RECs. RECs, also known as green certificates, green tags, or tradable renewable certificates, are financial products that are sold, purchased, and traded. These financial products allow the purchaser to pay for renewable generation without the need for physical or contractual delivery of electricity generated from qualifying energy sources.

Net metering allows electric utility customers to install qualifying renewable energy systems that are connected to an electric utility's distribution system (or grid) on their property to offset their use of electricity from the grid. The programs vary, but in general, electric utility customers are billed for the net amount of the grid-supplied electricity that they use. The net amount is the customer's total electricity consumption minus the amount that is generated by their renewable system. In some states, customers can sell excess electricity generated by their system to the utility. Forty-three states and the District of Columbia have statewide or district wide net metering programs that apply to all or certain types of electric utilities, such as solar panels.

Several states and individual electric utilities in the United States have established special rates for purchasing electricity from certain types of renewable energy systems. These rates, sometimes known as feed-in tariffs (FITs), are generally higher than retail electricity rates to encourage new projects of specific types of renewable energy technologies. Consumers in nearly every state can purchase green power, which represents electricity generated from specific types of renewable energy resources, such as wind power. Most of these programs generally involve the physical or contractual delivery of the generation resource to the customer or utility.

There are several federal and state requirements and incentives for the production, sale, and use of ethanol, biodiesel, and other fuels made from biomass. The federal Energy

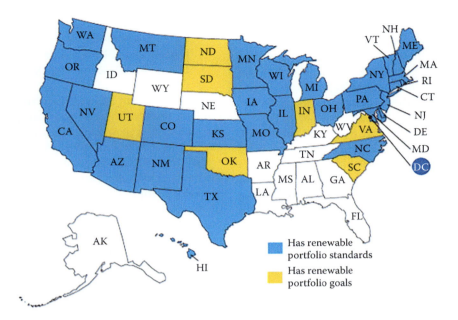

FIGURE 11.25 Statewide renewable portfolio standards and goals, 2015. (From EIA.)

Independence and Security Act of 2007 requires that 36 billion gallons of biofuels be used in the United States per year by 2022. Several states have their own renewable fuel standards or requirements. There are other federal programs that provide financial support and incentives for ethanol and other biofuels producers. Many states have their own programs that support or promote the use of biofuels (U.S. Energy Information Association, 2015).

The DOE and other federal government agencies, fund research and development of renewable energy technologies. Most of the research and development is carried out at the National Labs and in cooperation with academic institutions and private companies. The availability of these programs depends on annual appropriations from the United States Congress. Information on the portfolio standards—Database of State Incentives for Renewable Energy and Efficiency (DSIRE)—can be found online at: http://www.dsireusa.org/. With all the incentives available today, many are taking advantage of going greener in their personal lives—a wise choice indeed.

CONCLUSIONS

After a careful look at all the renewable energy options available today, coupled with the fact that prices are coming down in terms of being able to personally invest in renewable energy options, the benefits of renewable energy should be apparent on many levels. Renewable energy technologies are clean sources of energy that have a much lower environmental impact than conventional energy technologies and are forward-looking in nature, able to provide for future generations. In addition, it never runs out; whereas, nonrenewables are facing a finite future. Renewables are also a healthy approach for the job market and the economy. Most renewable energy investments

are spent on materials and workmanship to build and maintain the facilities, rather than on costly energy imports. Renewable energy investments are usually spent within the country, and more often than not—locally. This means your energy dollars stay home to create jobs and fuel local economies, rather than going overseas, keeping you and your community employed. Renewables also mean energy security—a very important issue with today's unstable political climate. In the past few decades, the United States and other countries have increased their dependence on foreign oil supplies instead of decreasing it. This impacts more than just national energy policies for those countries. This impacts those nations' energy securities, which continue to be threatened by their dependency on fossil fuels. Because these conventional energy sources are vulnerable to political instabilities, trade disputes, embargoes, and other disruptions, it puts countries that are currently caught in this situation at extreme risk constantly. The only way to eliminate this constant sense of tension is to break that tie and transition to renewables. The time for transition is now.

REFERENCES

Bhattarai, K., W. M. Stalick, S. McKay, G. Geme, and N. Bhattarai. 2011. Biofuel: An alternative to fossil fuel for alleviating world energy and economic crises. *Journal of Environmental Science and Health. Part A, Toxic/Hazardous Substances & Environmental Engineering* 46(12):1424–1442. (accessed September 28, 2016).

Renewable Energy World. 2016. Solar hot water. *Renewable Energy World.* http://www.renewableenergyworld.com/solar-energy/tech/solarhotwater.html (accessed October 8, 2016).

Renewable Energy World. June 2, 2010. Biomass generates 32% of all energy in Sweden. http://www.renewableenergyworld.com/articles/2010/06/biomass-generates-32-of-all-energy-in-sweden.html (accessed October 5, 2016).

Steer, A. and C. Hanson. January 29, 2015. Biofuels are not a green alternative to fossil fuels. *The Guardian.* https://www.the-guardian.com/environment/2015/jan/29/biofuels-are-not-the-green-alternative-to-fossil-fuels-they-are-sold-as (accessed September 28, 2016).

United States Energy Information Administration. 2016. Today in energy. https://www.eia.gov/todayinenergy/detail.php?id=26212 (accessed May 21, 2017)

U.S. Energy Information Association. 2015. *Annual Energy Outlook 2015, With Projections to 2040.* http://www.eia.gov/forecasts/aeo/pdf/0383 (2015).pdf (accessed October 7, 2016).

SUGGESTED READING

Boyle, G. 2012. *Renewable Energy: Power for a Sustainable Future,* 3rd ed. Oxford, U.K.: Oxford University Press, 584 p.

Brown, L. 2015. *The Great Transition: Shifting from Fossil Fuels to Solar and Wind Energy,* 1st ed. New York: W. W. Norton & Company, 192 p.

Department of Energy. 2016. Renewable energy. http://www.energy.gov/science-innovation/energy-sources/renewable-energy (accessed October 8, 2016).

Everett, B. et al. (Editors). 2012. *Energy Systems and Sustainability: Power for a Sustainable Future,* 2nd ed. Oxford, London: Oxford University Press, p. 672.

Milano, J., C.O. Hwai, H.H. Masjuki, W.T. Chong, M.K. Lam, P.K. Loh, and V. Vellayan. 2016. Microalgae biofuels as an alternative to fossil fuel for power generation. *Renewable and Sustainable Energy Reviews* 58:180–197. (accessed September 28, 2016).

National Renewable Energy Laboratory. 2016. Learning about renewable energy. U.S. Department of Energy. http://www.nrel.gov/workingwithus/learning.html (accessed October 8, 2016).

U.S. Energy Information Center, September, 2016.

Witte, G. October 15, 2016. In Scotland, gusts of wind usher in a quiet energy revolution. *The Washington Post.* https://www.washingtonpost.com/world/europe/in-scotland-gusts-of-wind-usher-in-a-quiet-energy-revolution/2016/10/15/e5da2f5a-8a57-11e6-8cdc-4fbb1973b506_story.html?utm_term=.fef571e9cf8b (accessed May 21, 2017).

12 Consequences of Energy Sources on the Environment

OVERVIEW

This chapter focuses on the consequences we face through the use of energy resources: consequences on the natural environment, on resources, economic consequences, and why oftentimes we must make informed choices that may not be ideal, but are the best choices for the bigger picture.

INTRODUCTION

For millennia, it seemed that humans had been able to live in their environments and make do with what they had. They were able to live off the land, use the resources available to them, and take care of what they had, because they realized the importance it held in their lives and the direct link it had to their survival. Then a significant change happened during the last few centuries where the focus was shifted to intensive agriculture and then the Industrial Revolution occurred. Discoveries were made, what was manual before became mechanized, and our world and the way we looked at it changed forever. During this time, populations worldwide began to multiply in size, and unparalleled demands on the earth's natural resources exploded like never before; and with it came consequences.

DEVELOPMENT

There are always consequences on a society from energy consumption, whether it is a large first-world country with cutting-edge technology, a third-world country just beginning to develop its technology and infrastructure, or a country developmentally somewhere in between. All energy uses and developments have impact. We are going to take a look at what those impacts are and how we, as a world civilization, can best approach the challenges they offer in order to make the best-informed decisions possible.

CONSEQUENCES OF ENERGY CONSUMPTION ON SOCIETY

During the Industrial Revolution and this time of notable transition, lifestyles and roles changed significantly. Populations grew rapidly, and with this came huge demands for energy and raw materials to produce goods and supplies. As civilizations advanced both technologically and socially, their wants and needs began to skyrocket. Enormous demands were placed on the environment and its resources. At the time, the spoils were just there for the taking—ideas of conservation were not generally considered and the landscape began to be transformed and degraded as one impact upon another was

launched at it. Before long, negative impacts came from many different directions at once through a combination of the following types of activities:

- More land was being cultivated.
- Chemical fertilizers, insecticides, and herbicides were used to increase production.
- Chemicals began polluting the air, soil, and water. Fertilizer run-offs caused toxic algal blooms that began killing aquatic animals.
- Choosing to raise just once crop (called monoculture) lowered an area's diversity and made it more vulnerable to disease.
- Traveling greater distances for food began abusing more of the environment.

With all these changes occurring while people searched for a better lifestyle and made demands on the environment to obtain it, their attention was shifted away from the traditional ideas of conservation and sustainable living. As per human nature, the more comforts people had, the more they wanted; and once they had it, the less they were willing to give up. As we all progressed together, our gas guzzling cars got more comfortable and convenient. Then the Age of Automation was born and before you knew it, everyone had central air, high-tech appliances and electronics, and all kinds of gadgets to make our lives much more comfortable. Today, it is occurring at almost breakneck speed. You might be asking about now, "what have humans done?"

The Human Condition and Mindset

Unfortunately, humans are the most polluting species on the planet. Earth is very good at recycling waste, but because of the size of our population and the amount we throw away, we are generating much more than earth can manage. It is important to keep in mind that pollution occurs at various levels and does not just impact our planet; it impacts all species, including humans, who dwell on it. We pollute in several ways: we impact earth's soil, water, and air.

Soil Pollution

Soil contamination, or soil pollution, as part of land degradation is caused by the presence of xenobiotic (human-made) chemicals or other alteration in the natural soil environment. It is typically caused by industrial activity, agricultural chemicals, or improper disposal of waste. The most common chemicals involved are petroleum hydrocarbons, polynuclear aromatic hydrocarbons (such as naphthalene and benzo(a)-pyrene), solvents, pesticides, lead, and other heavy metals.

Contamination is correlated with the degree of industrialization and intensity of chemical usage.

The concern over soil contamination centers mainly around the health risks, from direct contact with the contaminated soil, vapors from the contaminants, and from secondary contamination of water supplies within and underlying the soil (EPA, 2013). Mapping of contaminated soil sites and the resulting cleanup are time-consuming and expensive tasks, requiring extensive amounts of geology, hydrology, chemistry, computer modeling, and GIS skills in environmental science, as well as an appreciation of the history of industrial chemistry (George, 2016).

In North America and Western Europe, the extent of contaminated land is well documented, and legal frameworks are in place to deal with the problems. Developing countries tend to be less tightly regulated despite some of them having undergone significant industrialization. Pesticides, herbicides, large landfills, waste from food processing industries, and nuclear waste generated from nuclear reactors and weapons can deplete the soil of its nutrients and make it virtually lifeless and unproductive. According to the EPA:

> Usually, contaminants in the soil are physically or chemically attached to soil particles, or, if they are not attached, are trapped in the small spaces between soil particles (Patterson, 2016) (Figure 12.1).

Water Pollution

Water pollution is the contamination of water bodies (e.g., lakes, rivers, oceans, aquifers, and groundwater). This form of environmental degradation occurs when pollutants are directly or indirectly discharged into water bodies without proper treatment to remove harmful compounds. It affects the entire biosphere—all plants and organisms living in these bodies of water. In almost all cases, the effect is damaging not only to individual species and population but also to the natural biological communities.

FIGURE 12.1 Soil contamination caused by underground storage tanks containing tar. Encountered during remediation works at a disused gasworks. (Courtesy of Dumelow. Creative Commons.)

FIGURE 12.2 Garbage on the beach contributes to water pollution. It can get swept into the water and cause harm to birds, fish and other wildlife, which upsets the fragile marine ecosystem. (Courtesy of GreenLiving; Kerr.)

Effluence from industries, fertilizer run off, and oils spills all damage fragile ecosystems. According to the Water Project, "Nearly a billion people do not have access to clean and safe water in our world." Worldwatch Institute says:

> The 450 million kilograms of pesticides U.S. farmers use every year have now contaminated almost all of the nation's streams and rivers, and the fish living in them, with chemicals that cause cancer and birth defects (Cummings, 2013) (Figure 12.2).

Air Pollution

Air pollution is the introduction of particulates, biological molecules, or other harmful materials into Earth's atmosphere, causing diseases, allergies, death to humans, damage to other living organisms such as animals and food crops, or the natural or built environment. Air pollution may come from anthropogenic or natural sources. Pollutants are classified as primary or secondary. Primary pollutants are usually produced from a process, such as ash from a volcanic eruption. Other examples include carbon monoxide gas from motor vehicle exhaust or the sulfur dioxide released from factories. Secondary pollutants are not emitted directly. Instead, they form in the air when primary pollutants react or interact. Ground-level ozone is one well-known example of a secondary pollutant. Some pollutants may be both primary and secondary: they are both emitted directly and formed from other primary pollutants.

Burning fossil fuels and toxic gases produced in factories cause pollution. Air pollution infiltrates the environment and threatens the health of all who inhabit the earth. According to the United Nations:

> The estimations we have now tell us there are 3.5 million premature deaths every year caused by household air pollution and 3.3 million deaths every year caused by outdoor air pollution (GreenLiving, 2016) (Figure 12.3).

(a)   (b)

FIGURE 12.3 Two photos taken in the same location in Beijing in August 2005. (Courtesy of Bobak. Creative Commons.) The photograph (a) was taken after it had rained for two days. The photograph (b) shows smog covering Beijing in what would otherwise be a sunny day.

DID YOU KNOW?

- The generation of electric power produces more pollution than any other single industry in the United States.
- Burning fossil fuels such as coal or oil creates unwelcome by-products that pollute when released into our environment, changing the planet's climate and harming ecosystems.
- The electricity production industry in the U.S. is responsible for:
  - 62 percent of sulfur dioxide emissions that contribute to acid rain
  - 21 percent of nitrous oxides emissions that contribute to urban smog
  - 40 percent of carbon emissions that contribute to global climate change
- Nitrous oxide emissions also contribute to ground-level ozone, particulate matter (PM) pollution, haze in national parks and wilderness areas, brown clouds in major western cities, and acid deposition in sensitive ecosystems.
- Elevated ozone levels have led to the adverse health effects of smog and millions of dollars in agricultural damage.

Breaking Bad Habits

A major disadvantage of nonrenewable energy is the challenge of breaking humans of their habit of relying on it. The Union of Concerned Scientists reports that it is an uphill battle to sway consumers that the so-called "public goods" of renewable energy, such as reducing pollution for everyone, may not be enough to convince them to pay more for cleaner energy.

As countries disagree through wars and differences, the prices of nonrenewable energy such as oil have become a commodity, where price fluctuation is always eminent. The burning of fossil fuels continues to rise, producing high levels of carbon dioxide (CO_2), which climatologists believe is a major cause of climate change. Unfortunately, we have been witnessing this for a long time—a situation where sadly, only positive responses are seen when the effects of the problem seem to reach out and impact those who are not willing to make the changes. When it touches them personally, then they are willing to make sacrifices. Oftentimes, when the tension of the moment dies down, they revert right back to their previous behavior. The reality is, the bad habits will have to be not only broken but also be eventually replaced by sustainable ones.

CONSEQUENCES OF ENERGY USE ON THE ENVIRONMENT

Large or small, everything we do leaves its mark on the environment. This includes the energy choices we make. Because we cannot completely eliminate all traces of our interactions through them with the earth, it is important to understand them. It is important to understand that all energy sources have some impact on our environment. The fossil fuels—coal, oil, and natural gas—do substantially more harm than renewable energy sources by most measures, including air and water pollution, damage to public health, wildlife and habitat loss, water use, land use, and global warming emissions. It is also important, however, to understand the impacts associated with producing power from renewable sources such as wind, solar, geothermal, biomass, and hydropower. The exact type and intensity of environmental impacts vary depending on the specific technology used, the geographic location, level of construction and disturbance, and other factors. By understanding the current and potential environmental issues associated with each renewable energy source, it is possible to take measures to effectively avoid or minimize the impacts as they become a larger part of our energy supply in the future.

AIR POLLUTION AND PARTICULATES

PM, also called particle pollution, is the term for the mixture of solid particles and liquid droplets found in the air. Some particles—such as dust, dirt, soot, or smoke—are large or dark enough to be seen with the naked eye. Others are so small they can only be detected using an electron microscope. There are two types of particle pollution:

- PM10: inhalable particles, with diameters that are generally 10 micrometers and smaller
- PM2.5: fine inhalable particles, with diameters that are generally 2.5 micrometers and smaller (for comparison, a human hair is 30 times larger in diameter than these particles)

These particles come in many sizes and shapes and can be made up of hundreds of different chemicals. Some are emitted directly from a source, such as construction sites, unpaved roads, fields, smokestacks, or fires. Most particles form in the atmosphere as a result of complex reactions of chemicals such as sulfur dioxide and nitrogen oxides, which are pollutants emitted from power plants, industries, and automobiles. Therefore, they are associated with the use of energy sources; mostly, nonrenewable energy.

The concern about this PM is that it contains microscopic solids or liquid droplets that are so small that they can be inhaled and cause serious health problems when they travel deep into people's lungs. Particles less than 10 micrometers in diameter pose the greatest problems, because they can get deep into the lungs, and some may even get into the bloodstream. Therefore, exposure to such particles can affect both the lungs and heart. Numerous scientific studies have linked particle pollution exposure to a variety of health problems, including

- Premature death in people with heart or lung disease
- Nonfatal heart attacks
- Irregular heartbeat
- Aggravated asthma
- Decreased lung function
- Increased respiratory symptoms, such as irritation of the airways, coughing, or difficulty breathing

People with heart or lung diseases, children, and older adults are the most likely to be affected by particle pollution exposure.

Fine particles (PM2.5) are the main cause of reduced visibility (haze) in parts of the United States, including many of our treasured national parks and wilderness areas. One of the most basic forms of air pollution—haze—degrades visibility in many U.S. cities and scenic areas. Haze is caused when sunlight encounters tiny pollution particles in the air, which reduce the clarity and color of what we see, particularly during humid conditions. Since 1988, the federal government has been monitoring visibility in national parks and wilderness areas. In 1999, the EPA announced a major effort to improve air quality in national parks and wilderness areas.

Particles can be carried over long distances by wind and then settle on ground or water. Depending on their chemical composition, the effects of this settling may include

- Making lakes and streams acidic.
- Changing the nutrient balance in coastal waters and large river basins.
- Depleting the nutrients in soil.
- Damaging sensitive forests and farm crops.
- Affecting the diversity of ecosystems.
- Contributing to acid rain effects.
- Damaging materials.

PM can stain and damage stone and other materials, including culturally important objects such as statues and monuments. Some of these effects are related to acid rain effects on materials. EPA regulates inhalable particles. Particles of sand and large dust, which are larger than 10 micrometers, are not regulated by EPA (EPA, 2016).

EPA's national and regional rules to reduce emissions of pollutants that form PM will help state and local governments meet the Agency's national air quality standards.

It is important to reduce exposure to PM. This can be done by paying attention to air quality alerts, which are issued when PMs reach harmful levels. For example, the Air Quality Index (AQI) tells you how clean or polluted your outdoor air is, along with associated health effects that may be of concern. The AQI translates air quality data into numbers and colors that help people understand when to take action to protect their health.

ACID RAIN

Acid rain, or acid deposition, is a broad term that includes any form of precipitation with acidic components that fall to the ground from the atmosphere in wet or dry forms. This can include rain, snow, fog, hail, or even dust that is acidic.

The image illustrates the pathway for acid rain. It results when sulfur dioxide (SO_2) and nitrogen oxides (NO_x) are emitted into the atmosphere and transported by wind and air currents. The SO_2 and NO_x react with water, oxygen, and other chemicals to form sulfuric and nitric acids. These then mix with water and other materials before falling to the ground (Figure 12.4).

While a small portion of the SO_2 and NO_x that cause acid rain is from natural sources such as volcanoes, most of it comes from the burning of fossil fuels. The major sources of SO_2 and NO_x in the atmosphere are

- Burning of fossil fuels to generate electricity. Two thirds of SO_2 and one-fourth of NO_x in the atmosphere come from electric power generators.
- Vehicles and heavy equipment.
- Manufacturing, oil refineries, and other industries.

Winds can blow SO_2 and NO_x over long distances and across borders making acid rain a problem for everyone, and not just those who live close to these sources.

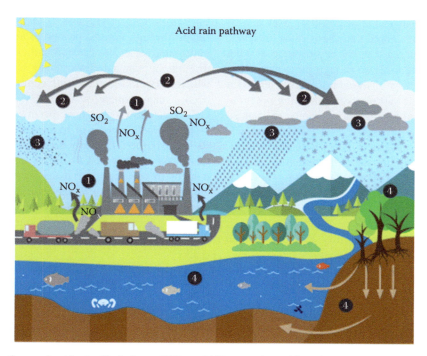

FIGURE 12.4 (1) The pathway of acid rain: Emissions of SO_2 and NO_x are released into the air, where (2) the pollutants are transformed into acid particles that may be transported long distances. (3) These acid particles then fall to earth as wet and dry deposition (dust, rain, snow, etc.), and (4) may cause harmful effects on soil, forests, streams, and lakes. State with extremely beautiful resources, unfortunately the locations of the tar sands and oil shale coincide with these areas of high aesthetic beauty, causing a cultural and economic conflict. (Courtesy of EPA, Washington, DC.)

There are two types of acid rain deposition: wet and dry. Wet deposition is what is most commonly thought of as acid rain. The sulfuric and nitric acids formed in the atmosphere fall to the ground mixed with rain, snow, fog, or hail. Acidic particles and gases can also deposit from the atmosphere in the absence of moisture as dry deposition. The acidic particles and gases may deposit to surfaces—water bodies, vegetation, buildings—quickly or may react during atmospheric transport to form larger particles that can be harmful to human health. When the accumulated acids are washed off a surface by the next rain, this acidic water flows over and through the ground, and can harm plants and wildlife, such as insects and fish.

The amount of acidity in the atmosphere that deposits on the earth through dry deposition depends on the amount of rainfall an area receives. For example, in desert areas, the ratio of dry to wet deposition is higher than an area that receives several inches of rain a year. When acid deposition is washed into lakes and streams, it can cause some to turn acidic.

CO$_2$ AND OTHER GREENHOUSE GASES

Greenhouse gases (GHGs) are a group of compounds that are able to trap heat—in the form of long wave radiation—in the atmosphere, keeping the earth's surface warmer than it would be if they were not present. These gases are what cause the natural greenhouse effect—a process necessary to sustain life on earth. As discussed in the climate portion of this book, any unnatural increases in the amount of GHGs in the atmosphere enhance the greenhouse effect and leads to climate change.

The more GHGs that are in the atmosphere, the more heat stays on earth.

There are two methods that enable additional GHGs to enter the atmosphere. One of them is through human activity. The main human sources of GHG emissions are fossil fuel use, deforestation, intensive livestock farming, use of synthetic fertilizers, and industrial processes. The other method is through biologic processes such as animal and plant respiration. The principal GHGs include the following:

- Carbon dioxide (CO_2)
- Methane (CH_4)
- Nitrous oxide (N_2O)
- Fluorinated gases
- Industrial gases

Once GHGs form in the atmosphere, they are extremely hard to get rid of because they have such a long half-life. In fact, their long atmospheric lifetime allows them to have a lasting effect on climate change. Because they do not react to changes in either temperature or air pressure, they do not get removed easily like water that condenses to become rain or snow. Carbon dioxide and the other forcing GHGs are the key gases within the earth's atmosphere that sustain the greenhouse effect and control its strength. The table illustrates various GHGs found in the atmosphere. Many are extremely potent—some can continue to reside in the atmosphere for thousands of years after they have been emitted. Some gases are 140 to 23,900 times more potent than CO_2 in terms of their ability to trap and hold heat in the atmosphere over a 100-year period. It is important to note that these gases and their effects will continue to increase in the atmosphere as long as they continue to be emitted. Even though these gases represent a very small proportion of the atmosphere—less than 2 percent of the total—their enormous heat-holding potential makes them significant and represents a serious addition to climate change.

In order to understand specific GHGs' potential impact, they are rated as to their global warming potential (GWP). The GWP of a GHG is the ratio of global warming—or radiative forcing—from one unit mass of a GHG to that of one unit mass of CO_2 over a period of time, making the GWP a measure of the "potential for global warming per unit mass relative to carbon dioxide." In other words, GHGs are rated on how potent they are compared to CO_2.

GWPs take into account the absorption strength of a molecule and its atmospheric lifetime. Therefore, if methane has a GWP of 23 and carbon has a GWP of 1 (the standard), this means that methane is 23 times more powerful than CO_2 as a GHG. The reference standards are what appear in the table. The higher the GWP value, the larger the infrared absorption and the longer the atmospheric lifetime. Based on this table, even small amounts of sulfur hexafluoride and HFC-23 can contribute a significant amount to climate change.

Global Warming Potential of Greenhouse Gases

Greenhouse Gas	Lifetime in the Atmosphere	GWP over 100 years (Compared to CO_2)
Carbon dioxide	50–200 years	1
Methane	12	23
Nitrous oxide	120	296
CFC 115	550	7,000
HFC-23	264	11,700
HFC-32	5.6	650
HFC-41	3.7	150
HFC-43-10mee	17.1	1,300
HFC-125	32.6	2,800
HFC-134	10.6	1,000
HFC-134a	14.6	1,300
HFC-152a	1.5	140
HFC-143	3.8	300
HFC-143a	48.3	3,800
HFC-227ea	36.5	2,900
HFC-236fa	209	6,300
HFC-245ca	6.6	560
Sulfur hexafluoride	3,200	23,900
Perfluoromethane	50,000	6,500
Perfluoroethane	10,000	9,200
Perfluoropropane	2,600	7,000
Perfluorobutane	2,600	7,000
Perfluorocyclobutane	3,200	8,700
Perfluoropentane	4,100	7,500
Perfluorohexane	3,200	7,400

Source: IPCC

The gases CO_2, CH_4, and N_2O are emitted to the atmosphere through natural processes and human activities. The fluorinated gases are created and emitted almost exclusively through human activities. Since the Industrial Revolution, which began in the eighteenth century, human activities have been a major source of all GHGs. Today, the growth of all GHG concentrations is now directly controlled by humans.

Fossil fuels consist of hydrogen and carbon. When fossil fuels are burned, the carbon combines with oxygen to create CO_2. The amount of CO_2 produced depends on the carbon content of the fuel. For example, for the same amount of energy produced, natural gas produces about half and petroleum produces about three-fourths of the amount of CO_2 produced by coal. If no new policies are introduced to limit GHG emissions, the U.S. Energy Information Administration (EIA) projects world energy-related carbon dioxide emissions to increase from 31.2 billion metric tons in 2010 to 36.4 billion metric tons in 2020, to 45.5 billion metric tons in 2040. Much of the growth in emissions is attributed to developing countries that are not members of the Organization for Economic Cooperation and Development (OECD) that continue to rely heavily on fossil fuels to meet their rapid growth in energy demand. Their projected portion of the increased emissions is projected to be 69 percent of the total (EIA, 2013).

Today, the primary sources of GHG emissions in the United States are as follows:

- Electricity production (30 percent of GHG emissions)—Electricity production generates the largest share of GHG emissions. Approximately 67 percent of our electricity comes from burning fossil fuels, mostly coal and natural gas (EIA, 2016).
- Transportation (26 percent of GHG emissions)—GHG emissions from transportation primarily come from burning fossil fuel for our cars, trucks, ships, trains, and planes. Over 90 percent of the fuel used for transportation is petroleum based, which includes gasoline and diesel.
- Industry (21 percent of GHG emissions)—GHG emissions from industry primarily come from burning fossil fuels for energy, as well as GHG emissions from certain chemical reactions necessary to produce goods from raw materials.
- Commercial and Residential (12 percent of GHG emissions)—GHG emissions from businesses and homes arise primarily from fossil fuels burned for heat, the use of certain products that contain GHG, and the handling of waste.
- Agriculture (9 percent of GHG emissions)—GHG emissions from agriculture come from livestock such as cows, agricultural soils, and rice production.
- Land Use and Forestry (offset of 11 percent of GHG emissions)—Land areas can act as a sink (absorbing CO_2 from the atmosphere) or a source of GHG emissions. In the United States, since 1990, managed forests and other lands have absorbed more CO_2 from the atmosphere than they emit.

From year to year in the U.S., emissions can rise and fall due to various factors, such as: changes in the economy; the price of fuel; cold winter conditions resulting in an increase in fuel demand, especially in residential and commercial sectors; an increase in transportation emissions resulting from an increase in vehicle miles traveled; and an increase in industrial production across multiple sectors.

At the global scale, the key GHGs emitted by human activities are listed as follows:

- Carbon dioxide (CO_2): Fossil fuel use is the primary source of CO_2. The way in which people use land is also an important source of CO_2, especially when it involves deforestation. CO_2 can also be emitted from direct human-induced impacts on forestry and other land use, such as through deforestation, land clearing for agriculture, and degradation of soils. Conversely, land can also remove CO_2 from the atmosphere through reforestation, improvement of soils, and other activities.
- Methane (CH_4): Agricultural activities, waste management, energy use, and biomass burning all contribute to CH_4 emissions.
- Nitrous oxide (N_2O): Agricultural activities, such as fertilizer use, are the primary source of N_2O emissions. Biomass burning also generates N_2O.
- Fluorinated gases (F-gases): Industrial processes, refrigeration, and the use of a variety of consumer products contribute to emissions of F-gases, which include hydrofluorocarbons (HFCs), perfluorocarbons (PFCs), and sulfur hexafluoride (SF_6).
- Black carbon is a solid particle or aerosol, not a gas, but it also contributes to the warming of the atmosphere.

Global GHG emissions can also be broken down by the worldwide economic activities that lead to their production (IPCC, 2014) (Figure 12.5).

- Electricity and Heat Production (25% of global GHG emissions): The burning of coal, natural gas, and oil for electricity and heat is the largest single source of global GHG emissions.
- Industry (21% of global GHG emissions): GHG emissions from industry primarily involve fossil fuels burned on site at facilities for energy. This sector also includes emissions from chemical, metallurgical, and mineral transformation processes not associated with energy consumption and emissions from waste management activities.
- Agriculture, Forestry, and Other Land Use (24% of global GHG emissions): GHG emissions from this sector come mostly from agriculture (cultivation of crops and livestock) and deforestation. This estimate does not include the CO_2 that ecosystems remove from the atmosphere by sequestering carbon in biomass, dead organic matter, and soils, which

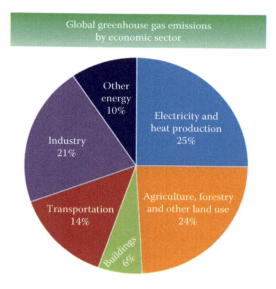

FIGURE 12.5 Global greenhouse gas emissions by economic sector. (From IPCC [2014]; based on global emissions from 2010. Details about the sources included in these estimates can be found in the Contribution of Working Group III to the Fifth Assessment Report of the Intergovernmental Panel on Climate Change.)

offset approximately 20 percent of emissions from this sector (FAO, 2014).
- Transportation (14% of global GHG emissions): GHG emissions from this sector primarily involve fossil fuels burned for road, rail, air, and marine transportation. Almost all (95 percent) of the world's transportation energy comes from petroleum-based fuels.
- Buildings (6% of global GHG emissions): GHG emissions from this sector originate from on-site energy generation and burning fuels for heat in buildings or cooking in homes.
- Other Energy (10% of global GHG emissions): This source of GHG emissions refers to all emissions from the energy sector, which is not directly associated with electricity or heat production, such as fuel extraction, refining, processing, and transportation.

In 2015, the top CO_2 emitters were China, the United States, the European Union, India, the Russian Federation, Japan, and Canada. This includes activities from fossil fuel combustion, cement manufacturing, and gas flaring. In areas such as the United States and Europe, changes in land use associated with human activities, such as agriculture and forestry, have the net effect of absorbing CO_2, partially offsetting the emissions from deforestation in other regions.

MINING

Whenever a resource must be mined in order to be acquired, the process is destructive and invasive. The type of mining, scale of the project, and methods used, determine how invasive it is for a particular area and determine the ultimate impact done to an area.

Habitat Destruction

The initial step of the mining operation is to clear the land. Often, clearing the land means deforestation, which immediately destroys wildlife habitats. As the land is cleared and low bushes and vegetation are burned and destroyed, wildlife either moves on or dies. Deforestation also affects the nesting habits and migratory patterns of birds, as well as the pollination of flowers and edible plant life. It destroys the homes of valuable insect life. This eliminates the natural process of converting CO_2 into oxygen.

In all ecosystems, the entire system must remain intact or it causes a ripple effect of negative consequences. When a tree is cut down, its loss can have an immediate and profound effect on the survival of other life in, around, and near it. Unfortunately, mining re-configures the land and its contours. Rain and subsequent groundwater is diverted. As equipment is sunk and the ground hollowed out, chemicals such as cyanide, mercury, methyl-mercury, and arsenic are forced through pipes. Tailings are left as waste piles. The water that runs off them goes into streams, creeks, rivers, and lakes and contaminates those resources.

Impacts to Urban Areas

Because of this, it is not hard to see how a large-scale mining process can affect the environment. If urbanization is within close proximity, then exposure to chemicals can affect inhabitants there and cause problems from conditions as minor as skin rashes to diseases as serious as cancer. Drinking water with lead and other chemicals can affect infants and cause birth defects. In addition, the growth of vegetation whether natural or planted can be affected by contaminated water. Fish are subject to being poisoned, and fish breeding grounds can be destroyed because of chemical poisoning and the diversion of the natural flow of clean water.

Stagnant water can also collect in any pools of water left standing in pits as the earth is dug out, becoming a breeding ground for mosquitoes and other microorganisms. Diseases such as malaria and meningitis are linked to these insects, which can cause serious health problems.

Fracking

In mining areas, where hydraulic fracturing, or "fracking" is occurring, other environmental issues are a concern. Fracking involves blasting huge amounts of water, sand, and chemicals deep into underground rock formations to access valuable oil and natural gas. While this is a form of alternative energy, it also has harmful environmental implications, causes local air pollution, may trigger earthquakes, and contaminate clean water supplies. Fracking's enormous consumption of water is especially concerning considering that much of the United States is currently suffering from drought. It requires more water than conventional gas drilling; but when natural gas is used in place of coal or nuclear fuel to generate electricity, proponents argue that it ends up saving water.

Those living near fractured wells are potentially at risk of health threats given the increased amount of volatile organic compounds (VOCs) and air toxins in the area. Again, proponents do counter that when natural gas replaces energy sources such as coal as a fuel for generating electricity, the benefits to air quality include lower CO_2 emissions than coal and almost none of the mercury, sulfur dioxide, or ash. In terms of global climate change, however, scientists are still unsure of what role fracking's, resulting toxins ultimately play in the GHG effect.

The issue most often associated with fracking is its impact on groundwater contamination. Avner Vengosh of Duke University, author of a study to determine the environmental effects of hydraulic fracturing, believes, "Wastewater disposal is one of the biggest issues associated with fracking" (Iacurci, 2014a).

In a report released by Nature World News, previous research has shown that 10–40 percent of the chemical mixture injected into the ground during fracking flows back to the surface during well development (Iacurci, 2014b).

It is clear that further information is still needed to fully determine fracking's role in groundwater contamination, climate change, and air pollution. The ultimate goal would be to develop practices that could optimize fracking's environmental cost–benefit balance.

RADIOACTIVE WASTE

Nuclear waste disposal—also called radioactive waste management—is an important part of nuclear power generation, and there are several strict guidelines that have to be followed by nuclear power plants and other companies to ensure that all nuclear wastes are disposed of safely. Fortunately, the amount of radioactive material that is left over from nuclear power plants is small compared to the waste produced by other methods of generating energy—such as burning coal; but it can be expensive and it must be done correctly.

When nuclear waste is disposed of, it is put into storage containers made of steel, and then placed inside a further cylinder constructed of concrete. These protective layers prevent the radiation from getting outside and harming the atmosphere or its immediate surroundings. It is a relatively easy and inexpensive method of containing very hazardous materials and does not need special transportation or to be stored in a specific place. The radioactive waste can be easily stored at an on-site reactor facility or adjacent to the source reactor. A counter-argument to this is that waste storage should be in a secure location rather than at the nuclear facility.

Other storage methods involve the selection of an appropriate geologic location for the storage of high-level radioactive waste. In this method, deep and stable geologic formations can be selected to store the nuclear waste long term. In this process, large and stable geologic locations are first located and then excavated to form long tunnels that extend several kilometers in length under the surface using conventional mining technology. The spent fuel and radioactive waste are then placed in the tunnels. Since the geologic formations chosen in this method are far from human population centers, the nuclear wastes are expected to be stably isolated from human interference for the long term. This technique is still under investigation and development. Several countries in the world (such as England and France) have considered this option. There are still many concerns about this geologic disposal technique, however, because the stored nuclear waste has potential to leak into the environment if any huge geologic change occurs, such as significant seismic activity. Any leakage at all would be catastrophic due to the extremely long half-lives of the waste. Because of this, many countries oppose the use of this potential disposal technique.

One of the biggest concerns with the disposal of nuclear waste is the effect the hazardous materials could have on animals and plant life. Although most of the time the waste is well sealed inside huge drums of steel and concrete, if an accident were to happen and a leak to occur, it could have drastic consequences on life around the area, causing cancer and genetic problems.

LANDFILLS

The environmental problems caused by landfills are numerous. While there are many problems with landfills, the negative effects most commonly occur in two areas: atmospheric effects and hydrologic effects.

Atmospheric Effects

According to the EPA, the methane produced by the rotting organic matter in unmanaged landfills is 20 times more effective than CO_2 at trapping heat from the sun. Not only does methane get produced by the various forms of rotting organic matter that find their way into landfills, but household cleaning chemicals often make their way there as well. The mixture of chemicals such as bleach and ammonia in landfills can produce toxic gases that can significantly impact the quality of air in the vicinity of the landfill.

Aside from the various types of gases that can be created by these landfills—such as methane, CO_2, nitrogen, ammonia, sulfides, and hydrogen—dust and other forms of nonchemical contaminants can make their way into the atmosphere. This contributes further to the air quality issue that plagues modern landfills.

Hydrologic Effects

Landfills also create a toxic soup of industrial and home-cleaning chemicals. People dispose of everything from industrial solvents to toilet bowl cleaners in landfills, and these chemicals accumulate and mix over time. There is an immediate concern for the welfare of the wildlife that comes into contact with these chemicals. In fact, it is common for animals to suffer painful deaths resulting from chemical contamination at landfills.

Aside from chemicals, electronic waste is also a large contributor to water quality issues near landfills. Consumer electronics contain everything from lead to cadmium. The EPA reports that in 2009, of the 2.2 million metric tons of electronic waste, less than 25 percent were recycled. In addition, these chemicals accumulate and are washed away periodically by rain, potentially bringing them toward municipal water supplies (EPA, 2011).

Additional Landfill Environmental Problems

Emissions are not the only types of problems associated with landfills. Other issues to contend with include fires and decomposition.

Landfill gases, and the shear amount of landfill waste, can easily ignite a fire. Fires can be difficult to put out and contribute to the pollution of the air and water. They can also potentially destroy habitats nearby if not brought under control fast enough. The most flammable gas that is most commonly produced by landfills is methane, which is extremely combustible. Firefighters will often use a fire-retardant foam to fight fires in landfills due to the presence of chemicals that would not be subdued by water, further adding to the chemical load of these landfills.

Occasionally, landfills are covered with earth, seeded with grass, and transformed into recreational areas. The management of gasses coming out of these sites is a constant issue, and creates an ongoing cost despite the new facade of the landfill. Products that are natural, such as wasted fruits and vegetables, will decompose within weeks while items like Styrofoam can take over a million years to decompose.

One Creative Solution

A number of landfills have been in use since long before the popularity of recycling. These landfills contain a wealth of mineral resources that are simply sitting there rotting away, and this has created a unique opportunity for "green" American mining. Miners have bought the rights to a number of different landfill facilities to conduct mining operations. With all of the precious metals and other minerals that are in electronic waste, more and more companies are looking at landfills as gold mines. This extra activity comes with larger atmospheric pollution via dust; however, this is generally offset by the amount of pollution that is not being generated by mining new materials and shipping them around the world.

IMPACTS OF ENERGY SOURCES

This section will discuss the environmental consequences of both nonrenewable and renewable energy resources.

Petroleum and Crude Oil

Great strides have been made to ensure that oil and gas producers make as little impact as possible on the natural environments in which they operate. These include drilling multiple wells from a single location to minimize damages to the surface, using environmentally sound chemicals to stimulate well production and restoring the surface as nearly as possible to pre-drilling conditions (as required by landowners and state or federal agencies, who often must approve the company's completion of restoration activities).

When many people think of oil and the environment, they think of oil spills. The reality is that the exploration and production of oil rarely create an oil spill. Most oil spills are the result of accidents at oil wells or with the pipelines, ships, trains, and trucks that move oil from wells to refineries. Oil spills contaminate soil and water and may cause devastating explosions and fires. The United States federal government and industry are involved in developing standards and regulations to reduce the potential for accidents and spills along with effective responses to clean up spills when they occur.

After the Exxon Valdez oil spill in Prince William Sound, Alaska, in 1989, the U.S. Congress passed the Oil Pollution Act of 1990, which required all new oil tankers built for use between U.S. ports to have a full double hull. This act led the International Maritime Organization to also establish double-hull standards for new oil tankers in 1992 in the International Convention for the Prevention of Pollution from Ships (MARPOL). The amount of oil spilled from ships dropped significantly during the 1990s partly because of these double-hull standards.

The Deep Horizon drilling rig explosion and oil spill in the Gulf of Mexico in 2010 caused the U.S. government and the oil industry to review drilling technologies, procedures, and regulations to reduce the potential for similar accidents to occur. The U.S. government also replaced the Minerals Management Service (MMS), which administered offshore oil and natural gas leases, with the Bureau of Ocean Energy Management (BOEM) and the Bureau of Safety and Environmental Enforcement (BSEE) to provide more effective oversight and enforcement of environmental regulations related to offshore energy development.

In response to several major accidents involving trains carrying crude oil, the U.S. Department of Transportation has proposed new standards for railroad tank cars, braking controls, and speed restrictions to reduce the potential for railroad accidents and oil spills. Unfortunately, a major reason for spills from pipelines in developing countries is civil unrest.

One unique restoration effort is putting offshore oil rigs to use. Oil wells are plugged when they are no longer economical, and the area around the well may be restored. Some old offshore oil rigs are tipped over and left on the sea floor in what is referred to as the Rigs-to-Reefs program. Within a year after a rig is toppled, it is often covered with barnacles, coral, sponges, clams, and other sea creatures. These artificial reefs attract fish and other marine life, and they have increased fish populations in addition to recreational diving opportunities (Figure 12.6).

Natural Gas

Natural gas is a fossil fuel, although the global warming emissions from its combustion are much lower than those from coal or oil. It emits 50–60 percent less carbon dioxide (CO_2) when combusted in a new, efficient natural gas power plant compared with emissions from a typical new coal plant (NETL, 2010). Considering only tailpipe emissions, natural gas also emits 15–20 percent less heat-trapping gases than gasoline when burned in today's typical vehicle (FuelEconomy.gov., 2013).

FIGURE 12.6 Fish at an oil platform, Gulf of Mexico. (Courtesy of U.S. Bureau of Ocean Energy Management.)

The drilling and extraction of natural gas from wells and its transportation in pipelines result in the leakage of methane, the primary component of natural gas that is 34 times stronger than CO_2 at trapping heat (Myhre et al., 2013). Whether natural gas has lower life cycle GHG emissions than coal and oil depends on the leakage rate, the GWP of methane over different time frames, and the energy conversion efficiency.

Cleaner burning than other fossil fuels, the combustion of natural gas produces very small amounts of sulfur, mercury, and particulates. It also produces nitrogen oxides (NO_x), which are the precursors to smog, but at lower levels than gasoline and diesel used for motor vehicles. The Department of Energy has stated that every 10,000 U.S. homes powered with natural gas instead of coal avoids the annual emissions of 1,900 tons of NO_x, 3,900 tons of SO_2, and 5,200 tons of particulates (Alvarez, 2012). Reductions in these emissions translate into public health benefits, as these pollutants have been linked with problems such as asthma, bronchitis, lung cancer, and heart disease for hundreds of thousands of Americans (California Environmental Protection Agency Air Resources Board, 2012).

Despite these benefits, however, unconventional gas development can affect local and regional air quality. Some areas where drilling occurs have experienced increases in concentrations of hazardous air pollutants, PM, and ozone. Exposure to elevated levels of these air pollutants can lead to adverse health outcomes, including respiratory symptoms, cardiovascular disease, and cancer (EPA, 2013).

Land Use and Wildlife

The construction and land disturbance required for oil and gas drilling can alter land use and harm local ecosystems by causing erosion and fragmenting wildlife habitats and migration patterns. When oil and gas operators clear a site to build a well pad, pipelines, and access roads, the construction process can cause erosion of dirt, minerals, and other harmful pollutants into nearby streams (UCS, 2016).

Water Use and Pollution

Unconventional oil and gas development may pose health risks to nearby communities through contamination of drinking water sources with hazardous chemicals used in drilling the wellbore, hydraulically fracturing the well, processing and refining the oil or gas, or disposing of wastewater (Colborn et al., 2011). Naturally occurring radioactive materials, methane, and other underground gases have sometimes leaked into drinking water supplies from improperly cased wells; methane is not associated with acute health effects but in sufficient volumes may pose flammability issues. The large volumes of water used in unconventional oil and gas development also raise water-availability concerns in some communities.

There have been documented cases of groundwater near oil and gas wells being contaminated with fracking fluids as well as with gases, including methane and VOCs (EPA, 2012d). One major cause of gas contamination is improperly constructed or failing wells that allow gas to leak from the well into groundwater.

Another potential avenue for groundwater contamination is natural or man-made fractures in the subsurface, which could allow stray gas to move directly between an oil and gas formation and groundwater supplies. Groundwater can become contaminated with fracking fluid. In several cases, groundwater has been contaminated from surface leaks and spills of fracturing fluid. Fracturing fluid also may migrate along abandoned wells, around improperly sealed and constructed wells, through induced fractures, or through failed wastewater pit liners (Vidic et al., 2013).

Unconventional oil and gas development also poses contamination risks to surface waters through spills and leaks of chemical additives, spills and leaks of diesel or other fluids from equipment on-site, and leaks of wastewater from facilities for storage, treatment, and disposal. Unlike groundwater contamination risks, surface water contamination risks are mostly related to land management and to on- and off-site chemical and wastewater management.

The EPA has identified more than 1,000 chemical additives that are used for hydraulic fracturing, including acids (notably hydrochloric acid), bactericides, scale removers, and friction-reducing agents. Only about dozen chemicals are used for any given well, but the choice of which chemicals is well-specific, depending on the geochemistry and needs of that well (EPA, 2012a). It is common to see large quantities—such as tens of thousands of liters for each well—of the chemical additives trucked to and stored on a well pad. If not managed properly, the chemicals could leak or spill out of faulty storage containers or during transport.

The growth of fracking and its use of huge volumes of water per well can pose a problem to local ground and surface water supplies, particularly in water-scarce areas. The amount of water used per well can vary because of differences in geologic formation, well construction, and the type of hydraulic fracturing process used. The EPA estimates that 70–140 billion gallons of water were used nationwide in 2011 for fracturing an estimated 35,000 wells (EPA, 2012c). Unlike other energy-related water withdrawals, which are commonly returned to rivers and lakes, most of the water used for unconventional oil and gas development is not recoverable. Depending on the type of well along with its depth and location, a single well with horizontal drilling can require 3–12 million gallons of water when it is first fractured—dozens of times more than what is used in conventional vertical wells (EPA 2013a). Similar enormous volumes of water are needed each time a well undergoes a "work over," or additional fracturing later in its life to maintain well pressure and gas production. A typical shale gas well will have about two work overs during its productive life span (Breitling Oil and Gas, 2012).

Earthquakes

Hydraulic fracturing itself has been linked to low-magnitude seismic activity—less than 2 moment magnitude (M) (the moment magnitude scale now replaces the Richter scale)—but these mild events are usually undetectable at the surface. The disposal of fracking wastewater by injecting it at high pressure into deep Class II injection wells, however, has been linked to larger earthquakes in the United States (National Research Council, 2013). At least half of the 4.5 M or larger earthquakes to strike the interior of the United States in the past decade have occurred in regions of potential injection-induced seismicity. Although it can be challenging to tie individual earthquakes to injection, in many cases, the association is supported by timing and location of the events (Van der Elst et al., 2013).

Coal

About 60 percent of coal in the United States is excavated from the earth in surface mines; the rest comes from underground mines. Surface coal mining can dramatically alter the landscape. Coal companies throughout Appalachia often remove entire mountain tops to expose the coal below. The wastes are generally dumped in valleys and streams. In West Virginia, for example, more than 121,400 hectares of hardwood forests (half the size of Rhode Island) and 1,610 kilometers of streams have been destroyed by this practice. Underground mining is one of the most hazardous of occupations, killing and injuring many in accidents, and causing chronic health problems.

Coal Transportation and Storage

A typical coal plant requires 40 railroad cars to supply 1.3 metric tons in a year, which equates to 14,600 railroad cars each year. Railroad locomotives, which rely on diesel fuel, emit nearly 900,000 metric tons of nitrogen oxide (NO_x) and 47,000 metric tons of coarse and small particles in the United States. Coal dust blowing from coal trains contributes PM to the air.

Coal burned by power plants is generally stored onsite in uncovered piles. Dust blown from coal piles irritates the lungs and often settles on nearby houses and yards. Rainfall creates runoff from coal piles. This runoff contains pollutants that can contaminate land and water.

Waste Maintenance

Waste created by a typical coal plant includes more than 114,000 metric tons of ash and 175,000 metric tons of sludge from the smokestack scrubber each year. Nationally, at least 42 percent of coal combustion waste ponds and landfills are unlined.

Toxic substances in the waste—including arsenic, mercury, chromium, and cadmium—can contaminate drinking water supplies and pose health issues by damaging vital human organs and the nervous system.

In coal power plants with once-through cooling systems, once the 265–680 billion liters of water have cycled through the power plant (for a typical 600-megawatt plant), they are released back into the lake, river, or ocean. This water is hotter (by up to 20°F–25°F) than the water that receives it, creating "thermal pollution" that can decrease fertility and increase heart rates in fish.

Typically, coal power plants also add chlorine or other toxic chemicals to their cooling waters to decrease algae growth. These chemicals are also discharged back into the environment.

Much of the heat produced from burning coal is wasted. A typical coal power plant uses only 33–35 percent of the coal's heat to produce electricity. The majority of the heat is released into the atmosphere or absorbed by the cooling water.

Air Pollution Issues

Coal plants are the nation's top source of carbon dioxide (CO_2) emissions, the primary cause of global warming. In 2011, utility coal plants in the United States emitted a total of 1.5 billion metric tons of CO_2. A typical coal plant generates 3.2 million metric tons of CO_2 per year.

Burning coal is also a leading cause of smog, acid rain, and toxic air pollution. Some emissions can be significantly reduced with readily available pollution controls, but most U.S. coal plants have not installed these technologies.

- *Sulfur dioxide (SO_2)*: Coal plants are the United States' leading source of SO_2 pollution, which takes a major toll on public health, including by contributing to the formation of small acidic particulates that can penetrate into human lungs and be absorbed by the bloodstream. SO_2 also causes acid rain, which damages crops, forests, and soils, and acidifies lakes and streams. A typical uncontrolled coal plant emits 13,000 metric tons of SO_2 per year. A typical coal plant with emissions controls, including flue gas desulfurization (smokestack scrubbers), emits 6,400 metric tons of SO_2 per year.
- *Nitrogen oxides (NO_x)*: NO_x pollution causes ground-level ozone, or smog, which can burn lung tissue, exacerbate asthma, and make people more susceptible to chronic respiratory diseases. A typical uncontrolled coal plant emits 9,300 metric tons

of NO_x per year. A typical coal plant with emissions controls, including selective catalytic reduction technology, emits 2990 metric tons of NO_x per year.
- *PM*: PM (also referred to as soot or fly ash) can cause chronic bronchitis, aggravated asthma, and premature death, as well as haze obstructing visibility. A typical uncontrolled plant emits 450 metric tons of small airborne particles each year. Baghouses installed inside coal plant smokestacks can capture as much as 99 percent of the particulates.
- *Mercury*: Coal plants are responsible for more than half of the U.S. human-caused emissions of mercury, a toxic heavy metal that causes brain damage and heart problems. Just 1/70th of a teaspoon of mercury deposited on a 25-acre lake can make the fish unsafe to eat. A typical uncontrolled coal plants emits approximately 77 kilograms of mercury each year. Activated carbon injection technology can reduce mercury emissions by up to 90 percent when combined with baghouses. This technology is only found in 8 percent of the U.S. coal holdings.

Other harmful pollutants emitted annually from a typical, uncontrolled coal plant include approximately:

- 52 kilograms of lead, 1.8 kilograms of cadmium, other toxic heavy metals, and trace amounts of uranium. Baghouses can reduce heavy metal emissions by up to 90 percent.
- 653 metric tons of carbon monoxide, which causes headaches and places additional stress on people with heart disease.
- 200 metric tons of hydrocarbons, VOCs, which form ozone.
- 102 kilograms of arsenic, which will cause cancer in one out of 100 people who drink water containing 50 parts per billion (Nescaum, 2011).

Nuclear

Nuclear power plants produce no air pollution or carbon dioxide, but they do produce by-products, such as nuclear waste and spent fuels. The main environmental concern related to nuclear power is the creation of radioactive wastes such as uranium mill tailings, spent (used) reactor fuel, and other radioactive wastes. These materials can remain radioactive and dangerous to human health for thousands of years. Radioactive wastes are subject to special regulations that govern their handling, transportation, storage, and disposal to protect human health and the environment. The U.S. Nuclear Regulatory Commission (NRC) regulates the operation of nuclear power plants in the United States.

Radioactive wastes are classified as low-level waste and high-level waste. The radioactivity in these wastes can range from just above natural background levels, such as in uranium mill tailings, to much higher levels, such as in used (spent) reactor fuel or in the parts inside a nuclear reactor. The radioactivity of nuclear waste decreases over time through a process called radioactive decay. The amount of time necessary to decrease the radioactivity of radioactive material to half its original level is called the radioactive half-life of the material. Radioactive waste with a short half-life is often stored temporarily before disposal to reduce potential radiation doses to workers who handle and transport the waste. This storage system also reduces the radiation levels at disposal sites. Most nuclear waste is low level (for example, disposable items that have come into contact with small amounts of radioactive dust), and special regulations are in place to prevent them from harming the environment. There is some spent fuel that is highly radioactive and must be stored in specially designed facilities. In addition to the fuel waste, much of the equipment in the nuclear power plants becomes contaminated with radiation and will become radioactive waste after the plant is closed. These wastes will remain radioactive for many thousands of years, which may not allow reuse of the contaminated land.

Nuclear power plants use large quantities of water for steam production and for cooling, affecting fish and other aquatic life. Likewise, heavy metals and salts can build up in the water used in the nuclear power plant systems.

Tar Sands and Shale Gas

Both the mining and processing of tar sands involve a variety of environmental impacts, such as GHG emissions, disturbance of mined land; impacts on wildlife; and air and water quality (Cathles, L.M., et al, 2012). The development of a commercial tar sands industry in the United States would also have significant social and economic impacts on local communities.

Tar sands greatly impact water supplies. For every gallon of gasoline produced by tar sands, about 22 liters of freshwater are consumed during the extraction, upgrading, and refining process. That is roughly three times as much as used for conventional oil. Much of this water is polluted by toxic substances harmful to human health and the environment. When surface mining is used, the wastewater ends up in toxic storage ponds. These ponds can cover over 78 square kilometers—making them some of the largest man-made structures on the planet.

In addition, a gallon of gasoline made from tar sands produces about 15 percent more CO_2 emissions than one made from conventional oil. This significant difference is attributable to the energy-intensive extraction, upgrading, and refining process. Also important to keep in mind, the carbon emissions associated with extracting tar sands could increase over time, as in-situ mining—which creates more emissions than surface mining—is used to extract the bitumen located deeper and deeper in the earth. When in-situ mining is used, wastewater is stored in the same well the bitumen is extracted from—risking contaminated groundwater if a leak were to occur. This choice of energy source requires careful consideration. The world's oil and gas companies have a responsibility to cut GHG emissions from their operations. It would make sense to assume that could include avoiding the dirtiest oil resources, such as tar sands.

Similar to tar sands, both the mining and processing of oil shale involve a variety of significant environmental impacts, such as GHG emissions, disturbance of mined land, disposal of spent shale, use of water resources, and impacts on air and water quality. The development of a commercial oil shale industry in the United States would also have significant social and economic impacts on local communities. Other impediments to development of the oil shale industry in the United States include the relatively high cost of producing oil from oil shale (currently greater than $60 per barrel), and the lack of regulations to lease oil shale. Of special concern in the relatively arid western United States is the large amount of water required for oil shale processing. Currently, oil shale extraction and processing require several barrels of water for each barrel of oil produced, though some of the water can be recycled. The process requires as much as five barrels of water—for dust control, cooling and other purposes—for every barrel of shale oil produced.

In addition, extracting operations destroy affected landscapes, forcing plants and animals out, with regeneration unlikely for decades. Oil shale extraction is also very energy intensive, so it does not offer a viable solution to the climate change problem. Researchers have found that a gallon of shale oil can emit as much as 50 percent more CO_2 than a gallon of conventional oil would over its given life cycle from extraction to tailpipe.

Due to these concerns and others, several environmental groups are fighting any development of the resource on sensitive arid lands in the western United States. The environmental groups have also brought up the fact that development of the oil shale resource would require the construction of ten new coal-fired power plants in order to get at and process the oil shale, significantly upping the carbon footprint of the entire region. The area is also near three national parks in one of the least developed parts of the United States (Figure 12.7).

Wind

Wind is a clean energy source. It produces no air or water pollution because no fuel is burned to generate electricity. The most serious environmental impact from wind energy may be its effect on bird and bat mortality. Wind turbine design has changed dramatically in the last couple of decades to reduce this impact. Turbine blades are now solid, so there are no lattice structures that entice birds to perch. Also, the blades' surface area is much larger, so they do not have to spin as fast to generate power. Slower-moving blades mean fewer bird collisions.

Land Use and Wildlife Habitat Issues

The land use impact of wind power facilities varies depending on the site: wind turbines placed in flat areas typically use more land than those located in hilly areas. Wind turbines do not occupy all of this land; they must be spaced approximately 5–10 rotor diameters apart. A rotor diameter is the diameter of the wind turbine blades. The turbines themselves and the surrounding infrastructure—including roads and

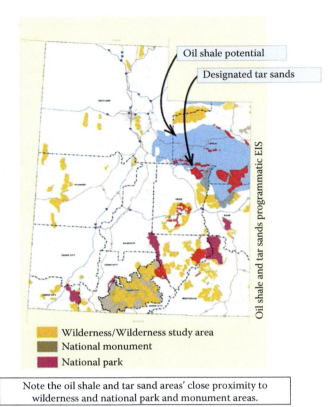

Oil shale potential

Designated tar sands

Oil shale and tar sands programmatic EIS

Wilderness/Wilderness study area
National monument
National park

Note the oil shale and tar sand areas' close proximity to
wilderness and national park and monument areas.

FIGURE 12.7 Oil shale and tar sand deposits in Utah. Two key problems with these deposits—which are predominantly located in Utah—are (1) lack of water in this sensitive arid area; and (2) many of the deposits are co-located with prime wilderness, recreation, scenic/historic, and national park areas. (From BLM.)

transmission lines—occupy only a small portion of the total area of a wind facility.

A survey by the National Renewable Energy Laboratory (NREL) of large wind facilities in the United States found that they use between 30 and 141 acres per megawatt of power output capacity (a typical new utility-scale wind turbine is about 2 megawatts). However, less than one acre per megawatt is disturbed permanently, and less than 3.5 acres per megawatt are disturbed temporarily during construction (Denholm, 2009). The remainder of the land can be used for many other productive purposes, such as livestock grazing, agriculture, highways, and hiking trails (NREL, 2012). As another option, wind facilities can be placed on abandoned or underused industrial land (or other commercial and industrial locations), which reduce concerns about conflicting land use (NREL, 2010).

Offshore wind facilities, which are currently not in operation in the United States but may become more common, especially off the East Coast, require larger amounts of space because the turbines and blades are bigger than their land-based counterparts. Depending on their location, such offshore installations may compete with a variety of other ocean activities, such as fishing, recreational activities, sand and gravel extraction, oil and gas extraction, navigation, and

aquaculture. Using best practices during the planning phase can help minimize potential land use impacts of offshore and land-based wind projects (Michel et al., 2007).

The impact of wind turbines on wildlife, most notably on birds and bats, has been extensively document and studied. A recent National Wind Coordinating Committee (NWCC) review found evidence of bird and bat deaths from collisions with wind turbines due to changes in air pressure caused by the spinning turbines, as well as from habitat disruption. The NWCC concluded that these impacts are relatively low and do not pose a threat to species populations (NWCC, 2010).

Research into wildlife behavior and advances in wind turbine technology have helped to reduce bird and bat deaths. Wildlife biologists have found that bats are most active when wind speeds are low. Using this information, the Bats and Wind Energy Cooperative concluded that keeping wind turbines motionless during times of low wind speeds could reduce bat deaths by more than half without significantly affecting power production (Arnett et al., 2010). Other wildlife impacts can be mitigated through more careful siting of wind turbines. The U.S. Fish and Wildlife Services have taken a leadership role in this effort by convening an advisory group including representatives from industry, state and tribal governments, and nonprofit organizations that made comprehensive recommendations on appropriate wind farm siting and best management practices (FSW, 2012).

Offshore wind turbines can have similar impacts on marine birds, but as with onshore wind turbines, the bird deaths associated with offshore wind are minimal. Wind farms located offshore will also impact fish and other marine wildlife. Some studies suggest that turbines may actually increase fish populations by acting as artificial reefs. The impact will vary from site to site, which means that proper research and monitoring systems are needed for each offshore wind facility (Michel et al., 2007).

Community and Water Use Issues

Sound and visual impact are the two main public health and community concerns associated with operating wind turbines. Most of the sound generated by wind turbines is aerodynamic, caused by the movement of turbine blades through the air. There is also mechanical sound generated by the turbine itself. Overall sound levels depend on turbine design and wind speed.

Some people living close to wind facilities have complained about sound and vibration issues, but industry and government-sponsored studies in Canada and Australia have found that these issues do not adversely impact public health (Chief Medical Officer of Heath of Ontario, 2010; National Health and Medical Research Council, 2010). It is important for wind turbine developers to take these community concerns seriously, however, by following "good neighbor" best practices for siting turbines and initiating open dialogue with affected community members. With continuing technological advances, such as minimizing blade surface imperfections and using sound-absorbent materials, this should also help reduce wind turbine noise (Bastasch et al., 2006).

Under certain lighting conditions, wind turbines can create an effect known as shadow flicker. This annoyance can be minimized with careful siting, planting trees or installing window awnings, or curtailing wind turbine operations when certain lighting conditions exist (NREL, 2012).

The Federal Aviation Administration (FAA) requires that large wind turbines, like all structures over 200 feet high, have white or red lights for aviation safety. The FAA recently determined, however, that as long as there are no gaps in lighting greater than a half-mile, it is not necessary to light each tower in a multi-turbine wind project. Daytime lighting is unnecessary as long as the turbines are painted white.

When it comes to aesthetics, wind turbines can elicit strong reactions. To some people, they are graceful sculptures; to others, they are eyesores. Whether a community is willing to accept them in return for cleaner power does need to be discussed as part of the process prior to implementing a project if it is going to be successful.

There is no water impact associated with the operation of wind turbines. As in all manufacturing processes, some water is used to manufacture steel and cement for wind turbines.

Life Cycle Global Warming Emissions

While there are no global warming emissions associated with operating wind turbines, there are emissions associated with other stages of a wind turbine's life cycle, including materials production, materials transportation, on-site construction and assembly, operation and maintenance, and decommissioning and dismantlement. Estimates of total global warming emissions depend on a number of factors, including wind speed, percent of time the wind is blowing, and the material composition of the wind turbine (National Academy of Sciences, 2010). Most estimates of wind turbine life cycle global warming emissions are between 0.09 and 0.18 hectograms of carbon dioxide equivalent per kilowatt-hour. To put this into context, estimates of life cycle global warming emissions for natural gas generated electricity are between 2.7 and 9 hectograms of carbon dioxide equivalent per kilowatt-hour and estimates for coal-generated electricity are 6.4 and 16.3 hectograms of carbon dioxide equivalent per kilowatt-hour (IPCC, 2011).

Solar

The sun provides an unlimited resource for generating clean and sustainable electricity without toxic pollution or global warming emissions. The potential environmental impacts associated with solar power, such as land use and habitat loss, water use, and the use of hazardous materials in manufacturing, depend on the technology used—whether it is photovoltaic (PV) solar cells or concentrating solar thermal plants (CSP). The size of the system also is a factor; a rooftop system is going to play a lesser role than a large utility-scale system. The following sections summarize the impacts in terms of land use, water use, hazardous materials, and global warming emissions.

Land and Water Use Impacts

Depending on the location, larger utility-scale facilities have raised concerns about land degradation and habitat loss.

FIGURE 12.8 The Ivanpah solar power facility is a concentrated solar thermal plant in the California Mojave Desert 64 km southwest of Las Vegas, with a gross capacity of 392 MW. It deploys 173,500 heliostats, each with two mirrors, focusing solar energy on boilers located on three centralized solar power towers. It is currently the world's largest solar thermal power station. (Courtesy of Craig Butz, Creative Commons.)

Depending on the technology utilized, the total land area required varies. The topography of the area also plays a role, as it determines how the site must be configured. The intensity of the solar resource is also a consideration in total land use. Estimates for utility-scale PV systems range from 1.4 to 4 hectares per megawatt, while estimates for CSP facilities are between 1.6 and 6.7 hectares per megawatt (UCS, 2016).

Unlike wind facilities, solar projects lack the opportunity to share land with agricultural uses. Land impacts from utility-scale solar systems can be minimized, however, by siting them at lower-quality locations such as brownfields, abandoned mining land, or existing transportation and transmission corridors (NREL, 2012). Some CSP projects, such as the Ivanpah Solar Power Facility, are located in desert areas. Smaller scale solar PV arrays, which can be built on homes or commercial buildings, also have minimal land use impact (Figure 12.8).

In October 2016, California-based energy company SolarReserve announced plans for a massive CSP plant in Nevada that will be the largest of its kind once it is built. A $5 billion project, it will generate between 1,500 and 2,000 megawatts of power, which is enough to power about one million homes. That amount of power is the equivalent of a nuclear power plant or the Hoover Dam. The project will feature at least 100,000 mirrored heliostats that will capture the sun's rays and concentrate it onto 10 towers that are equipped with a molten salt energy storage system. When the molten salt is heated to more than 1,000°, it boils water and creates a steam turbine that continuously drives generators. The advantage of this system is that it can be used after sunset, which makes it a true on-demand, 24-hour-a-day system. The project is expected to be completely emission free—running without the requirement for natural gas or oil back up. The system will be sited on 6,500 hectares. Sites of least impact are currently being assessed. The project is expected to begin construction in 2019 (Chow, 2016).

Solar PV cells do not require water for generating electricity. Water is required, however, to manufacture solar PV components. CSPs, like all thermal electric plants, do require water for cooling. Water use depends on the plant design, plant location, and the type of cooling system. Cooling system types include wet-recirculating—or closed-loop—technology, once-through systems, and dry cooling systems.

Wet-recirculating—or closed-loop—systems reuse cooling water in a second cycle rather than immediately discharging it back to the original water source. Most commonly, wet-recirculating systems use cooling towers to expose water to the ambient air. Some of the water evaporates, and the rest is then sent back to the condenser in the power plant. Because wet-recirculating systems only withdraw water to replace any water that is lost through evaporation in the cooling tower, these systems have much lower water withdrawals than once-through systems, but usually have higher water consumption. In the western United States, wet-recirculating systems are most commonly used.

Once-through systems take water from nearby sources, such as rivers, lakes, aquifers, or even the ocean. They then circulate it through pipes to absorb heat from the steam in systems called condensers and discharge the now warmer water to the local source. Once-through systems used to be the most popular because of their simplicity, low cost, and the possibility of siting power plants in places with abundant supplies of cooling water. This type of system is common in the eastern United States. Today, very few new power plants use this type of cooling because of the disruptions they cause to local ecosystems due to the heavy water withdrawals involved and because of the increased difficulty in siting power plants near available water sources.

Dry-cooling systems use air instead of water to cool the steam exiting a turbine. Dry-cooled systems use no water and can decrease total power plant water consumption by more than 90 percent. There are tradeoffs to these water savings, however, in the form of higher costs and lower efficiencies.

Given the different cooling system options, the CSP that use wet-recirculating technology with cooling towers withdraw between 2,270 and 2,460 liters of water per megawatt-hour of electricity produced. CSP with once-through cooling technology have higher levels of water withdrawal, but lower total water consumption (because water is not lost as steam). Dry-cooling technology can reduce water use at CSP by approximately 90 percent (NREL, 2012). The overall tradeoffs to these water savings are higher costs and lower efficiencies. Dry-cooling technology is significantly less effective at temperatures above 38°C. Many regions in the United States that have the highest potential for solar energy also happen to be those with the driest climates, so careful consideration of these water tradeoffs is critical.

Hazardous Materials and Life Cycle Global Warming Emissions

The PV cell manufacturing process includes a number of hazardous materials, most of which are used to clean and purify the semiconductor surface. These chemicals, similar to those used in the general semiconductor industry, include hydrochloric acid, sulfuric acid, nitric acid, hydrogen fluoride, 1,1,1-trichloroethane, and acetone. The amount and type of chemicals used depend on the type of cell, the amount of cleaning that is needed, and the size of silicon wafer (NREL, 2012). Industry employees also face risks associated with inhaling silicon dust. Because of this, PV manufactures must pay strict adherence to safety procedures to ensure employees are not exposed to harmful chemicals and that manufacturing waste products are disposed of properly.

Thin-film PV cells contain a number of more toxic materials than those used in traditional silicon photovoltaic cells, including gallium arsenide, copper–indium–gallium–diselenide, and cadmium telluride (NREL, 2012). If these are not handled and disposed of properly, they could pose serious environmental or public health problems. Financial incentives have been put in place, however, to ensure that manufacturers will ensure that these highly valuable and often rare materials are recycled rather than thrown away.

Even though there are no global warming emissions associated with generating electricity from solar energy, emissions associated with other stages of the solar life cycle, including manufacturing, materials transportation, installation, maintenance, decommissioning, and dismantlement, must be accounted for. Most estimates of life cycle emissions for photovoltaic systems are between 0.03 and 0.08 kilograms of carbon dioxide equivalent per kilowatt-hour.

Most estimates for concentrating solar power range from 0.04 to 0.09 kilograms of carbon dioxide equivalent per kilowatt-hour. In both of these cases, it is still far less than the life cycle emission rates for natural gas (0.6–2 lbs of CO_2E/kWh) and coal (1.4–3.6 lbs of CO_2E/kWh) (IPCC, 2011).

Biofuels: Biomass, Ethanol, and Biodiesel

Biofuels look like an ideal energy solution. Using biomass for energy can have both positive and negative impacts on the environment. Using biomass for energy provides an alternative to using fossil fuels such as coal, petroleum, or natural gas. While burning fossil fuels and burning biomass both release CO_2, the rationale in favor of biomass is that when a plant (biomass) is the source of the fuel, it has already captured the CO_2 through photosynthesis, and is subsequently releasing it, making it a net zero gain of CO_2.

Like most things, however, they do come with their drawbacks. It takes a tremendous amount of energy to grow crops, produce the fertilizers and pesticides necessary to ensure their growth, and process the grown plants into fuel. Today there is an ongoing debate about whether ethanol produced from corn provides more energy than it uses for growing and processing the plants. Also, those opposing biofuels claim that fossil fuels provide much of the energy in biofuels production, so biofuels may not replace as much oil as they use.

It is important to be aware of the fact, however, that while biomass can create harmful emissions such as CO_2 and sulfur when it is burned, it still causes less pollution than fossil fuels. Even burning wood in a fireplace or stove can create pollutants such as carbon monoxide. Burning municipal solid waste, or garbage that would otherwise go into a landfill, can also cause potentially dangerous emissions. Combustion of these materials must be carefully controlled. Disposing of the resulting ash can pose a problem, as it may contain harmful metals such as lead and cadmium. Waste-to-energy plants produce air pollution when MSW is burned to produce steam or electricity.

The EPA enforces strict environmental rules to waste-to-energy plants, and it requires that waste-to-energy plants use air pollution control devices, such as scrubbers, fabric filters, and electrostatic precipitators to capture air pollutants.

Scrubbers are utilized to clean emissions from these facilities by spraying a liquid into the chemical gas to neutralize the acids present in the stream of emissions. Fabric filters and electrostatic precipitators also remove particles from the combustion gases. The particles—called fly ash—are then mixed with the ash that is removed from the bottom of the waste-to-energy plant's furnace. A waste-to-energy furnace burns at high temperatures (982°C to 1,093°C) that make complex chemicals break down into simpler, less harmful compounds.

When disposing of ash from waste-to-energy plants, ash can contain high concentrations of various metals that were present in the original waste. Textile dyes, printing inks, and ceramics, for example, may contain lead and cadmium. Separating waste before combustion can solve part of the problem. Because batteries are the largest source of lead and cadmium in municipal waste, they should not be included in regular trash. Florescent light bulbs should also not be included in regular trash because they contain small amounts of mercury.

The EPA tests ash from waste-to-energy plants to make sure that it is not hazardous. The test looks for chemicals and metals that would contaminate ground water. Ash that is considered safe is used in municipal solid waste landfills as a cover layer. Ash is also used to build roads and to make cement blocks.

Utilizing wood, and charcoal made from wood, for heating and cooking can replace fossil fuels and may result in lower CO_2 emissions. Wood may be harvested from forests or woodlots that have to be thinned, or it may come from urban trees that fall down or that have to be cut down. Wood smoke contains harmful pollutants such as carbon monoxide and PM. Burning wood in an open fireplace for heating is an inefficient way to produce heat; however, it can also produce air pollution. Modern wood burning stoves and fireplace inserts are designed to reduce the amount of particulates emitted by the appliance. Wood and charcoal are major cooking and heating fuels in less-developed countries, and the wood may be harvested faster than trees can grow, which results in deforestation. Planting fast-growing trees for fuel and using fuel-efficient cooking stoves can help slow deforestation and improve the environment.

Ethanol is often added to gasoline, and while these mixtures burn cleaner than pure gasoline, they also have higher "evaporative emissions" from dispensing equipment and fuel tanks. These emissions do contribute to ozone problems and smog. Burning ethanol also creates carbon dioxide. Biodiesel creates less sulfur oxides, PM, carbon monoxide, and hydrocarbons when burned than traditional petroleum diesel. But biodiesel does create more nitrogen oxide than petroleum diesel.

Hydropower

Hydroelectric power includes energy forms from the massive hydroelectric dams to the small run-of-the-river hydro plants. Large-scale hydroelectric dams continue to be built in many parts of the world—such as China and Brazil. It is not expected, however, that any will be built in the United States in the foreseeable future. Plans for the United States include increasing the capacity of current reservoirs and possible construction of new run-of-the-river projects. Key impacts for hydropower are principally in land use issues, wildlife impacts, and life cycle global warming emissions. We will discuss these in the following sections.

Land Use and Wildlife Impacts

The size of the reservoir created by a hydroelectric project can vary widely, depending largely on the size of the hydroelectric generators and the topography of the land. Hydroelectric plants in flat areas tend to require much more land than those in hilly areas or canyons where deeper reservoirs can hold a greater volume of water in a smaller space.

When a hydroelectric reservoir is constructed, the land must be flooded. Once covered, it is taken for other uses permanently. This can have an extreme environmental impact. Impacts include forest destruction; loss of wildlife habitat; destruction of agricultural land; loss of scenic lands; and loss of cultural assets, such as archaeological sites, to name a few. In many instances, such as the Three Gorges Dam in China, entire communities have also had to be relocated in order to construct a reservoir (Yardley, 2007).

Dammed reservoirs are used for multiple purposes, such as agricultural irrigation, flood control, and recreation, so not all wildlife impacts associated with dams can be directly attributed to hydroelectric power. Hydroelectric facilities, do however, have a major impact on aquatic ecosystems. For example, although several methods exist to minimize the environmental impact to wildlife when reservoirs are constructed—including building fish ladders and in-take screens—fish and other organisms can still be injured and killed by turbine blades. There can also be wildlife impacts both within the dammed reservoirs and downstream from the facility. Reservoir water is usually more stagnant than normal river water. As a result, the reservoir will have higher than normal amounts of sediments and nutrients, which can cultivate an excess of algae and other aquatic weeds. These weeds can crowd out other river animal and plant life, and they must be controlled by manual harvesting or through the introduction of fish that eat the invasive algae and weeds (NREL, 2012). In addition, water is lost through evaporation in dammed reservoirs at a much higher rate than in flowing rivers. Where reservoirs are constructed in highly arid regions, any water loss is a serious issue.

In addition, if too much water is stored behind the reservoir, segments of the river downstream from the reservoir can dry out. Therefore, most hydroelectric operators are required to release a minimum amount of water at certain times of year. If not released appropriately, water levels downstream will drop and animal and plant life can be harmed. In addition, reservoir water is typically low in dissolved oxygen and colder than normal river water. When this water is released, it could have negative impacts on downstream plants and animals. To mitigate these impacts, aerating turbines can be installed to increase dissolved oxygen, and multi-level water intakes can help ensure that water released from the reservoir comes from all levels of the reservoir, rather than just the bottom—which is the coldest and has the lowest dissolved oxygen levels.

Life Cycle Global Warming Emissions

Global warming emissions are produced during the installation and dismantling of hydroelectric power plants, but recent research suggests that emissions during a facility's operation can also be significant. Such emissions vary greatly depending on the size of the reservoir and the nature of the land that was flooded by the reservoir.

Small run-of-the-river plants emit between 0.005 and 0.014 kilograms of CO_2 equivalent per kilowatt-hour. Life cycle emissions from large-scale hydroelectric plants built in semi-arid regions are also modest: approximately 0.03 kilograms of carbon dioxide equivalent per kilowatt-hour. However, estimates for life cycle global warming emissions from hydroelectric plants built in tropical areas or temperate peatlands are much higher. After the area is flooded, the vegetation and soil in these areas decompose and release both CO_2 and methane. The exact amount of emissions depends on site-specific characteristics, but current estimates suggest that life cycle emissions can be over 0.2 kilograms of CO_2 equivalent per kilowatt-hour (IPCC, 2011b).

Tidal/Waves/Ocean Currents

Wave and tidal power—also referred to as hydrokinetic energy—encompasses an array of energy technologies, several of which are still in the experimental stages or in the early stages of deployment.

While actual impacts of large-scale operations have not been observed, a range of potential impacts can be projected. For example, wave energy installations can require large expanses of ocean space, which could compete with other uses—such as fishing and shipping—and cause damage to marine life and habitats (UCS, 2016). Some tidal energy technologies are located at the mouths of ecologically sensitive estuary systems, which could cause changes in hydrology and salinity that negatively impact animal and plant life.

In addition, while estimates for life cycle global warming emissions for wave and tidal power are preliminary, current research suggests that they would be below 0.02 kilograms of CO_2 equivalent per kilowatt-hour (IPCC, 2011c). Upcoming research over the next decade will give scientists a better idea as to the potential of these technologies.

Geothermal

The most widely developed type of geothermal power plant—called hydrothermal plants—are located near geologic "hot spots" where hot molten rock is close to the earth's crust and produces hot water. In other regions, enhanced geothermal systems—also called hot dry rock geothermal—which involve drilling into earth's surface to reach deeper geothermal resources, are able to allow broader access to geothermal energy.

Geothermal plants differ in terms of the technology they use to convert the resource to electricity (direct steam, flash, or binary) and the type of cooling technology they use (water-cooled and air-cooled). Environmental impacts differ depending on the conversion and cooling technology used. When looking at the impacts of geothermal, it is necessary to look at water quality and use, air emissions, land use, and life cycle global warming emissions. The following sections discuss this (Figure 12.9).

Water Quality and Use

Geothermal power plants can have impacts on both water quality and consumption. Hot water pumped from underground reservoirs often contains high levels of sulfur, salt, and other minerals. Most geothermal facilities have closed-loop water systems, in which extracted water is pumped directly back into the geothermal reservoir after it has been used for heat or electricity production. In these types of systems, the water is contained within steel well casings that are cemented to the surrounding rock (Kagel, 2008). There have been no reported cases of water contamination from geothermal sites in the United States (NREL, 2012a).

Water is also used by geothermal plants for cooling and re-injection. All geothermal power facilities in the United States use wet-recirculating technology with cooling towers (refer to explanation in the solar section). Depending on the cooling technology used, geothermal plants can require between 6,435 and 15,140 liters of water per megawatt-hour. Most geothermal plants, however, can use either geothermal fluid or freshwater for cooling. If the plant uses geothermal fluids instead of freshwater, it significantly reduces the plant's overall water impact (Macknick et al., 2011).

The majority of geothermal plants re-inject water into the reservoir after it has been used to prevent contamination and land subsidence problems. In most cases, however, not all water removed from the reservoir is re-injected because it is lost as steam. In order to maintain a constant volume of water in the reservoir, outside water is used. The amount of water needed depends on the size of the plant and the technology used; but because reservoir water is considered "dirty," it is usually not necessary to use clean water for this purpose. As an example, the Geysers geothermal site

FIGURE 12.9 The Sonoma Calpine 3 geothermal power plant at the geysers field in the Mayacamas Mountains of Somona County, Northern California. (Courtesy of Stepheng3, Creative Commons.)

in California injects nonpotable-treated wastewater into its geothermal reservoir (Kagel, 2008).

Air Emissions and Land Use

Any air emission issues faced will be determined by whether the system is an open- or closed-loop. In closed-loop systems, gases removed from the well are not exposed to the atmosphere and are injected back into the ground after giving up their heat, so air emissions are minimal. Open-loop systems emit hydrogen sulfide, carbon dioxide, ammonia, methane, and boron. Hydrogen sulfide, which has a distinctive "rotten egg" smell, is the most common emission encountered with these systems (Kagel, 2007).

In the atmosphere, when the hydrogen sulfide changes into sulfur dioxide (SO_2), which contributes to the formation of small acidic particulates, it is absorbed by the bloodstream, causing heart and lung diseases (NRC, 2010). Sulfur dioxide also causes acid rain, which damages crops, forests, and soils, and acidifies lakes and streams. However, SO_2 emissions from geothermal plants are approximately 30 times lower per megawatt-hour than from coal plants, which is the nation's largest SO_2 source.

Some geothermal plants also produce small amounts of mercury emissions, which must be mitigated using mercury filter technology. Scrubbers can reduce air emissions, but they produce a watery sludge composed of the captured materials, including sulfur, vanadium, silica compounds, chlorides, arsenic, mercury, nickel, and other heavy metals. This toxic sludge often must be disposed of at hazardous waste sites (Kagel, 2007)

The amount of land required by a geothermal plant varies depending on the following:

- Properties of the resource reservoir
- The amount of power capacity
- The type of energy conversion system
- The type of cooling system
- The arrangement of wells and piping systems
- The substation and auxiliary building needs

The geysers, which is the largest geothermal plant in the world, has a capacity of approximately 1,517 megawatts, and the area of the plant is approximately 78 square kilometers, which translates to approximately 5 hectares per megawatt. Similar to the geysers, many geothermal sites are located in remote and sensitive ecological areas, making it critical for project developers to take this into account during the planning process. Land subsidence is sometimes caused by the removal of water from geothermal reservoirs. Most geothermal facilities address this risk by re-injecting wastewater back into geothermal reservoirs after the water's heat has been captured.

Hydrothermal plants are sited on geological "hot spots," which tend to have higher levels of earthquake risk. There is some evidence that hydrothermal plants can lead to an even greater earthquake frequency (NREL, 2012). Enhanced geothermal systems (hot dry rock) can also increase the risk of small earthquakes. In this process, water is pumped at high pressures to fracture underground hot rock reservoirs similar to technology used in natural gas hydraulic fracturing (fracking). Earthquake risk associated with enhanced geothermal systems can be minimized by siting plants an appropriate distance away from major fault lines.

Life Cycle Global Warming Emissions

In open-loop geothermal systems, approximately 10 percent of the air emissions are CO_2, and a smaller amount of emissions are methane, a more potent global warming gas. Estimates of global warming emissions for open-loop systems are approximately 0.05 kilograms of CO_2 equivalent per kilowatt-hour. In closed-loop systems, these gases are not released into the atmosphere, but there are a still some emissions associated with plant construction and surrounding infrastructure.

Enhanced geothermal systems, which require energy to drill and pump water into hot rock reservoirs, have life cycle global warming emission of approximately 0.09 kilograms of carbon dioxide equivalent per kilowatt-hour (IPCC, 2011a).

CONCLUSIONS

As you have seen in this chapter, all energy sources have some impact on our environment. Each is unique with its own drawbacks and limitations, just as the impact of each energy project is dependent on the location, scale, size, configuration, and specific characteristics of the individual site under consideration. As technologies continue to rapidly advance and new knowledge is gathered, we will continue to progress in our use of existing and new energy resources. What we must not lose sight of is staying mindful of the health of the environment and the populations that live within its bounds.

REFERENCES

Alvarez, R. A., S. W. Pacala, J. J. Winebrake, W. L. Chameides, and S. P. Hamburg. 2012. Greater focus needed on methane leakage from natural gas infrastructure. *Proceedings of the National Academy of Sciences* 109:6435–6440.

Arnett, E. B., M. M. P. Huso, J. P. Hayes, and M. Schirmacher. 2010. *Effectiveness of Changing Wind Turbine Cut-in Speed to Reduce Bat Fatalities at Wind Facilities*. A final report submitted to the Bats and Wind Energy Cooperative. Austin, TX: Bat Conservation International.

Bastasch, M., J. van Dam, B. Søndergaard, and A. Rogers. 2006. Wind turbine noise–An overview. *Canadian Acoustics* 34(2):7–15.

Breitling Oil and Gas. 2012. US shale faces water, transparency complaints. October 4.

California Environmental Protection Agency Air Resources Board. 2012. Health Effects of Air Pollution. https://www.arb.ca.gov/research/health/health.htm (accessed May 21, 2017).

Cathles, L. M., L. Brown, M. Taam, and A. Hunter. 2012. A commentary on "The greenhouse gas footprint of natural gas in shale formations" by R.W. Howarth, R. Santoro, and A. Ingraffea. *Climatic Change*. doi:10.1007/s10584-011-0333-0.

Chief Medical Officer of Heath of Ontario. 2010. The potential health impact of wind turbines. Toronto, Canada: Ontario Ministry of Health and Long Term Care.

Chow, L. October 12, 2016. World's largest solar project would generate electricity 24 hours a day, power 1 million U.S. Homes. *EcoWatch*. http://www.ecowatch.com/worlds-largest-solar-project-nevada-2041546638.html (accessed October 14, 2016).

Colborn, T., C. Kwiatkowski, K. Schultz, and M. Bachran. 2011. Natural gas operations from a public health perspective. *Human and Ecological Risk Assessment: An International Journal*. 17(5):1039–1056.

Cummings, C. H. 2013. Ripe for change: Agriculture's tipping point. *Worldwatch Institute*. http://www.worldwatch.org/node/4119 (accessed October 12, 2016).

Denholm, P., M. Hand, M. Jackson, and S. Ong. 2009. *Land-use Requirements of Modern Wind Power Plants in the United States*. Golden, CO: National Renewable Energy Laboratory.

Energy Information Administration (EIA). 2013. International energy outlook 2013. http://www.eia.gov/forecasts/ieo/pdf/0484 (2013).pdf (accessed October 14, 2016).

Energy Information Administration (EIA). 2016. Electricity explained. *U.S. Energy Information Administration*. http://www.eia.gov/energyexplained/index.cfm?page=electricity_in_the_united_states (accessed October 14, 2016).

Environmental Protection Agency EPA, May 2011. Electronics waste management in the United States through 2009; executive summary. https://nepis.epa.gov/Exe/ZyNET.exe/P100BKLY.TXT?ZyActionD=ZyDocument&Client=EPA&Index=2011+Thru+2015&Docs=&Query=&Time=&EndTime=&SearchMethod=1&TocRestrict=n&Toc=&TocEntry=&QField=&QFieldYear=&QFieldMonth=&QFieldDay=&IntQFieldOp=0&ExtQFieldOp=0&XmlQuery=&File=D%3A%5Czyfiles%5CIndex%20Data%5C11thru15%5CTxt%5C00000001%5CP100BKLY.txt&User=ANONYMOUS&Password=anonymous&SortMethod=h%7C-&MaximumDocuments=1&FuzzyDegree=0&ImageQuality=r75g8/r75g8/x150y150g16/i425&Display=hpfr&DefSeekPage=x&SearchBack=ZyActionL&Back=ZyActionS&BackDesc=Results%20page&MaximumPages=1&ZyEntry=1&SeekPage=x&ZyPURL (accessed October 14, 2016).

Environmental Protection Agency (EPA). 2012a. Study of the potential impacts of hydraulic fracturing on drinking water resources. Progress report. EPA 601/R-12/011. December.

Environmental Protection Agency (EPA). 2012b. What are the six common air pollutants? April 20.

Environmental Protection Agency (EPA). 2012c. Study of the potential impacts of hydraulic fracturing on drinking water resources. Progress report. EPA 601/R-12/011. December.

Environmental Protection Agency (EPA). 2012d. Study of the potential impacts of hydraulic fracturing on drinking water resources. Progress report. EPA 601/R-12/011. December.

Environmental Protection Agency (EPA). 2013. Risk assessment guidance for superfund, human health evaluation manual, office of emergency and remedial response, U.S. Environmental Protection Agency, Washington D.C. 20450.

Environmental Protection Agency (EPA). 2013a. Natural gas extraction—hydraulic fracturing. July 12.

Environmental Protection Agency (EPA). 2016. Health and environmental effects of particulate matter (PM). https://www.epa.gov/pm-pollution/health-and-environmental-effects-particulate-matter-pm (accessed October 14, 2016).

FAO. 2014. Agriculture, forestry and other land use emissions by sources and removals by sinks (89 pp, 3.5 M, About PDF) Exit Climate, Energy and Tenure Division, FAO.

Fish and Wildlife Service (FSW). 2012. Recommendations of the wind turbine guidelines advisory committee.

FuelEconomy.gov. 2013. Find a car: Compare side-by-side. U.S. Department of Energy.

George, R., J. Varsha, S. Aiswarya, and P. A. Jacob. 2016. Treatment methods for contaminated soils–translating science into practice. *International Journal of Education and Applied Research*. http://ijear.org/vol4.1/rebecca.pdf (accessed October 11, 2016).

Iacurci, J. June 26, 2014a. Fracking flowback may lead to contaminated groundwater. *Nature World News*, June 26. http://www.natureworldnews.com/articles/7777/20140626/fracking-flowback-may-lead-to-contaminated-groundwater.htm (accessed October 3, 2016).

Iacurci, J. September 12, 2014b. The pros and cons of fracking. *Nature World News*, September 12. http://www.natureworldnews.com/articles/9011/20140912/the-pros-and-cons-of-fracking.htm (accessed October 15, 2016).

IPCC, 2011. IPCC special report on renewable energy sources and climate change mitigation. Prepared by Working Group III of the Intergovernmental Panel on Climate Change, O. Edenhofer, R. Pichs-Madruga, Y. Sokona, K. Seyboth, P. Matschoss, S.

Kadner, T. Zwickel, P. Eickemeier, G. Hansen, S. Schlömer, and C. von Stechow (eds). Cambridge, UK: Cambridge University Press, 1075 pp (Chapters 7 and 9).

IPCC, 2011a: IPCC special report on renewable energy sources and climate change mitigation. Prepared by Working Group III of the Intergovernmental Panel on Climate Change, O. Edenhofer, R. Pichs-Madruga, Y. Sokona, K. Seyboth, P. Matschoss, S. Kadner, T. Zwickel, P. Eickemeier, G. Hansen, S. Schlömer, and C. von Stechow (eds). Cambridge, UK: Cambridge University Press, 1075 pp (Chapter 4 and 9).

IPCC, 2011b: IPCC special report on renewable energy sources and climate change mitigation. Prepared by Working Group III of the Intergovernmental Panel on Climate Change, O. Edenhofer, R. Pichs-Madruga, Y. Sokona, K. Seyboth, P. Matschoss, S. Kadner, T. Zwickel, P. Eickemeier, G. Hansen, S. Schlömer, and C. von Stechow (eds). Cambridge, UK: Cambridge University Press, 1075 pp (Chapter 5 and 9).

IPCC, 2011c: IPCC special report on renewable energy sources and climate change mitigation. Prepared by Working Group III of the Intergovernmental Panel on Climate Change, O. Edenhofer, R. Pichs-Madruga, Y. Sokona, K. Seyboth, P. Matschoss, S. Kadner, T. Zwickel, P. Eickemeier, G. Hansen, S. Schlömer, and C. von Stechow (eds). Cambridge, UK: Cambridge University Press, 1075 pp (Chapter 6 and 9).

IPCC (2014). Climate change 2014: Mitigation of climate change. Exit Contribution of Working Group III to the Fifth Assessment Report of the Intergovernmental Panel on Climate Change, Edenhofer, O., R. Pichs-Madruga, Y. Sokona, E. Farahani, S. Kadner, K. Seyboth, A. Adler, I. Baum, S. Brunner, P. Eickemeier, B. Kriemann, J. Savolainen, S. Schlömer, C. von Stechow, T. Zwickel, and J. C. Minx (eds.). Cambridge, UK: Cambridge University Press.

Kagel, A. 2007. *A Guide to Geothermal Energy and the Environment*. Washington, DC: Geothermal Energy Association.

Kagel, A. 2008. *The State of Geothermal Technology. Part II: Surface Technology*. Washington, DC: Geothermal Energy Association.

Office of Energy Efficiency and Renewable Energy. 2010. Wind turbine interactions with birds, bats, and their habitats: A summary of research results and priority questions. https://energy.gov/eere/wind/downloads/wind-turbine-interactions-birds-bats-and-their-habitats-summary-research-results (accessed May 22, 2017).

Macknick et al. 2011. *A Review of Operational Water Consumption and Withdrawal Factors for Electricity Generating Technologies*. Golden, CO: National Renewable Energy Laboratory.

Michel, J., H. Dunagan, C. Boring, E. Healy, W. Evans, J. Dean, A. McGillis, and J. Hain. 2007. Worldwide synthesis and analysis of existing information regarding environmental effects of alternative energy uses on the outer continental shelf. MMS 2007-038. Prepared by Research Planning and ICF International. Herndon, VA: U.S. Department of the Interior, Minerals Management Service.

Myhre, G. et al. 2013. Anthropogenic and natural radiative forcing. In *Climate change 2013: The physical science basis: Contribution of Working Group I to the fifth assessment report of the Intergovernmental Panel on Climate Change*, T. F. Stocker, D. Qin, G.-K. Plattner, M. Tignor, S. K. Allen, J. Boschung, A. Nauels, Y. Xia, V. Bex, and P. M. Midgley

(eds). Cambridge, UK: Cambridge University Press, 659–740. Available at www.climatechange2013.org/images/report/WG1AR5_Chapter08_FINAL.pdf.

National Academy of Sciences. 2010. Electricity from renewable resources: Status, prospects, and impediments.

National Energy Technology Laboratory (NETL). 2010. Cost and performance baseline for fossil energy plants, Volume 1: Bituminous coal and natural gas to electricity. Revision 2. November. DOE/NETL-2010/1397. United States Department of Energy.

National Health and Medical Research Council (NHMRC). 2010. Wind turbines and health: A rapid review of the evidence. Canberra, Australia: National Health and Medical Research Council.

National Renewable Energy Laboratory (NREL). 2000. Life cycle assessment of a natural gas combined-cycle power generation system.

National Renewable Energy Laboratory (NREL). June 14, 2010. Brownfields' bright spot: Solar and wind energy, June 14. Available at http://www.nrel.gov/news/features/feature_detail.cfm/feature_id=1530

National Renewable Energy Laboratory (NREL). 2012. Renewable electricity futures study. M. M. Hand, S. Baldwin, E. DeMeo, J. M. Reilly, T. Mai, D. Arent, G. Porro, M. Meshek, D. Sandor (eds). 4 vols. NREL/TP-6A20-52409. Golden, CO: National Renewable Energy Laboratory.

National Renewable Energy Laboratory (NREL) 2012a. Renewable Electricity Futures Study. M. M. Hand, S. Baldwin, E. DeMeo, J. M. Reilly, T. Mai, D. Arent, G. Porro, M. Meshek, D. Sandor (eds). 4 vols. NREL/TP-6A20-52409. Golden, CO: National Renewable Energy Laboratory.

National Research Council (NRC). 2010. Hidden costs of energy: Unpriced consequences of energy production and use. Washington, DC: National Academies Press. Available at http://www.nap.edu/catalog.php?record_id=12794.

National Research Council (NRC). 2013. Induced seismicity potential in energy technologies. Washington, DC: The National Academies Press.

Nescaum. 2011. Control technologies to reduce conventional and hazardous air pollutants from coal-fired power plants. March 31, 2011.

Patterson, S. 2016. How do humans affect the environment? *GreenLiving*. http://greenliving.lovetoknow.com/How_Do_Humans_Affect_the_Environment (accessed October 14, 2016).

The National Academies of Sciences, Engineering, and Medicine. 2010. *Electricity from Renewable Resources: Status, Prospects, and Impediments*. Washington, DC: The National Academies Press, 386 p.

Union of Concerned Scientists. 2016. Environmental impacts of renewable energy technologies. http://www.ucsusa.org/clean_energy/our-energy-choices/renewable-energy/environmental-impacts-of.html#.WAAkzfkrLZ7 (accessed October 13, 2013).

United States Environmental Protection Agency. May 2011. Electronics waste management in the United States through 2009, EPA 530-R-11-002. http://www.epa.gov/wastes/conserve/materials/ecycling/docs/fullbaselinereport2011.pdf (accessed May 21, 2017).

Van der Elst, N. J. et al. 2013. Enhanced remote earthquake triggering at fluid-injection sites in the Midwestern United States. *Science* 341:164–167.

Vidic, R. D., S. L. Brantley, J. M. Vandenbossche, D. Yoxtheimer, and J. D. Abad. 2013. Impact of shale gas development on regional water quality. *Science* 340(6134). doi:10.1126/science.1235009.

Yardley, J. 2007. Chinese dam projects criticized for their human costs. *New York Times*, November 19.

SUGGESTED READING

American Wind Energy Association (AWEA) and the Canadian Wind Energy Association (CanWEA). 2009. Wind turbine sound and health effects: An expert panel review.

Argonne National Laboratory (ANL). 2012. GREET 2 2012 rev1. U.S. Department of Energy.

Harrison, R. W., D. Shindell, R. Santoro, A. Ingraffea, N. Phillips, and A. Townsend-Small. 2012. Methane emissions from natural gas systems. Background paper prepared for the National Climate Assessment. Reference number 2011-0003.

Harrison, S. S. 1983. Evaluating system for ground-water contamination hazards due to gas-well drilling on the glaciated Appalachian Plateau. Groundwater 21(6):689–700.

Harvey, S., V. Gowrishankar, and T. Singer. 2012. Leaking profits: The U.S. oil and gas industry can reduce pollution, conserve resources, and make money by preventing methane waste.

The Royal Society and The Royal Academy of Engineering. 2012. Shale gas extraction in the UK: A review of hydraulic fracturing. June.New York: Natural Resources Defense Council.

International Energy Agency (IEA). 2012. Golden rules for a golden age of gas: World energy outlook special report on unconventional gas. Paris, France. Available at http://www.worldenergyoutlook.org/media/weowebsite/2012/goldenrules/WEO2012_GoldenRulesReport.pdf.

Lyman, S. and H. Shorthill, 2013. 2012 Uintah Basin winter ozone & air quality study. Final report. Document no. CRD13-320.32. Commercialization and Regional Development. Utah State University. February 1.

National Energy Technology Laboratory (NETL). 2009. Modern shale gas development in the United States: A Primer. United States Department of Energy. April.

National Renewable Energy Laboratory (NREL). 1999. Life cycle assessment of coal-fired power production

McKenzie, L. M., R. Z. Witter, L. S. Newman, and J. L. Adgate. 2012. Human health risk assessment of air emissions from development of unconventional natural gas resources. Science of the Total Environment 424: 79–87. doi:10.1016/j.scitotenv.2012.02.018.

Ohio Department of Natural Resources, Division of Mineral Resources Management. 2008. Report on the investigation of the natural gas invasion of aquifers in Bainbridge Township of Geauga County, Ohio. September 1.

Skone, T. 2012. Role of alternative energy sources: Natural gas power technology assessment. DOE/NETL-2011/1536. National Energy Technology Laboratory

The Royal Society and The Royal Academy of Engineering. 2012. Shale gas extraction in the UK: A review of hydraulic fracturing. June.

Tollefson, J. 2013. Methane leaks erode green credentials of natural gas. Nature 493. doi:10.1038/493012a.

Wigley, T. M. L. 2011. Coal to gas: The influence of methane leakage. Climatic Change 108:601–608.

Williams, H. F. L., D. L. Havens, K. E. Banks, and D. J. Wachal. 2008. Field-based monitoring of sediment runoff from natural gas well sites in Denton County, Texas, USA. Environmental Geology 55:1463–1471.

Wiseman, H. J. 2013. Risk and response in fracturing policy. University of Colorado Law Review 84:758–761, 766–770, 788–792.

Cook, J. L. 2013. *Reliance on Fossil Fuels in the U.S.: Impacts to Culture, Health, Wildlife, and Environment.* LLC (Amazon.com): On Demand Publishing, 72 p.

McBroom, M. 2016. *The Effects of Induced Hydraulic Fracturing on the Environment.* Waretown, NJ: Apple Academic Press, 362 p.

13　The Geopolitics of Energy

OVERVIEW

This chapter presents the various geopolitical energy policies from different regions of the world to compare and contrast the differences and similarities in this sometimes rapidly changing stage. Interestingly, especially in the context of renewable energy and sustainably, is the activity that is taking place in North America, further illustrating the extreme obstacles being faced today in the attempt to create a sustainable world. This chapter covers the energy activities of Asia and India, the European Union (EU), the Middle East, North America, and Russia in order to illustrate how varied, rapidly changing, and volatile it can be today.

INTRODUCTION

To say the geopolitics of energy for the world's countries is a complex matter would be a gross understatement. Policies are constantly in flux and often intertwined with delicate relationships between other countries. Each country has their own set of policies and those policies can often set the tone of the relationship between the respective nations. Energy seems to be one of the most important commodities in this age, requiring negotiations to be conducted very carefully, as they can have far-reaching ramifications.

DEVELOPMENT

Geopolitics is the battle for space and power, played out in a geographical setting. Just as there are diplomatic geopolitics—what we are most used to hearing on the news—which involve the policies of countries, economic geopolitics, and military geopolitics, likewise, there are energy geopolitics. Throughout history, every international entity (e.g., empire, dynasty) has been based on an energy resource of some type, so the geopolitics of energy is not new. For example, during the eighteenth and nineteenth centuries, the British Empire was dominant during the Age of Coal and Steam. From the nineteenth to the early twenty-first centuries, the American Empire has dominated under the Age of Petroleum. Currently, as the Age of Petroleum has been in decline, vast shale gas discoveries are giving speculation to a possible new age of dominance in the United States as the upcoming Age of Natural Gas. Speculation is the keyword, however, in light of climate change and the crucial introduction of renewable energy. The bottom line remains, nonetheless: energy equates to power and world dominance.

THE GEOPOLITICS OF ENERGY—A NEW PLAYING FIELD

It cannot be argued that the world of energy and power is in flux. Centers of dominance are shifting from traditional playing fields, new trends are emerging, energy security is becoming more unbalanced, new markets are being shaped daily, new challenges and opportunities are appearing, times are more uncertain, landscapes are changing, and there is a growing list of winners and losers in this volatile landscape.

Mohan Malik, a professor at the Asia-Pacific Center for Security Studies in Honolulu, an expert in the geopolitics of energy, has created a conceptual new world map now dominated by a growing consumer market for energy in Asia and a growing market for production in the United States—a very different landscape from the traditional producer/consumer market of the past four decades. Says Malik, "Asia has become 'ground zero' for growth" as far as the consumption of energy is concerned (Forbes, 2014). His research indicates that over the next 20 years, 85 percent of the growth in energy consumption will come from the Indo-Pacific region. Currently, nearly 25 percent of the world's liquid hydrocarbons are consumed by China, India, Japan, and South Korea. China will account for 40 percent of the growing consumption until 2025, and after that India will emerge as the largest single source of increasing demand (World Energy Outlook, 2014). The rate of energy consumption growth for India will increase to 132 percent; in China and Brazil the demand will grow by 71 percent, and in Russia it will increase by 21 percent. Malik believes that this incredible increase in demand for gas will overtake the demand for oil and coal combined. He also believes that the Indo-Pacific region will become more and more reliant on the Middle East for its oil, stating that by 2030, 80 percent of China's oil will come from the Middle East, as will 90 percent of India's. Japan and South Korea are already totally dependent on oil imports.

While the Indo-Pacific region is becoming increasingly more energy dependent on the Middle East, in the other hemisphere the United States is also currently becoming a global energy producing giant. Malik states that U.S. shale oil production will more than triple between 2010 and 2020. Added to that, if the United States opened up its Atlantic and Pacific coastlines to drilling, oil production in the United States and Canada could eventually equal the consumption in both China and India (already, within a decade, shale gas has risen from 2 percent to 37 percent of U.S. natural gas production). The United States has now overtaken Russia as the world's largest natural gas producer. Some estimates put the United States as overtaking Saudi Arabia as the world's largest oil producer by the end of the current decade, although Malik believes this is unlikely.

Malik also makes the point, that if this were to happen, it would put America in a position of world energy dominance; and if those resources were combined with Canadian oil sands and Brazil's oil lying beneath salt beds, these would hypothetically have the potential to transform the Americas into the "new Middle East" of the twenty-first century. This represents a major shift in the geopolitics of energy and opens the door to possible ramifications. But now let us factor in Russia, another energy-rich country.

Russia is also shifting its focus of energy exports to East Asia. China may well become Russia's biggest export market for oil before the end of the decade, even while Russian energy firms are simultaneously developing closer relationships with Japan in order to protect themselves against their growing emphasis on China. (Chilcoat, 2016)

Add to that the rising maritime tensions in the South China Sea and adjacent East China Sea. As the Indo-Pacific waters (Greater Indian Ocean and South China Sea) are slated to become the world's new energy interstate with these new trade agreements, territorial tensions over which country owns what geographical feature in those waters is not only being driven by potential energy reserves and fish stocks in the vicinity, but also by idea that these areas are of geopolitical importance because of the changing world energy market.

What we can expect to see is everyone moving in that direction, which means we are already witnessing it being put into play now. We are seeing energy routes being established to the Indo-Pacific region. The Middle East will be exporting more and more hydrocarbons there. Russia is already exporting a lot of hydrocarbons to East Asia. And North America will soon be looking to the Indo-Pacific region to export its own energy, especially natural gas. Europe, because of its aging population, is not expected to grow in relative importance in world energy markets.

With economic importance often comes cultural and political importance. In this case, if that were to happen, then the current tension between an economically and demographically stagnant EU and a troubled, autocratic Russia—even if somewhat energy rich, it could push Europe into a state of decline, while North America and the countries near the Indian Ocean world become the new thriving centers of commerce. It could also be possible to see an alliance of some sort between Russia and China because of their energy ties; which could cause conflict and competition with the democratic West.

THE CONCEPT OF SECURITY VERSUS SURVIVAL

Ever since the Industrial Revolution in the eighteenth century, the geopolitics of energy—including who supplies it and who can get reliable access to supplies—has been a driving factor in global prosperity and security. The country that has control over it has always had the upper hand, and the countries that had good enough relationships with that country were able to get it in times of need fared well. Over the next few decades, energy politics is likely to determine which countries fare well (survive and prosper) and those that do not.

At no other time is the political nature of energy more prominent than during a crisis, especially when unstable oil markets drive up the prices and politicians hear their constituents protest. But in the recent past, it has become even more complicated. Transportation systems (especially in the United States) have become reliant on oil, which means that any disruption of the oil market can bring a great power to a standstill. Access to energy is critical to sustaining growth in countries (especially applicable now to China and India), not only to lift countries out of poverty, but also to keep up with the exploding populations. If it were taken away, it could unravel a country whether they have an authoritarian governed country or a democratic one. And it is these very factors that have turned the market power of energy suppliers into political power. Importers have come to compete for supplies, driving up prices, supplier wealth, and the capacity to play roles in regional and international politics that go far beyond the GDP of countries, such as Russia, Venezuela, and Iran.

These traditional geopolitical considerations have become even more complex with global climate change. The United Nations' IPCC has irrefutably documented that the use of fossil fuels is the principal cause of greenhouse gases that are driving up the temperature of the planet. Climate change will create severe flooding and droughts which will devastate many countries' food production, lead to the spread of various illnesses, and cause hundreds of thousands of deaths per year, particularly for those living in the developing world. Nearly two billion people were affected by climate-related disasters in the 1990s and that rate may double in the next decade. At the very same time countries that are competing for energy must radically change how they use and conserve energy. The politics of that debate, particularly how to pay for the costs and dissemination of new technologies, and how to compensate those who contribute little to climate change but will most severely experience its tragedies, are emerging as a new focal point in the geopolitics—as they should.

EMERGING TRENDS AND UNCERTAIN TIMES—ONE ANALYSIS

In a report issued by the Center for Strategic and International Studies Energy and National Security Program in 2010, they described the recently changing energy landscape as inherently complex and identified five major trends currently occurring:

- Shifting demand dynamics
- A changing resource base, supply choices, and delivery requirements for petroleum
- Investment, price volatility, and alternative fuels

- Key players and evolving rules
- Climate change and efforts to impose carbon constraints on a fossil fuel-dependent world

Shifting Demand Dynamics

Beginning in the early 2000s, before the economic crisis in the United States, the energy markets were already beginning to change. The oil stockpiles that existed were dwindling so that when any geopolitical or weather-related supply interruption occurred it caused a spike in the price (shortages cause the prices to go up). It worsened matters and caused supply problems when other complications occurred such as infrastructure problems, capability limitations, heightened geopolitical and investment risks, and volatile costs. Added to all of this was the growing environmental movement over the past decade, looking for alternative sources of renewable energy and urging the market to veer away from fossil fuels.

Also since the turn of the millennium, there has been the emergence of new global players that have entered the market and wanted to play by their own rules. This has posed threats to the United States' position, and created an increasingly unsustainable world energy balance along with the need for reform.

Even when adjusting for the impact of the economic downturn in the early 2000s, as a consequence of continued population growth and improved living standards for large portions of the world's population, most predictions—including those by the U.S. Energy Information Administration (EIA), the International Energy Agency (IEA), and the Organization of the Petroleum Exporting Countries (OPEC), project that global energy demand will increase by more than 45 percent between now and 2035. Most of this growth will be from China and India. Energy demand from developing economies has already overtaken energy demand requirements in the developed world. This change presents significant challenges and uncertainty that can affect both investments and policy choices for the future (Figure 13.1).

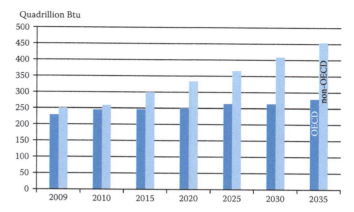

FIGURE 13.1 Energy consumption in the OECD and non-OECD Countries, 2009 to 2035. (From Energy Information Administration, International Energy Outlook 2010.)

Interestingly, this study believes that by 2035, fossil fuels will still comprise more than 80 percent of energy supplies while alternative and renewable fuels continue to increase (Figure 13.2). The most dramatic change will be with the world's developing and emerging economies. It is projected that the Non-Organisation for Economic Co-operation and Development (OECD) countries—including China, India, and also portions of the Middle East—will increase by 78 percent between now and 2035, which accounts for 75 percent of the total increase that is projected in future global demand. This equates to the majority of expected increase in energy-related CO_2, that which will contribute to climate change. These non-OECD nations presently derive nearly 90 percent of their energy needs from fossil fuels, with coal providing the largest share.

Of the total energy consumed worldwide, roughly 40 percent is used for power generation needs and 30 percent goes for transportation. Half the world's oil—43 million barrels a day—is dedicated to fueling transportation needs. The only way to curb this consumption at this point is through improved efficiency in the transportation sector or increasing oil prices. It is felt that the use of alternative

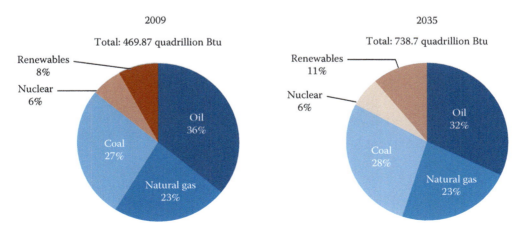

FIGURE 13.2 Global primary energy consumption by fuel, 2009 and 2035. (From EIA, IEO 2010.)

transportation fuels and technologies, such as ethanol, bio-fuels, compressed natural gas (CNG), electric vehicles, and batteries, is welcome and helpful, but is several decades away from being developed enough to really make a difference in order to be able to replace oil. Because of these issues, the future energy needs and technological situations seem daunting.

A Changing Resource Base

The study acknowledges that earth is rich with energy resources, both conventional and unconventional. They are, however, becoming increasingly difficult to access, produce, convert, and deliver to where they are needed in a cost-effective, secure, and environmentally friendly way. Most of the energy resources that are more easily acquirable have already been recovered. Because today's energy resources are largely geographically, geologically, technologically, environmentally, and financially more challenging to get to, it is going to be a serious challenge to make use of them.

Geographically, most of the known remaining conventional oil and natural gas resources are located in the Middle East, Africa, and Eurasia. The unconventional fuels, such as oil sands, oil shale, unconventional gas, and extra-heavy oil deposits are located in the Western Hemisphere (Figure 13.3). The problem with unconventional fuels is that their extraction and refining present considerable environmental challenges. The unconventional oil resources cannot serve as direct replacements yet—some of the extraction and processing technology is still being researched and developed.

Projected future oil demand exceeds current supply levels by more than 25 percent and requires more than 20 million barrels a day of additional supply above current levels. A challenge faced is that given projected rates of oil field decline on the order of 4 to 7 percent per year, meeting even more

modest 2030 oil needs will still require the addition of 60 million barrels per day of new production (a level comparable to current output). In order to achieve this, it will require sizable contributions from both OPEC and non-OPEC sources (Verrastro et al., 2010). Yet, non-OPEC output is expected to reach a plateau sometime in the middle to end of the next decade. In the short-term it may be possible that increases in non-OPEC needs could be met from Canada (oil sands), the United States' (offshore), Brazil, Russia, and the Caspian Sea region (ocean), but not in the amounts expected—only partially. The trend exists that when supplies are tight, other dimensions of the market take on new significance. Spare capacity, commercial stock levels, strategic stock policies and practices, domestic fuel subsidies, fuel switching, political instability, investment decisions, and the role of nontraditional investors are more carefully scrutinized when markets are tight like this.

Price Volatility

Continued demand and short supplies, along with increasing equipment and materials costs, resulted in soaring energy prices in 2007 and 2008. During the five year period from 2003 to 2008, energy prices rose drastically. In fact, oil prices rose by more than $100 per barrel, doubling in just one year alone. Then, from July 2008 to July 2009, prices declined by more than 50 percent, then rose again. In view of the fact that much of the world's economy was built on cheap energy, this caused havoc. In the United States, homes, cars, transportation habits, and heating and cooling preferences were all based on habits learned in a world where energy was relatively inexpensive. The experience of the past several years, with uncertainty, had now left serious uncertainty, making future investment decisions difficult.

In an environment now of uncertainty, there are estimates that industry and governments will need to invest

FIGURE 13.3 Global oil reserves and aggregate consumption by region, 2009–2030. (From EIA, IEO 2009.)

approximately $26 trillion between now and 2030 to meet the forecasted energy demand. This figure does not take into account the investment necessary to shift the global energy system from its current state to a lower carbon alternative (Verrastro et al., 2010). The inability to access lowest cost reserves (Middle East), combined with new demand for materials and labor, has increased project development costs considerably. In addition, investment decisions about particular fuels are dependent on a wide variety of considerations, including:

- The cost of fuel
- Projected long-term supply
- The investment framework
- Material costs
- The regulatory environment
- Reliability
- Other resources (e.g., water availability)
- Environmental impacts

These considerations will need to be looked at in order of importance. During times of high prices and perceived resource constraints, for example, abundant and cheap fuels are the most attractive. When environmental opposition to more carbon-intensive fuels or environmentally invasive extraction or production practices is predominant, more expensive and cleaner fuels are held at a premium. Usually a fuel hierarchy is created and used in different parts of the world, which affects the investment decisions that come up and helps make the decisions on which to focus on. Choices of fuels could be coal, nuclear, natural gas, or renewables.

Renewable energy sources have seen considerable growth in many parts of the world, as environmental concerns and requirements drive investors to cleaner energy options and seek greater security and diversity in their fuel mix. They have increased their investments in wind and solar energy. The total of all renewables—including wind, solar, geothermal, biomass, and hydroelectric power—accounts for less than 10 percent of global energy usage. It is expected to only reach 11 percent by 2035. The principal factors that are driving investment in clean energy technology are the concerns over climate change, increasing demand for energy, and energy security. Wind and solar power have been the largest recipients for financing, with biofuels coming in third. The United States and the EU are the principal investors, with China, India, and Brazil also beginning to invest. Recently, natural gas has entered the competition against renewable energy.

It is expected that fossil fuels will meet the majority of global energy demand for decades to come. Even when renewables begin to take over a fair share of the market, fossil fuels will still remain important due to the global energy system's sheer size. This makes it important for policymakers to find ways to ensure that the current energy system is well maintained, even as it eventually transitions to low-carbon alternatives.

Key Players and Evolving Rules

The energy playing field is also changing because new players with new agendas are entering now, and so are new investors, as well as new alliances. These introductions are challenging the old alliances while threatening change and a sense of the unknown. New priorities and interests are being introduced. As demand from developing countries grows and incremental supply from all countries becomes increasingly important for meeting demand even as significant reserves still lie in a handful of regions, the interaction between countries—both old and new—becomes very important. The decisions that are made are driven by three general things:

- Internal interests and drivers
- Geopolitical dynamics
- Geoeconomics

Internal Interests and Drivers

Many of the newly emerging economies have large populations with a wide divide between the wealthy and poor of their society. To stay in power and be effective, their political leaders must be able to bring their poor out of that condition and maintain economic and political stability. Because energy is the lifeblood of their economy, they must be able to make energy more affordable for their people by developing resources abroad and ensure adequate supplies. Simultaneously, energy suppliers are trying to figure out how best to provide energy for markets, without depleting their reserves. Some countries have become cautious with their resources, now more cautious with the processing of them, not willing to waste any or be involved in any unproven speculation. This means they must do more exploring and locate new reserves. Another challenge is how to deal with the reemergence of security-inspired, politically-driven foreign investment. Countries opting to exploit their newly-acquired leverage are in subtle and not-so-subtle ways redefining market competition. The United States and other countries that have been in the game longer under different rules have been slow to recognize and adjust to the dynamics of these new business relationships.

Geopolitical Dynamics

Geopolitical alliances continue to have a strong impact on energy production and trade. Now, however, new countries are coming onto the scene and upsetting the traditional alliances. They are disrupting the status quo and marking the beginning of a "new game" in the geopolitics of oil and other energy resources. Those that are resource-rich are entering the picture, some of them with the intent of using their oil resources to further their foreign policy objectives; exploiting their commodities in an effort to gain political power. The national oil companies also have a rapidly rising role. They now control the majority of conventional resources and account for more than half of current oil and natural gas production around the globe. As new energy-consuming countries emerge on the scene—such as China, India, and others in the developing world—these companies want to reach out and control these new energy-consuming giants right from the beginning, whose growth is concentrated in the industrial sectors but who, as living standards increase, are also likely to have an increasing impact on transportation fuels growth, and in the geopolitical alignments that accompany these changes.

The bottom line is that the rules of engagement are changing. The old rules can no longer be taken for granted—they are being challenged by the non-OECD countries. In addition, the new developing-world governments are bringing with them other political commitments and agreements that operate apart from traditional agreements and that is worrying the OECD members. One major worry is that actions taken by either side—traditional or nontraditional—can affect both sides of the supply/demand balance. Examples include restrictions on access to resource-rich areas by countries, changing regulations, and threats to security of both people and assets and sabotage by radical organizations. Geopolitical activity by nations can also have an effect, as seen in the rise of China's and India's desires to form alliances with producer governments to solely secure supplies in order to meet their projected energy needs. All these actions put the geopolitical arena in a state of upheaval.

Geoeconomics

As oil prices have risen, an enormous amount of capital wealth has been transferred from energy consumers to a small number of energy-producing nations. The energy-producing nations, however, do not have adequate safeguards to protect against corruption and inappropriate use of the revenue generated from their oil income. This allows oil-producing nations to self-finance development projects in their own countries and in other nations without disclosing their actions, leading to the degradation of international standards created by Western economies. This lack of transparency leads to instability and the fact that the United States and the EU, who are dependent on these resources could potentially lead to disaster.

In addition, because emerging economies are able to finance their own development or seek partnerships with one another rather than adopt the rules of existing global institutions, the traditional rules are further undermined, leaving the Western world even more vulnerable. Added to this, world opinion of Western-based alliances has declined in recent years as public opinion of the United States has taken a negative turn, putting the Western world in a very vulnerable position. The IEA is trying to recruit China and India and bring them on board, but so far they have not obligated and have not stated whether the benefits of membership are worth the obligations of belonging to the organization.

The IEA is an autonomous organization which works to ensure reliable, affordable, and clean energy for its 29-member countries. It has four main areas of focus:

- *Energy security*: Promoting diversity, efficiency, and flexibility within all energy sectors
- *Economic development*: Supporting free markets to foster economic growth and eliminate energy poverty
- *Environmental awareness*: Analyzing policy options to offset the impact of energy production and use on the environment, especially for combating climate change
- *Engagement worldwide*: Working closely with partner countries, especially major economies, to find solutions to shared energy and environmental concerns

Founded in 1974, the IEA was initially designed to help countries coordinate a collective response to major disruptions in the supply of oil, such as the crisis that occurred in 1973–1974.

While this remains a key aspect of its work, the IEA has evolved and expanded. It is a source of global information on energy and provides informative statistics and analysis. It analyzes the full spectrum of energy issues and advocates policies that will enhance the reliability, affordability, and sustainability of energy.

Climate Change and Carbon Constraints

Governments are attempting to get a handle on climate change by offering a fuel type that is low in CO_2 in order to combat the negative effects of climate change, such as warming temperatures, changes in rainfall, glacial melting, and rising sea levels. They feel these are the effects that have the greatest potential to transform the global energy system and cause the most damage (Figure 13.4). The use of fossil fuels and the acknowledgment by the IPCC that it is the leading cause of greenhouse gases (GHGs) in the atmosphere that lead to climate change is acknowledged by energy policymakers. They acknowledge that the world relies on fossil fuels for more than 80 percent of its energy needs and that reducing this dependence will require significant new investment, technology improvements, and massive-scale deployment sustained over a long period. Transitioning to a low-carbon energy future will require a complete transformation of the energy system that the world has relied on for a century, as they move toward a more sustainable one—one they note that is likely to be expensive and has not been tested at the scale yet that it will ultimately be required in order to serve the entire population. It is also understood to be inevitable because we cannot continue with fossil fuels due to their unsustainability. Many feel that the transformation to renewables has already begun globally.

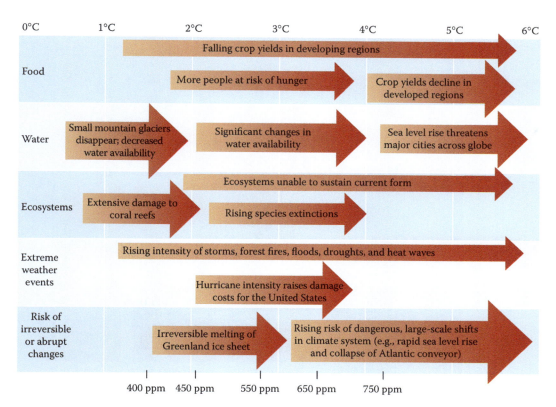

FIGURE 13.4 Projected climate impacts of rising temperatures and increasing CO_2 concentrations. (Adapted from Stern Review on the Economics of Climate Change, Report to the Prime Minister and Chancellor of the Exchequer, October 30, 2006.)

It is also understood that global population will reach about 9 billion by 2050, making another strong case for renewable energy and better developed self-sufficiency. In addition, given recent food shortages in some areas of the world partly due to climate change, being more self-sufficient with energy resources makes better sense from an energy security perspective than being so dependent on other countries for an energy supply. That said, the issue is still being worked out. Perhaps with the 2016 Paris Agreement, those details will get solved.

With agreements in existence between countries today, and the individual commitments to them, current projections are that the individual commitments are not enough to keep the temperature rise below the target 2°C (a stabilization target of about 400 ppm CO_2-equivalent in the earth's atmosphere). The more realistic amount looks like it will be somewhere between 2°C and 3°C (upward of 450 ppm). At any rate, even as nations try to advance technological solutions, we must be ready to adapt to the possibly of unavoidable impacts of climate change (unavoidable impacts of changing patterns of rainfall and drought, increasingly severe weather conditions and higher temperatures). The climate changes that are already set in motion are now inevitable; there is nothing we can do to stop them (Figure 13.5). Society must be prepared to adapt to changing rainfall patterns, drought, increasingly severe weather conditions, shifting seasons, and hotter temperatures. For a unified, global approach with security and transparency, policies should be written and enforced from the top down,

promoting renewable energy, establish emissions reduction targets, and invest in technology research and development. Climate change also needs to be taken into consideration when making investment decisions. With upcoming technology, it would also be wise to leave the door open on working with countries we do not currently work with—maintaining amenable diplomatic relations. There may come a time when we use other commodities for energy, such as indium (found in photovoltaic cells), platinum (catalysts), lithium and lanthanum (used in batteries), rhenium (used for nickel-based super alloys), and rhodium (an anticorrosive material used in high-temperature coatings) and need good working relationships with countries we have not worked with before.

GEOPOLITICAL ENERGY PERSPECTIVES FROM THE EU

The EU has been working on an initiative for an EU-wide Energy Union, with the European Commission's adoption of a "Framework Strategy for a Resilient Energy Union with a Forward-Looking Climate Change Policy" (European Commission, 2015). The new strategy document outlines the Commission's desire to attain "secure, sustainable, competitive, affordable energy for every European." The initiative is intended to transform energy markets and energy/climate policy across the EU. Its purpose is to facilitate coordination and integration in energy security, trade, regulation and efficiency, as well as in low-carbon development and research

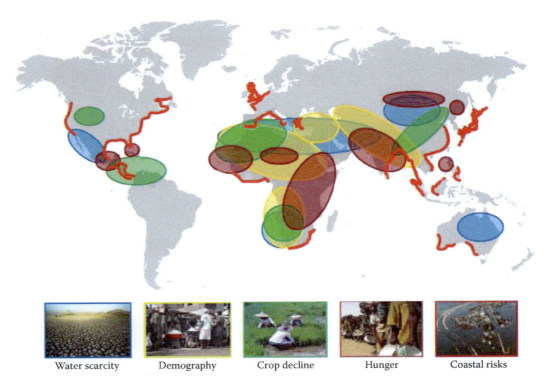

| Water scarcity | Demography | Crop decline | Hunger | Coastal risks |

FIGURE 13.5 Climate change as a threat multiplier. (From World Meteorological Organization [WMO/CSIS], Washington, DC.)

and innovation between the member countries. Two main processes are driving climate and energy policy across Europe: (1) the 2030 Climate and Energy Package, which formed the basis of the EU's contribution at the United Nations Framework Convention on Climate Change (UNFCCC) negotiations and (2) the development of a new energy governing body—the Energy Union (World Nuclear Association, 2016).

During a conference in 2015 of the Friends of Europe, a leading energy think-tank that specializes in creating solutions to global and European problems met over the geopolitics of energy. The event, termed "The New Geopolitics of Energy—Winners and Losers," focused on a select group of policymakers, business leaders, and industry experts to discuss both global and European outlooks on the world's dynamic and changing energy landscape. Their observations about the topic, while in some aspects echoed the views of the United States, they also gave some insight into the topic from the perspective they share uniquely with the world.

They believe we are living in a time of great change in both global and European energy markets. As Mark Lewis, a Strategic Advisor at Friends of Europe, put it: "If there is one word that describes the current state of energy affairs in the world, it is *disruption*."

Europe's consensus is that climate change, the crisis in Ukraine, instability in the Middle East, plummeting oil prices, the emergence of North America as a major energy producer, and the arrival of new renewable energy technologies have all triggered a massive amount of change in global energy diplomacy and market balances. Elmar Brok MEP, Chair of the European Parliament Committee on Foreign Affairs believes

that because of all this upset, the EU needs to create and implement its Energy Union as soon as possible. As global markets increase in size, the EU must strive to maintain its presence and weight in global political circles, especially in energy issues (Figure 13.6).

FIGURE 13.6 The member states of the European Union are about to enter into a new European energy union to reform how Europe produces, transports, and consumes energy, focusing on energy security and decarbonizing the continent's economies (World Nuclear News, 2015). (From Radu Gherasim GFDL.)

MOSCOW, UKRAINE, AND EUROPE

One of the key tools in Russia's "hybrid war" taking place within Europe is its energy component. Hybrid warfare is a military strategy that blends conventional warfare, irregular warfare, and cyber warfare. Russia is a state heavily reliant on energy and raw materials, where hydrocarbons are not just a commodity, but also a tool for achieving geopolitical objectives. To achieve its political goals, Moscow has used energy as a manipulative tool against Ukraine on three levels:

- Political
- Economic
- Informational

Ukrainian energy infrastructure is of significant interest to the Kremlin, since its occupation or destruction not only cause significant economic losses to Ukraine, but also threatens the European countries' energy security.

In the Russian "gas aggression" that occurred against Ukraine in 2006 and 2009, Russia—in 2006—conducted so-called "punishment actions," cutting off the gas supply to Ukraine and reducing the volume of gas transit via Ukraine to the EU. As a result, Ukraine was punished for its 2004 Orange Revolution and Europe was punished for having supported Ukraine. The crisis in 2009 also had a specific purpose. It was meant to provoke an internal political crisis in Ukraine and worsen the country's relationship with the EU. Moscow's idea was that in case the gas supply to Ukraine would completely stop, Kyiv's (the capital of Ukraine) authorities would be unable to deliver gas from western storage to the main industrial centers in the east and south of Ukraine, thereby leaving them without heat. This was meant to provoke a "social explosion" within Ukraine.

Because of Russia's official declaration about using energy resources and infrastructure in order to "address national and global problems," some experts think that Russia is using their post-Soviet space phenomenon of "energy wars" (using energy as a weapon to further control) on the EU and NATO now. Russia has used the energy weapon against European countries before: such as in the reduction of oil supplies to the Czech Republic in 2008 when Prague signed an agreement regarding the deployment of American missile defense radars on its territory. Before that, in 2007, Russia suspended the supply of oil and coal to Estonia for one month due to the transfer of Soviet soldiers, explaining that it was a technical problem related to logistics. Recently, in 2015, Russia reduced the oil transit through Lithuanian ports by 20 percent without giving a reason for it.

Because presently there is no integrated gas infrastructure in the EU, the Kremlin can manipulate the volumes, directions, and prices of gas. If there were a deterioration of relations between Russia and NATO, Moscow would be in a position to reduce gas supplies to European countries in combination with waging an information and psychological warfare campaign and, perhaps, cyber-attacks. The Kremlin's new pipeline projects will allow for a selective reduction or suspension of gas supply to Germany, Poland, Hungary, Romania, and others, impacting these countries' energy security in the process.

The active promotion of the Nord Stream 2 pipeline has created some divisions between EU member states and the European Commission. Without a doubt, the main beneficiary of such divisions is Moscow. The Nord Stream 2 pipeline constitutes an important and dangerous barrier to the EU's Energy Union strategy and an obstacle in the plan to boost liquefied natural gas (LNG) imports.

The energy component has become a tool of Russian propaganda to exert psychological pressure on Ukrainian society as well as the international community. At the beginning of the active phase of the Ukrainian-Russian confrontation, Moscow, without any evidence, repeatedly accused Kyiv of illegal extraction of transit gas intended for European consumers. The purpose of those false accusations was to create the image of Ukraine as an unreliable supplier in the eyes of its European partners and to cultivate distrust toward Ukrainian authorities.

The launch of stable reverse gas supplies to Ukraine from its European partners and the country's refusal in 2015 to purchase gas from Russia became the main talking point centered on the question of whether Ukraine would survive the 2015/2016 heating season. A number of experts from the Kremlin predicted an "energy Armageddon" for Ukraine if it does not buy Russian gas.

Furthermore, after Russia annexed Crimea, the Ukraine faces an unprecedented energy crisis after the loss of coal mines and shale gas fields in war-torn regions, and Russia's push to bypass it as a transit country for natural gas to Europe. The EU was warned by Russia not to attempt to block any new pipelines.

The European Commission launched an initiative to create a European Energy Union to reform how Europe produces, transports, and consumes energy. On February 4, 2015, the European Commission began debates on the formation of the Energy Union which included the following objectives:

- Diversifying of energy sources currently available to the Member States
- Helping EU countries become less dependent on energy imports
- Making EU a world leader in renewable energy and leading the fight against global warming

The Energy Union has five stated goals:

- Ensuring security of supply
- Building a single internal energy market
- Raising energy efficiency
- Decarbonizing national economies
- Promoting research and innovation

There are some concerns about the plan, however, such as building a single internal energy market; specifically that it will include enforcement and decarbonizing of everyone's economies, which could cause problems if that definition of decarbonizing differs from Germany's.

Germany has a deep decarbonization plan called the Deep Decarbonization Pathways Project (DDPP) which was co-founded in 2013 by the United Nations' Sustainable Development Solutions Network (SDSN) and the Institute for Sustainable Development and International Relations (IDDRI). The DDPP is a collaborative global initiative that aims to demonstrate how individual countries can transition to a low-carbon economy consistent with the internationally agreed target of limiting the anthropogenic increase in global mean surface temperature to less than 2°C compared with pre-industrial times. In order to achieve this target, global net GHG emissions would need to approach zero by the second half of the century. This would require a profound transformation of energy systems by mid-century, and would have to be achieved through steep declines in carbon intensity in all sectors. This is what is referred to as "deep decarbonization." The German government's plan is geared to reduce domestic GHG emissions by 80–95 percent by 2050 and is designed as a pathway to a GHG-free future in the following decades (Hillebrandt et al., 2015).

With the implementation of the Energy Union, several questions have been raised, such as:

- Will England insist on more wind in Ireland, against their will?
- Will Germany push for greater dependence on Russian natural gas?
- Will Germany try to shut down France's nuclear energy, even though they produce more carbon-free electricity than all other low-carbon sources in Europe combined?

These are the questions being raised now, and show the wide variety of issues that need answers within just the European nations alone. Some EU countries want to exploit domestic fossil fuels. Besides Germany's expansion of dirty coal, the United Kingdom stresses the need to use "all lower carbon sources at our disposal," which means they are referring to using "natural gas." The United Kingdom has a reasonable amount of shale gas.

It has been suggested than an Energy Union might be good if they stick to things like ensuring energy supply,

raising efficiency, and promoting research and innovation. Specifically, it has been suggested that they make an EU-wide smart grid; build more efficient distribution and pipeline delivery options; encourage new technologies, efficiency, and conservation; and provide cheap energy to those who are too poor to buy it on the retail market. But skeptics do not believe this is likely, due to the importance the EU's Commission has placed on its Climate Action and Energy plan. The Climate Action and Energy plan's goal is to:

- Formulate and implement climate policies and strategies
- Take a leading role in international negotiations on climate
- Implement the EU's Emissions Trading System
- Monitor national emissions by EU member countries
- Promote low-carbon technologies and adaptation measures

Energy security is currently high on Europe's political agenda, and was a powerful motivator for the climate agreement in Paris by the end of 2015. Although low oil prices have actually helped Europe against Putin's energy control, the low oil prices are seen as only the temporary fallout of a battle between Saudi Arabia and the United States to bankrupt the American oil shale industry; and that battle is seen as temporary, until one side or the other backs off.

In response to Europe's disparity, vice president of the Energy Union Maroš Šefčovič commented, "We will work to ensure a coherent approach to energy across different policy areas, to create more predictability."

But that is what many see as the problem. When the topic is energy—a commodity surrounded by need, diverse regulatory and market trading systems, money, and power—it is complicated and unpredictable. It is made even more complicated when it must include several countries with very different interests and value systems.

The IEA Executive Director Maria van der Hoeven, points out further complications: "In the coming decades the EU is expected to retire half of its electricity capacity. Nuclear plants are aging, with half of the EU's existing nuclear capacity to be retired by 2040. Environmental rules also require the phase-out of old coal-fired power plants" (Conca, 2015).

Europe's largest energy companies are not pleased with Europe's massive decarbonization plan. The CEOs of Germany's E.ON AG, France's GDF Suez SA, and Italy's Eni SpA have stated that the stability of Europe's electricity generation is at risk from the market structures caused by skyrocketing renewable energy subsidies that have been introduced across the continent over the last decade (Amiel 2013).

Since renewables and their subsidies are a major component of the new European Energy Union's strategy, it is expected that there will be some resistance from

Europe's industry who already think that many European governments' made an unwise decision at the turn of the millennium to promote renewable energy by any means possible (Capgemini, 2013).

The pricing of energy is currently a problem and the EU pricing system does need to be re-designed. For example, in France, nuclear power wholesales for about €40/MWhr ($54/MWhr), but electricity generated from renewables in most of the EU is guaranteed at about €80/MWhr ($108/MWhr), regardless of demand. Retail prices, however, are generally higher than either (EU report on Energy), ranging from €100/MWhr ($135/MWhr) to over €200/MWhr ($270/MWhr). In order for an Energy Union to be successful, this would all have to be aligned fairly for everyone.

The Energy Union also focuses on security. Michael Rühle, Head of the Energy Security Section at NATO says, "In the world today, energy can be a matter of national security, whether we like it or not, and the winners of yesterday can be the losers of tomorrow. It's not just about economics. Policymakers, industry, and consumers need greater awareness of the nexus between energy, resources, and security" (FOE, 2015).

Stable and diversified access to energy is a critical issue for policymakers and it is becoming more important since many of the regions that supply energy, or are transit regions, are suffering from major geopolitical tensions. The goal of acquiring or controlling territories rich in fossil fuel resources has motivated conflicts such as the spread of Islamic State militants into Iraq's oilfields, attacks against Libya's oil production and transportation capabilities, and political and economic sanctions against Russia following the crisis in Ukraine.

Marietje Schaake MEP, Vice-Chair of the European Parliament Delegation for relations with the United States believes that such security issues should not be underestimated by the EU, as instability outside Europe has an effect not only on energy prices but also on immigration flows and the EU's internal political, social, and economic stability (FOE, 2015).

Joan MacNaughton, Executive Chair of the World Energy Trilemma at the World Energy Council makes an interesting point when discussing energy and resource issues that are increasingly mixed with territorial disputes and fluctuating oil prices which then raise the question of instability and risks to global security. She says, "A low-price scenario has real geopolitical implications for instability in the Middle East and Sub-Saharan Africa in particular. With a third of the world's newly discovered fossil fuel deposits over the last 5 years in Sub-Saharan Africa, countries such as Angola, Namibia, and Nigeria are investing heavily in exports, leaving their economies and regional stability open to risks from fluctuating oil prices. In order to resolve global energy supply and security issues, policymakers need to prioritize three goals in particular. All nations need to look at (1) energy access, (2) affordability, and (3) sustainability in order to find the best energy mix and quality of life for their populations." She added that the impact of climate change in Sub-Saharan Africa and the economic deprivation from lack of access to energy are two strong drivers of instability. One manifestation of this instability is migration, which is quickly becoming a major disruptive element for Europe. It is in the best interests of the EU and its partners to address these climate-driven challenges.

The European Nations are also mindful of the shifts in the geopolitical energy supply and demand. While OECD countries are reducing dependence on fossil fuels, demand from non-OECD countries is increasing significantly, along with the risks associated with variable prices. The energy demand is transferring from the West to the East and is a reflection of a shifting of wealth and power in the world.

"The turmoil in energy markets and the discussion on energy security are nothing new," says Marat Terterov, Principal Coordinator of the Energy Charter Secretariat and Principal Director of the European Geopolitical Forum (FOE, 2015). Even when faced with increasing disruption, members of the OPEC are producing 31 million barrels a day.

THE MIDDLE EAST POSITION

While one in every three barrels of exported crude petroleum still originates from the Middle East, there are significant shifts happening in this sector of the economy which are significantly changing the long-held geopolitical situation of the region. According to a recent report from Anja Kasperson, Former Head of Geopolitics and International Security, World Economic Forum, the Middle East decline from the arena they once held such prominent power over is because of the following reasons:

- The United States does not depend on the region as it once did; it now depends more on its own reserves, including banking on its shale gas.
- Iran's diplomatic efforts are not currently successful with the United States.
- Energy prices are volatile, frequently hitting extreme lows, which are felt hard in the Middle East's economy.
- Two former large producers, Libya and Iraq, are in deep crises and producing well below potential; and there is no easy solution in sight.
- Extremists control vast regions of territory across Syria and Iraq and sell crude oil at steep discounts. Though not enough to significantly impact the oil market overall, this factor is sufficient to ensure a steady influx of money that will continue to create havoc in the region and slow down trade and pipelines.
- Saudi Arabia has found itself in a tight spot in this new context and its allegiances are becoming more muddled by the day—over ISIS, over Syria, over Iran, and over the global oil regime by using oil prices to defend their market share and position themselves politically.

Because of all the turmoil in the region, energy policies are going to have to adapt. In order to have strong, workable policies there, they will have to take into consideration the local and regional environment, specifically the individual policies of regional powers. By having a forward-looking energy policy that does this, they will be in a better position for uninterrupted energy access. They must also change their mindset and abandon their former idea that strict security measures and state control are necessary for sustainability. Instead, they need to realize that socioeconomic sustainability and working together is critical for improving the security of the region. At times, oil-dependent governments have suggested economic reforms when the commodity price has been low, but when prices recover and raise again, implementation is either put on the back burner or halted so that no progress is made. But reform is needed now more than ever before—specifically transparent governance and physical security in the region. De-politicization of the energy policies in the region is critical in order to obtain stability in the Middle East. In order for this to happen, the role of government and the private sector is going to have to be reconfigured. Currently, governments still control most of the energy sector in order to manage their economies. In reality, however, this is inefficient. Based on the opinion of the World Economic Forum, if governments limited their roles to clear and fair regulation and allowed the private sector to maximize the energy industry's performance, it would provide governments more wealth to invest in essential public service entities and projects (Kaspersen, 2015).

What occurs globally will have significant impacts on the Middle East's future energy outlook. The politicization of oil prices and the current drop in value could negatively impact the shale oil boom that is currently occurring in the United States and Canada. China's growing energy consumption, however, will still continue to grow, which means that the majority of the Middle East's wealth will still be from petroleum, which further reinforces the need for transparency (Kaspersen, 2015).

NORTH AMERICA: THE NEW MIDDLE EAST?

Within the past few years, North America has become one of the most rapidly growing oil and gas producing regions in the world. Production of shale oil and gas and oil sands (also called tar sands) are increasing as new infrastructures are developed and new technologies are used, enabling more resources to be located. These deposits could be so vast that it could change North America into the new Middle East by 2020, according to a study published by ABC Investments (2013).

In 2011, the U.S. government estimated that the world has proven reserves of approximately 1.35 trillion barrels of oil and that geological reports substantiate that North America has more oil than anywhere else worldwide. The U.S. Energy Information Administration ranked the United States as 13 of the top 15 countries in the world with the largest oil reserves. As of 2010, it was listed as having proved oil reserves of

FIGURE 13.7 Oil sands deposits of Alberta, Canada. (From Energy Resources Conservation Board [ERCB], Edmonton, Canada.)

20.68 billion barrels with a total production of 9.68 million barrels per day (b/d). Canada is ranked third, with proved oil reserves of 175.2 billion barrels for a daily production of 3.48 million barrels.

The wealth in oil lies in oil sands. One of the largest deposits is the oil sands of Alberta (Figure 13.7). The deposit lies in the Athabasca region, encompassing 140,000 square kilometers, making it the largest oil deposit in North America. It is not without its challenges, however. It's thick, soaked sands make it extremely difficult for the crude oil to be extracted from it. In the past, experts believed that only a fraction of the oil could ever be recovered. Today, technology has improved enough that it is estimated that 175 billion barrels are recoverable. In addition, as mentioned in the Summary of Alberta's resource potential of the Energy Resources Conservation Board, several shale and siltstone formations were examined (the Duvernay, Muskwa, Montney, basal Banff/Exshaw, Wilrich, north Nordegg, and Rierdon) and showed significant resources, as well. It is estimated there are 3424 Tcf (trillion cubic feet) of natural gas, 58.6 billion barrels of natural-gas liquids, and 423.6 billion barrels of oil.

The Bakken Formation is another major discovery. It is a natural geological phenomenon that consists of a rock unit occupying about 518,000 square kilometers of the subsurface of the Williston Basin, underlying parts of Montana, North Dakota, and Saskatchewan (Figure 13.8). It is the second-largest oil source in North America. Proved oil reserves are 24 billion barrels and estimated total reserves are huge with

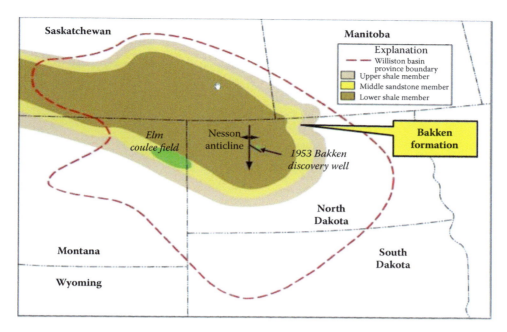

FIGURE 13.8 The Bakken formation. (From United States Geological Survey [USGS].)

501 billion barrels. Geologists estimate that anywhere from 3 to 50 percent of it is recoverable by currently available technology.

The challenge with the Bakken formation is that oil is encased in sheets of nonporous shale. Traditional drilling methods yield little usable oil compared to the expense required to retrieve it. In 2000 as oil prices began to raise, Montana-based geologist named Dick Findley developed a new method to cope with the flat shape of the deposits. He promoted the concept of horizontal drilling. To get the oil out of the rock, he came up with fracturing by pumping sand at high pressure into the well, collapsing the oil-rich rock, and allowing the oil to flow back up.

The quality of the oil obtained from the formation is excellent—free of water and other impurities. There are currently no reliable estimates as to how much oil is actually on the Canadian side of the Bakken. About 25 percent of the Bakken is in Saskatchewan, which is already producing 5 million barrels a year.

North America is what some refer to as a shale gas revolution. Shale is a reservoir—a source rock—with natural gas trapped within it. The natural gas found in these rocks is considered unconventional, similar to coal bed methane. Current technology makes this type of gas more accessible and economical today than it was 30 years ago, however (Figure 13.9). Recent abundant natural gas discoveries in North America have triggered an Industrial Revolution in energy intensive industries. Demand for this type of energy has increased; because natural gas is cheaper than oil, its production was the starting point of the now game-changing shale revolution that is taking place. Gas-fired power generation, natural gas vehicles, and LNG exports are some examples of the new applications of this energy source.

Currently, daily production rates of shale gas in the United States are low, about 50–250 Mcf (thousand cubic feet) per day compared to conventional reservoirs. By 2020, however, U.S. shale liquids growth projections could increase up to 21 Mcf per day. Recoverable reserves are estimated at 500 to 1,000 trillion cubic feet. Shale gas reservoirs have wells that may produce for as long as 80 years and a geographic extent that may exceed thousands of square kilometers.

Based on a 2012 assessment by the U.S. Geological Survey, estimated volumes of technically recoverable crude oil resources resulting from reserve growth for discovered fields outside the United States that have reported in-place oil and gas volumes are 500 million barrels of oil equivalent or more. The mean volumes were estimated at 665 billion barrels of crude oil, 1,429 trillion cubic feet of natural gas, and 16 billion barrels of natural gas liquids, which constitutes a significant portion of the world's oil and gas resources. These shale plays, accessed by hydraulic fracturing (fracking) and horizontal drilling technology, may allow a way to extract shale gas for operators on site. It is possible that this relatively new way of extraction could transform the United States and Canada into energy-independent countries in the future. There is controversy, however, on the environmental effects and safety of fracking as mentioned previously.

The politics of energy are currently receiving key attention with regulatory and environmental concerns as the North American oil supply network undergoes a massive transformation. That puts pressure on existing transportation infrastructure by raising obstacles with restrictions on crude oil exports. Pipeline infrastructure build out has also been slowed or halted. One example is the Keystone XL Pipeline Project, which has been delayed due to environmental concerns (Figure 13.10). The Keystone XL pipeline is a proposed 1,897-kilometer pipeline that would run from the oil sands in

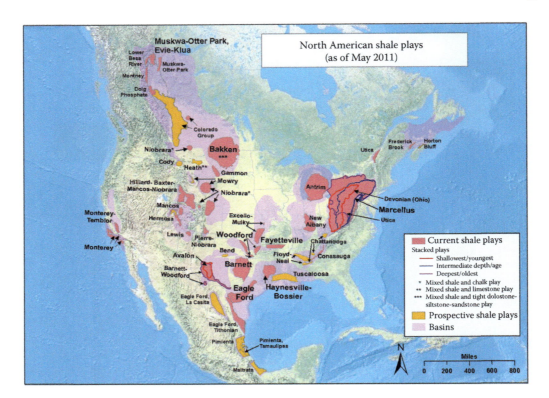

FIGURE 13.9   Shale gas reserves. (From U.S. Energy Information Administration.)

FIGURE 13.10   Proposed pipeline infrastructure. (From CAPP, CIRA.)

Alberta, Canada, to Steele City, Nebraska, where it could join an existing pipe (Figure 13.11). It could carry 830,000 barrels of oil each day. The proposed XL pipeline has the same origin and destination as an operational pipe, also called Keystone, which was granted presidential permit in 2008 by President George W. Bush; but this one takes a more direct route. The XL pipeline would allow for an increased supply of oil from Canada.

A section running south from Cushing in Oklahoma to the Gulf opened in January 2014. At the coast there are additional refineries and ports from which the oil can be exported. The pipeline would be a privately-financed project, with the cost of construction shared among TransCanada, an energy company based in Calgary, Alberta, and other oil shippers. U.S.-produced oil, although less than Canadian, would also be transported by Keystone XL.

FIGURE 13.11 Route of the Keystone pipeline. (Courtesy of Meclee, Creative Commons.)

Presently, Canada sends 550,000 barrels of oil per day to the United States via the existing Keystone Pipeline. The oil fields in Alberta are landlocked and as they are further developed, require means of access to international markets. Many of North America's oil refineries are based in the Gulf Coast, and industry groups on both sides of the border want to benefit. Arguments for the pipeline argue that an increased supply of oil from Canada would mean a decreased dependency on the Middle Eastern market. According to market principles, the more oil in the market, the lower the price for consumers. In addition, the U.S. State Department estimates the project would create 42,000 jobs over a two year construction period—35 of which would remain after the pipeline is built.

To date, the Canadian National Energy Board approved the Keystone XL pipeline in March 2010. Because the XL pipeline crosses the U.S./Canadian border, the project requires presidential permit prior to construction. The assumption was that President Obama would approve the project at that time, but Congress demanded action within 60 days, and the U.S. President turned it down, citing an inadequate environmental assessment. In the meantime, the Environmental Protection Agency (EPA) assessed the project, encouraging President Obama not to approve the pipeline. EPA regulations themselves would not block construction, but litigators could use the EPA to stop construction, even after it is approved by the president.

In February 2015, Congress voted to begin construction immediately, but President Obama vetoed the bill, saying it undermined the necessary review process. Shortly before President Obama rejected the application, the firm hoping to build the pipe, TransCanada asked the U.S. government to put its review of the project on hold. The Keystone XL pipeline has become a controversial project for several reasons. While the U.S. State Department initially said in 2011 Keystone XL would not have significant adverse effects on the environment, that same year the same department determined TransCanada would need to assess alternative routes in Nebraska because the Sandhills region is a fragile ecosystem. Beyond the risks of spillage, the pipeline represents a commitment to develop Alberta's oil sands (BBC News, 2015).

Despite the recent push to find renewable sources of energy and move away from fossil fuels, the amount of oil produced in northern Alberta is projected to double by 2030. It is argued by environmentalists that by developing the oil sands, fossil fuels will be readily available and the trend toward warming of the atmosphere will not be curbed. President Obama's decision to approve or refuse the pipeline is therefore held up as symbolic of America's energy future (Figure 13.12a and b).

In addition, it has been determined that more energy is required to extract oil from the Alberta oil sands than in traditional drilling, and Environment Canada says it has found industry chemicals seeping into ground water and the Athabasca River. This risk to local communities is one of the reasons so many have been opposed to the project. A group called the Cowboy-Indian Alliance marched on the U.S. Capitol earlier this year, citing destruction of local environments.

First Nations groups in Northern Alberta has even gone so far as to sue the provincial and federal government for damages from 15 years of oil sands development they were not consulted on, including treaty-guaranteed rights to hunt, trap, and fish on traditional lands. With the shift in power in the United States, the future remains to be seen.

RUSSIA'S ENERGY POLICY

Russia has an Energy Strategy policy which was approved in 2000 and has been outlined up until 2020. The policy identifies several key components, including:

- Energy development
- Technical development
- Energy efficiency
- Sustainable development
- The environment
- Improved effectiveness and competitiveness

In July, 2008, Russian President Dmitry Medvedev allowed the allocation of oil and gas deposits on the continental shelf without an auction. They signed an agreement with China on February 17, 2011, that in exchange for $25 billion in Chinese loans to Russian oil companies, Russia would supply China

(a)

(b)

FIGURE 13.12 (a) Keystone XL demonstration at the White House, August 23, 2011. (Courtesy of Josh Lopez, Creative Commons.) (b) An estimated crowd of 35,000 to 50,000 gathers near the Washington monument on February 17, 2013 to protest the Keystone XL pipeline and support action on climate change. (Courtesy of Jmcdaid, Creative Commons.)

with large quantities of crude oil via new pipelines for the following 20 years (Zhu, 2015).

Russia has considerable oil reserves and it comprises a significant portion of their economy. Oil and gas comprise more than 60 percent of their exports and account for more than 30 percent of their gross national product (GNP). As per their energy policy, they pump 10.6 million barrels of oil a day (4 billion barrels a year), and have 100 billion barrels of proven oil reserves—enough projected to supply their needs for 30 years. Because of their finite resource amount and the fact they are depleting it while oil is at historic low prices; it is having negative ramifications

for their economy. Currently, they are highly dependent on the export of these natural resources and they are also using them to their political advantage. The situation mentioned with the EU is one example, and the United States and other Western countries have stepped in to lessen the dependence of the EU on Russian resources. One of the worst uses of oil and gas for political gain was the situation in the Ukraine. Beginning in the mid-2000s, Russia and Ukraine had several disputes and Russia then threatened to cease the supply of gas (Figure 13.13). Russia plans to completely abandon gas supplies to Europe through Ukraine after 2018.

FIGURE 13.13 Major Russian gas pipelines to Europe. (Courtesy of Samuel Bailey, Creative Commons.)

RUSSIA–UKRAINE GAS DISPUTES

The Russia–Ukraine gas disputes refer to a number of serious disputes between Ukrainian oil and gas company Naftogaz Ukrayiny and Russian gas supplier Gazprom over natural gas supplies, prices, and debts. These disputes have escalated beyond simple business disputes into transnational political issues—involving political leaders from several countries—that threaten natural gas supplies in numerous European countries dependent on natural gas imports from Russian suppliers, which are transported through Ukraine. Russia provides approximately one-fourth of the natural gas consumed in the EU; of which approximately 80 percent of those exports travel through pipelines across Ukrainian soil prior to arriving in the EU.

A serious dispute began in March 2005 over the price of natural gas supplied and the cost of transit. During this conflict, Russia claimed Ukraine was not paying for gas, but diverting that which was intended to be exported to the EU from the pipelines. Ukrainian officials at first denied the accusation, but later Naftogaz admitted that natural gas intended for other European countries was retained and used for domestic needs. The dispute escalated to a critical point on January 1, 2006, when Russia cut off all gas supplies passing through Ukrainian territory. On January 4, 2006, a preliminary agreement between Russia and Ukraine was achieved, and the supply was restored. The situation calmed until October 2007 when new disputes began over Ukrainian gas debts. This led to reduction of gas supplies later in March 2008. During the last months of 2008, relations once again became tense when Ukraine and Russia could not agree on the debts owed by Ukraine.

In January 2009, the disagreement resulted in supply disruptions to several European nations, with 18 European countries reporting major drops in or complete cut-offs of their gas supplies transported through Ukraine from Russia. In September 2009 officials from both countries stated they felt the situation was under control and that there would be no more conflicts over the topic, at least until the Ukrainian 2010 presidential elections (McBride, 2009). In October 2009, however, another disagreement arose about the amount of gas Ukraine would import from Russia in 2010. Ukraine intended to import less gas in 2010 as a result of reduced industry needs because of its economic recession; but Gazprom insisted that Ukraine fulfill its contractual obligations and purchase the previously agreed upon quantities of gas. Then, on June 8, 2010, a Stockholm court of arbitration ruled Naftogaz of Ukraine must return 12.1 billion cubic meters of gas to RosUkrEnergo, a Swiss-based company in which Gazprom controls half the interest. Russia accused the Ukrainian side of diverting gas from pipelines passing through Ukraine in 2009. Several high-ranking Ukrainian officials stated the return "would not be quick," further illustrating the extreme geopolitical tension in this area. Those hurt the most during these tensions were the people in the areas where the gas was banned during the cold winter months who nearly froze to death.

Under Putin, efforts have been made to attempt control over European energy. Russia played a major role in canceling the construction of the Nabucco pipeline, which was to have transported gas from the Caspian Sea to Europe in order to bypass Russia, overriding the South Stream pipeline. The pipeline was sponsored by the EU and was slated to run from Azerbaijan across Georgia and Turkey to the Bulgarian border. Russia's South Stream project was begun in 2007, and was designed to export gas from Russia under the Black Sea and through the Balkans to Western Europe. Work on South Stream began in December 2012 and was expected to be completed by 2018 (Weiss, 2013). The pipeline was designed to be capable of supplying 63 billion cubic meters of gas per year. The failure of the Nabucco pipeline project was viewed in terms of lack of coordination on the EU side as well as Russian aggression. In Eastern Europe, where the Nabucco pipeline was to have supplied gas from the Caspian region to countries that are highly dependent on Russian gas, governments are seeking to develop the production of shale and liquefied gas. This situation brings up the issue supporting why groups of countries, such as the EU, need to band together and have coordinated energy plans with a future vision so that they are not in a position to be controlled by the political interests and agendas of other major political powers.

Then in December 2014, more than 7 years after the South Stream pipeline project was first announced, Putin suddenly announced during a visit to Turkey that the project was cancelled, stating that the European Commission had been unconstructive. According to Putin, "It's not that the European Commission has helped to implement this project—it's that we see obstacles being created in its implementation." In reality, Russia expected their international law to govern EU-Russian relations, not the European Commission's Intergovernmental Agreements (IGAs) to be the ruling body, so they shut the project down (Behrens, 2014; Reuters, 2014).

The cancellation does cause an inconvenience for Brussels and finding an alternative source of gas for southeastern Europe, however. The South Stream would have provided energy for Serbia, Bulgaria, and Hungary because it offered a supply of gas that did not pass through Ukraine, which meant that there was less of a risk for supply disruption. Vaclav Bartuska, the Czech government's ambassador for energy security noted, "The decision shows that the outside world does have an impact on Russia, and probably a bigger one than Russia would like to admit."

Countries further south which depend nearly entirely on imported Russian gas shipped to them via Ukraine were even more dismayed at the cancellation of the pipeline's construction. They were victims of the January 2009 pricing dispute when the shipment of gas stopped moving through Ukraine and they were left with fuel shortages in sub-zero winter weather.

Then, to show how volatile the political situation is, in June 2016, Putin stated that Russia had not "definitively" canceled either the South Stream or Turkstream European gas export pipeline projects but needed a clear position on them from Europe (Dyomkin, 2016). The Turkstream project was slated to increase Russian gas exports to southern Europe via Turkey, but talks were frozen on that after a Sukhoi Su-24 attack aircraft was shot down by a Turkish jet near the Syrian-Turkish border in November, 2014.

Russia has also been involved in talks with Israeli Prime Minister Benjamin Netanyahu concerning taking part in the development of Israel's offshore gas fields; and also stated that they would look for other markets—most likely Asia—if Poland stopped buying Russian gas after their contract expired, as further illustration of the volatile, rapidly changing energy landscape (Klimentyev, 2014).

CONCLUSIONS

As we have seen, many countries worldwide find the fossil fuel market ranking high on their GNP, and it certainly plays a highly significant role in everyone's daily life. Those in the developed world cannot get through a single day comfortably without the use of fossil fuels—and even if every country in the world pledged to make the switch to renewable energy right now, it would still take us years to rebuild our infrastructure to completely make the switch. So, the bottom line is that we are stuck with fossil fuels–even if it is temporary, for a few years yet. But whatever that time period may be, is partially dependent on technology, and partly dependent on us. A big part of the equation is public attitude and acceptance of renewables versus the comfortable position of fossil fuels. And even more important: each country's respective government. Without government support, it will be virtually impossible to make the switch. Governments must be fully supportive and on board before transitions can be made. If governments are still leaning and negotiating for fossil fuels, then it slows the transition to renewables down. It is back to the age old question: how do we make this transition to sustainable energy (which we must have) and eventually phase fossil fuels out (which the majority of the world wants to keep)?

REFERENCES

ABC Investments. February 17, 2013. Deliverance day: The new middle east of America by 2020? *Seeking Alpha.* http://seekingalpha.com/article/1199701-deliverance-day-the-new-middle-east-of-america-by-2020 (accessed November 11, 2016).

Amiel, G. October 11, 2013. Energy bosses call for end for subsidies for wind, solar power. *Wall Street Journal.* http://stream.wsj.com/story/latest-headlines/SS-2-63399/SS-2-352276/ (accessed November 10, 2016).

BBC News. November 6, 2015. Keystone XL pipeline: Why is it so disputed? *BBC News*, November 6. http://www.bbc.com/news/world-us-canada-30103078 (accessed November 11, 2016).

Behrens, A. December 5, 2014. The declared end of south stream and why nobody seems to care. *CEPS Commentary.* https://www.ceps.eu/system/files/AB%20Southstream%20Pipeline.pdf (accessed November 13, 2016).

Capgemini. October 10, 2013. European energy markets observatory 2013. *Capgemini.* https://www.capgemini.com/resources/european-energy-markets-observatory-2013-full-study (accessed November 10, 2016).

Chilcoat, C. March 10, 2016. Energy relations are reshaping geopolitics. *Oil Price.* http://oilprice.com/Energy/Energy-General/How-Russia-Japan-Energy-Relations-Are-Reshaping-Geopolitics.html (accessed November 13, 2016).

Conca, J. 2015. The role of the new European energy union. *Forbes.* http://www.forbes.com/sites/jamesconca/2015/02/09/the-role-of-the-new-european-energy-union/#6396959b387e (accessed November 10, 2016).

Dyomkin, D. June 7, 2016. Putin says Russia hasn't canceled south stream, turkstream gas projects. *Commodities.* http://www.reuters.com/article/us-russia-gas-exports-idUSKCN0YT228. (accessed November 13, 2016).

European Commission. 2015. Energy union: Secure, sustainable, competitive, affordable energy for every European. Press release, Brussels. February 25, 2015. http://europa.eu/rapid/press-release_IP-15-4497_en.htm (accessed November 13, 2016).

Friends of Europe. July 2, 2015. The new geopolitics of energy—winners and losers. *Friends of Europe.* http://www.friendsofeurope.org/event/new-geopolitics-energy-winners-losers/ (accessed November 7, 2016).

Hillebrandt, K. et al. 2015. Pathways to deep decarbonization in Germany. *SDSN–IDDRI.* http://deepdecarbonization.org/wp-content/uploads/2015/09/DDPP_DEU.pdf (accessed November 10, 2016).

Kaplan, R. D. April 4, 2014. The geopolitics of energy. *Forbes.* http://www.forbes.com/sites/stratfor/2014/04/04/the-geopolitics-of-energy/#d801359547c6 (accessed November 7, 2016).

Kaspersen, A. June 25, 2015. The ghanging geopolitics of oil in the middle east. *World Economic Forum.* https://www.weforum.org/agenda/2015/06/the-changing-geopolitics-of-oil-in-the-middle-east/ (accessed November 11, 2016).

Klimentyev, M. December 2, 2014. Cancellation of south stream pipeline gives EU an economic headache. *Newsweek.* http://www.newsweek.com/cancellation-south-stream-pipeline-gives-eu-economic-headache-288516 (accessed November 11, 2016).

McBride, J. 2009. UPDATE 2-Gazprom sees political risk to Ukraine gas payments. *Reuters.* http://www.reuters.com/article/gazprom-ukraine-idUSLC31093620090912?pageNumber=1&virtualBrandChannel=0 (accessed November 11, 2016).

Reuters. December 2, 2014. Cancellation of south stream pipeline gives EU an economic headache. *Newsweek.* http://www.newsweek.com/cancellation-south-stream-pipeline-gives-eu-economic-headache-288516 (accessed November 12, 2016).

Verrastro, F. A., S. O. Ladislaw, M. Frank, and L. A. Hyland. October 2010. The geoppolitics of energy: Emerging trends, changing landscapes, uncertain times. *Center for Strategis & International Studies.* https://csis-prod.s3.amazonaws.com/s3fs-public/legacy_files/files/publication/101026_Verrastro_Geopolitics_web.pdf (accessed November 7, 2016).

Weiss, C. July 13, 2013. European union's nabucco pipeline project aborted. *International Committee of the Fourth International.* https://www.wsws.org/en/articles/2013/07/13/nabu-j13.html (accessed November 11, 2016).

World Energy Outlook 2014. International energy agency. http://www.worldenergyoutlook.org/weo2014/ (accessed November 7, 2016).

World Nuclear Association. October 2016. Nuclear power in the European Union. *World Nuclear Association.* http://www.world-nuclear.org/information-library/country-profiles/others/european-union.aspx (accessed November 10, 2016).

World Nuclear News. February 6, 2015. Europe starts work on energy union. *World Nuclear News.* http://www.world-nuclear-news.org/EE-Europe-starts-work-on-Energy-Union-0602154.html (accessed November 10, 2016).

Zhu, C. 2015. China, Russia ink oil loan agreement. *Baijing Online.* http://english.caijing.com.cn/2009-02-18/110070270.html (accessed November 12, 2016).

SUGGESTED READING

Goodell, J. 2007. *Big Coal: The Dirty Secret Behind America's Energy Future.* New York: Mariner Books. 352 p.

Hauter, W. 2016. *Frackopoly: The Battle for the Future of Energy and the Environment.* New York: The New Press. 384 p.

Hillebrand, E. 2015. *Energy, Economic Growth, and Geopolitical Futures: Eight Long-Range Scenarios.* Cambridge, MA: MIT Press. 248 p.

Open University. 2016. *Future Energy Demand and Supply.* Kindle Edition. https://www.amazon.com/Future-energy-demand-supply-University-ebook/dp/B01D8×6QPI/ref=sr_1_8?s=books&ie=UTF8&qid=1479087727&sr=1-8&keywords=the+future+of+energy (accessed November 13, 2016).

Pascual, C. January 8, 2008. The geopolitics of energy: From security to survival. *Brookings.* https://www.brookings.edu/research/the-geopolitics-of-energy-from-security-to-survival/ (accessed November 7, 2016).

Pascual, C. September 2015. The new geopolitics of energy. Columbia/SIPA Center on Global Energy Policy. http://energypolicy.columbia.edu/sites/default/files/energy/The%20New%20Geopolitics%20of%20Energy_September%202015.pdf (accessed November 7, 2016).

Pascual, C., J. T. Mathews, D. Gordon, and B. Jones. July 24, 2014. The new geopolitics of energy: Challenges and opportunities. *Carnegie Endowment for International Peace.* http://carnegieendowment.org/2014/07/24/new-geopolitics-of-energy-challenges-and-opportunities/hfua (accessed November 7, 2016).

Schwendiman, G. 2015. *The Future of Clean Energy: Who Wins and Who Loses as the World Goes Green.* Bloomington, IN: AuthorHouse. 216 p.

14 Applying Technology to Sustainability, Part I
Urban Planning and Green Buildings

OVERVIEW

This chapter opens the discussion about utilizing technology and various cutting-edge applications to promote and encourage sustainability. In this first of a two-part installment, we will be taking a close look at urban planning techniques and green building strategies. This is especially important today as we attempt to wean our societies off of traditional energy sources and make the transition to more sustainable options. It is also timely as we address critical issues, such as climate change, and look for new ways to deal more effectively with the natural environment around us in an effort to conserve resources and provide for current populations and beyond.

This chapter looks first at urban planning and smart growth. It discusses the basic tenets of what smart growth is and why it is important to move in that direction for the future. Next, it introduces the concept of the Smart Grid and what its implementation will mean for the country as a whole, and what opportunities it will provide for every individual. Following that, we will transition to green building and how the concept of cogeneration can be a logical means for providing energy to heat some buildings. Green building benefits, ideas, strategies, and goals will be presented and examples of those leading the way will be shared as inspiration for others to follow.

INTRODUCTION

Green spaces are a great benefit to the environment. They filter pollutants and dust from the air, they provide shade and lower temperatures in urban areas, and they reduce erosion of soil into waterways. Those are some of the environmental benefits that green spaces provide. But there are so many more: there are economic benefits, lifestyle benefits, community benefits, recreational benefits, health benefits, and aesthetic benefits. And the list just goes on. Green spaces add value in multiple dimensions. Yet to have them, they have to be well thought out, planned for, and designed properly. That is where urban planning comes in. A rapidly growing field today, urban planning provides the blueprint—or "greenprint"—for sustainable cities of the future. Green buildings follow the same guidelines. They must also be planned for sustainability. Fortunately, the trend today is to now think these aspects through and take into account the environment and sustainability. That is where the fields of sustainable urban planning

and green architecture come into play. Let us take a look at how they are benefitting millions across the planet.

DEVELOPMENT

Development decisions affect many of the things that touch people's everyday lives, such as their homes, their health, the schools their children attend, the taxes they pay, their daily commute, the natural environment around them, economic growth in their community, and opportunities to achieve their dreams and goals. What, where, and how communities build will affect their residents' lives for generations to come. In order to make a lasting difference in the way we use natural resources and fight climate change, it will be necessary to make critical lifestyle choices that may be very different from those we are traditionally used to. A new approach is now catching on and taking hold that seems to solve these problems in a way that benefits both human and natural resources now and in the long term.

URBAN PLANNING AND SMART GROWTH

Today, communities of all sizes across the country are using creative strategies to develop in ways that preserve natural lands and critical environmental areas, protect water and air quality, and reuse already-developed land. They are conserving resources by reinvesting in existing infrastructure and rehabilitating historic buildings. By designing neighborhoods that have homes near shops, offices, schools, houses of worship, parks, and other amenities, communities are now giving residents and visitors the option of walking, bicycling, taking public transportation, or driving as they go about their business. In addition, by offering a varied selection of housing types, it is now making it easier for senior citizens to stay in their neighborhoods as they age, young people to afford their first home, and families at all stages in between to find a range of homes that work for them that they can afford. This is *smart growth*.

Smart growth approaches are uniquely designed to enhance neighborhoods and involve residents in the development decisions. Because of this innovative approach, these communities are creating vibrant places to live, work, and enjoy recreational opportunities. The resulting high quality of life makes these communities economically competitive, creates business opportunities, and strengthens the local tax base. Based on the experience of communities around the nation that have used smart growth approaches to create and maintain appealing neighborhoods, the Smart Growth Network developed a set of ten basic principles to guide smart growth strategies:

- Utilize mixed land uses.
- Take advantage of compact building design.
- Create a range of housing opportunities and choices.
- Create walkable neighborhoods.
- Foster distinctive, attractive communities with a strong sense of place and belonging.
- Preserve open space, farmland, natural beauty, and critical environmental areas.
- Strengthen and direct development toward existing communities.
- Provide a variety of transportation choices.
- Make development decisions predictable, fair, and cost-effective.
- Encourage community and stakeholder collaboration in developmental decisions.

The Smart Growth Network is a partnership of government, business, and civic organizations that support smart growth. The EPA is one of the founding partners of the network. Created in 1996, the network has become a central collection site of information for individuals, businesses, and government institutions interested in learning about smart growth strategies. The Smart Growth Network website, Smart Growth Online (www.smartgrowth.org), features many smart growth-related news, events, information, research, presentations, and publications on the topic for anyone interested in learning more. The partners in the endeavor include the following groups:

American Farmland Trust (www.farmland.org)
American Institute of Architects, Center for Communities by Design (http://www.aia.org/about/initiatives/AIAS075265)
American Planning Association (https://www.planning.org/)
American Public Health Association (http://www.apha.org)
American Society of Landscape Architects (https://www.asla.org/)
Association of Metropolitan Planning Organizations (http://www.ampo.org/)
Association of State and Territorial Health Officials (http://www.astho.org/)
Center for Neighborhood Technology (http://www.cnt.org/)
Congress for the New Urbanism (https://www.cnu.org/)
Conservation Fund (http://www.conservationfund.org/)
Delaware Valley Smart Growth Alliance (http://www.delawarevalleysmartgrowth.org/)

Enterprise (http://www.enterprisecommunity.com/)
Environmental Finance Center Network (https://www.epa.gov/envirofinance/efcn)
Environmental Law Institute (http://www.eli.org/)
Florida Department of Health (http://www.floridahealth.gov/)
Forterra (http://forterra.org/)
Funders Network for Smart Growth and Livable Communities (http://www.fundersnetwork.org/)
Institute of Transportation Engineers (http://www.ite.org/)
International City/County Management Association (http://icma.org/en/icma/home)
Local Government Commission (https://www.lgc.org/)
Local Initiatives Support Corporation (http://www.lisc.org/)
National Association of Counties (http://www.naco.org/)
National Association of Conservation Districts (http://www.nacdnet.org/)
National Association of Local Government Environmental Professionals (http://www.nalgep.org/)
National Association of Realtors (http://www.realtor.org/)
National Center for Smart Growth Research and Education (http://www.smartgrowth.umd.edu/)
National Multifamily Housing Council (http://www.nmhc.org/)
National Oceanic and Atmospheric Administration (http://www.noaa.gov/)
National Trust for Historic Preservation (http://www.preservationnation.org/)
Natural Resources Defense Council (https://www.nrdc.org/)
Northeast-Midwest Institute (http://www.nemw.org/)
Project for Public Spaces (http://www.pps.org/)
Rails-To-Trails Conservancy (http://www.railstotrails.org/)
Scenic America (http://www.scenic.org/)
Smart Growth America (https://smartgrowthamerica.org/)
State of Maryland (http://www.mdp.state.md.us/)
Surface Transportation Policy Partnership (http://www.mdp.state.md.us/)
Sustainable Community Development Group (http://sustainablecommunitydevelopmentgroup.org/)
Trust for Public Land (http://www.tpl.org/)
Urban Land Institute (http://uli.org/)
U.S. Environmental Protection Agency (https://www.epa.gov/smartgrowth)
U.S. Forest Service (http://www.fs.fed.us/)
Virginia Tech Metropolitan Institute (http://www.spia.vt.edu/faculty-research/mi/)

Basic Objectives of Smart Growth Urban Planning

There are several considerations and goals of smart growth urban planning. This section will address many of them, such as the importance of sustainable urban development, fighting against sprawl, why conserving open spaces is necessary, the role of sustainable zoning, the necessity of investing in downtowns, why creating sustainable landscapes and buildings is key, the benefits of redeveloping brownfields, and how smart growth improves both air and water quality, thereby addressing the issue of climate change.

Sustainable Urban Development

When we refer to smart growth, sustainable urban development should be guided by a sustainable planning and management vision that incorporates three critical components. They are

- Interconnected green space
- A multimodal transportation system
- Mixed-use development

Diverse public and private partnerships should be used to create sustainable and livable communities that protect historic, cultural, and environmental resources. In addition, policymakers, regulators, and developers should support sustainable site planning and construction techniques that reduce pollution and create a balance between built and natural systems (Figure 14.1a).

New sustainable urban developments or re-developments should provide a variety of commercial, institutional, and educational uses as well as housing styles, sizes, and prices. The provision of sidewalks, trails, and private streets, connected to transit stops and an interconnected street network

(a)

(b)

(c)

(d)

FIGURE 14.1 (a) The smart growth concept is a planned economic and community development that attempts to curb urban sprawl. (From Federal Highway administration/Department of Transportation.) (b) Murray City, Utah, smoke stacks. This had been an industrial area for years. Even after the industrial area shut down, the smokestaks, part of a smelter in the silver industry, had stood abandoned for decades when the city finally decided to demolish them, clean the site up, and redevelop the land. (Courtesy of Wikipedia Commons.) (c) Today, the Intermountain Medical Center occupies the land the smelter previously occupied. It is surrounded by a busy shopping center. (Courtesy of Wasatch Images.) (d) Redeveloped brownfield. Looking northeast from near Chapel Avenue at Port Liberte town houses with Waterfront Walkway along canal on a sunny early afternoon. (Courtesy of Jim Henderson, Creative Commons.)

within these mixed-use developments should provide mobility options and help reduce pollution by reducing vehicle trips. Walking, bicycling, and other mobility options should be encouraged throughout the urban mixed-use core and mixed-use neighborhoods with easily accessed and well-defined centers and edges.

Fighting Sprawl

The current use of land-use-based zoning, real-estate tax laws, and highway design regulations has created automobile-dominated sprawl conditions where cars are required for almost all activities. This has led to the ever-increasing congestion and longer commute times that most cities today, unfortunately, have to deal with. This decentralized suburban expansion has created developments with no sense of place, which consume exorbitant amounts of land, and make it necessary to have huge infrastructure commitments. This often contributes to the deterioration of urban centers, leaving them vacant and in economic decline; where they were once bustling centers of commerce.

Development patterns that result in sprawl are not in the long-term best interest of cities, small towns, rural communities, or agricultural lands. As communities plan for growth and change, in-fill and redevelopment should be utilizing the existing infrastructure—not spreading out and consuming new resources. Public agencies should be promoting and overseeing processes for remediation of urban sites to relieve the pressure to develop at the urban fringe. Plans should identify open lands that can either be sustainably developed into green spaces or left alone. By doing this, preserving the open green spaces in cities would eliminate sprawl because these spaces would provide important outlets within the city instead.

Sustainable urban development means responsible growth and development strategies that are broader in vision and more regional in scale. Several sustainable growth strategies currently exist. Some of them include

- Urban in-fill: new development that is sited on vacant or undeveloped land within an existing community and that is enclosed by other types of development
- Suburban redevelopment: reinventing infrastructure for compact development, thereby removing the necessity to travel extensive distances outside residential areas
- Open-land development: sustainably developing open lands into green spaces

These can all lead to more diverse housing styles and multimodal transit. Because each community is unique, the solution for each location must be tailored to fit its particular needs. Responsible and innovative development strategies at the federal, state, and local levels are needed to guide private development.

Conserving Open Spaces

Preserving natural lands and encouraging growth in existing communities protect farmland, wildlife habitat, outdoor recreation, and natural water filtration, which ensures clean drinking water. Communities should take advantage of government and private initiatives, such as conservation districts and open land trusts in order to preserve open space. Open space can help curb scattered development, protect watersheds and natural habitat, maintain historic and cultural assets, and provide diverse recreational opportunities. Lack of proper planning at the onset can lead to the fragmentation of conserved lands. Therefore, it is critical to take open spaces into consideration during the planning process.

Sustainable Zoning

The replacement of conventional zoning codes that control land use with those that control physical form (Form-Based Code, FBC) can benefit sustainable growth and development. A FBC is a way of regulating land development to achieve a specific urban form. FBCs focus on the physical form of an area—rather than separation of uses, as conventional codes do today—as the organizing principle, with less of a focus on regulated land use. FBCs offer a powerful alternative to the conventional zoning regulation that has been traditionally used.

FBCs are viewed as a new response to the modern challenges of urban sprawl, deterioration of historic neighborhoods, and neglect of pedestrian safety in new developments. Because current zoning regulations typically discourage compact, walkable urbanism, FBCs are a tool to address these problems and provide local governments the regulatory means to achieve better development objectives. Implementing FBCs can result in communities that fit their place and time; have a mix of uses that are appropriately scaled; enjoy pedestrian friendly, well-defined public realms; and are overall more sustainable.

Investing in Downtowns

Encouraging businesses, nonprofits, governments, and cultural institutions to locate their offices and other facilities within the urban core as opposed to suburban or fringe locations can support sustainable urban development. Investing in an urban core can help support urban core revitalization efforts and attract and retain businesses and services. Often, tax credits or other incentives are needed to encourage the preservation or rehabilitation of historic properties or green spaces within the urban core.

Creating Sustainable Landscapes and Buildings in an Urban Context

Decreasing impervious pavement areas, providing abundant and usable interconnected greenways and open space, implementing sustainable stormwater techniques, and planting or preserving vegetation all help mitigate greenhouse gas (GHG) emissions from urban areas. These are all important steps in fighting climate change. In addition, siting buildings to maximize passive heating/cooling, using energy-efficient building technologies, including green or cool roofs can also help lower building emissions—another plus in dealing with rising atmospheric temperatures. Communities should move toward energy conservation and nonoil and coal-based alternatives, such as solar, wind, thermal, and biomass, which can reduce dependency on nonrenewable resources, as well as minimize air, water, and thermal pollution. In creating sustainable landscapes, it provides a proactive avenue for realistic long-term solutions.

Redeveloping Brownfields

Brownfields are abandoned, idled, or underused industrial and commercial properties where redevelopment is complicated by real or perceived environmental contamination. Cleaning up and redeveloping a brownfield can remove blight and environmental contamination, promote neighborhood revitalization, lessen development pressure on undeveloped land, and use existing infrastructure. The redevelopment of brownfield sites enables communities to reuse abandoned areas that are often located in urban centers with existing infrastructure

There are several benefits of utilizing brownfields. Brownfield redevelopment ensures that contaminated land is cleaned up and restored. Many existing brownfield sites are contaminated as a result of past industrial or commercial uses. Depending on what these sites were formerly used for, contaminants may include a range of toxins, such as petroleum, metals, asbestos, pesticides, PCBs, and solvents. These contaminants may create significant health and safety risks for those who live and work close to brownfield properties. When abandoned brownfield lands are left neglected, contaminants may migrate off-site, creating hazards for others nearby. Even where soil and groundwater at a brownfield site are not contaminated, deteriorating buildings and surface debris on these sites may still pose health and safety risks and have a negative impact on property values and on a neighborhood's image. Cleaning up these sites helps to improve the quality of the environment in the community and removes real and perceived threats to health and safety.

Cleaning up brownfields can be expensive, so contamination is unlikely to be removed unless a site is intended for a new use that makes the cleanup financially worthwhile. Municipalities have begun to see the potential for brownfields to improve communities and make them more sustainable as the population grows, and may offer incentives to promote brownfield redevelopment (Figure 14.1b–d).

Communities may experience many environmental, social, and economic benefits from brownfield redevelopment, such as:

- Removing actual and potential sources of land, water, and air contamination
- Recovering desirable locations, allowing for smarter growth through urban intensification
- Removing or renovate abandoned and derelict buildings, decreasing the risk of injury, vandalism, and arson
- Preserving historical landmarks and heritage architecture
- Beautifying urban landscapes
- Reviving older urban communities and surrounding areas
- Locating new development in areas where better use can be made of existing municipal infrastructure and services such as transit

- Increasing property assessment values and the resulting tax base, leading to increased revenue for governments
- Reducing urban sprawl
- Preserving open green land, which may be productive farmland or environmentally significant land

Communities should take advantage of programs that focus on facilitating the cleanup and reuse of these areas by awarding grants, capitalizing loan funds, and providing technical assistance (cielap.org, 2016).

Improving Air and Water Quality

Compact communities with a mix of uses and transportation options make it much easier for people to choose to walk, bicycle, or take public transit instead of driving. Those who choose to drive generally can drive shorter distances, cutting down drastically in GHG emissions. Less travel by motor vehicles reduces air pollution by smog-forming emissions and other pollutants (National Household Travel Survey).

Compact development and open space preservation also help protect water quality by reducing the amount of paved surfaces and by allowing natural lands to filter rainwater and runoff before it reaches drinking water supplies. Green infrastructure techniques, which mimic the natural processes of capturing, holding, absorbing, and filtering stormwater, can be incorporated into streets, sidewalks, parking lots, and buildings.

Addressing Climate Change

Smart growth offers transportation options and land use patterns that reduce air pollution, which also cut the emission of GHGs that contribute to climate change. By using energy-efficient, green building techniques, these also reduce GHG emissions from energy use. Smart growth strategies are beneficial because they enable communities to be forward-thinking and prepare for the impacts of climate change.

A NEW WAY TO PLANT URBAN TREES

Urban trees are a critical part of green infrastructure systems and provide a range of benefits. They reduce the urban heat island effect, manage stormwater, and provide shade that lengthens the life of materials. In the summer, shadier streets also mean lower neighborhood temperatures. The larger the tree, the more valuable it is in an urban setting. Peter MacDonagh, an urban landscape architect of Kestrel Design Group, states that "older trees are far more valuable than younger ones, so work needs to be done to preserve these." A 30-inch diameter breast height (DBH) tree provides 70 times the ecological benefits of a 3-inch DBH tree. For example, a large tree intercepts 79 percent of rain hitting the ground, providing the "best green infrastructure you can find."

He stresses that the key to attaining the mature trees necessary to benefit a green urban environment lays in strict adherence to the soil type the trees are planted in. They require an exacting composition of loam, sand, compost, and clay silt due to their ability to provide the correct nutrients for the vegetation while filtering out heavy metals, phosphorous, and nitrogen. Nitrogen is especially harmful because its runoff can cause algae blooms and kill other life if it is allowed to get to the watershed in large amounts.

New techniques are also being used with the way the trees are being planted. A new technique utilizing suspended pavements is being tested. These utilize 48-inch deep silva cells—which resemble rubber packing crates. Within the loam and silva cells are irrigation systems that move water to the trees. Tested in a pilot project in Queens Quay, Toronto, this approach has proven successful (Green, 2010).

In another project on the Lincoln Center roof in New York City, 30 trees were planted on a roof deck. Landscape architect, James Urban observed that "There were lots of obstacles—everything is going on in the built urban environment." Urban navigated the shallow roof, elaborate lighting systems, and thin paving on top of the deck. He ended up adding in "geogrids" and gravel that helped ensure the new platform could provide a safe growing environment for trees and also bear the weight of a light pick-up truck or ambulance. Able to solve those obstacles, the one concern he was left with was the CO_2 that was emitted during the construction of the project; cautioning that those emissions needed to be taken into account and a solution necessary in order to make projects like this as green as possible (Green, 2010).

In a project completed in Normal, Illinois, funded by federal, state, and local governments, a "people space in the center of a roundabout" was created. The area features a set of urban trees, outer lawn, bog water infiltration system, and circular stream filled with cleansed water. According to landscape architect Peter Schaudt, it represents the "park of the future." While there were initial concerns about the completed project being safe for a public gathering place, it has resulted in a safe space mostly made so due to the trees, which separate the cars from the social space. Trees in the traffic circle and nearby streets were supported by silva cells, loam and drip irrigation (similar to those used in Toronto). Schaudt adds, "The circle's trees were set up to live a long time." He believes in planning for the "4th dimension—time." And likes to show those he works with what the site will look like in 25 years.

Architect Peter MacDonough stresses that well-planted trees are not only more cost-efficient, they also provide more ecosystem benefits. As an example, the government in Minneapolis conducted research and discovered that they could either spend $3.5 million on new stormwater conveyance pipes to handle current runoff or invest $1.5 million on silva cell systems (Green, 2010). Looking toward the future and the benefits of investing in a green future can pay big dividends.

Green Infrastructure in Urban Settings

The future of wildlife and wildlife habitat in rural, suburban, and urban settings depends on an environmentally responsible strategy of land management that emphasizes a mix of spaces for both people and wildlife. The use of ecological information during the design process creates a more compatible combination between land use and the natural environment (Figure 14.2a–c). This would also increase public awareness of wildlife, wildlife habitats, and their value to human welfare. Wildlife and wildlife habitat values need to be considered early in any sustainable urban development process.

If land-use decisions are implemented independently, rather than coordinated on a multiple-use level, this will result in a landscape with fragmented wildlife habitat. While both human and wildlife needs will be compromised, wildlife will see the biggest impact in terms of climate change impacts because they cannot easily migrate to new areas—especially when those areas may already be urbanized and destroyed. The concept of landscape ecology is to design landscape mosaics—matrices of patches, corridors, and ecological edges—to serve a broader and more diverse range of ecological uses. If comprehensive land-use planning and design are done keeping the future of wildlife in mind, it will make sure they are protected from the beginning and that their interests will be taken into consideration so that they will not face endangerment or extinction (Figure 14.2d).

What Are Wildlife Corridors?

Corridors are habitats that are typically long relative to their width, and they connect fragmented patches of habitat. They can vary greatly in size, shape, and composition. The main goal of corridors is to facilitate movement of individuals, through both dispersal and migration, so that gene flow and diversity are maintained between local populations. By linking populations throughout the landscape, there is a lower chance for extinction and greater support for species richness.

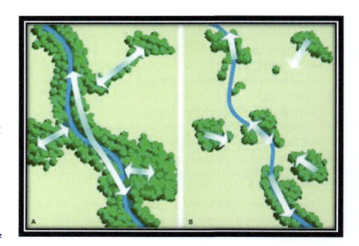

FIGURE 14.2 Qunli stormwater wetland park by Turenscape in Harbin, China. (a) Native wetland grasses grow in the ponds at various depths, and groves of native silver birch trees (Betula) create a dense forest setting. (From http://inhabitat.com/qunli-stormwater-wetland-park-stores-rainwater-while-protecting-the-environment-from-urban-development/turenscape-qunli-stormwater-wetland-park-harbin-china-10/). (b, c) The urban landscape creates a stark contrast with the lush green wetland that stretches out beyond it. (b: From http://inhabitat.com/qunli-stormwater-wetland-park-stores-rainwater-while-protecting-the-environment-from-urban-development/turenscape-qunli-stormwater-wetland-park-harbin-china-6/; c: From http://inhabitat.com/qunli-stormwater-wetland-park-stores-rainwater-while-protecting-the-environment-from-urban-development/turenscape-qunli-stormwater-wetland-park-harbin-china-5/). (d) Overpasses such as this one allow for traffic to continue for human convenience, while allowing wildlife to pass unharmed beneath from place to place. (Courtesy of Radomil, Creative Commons.)

How Do Wildlife Corridors Work?

Corridors work by increasing connectivity between patches that are isolated by human habitat fragmentation, caused primarily by urbanization, agriculture, and forestry. Plants and animals can use corridors for both dispersal and migration, which are two key movement patterns for species. Human-dominated habitats surrounding natural areas present barriers that plants and animals are unable or highly reluctant to move through, which is why corridors are necessary. These inhospitable places may have higher abundances of predators, lower resource availability, or reduced shelter. When a corridor is present, however, it provides an unbroken path of suitable habitat that can provide safe passage for animals or plants without being hindered as they travel through agricultural or urban landscapes. This connectivity is critical to a species' survival, as it promotes gene flow between populations and supports higher species diversity.

What Types of Corridors Exist?

Today, there are many kinds of corridors, and the differences are often due to the purpose of the corridor. Corridors can exist naturally, such as riparian corridors that link two different populations dependent on isolated wetlands. They can also be constructed through management practices, such as preservation of multiple land parcels to create a wildlife corridor for large mammals. Corridors can be artificially constructed, such as overpasses or underpasses on highways, for the sole purpose of funneling individuals away from human threats. Stream corridors can consist of a network of protected watersheds that allow fish to migrate long distance without being hindered by a road blockage or dam. Corridors can be large, as is typical in wildlands that follow mountain chains; or small, as is typical of greenways or wildlife overpasses in urban landscapes. The different types of corridors, however, all focus on one goal: to ensure connectivity between isolated habitat patches so that one or many species can move freely throughout the landscape.

What Are "Corridors" versus "Connectivity"?

Landscape connectivity refers to how plants or animals are able to move through a landscape. More connectivity means fewer barriers to dispersal or migration. Corridors are the clearest way to increase connectivity, as they provide structural connections between habitats in the landscape. There are other ways to increase connectivity, however, where strict conservation is not possible. It may be possible to reduce the distance between conservation areas, or to manage lands in ways that are less harsh for wildlife. Plant and animal dispersal and migration are not always incompatible with human uses of the landscape. Conservation goals often include both benefits to humans and supporting biodiversity.

What Are Some Examples?

There are many examples of both natural corridors and man-made corridors in the United States and around the world, including ones in South Carolina, Florida, Australia, Germany, the Netherlands, Canada, Nepal, and France.

The Netherlands is home to over 600 wildlife crossings known as ecoducts. The longest of these is Natuurbrug Zanderij Crailoo, which spans a distance of 800 meters long and 50 meters wide.

(http://www.atlasobscura.com/places/natuurbrug-zanderij-crailoo)

Red crabs climb over an overpass to cross a road on Christmas Island during their migration. (Christmas Island National Park, Australia) (Source: https://envirothink.wordpress.com/2015/06/09/new-wildlife-corridor-to-be-built-in-washington-state/)

A green bridge near Grevesmühlen, Germany. (Source: https://envirothink.files.wordpress.com/2015/06/green-bridge-over-the-a20-near-grevesmc3bchlen-germany.jpg)
(Information source: http://conservationcorridor.org/the-science-of-corridors/)

With rapid global population growth and increasing urban and suburban density, there are those who consider green spaces a luxury. High-performing green spaces, or green infrastructure, however, provide genuine economic, ecological, and social benefits. Integrating green infrastructure into the built environment must be made a priority—even if it requires a change in mindset—for successful sustainable living.

Green infrastructure represents a paradigm shift in looking at the human/environment interface in a new way. It should be seen as a way to better understand the valuable services that nature naturally provides the human environment. At both the regional and national levels, interconnected networks of park systems and wildlife corridors preserve ecological function, manage water, provide wildlife habitat, and create a balance between built and natural environments (ASLA, 2016). There is no substitute of the many natural and aesthetic benefits that green spaces provide. Their value has long been seen in the intrinsic values that are part of the national park system in the United States, for example. The public has long valued those places, and the green spaces in this context have similar value and worth.

At the urban level, urban forestry and other green approaches are critical for reducing energy usage costs, managing stormwater runoff, and creating clean, temperate air. Transportation networks can be made green with the addition of artful bioretention systems. A bioretention system is a system that manages stormwater through the use of filtration that is modeled after the biological and physical characteristics of an upland terrestrial forest or meadow ecosystem. They use vegetation, such as trees, shrubs, and grasses, to remove pollutants from stormwater runoff. Sources of runoff are diverted into bioretention systems directly as overland flow or through a stormwater drainage system. They can also be constructed directly in a drainage channel or swale.

Green roofs, walls, and other infrastructure within or on buildings are also beneficial. They reduce energy consumption and dramatically decrease stormwater runoff. At all scales, the green approach provides genuine lasting ecological, economic, and social benefits. It has proven to be effective and cost-efficient for the following:

- Absorbing and sequestering atmospheric CO_2
- Filtering air and water pollutants
- Stabilizing soil to prevent or reduce erosion
- Providing wildlife habitat
- Decreasing solar heat gain
- Lowering the public cost of stormwater management infrastructure and providing flood control
- Reducing energy usage through passive heating and cooling

Because of these benefits, green infrastructure is crucial to combating climate change, creating healthy built environments, and improving the quality of life (ASLA, 2016).

Forests and Nature Reserves Within Urban Settings

Forests and nature reserves—which can include areas as diverse as beaches, wetlands, prairies, meadows, and desert landscapes—have long been recognized as places for social interaction, psychological renewal, recreation, and educational opportunities. Recently, scientists have gained a better understanding of how natural areas provide ecosystem services such as carbon sequestration, air and water purification, and large-scale nutrient cycling for entire regions. Preserving and restoring natural areas are the best ways to ensure that areas will stay healthy, agricultural lands will remain productive, and the effects of climate change can be kept to a minimum. In order for this to happen, however, governments, organizations, and communities must be willing to invest in forests and nature reserves and ensure their protection. It must be a well-planned effort. It would also be beneficial to put into effect multiple-use trails; rails-to-trails; and both "wild and scenic" and recreational river and greenways programs, which all provide educational opportunities that enable people to learn about the value of forests and reserves. Populations always seem to be more willing to protect resources when they understand their true value and have a vested interest in its success.

As urbanization continues, it is critical for communities to also work on a *regional level* to coordinate planning and design for maintaining forests and reserves. The "big picture" approach is important to avoid segmented landscapes, both for humans and wildlife. Forest and nature reserve benefits include

1. Water Management Benefits: Watersheds with more forest cover have been shown to have higher groundwater recharge, lower stormwater runoff, and lower levels of nutrients and sediment in streams than areas dominated by urban and agricultural uses. A U.S. Geological Survey study of nutrients in undeveloped watersheds (mostly forested) found that forests "produced the best water quality in the country." Maximizing forest cover is beneficial for the following reasons:

   a. Forests delay water through greater interception in their multilayered leaf canopy.

   b. Forests have sturdy, long-lived roots that help to anchor soil against erosion. Deep roots also promote greater evapotranspiration (drawing water from soil and releasing it as water vapor into the atmosphere) and thus create soil water deficits, allowing forest soils to absorb more water during storms.

   c. The forest litter layer further promotes infiltration of water into the soil, slows water flow above-ground, and decreases erosion.

   d. Sub-surface water flow minimizes pollutant and sediment contamination of waterways and gives ample opportunity for plants and microbes to take up nutrients.

   e. Nitrogen loading downstream decreases as the percentage of forested land increases (World Resources Institute, 2016).

2. Health, Energy, and Livability Benefits: A "greening" project was completed in Portland in 2012, and one of the things they wanted to define was an estimate of the magnitude of "other benefits" for the city and its occupants in terms of health, community, livability, and energy. The project encompassed several goals: within the project area it improved evapotranspiration, reduced stormwater flows, and increased surface infiltration and groundwater recharge. It restored or preserved aquatic and terrestrial habitat and improved water quality in terms of reduced metals and total suspended solids. The improvements implemented in the study were classified into three main categories for metric tracking: health, energy and carbon sequestration, and community livability.

The following lists the green infrastructure that was involved in the project along with the associated benefits as a result:

- *Ecoroofs*: Ecoroofs replaced the conventional roofing materials with a living, breathing vegetated roof system. The ecoroofs consisted of a layer of vegetation over a growing medium on top of a waterproof membrane. These roofs were able to significantly decrease stormwater runoff by detaining and evaporating the stormwater directly at the site. They also saved energy, filtered pollutants, cooled urban heat islands, provided habitat, created green spaces, improved community livability, and provided educational opportunities.

- *Benefits*: Improved air quality, improved aesthetic value, increased energy savings, reduction of CO_2 emissions, improved mental health.
- *Green Streets*: Green streets are vegetated curb extensions, streetside planters, or infiltration basins (rain gardens) that collect stormwater runoff from streets. Green streets reduced stormwater flow to sewers, reduced pollutants and limited erosion in urban streams, provided wildlife habitat and neighborhood green spaces, and refreshed groundwater supplies.
  - *Benefits*: Improved air quality, increased home value, increased energy savings, reduced CO_2 emissions, and increased social capital.
- *Trees*: Trees protect watershed health by absorbing rain (which restores hydrology) and preventing erosion (which protects water quality and habitat). In this way, trees are a vital, long-term, and low-cost component of Portland's green infrastructure for managing stormwater. Trees also cleaned the air, created restorative spaces, and provided cooling shade and wildlife habitat. Street trees can improve property values and slow traffic, making streets safer for pedestrians, bike riders, and motorists.
  - *Benefits*: Improved air quality, increased property value, increased energy savings, reduced CO_2 emissions, increased social capital, improved mental health, and increased environmental equity.
- *Invasive Removal and Revegetation*: Invasive plants have an impact on water quality, biodiversity, fish and wildlife habitat, tree cover, and fire risk and costs. Increasing efforts to prevent and control invasions in high quality natural areas is the most cost-effective and ecologically successful approach. Removing invasive vegetation and restoring native plants reduced stormwater volume, filtered stormwater pollutants, provided habitat diversity, and cooled the air, pavement, and streams.
  - *Benefits*: Possible increased greenness.
- *Culvert Removal*: Culverts contribute to flooding and erosion. Associated streambed restoration projects are planned through partnerships.
  - *Benefits*: Improved aesthetics.
- *Land Purchase*: Development on steep slopes and drainage ways can cause landslides and erosion, increase flooding problems, and harm water quality and habitat. Public acquisition of natural areas protected these areas from development and preserved watershed and floodplain functions.
  - *Benefits*: Improved air quality and improved property value.
- *Planting in Natural Areas*: All invasive species are removed and native species are (re)established

in order to create a healthier environment. This includes native trees, shrubs, grasses, and wildflowers. Creating healthy native ecosystems helped reduce stormwater volume, filtered pollutants, provided cooling shade for streams, and provided diverse habitat to support native wildlife.

- *Benefits*: Improved air quality, increased energy savings, improved mental health, improved property values.

They concluded from the project that there were many positive relationships between human health and green infrastructure. Natural areas and increased levels of vegetation can directly improve health by improving air quality and reducing air quality related illnesses. Additionally, by increasing the level of greenery in the urban environment, green infrastructure can indirectly enhance both physical health (by increasing an individual's desire to get outside in the fresh air and exercise) and mental health (by improving the visual quality of the environment, which in turn reduces stress and mental fatigue).

From an energy and GHG perspective, there was a net decrease in energy use and associated CO_2 and GHG emissions. Energy use is less because of reduced stormwater generation, which results in less energy use needed for stormwater pumping and treatment. Increased urban vegetation leads to cooler summertime ambient air temperatures, which creates a lower heat island effect, and decreases energy use for summer cooling. The increased vegetation also increases the insulation and shading of buildings, which reduces the energy use in buildings. Reduced energy use in these areas will decrease CO_2 and other GHG emissions as less power generation from fossil fuels is required. In addition, the planting of trees and other vegetation will increase carbon sequestration in the city.

In terms of community livability, having an increase in vegetation and green spaces in urban areas is linked to a wide variety of community benefits, such as increased access to nature, social connections, recreational opportunities, walkability, and reduced crime rates. Trees, vegetation, and green spaces do have an overwhelmingly positive effect on the community as a whole.

Retrofitting and Repurposing

Retrofitting an existing building can often be more cost-effective than building a new facility. Since buildings consume a significant amount of energy, particularly for heating and cooling, and because existing buildings comprise the largest segment of the built environment today, an important step toward energy conservation is to look at retrofits to reduce energy consumption and the cost of heating, cooling, and lighting buildings. But conserving energy is not the only reason for retrofitting existing buildings. The goal should also be to create a high-performance building by applying the integrated, whole-building design process, to the project during the planning phase that ensures all key design objectives are met. For example, an integrated project team could develop a single design strategy that would meet multiple design objectives. Doing so would ensure that the building ultimately costs less to operate; lasts longer; and contributes to a better, healthier, more acceptable environment for people to live and work in. It also increases the value of the building.

There are several indoor benefits to consider, as well. By improving indoor environmental quality, decreasing moisture penetration, and reducing mold, the occupant health and productivity will be improved. It is a good rule of thumb that when considering a retrofit, also consider upgrading for accessibility, safety, and security at the same time.

The unique aspects for retrofit of historic buildings must be given special consideration. Designing major renovations and retrofits for existing buildings to include sustainability initiatives will reduce operation costs and environmental impacts and can increase building adaptability, durability, and resiliency. When remodeling an old building to make it functional in a sustainable world, it is important to remember the three R's: reuse, repurpose, refurbish.

Consider the twist to this old adage:

Out with the new, in with the old.

Although it may seem counterintuitive, businesses can avoid making substantial new purchases and still reduce their environmental impact while strengthening their bottom line. All it takes is some imagination, determination, and commitment.

The Electric Grid/Smart Grid

"The Grid" that people often talk about in the United States refers to the country's electric grid—a network of transmission lines, substations, and transformers that deliver electricity from power plants to homes and businesses. It is what everyone plugs into when they turn on their light switches or turn on a computer. The current electric grid in the United States was built in the 1890s and improved upon as technology advanced through each decade. Today, it consists of more than 9,200 electric generating units with more than one million megawatts of generating capacity connected to more than 483,000 kilometers of transmission lines (Smartgrid, 2016). Although the electric grid is considered an engineering marvel, the way that it evolved has made it a patchwork-style construction and the demand currently being placed on it is pushing it to its limits. In order to progress, a new type of electric grid is necessary—one that is built from the bottom up in a logical, cohesive way in order to accommodate the sheer volume and demand of digital and computerized equipment and technology that is completely dependent on it. A grid is currently needed that can automate and manage the increasingly complex nature of the system and accommodate the future requirements of this century and beyond.

What makes a grid "smart" is the implementation of digital technology that would allow for two-way communication between the utility and its customers, and data that can be sensed along the transmission lines. Like the Internet, the Smart Grid envisioned for the country will consist of controls, computers, automation, and new technologies and equipment all working together. In the Smart Grid, these technologies will work with the electrical grid to respond digitally to any rapidly changing electric demand (Figure 14.3).

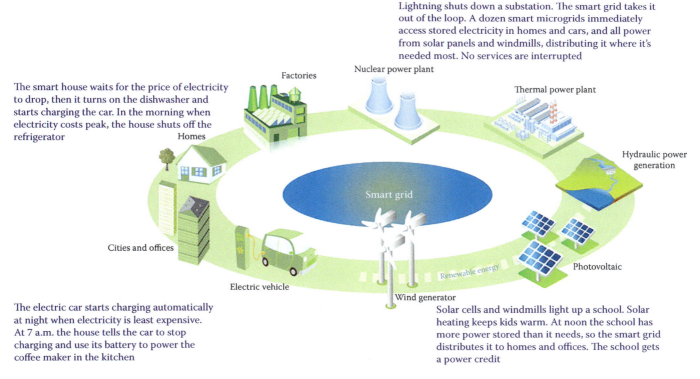

FIGURE 14.3 Example of how a Smart Grid operates. (Tsao, M.)

The Smart Grid represents a unique opportunity to boost the energy industry into a new era of reliability, availability, and efficiency that will contribute to both the country's economic and environmental health. The expected benefits associated with the Smart Grid include

- More efficient transmission of electricity.
- Quicker restoration of electricity after power disturbances.
- Reduced operations and management costs for utilities and ultimately lower power costs for consumers.
- Reduced peak demand, which will also help lower electricity rates.
- Increased integration of large-scale renewable energy systems.
- Better integration of customer-owner power generation systems, including renewable energy systems.
- Improved security.

Today, when an electricity disruption, such as a blackout, occurs, it can have a domino effect—a series of failures that can affect businesses, communications, traffic, and security. This is especially a threat during the winter, when homeowners can be left without heat. A Smart Grid would add resiliency to the electric power system and make it much better able and prepared to address emergencies such as severe storms, earthquakes, large solar flares, and terrorist attacks. Because of its two-way interactive capacity, the Smart Grid would allow for automatic rerouting when equipment failures or outages occur. This would minimize outages and the effects when they do happen. When a power outage did occur, Smart Grid technologies would be able to detect and isolate the outages, containing them before they became large-scale blackouts. The new technologies, once implemented, will also help ensure that electricity recovery resumes quickly and strategically after an emergency—routing electricity to emergency services first, for example. In addition, the Smart Grid would take greater advantage of customer-owned power generators to produce power when it is not available from utilities. By combining these "distributed generation" resources, a community could keep its health centers, police departments, traffic lights, phone systems, and grocery stores operating during emergencies. In addition, the Smart Grid is a way to address an aging energy infrastructure that needs to be upgraded or replaced. It is a way to address energy efficiency, to bring increased awareness to consumers about the connection between electricity use and the environment. And it is a way to bring increased national security to our energy system—drawing on greater amounts of local electricity that is more resistant to natural disasters and attack.

The Smart Grid is not just about utilities and technologies, it is about giving the public the information and tools they need to make choices about energy use. In a similar way that they can already manage activities such as personal banking from home computers, they will also be able to manage their electricity use. A smarter grid will enable an unprecedented level of consumer participation. It will be possible to monitor electrical use. "Smart meters," and other mechanisms, will allow home owners to see how much electricity is used and when. Combined with real-time pricing, this will allow homeowners to save money by using less power when electricity is most expensive. Those on the Smart Grid will be able to also generate their own power.

The Smart Grid will consist of millions of components—controls, computers, power lines, and new technologies and equipment. It will take time for all the technologies to be perfected, equipment installed, and systems tested before it comes fully on line. Its installation will evolve over the next decade or so. Once it is installed and functional, the Smart Grid will likely bring the same kind of transformation that the Internet has already brought to the way we live. It will offer many opportunities for consumers to save energy and for utilities to operate the grid in a more efficient, effective, and reliable way. But it will also probably require people to do things differently, such as wait to run the dishwasher at a time when it is not a peak energy demand; and the benefit of changing that part of their lifestyle will be savings in energy costs. Even better—for people who generate their own power, it can even result in the utility company sending a check to the customer.

Many utilities already offer their customers ways to save extra money on their utility bills. For example, for some customers who have central air conditioning systems, their utility company will place a remote-control switch on the air conditioner to cycle the air conditioner on and off during times of peak power demand. In return, those customers receive a credit on their electrical bill. The Smart Grid will be able to offer programs, resulting in greater energy savings with less inconvenience to businesses and homeowners. For example, they may offer time-of-use pricing, net metering, and compensation programs for plug-in electric vehicles (PEVs).

Time-of-Use Programs

One of the most important ways the public can get involved with the Smart Grid is to take advantage of time-of-use programs. Throughout the day, the demand for energy changes: it is usually lowest in the middle of the night and highest from about noon to 9 p.m., but it can vary according to weather patterns and what is happening during that time. Power plants and utilities have to work harder to meet the needs of electric consumers when the demand is highest. During peak energy usage, utilities sometimes have to bring less-efficient—and often more-polluting—power generation facilities on line or purchase power from neighboring utilities at a higher cost. In the worst cases, utilities may have to implement rolling blackouts or reduce the voltage of the system—a technique called a *brownout.*

Time-of-use rates encourage electricity users to use energy when the demand is low by giving them a lower price for electricity during those times. Distributing the demand more evenly ensures that a steady and reliable stream of electricity is available for everyone. Home energy management systems are often used to help homeowners make the most of time-of-use pricing. They can be easily accessed with a home computer or hand-held mobile device to see when the prices are the highest, which appliances used the most electricity, and when the prices rise during the day, so that the consumer can remotely shut the power off at their home.

Net Metering

For people who generate their own power at home—using a rooftop solar power system, for instance—net metering is an option already available in many states. In general, net metering involves the use of a meter that can record when power flows back into the grid as a credit. Some mechanical meters will literally spin backward, although most utilities are using digital meters for net metering.

Most people with net-metered systems are allowed to accumulate credits for excess power generation—that is, power fed into the grid from their home power systems—on a monthly basis. At the end of the year, home power generators usually have the option of carrying over credits into the next year or receiving a check for their excess power generation. That is correct: your utility can pay you for power.

The Smart Grid will enable enhancements to these net metering programs. For instance, a utility might pay more for customer-generated power during times of peak power demand, while paying less for off-peak power. Like time-of-use pricing, such a pricing structure, will encourage home generators to minimize their energy use during times of peak demand so they can maximize the amount of power fed onto the grid.

Other Financial Incentives

The Smart Grid will open up countless new ways for consumers and their utilities to interact with energy. Many of these new capabilities will offer energy savings, but some may also include minor sacrifices or inconveniences, and for those, utilities are likely to offer financial incentives to encourage participation. For instance, smart appliances could offer countless subtle ways for utilities to shift electrical demand to off-peak hours. Your dishwasher could defer running until later in the evening, or your refrigerator could defer its defrost cycle. Your air conditioner could slightly extend its cycle time to help lower your power demand during peak hours. Unneeded lights or electrical devices could even power off.

For these programs, utility companies may offer some financial incentive for participation. The result could be a double savings: a direct energy savings on the electricity bill and the extra benefit of earning an incentive payment from the utility.

One future area of potential financial incentives involves the use of PEVs as sources of stored energy for the utility. For those who own a PEV, their utility might pay an incentive for occasionally drawing power from the battery pack. Because extra cycling of the battery pack could shorten its useful life (at least with today's battery packs), the utility could provide extra compensation to help account for the slight degradation in the useful service life of the batteries.

Overall, the Smart Grid will open the door to many new possibilities for utilities and their customers to reach

agreements on ways to save energy. The financial incentives available could encourage a wide range of new consumer options. Consumers may be willing to pay slightly more for a smart appliance, for instance, if it can also become a new source of revenue. And utility incentives could also encourage customers to install a home generation system, such as a small wind turbine or solar power system. The result is a win-win-win: a win for the customer, for the power provider, and for the community.

Grid Operation Centers, Blackouts, and Solutions

Today's electrical transmission system—including the giant power lines and transmission towers that snake across the landscapes—operates much like a system of interconnected streams. Power flows through the transmission system along the path of least resistance, finding multiple paths between the power plants and the cities that are demanding the power. Grid operators actually have very little control over today's system. Their primary task is to make sure that as much power is being generated as is being used—if not, the grid's voltage could drop, causing the grid to become unstable. Operators generally know which lines are in service and when relays have opened to protect lines against faults, but they have limited control capabilities. Unfortunately, like water in a bathtub, power can "slosh around" within the grid, developing oscillations that, under the worst of conditions, could lead to widespread blackouts. To compound the problem, grid operators also have limited information about how the power is flowing through the grid. The Smart Grid will help solve this problem by adding new capabilities for measurement and control of the transmission system. These technologies will make the grid much more reliable and will minimize the possibility of widespread blackouts.

August 14, 2003, is a date that few grid operators in North America will ever forget. Although power demand was high that day, the situation seemed under control until a relatively insignificant power line in Ohio overheated and tripped offline. Like throwing a rock in a pond, the power line failure triggered oscillations in the transmission systems that lined the shores of the Great Lakes and provided power to the Northeast and parts of Canada. Those oscillations eventually overloaded the system, causing a massive blackout that stretched from Michigan and Ohio, through the Northeast, and into Canada. With subways down, many commuters in New York City struggled to find a way home. The blackout affected an estimated 10 million people.

Smart Grid technologies offer a new solution to the problem of monitoring and controlling the grid's transmission system. New technologies called Phasor Measurement Units (PMU) sample voltage and current many times per second at a given location, providing a snapshot of the power System at work. PMUs provide a new monitoring tool for the Smart Grid. In the current electric grid, measurements are taken once every 2 or 4 seconds, offering a steady-state view into the power system behavior. Equipped with Smart Grid communications technologies, measurements can be taken many times a second, offering dynamic visibility into the power system. This makes it easier to detect the types of oscillations that led to that infamous 2003 blackout.

The "Self-Healing" Grid

Smart Grid technologies also offer a new means of controlling the transmission system. New high-power electronics function essentially as large-scale versions of transistors, adding a new level of control to the transmission system. New technologies can help control unwanted power oscillations and avoid unproductive flows of current through the grid that waste energy. New measurement technologies enable a new automated approach to control the grid. Software can potentially monitor the grid in real time for disturbances that could lead to blackouts and take action against it. Such monitoring software could cut down the oscillations in the power grid, or even reroute power through the grid to avoid overloading a transmission line. If a power-line needed to be removed from service, control software could reroute it in the least disruptive way. This approach is often referred to as the "self-healing" grid because it acts in its own self-interest. These distribution system measures are sometimes referred to as *distribution intelligence.*

Distribution intelligence can also incorporate more sophisticated ground-fault detectors to minimize the possibility of people getting accidently shocked or electrocuted when encountering downed power lines. Most utilities are only starting on the road to true distribution intelligence, but the market is expected to boom in the coming years.

ENERGY EFFICIENCY: GREEN BUILDINGS

One way a real different can be made is through green building. Green building, or sustainable design, is the practice of increasing the efficiency with which buildings and their sites use energy, water, materials, and reducing building impacts on human health and the environment over the entire life cycle of the building. The growth and development of our communities have a large impact on our natural environment. The manufacturing, design, construction, and operation of the buildings in which we live and work are responsible for the consumption of many of our natural resources. Lowering this impact is an important way we can help the environment, use less energy, and slow the effects of climate change. This section discusses how to accomplish that.

Combined Heat and Power—Cogeneration

CHP, also known as cogeneration, is the simultaneous production of power (electricity) and heat from natural gas (dominantly), coal, oil, biomass, biogas and biofuel, solar (PV), and waste heat (recovery). Whereas conventional energy production just treats waste heat as an inevitable loss,

CHP treats it as productive thermal energy which can be used. This offers major economic and environmental benefits because it turns otherwise wasted heat into a viable energy source. For power stations alone, this greater efficiency can mean CO_2 emissions are cut by up to two-thirds when compared with conventional coal-fired power stations. Waste heat can be heat from waste incineration, waste heat from power production, and/or industrial/commercial/even residential waste heat. Fuel sources vary from project to project and country to country.

For example, while boiling water to generate electricity, the low-pressure unused steam can be sold for industrial processes or space heating (Figure 14.4). Cogeneration is a high-efficiency energy system that produces either electricity (or mechanical power) and valuable heat from a single fuel source. For example, a hospital cogeneration plant could produce some of the power and all the hot water needed for its laundry and hot water system from the waste heat it generates. Similarly, office buildings can produce power for electricity and air conditioning from the waste heat generated by its air conditioning engines. Iceland is a good example of

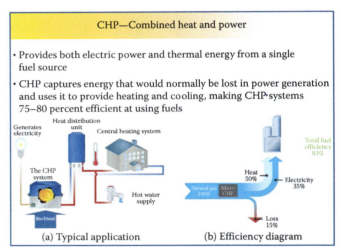

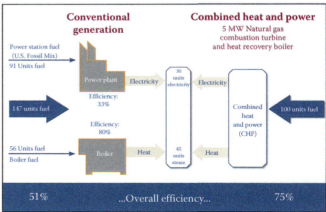

FIGURE 14.4 Cogeneration, or combined heat and power (CHP), is the utilization of boilers, turbines, and/or engines to simultaneously generate electricity power and heat that can be useful in several ways such as for hot water, steam, district heating, and water desalination. (From EPA, Washington, DC.)

utilizing CHP. Their dominant source of energy for its utilization is geothermal and from it they successfully operate district heating.

In general, businesses, government agencies, and facilities most likely to benefit from cogeneration are those that use large quantities of hot water, heat, steam, or chilling such as hospitals, large maintenance facilities, aquatic centers, and laundries. Others who could benefit from cogeneration include hotels; office buildings; food and beverage processors; chemicals and plastics producers; pulp, paper, and fiber board manufacturers; metals processors; textile producers; shopping centers; and universities.

CHP can be used in several energy technologies. Often, systems are developed exclusively for onsite generation of electrical and/or mechanical power, in addition to HVAC and water heating. CHP is most often developed with a gas turbine and a heat recovery unit or a steam boiler with a steam turbine.

There are several benefits of cogeneration. The most significant benefits include the following:

- It can significantly reduce energy costs and greenhouse emissions by up to two-thirds.
- Local air quality benefits can be achieved through the replacement of older coal-fired boilers.
- Cogeneration increases the utilization of resources.
- It requires less fuel to produce a given energy output
- It can produce a given energy output on site, avoiding the grid and associated energy losses.
- Consumers save money on their utility bills, offering a reliable source of high-quality energy and HVAC.

When waste heat is captured and used a second or even third time, the energy savings can be significant, because one fuel is being used for several purposes. This significantly reduces the carbon footprint of the facility. The amount saved varies from building to building because each application of CHP is unique, as well as dynamic in nature. Reductions in energy use of up to 25 percent have been reported.

Energy savings are not the only reason to consider CHP. It provides seamless reliability and resilience to the building. By having a system up and running continuously, all year long, producing utility quality electric power, all critical operations and services in that building will be up and running when the power grid is down. Additionally, when CHP waste heat is used in conjunction with absorption chillers, air conditioning is always available, which is not true with just a back-up generator alone. This feature would prove critical to have in the event of catastrophic weather events, such as hurricanes. In summary, CHP provides the services of a back-up generator with day-to-day services year-round, and is always up and running.

There are three types of CHP systems available—the Combustion Engine, Gas Turbine, and Fuel Cell. All three systems use clean burning and plentiful natural gas:

- Combustion Engines resemble, in many ways, a car engine, and can be sized from 10 to 100 kilowatts per unit. Combinations of several 100 kilowatts units are often used in tandem to create systems large enough for multifamily housing or commercial use.
- Gas Turbine units can be as small as 50 kilowatts and go up from there to megawatt unit systems.
- Fuel Cell systems are incredibly efficient in their use of natural gas, and as such, produce virtually no emissions, making these units very environmentally friendly. The most common type of fuel cell is the Solid Oxide Fuel Cell. With a Solid Oxide Fuel Cell, electricity and heat are produced as natural gas (methane) that is passed over a series of plates, creating an electrochemical reaction. The heat is very high quality and is around 538°C. The extremely efficient use of the fuel is the reason for the virtual emission-free output.

In order to use CHP properly and get the full benefits from it, the facility must be fully evaluated as to the ability to use a sufficient amount of the waste heat from the electric generation side of the CHP.

Waste Heat Utilization—A Case in Point

Recycling heat is fairly common in Europe. Denmark gets roughly half of its electricity from recycled heat, followed by Finland at 39 percent, and Russia at 31 percent. In the United States, it is currently only 12 percent. According to a report by Lawrence Livermore National Laboratory and the Department of Energy, the U.S. wastes more than half of the total energy it produces—mostly as heat, but also as gas, biomass, and methane. Using that waste could reduce CO_2 emissions by 17 percent. "It's free energy, essentially," says Brendan Owens, vice president of Leadership in Energy and Environmental Design (LEED) at the U.S. Green Building Council (Hughes, 2014) (Figure 14.5).

FIGURE 14.5 Waste heat that exits processing plants represents free energy if it was utilized. (Courtesy of Jorge Royan, Wikimedia Commons.)

Recycling heat is a fairly simple process. New buildings often have condensing water heaters, which use gas burners not only to warm up water (just as other heaters do) but also capture the heat in the combusted gas that is going out the flue. This occurs on a larger scale, too. In 1882, when Thomas Edison built the world's first commercial power plant in Manhattan, he sold its steam to heat nearby buildings. Today, these are the types of plants that are known as CHP, or cogeneration, plants. Edison's old plant eventually became the massive Con Edison, whose operations today produce 19.7 billion pounds of steam a year. The ArcelorMittal steel mill in East Chicago, Indiana, is another good example. It uses extra blast furnace heat to make steam that then generates electricity for the mill, saving approximately $20 million a year and preventing 308,442 metric tons of CO_2 emissions—the equivalent of taking 62,000 cars off the road.

So why do not we have more projects like these? It is partly logistics: trapped heat cannot travel far, and centralized power plants of the U.S.'s tend to be farther from urban centers than European ones. Even in ideal circumstances, large energy projects usually have big, upfront costs that may not be recouped for years. And U.S. government tax breaks on heat recycling are less than those on other clean technologies. Corporations get a 30 percent tax rebate for solar and wind power, but just 10 percent for CHP.

Energy regulations in the United States have unintended consequences that create other barriers to recycling heat. Although some electric generation has been deregulated since the 1970s, energy transmission and distribution are still monopolistic. In many states, it is actually illegal for non-utilities to sell excess electricity to their neighbors. Because of the Clean Air Act, factories must limit their emissions—but the law does not give credits for energy efficiency or emissions prevention. It effectively penalizes companies that invest in waste-heat systems. What is more, the act's stricter rules only apply to new sources of pollution, grand-fathering in older, dirtier plants and giving incentives for companies to hold onto them. Unfortunately, energy policy is very complicated in the United States, and there is no single solution. But time is long overdue to start thinking of ways to start effectively dealing with recycled heat. The energy potential is enormous.

District Heating

Introduced in the 1980s, district energy systems produce steam, hot water, or chilled water at a central plant. The steam, hot water, or chilled water is then piped underground to individual buildings for space heating, domestic hot water heating, and air conditioning. As a result, individual buildings served by a district energy system do not need their own boilers or furnaces, chillers, or air conditioners.

A typical district heating installation consists of a highly insulated "heat main" of flow and return pipes distributing hot water (or steam) past all buildings which are connected. A junction point allows easy connection to each building, from which hot water can be taken from the main to a heat

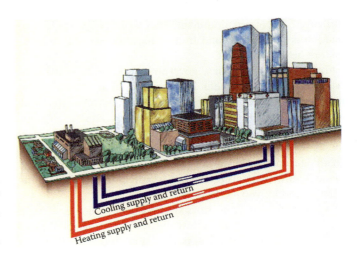

FIGURE 14.6 Schematic of district heating. (From International District Energy Association, Westborough, MA.)

exchanger (heat substation) within each building. The heating circuit within the building is isolated from the heat main (Figure 14.6).

Temperature measurements of the flow and return lines, plus a flow meter (together forming a heat meter), allow the actual heat usage within each building, or even apartment, to be separately measured, delivered and billed for accordingly. Remote meter reading, by a modern, secure web interface, or a drive-by, is both usually possible, as are remote diagnostics to ensure reliable operation.

The district energy system does that work for them, providing valuable benefits including the following:

- Improved energy efficiency
- Enhanced environmental protection
- Fuel flexibility
- Ease of operation and maintenance
- Reliability
- Comfort and convenience for customers
- Decreased life cycle costs
- Decreased building capital costs
- Improved architectural design flexibility

When steam, hot water, or chilled water arrives at a customer's building, they are ready to use. They are 100 percent efficient "at the door," compared with 80 percent efficient or less when burning natural gas or fuel oil at a building.

With district energy, building managers no longer need to burn fuels or store or use refrigerants on site, so the site is safer and more environmentally sound. It also eliminates the need to build unsightly smokestacks. Instead, fuel and refrigerants are used at district energy plants. These systems employ stringent emission controls—more so than individual buildings—and this provides air-quality benefits.

The beauty of a district energy system is that since it serves so many customers from one location, it can accomplish things individual buildings usually cannot. For example, district energy systems can use a variety of conventional fuels such as coal, oil, and natural gas, whichever fuel is most competitive at the time. And because of a district energy system's size, the district energy plant can also transition to use renewable fuels, such as various forms of biomass (wood and food processing waste, geothermal heat, and CHP).

Because boilers or chillers are not necessary, there is less maintenance, monitoring, and equipment permitting. District energy customers also eliminate the need for fuel deliveries, handling, and storage, so there are fewer safety and liability concerns for employees and building occupants. The use of district energy service frees up valuable building space by eliminating the need for mechanical rooms. These systems are very reliable, generally operating at a reliability 99.99 percent. District energy is available whenever a building needs heating or cooling. So even if there are unusually warm days in January, a building can receive chilled water or steam for air conditioning, without starting up its own chillers.

The cost to employ this heating method is also less expensive than traditional methods and there is more architectural design flexibility. Because there are no boilers or furnaces and roofs free of smoke stacks and cooling towers, this enables greater building design flexibility. Because of this, older buildings can be renovated fairly easily.

District heating has become the favored method of heating of many cities in Europe. It has also risen in popularity and use throughout much of the rest of the world. The concept was actually invented more than a century ago. It was introduced in 1903 in Moscow, Frederiksberg and Copenhagen—in all three locations the same year.

Green Buildings

The ideal green building would be a building project that would allow the builder to preserve most of the natural environment around the project site, while still being able to produce a building that is going to serve its purpose. The construction and operation are designed to promote a healthy environment for all involved, and it will not disrupt the land, water, resources, and energy in and around the building.

The EPA provides this definition: "Green building is the practice of creating structures and using processes that are environmentally responsible and resource efficient throughout a building's life cycle from siting to design, construction, operation, maintenance, renovation, and deconstruction. This practice expands and complements the classical building design concerns of economy, utility, durability, and comfort. Green building is also known as a sustainable or high performance building."

Let us look at why it is so important to go green. Most people will find when going green that they are able to reduce their carbon footprint and actually contribute in a significant way to the environment. Green buildings are designed in such a way as to reduce the overall impact on the environment and human health by

- Reducing trash, pollution, and degradation of environment.
- Efficiently using energy, water, and other resources.
- Protecting occupant health and improving productivity.

Does Going Green Cost a Lot More?

It is a common misconception that going green costs a lot more money than remaining traditional. While some things may cost a bit more up front to go green—because green materials and products can be more costly—it is important to consider the type of savings that will ultimately be realized. It also enables the investor to save on energy costs, because going green also means conserving energy. Green building should be considered an investment—one that will eventually save money, one that saves energy, and one that helps the environment. It is a win-win situation all around.

Benefits and Goals of Green Building

With new technologies constantly being developed to complement current practices in creating greener structures, the benefits of green building can range from environmental to economic to social. By adopting greener practices, it is possible to take maximum advantage of environmental and economic performance. Green construction methods, when integrated with design, provide the most significant benefits. Benefits of green building include:

Environmental benefits:

- Reduction of water waste
- Conservation of natural resources
- Improvement of air and water quality
- Protection of biodiversity and ecosystems

Economic Benefits:

- Reduction of operating costs
- Improvement of occupant productivity
- Creation of green market products and services

Social Benefits:

- Improvement of quality of life
- Minimization of strain on local infrastructure
- Improvement of occupant health and comfort

It is imperative to consider the goals of green building. Naturally, one of the main goals is to make the earth more sustainable, but it really goes deeper than that. Once the decision is made to go green, the real goal becomes to help sustain the environment without disrupting the natural habitats around it. When a building project is begun, and the natural habitats around it are disrupted, it can actually make an impact on the wildlife and environment that can have a ripple effect. Even the smallest positive changes made will help to promote a better planet earth and a better place for us all to live—not just us humans but also the plants and wildlife that take up their residence here on earth as well.

Green buildings are something that everyone can get involved in. If someone does not plan on rebuilding their home, then they may just want to make a few green changes within their home to ensure that they are able to accomplish some green goals. For example, they cut down on energy usage in order to positively impact the environment. It is not as hard as some people make it out to be, and contributing toward the effort in a positive way is a great feeling (Kukreja, 2016).

GREENING YOUR PLACE OF WORK

Let us consider the workplace. There are several things that can be done to green it up. Here are a few tips that can make a real difference:

- Make meetings paper-free: Avoid printing out lengthy documents or even the typical meeting agenda. In this day of computer technology, it may not even be necessary to bring any paper to a meeting at all.
- Limit paper mail: Paper mail can largely be eliminated these days. For those who are on paper mailing lists where there is an electronic (online) option, opt for that instead.
- Change the copier settings: The first rule of thumb is to reduce the amount of copying and printing that is done. For printing that cannot be avoided, always print on both sides of the paper, eliminating half the paper usage right off the top.
- Limit what is purchased in advance: If dated, or perishable, items that are purchased are limited, it can save in the long run. For items like printer cartridges that have a shelf life, having them expire before they can be used is wasteful. If an item does have a long shelf life, however, the bulk rule still applies, and it is good to stock up and use less packaging.
- Purchase Energy Star electronics: When it is time to purchase new equipment, such as refresh cycles for electronics, always purchase Energy Star technology. This includes items such as computers, heating and cooling systems, printers, lights, copiers, microwaves, refrigerators, coffee makers, and so forth.
- Get into the recycling habit: There are recycling companies that contract with businesses and offer curbside pickup once a week from bins that they supply. Bins can be placed in convenient locations where they will be sure to be used, such as next to copier machines. This represents a great opportunity to educate fellow coworkers.

- Purchase nontoxic materials: Nontoxic materials such as nontoxic cleaning supplies are less expensive than chemical materials, and also eliminate the need of restricted disposal.
- Lease only what is necessary: If a large business space is not absolutely necessary, do not lease it; lease only the square footage necessary in order to eliminate wasteful lighting, heating, or maintaining unused space.
- Offer incentives for going green on the way to work: For those in a position to offer green transportation incentives, encourage fellow employees to carpool, take public transportation, bike, or walk to work. Some businesses offer discounted carpool parking or bus passes.

The bottom line is that by rewarding employees for going green and encouraging responsible behavior, it is possible to impress customers with healthy attitudes and innovative approaches to business. It also sets a great example for other businesses to follow.

In the past, traditional buildings have not been designed or constructed with conservation in mind. Many have had a negative impact not only on the environment but also on their occupants. Non-green traditional buildings were not only designed without the environment in mind but they were also expensive to operate, contributed to excessive resource consumption, generated excess waste, and added to pollution levels.

Today, a greater percentage of the public is looking at the concept of sustainable green energy projects—either as renovations of existing structures or as newly constructed structures.

Many decisions are made regarding the construction and design of a sustainable building long before the foundation is even dug. Questions such as the following have to be answered:

- What kinds of materials (sustainable) should the building be made of?
- Where and how does certified (sustainable) wood get purchased?
- Which materials can be purchased locally, and which cannot? This is important in reducing waste and shipping costs.
- Where can carbon emissions be reduced?
- How many tons of greenhouse gases will the construction of the building put into the atmosphere?
- Are there any carbon offsets that can be acquired to build the building?

This is where the LEED Certification Program comes into play.

The LEED Certification Program

The LEED certification program has available a set of standards of sustainable design and development that can be used by developers, architects, engineers, real estate professionals, and others who are interested in sustainable construction. These standards apply to virtually all constructions, including new construction, existing buildings, commercial buildings, residential homes, and neighborhoods as a whole. The LEED rating that a project receives is based on the following criteria:

- Site selection
- Water and energy efficiency
- Materials used
- Indoor environmental quality

The voluntary certification program, developed by the U.S. Green Building Council (USGBC), provides building owners and operators with detailed guidance on how to build any type or style of green building. LEED also provides for awareness and education. Its philosophy is that a green home is truly green only if the people who live in it use the green features to maximum effect. Therefore, they encourage home builders and real estate professionals to provide homeowners, tenants, and building managers with the education and tools they need to understand what makes their home green and how to make the most of those features.

The following are some of the techniques that LEED recognizes in sustainable construction:

- The use of low volatile organic compounds (VOCs) paint.
- Plywood processed without the use of formaldehyde.
- Use of large windows that provide plenty of fresh air and natural light.
- Installation of energy- and water-efficient appliances (Energy Star).
- Installation of low-emitting carpet.
- Sites must not be built on vulnerable areas such as floodplains, prime farmland, habitats of threatened or endangered animals, land close to wetlands.
- Must be within walking distance of at least 10 basic services.
- Must provide space for storage and collection of recyclables.
- Establishes a minimum level of indoor air quality performance.
- Minimizes cigarette smoke.
- Is located near alternative transportation.
- Reuses or recycles construction materials when possible.

Several businesses that have remodeled to become green have determined that it has had a positive effect in increased worker productivity and better economic returns. According to the DOE and the Rocky Mountain Institute, a study of the effect of office design on productivity found a direct correlation between specific changes in the physical environment and worker productivity. Some of the specific cases are summarized as follows:

Leaders Who Have Stepped Forward

Several businesses recently have stepped forward and gone green, some making small improvements, some larger ones; others retrofitting existing structures, while still others beginning with completely new construction. The following a few of the recent examples.

Boeing Turns on the Green Light

Boeing—the manufacturer of aircraft—participates in the EPA's voluntary Green Lights program to promote energy-efficient lighting. It recently retrofit its 743,000,000 square meter facility, reducing electricity use by up to 90 percent in some of its plants. Boeing calculates its overall return on investments in the new lighting to be 53 percent—the energy savings paid for the lights in just 2 years.

The Boeing Company has also discovered some other interesting results. With its more efficient lighting, the employees have noticed that the glare inside the work area has been reduced. Friedman said, "The things the employees tell us are almost mind-boggling. One woman who puts rivets in 9-meter wing supports had been relying on touch with one part because she was unable to see inside. Now, for the first time in 12 years, she could actually see inside the part." Friedman also said that "Most of the errors in the aircraft interiors that used to slip through were not being picked up until installation in the airplane, where it is much more expensive to fix. Even worse, some imperfections were found during the customer walk-through, which is embarrassing and costly. Although it is difficult to calculate the savings from catching errors early, a manager estimated that they exceeded the energy savings for that building."

Improving Morale and Energy Efficiency

The Pennsylvania Power and Light Company's older lighting system was causing glare from the work surfaces to shine into employees' eyes, making their work less efficient—it took them longer to complete tasks and caused an increase in the number of errors they made. Russell Allen, superintendent of the office complex, remarked, "Low-quality seeing conditions were also causing morale problems among employees. In addition to the (glare and lighting) reflections, workers were experiencing eye strain and headaches that resulted in sick leave."

The power company decided to invest in modern, energy-efficient, nonglare track lighting that gave control to each workstation so that each employee could adjust his or her own lighting for comfort and efficiency. The results were noticed immediately.

Allen noted that "as lighting quality is improved, lighting quantity can often be reduced, resulting in more task visibility and less energy consumption."

When Allen did a cost analysis on the effects of the lighting change, he was surprised. He reported that the lighting energy use dropped by 69 percent, and total operating costs fell 73 percent. The annual savings alone from the reductions in energy operating costs completely paid for the lighting system in less than 4 years—a 25 percent return on investment. In addition, the newer lighting lowered heat loads (because of better efficiency), resulting in lower space-cooling costs. Employee productivity also rose 13.2 percent. The savings in salary due to the increase in employee productivity paid for the new lighting in just 69 days. According to Allen, "Not only is this an amazing benefit, it is only one of several."

Before the upgrade, employees used an average of 72 hours of sick leave a year. Because the new lighting relieved eye fatigue and headaches, as well as boosted morale, the absenteeism dropped to 25 percent. The improved lighting also reduced the number of errors employees made, producing overall higher-quality products. Allen concluded, "Personally, I would have no qualms in indicating that the value of reduced errors is at least $50,000 a year. If this estimate was included in the calculation, the return on investment would exceed 1,000 percent."

CONCLUSIONS

There are multiple benefits from sustainable urban planning. Green cities have the ability to improve quality of life, expand opportunities, address the issue of climate change, reduce CO_2 emissions, reduce our dependence on fossil fuels, and make us more environmentally conscious of the world around us. They present us with different lifestyle opportunities, unique recreational areas and ways to enjoy the outdoors, pedestrian streets, clean water sources, and integrated public transportation options. They show us opportunities like we have not seen before and provide us with ways to adapt to a greener, cleaner tomorrow—and they do it beautifully and with innovation.

Green buildings follow the same philosophy. This is truly the age of invention; the age of new ideas and adaptation. With the acceptance of new designs and new ways to look at old buildings and resources, we can not only benefit the environment but also improve our social outlook and human health—the physical, mental, and social aspects. If we are willing to take on the challenge and be open to doing things a little bit differently in the future, the future is ours.

REFERENCES

American Society of Landscape Architects (ASLA). 2016. Green infrastructure. *American Society of Landscape Architects: Professional Practice* https://www.asla.org/ContentDetail.aspx?id=43532 (accessed October 1, 2016).

Canadian Institute for Environmental Law and Policy (CIELAP). 2016. Brownfield redevelopment: What you should know. http://cielap.org/brownfields/benefits.html (accessed October 16, 2016).

eInstruments. 2016. Using portable emissions analyzers for optimizing the performance of a cogeneration plant. Application Note #IA-13-0801. http://www.e-inst.com/cogeneration/ (accessed October 18, 2016).

Green, J. November 18, 2010. A new way to plant urban trees. *Uniting the Built and Natural Environments. American Society of Landscape Architects.* https://dirt.asla.org/2010/11/18/a-new-way-to-plant-urban-trees/ (accessed October 15, 2016).

Hughes, V. March 13, 2014. Waste heat is free energy. So why aren't we using it? *Popular Science.* http://www.popsci.com/article/science/waste-heat-free-energy-so-why-arent-we-using-it (accessed October 5, 2016).

International District Energy Association. 2016. What is district energy? *Energy and Power Solutions.* http://www.districtenergy.org/what-is-district-energy (accessed October 16, 2016).

Kukreja, R. 2016. What is a green building? *Sunpower.* https://us.sunpower.com/what-green-building/ (accessed October 19, 2016).

National Household Travel Survey. March 2009. The carbon footprint of daily travel. U.S. Department of Transportation. http://nhts.ornl.gov/briefs/Carbon%20Footprint%20of%20Travel.pdf (accessed October 20, 2016).

Smartgrid.gov 2016. What is the smart grid? https://www.smartgrid.gov/the_smart_grid/smart_grid.html (accessed October 17, 2016).

Tsao, M. May 22, 2011. Living off the grid: Smart grids are current technology at its best. Michigan Engineering, University of Michigan. http://www.engin.umich.edu/college/about/news/stories/2011/may/living-off-the-grid-smart-grids-are-current-technology-at-its-best (accessed October 18, 2016).

World Resources Institute. Edited by Todd Gartner et al. 2016. Natural infrastructure: Investing in forested landscapes for source water protection in the United States. *World Resources Institute.* http://www.wri.org/sites/default/files/wri13_report_4c_naturalinfrastructure_v2.pdf (accessed October 1, 2016).

SUGGESTED READING

Dodson, J. and N. Sipe. 2016. *Planning After Petroleum: Preparing Cities for the Age Beyond Oil.* New York: Routledge. 272 p.

EPA. 2016. U.S. Greenhouse gas inventory report: 1990–2014. *U.S. Environmental Protection Agency.* https://www.epa.gov/ghgemissions/us-greenhouse-gas-inventory-report-1990-2014. (accessed October 1, 2016).

Farr, D. 2007. *Sustainable Urbanism: Urban Design with Nature.* Hoboken, NJ: Wiley. 352 p.

Gould, K. A. and T. L. Lewis. 2016. *Green Gentrification: Urban Sustainability and the Struggle for Environmental Justice.* New York: Routledge. 192 p.

Harnik, P. 2010. *Urban Green: Innovative Parks for Resurgent Cities,* 2nd ed. Washington, DC: Island Press. 208 p.

Levy, J. M. 2016. *Contemporary Urban Planning,* 11th ed. New York: Routledge. 476 p.

Moore, C. 2016. *Green Roofs and Living Walls for Architects.* Kindle Edition. ISDM.

Sullivan, M. June 30, 2007. World's first air-powered car: Zero emissions by next summer. *Popular Mechanics.* http://www.popularmechanics.com/cars/a1665/4217016/ (accessed October 20, 2016).

15 Applying Technology to Sustainability, Part II

Smart Transportation, Alternative Fuels, and New Fuel Technology

OVERVIEW

In the second installment of Applying Technology to Sustainability, we will be focusing on the issues of smart transportation, alternative fuels, advanced fuels, and new fuel technology. These represent several exciting new areas of technology in the fast-paced world of sustainability. In the smart transportation arena, we will look at smart street design and its growing importance in urban growth today, transit-oriented development, and the ins and outs of sustainable transportation planning. From there, we will next take a look at the explosion of alternative fuels and transport options that are either available or on the horizon, including fuel cells, hybrids, flex fuels, plug-ins, air powered transportation, and cars of the future. We will look at alternative fuels and advanced fuels, and then delve into new fuel technology. Finally, we will explore the latest innovations that have been in the news recently. You may be surprised at what you will find, and the commodities scientists have found to power the future.

INTRODUCTION

We have already seen how the cities of the future will benefit from smart growth through sustainability and urban planning. They will also benefit when those same smart growth strategies are applied to transportation systems. They will better serve the growing populations while providing economic growth for both businesses and communities. Such strategies might include creating transit options, such as buses, trolleys, subways, light rail, street cars, and ferries, which can accommodate more travelers in the same space while creating better options for getting from one location to another. Streets can be designed for multiple types of applications, making neighborhoods more functional and appealing.

People want more transportation choices and options. These smart growth techniques can provide all that, while giving each person the choice to select the best way to get around. This chapter takes a look at just how those plans can be taken from ideas and brought to fruition to form a healthier tomorrow.

DEVELOPMENT

Transportation and land use patterns that are closely connected. Transportation facilities and networks have the power to shape development, influence property values, and determine a neighborhood's character and quality of life. In addition, transportation networks have important consequences for the environment, including air and water quality, climate change, and open space preservation. How communities develop also affects how convenient and appealing public transportation, bicycling, and walking are for their residents. It is this development phase that gives each person a sense of ownership and belonging.

TRANSPORTATION EFFICIENCY IN AN URBAN SETTING

If transportation has been carefully integrated and land use planning concepts have been carefully thought out, it gives people more choices for getting around their town and region. When homes, offices, stores, and civic buildings are near transit stations and close to each other, it makes it convenient to walk, bicycle, or take transit. When there are more choices, it becomes easier—and more enjoyable—to incorporate physical activity into daily routines, reduces transportation costs, and allows more freedom and mobility to low-income individuals, senior citizens, disabled persons, and others who cannot or choose not to drive or own a car.

By providing a range of transportation choices and walkable neighborhoods that support them, it can also help improve air quality and reduce greenhouse gas (GHG) emissions. According to EPA's 1990–2014 Emissions Inventory, roughly 16 percent of U.S. GHG emissions comes from cars and light-duty trucks (including pickup trucks, Sports Utility Vehicles [SUVs], and minivans) (EPA, 2016). Developing more compactly and investing in public transit as well as other transportation options, makes it easier for people to drive less, which lowers GHG emissions. Keeping cars off the road and lessening total miles traveled, also reduces other harmful pollutants found in the atmosphere, such as carbon monoxide, sulfur dioxide, and particulate matter, which are also emitted by motor vehicles. There are four approaches that are commonly used to enhance the quality of life and protect human health along with the environment. They are smart street design, transit-oriented development, parking management, and sustainable transportation planning.

Smart Street Design

In the past, transportation planners have overlooked the important role streets play in shaping neighborhoods. It used to be that decisions about street size and design in many communities focused on getting as many cars as possible through the streets as quickly as possible. Street design now determines whether an area will be safe and inviting for pedestrians, bicyclists, and transit users, which also affects the success of certain types of retail, influences land values and tax receipts, and influences overall economic strength and growth. It is not just about cars anymore.

Street design also has important environmental impacts. It can determine the viability of less-polluting modes of transportation, affecting air quality and climate change. It also influences the volume of stormwater runoff, the water quality of that runoff, and the magnitude of the heat island effect.

Transit-Oriented Development

The United States is in the process of a demographic shift that will have major effects on the nation's housing market and development patterns in the near future. The fastest-growing demographic groups are the older, single-person households; and nonwhite households. They prefer homes within walking distance of transportation alternatives, shopping, restaurants, parks, and cultural amenities.

Market surveys and research have consistently shown that at least one-third of homebuyers prefer homes in smart growth neighborhoods, and this share is growing. Transit-oriented development (TOD) creates walkable communities for people of all ages and incomes and provides more transportation and housing choices. Because of these preferences, TOD is compact development built around a transit station or within easy walking distance (typically a half-mile) of a station and containing a mix of land uses such as housing, offices, shops, restaurants, and entertainment.

TOD can help lower household transportation costs, boost public transit ridership, reduce GHG emissions and air pollution, encourage economic development, and make housing more affordable by reducing developer expenditures on parking and allowing higher-density zoning.

Parking Management

Parking requirements can be an obstacle to compact development. The parking requirements found in many conventional zoning codes often call for off-street parking based on generic standards, not on individual sites' needs and context, and require too much parking to be provided on the development site. With their high costs and space requirements, conventional parking regulations can discourage

compact, mixed-use development and redevelopment in older neighborhoods. Large expanses of surface parking and stand-alone parking structures can discourage walking and make driving the only viable transportation between destinations. Better-managed parking can support active, economically strong, mixed-use districts; encourage walking and transit use; and reduce the costs of redevelopment and infill projects.

Sustainable Transportation Planning

Transportation planning will get the best results for communities when it is part of a comprehensive approach that includes land use and environmental planning at both the local and regional levels. Transportation planning and design choices have a direct influence on development patterns, choices of travel mode, the cost of infrastructure, and potential redevelopment.

This integrated approach requires transportation and land use planners to

- Examine the effects of transportation projects on future growth, development, and long-range economic goals.
- Assess each project's effects on air and water quality and other environmental resources.
- Determine whether transportation and other infrastructure can be built on a timetable that is compatible with development or redevelopment projects.

It is important that transportation configurations meet residents' needs and expectations in order for the entire community design to be accepted. This phase of the planning is one of the first that must be presented and accepted to ensure the project's ultimate success.

ALTERNATIVE FUELS AND TRANSPORT OPTIONS

It is often said that Americans have been in love with their cars ever since Henry Ford turned the car into a "must-have" item. Interestingly, when Ford's Model T was introduced in 1908, it got 12-kmL (28.5-mpg) fuel efficiency. Since then, even though technology has improved and cars today have a wide selection of features and go farther and faster, when it comes to fuel efficiency, technology has digressed. In response to the oil shortage crisis in the 1970s, the United States was determined to begin producing cars that doubled the mileage of existing cars. Since then, however, with the pressure off, the worry about fuel efficiency went by the wayside and the focus shifted instead to performance cars such as SUVs. By 2005, most of the cars on the highways in the United States were less fuel efficient than those on the road in the 1980s.

According to the Environmental Defense Fund (EDF), if the exhaust coming from a car had an actual weight, an average household with two medium-sized sedans would emit more than 9,072 kilograms of CO_2 a year (NHTS, 2009). Even worse, SUVs were emitting up to 40 percent more than smaller cars. Today, fuel efficiency has once again become a concern for two reasons:

- The impacts of climate change
- U.S. dependence on foreign sources of oil

A gallon of gasoline weighs slightly more than 2.5 kilograms. When it is burned as fuel in a vehicle, the carbon in it combines with oxygen and produces approximately 8.5 kilograms of CO_2. Add to this the energy that was expended in making and distributing the fuel, the total global warming pollution is about 3 kilograms per liter (25 pounds of CO_2 per gallon).

To illustrate the impact, a car that gets 21 mpg (9 kilograms per liter) and is driven 30 miles (48 kilometers) a day uses 1.5 gallons (5.5 liters) each day and emits 35.5 pounds (1.5 kilograms) of CO_2 every day. When multiplied by the millions of cars that are driven in the United States each day, this adds up. This means that one million cars emit the equivalent of 35.7 million pounds (16.2 million kilograms) of CO_2 every day, 2 million cars contribute 71.4 million pounds (32.5 million kilograms) of CO_2 every day, and 3 million cars contribute 107.1 million pounds (48.5 million kilograms) of CO_2 every day.

The table illustrates what the true costs of lower fuel efficiency add up to and why it is important to take action now toward a greener tomorrow.

Population growth and the spread of suburban areas have put even more cars on the road and more miles driven each day. These trends combined with inefficient mileage spell environmental disaster. One way to combat this dangerous trend is with green technology in the auto industry—the advent of the hybrid cars, electric vehicle options, flexible fuels, fuel cells, plug-in cars, and other forms of transportation technology is currently developing.

Fuel Cells

There are currently several types of fuel cells. The type most often used in vehicles is polymer electrolyte membrane (PEM) fuel cells—also called proton exchange membrane fuel cells. A PEM fuel cell uses hydrogen fuel and oxygen from the air to produce electricity (Figure 15.1).

Hydrogen fuel is channeled through field flow plates to the anode on one side of the fuel cell, while oxygen from the air is channeled to the cathode on the other side of the cell. At the anode, a platinum catalyst causes the hydrogen to split into positive hydrogen ions (protons) and negatively charged electrons. The PEM allows only the positively charged ions to pass through to the cathode. The negatively charged electrons must travel along an external circuit to the cathode, creating an electrical current. At the cathode, the electrons and positively charged hydrogen ions combine with oxygen to form water, which flows out of the cell.

Most fuel cells designed for use in vehicles produce less than 1.6 volts of electricity—not nearly enough to power a vehicle. Because of this, it is necessary to place multiple

The Annual Cost of Lower Fuel Efficiency

Average Gas Mileage	Average Fuel Used (Based on 12,000 Miles per Year)	Approximate Greenhouse Gas Pollution	Approximate Cost (Based on $2.30/Gallon)
50 MPG	240 gallons	2.7 tons/year	$552
40 MPG	300 gallons	3.4 tons/year	$690
30 MPG	400 gallons	4.5 tons/year	$920
25 MPG	480 gallons	5.4 tons/year	$1,104
20 MPG	600 gallons	6.8 tons/year	$1,380
15 MPG	800 gallons	9 tons/year	$1,840
10 MPG	1,200 gallons	13.6 tons/year	$2,760

Source: Argonne National Laboratory

For the average American who owns a car, driving is one of the top two daily pollution-causing activities (electricity use is the other one). Because of this, by choosing a greener vehicle, it is one way that a person can make a significant difference in the fight against climate change. According to the EDF, vehicle choice is one of the most powerful decisions a person can make, and there is a triple benefit associated with it:

- It protects the climate.
- It reduces the country's dependence on oil.
- It saves money at the pump.

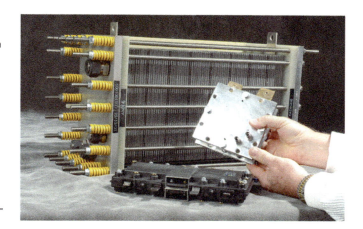

FIGURE 15.1 A hydrogen fuel cell generates electricity through an electrochemical reaction using hydrogen and oxygen. Hydrogen is sent into one side of a proton exchange membrane (PEM). The hydrogen proton travels through the membrane, while the electron enters an electrical circuit, creating a DC electrical current. On the other side of the membrane, the proton and electron are recombined and mixed with oxygen from room air, forming pure water. Because there is no combustion in the process, there are no other emissions, making fuel cells a clean and renewable source of electricity. (Courtesy of Matt Etiveson, Department of Energy/National Renewable Energy Laboratory [DOE/NREL], Golden, CO.)

fuel cells onto a fuel cell stack. The potential power generated by a fuel cell stack depends on the number and size of the individual fuel cells that comprise the stack and the surface area of the PEM. Although there are not many fuel cell vehicles on the road today, researchers believe they will 1 day revolutionize the transportation sector. The technology has the potential to significantly reduce energy use and harmful emissions as well as the United States' dependency on foreign oil. A radical departure from current vehicles, they completely eliminate the conventional internal combustion engine. Like battery electric vehicles, FCVs are propelled by electric motors. But while battery electric vehicles use electricity from an external source, FCVs create their own electricity. The fuel cell creates the electricity through a chemical process using hydrogen fuel and oxygen. FCVs can be fueled with pure hydrogen gas stored onboard in high-pressure tanks. They can also be fueled with hydrogen-rich fuels such as methanol, natural gas, or even gasoline, but these fuels must first be converted into hydrogen gas by an onboard device called a reformer (Figure 15.2).

FCVs that are fueled with pure hydrogen emit no pollutants—only water and heat. FCVs using hydrogen-rich fuels and a reformer produce only small amounts of air pollutants. FCVs are twice as efficient as today's similarly sized conventional models.

Research still needs to be completed in order to begin producing FCVs for mainstream America. Effective and efficient ways to produce and store hydrogen must be determined. Currently, extensive research is underway to solve the limitations so that FCVs will become the cars of tomorrow. Partnerships with private research and the government are underway right now to make this happen with projects such as FreedomCAR (a DOE initiative) and the California Fuel Cell Partnership (a California initiative).

Hybrids

Hybrids combine a small combustion engine with an electric motor and battery. The two technologies can be combined to reduce fuel consumption and tailpipe emissions. Most of the hybrids on the road today compliment their gas engines by charging a battery while braking—a concept called *regenerative braking*.

Engines that run on diesel or other alternative fuels can also be used in hybrids. A hybrid drive is fully scalable, which means that the drive can be used to power everything from small commuter cars to large buses and watercraft. The technology can even work on locomotives. Hybrids get more miles per gallon than most nonhybrids; they also usually have very low tailpipe emissions (Figure 15.3).

Hybrids are able to reduce smog pollution by 90 percent compared with the cleanest conventional vehicles on the road today. For example, the Toyota Prius, introduced in 1997, achieves a 90 percent reduction in smog-forming pollutants over the current national average. Because hybrids do have an internal combustion engine, however, they will never be able to achieve zero emissions. They do consume much less fuel. They do cut global warming emissions by a third to a half. Models currently under development should be able to cut even more.

FIGURE 15.2 Zero emission hydrogen fuel cell bus at a bus stop outside a parking garage in the Connecticut transit district. This bus is nonpolluting and has more than twice the fuel economy of a standard diesel bus. (Courtesy of Leslie Eudy, DOE/NREL, Golden, CO.)

FIGURE 15.3 Alternative fuel vessels are doing their part to cut traffic congestion and emissions in Fort Lauderdale, Florida. The hybrid-electric waterbus runs on electricity and B20 biodiesel, consuming just half the fuel per day of a similar diesel-powered boat. (Courtesy of DOE/NREL, Golden, CO.)

By combining gasoline and electric power, hybrids have the same or greater range than traditional combustion engines. For example, the 2016 Toyota Prius gets 54 mpg in highway driving and 50 mpg in city driving, with a total range of about 570 miles (917 kilometers) on a full tank of gas. The 2016 Hyundai Hybrid is rated at 42 mpg in highway driving and 39 in city driving, with a total range of 600 miles (966 kilometers).

Hybrids have the same or better performance than their traditional counterparts. As public support catches on, more hybrids will be introduced. The initial cost of hybrids is slightly higher, but with fuel savings over the length of the car's lifetime, the costs are competitively priced.

Flex Fuels

Flexible fuel vehicles (FFVs) are designed to run on gasoline or a blend of up to 85 percent ethanol. They are similar to gasoline models and have only a few engine and fuel system modifications. FFVs have been produced since the 1980s. The opinion of the Union of Concerned Scientists (UCS) is that while there may be potential benefits from getting more FFVs out on the road, the benefits are not worth any increase in oil dependency.

A better alternative is the capability of driving cars that run on alternative fuels as long as the main fuel choices are not fossil fuels—they need to be biomass fuels, domestic fuels, and advanced, alternative fuels that are good for the environment, national security, and the economy.

One of the main concerns about the FFV program is what it refers to as the dual-fuel loophole. The dual-fuel loophole allows manufacturers to earn credits toward meeting federal fuel economy standards by producing vehicles that are able to run on both petroleum and an alternative fuel, even if they never actually use alternative fuel. The best approach to this dilemma is to make sure that auto manufacturers are ensuring that the fuel tank designed to operate on E85 percent ethanol fuel is actually being run on E85 ethanol fuel.

Plug-in Hybrid Vehicles

Plug-in hybrid technology allows gasoline-electric hybrid vehicles to be recharged from the electrical power grid and run many miles on battery power alone. Because electric motors are far more efficient than internal combustion engines, vehicles that use electricity almost always produce less climate change pollution than gasoline vehicles, even when the electricity used to fuel them is generated from coal. The benefits are even greater when vehicles are fueled with renewable-generated electricity.

A gas engine provides additional driving range as needed after the battery power is gone. Plug-in hybrids may never need to run on anything but electricity for shorter commutes. The combination of gas and electric driving technologies can already achieve up to 150 mpg (63.5 kilograms per liter).

FIGURE 15.4 NREL's plug-in hybrid electric vehicle (PHEV) at the National Wind Technology Center (NWTC). This car gets 100 mpg (42.5 kilograms per liter). (Courtesy of Mike Linenberger, DOE/NREL, Golden, CO.)

In June 2008, the DOE National Renewable Energy Laboratory (NREL) modified a 2006 Toyota Prius sedan to achieve an amazing 100 mpg (42.5 kilograms per liter). The experimental plug-in runs the initial 60 miles (96.5 kilometers) mostly on battery, with the remainder achieved under engine power. According to NREL, the sedan's performance is more than doubles the fuel economy of a standard Prius, which is rated 54/50 mpg. In addition, it is a five-fold improvement over the 20 mpg averages that passenger cars and light trucks in the United States formerly achieved. The plug-in hybrid runs on electricity at low speeds, then the batteries and the gasoline engine share the work. The batteries recharge automatically as the car is running (Figure 15.4).

NREL researchers added several features to the plug-in Prius to break the 100-mpg barrier, including:

- A plug to recharge its batteries directly from the utility grid using a standard 110-volt electrical outlet
- A larger lithium-ion battery that allows the car to operate on electricity for longer trips at speeds up to 35 mph (56.5 kph)
- A rooftop solar panel that charges the battery while the car is driving or parked outdoors, adding 5 miles (8 kilometers) to the vehicle's range

"The stored power in the battery does a great job of displacing petroleum," said Tony Markel, a senior engineer with the Vehicle Systems Analysis Group. "Most people's daily commute is about 30 miles, so this car would run virtually on battery for their entire drive."

The NREL Prius is a unique research prototype and is not available to the public. It costs about $70,000—the cost of the standard Prius plus $42,500 for the modifications.

Electric Vehicles

A battery-electric vehicle (BEV) uses electricity stored in its battery pack to power an electric motor that turns its wheels. The battery pack is recharged by connecting it (plugging it in) to a wall socket or other electrical source, such as a solar panel.

Because these vehicles use electricity as the fuel source, there are no emissions from a tailpipe when recharging the electricity, and it costs pennies (compared to dollars at the pump for gasoline-powered vehicles) to operate.

Since BEVs do not have tailpipes, they do not produce any tailpipe emissions. They do recharge, however, using electricity generated at power plants that emit global warming and smog-forming pollutants. If an electric vehicle is charged at a facility that strictly uses renewable energy such as solar power, hydropower, or wind power, then electric car technology is completely green. If the vehicle is recharged at a facility with electricity that is generated from power plants using fossil fuels, they are still up to 99 percent cleaner than conventional vehicles and can cut climate change-causing emissions by as much as 70 percent. They are extremely energy efficient. The electric motors convert 75 percent of the chemical energy from the batteries to power the wheels. As a comparison, internal combustion engines convert 20 percent of the energy stored in gasoline (Figure 15.5).

BEVs are more expensive to purchase than standard cars. This is largely attributed to the fact that their advanced battery packs are expensive to produce. The positive side is that it costs only about one-third the price of refueling a gasoline-powered vehicle. BEVs also offer quiet, smooth driving experiences. They can travel 50–100 miles (80.5–161 kilometers) per charge depending on the battery type and driving conditions. Electric vehicles reduce energy dependency because the energy supply is solely domestic. It takes four to eight hours to fully recharge the battery.

Air Powered

Air-powered cars are vehicles that are being developed to run only on compressed air. This zero-emission fuel is believed to hold some promise for future car models and is being explored in Europe, Asia, and the United States. Air power can be substantial as seen in pneumatic air powered tools (Figure 15.6).

According to an article in *Popular Mechanics*, a car powered by air is a reality. In fact, compressed air vehicles—commonly referred to as air cars—have been running around for several years.

Compressed air is used every day to perform difficult tasks. For instance, mechanics rely on air-driven pneumatic tools to turn nuts and bolts. Pneumatic tools are powerful—even at a relatively low psi pressure setting. Compressed air is a force that with enough power can propel a wheel-driven car (Sullivan, 2007).

Guy Negre of Motor Development International (MDI) has developed just such an air-powered vehicle. The vehicle has no combustion—its power comes solely from compressed air run by electricity from the grid. To make the vehicle go, a pair of air-driven pistons turns a crankshaft that produces a rotational force. The technology can potentially be paired in two-, four-, or six-cylinder engine configurations. While there is no combustion, the only engine heat comes from friction, enabling it to be constructed of lightweight aluminum.

In 2007, Tata Motors licensed the rights from MDI for $28 million to build and sell Tata air cars in India. The Nano is Tata's scooter replacement. It is very popular in developing countries, but it does not meet U.S. federal emissions and safety requirements.

Air cars have not escaped the attention of U.S. automakers. Ford, for instance, has worked with an engineering team at UCLA to develop an air hybrid. Interestingly, air-powered

FIGURE 15.5 Electric car. (Courtesy of Tom Brewster, DOE/ NREL, Golden, CO.)

FIGURE 15.6 Air-powered vehicle. (Courtesy of DOE/NREL, Golden, CO.)

cars are not a new idea; the concept predates the internal combustion engine. Author Jules Verne, in his book *Paris in the twenty first Century*, describes a transportation system utilizing compressed air. Today, scientists are turning his concept into reality.

The Top Environmentally Friendly Cars Today

In light of climate change and other environmental issues, automobile manufacturers worldwide have been working to make vehicles more fuel efficient and environmentally friendly.

Mike Marshall, director of automotive emerging technologies, said, "High gas prices, coupled with consumers becoming more familiar with alternative powertrain technology, are definitely increasing consumer interest in hybrids and flexible fuels. However, the additional price premiums associated with hybrid vehicles, which can run from \$3,000 to \$10,000 more than a comparable nonhybrid vehicle, remain the biggest concern among consumers considering a hybrid."

There is high consumer interest in hybrids and vehicles that run on alternative fuels such as diesel or E85 (ethanol). Prices are still somewhat restrictive for many drivers, however. Fortunately, many standard models and makes now offer good mileage as they are. Thirty of some of the most environmentally friendly cars include the following:

Thirty of the Most Environmentally Friendly Cars of 2016

Smart for Two Electric Drive Convertible/Coupe

Chevrolet Spark EV	Volkswagen Jetta Hybrid	Hyundai Sonata Plug-In Hybrid
Fiat 500e	Nissan Versa Note	Toyota Prius Two Eco
Toyota Prius Eco	BMW X3	Ford Focus Electric
Volkswagen E-Golf	Ford Fusion Hybrid	Audi A3 E-Tron Ultra
Nissan Leaf	Lexus CT 200h	Fiat 500e
Kia Soul Electric	Mazda CX-5	Mini Cooper
Toyota Prius C	Mazda Mazda3	Subaru BRZ
Toyota Prius	Toyota Avalon Hybrid	Tesla Model S
Ford Focus Electric	Toyota Highlander Hybrid	Ford C-MAX Hybrid SEL
Chevrolet Volt	Toyota Prius V	

Based on studies conducted by Forbes, Cars.com, thrillist.com.

Another major focus of the automobile industry in the United States is on reducing harmful GHGs and the United States' dependence on foreign oil. These are two of the biggest hurdles facing the nation today. Accomplishing both tasks will require the same course of action: reducing the amount of fossil fuel that is burned every day. According to the National Resources Defense Council, the United States spends more than \$200,000 per minute on foreign oil—\$13 million per hour—and two-thirds of that is used for transportation.

By increasing the efficiency of cars, trucks, and SUVs, many environmental problems can be solved. Driving more fuel-efficient cars is one way to achieve this.

To date, Honda is known as the greenest automaker. It has the best overall smog performance in four out of five classes of vehicles and better than average global warming scores in every class. As more consumers become aware and educated, automobile manufacturers will be pressured into meeting the demands for greener technology.

Cars of the Future

Cars of the future, no doubt, will evolve to run cleaner, faster, and more efficiently than ever before because technology is constantly being redefined and improved by automakers and engine manufacturers, and watched over by the scientific mind in a quest to fight the very real effects of climate change. As new technologies today leap from the concept level to the drawing board, and into reality—such as hybrid cars and hydrogen fuel cells—they will continue to push the edges of today's automobile technology and redefine what is possible.

There will, no doubt, be many new innovations introduced in the future as new discoveries are made—new elements, new technology, new mediums, new types of motion—which will also contribute to better fuel economy, wiser environmental management, and lower GHG emissions. As the general public becomes more aware of the pertinent issues concerning global warming, the public demand will also play a significant role in what manufacturers supply in terms of efficiency.

Advanced materials—such as metals, polymers, composites, and intermetallic compounds—also play an important role in improving the efficiency of transportation engines and vehicles. Weight reduction is one of the most practical ways to increase the fuel economy of vehicles while reducing exhaust emissions. The less a car weighs, the better mileage efficiency it can achieve.

The use of lightweight, high-performance materials will contribute to the development of vehicles that provide better fuel economy but are comparable in size, comfort, and safety to today's vehicles. This way, making the change to increase energy efficiency will not impact the level of comfort previously enjoyed.

The development of propulsion materials and technologies will help reduce costs while improving the durability and efficiency. Regardless of what future surprises technology has in store, one thing is certain: the issue of climate change must be addressed and acted upon immediately if there is to be a future with choices.

Sustainable Mass Transit Options

Sustainable mass transit should offer many benefits, such as:

- Reducing CO_2 emissions
- Helping produce numerous financial benefits
- Creating urban centers of productivity
- Reducing dependence on foreign oil

FIGURE 15.7 Mass transportation helps lower greenhouse gas emissions, especially with alternative fuel models, such as this hybrid electric bus. Using public transportation promotes sustainable living. (Courtesy of Wasatch Images.)

FIGURE 15.8 Light rail public transportation is one low-cost system that helps commuters adopt a green lifestyle and help the environment. This system in Salt Lake City, Utah, was built in 2002 and is so popular that since the initial route was built, four additional extensions have been added. (Courtesy of Wasatch Images.)

As far as carbon-based transit, the amount of oil used per passenger is greatly reduced with the use of any type of sustainable mass transit when compared to the fossil fuels used when people choose to drive their personal vehicles. New fuel technologies such as biofuel and hydrogen, in addition to electric vehicles, represent a new trend in environmentally sound and economically beneficial mass transit. Many metropolitan areas today are utilizing buses that use alternatives to fossil fuels (Figure 15.7). The higher the quantity and quality of public transportation, the greater the incentive will be for the public to choose it as a transportation option. Convenience is another factor that must be considered. An appropriate number of routes and times per day of pickup and drop-off are essential for public participation. If it is not convenient, then people will take their own cars instead.

One of the more promising modes of public transport is rail, particularly electric urban rail. Light rail provides various benefits relating to cost effectiveness; the low maintenance needs and low energy demands of light rail make this form of transportation highly efficient. Environmental benefits include the reduction of CO_2, as well as the reduction of carbon monoxide and nitrogen oxide (Figure 15.8).

Traffic congestion is greatly reduced in any urban environment, and ultimately light rail systems can replace highways. Light rail creates jobs both by producing a new source of capital and by creating new, busy economic centers. Not only does light rail replace the use of fossil fuels, but the development cost of light rail systems is about half that of building freeways.

ALTERNATIVE FUELS

An alternative fuel is any material or substance that can be used as a fuel other than that derived from the standard fossil fuels (coal, oil, or gas). There are several types of alternative fuels currently in use and many more being researched and developed. The fuels that are most commonly used today are biodiesel, bioalcohols (methanol, ethanol, and butanol), chemically stored electricity (such as batteries and fuel cells), nonfossil hydrogen, nonfossil methane, nonfossil natural gas, vegetable oil, and other sources produced from biomass. Each of these fuels has its advantages and disadvantages. The advantages of the alternative fuels currently being used are that less pollution is emitted from them compared to fossil fuels. The obvious disadvantage, of course, is that some fossil fuels are still being used in them—but we are taking a step in the right direction.

Biofuels and Clean Vehicles

It is well understood that one of the biggest contributors to climate change is the burning of fossil fuels, and the United States is one of the largest contributors to the problem. Transportation-related emissions are responsible for 40 percent of the United States' total global warming pollution. Using new technology in the area of transportation—more efficient vehicles and lower carbon fuels (fuels that generate far less heat-trapping gases per unit of energy)—is one area where people can make a significant difference. Hydrogen, electricity, and biofuels all have this capability. This also helps national security by reducing the country's dependence on foreign oil.

There are a wide range of characteristics among each alternative fuel type and their unique environmental emissions, if any, and their environmental impacts. Standards are currently being developed that will require fuel providers to account for and reduce the heat-trapping emissions associated with both the production and use of fuel.

California, the nation's largest market for transportation fuel, is developing a low-carbon fuel standard that will require fuel providers to verify there is a reduction in global warming emissions per unit of energy delivered.

When carbon emissions are calculated, all emissions are accounted for during the fuels' entire life cycle. The accounting system is very specific and addresses any uncertainties. It also allows for changes over time as technology improves for assessments, as well as products, to become more refined.

Through being able to keep track of the exact performance of each type of new fuel, it allows researchers to be able to assess just how effective the alternative fuels are toward making headway against climate change.

Low-Carbon Fuel Standards

Smart fuel policies, such as California's, which took effect in 2010, are important because they promote carbon reduction through the entire fuel manufacturing process. According to the UCS, low-carbon fuel standards (LCFSs) also create market certainty for cleaner fuels, making sure the fuel industry does its part along with the automakers and consumers to reduce transportation emissions that relate to climate change. Other states that are also considering developing low-carbon fuel standards are Arizona, Minnesota, New Mexico, Oregon, and Washington.

LCFSs are designed to work in tandem with the Obama administration's auto fuel standards, which include tailpipe emissions to be cut by more than 30 percent. What makes the LCFS unique is that it deals with the entire life cycle emissions of fuels on an average per-gallon basis. Instead of dictating specific technologies or fuel types, it allows suppliers lateral freedom to decide which methods they will use to meet the reduced emissions targets. The key aspect is the life cycle component, which requires *every* aspect involved in the fuel's retrieval, creation, and use to be accounted for in the way it contributes to climate change—a concept called "well-to-wheels," which must be taken into account.

In "wells-to-wheels," the following are taken into account:

- The emissions generated at the extraction source
- The refinery process
- The tailpipe emissions

For example, where the fuel is grown as an agricultural crop (instead of mined from the ground), all the emissions from tractors and fertilizers used to grow the crop, all the energy used to convert the crop to a fuel, and any other indirect sources of pollution—including any emissions generated from changes in land use as a result of biofuel production—must be accounted for.

Because of all the interrelationships involved, accounting for all these inputs can become very complex. The UCS stresses that these issues must be openly discussed and a consensus reached without any political lobbying or interference, to keep the technical issues and results reliable in order to avoid a distorted analysis. Otherwise, it will keep the LCFS from being able to deliver the needed low-carbon fuels.

Under the LCFS program, fuel suppliers are not mandated to try particular technologies or specific fuels; they are free to choose how they meet their emissions targets. For instance, they can choose to blend lower-carbon biofuels into gasoline to lower the carbon content; they can choose to reduce emissions from the refining process; or they can sell natural gas (which has a lower carbon content) for use as a transportation fuel. They also have the option of using trading credits, which provide even more flexibility and lower the loss of compliance if other methods, such as switching technologies, prove to be too costly. In this case, for example, fuel suppliers could purchase credits from electric utilities that supply low-carbon electricity to plug-in hybrids.

An important aspect in the LCFS concept is that an accurate life cycle accounting must be kept, including any indirect emissions. Indirect emissions include situations such as when land must be cleared to grow crops that will be used for biofuels. If forested areas are removed, the lost carbon storage must be accounted for. If all the direct and indirect impacts to net carbon balance are not accurately accounted for, then calculations of carbon reductions cannot be accurate. Otherwise, a study may declare a decrease in fuel emissions; yet actually contribute to an increase in global warming pollution.

A Complicated, but Innovative Way to Reward

The LCFS is important for three major reasons:

- Although it may sound convoluted at times, it promotes improvements in the supply chain.
- It protects against high-carbon fuels.
- It creates choices and spurs innovation.

When the full life cycle is considered, it offers the fuel provider opportunities to improve and lower the carbon content anywhere along the supply chain. It protects against high-carbon fuels by offering a clear incentive to use clean fuels over polluting fuels. For example, the coal to liquids technology has a life cycle global warming pollution almost double that of petroleum. It is a bigger incentive to choose a low-carbon fuel to begin with. This concept makes the dirtier fuel businesses pay the price for their higher pollution.

Because the LCFS does not mandate using a specific approach or a certain technology, it opens the playing field and focuses only on companies' abilities to deliver cost-effective low-carbon fuel. This way it is not encumbered by government mandates that could slow new developments. Its design, instead, is to focus on who can supply the lowest-carbon fuels. According to the UCS, the investors and the marketplace (public) will be the ones who will decide on the ultimate winners.

President Obama has also called for a nationwide low-carbon fuel standard to help meet his goal of cutting GHG emissions more than 80 percent by mid-century. California's LCFS would require refineries, producers, and importers of motor fuels sold in California to reduce the carbon intensity of their products by 10 percent by 2020, with greater cuts thereafter.

At the national level, work is being done to encourage the information of heat-trapping emission requirements into the current renewable fuel standard. Several bills have also been introduced in congress that would establish low-carbon fuel standards. Advanced clean vehicle technologies are available now. Fuel cells have come a long way in research. Various cities in the world have demonstrated fuel cell bus programs, such as Chicago, Illinois, and Vancouver in Canada. In California, a partnership among automakers, the government, and fuel cell manufacturers is testing fuel cell technology and is expected to produce over 60 demonstration vehicles in the next few years.

In addition, the California zero emission vehicle (ZEV) program requires auto manufacturers to sell increasing numbers of zero emission vehicles over the next decade to further promote fuel cell vehicles. The trend has been set, and the public is voicing its eagerness to have these vehicles available. The majority of automobile manufacturers have announced their plans to begin selling fuel cell passenger vehicles in the next few years.

There are two different concepts when referring to advanced clean vehicles:

- A fuel-efficient car
- A low-emitting car

A higher fuel efficiency results in less climate change pollution. A low-emitting vehicle releases fewer smog-forming pollutants. The amount of fuel that a car burns determines how much CO_2 it releases. Air pollution-control devices on cars reduce other pollutants, such as CO, or smog-forming pollutants, such as NO_x and volatile organic compounds (VOCs). All vehicles with high fuel efficiencies do not necessarily reduce urban smog, such as diesels. The vehicle must also be one that has low emissions. Truly green cars address all the issues:

- Climate change
- Air pollution
- America's dependence on oil

ALTERNATIVE AND ADVANCED FUELS

According to the DOE, there are more than a dozen alternative and advanced fuels in production or being used today. They include the following:

- Biodiesel
- Electricity
- Ethanol
- Methanol
- Hydrogen
- Natural gas
- Propane

As the general population becomes more educated and aware of their existence and availability and as the price of gasoline

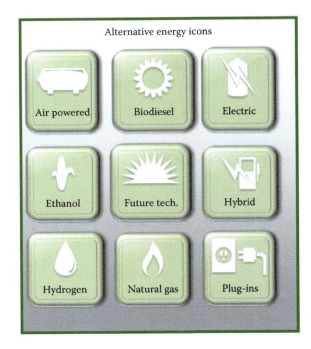

FIGURE 15.9 Icons signifying renewable energy. (From Gines, J. K., *Climate Management Issues: Economics, Sociology, and Politics*, CRC Press, Boca Raton, FL, 2012.)

at the pumps skyrockets, there is a growing interest in using them as the green revolution gains momentum. These fuels provide numerous benefits, including that they are environmentally friendly and they reduce the United States' dependence on foreign oil sources.

Alternative fuels have been defined by the Energy Policy Act (EPAct) of 1992. The EPAct was passed by congress to reduce the nation's dependence on imported petroleum by requiring specific vehicle fleets to acquire alternative fuel vehicles that are capable of operating on nonpetroleum fuels (Figure 15.9).

Biodiesel

Biodiesel is a renewable alternative fuel that is made from soybeans, biomass, vegetable oil, animal fats, and recycled restaurant greases. It can be used in its pure state (B100) or blended with petroleum diesel. B2 is 2 percent biodiesel. There is also B5 and B20 blends. The liquid fuel is comprised of fatty acid methyl esters (FAMEs), or long-chain alkyl esters.

Biodiesel is produced from renewable sources such as new and used vegetable oils and animals fats. It is also a much cleaner burning fuel than the traditional petroleum diesel, although it has physical properties similar to those of traditional diesel. Biodiesel is also nontoxic and biodegradable. It can be used in most diesel cars without modification, making it an attractive choice of alternative fuel. The following table illustrates the advantages and disadvantages of biodiesel as identified by the DOE.

Advantages and Disadvantages of Biodiesel

Advantages	Disadvantages
Domestically produced from nonpetroleum renewable resources	Use of blends above 85 percent not yet warranted by automakers
Can be used in most diesel engines, especially newer ones	Lower fuel economy and power (10 percent lower for B100, 2 percent for B20)
Less air pollutants (other than nitrogen oxides) and greenhouse gases	Currently more expensive
Biodegradable	More nitrogen oxide emissions
Nontoxic	B100 generally not suitable for use in low temperatures
Safer to handle	Concerns about B100s impact on engine durability

The U.S. biodiesel industry is still relatively small, but it is growing rapidly. The bulk of biodiesel manufacturing comes from industries involved in making products from vegetable oil or animal fat. One example is the detergent industry. The soy industry has been one of the major driving forces behind biodiesel commercialization because of overproduction of soy oils and falling prices in the market the past few years. According to the DOE, there is enough virgin soy oil, recycled restaurant grease, and other acceptable feedstock available in the United States to provide quality feedstock for approximately 1.5 billion gallons (5.5 billion liters) of biodiesel per year, which represents about 5 percent of the diesel used in the United States.

Electricity

Electricity can be used to power both electric and plug-in hybrid electric vehicles directly from the power grid. An important aspect of electric vehicles is that they do not produce any tailpipe emissions. The only emissions that can be attributed to electricity are those generated in the production process at the power plant that generates the electricity. The electricity option works well for short-range driving.

Electricity is used as the transportation fuel to power battery electric vehicles (BEVs). These vehicles store the electricity in a battery, which then powers the vehicle's wheels through an electric motor. The batteries have a limited storage capacity, and once the charge has been used up, they must be recharged by being plugged into an electrical source. Examples of these vehicles are the battery-powered escort courtesy cars common in airport terminals.

Ethanol

Ethanol is a renewable fuel made from biomass (various plant materials). Ethanol contains the same chemical compound (C_2H_5OH) found in alcoholic beverages, which is why it is also known as ethyl alcohol or grain alcohol. Presently, nearly half of the gasoline produced in the United States contains ethanol in a low-level blend to oxygenate the fuel and reduce air pollution. According to the DOE, studies have estimated that ethanol and other biofuels could replace 30 percent or more of the U.S. gasoline demand by 2030.

It takes several steps to produce ethanol. Initially, biomass feedstock must be grown. Ethanol is a clear, colorless liquid and can be produced from either starch- or sugar-based feedstock such as corn grain or sugarcane. In the United States, ethanol is primarily produced from corn grain. In countries, like Brazil it is produced from sugarcane. It can also be produced from cellulosic feedstock such as grass, wood, crop residues, or old newspapers, but the process is more involved and complicated.

Plants contain the cellulosic materials cellulose and hemicellulose. These complex polymers are what form the structure of plant stalks, leaves, trunks, branches, and husks. They are also found in products made from plants, such as sugar. In order to make ethanol from cellulosic feedstock the materials must be broken down into their component sugars for fermentation to ethanol in a process called biochemical conversion. Cellulosic feedstock can also be converted into ethanol in a process called biochemical conversion. Cellulosic ethanol conversion processes are a major focus of current DOE research.

According to scientists at the DOE, ethanol works very well in internal combustion engines. Historically, Henry Ford and other early automakers believed ethanol would be the world's primary fuel source (before gasoline became so readily available).

As a comparison, a pure gallon of ethanol contains 34 percent less energy than a gallon of gasoline. It is also a high-octane fuel. Low-octane gasoline can be blended with 10 percent ethanol to achieve the standard 87-octane requirement. Ethanol is the principal component today of 85 percent ethanol and 15 percent gasoline (E85).

The DOE has identified several benefits of ethanol. It is a renewable, largely domestic transportation fuel. When it is used as a low-level blend such as E10 (10 percent ethanol, 90 percent gasoline), or high-level such as E85 (85 percent ethanol, 15 percent gasoline), ethanol still helps reduce the United States' dependence on foreign oil and combats climate change. In the United States, ethanol is produced almost entirely from domestic crops.

Another major positive is that the CO_2 released when ethanol is burned is balanced by the CO_2 captured when the crops are grown to make ethanol. Based on data from Argonne National Laboratory, on a life cycle basis, corn-based ethanol production and use reduces GHG emissions by up to 52 percent compared to gasoline production and use. They project that cellulosic ethanol use could reduce GHGs by as much as 86 percent. Argonne National Laboratory is DOE's largest research center, and is located in Chicago.

Ethanol is also completely biodegradable and, if spilled, poses much less of an environmental threat than does petroleum to surface and groundwater. As an example of this, after the sinking of the Bow Mariner off the Virginia coast in February 2004, U.S. officials noted the 3-million–gallon (11-million-L) cargo of industrial ethanol had dissipated quickly and did not pose an environmental threat to humans or marine life.

For more than 20 years, the DOE has been heavily involved in research and development to identify and develop promising energy crops. Its prominent research areas include an integrated analysis of biomass resources, feedstock sustainability, and feedstock systems engineering.

The U.S. Department of Agriculture (USDA) is also heavily involved in ethanol feedstock research and development. Three specific fields it is currently focused on include the following:

- Selective breeding and genetic engineering to improve feedstock crop yields
- Developing sustainable approaches to feedstock production
- Investigating the environmental and economic impacts of feedstock crops on farmland

The Biomass Research and Development Initiative is a DOE/USDA effort to coordinate and accelerate all federal bio-based products and bioenergy research and development. The initiative funds ethanol feedstock production research and development.

The DOE Office of Science also supports fundamental research on ethanol feedstock, including the development of optimized energy crops. It has formed three bioenergy research centers to accelerate basic research on cellulosic ethanol and other biofuels.

Methanol

Methanol, also known as wood alcohol, is considered an alternative fuel under the Energy Policy Act of 1992. Currently, most of the methanol production is accomplished through a process that utilizes natural gas as a feedstock. Methanol can be used to make methyl tertiary-butyl ether (MTBE), an oxygenate that is blended with gasoline to enhance octane and create cleaner burning fuel. MTBE production and use has declined in recent years, however, because it was responsible for some groundwater contamination.

As an engine fuel, methanol has similar chemical and physical characteristics as ethanol. Methanol is methane with one hydrogen molecule replaced by a hydroxyl radical. In the formation process, steam reforms natural gas to create a synthesis gas, which is then fed into a reactor vessel in the presence of a catalyst. This process produces methanol and water vapor. Even though a variety of feedstock can be used to create methanol, with today's economy, the use of natural gas is the preferred production method (Figure 15.10).

There are several advantages to methanol's physical and chemical characteristics for use as an alternative fuel:

- It has a relatively low production cost.
- It has a lower risk of flammability compared to traditional gasoline.
- It can be manufactured from a variety of carbon-based feedstock.
- It can help reduce the United States' dependency on foreign oil.
- It can be converted into hydrogen.

FIGURE 15.10 Ethanol and methanol fuel pumps for refueling alternative-fueled vehicles. (Courtesy of Warren Gretz, DOE/NREL, Golden, CO.)

Researchers are currently trying to find a way to utilize methanol in fuel cell vehicles for use in the future.

Hydrogen

Hydrogen is the latest energy source that science is looking toward with the expectation that it could completely revolutionize not only the alternative fuel industry, but also possibly the entire energy system.

Hydrogen is an energy carrier, not an energy source. Energy is required to separate it from other compounds. Once produced, hydrogen stores energy until it is delivered in a usable form, such as hydrogen gas delivered into a fuel cell. Hydrogen is the most abundant element in the universe. It makes up roughly 90 percent of the universe by weight. Hydrogen is the lightest of the gases and when it is burned, its only waste product is water. Because of its abundance, simplicity, efficiency, and lack of toxic emissions, it is viewed as the perfect fuel in the face of climate change.

Hydrogen can be produced from fossil fuels, nuclear energy, biomass, and by electrolyzing water. The environmental impact and energy efficiency of hydrogen depends on how it was produced. If hydrogen is produced with renewable energy and used in fuel cell vehicles it is the one alternative fuel that holds the promise of eventually obtaining a virtually pollution-free transportation system network and a long-awaited freedom from the United States' crippling dependence on foreign oil in a politically unbalanced world.

At the earth's surface temperature and pressure, hydrogen is colorless. It is rarely found alone in the natural world, however. It is usually bonded with other elements. Only small amounts are actually present in the atmosphere. Hydrogen exists in enormous quantities in water, in hydrocarbons, and in other organic matter. Efficiently producing hydrogen from existing compounds is the major hurtle that researchers are facing today.

According to the DOE, using steam to reform natural gas accounts for about 95 percent of the approximately 9 million tons (907 thousand metric tons) of hydrogen produced in the

United States each year. This level of hydrogen production could fuel more than 34 million cars. The major hydrogen-producing states today are California, Louisiana, and Texas. Almost all the hydrogen produced in the United States is used for refining petroleum, treating metals, producing fertilizer, and processing foods. NASA has also used it extensively since the 1950s for space flight.

Hydrogen is used to fuel internal combustion engines and fuel cells, which can then power low- or zero-emission vehicles such as fuel cell vehicles. Today, there are major research and development efforts going on to make this a viable and conveniently accessible technology to the public. For years, hydrogen technology has been looked at as the ideal technology to supply the world with energy, and still win the battle against climate change.

Hydrogen is considered an alternative fuel under the Energy Policy Act of 1992. Its chief interest as an alternative transportation fuel is attributed to three factors:

- It is clean burning.
- It can be produced domestically, reducing or eliminating dependence on foreign countries.
- It has an extremely high-efficiency potential (up to three times more efficient than today's gasoline-powered vehicles).

There is still research that needs to be done, however. The energy in 2 pounds (1 kilogram) of hydrogen is equivalent to the energy in 1 gallon (3.5 liter) of standard automobile gasoline. For a vehicle to travel 300 miles (483 kilometer), a fuel cell would have to store 11 to 29 pounds (5 to 13 kilogram) of hydrogen. Because hydrogen has a low volumetric energy density (a small amount of energy by volume compared with fuels such as gasoline), it would require a tank larger than a car's trunk to store it all. Therefore, it is not quite practical yet; advanced technologies are needed to reduce storage space and weight.

Hydrogen storage technologies being researched currently encompass high-pressure tanks with gaseous hydrogen compressed up to 10,000 pounds per square inch (4,536 kilogram per 6.5 square centimeter), cryogenic liquid hydrogen cooled −423°F (−253.5°C) in insulated tanks, and chemical bonding of hydrogen with another material (such as metal hydrides).

Natural Gas

Natural gas is a mixture of hydrocarbons, mostly methane. When it is delivered through the pipeline system in the United States, it also contains hydrocarbons such as ethane and propane, and other gases such as helium, carbon dioxide, nitrogen, hydrogen sulfide, and water vapor.

Natural gas has a high octane rating and works well in spark-ignited internal combustion engines. It is one of the cleaner fossil fuels, is nontoxic, noncorrosive, and noncarcinogenic. It also has the distinction of not polluting soil, groundwater, or surface water. The bulk of usable natural gas is obtained from oil and gas wells. Smaller amounts can be obtained from other sources, such as synthetic gas, landfill gas, other biogas resources, and gas by-products from coal.

Roughly one-fourth of the energy used in the United States comes from natural gas. Of that amount one-third of it goes to commercial and residential uses, one-third to industrial uses, and one-third to electrical power production. Only about one-tenth of 1 percent is currently used for transportation fuel. Due to the fuel's gaseous nature, it has to be stored on board the vehicle in either a compressed gaseous (CNG) or liquefied (LNG) state.

CNG must be stored on board a vehicle in tanks at high pressure—up to 3,600 pounds (1,663 kilogram) psi. A CNG-powered vehicle gets about the same fuel economy as a conventional gasoline vehicle on a gasoline gallon equivalent (GGE) basis. A GGE is the amount of alternative fuel that contains the same amount of energy as a gallon of gasoline. A GGE equals about 5.5 pounds (2.5 kilogram) of CNG.

To store more energy on board a vehicle in an even smaller volume, natural gas can be liquefied. To produce LNG, natural gas is purified and condensed into liquid by cooling it to −260°F (−162°C). At atmospheric pressure, LNG occupies only 1/600 the volume of natural gas in its vapor form. A GGE equals about 1.5 gallons (5.5 liter) of LNG. Because it must be kept at such cold temperatures, LNG is stored in double-walled, vacuum-insulated pressure vessels. LNG fuel systems are mainly used with heavy-duty vehicles. Natural gas vehicles and infrastructure development (the filling stations, etc.) can also facilitate the transition to hydrogen technology and fuel cell vehicles. According to the DOE, with the highest hydrogen-to-carbon ratio of any energy source, natural gas is an efficient source of hydrogen—it is the number one source of commercial hydrogen used in the United States.

The vast network of natural gas transmission lines offers the potential for convenient transportation of natural gas to future refueling stations that reform hydrogen from the gas. The DOE has identified important similarities between natural gas and hydrogen technologies that make the lessons learned from natural gas technology an aid to the future transition from conventional liquid fuels to gaseous hydrogen fuel. Their similarities include fuel storage, fueling methods, station site locations, facilities, and public acceptability.

Propane

Propane—also known as liquefied petroleum gas (LPG), or autogas in Europe—is a three-carbon alkane gas. Stored under pressure inside a tank, propane turns into a colorless, odorless liquid. As pressure is released, the liquid propane vaporizes and turns into gas that is used for combustion. Propane has a high octane rating. It is nontoxic and presents no threat to soil, surface water, or groundwater.

Propane is produced as a by-product of natural gas processing and crude oil refining. It accounts for only about 2 percent of the energy used in the United States and is used to heat homes (usually in more remote or rural areas where gas lines have not been run), cook, refrigerate food, dry clothes, power farm and industrial equipment, dry corn, and heat barbeques. The chemical industry uses propane as a raw material for making plastics and other compounds. Less than 2 percent of propane consumption is used for transportation fuel.

The idea of using liquefied petroleum gas as an alternative transportation fuel comes mainly from the fact that it has domestic availability. Because of its high-energy density and its clean-burning qualities, it is also the most commonly used alternative transportation fuel and the third most used vehicle fuel (behind gasoline and diesel).

When propane is used as a vehicle fuel, it can be run as a mixture of propane with smaller amounts of other gases. According to the Gas Processors Association's HD-5 specification for propane as a transportation fuel, it must consist of 90 percent propane, no more than 5 percent propylene, and 5 percent other gases, mainly butane and butylene.

Propane—a gas at normal pressure and temperature—is stored on board vehicles in a tank pressurized at 300 pounds (136 kilogram) psi, which is equal to a pressure about twice that of an inflated truck tire. A gallon of propane has about 25 percent less energy than a gallon of gasoline.

And that is just the beginning. In the next section, we will see where researchers are in their work on new fuel technology.

New Fuel Technology

One of the top tasks on researchers' lists today is to identify new alternative fuels for our vehicles and for other applications. Several fuels are just in the beginning stages of development. According to the DOE, each one of them promises benefits in the forms of increased energy, increased national security, reduced emissions, higher performance, and economic stimulation.

Most of the fuels discussed in this section are considered alternative fuels under the Energy Policy Act of 1992. They include the following:

- Biobutanol
- Biogas
- Biomass to liquids
- Coal to liquids
- Gas to liquids
- Hydrogenation-derived renewable diesel
- P-series fuels
- Ultralow sulfur diesel
- Green charcoal

Biobutanol

Biobutanol is a 4-carbon alcohol (butyl alcohol) produced from biomass feedstock. Its primary use today is as an industrial solvent in products like enamels and lacquers. Similar to ethanol, biobutanol is a liquid alcohol fuel that can be used in today's gasoline-powered internal combustion engines. Its chemical properties make it a good fuel to blend with gasoline. It is also compatible with ethanol blending and can improve the blending of ethanol with gasoline. The energy content of biobutanol is 10–20 percent lower than that of gasoline.

According to the EPA, biobutanol can be blended as an oxygenate with gasoline in concentrations up to 11.5 percent by volume. Blends of 85 percent or more biobutanol with gasoline are required to qualify as an alternative fuel. Today, a company called Butyl Fuel, LLC, through a DOE Small Business Technology Transfer grant, is working on developing a process aimed at making biobutanol production economically competitive with petrochemical production processes. Butyl Fuel's current plans are to market its biobutanol as a solvent first, and then market it as an alternative fuel in the future. To date, there is no infrastructure in place for fueling vehicles with biobutanol; but because biobutanol does not corrode pipes or contaminate water as ethanol can, researchers expect that biobutanol will be able to be distributed through existing gasoline infrastructure, including existing pipelines.

The DOE has identified the following benefits of biobutanol:

- It can be produced domestically from several home-grown types of feedstock.
- Its domestic production can create more jobs in the United States.
- GHG emissions are reduced because the CO_2 captured when the feedstock crops are grown balances the CO_2 released when the biobutanol is burned.
- It is easily blended with gasoline for use in vehicles.
- Its energy density is only 10–20 percent lower than gasoline's.
- It is compatible with the current gasoline distribution infrastructure and would not require new or modified pipelines, blending facilities, storage tanks, or retail station pumps.
- It is compatible with ethanol blending and can improve the blending of ethanol with gasoline.
- It can be produced using existing ethanol production facilities with relatively minor modifications.

Biobutanol is also being currently researched by other industries and government groups. The USDA Agricultural Research Service (ARS) is studying biobutanol production as a part of a study in its bioprocess technologies for production of biofuels from lignocellulosic biomass.

Biogas

Biogas is a gaseous product of the anaerobic digestion of organic matter from sources such as sewage sludge, agricultural wastes, industrial wastes, animal by-products, and municipal solid wastes. In landfills, anaerobic digestion of wastes occurs naturally. Gas collection is viable in landfills that are at least 12 meters deep and contain at least 907,000 metric tons of waste. It consists of 50–80 percent methane; 20–50 percent CO_2; and traces of hydrogen, nitrogen, and carbon monoxide. It is sometimes referred to as swamp gas, landfill gas, or digester gas. When its composition is upgraded to a higher standard of purity, it is referred to as renewable natural gas.

After biogas is produced and extracted, it has to be upgraded for pipeline distribution or use as a vehicle fuel. This process requires that the methane proportion be increased and the CO_2 contaminants be decreased. The International Energy Agency (IEA) estimated that in 2005, 185 anaerobic digestion plants had the capacity to process 5 million metric tons of municipal solid and organic industrial waste to generate 600 megawatts (MW) of electricity. The CIVITAS Initiative issued a report that estimated that European biogas production could satisfy up to 20 percent of Europe's natural gas consumption. A report titled "Natural Gas Vehicles for America" cites a 1998 study estimating that the biogas potential at that time from landfills, animal waste, and sewage is equivalent to 6 percent of U.S. natural gas consumption or 38 billion liters equivalents of transportation fuel. This calculates into about 7 percent of the 2006 U.S. gasoline consumption.

Biogas is used for many different purposes. In rural communities it is used for household cooking and lighting. Large-scale digesters provide biogas for heat and steam, electricity production, chemical production, and vehicle fuel. Once biogas is upgraded to the required level of purity and compressed or liquefied, biogas can be used as an alternative vehicle fuel in the same form as conventionally derived natural gas (CNG or LNG).

A 2007 DOE report estimated that 12,000 vehicles worldwide are using biogas. It also predicted that by 2012 there would be 70,000 to 120,000 on the road. The majority of these vehicles exist in Europe, with Sweden claiming that more than half of the gas used in its 11,500 natural gas vehicles is biogas. Biogas vehicle activity has not been as high in the United States. DOE's research and development-sponsored projects have been working on further development of biogas technologies. The DOE has identified the following benefits of biogas as an alternative fuel:

- It represents increased energy security for the nation.
- It serves as a conduit to pave the way for fuel cell vehicles in the future.
- It improves public health and the environment through reduced vehicle emissions.
- It offsets the use of nonrenewable resources such as coal, oil, and fossil fuel natural gas.
- It reduces GHG emissions.
- Its production creates jobs and benefits for the local economy.
- It helps treat waste disposal naturally, requiring less land area, which reduces the amount of material that must be landfilled.

Biogas is expected to become a bigger player in the United States in the future as alternative fuels gain momentum. In Finland in 2015, biogas consumption increased to 35 percent in their transportation sector. In the past decade, biogas as a vehicle fuel has grown to 1,200 percent. The country now has 24 biogas filling stations, illustrating its potential as a promising new fuel.

Biomass to Liquids

Biomass to liquids (BTL) describes processes for converting biomass into a range of liquid fuels such as gasoline, diesel, and petroleum refinery feedstock. These processes are different from the enzymatic/fermentation processes and processes that use only part of a biomass feedstock, such as the processes that produce ethanol, biobutanol, and biodiesel.

Currently, the major biomass to liquids production processes are:

- *Gas to liquids*: Involves the conversion of biomass into gas and then into liquids
- *Pyrolysis*: Involves the decomposition of biomass in the absence of oxygen to produce a liquid oil

Biomass to liquids processes have the potential to produce a wide range of fuels and chemicals. These fuels include gasoline, diesel, and ethanol. A major benefit of these fuels is their compatibility with currently existing vehicle technologies and fuel distribution systems. Biomass-derived gasoline and diesel could be transported through existing pipelines, dispensed at existing fueling stations, and used to fuel today's gasoline- and diesel-powered vehicles. Another major benefit is that these fuels reduce regulated exhaust emissions from a variety of diesel engines and vehicles, and their near-zero sulfur content enables the use of advanced emission control devices.

Coal to Liquids

Coal to liquids (CTL) is the process of converting coal into liquid fuels, such as diesel and gasoline. The principal method available today is called the Fischer-Tropsch process, which is a two-step process: it first converts the coal into gas, and then into liquid. There are several other processes that are able to directly convert coal into liquids, including a process called liquefaction, but they are not as common.

Coal to liquids has the ability to produce a number of useful fuels and chemicals, including transportation fuels. In addition to diesel and gasoline, methanol can also be produced. The largest benefit of this technology is that the resulting fuel is compatible with currently existing vehicle technology and fuel distribution systems. In addition, coal-derived gasoline and diesel could be transported through the existing pipeline infrastructure, dispensed at existing fueling stations, and used in vehicles without any modifications.

The DOE has identified the following benefits to pursuing research and development with the coal to liquid technology:

- The fuel can be used directly in existing vehicles.
- The fuel is compatible with existing infrastructure (pipelines, storage tanks, and retail station pumps).
- The fuel provides vehicle performance similar to or better than conventional diesel.

- Fischer-Tropsch diesel reduces regulated exhaust emissions from a variety of diesel engines and vehicles. The near-zero sulfur content of the fuels allows the utilization of advance emission control devices.

The EPA has run its own tests on the Fischer-Tropsch diesel and determined the coal to liquids diesel has benefits over traditional diesel, including the fact that it is cleaner burning, with less nitrogen oxide, little to no particulate emissions, and is lower in hydrocarbon and carbon monoxide emissions.

Gas to Liquids

Gas to liquids is the process of converting natural gas into liquid fuels, such as diesel, methanol, and gasoline. The principal production method today is the Fischer-Tropsch process. This conversion process can produce a range of fuels and chemicals. Similar to coal to liquids, a major benefit of the gas to liquids is its compatibility with existing vehicle technologies and fuel distribution infrastructure. To date, tests that have been run on diesel using the Fischer-Tropsch method provide similar or even better vehicle performance results than conventional diesel does. These fuels can also be produced using natural gas reserves, such as stranded reserves, that are uneconomical to recover using other methods.

According to a report in the Petroleum Economist in May 2002, there are about 800 small, undeveloped fields (stranded reserves) that are potential candidates for Fischer-Tropsch gas to liquids projects of up to around 10,000 barrels a day. If stranded reserves are used to produce liquid fuels, it reduces the need to flare natural gas, which results in reduced GHG emissions.

Hydrogenation-Derived Renewable Diesel

Hydrogenation-derived renewable diesel (HDRD) is the product of fats or vegetable oils—either alone or blended with petroleum—have been refined in an oil refinery. HDRD that is produced this way is sometimes referred to as second-generation biodiesel. Not much work has been done on it, but researchers expect HDRD will be able to either substitute directly for or blend in any proportion with petroleum-based diesel without modification to vehicle engines or the fueling infrastructure. HDRD's ultralow sulfur content and high cetane number (a measure of the combustion quality of diesel fuel) will provide both vehicle performance and emissions benefits. The technology is not widely available today, but it is gaining recognition.

Researchers see many benefits resulting from its development and future implementation:

- It should be able to be produced domestically, creating U.S. jobs.
- It reduces GHG emissions because the CO_2 is captured when the feedstock crops are grown, balancing the CO_2 that is released when the fuel is burned.
- It should be able to be used in today's diesel-powered vehicles.

- It should be fully compatible with current infrastructure (pipelines, fueling stations, and storage).
- It can be produced using existing oil refinery capacity and does not require extensive new facilities.
- It should provide similar, or even better, performance over commercial diesel.
- It has an ultralow sulfur content and should be able to work with advanced emission control devices on all vehicles.

P-Series

P-Series fuel is a blend of natural gas liquids, ethanol, and the biomass-derived cosolvent methyltetrahydrofuran (MeTHF). P-series fuels are clear, colorless, 89 to 93 octane, liquid blends that are formulated to be used in flexible-fuel vehicles. P-series fuel can be used alone or mixed with gasoline in any proportion inside a flex-fuel vehicle's fuel tank. Currently, the P-series is not yet being produced in large quantities and is not widely used.

Ultra-Low Sulfur Diesel

Ultra-low sulfur diesel (ULSD) is diesel fuel with a 15 ppm or lower sulfur content. The low sulfur content enables the use of advanced emission control technologies on both light-duty and heavy-duty diesel vehicles. Most of the highway diesel fuel refined in or imported into the United States is required to be ultralow sulfur diesel. Today, most of the ultralow sulfur diesel is produced from petroleum. In the future, it will be possible to make it from biomass to liquids, coal to liquids, and gas to liquids. The EPA has identified the following major benefits:

- Ultralow sulfur diesel can use catalytic converters and particulate traps that nearly eliminate emissions of nitrogen oxides and particulate matter—all pollutants related to serious health problems.
- Emission reductions from the use of clean diesel will be equivalent to removing the pollution from more than 90 percent of today's trucks and buses when the current heavy-duty vehicle fleet is completely replaced in 2030.
- Diesel engines are 20–40 percent more efficient than comparable gasoline engines.
- Ultralow sulfur diesel uses existing fueling infrastructure and works with existing engine and vehicle technologies.
- Replacing some gasoline vehicles with diesel vehicles will result in reduced U.S. petroleum fuel use and GHG emissions.

The Energy Policy Act of 1992 was passed by congress to reduce the nation's dependence on imported petroleum by requiring certain fleets to acquire alternative fuel vehicles, which are capable of operating on nonpetroleum fuels. Ultralow sulfur diesels will assist in working toward that goal.

Green Charcoal

The development of cooking fuels using biomass has created another type of fuel—charcoal. This involves a continuous process of pyrolysis of vegetable waste (agricultural residues, renewable, wild-growth biomass) and transforms it into a product referred to as green charcoal.

This newly created domestic fuel performs the same as charcoal made from wood but at half the cost. Used in Africa, this fuel has many more uses than cooking; it also represents a freedom from being held hostage to a scarcity of resources and the long distances to and great costs of available fuels in Africa (Figure 15.11).

Today, more than 2 million people worldwide face domestic energy shortages. In many parts of Africa, Latin America, and Asia, wood is becoming scarce, and modern energy

(a)

(b)

FIGURE 15.11 (a, b) Green charcoal briquettes are made from biomass, sawdust, and shredded and soaked paper in Bukava and the Virunga Mountains in Congo. This alternative fuel is used instead of traditional charcoal to protect the mountain gorillas of the region (obtaining charcoal in the region endangers the gorilla). The biomass briquettes burn charcoal. They are mixed and patted into molds until they dry into briquettes (image [a]). Once they are ready to burn, their heat output is similar to charcoal, enabling the villagers to cook their food efficiently and sustainably. (From Gines, J. K., *Climate Management Issues: Economics, Sociology, and Politics*, CRC Press, Boca Raton, FL, 2012.)

supplies are difficult to come by or nonexistent. In the Sahel region, inhabitants have to walk about 19.5 kilometers a day just to find a household supply of wood. In the small villages, families are forced to spend up to one-third of their income on wood or charcoal. As the villagers gather wood, they steadily deforest the area, increasing the ill effects of drought and deforestation, which then leads to climate change.

In an effort to halt deforestation in the savannah zones, a renewable replacement substance to effectively use as an energy source was introduced to the people in the region. A process was developed to carbonize vegetative material into pellets or briquettes. Savannah weeds, reeds, and various types of straw (wheat, rice, or maize), cotton stems, rice husk, coffee husk, bamboo, or any plant with sufficient lignin content can be used to produce green charcoal.

The advantage to using green charcoal is that it preserves forests. Green charcoal also eliminates methane emissions. The process of producing the briquettes does not release any GHGs.

Another reason for using green charcoal is to avoid the buildup of soot. In an April 16, 2009 article in the *New York Times*, Elisabeth Rosenthal reported that research conducted by Dr. Veerabhadran Ramanathan, professor of climate science at Scripps Institute of Oceanography and one of the world's leading climate scientists showed that soot (also called black carbon) builds up in tens of thousands of villages in the developing world. It is actually a pollutant and a major (although previously ignored) source of global climate change.

Ramanathan says that black carbon has recently emerged as a serious contributor—responsible for 18 percent of the global warming that has occurred, compared to a 40 percent contribution from CO_2. Focusing on black carbon emissions and reducing them is currently being viewed by scientists such as Ramanathan as an inexpensive, easy, relatively fast way to slow global warming, especially in the near term. One way to accomplish this goal is to replace the common primitive cooking stoves found in most homes in developing countries with a modern version that emits much less soot.

According to Ramanathan:

> it is clear to any person who cares about climate change that these [replacement stoves] will have a huge impact on the global environment. In terms of climate change, we're driving fast toward a cliff, and this could buy us time.

Another advantage to this is that because soot has a fairly short life span (a few weeks), decreasing the input of soot in the atmosphere would have an immediate effect in combating climate change.

The discovery of black carbon's influence is so new that it was not even mentioned in the IPCC's assessment reports until 2014. It has been through recent research by institutions such as Scripps and NASA that black carbon has become better understood. It is now believed that black carbon could account for as much as half of the current Arctic warming.

While soot does not travel globally like CO_2, it does travel. Soot from India has been found in the Maldives and on the Tibetan Plateau. From the United States, it travels to the Arctic. Professor Syed Iqbal Hasnain, a glacier specialist in India, has predicted that the Himalayan glaciers will most likely lose 75 percent of their mass by 2020 because of black carbon deposition.

In an effort to take action on this, the U.S. congress introduced a bill in March 2009 that would require the EPA to specifically regulate black carbon and direct aid to black carbon reduction projects abroad. This major effort would include providing modern cook stoves to 20 million homes in developing countries, where black carbon pollution poses the biggest problems. These new stoves cost about $20 and use solar power, making them much more efficient while reducing soot levels more than 90 percent. The solar stoves do not use wood or biomass. Other new stove options burn cleaner.

In March 2009, a cook stove project called Surya was initiated. It began market testing six different styles of cook stoves in small villages in India in an effort to begin the conversion of stoves to reduce black carbon pollution.

Research scientists are busy developing alternative and advanced fuels, and are also busy looking for options beyond fossil fuels to supply energy for the future. If significant advances are not made within the next decade, there will be little hope of offsetting the permanent destruction from climate change. Research scientists and policymakers are up against the clock, which is ticking fast.

On the Cutting Edge

In a recent study released by the Fuel Freedom Foundation, they believe there is promise in six new fuel technologies that, as they say "could be game changers" in the field. Innovation is the name of the game in this study. According to their research, here are some new options under scrutiny as possible sources for new transportation fuels:

- Whiskey
- Microalgae-biomass
- The "artificial leaf"
- Cactus
- Biofuel from grapes
- Plastic waste

The use of whiskey is showing promise in this exciting field. A Scottish professor developed a new technique that takes waste from whiskey-making factories and turns it into biobutanol. Due to its success the professor's organization—Celtic Renewables—has been awarded nearly $17 million from the U.K. government to build a facility with the sole purpose of converting whiskey waste to biofuel. The facility is scheduled to begin operation by the end of 2018 and plans on producing approximately one million liters of fuel each year (Taft, 2015).

For years, microalgae-biomass has been considered as a possible option for an alternative fuel. Efforts to produce it in industrial quantities, however, have continually been put on the back burner either due to high startup costs or issues with efficiency. An Israeli company, UniVerve Ltd., is now looking

at the issue in a new light with a new approach that could solve all the prior issues. They have developed a new technique called the Hanging Adjustable V-shaped Pond (HAVP) cultivation system which allows light to hit the algae from all sides by suspending the algae in a unique triangular structure. This design significantly increases output without increasing costs. The fuel produced by the HAVP system is projected to cost around $50 per barrel, and it is being promoted as the new competitor for crude oil to contend with.

The "artificial leaf" is cutting-edge innovation. Developed by Harvard engineer Daniel Nocera in 2011, he and his team claim that the "leaf" can produce hydrogen fuel. However, when it became clear that hydrogen was not going to become a viable transportation fuel for some time, Nocera got back to work and genetically modified a strain of bacteria to produce a burnable liquid fuel from hydrogen. The new bacteria, called *Ralstonia eutropha* has already proven to be more efficient than plants at turning energy into a useable liquid fuel. Nocera is still innovating and improving his invention to increase its performance. It is certainly an up-and-coming innovation to keep focused on for the future.

Cacti are earning plenty of attention in the United Kingdom these days. Researchers have been improving the process of making biofuel from cactuses and similar vegetation. Known as crassulacean acid metabolism (CAM) plants, these cactuses can be grown on arid or semi-arid land and could produce as much energy as natural gas if planted on even 4 to 12 percent of the available semi-arid land in the world.

Australian scientists are also making breakthroughs in fuel technology. Grape marc, the waste product left over from the wine-making process, could produce up to 4.9 billion liters of biofuel annually. Australian researchers have successfully developed a fermentation process that yields 397 liters of bioethanol per metric ton of grape marc. This development has attracted multiple investors, with companies like United Airlines investing $30 million toward commercial development of the system.

New fuel technology may even have a viable answer for all the plastic waste harming marine wildlife. With 8.1 million metric tons of plastic waste having entered the oceans in 2015 alone, it is clear the environmental problem needs to be dealt with—and a teenager from Egypt is attempting to change the monumental problem. Azza Faiad's new catalyst, known as aluminisilicate, converts plastic into feedstocks for biofuels. Faiad estimates that if approximately $78 million worth of biofuel were produced per year in Egypt alone, and if spread worldwide, the process could provide an answer to the plastic waste problem, while simultaneously producing large quantities of clean biofuel to displace gasoline.

In the News

Recently, there seems to be no shortage of studies being conducted, searching for new and innovative sources of clean, renewable energy. The following sections present a few of them to ignite the imagination as to the possibilities.

New Fuel Cells

Hydrogen has always been attractive as an energy source because it stores about three times as much energy by weight as gasoline. When it is used in a fuel cell to produce electricity, its by-products are water that is pure enough to drink, and heat that can be used for other purposes. But there is a down side: hydrogen is expensive and difficult to isolate. Additionally, the atoms in hydrogen gas are so sparse that it is difficult to pack enough of them in a small space—such as a fuel tank—to do much good, and it can be highly flammable. Added to that—hydrogen fueling stations are not common, so fueling hydrogen-powered cars is logistically difficult. This has hampered the development of hydrogen-fueled cars because people did not like to buy an expensive car that is difficult to fuel, and fueling stations are not going to invest in hydrogen unless there are a lot of hydrogen-car owners. It is a Catch-22.

That has led to a compromise. One avenue of research these days is directed at using the hydrogen that is already available in commonly used fuels (like gasoline, methanol, or diesel) to power fuel cells. But that requires a "reforming" of the fuel, either of the car's engine or at the filling station, to extract the hydrogen and feed it to the fuel cell. The process has turned out to be expensive and challenging, but industry leaders like DaimlerChrysler and Ford have finally proved that it can be done. In the meantime, researchers at the University of Pennsylvania have approached the problem from a different perspective. Instead of reforming standard fuels, they have developed a fuel cell that can run directly on ordinary diesel fuel.

"There used to be a saying that you could run a fuel cell on any fuel as long as it is hydrogen," says Raymond J. Gorte, professor of chemical engineering at Penn and the lead author of a paper reporting the findings in the *Journal of the Electrochemical Society* (Dye, 2015).

The researchers at Penn figured out how to get around the very problem that has stumped so many other researchers. Other attempts to use fuels other than pure hydrogen have worked, but only briefly. The problem was that previous systems used nickel as a catalyst, causing hydrogen and oxygen to combine and freeing up electrons, which provided electricity.

"But nickel is also an excellent catalyst for making graphite out of hydrocarbons, and that's a problem," John M. Vohs, chairman of chemical engineering at Penn, says. The graphite "plugged up the whole fuel cell, shutting it down." The Penn scientists switched to a composite of copper and ceria, which is used in automotive catalytic converters, and the problem was solved.

Running fuel cells on diesel fuel would not eliminate the need for hydrocarbons, but it could reduce it. "An internal combustion engine is 15 percent efficient in turning the fuel into power, while in a fuel cell you could probably at least double that," Vohs says.

What may be even more important with this discovery is that fuel cells that could run on standard fuels could help during a *transition period*. The infrastructure is already in place to support fuel cells that can run on diesel fuel. So for now, it would be possible to fill up with diesel, paving the way for the day when hydrogen might be readily available. It could serve as a bridge to future hydrogen fuel cells.

What seems even more likely in the near term is that we will see a growing use of fuel cells in various applications before they become common in vehicles. "To be honest with you, transportation is the hardest application in that you have to have a system that's mobile and robust," Vohs says. "You have to be able to hit potholes in the road and have everything survive."

He believes an ideal application would be to use fuel cells as individual power plants at our homes. It takes a lot of power to run a car—possibly ten times more than it takes to supply all the electricity needed to run a house—so a small fuel cell down in the basement could do the job. That would cut down on the enormous loss of power through transmission lines, because the power would be produced where it is needed, thus saving additional energy. Vohs thinks that application could be available within five years, using natural gas as the fuel (Dye, 2015).

Spinach Power

Using a simple membrane extract from spinach leaves, researchers from the Technion-Israel Institute of Technology have developed a bio-photo-electro-chemical (BPEC) cell that produces electricity and hydrogen from water using sunlight.

The unique combination of a human-made BPEC cell and plant membranes, which absorb sunlight and convert it into a flow of electrons opens the door for the development of new technologies in the creation of clean fuels from renewable sources: water and solar energy.

The BPEC cell developed by the researchers is based on the naturally occurring process of photosynthesis in plants, in which light drives electrons that produce storable chemical energetic molecules.

In order to utilize photosynthesis for producing electric current, the researchers added an iron-based compound to the solution. This compound enables the transfer of electrons from the biological membranes to the electrical circuit, enabling the creation of an electric current in the cell. The electrical current can also be channeled to form hydrogen gas through the addition of electric power from a small photovoltaic cell that absorbs the excess light. The solar energy to be converted into chemical energy is then stored as hydrogen gas formed inside the BPEC cell. This energy can then be converted into heat and electricity by burning the hydrogen, in the same way hydrocarbon fuels are used.

Unlike the combustion of hydrocarbon fuels—which emit CO_2 into the atmosphere and pollute the environment—the product of hydrogen combustion is clean water. A closed cycle that begins with water and ends with water, it allows the conversion and storage of solar energy in hydrogen gas, which could be a clean and sustainable substitute for hydrocarbon fuel (Pinhassi et al., 2016).

Harvesting Sunlight to Split CO_2 into Alcohol Fuels

Chemists at the University of Texas at Arlington have been the first to demonstrate that an organic semiconductor polymer called polyaniline is a promising photocathode material for the conversion of CO_2 into alcohol fuels without the need for a co-catalyst.

"This opens up a new field of research into new applications for inexpensive, readily available organic semiconducting polymers within solar fuel cells," said principal researcher Krishnan Rajeshwar, UTA distinguished professor of chemistry and biochemistry and Co-Director of UTA's Center for Renewable Energy, Science and Technology.

> These organic semiconducting polymers also demonstrate several technical advantages, including that they do not need a co-catalyst to sustain the conversion to alcohol products and the conversion can take place at lower temperatures and use less energy, which would further reduce costs, Rajeshwar added.

In this proof-of-concept study, the researchers provide insights into the unique behavior of polyaniline obtained from photoelectrochemical measurements and adsorption studies, together with spectroscopic data. They also compared the behavior of several conducting polymers.

The stationary currents recorded after two hours during testing suggested that the polyaniline layer maintained its photoelectrochemical ability for the studied time period. While in the gas phase, only hydrogen was detected, but potential fuels such as methanol and ethanol were both detected in the solution for carbon dioxide-saturated samples.

"Apart from these technical qualities, as a polymer, polyaniline can also be easily made into fabrics and films that adapt to roofs or curved surfaces to create the large surface areas needed for photoelectrochemical reduction, eliminating the need for expensive and dangerous solar concentrators," Rajeshwar added. This is seen as important research toward global environmental impact and the need to reduce the impact of CO_2 emissions. By finding an inexpensive, readily available photocathode material, it is hoped that it will open up new options to create cheaper, more energy-effective solar fuel cells (Hursán, 2016).

CONCLUSIONS

As this chapter has illustrated, there are many options and choices available for each person in a sustainable community. As new discoveries are made, it opens and expands that innovative horizon just a bit more. These are exciting times to be living—and planning—in. Options face us at nearly every turn. As we continue to develop and grow our new sustainable communities, the good choices we make will be felt for generations to come.

REFERENCES

Dye, L. September 27, 2015. New fuel cells seen in near future. *ABC News.* http://abcnews.go.com/Technology/story?id=98252&page=1 (accessed October 21, 2016).

eInstruments. 2016. Using portable emissions analyzers for optimizing the performance of a cogeneration plant. Application Note #IA-13-0801. http://www.e-inst.com/cogeneration/ (accessed October 18, 2016).

EPA. 2016. U.S. greenhouse gas inventory report: 1990–2014. *U.S. Environmental Protection Agency.* https://www.epa.gov/ghgemissions/us-greenhouse-gas-inventory-report-1990-2014. (accessed October 1, 2016).

Gines, J. K. 2012. *Climate Management Issues: Economics, Sociology, and Politics.* Boca Raton, FL: CRC Press.

Green, J. November 18, 2010. A new way to plant Urban trees. *Uniting the Built and Natural Environments. American Society of Landscape Architects.* https://dirt.asla.org/2010/11/18/a-new-way-to-plant-urban-trees/ (accessed October 15, 2016).

Hursán, D., A. Kormányos, K. Rajeshwar, and C. Janáky. 2016. Polyaniline films photoelectrochemically reduce CO_2 to alcohols. *Chemical Communications* 52(57):8858. DOI: 10.1039/C6CC04050K.

National Household Travel Survey. March 2009. The carbon footprint of daily travel. U.S. Department of Transportation. http://nhts.ornl.gov/briefs/Carbon%20Footprint%20of%20Travel.pdf (accessed October 20, 2016).

Sullivan, M. June 30, 2007. World's first air-powered car: Zero emissions by next summer. *Popular Mechanics.* http://www.popularmechanics.com/cars/a1665/4217016/ (accessed October 20, 2016).

Taft, N. October 7, 2015. 6 new fuel technologies that could be game changers. *Fuel Freedom Foundation.* https://www.fuelfreedom.org/6-new-fuel-technologies-that-could-be-game-changers/ (accessed October 21, 2016).

SUGGESTED READING

Duany, A. and J. Speck. 2009. *The Smart Growth Manual.* New York: McGraw-Hill, 240 p.

Halderman, J. D. 2015. *Hybrid and Alternative Fuel Vehicles.* Prentice Hall, NJ: Pearson, 384 p.

Laughjlin, R. B. 2013. *Powering the Future: How We Will (Eventually) Solve the Energy Crisis and Fuel the Civilization of Tomorrow.* New York: Basic Books, 240 p.

Lee, S. and I. Samuel. 2015. *Street Smart: The Rise of Cities and the Fall of Cars.* New York: Public Affairs, 312 p.

Montgomery, C. 2014. *Happy City.* New York: Farrar, Straus, and Giroux, 368 p.

Speck, J. 2013. *Walkable City: How Downtown Can Save America, One Step at a Time.* San Francisco, CA: North Point Press, 320 p.

16 Sustainable Agriculture and Industry

OVERVIEW

In this chapter, we will take a look at two important components in society today: agriculture and industry. Both important aspects of community, they are seeing trends in the color green these days. We will take a look first at agriculture and some of the positive environmental steps it is taking, such as employing sustainable practices, organic farming, community-based approaches, healthy land management practices, and green conservation measures. Following that, we will focus on industry and examine some of the exciting new trends emerging from that sector, including food security, ecologically sustainable development concepts, and eco-industrial parks (EIPs).

INTRODUCTION

One of the most critical economic sectors is agriculture. No one can survive without it, making it one key segment in every country's present, and future, that must be healthy and sustainable. If it is not, then opportunities will be lost at the expense of society, natural resources, and the future. It is critical that as we plan for both the present, and future, we critically evaluate our actions, and make sure that as stewards of the planet's food resources, we maintain the resource responsibly.

DEVELOPMENT

Agriculture has several sustainable options available today. For example, the demand and increased revenue generated from organic products have changed the way many farmers are doing business. The U.S. Department of Agriculture says that organic farming has been one of the fastest growing segments of U.S. agriculture for more than a decade. For comparison, in 1990, there was less than 4,047 square kilometers of certified organic farmland. By 2002, that area had doubled, and then it had doubled again between 2002 and 2005. In 2011, the area had grown to 21,853 square kilometers (USDA, 2016).

Organic livestock sectors have grown even faster. In 2008, U.S. producers dedicated approximately 19,425 square kilometers of farmland to organic production, 10,926.5 square kilometers to cropland, and 8,498.5 square kilometers to rangeland and pasture, and the trend has continued to skyrocket since.

Alternative farming is another rapidly growing concept. Alternative farming is a general term that represents many different practices and agricultural methods that all share similar goals. Alternative agriculture places additional emphasis on conservation of the land and preserving its resources. Practices that are emphasized in alternative farming include the following:

- Building new topsoil (composting)
- Using natural biological approaches instead of chemical pesticides for controlling insects
- Conserving soil by rotating crops, letting unused vegetation recycle back into the ground, plowing the land relative to its needs, and reducing the amount of tilling of the soil

Three important areas in the growing green trend in the agricultural sector are

1. Sustainable agriculture
2. Organic farming
3. Community-based farming

The following sections discuss these areas.

SUSTAINABLE AGRICULTURE AND ECOLOGICAL SYSTEMS

Sustainable agriculture looks at the farming cycle as a complete system. This places an emphasis on working in harmony with entire ecological systems. As farmers see shifts take place in our ever-changing environment, their goal is to change their farming methods to stay in harmony with the environment.

The following are some advantages of organically grown food:

- Organic food is rich in vitamins, minerals, and enzymes, which help people fight off infections more effectively.
- Because they are pesticide-free, organic food reduces the risk of cancer and certain birth defects.
- Organic products tend to taste better than nonorganic ones. For instance, a study by Washington State University claimed that organic apples were sweeter and had a firmer texture compared to those of conventional farming.
- Organic farming is better for the environment because it does not use toxic chemicals that run off to pollute soil and water. It also promotes a use of the land that will maintain richness in the soil for the next generations.

A Living, Changing System

Farmers practicing sustainable agriculture look at the agricultural system as a living thing with different needs for different areas within the farm. Instead of applying the same practices evenly over the entire farm, each area on the farm is assessed and treated according to its own needs.

Another important distinction is that sustainable agriculture seeks to balance farm profit over the long term with needs for good soil and clean water, a safe and abundant food supply, and rural communities that are rewarding to live in. Farmers of sustainable agriculture look at agriculture and ecology together, referring to it as agroecology.

In traditional methods, soil scientists study soils, hydrologists study water, and agronomists study crops. Studying these different components separately, however, can result in a lack of understanding or appreciation of how the entire system fits and works together. In agroecology, the entire system (soil, water, sun, plants, air, animals, microorganisms, and people) is studied together.

Farmers who practice sustainable agriculture strive to understand the complex relationships among all parts of the agroecosystem. Sustainable agricultural practices include the following components:

- Refraining from taking or using chemicals, or using very few chemicals, to reduce pest damage to crops
- Using fewer herbicides (weed killers)
- Minimizing runoff in order to reduce soil erosion
- Testing the soils to determine which nutrients are available
- Rotating crops so that the land is not overused or depleted of critical nutrients (crop rotation also reduces insects and weeds)
- Minimizing soil erosion by using contour plowing (keeping plowing patterns compatible with the slope of the land), covering crops to protect the soil, using no-till methods, and utilizing perennial plants (plants that bloom each year without having to be reseeded)
- Improving and protecting wildlife habitat
- Monitoring grazing practices to ensure that the land is not being abused and overgrazed

People often have definite opinions about this type of farming. Some have argued that sustainable farming is not as productive and is more expensive than traditional farming. Others believe that even if that is the case, they are willing to pay more for the food produced from this system. Still others maintain that it does not represent lower productivity but instead builds on current agricultural achievements and can produce large volumes of crops without harming the land.

Regardless of the various opinions concerning sustainable farming, it has become an important part of agriculture in the past few years. Many farmers and ranchers have chosen to use more conservative practices on their lands, and they see it as the future of a successful, long-term relationship with the ecosystem.

ORGANIC FARMING

The organic produce section at grocery stores has become very popular. Some stores even divide their produce into separate sections—organic versus nonorganic—for shoppers' convenience. Some stores offer nothing but organic items. Organic does not refer to the food itself, but as to how the food is produced. Organic foods are produced without using any synthetic pesticides or fertilizers. They are also not given any ionizing radiation. Organic crops are grown on soil that has been chemical-free for at least 4 years. Organic farming is also meant to maintain the land and keep the surrounding ecosystems healthy. Organic livestock cannot be fed nonorganic feed or given any type of growth hormone or antibiotic. Before a product can be labeled "organic," a government-approved certifier must inspect and certify it.

Most organic farmers strive to make the best use of land, animal, and plant interactions; preserve the natural nutrients; and enhance biodiversity. They practice soil and water conservation to keep erosion down; they use organic manure and mulch to improve soil structure. They also use natural pest controls, such as biological controls (using an insect's natural predators rather than chemicals), as well as plants with pest-control properties. They rotate their crops to keep production and fertility higher.

When properly managed, organic farming reduces or eliminates water pollution and helps conserve water and soil on farms. Today, organic farming represents only a small section of agriculture, but it is gaining quite a following. Because it does not require expensive chemicals, many developing countries are able to produce organic crops to export to other countries, helping their economies.

COMMUNITY-BASED FARMING

One form of alternative agriculture is community-based farming—or community-supported agriculture (CSA). CSA consists of many participants in a local community working together to cultivate and care for an area of land, which will produce food for them to eat.

CSA first began in Europe and Japan and was developed as a way to have a different social and economic system. Farming practices like this also exist in Israel on kibbutz farms. CSA in America provides an opportunity for nonfarmers and farmers to join together to advance the science of agriculture. Many people participate in CSA so that they can have a direct connection to their personal food supply and because they are concerned about the widespread use of pesticides in conventional agriculture. They also want to participate in a stewardship role for the land and its future.

Participants usually purchase their share of the harvest ahead of time. Then, as the crops are grown and cultivated, from late spring to early fall, the participants receive a supply of the crops that are grown, such as fruits, vegetables, and herbs. An organic farming approach is often taken.

Recycling, Reducing, and Reusing in Sustainable Farming

Only through active measures of conservation can natural resources be protected now and into the future. For farmers to conserve resources, it is important to practice sustainable farming. Farmers must use the land responsibly so that it can be productive in the present as well as the future. Because side effects of traditional agriculture can include loss of biodiversity and habitat, erosion and soil loss, soil contamination, possible degradation of water quality, and reduction of water quantity, conservative measures must be practiced.

Recycling in agriculture can encompass managing the land in such a way that as one crop depletes certain nutrients, another can be planted afterward that replaces the lost nutrients back into the soil. Crop rotation is an example of this type of practical recycling.

Reducing impacts is also critical. By controlling erosion, controlling the amount of irrigation water used (by using water-efficient methods), or by controlling the amount of pesticides or fertilizers used, it is possible to reduce adverse environmental effects. Reusing resources is also environmentally responsible. Manure can be collected and reused as fertilizer instead of being left to seep into water sources and contaminate them. Composting is another method of reusing organic components to increase the fertility of agriculture.

Not only are recycling, reducing, and reusing important on the farm, they are also important downwind or downstream from the agricultural operation. Effects can often be far-reaching—they can deplete aquifers or cause widespread impacts on habitats.

Conservation measures are important everywhere, but some areas are more at risk than others. The ability of a particular soil to erode depends on soil type, topography (slope of the land), organic matter, local geologic and erosive processes, and climate. Plowing destroys plant roots, which would otherwise help stabilize the soil. Soils that are disturbed by plowing and cultivation are prone to erosion by water runoff and wind.

Some Possible Solutions

Soil erosion can be reduced if farmland is allowed to recover and remain fallow, letting natural succession take place, or by restoring native vegetation. Another viable solution is to remove highly erodible land from production. If land cannot be retired, there are many plowing and cultivation techniques, such as contour plowing, that help conserve the land. Many of these techniques will be discussed in the section on conservation measures. The success of various conservation techniques depends on local soil conditions, topography, and the type of crop that is being planted.

Managing Plant Nutrients

Another way that farmers can practice responsible conservation is to manage plant nutrients. Nutrients naturally cycle among water, air, soil, and biota (living things). Agriculture can disrupt this natural cycling because it redistributes nutrients, depletes soils of some nutrients, and concentrates nutrients in eroded sediments and waterways.

In addition, many crops need great amounts of nutrients (such as nitrogen and phosphorus), and they deplete the soil of them faster than the native plants in the area do. This necessitates the application of fertilizer, which must be carefully monitored.

Harvesting crops also creates a nutrient sink. When the crop is completely removed, it cannot be recycled back into the complete system (soil). It is for these reasons that farmers must pay close attention to the health of the soil. Soil surveys are conducted by the Natural Resources Conservation Service (NRCS) for the entire country. These surveys contain a wealth of information about local soil types and suggested land uses based on soil type in order to promote conservation and sustainable farming.

Overgrazing on Rangeland

Another area of environmental concern is the effect of overgrazing on rangeland. In the United States, about 40 percent of the land is considered rangeland. If conservation measures are not practiced, overgrazing by livestock can alter plant communities by removing some species and allowing inedible species, invasive species, and noxious weeds, to take over. In many areas of the world, forests have been cleared and converted to rangeland or pasture, which has resulted in a significant loss of biodiversity and wildlife habitat. In some places, it has changed the structure and function of the ecosystem.

Proper Pest Management

Pest management is also an issue in proper conservation measures. Using broad-spectrum pesticides (pesticides that affects a wide range of species, not just a specific one) can eliminate the beneficial species as well. For example, it can eliminate a pest's natural predator or plant pollinators.

In addition, some pesticides are not readily biodegradable. Long after they have been used, the pesticides can collect in the tissues of plants and animals or collect in sediments. They can then reenter the food web and cause widespread problems. This is why many farmers who practice sustainable farming use some form of integrated pest management (IPM)—a combination of several pest control techniques. In IPM, if pesticides are used, the farmer carefully determines which specific pesticide will affect the pest, determine how much to use (so that it does not reside long term in the environment), and when the best time frame exists in which to apply it most effectively. These types of conservation measures will enable a greener long-term management of the land.

Green Conservation Measures

There are several agricultural conservation measures that can be used in order to protect natural resources, including the following:

- Crop rotation
- No-till, conservation tillage, and other crop residue management
- The use of cover crops

- Terracing, grassed waterways, contour strip cropping, and contour buffer strips
- Erosion control and grade stabilization
- Composting

Crop rotation is the practice of changing the crops grown in a field, usually in a planned sequence. Crops need nutrients in order to grow. Grass plants, such as wheat, oats, and corn, use nitrogen to grow. Legumes, such as soybeans and alfalfa, have a symbiotic relationship with the nitrogen-fixing bacteria in the soil. The legumes and bacteria together create a form of nitrogen that is usable by the grass crops. Because of this basic relationship, many farmers rotate their fields between grass and legume crops to keep a supply of nitrogen in the soil. Cover crops, such as clover and hay, are also cycled into the rotation in order to add organic material to the soil.

Crop rotation is also an effective way to control weeds, insects, and disease, because it naturally breaks the cycles of these different pests. It reduces soil erosion and saves fertilizer costs because the nitrogen that the grasses deplete is naturally added back into the soil by the legumes. It also reduces the potential for nitrate leaching to groundwater because the nitrogen is being actively produced and used from one planting to the next. This way, it does not build up.

No-Till, Conservation Till, and Crop Residue Management

No-till agriculture is the practice of not plowing or disturbing the field. Because the soil is not disturbed, it minimizes the erosion and deposition of sediments into nearby water, a negative side effect of plowing. Instead, no-till agriculture ensures the soil is anchored to the plant root systems.

Conservation tillage is a management practice that minimizes soil disturbance. It provides long-term crop residues and vegetation on croplands. It reduces erosion and surface runoff of pesticides and heavy metals.

Both no-tilling and conservation tilling methods retain the crop residue and vegetative cover, which keeps the soil cooler longer and allows it to retain more moisture. It also enhances the fertility of the soil. Crop residue management is any tillage method that leaves crop residue on the surface to reduce the effects of erosion. The residue acts as a protective layer that shields the soil from wind and rain until the emerging crops are able to provide their own protective canopy. Crop residue improves soil tilth and adds organic matter to the soil. Less tillage also reduces soil compaction.

The Importance of Cover Crops

A cover crop is a close-growing crop that temporarily protects the soil before the next crop is established. Commonly used cover crops include oats, winter wheat, and cereal rye. Legume cover crops add nitrogen to the soil and provide low-cost fertilizer for future grain crops. They are planted as soon as possible after the harvest in fields where residue cannot sufficiently protect the soil from water and wind erosion during the winter and spring months.

A cover crop can also be planted after the last cultivation in order to provide a longer growing period. In sandy soils, they can be used to reduce nitrate leaching. Cover crops are helpful when crops are grown that do not provide a lot of residue, such as soybeans or corn. They are also beneficial on land that is easily erodible.

DID YOU KNOW THIS ABOUT SOIL?

- There are more living organisms in 1 cubic foot (0.02 cubic meter) of soil than there are people in the United States
- It can take 100 to 500 years to create 1 inch (2.5 centimeter) of soil
- The average quarter acre of lawn contains 50 to 250 earthworms
- The best china dishes are made from soil
- Throughout history, civilizations rose or fell depending on the fertility of their topsoil
- Almost all of the antibiotics we take to help us fight against infections were obtained from soil microorganisms

Terracing

A terrace is an earthen embankment that follows the contour of a hillside. It breaks a long slope into shorter segments like a set of stair steps, which greatly reduces erosion. Terraces also intercept the flow of water by serving as small dams on a hillside. The terraces intercept runoff water and either help guide it to a safe outlet or are designed to collect the water and temporarily store it until it can filter into the ground. Some terraces are designed to serve as a channel to slow runoff and carry it to a designated outlet, such as a grassed waterway.

Utilizing Grassed Waterways

A grassed waterway is a natural drainage way that is established with grass in order to prevent gullies from forming in fields. The natural drainage way is graded and shaped to form a smooth, shallow channel. It is then planted with grass so that a thick sod covers the drainage way.

The grass serves other purposes, as well, such as follows:

- It traps sediment washed from the cropland.
- It absorbs chemicals, heavy metals, and nutrients contained in the runoff water.
- It provides cover for small birds and animals.

Each grassed waterway is slightly different; the design depends on the nature of the field it drains (Figure 16.1).

Contour Strip Cropping

Contour strip cropping is the practice of planting various row crops and hay (or small grains) in alternating strips planted side by side. Tilling and planting carry across the slopes, following the contours of the land. When farming is done on the

FIGURE 16.1 An aerial view of a field buffer and grassed water system. (Courtesy of Tim McCabe/United States Department of Agriculture USDA.)

contour, it creates small ridges that slow runoff water. When the water is slowed, it has time to infiltrate the ground and filter the sediments. Farming on the contour, rather than up or down the slope, also reduces wear and tear on the farm equipment and reduces fuel consumption (Figure 16.2).

Crop rotation is practiced with this farming method. Over successive years, the hay strips are rotated with the grain and crops. Rotating strips from corn to legumes allows the corn to use the nitrogen added to the soil by the legumes. It also reduces soil loss by 50 percent because the different crops alternating in different areas of the soil strengthen the soil characteristics over time. This way, the soil resources are not being depleted as they would be if only one crop was grown on that area of land (there would be no nutrient exchange and natural balancing).

The ends of the rows are often planted with grasses to reduce erosion and make it easier to turn equipment. In areas where runoff is concentrated, grassed waterways can be used. Besides achieving decreased runoff and erosion with strip

FIGURE 16.2 A field showing contour strip cropping. (Courtesy of USDA.)

cropping, contour strip cropping also increases the stability of the local soil.

Contour Buffer Strips

Contour buffer strips are strips of grass or other permanent vegetation in a contoured field that help trap sediment, pollutants, and nutrients (such as from fertilizers). Sediments can be kept from moving within and from farm fields. It is similar to contour strip cropping except that the permanent grass strips are narrower than the hay or grain strips used in contour strip cropping.

This conservation solution works well because the buffer strips are established along the contour of the land (they logically follow the direction of the slopes instead of run up or down the slopes sideways). Because of this, runoff flows slowly and evenly across the grass strips greatly reducing erosion.

Vegetation is usually kept tall during spring runoff to slow it down and further control erosion. The vegetation is conservation oriented because it can provide wildlife habitat for small birds and other animals. Buffer strips are an inexpensive substitute for terraces.

Buffer strips can also be used in urban settings for the landscaping of yards on hills. There are also other advantages. They

- Improve soil, air, and water quality.
- Restore biodiversity.
- Create scenic landscapes.
- Protect livestock from harsh weather.
- Protect buildings from wind damage.

Recycling Agricultural Wastes to Produce Hot Water

Composting is the time-honored process of converting agricultural or gardening wastes into fertilizer. During the composting process, an important by-product is heat. This production of heat, if tapped, can be used to supply hot water for a home. If the compost pile is large enough and produces enough heat, the heat can be captured by a simple heat exchanger. The heat exchanger can be a coil of flexible plastic pipe embedded in the interior of the compost heap.

Heat from decomposition penetrates the pipe and heats the water, which is circulating inside. The cold water, initially put into the pipe, warms up while it is in the portion of the pipe within the compost heap. When a faucet is opened at the other end, hot water emerges until the incoming cold water replaces the heated water. When the stored hot water is used up, repeating the process regenerates it.

Seed Hybridization and Genetic Engineering of Crops

For generations, farmers have created hybrids by cross-pollinating plants with different genetic traits. Because of this, the most common understanding of the term "hybrid" is to mean that the seed was commercially produced. It is important to note, however, that there is also a naturally occurring, ongoing process of hybridization, as well. In our usage of it here, we will refer to the human influenced creation of hybrids.

Because the goal of hybridization is to ultimately create a superior product, an important factor is the heterosis (the tendency of a cross-bred individual to show qualities superior to those of both parents) or combining ability of the parent plants. Crossing any particular pair of inbred strains may or may not result in better, stronger, or enhanced offspring. Because of this, the parent strains used are very carefully selected in order to achieve the uniformity that comes from the uniformity of the parents, and the superior performance that comes from heterosis.

In agriculture and gardening, hybrid seeds are those produced by cross-pollinated plants. They are seeds and plants that are the result of first-generation cross-breeding raised for commercial purposes. Hybridization can occur naturally when wind or bees or other insects carry pollen from one plant to another resulting in plants that can best survive in the particular environmental conditions where they exist. In this sense, all plants are hybrids, having evolved genetically over centuries to obtain their particular characteristics. When gardeners talk about hybrids, however, they usually mean F1 hybrids, plants grown from seeds that are cross-pollinated with help from humans. To explain it simply, hybridization is simply plant breeding sped up. It is the process whereby the pollen from one plant with desirable characteristics is rubbed on the stigma of another plant with desirable characteristics. This flower produces a seed that, when grown, exhibits a combination of desirable characteristics. Controlled hybrids provide very uniform characteristics because they are produced by crossing two inbred strains. Hybrids are designed to improve the characteristics of the resulting plants, such as designing them to have better yields, possess greater uniformity, show improved color, or be more disease resistant.

Farmers had experimented with this idea for a long time before Gregor Mendel began experimenting with it in 1865 and presented his breakthrough paper "Experiments in Plant Hybridization," which summarized his observations of the process with pea plants. He discovered dominant and recessive genes and how they could be shared. Now considered the father of genetics, he laid the groundwork for the plant breeding done today—and genetic engineering (GE) in general. Hybridizing did not become popular, however, until the 1930s when the process of growing corn with "hybrid vigor" became well known. Controlling the pollination of corn became a key turning point in the agricultural industry.

Home gardening has also been strongly influenced by hybridization. The turning point for it was the introduction of Burpee's Big Boy Hybrid in 1949—which was the beginning in a long series of hybrid tomato types. Tomato hybrids that have since been introduced carry traits such as early maturity, juiciness, size, compactness, bush-type plant growth, various acid-sweetness balances (flavor), shipability, and storage traits.

Today, seed companies introduce new hybrids annually. Some become favorites while others do not and then just fade away. But they all have one thing in common, they are F1 hybrids, which means they are the first generation of their particular hybrid. It is important to note that if their seeds are not grown until the two year, the resulting plant will not be the same as the F1 hybrid was designed to be. It will be a second generation, F2 hybrid and will have a small percentage of the true plants, but the majority of the plant that grows from it will be something different, and probably not desirable. The hybrid breeding process has some very specific guidelines.

EFFECT OF HYBRIDIZATION ON BIODIVERSITY AND FOOD SECURITY

In agriculture and animal husbandry, the green revolutions' use of conventional hybridization has increased yields by breeding "high-yielding varieties." The replacement of locally indigenous breeds, compounded with unintentional cross-pollination and cross-breeding (genetic mixing), has reduced the gene pools of various wild and indigenous breeds resulting in the loss of genetic diversity. Since the indigenous breeds are often better adapted to local extremes in climate and have immunity to local pathogens this represents a significant genetic erosion of the gene pool for future breeding. Newer, genetically engineered varieties can be a problem for local biodiversity. Some of these plants contain designer genes that would be unlikely to evolve in nature, even with conventional hybridization. These may pass into the wild population with unpredictable consequences and may be detrimental for the success of future breeding programs. Seed producers can patent the parents and process of their hybridization and "own" the seed. This has become a controversial issue in the past years. Genetically engineered seeds, or GMOs, are a different issue. They are not created by pollination, but by genetic manipulation.

Genetic Engineering of Food

GE is the modification of an organism's genetic composition by artificial means, often involving the transfer of specific traits, or genes, from one organism into a plant or animal of an entirely different species. When gene transfer occurs, the resulting organism is referred to as a transgenic or a genetically modified organism (GMO).

GE is different from traditional cross-breeding, where genes can only be exchanged between closely related species. With GE, genes from completely different species can be inserted into one another. For example, scientists in Taiwan have successfully inserted jellyfish genes into pigs in order to make them glow in the dark.

All life is made up of one or more cells. Each cell contains a nucleus, and inside each nucleus are strings of molecules called deoxyribonucleic acid (DNA). Each strand of DNA is divided into small sections called genes. These genes contain a unique set of instructions that determine how the organism grows, develops, looks, and lives. During GE processes, specific genes are removed from one organism and inserted into another plant or animal, thus transferring specific traits.

Genetically Engineered Crops

Genetically engineered, or modified, crops are plants utilized in agriculture, which have had their DNA modified through the use of GE techniques. Generally, the goal is to introduce a new trait into a plant, which does not occur naturally in the species so that it can serve some specific purpose.

For example, some food crops have been genetically modified to be more resistant to certain pests, diseases, or a specific environmental condition. They may have been modified to stay fresh longer or have less of a resistance to a specific herbicide. Some crops are altered to make them more nutritious. Some nonfood crops are modified for their pharmaceutical aspects, their use as biofuels, for use in bioremediation, or to serve some other industrial purpose.

Let us look at bioremediation as an example. Bioremediation is a waste management technique that utilizes organisms to remove or neutralize pollutants from a contaminated site. They are used to help clean up spills in hazardous materials (hazmat) incidents. Typically, the bioengineered substance breaks down the hazardous substance into less toxic or nontoxic substances in the clean-up process. The substance being used in the clean-up effort can be capable of bioremediation on its own or have had a component added to it to help encourage the bioremediation process, such as fertilizer or oxygen that enhance the growth of the pollution-eating microbes found within the medium. This is a process called biostimulation.

Some researchers feel that these techniques have great potential, that it can be utilized to clean up very hazardous waste sites (Lovley, 2003). For example, the most radioresistant organism known, called bacterium *Deinococcus*, was successfully modified to consume and digest toluene and ionic mercury from extremely dangerous radioactive nuclear waste. A major concern, however, is that releasing genetically augmented organisms into the natural environment may not be wise because it could prove extremely difficult to track, monitor, and control them.

Using genetic modification technology on crops, however, has been around and practiced for over 30 years. The first genetically modified crop plant—an antibiotic-resistant tobacco plant—was produced in 1982. The first field trials occurred in France and the United States in 1986, when tobacco plants were engineered to be herbicide resistant. The following year, Plant Genetic Systems was the first company to genetically engineer insect-resistant tobacco plants by incorporating genes that produced insecticidal proteins.

Nearly 400 million acres of farmland worldwide are now used to grow GE crops such as cotton, corn, soybeans, and rice. In the United States, GE soybeans, corn, and cotton make up 93, 88, and 94 percent of the total acreage of the respective crops. The majority of genetically engineered crops grown today are engineered to be resistant to pesticides and/or herbicides so that they can withstand being sprayed with weed killer while the rest of the plants in the field die (Grace Communications Foundation, 2016).

In 2014, in the largest review done to date on the effects of genetically modified crops were reported in a meta-analysis as positive. The meta-analysis considered all published English-language examinations of the agronomic and economic impacts between 1995 and March 2014 for three major GM crops: soybean, maize, and cotton. The study found that herbicide-tolerant crops have lower production costs, while for insect-resistant crops the reduced pesticide use was offset by higher seed prices, leaving overall production costs about the same as the costs for non-GM crops.

Yields increased 9 percent for herbicide tolerance and 25 percent for insect-resistant varieties. Farmers who had adopted GM crops made 69 percent higher profits than those who did not. The review found that GM crops help farmers in developing countries, increasing yields by 14 percentage points (GMC, 2014).

GM crops grown today, or under development, have been modified with various traits. These traits include improved shelf life, disease resistance, stress resistance, herbicide resistance, and pest resistance. They have been modified to become useful produces such as biofuel or drugs or have the ability to absorb toxins and for use in the bioremediation of pollution.

Others believe that not enough adequate research has been carried out to identify the effects of eating animals that have been fed genetically engineered grain nor have sufficient studies been conducted on the effects of directly consuming genetically engineered crops such as corn and soy. Yet despite our lack of knowledge, GE crops are widely used throughout the world as both human and animal food.

Genetically Engineered Animals

Scientists are currently working on ways to genetically engineer farm animals. Atlantic salmon has been engineered to grow to market size twice as fast as wild salmon, chickens have been engineered so that they cannot spread H5N1 avian flu to other birds, and research is being conducted to create cattle that cannot develop the infectious prions that can cause bovine spongiform encephalopathy (also known as mad cow disease). At this point, no GE animals have been approved by the FDA to enter the food supply. GE experiments on animals do, however, pose potential risks to food safety and the environment.

In 2003, scientists at the University of Illinois were conducting an experiment that involved inserting cow genes into female pigs in order to increase their milk production. They also inserted a synthetic gene to make milk digestion easier for the piglets. Although the experimental pigs were supposed to be destroyed, as instructed by the FDA, 386 offsprings of the experimental pigs were sold to slaughterhouses, where they were processed and sent to grocery stores as pork chops, sausage, and bacon.

University of Illinois representatives claimed that the piglets did not inherit the genetic modifications made to their mothers, but there was still a clear risk to the people who purchased products made from the 386 piglets. Since no genetically engineered animal products have ever been approved by the FDA, the pork products that reached supermarket shelves were technically illegal for human consumption. As a result of the accident, the FDA sent letters in May 2003 to all land-grant universities, reminding researchers that their work "may require" licensing under the animal drug law.

What Are the Concerns over Genetically Engineered Food?

Many concerns have been raised over the inadequate testing of the effects of GE on humans and the environment. GE is still an emerging field, and scientists do not know exactly what can result from putting the DNA of one species into another. The introduction of foreign DNA into an organism could trigger other DNA in the plant or animal to mutate and change. In addition, researchers do not know if there are any long-term or unintended side effects from eating GE foods.

Critics of GE believe that GE foods must be proven safe before they are sold to the public. Specific concerns over GE include the following:

- *Allergic reactions*: There are two concerns regarding allergic reactions. The first is with known allergens. For example, if genes from nuts are inserted into other foods, it could cause severe reactions in people with nut allergies. Therefore, there is concern that people with known allergies will not be aware that the genetically engineered food they are eating contains substances to which they are allergic. The second concern is that new allergies might be created, since new combinations of genes and traits have the potential to cause allergic reactions that have never existed before.
- *Gene mutation*: Scientists do not know if the forced insertion of one gene into another gene could destabilize the entire organism and encourage mutations and abnormalities.
- *Antibiotic resistance*: Almost all GE foods contain antibiotic resistance marker genes that help producers know whether the new genetic material was transferred to the host plant or animal. GE food could make disease-causing bacteria even more resistant to antibiotics, which could increase the spread of disease throughout the world.
- *Loss of nutrition*: GE may change the nutritional value of food.
- *Environmental damage*: Insects, birds, and wind might carry genetically altered pollen to other fields and forests, pollinating plants and randomly creating new species that would carry on the genetic modifications.

Gene Pollution Cannot Be Cleaned Up

Once released into the environment, genetically engineered organisms cannot be cleaned up or recalled. So, unlike chemical and nuclear contamination, which can at least be contained, genetic pollution cannot be isolated and separated from the environment in which it is spreading.

Superweeds

GE crops can cross-pollinate-related weed species, passing on their ability to survive the application of weed killers. Even without passing on that specific genetic trait, the widespread adoption of GE crops that are resistant to herbicides such as Roundup® has led to dramatic increases in the use of this weed killer, and weeds have gradually developed resistance to the herbicide. This leads to the evolution of superweeds that are very difficult to control. Already, superweeds

have infested 4.8 million hectares in the United States. At least 20 weed species worldwide are resistant to Roundup®, including aggressive weeds such as ragweed, pigweed, and waterhemp.

Terminator Seeds

Some GE seeds are engineered so that plants cannot reproduce their seeds. In many parts of the world, saving seeds from season to season is the only way farmers are able to survive and continue growing food. However, with GE technology, seeds can be sterile, forcing farmers to rely on seed companies for their livelihood, an expense they may not be able to bear.

What are some genetically engineered foods that have been approved for commercial use?

Alfalfa
Cherry Tomato*
Chicory (Cichorium intybus)
Corn
Cotton
Flax*
Papaya
Potato*
Rapeseed (Canola)
Rice*
Soybean
Squash
Sugar beet
Tomato*

* These GE crops were approved by the federal government but are not known to be commercially available.

Food Security and the Future

One important issue in agriculture and the future—and where green living comes into play—is food security. What this refers to is the availability of food and one's access to it. Here are the definitions that both the United Nation's Food and Agriculture Organization (FAO) and the U.S. Department of Agriculture (USDA) have given it:

- Food security exists when everyone at all times has physical, social, and economic access to sufficient, safe, and nutritious food to meet their dietary needs and food preferences for an active and healthy life.
- Food security on a household basis means that all members of a family have at all times access to enough food for an active, healthy life.
- The USDA states that as a minimum, it is the ready availability of nutritionally adequate and safe foods, and an assured ability to acquire acceptable foods in socially acceptable ways, such as without having to rely on emergency food supplies, scavenging, stealing, or other coping strategies.

In its most basic form, being food secure means not living in hunger or in fear of starvation. This concept exists on many levels—a personal level, family-unit level, local level, national level, and world level.

It has long been recognized that for a country to become advanced—or industrialized—it must first be able to feed itself; otherwise, it is not self-sufficient. With the push for industrialization that so many countries are focusing on, however, being self-sufficient may be in jeopardy. Perhaps, we are at a crossroads between food security and industrialization, and this is one area where green living can help.

The following lists some relevant facts about food security and hunger:

- According to the FAO of the United Nations, world-wide about 852 million people are chronically hungry due to extreme poverty, while up to 2 billion people lack food security intermittently.
- Every year, 6 million children die of hunger, which translates to 17,000 every day.
- There are many situations that affect food security, such as the fluctuation of world oil prices, climate change, global population growth, and the world-wide push to industrialize, which means that valuable agricultural lands are being converted to industrial areas (many rural areas are being converted into urban areas).
- In many areas of the world, national governments do not allow individuals to own land, therefore, people do not improve the land for farming purposes.
- Currently in the United States, there are approximately 2,000,000 farmers, which translates to less than 1 percent of the population.

A major dilemma the world faces is the loss of farmland and food security due to the many reasons mentioned. This may be a key opportunity for the green lifestyle to help. As countries reach out, research this complex topic, and attempt to find solutions, they will need to pull from current knowledge of farming techniques, green living, and the ongoing research of practices such as seed hybridization and the GE of crops (larger seeds and stalks, larger seeds, improving individual characteristics of seeds that will grow in a greater range of climates, and so forth).

These are all relevant questions, issues, and concerns, and this may be a key area for those currently pursuing a green career to become actively involved in the crucial research, development, and application of new techniques and technology in order to solve this global issue.

THE INDUSTRIAL SECTOR

As we have already touched on in several chapters so far, greening up in the industrial sector is going to be crucial over the next few decades as major changes occur globally with the earth's climate and ecosystems. Of particular importance to focus on are the concepts of ecologically sustainable development and eco-industrialism. These concepts are concerned directly with sustainability—the upcoming green wave of the future.

Ecologically Sustainable Development

Ecologically sustainable development is a key aspect that will be required for all future businesses and economic ventures. Similar to sustainable development, where development is allowed to occur but not at the expense of future generations' quality of life, ecologically sustainable development is simply the environmental component of sustainable development.

Because we, as a society, are responsible for our decisions as they affect the future, such as environmental degradation, we must fully consider the consequences of all our actions so that we do not compromise future generations and leave them with long-term damage that will negatively affect their existence, growth, and quality of life. In fact, through green living, it is our responsibility to ensure that the health, diversity, and productivity of the environment are maintained—or even enhanced, if possible—for the benefit of future generations.

It follows that any decision we make for ourselves now should include a future component to it in order to explore the consequences of our actions and decisions as they relate to the quality of life for generations down the road. Putting it even more simply, currently, we are our earth's stewards and need to make sure we plan for future generations as well as our own. This concept brings to life the fact that all the economic planning decisions we make will need to be thought through not only in terms of how what we do affects today's populations but how it affects all future ones, as well. Everything we do, every actions we take, has a long-term ripple effect. The bottom line is that any economic development cannot sacrifice the future quality of life for short-term gains and must help maintain the ecological processes on which all life depends.

ECO-INDUSTRIAL PARKS

EIPs are a new concept centered on the green industry. Basically, several industries team up and all build on a common site (Figures 16.3 and 16.4). The following characteristics apply to EIPs. They

- Become a sustainable community.
- Are designed to fit into their natural setting in order to minimize both environmental impacts and operating costs. For example, instead of landscaping, they use the natural vegetation.
- Utilize renewable resources in their design.
- Strive to minimize their environmental impact.
- Become a community of companies.
- Build and design ecologically appropriate buildings.
- Are designed to be able to adapt to change.
- Are focused on improving environmental performance.

The beauty of these systems is that each business co-located on a common property does not have to individually tackle the very same environmental issues and hurtles. They all combine

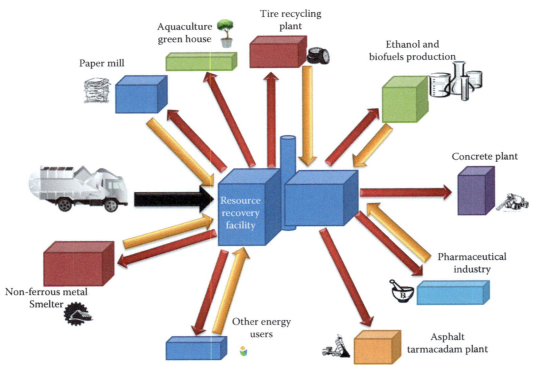

FIGURE 16.3 Eco-industrial part schematic. (From Department of Energy.)

FIGURE 16.4 An eco-industrial park in Oswego County, New York. (Courtesy of Oswego County Business.)

resources and efforts jointly to reduce the impact on the environment. EIPs are part of a sustainable community development idea, and they offer a new method for addressing future business and industry questions. They involve a network of firms and organizations that work together to improve both their economic and environmental performance—an industrial ecosystem of sorts. The businesses work together as a community that coordinates the use of energy, water, and materials. By creating and sharing resources, they all benefit, while reducing additional stress on the environment that would normally have occurred if they had worked alone.

An example of how they piggyback and share resources is the sharing of energy. They cut costs and reduce the load on the environment by designing connected energy systems. This can include flows of steam or heated water that travel from one plant to another, where they all share and benefit; or perhaps several nearby industries all tap into renewable solar or wind energy systems together.

They can also set up cascading material flows, where one company's waste may be used by another industry as part of its manufacturing process—the creation of a synergy of industries. Waste products may be reused internally by another business within the industrial park or perhaps marketed to someone else through a network set up by the industrial park.

Waste management is one area specifically focused on, where the entire eco-park may even support the establishment of a comprehensive waste reduction strategy, such as an integrated methodology of recycling, reuse, remanufacturing, and composting. In a similar manner, by-products generated from one industry in the park may be used by another in its industrial processes.

One well-known example of an EIP is in Kalundborg, Denmark, which is an industrial community consisting of a coal-fired power plant, a refinery, a pharmaceutical and industrial enzyme plant, a wallboard company, and a heating facility. This EIP has been referred to as an industrial symbiosis because of the way the companies share processes and work together (Figure 16.5).

FIGURE 16.5 Kalundborg eco-industrial park, Denmark. (Courtesy of MR3641/Wikimedia Commons.)

The Kalundborg EIP—also known as the Kalundborg Symbiosis—is the world's first working industrial symbiosis. The Kalundborg Symbiosis is an industrial ecosystem, where the by-product residual product of one enterprise is used as a resource by another enterprise, in a closed cycle. An industrial symbiosis is a local collaboration where public and private enterprises buy and sell residual products, resulting in mutual economic and environmental benefits.

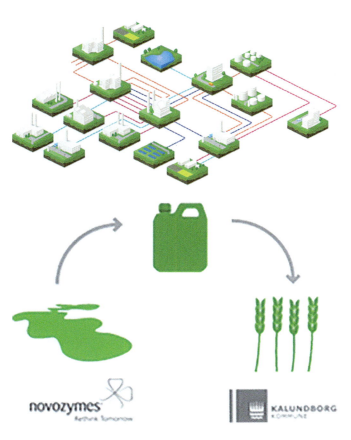

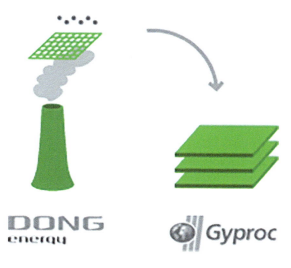

Result examples:

- They have saved 3 million cubic meters of water through recycling and reuse.
- Biogas is made out of yeast slurry from the production of insulin.
- Recycling of 150,000 tons of gypsum from desulphurization of flue gas (SO_2) has replaced the import of natural gypsum ($CaSO_4$) (see: http://www.symbiosis.dk/en.)

In the development of the Kalundborg Symbiosis, the most important element has been healthy communication and good cooperation between the participants. The symbiosis has been founded on human relationships, and fruitful collaboration between the employees that have made the development of the symbiosis-system possible. On the right, you can read about some of the most important lessons in creating and maintaining successful symbiosis that we have learned over the last four decades.

EIPs exist worldwide and appear, or are being developed, in locations such as Canada, Hong Kong, the Netherlands, Austria, Spain, Costa Rica, Mexico, Namibia, South Africa, Australia, in several U.S. cities, and the U.S. military. The beauty of these facilities is that they eliminate a lot of the redundant construction and maintenance costs and impacts. Because of their improved performance, they are being viewed more and more as the industrial model of the future.

CONCLUSIONS

We have learned that sustainable agriculture is the successful production of plant and animal products, including food, in a way which uses farming techniques that protect the environment, the welfare of animals, public health, and communities. Sustainable agriculture enables the production of healthy food without taking away from the ability of future generations to have the same thing. One of the key concepts to obtaining sustainable agriculture is finding the right balance between the need for food production and the preservation of natural environmental ecosystems. It is also important for sustainable agriculture to be able to find the right economic stability for farms if order to help farmers achieve the best quality of life possible. Because agriculture is the biggest employer in the world—approximately 40 percent of the world's population make their living in it—it is one field we need to make sure we are keeping sustainable in the long term.

The industrial picture presents a similar story. Because the industrial sector will only continue to grow, it will require the best minds and innovators to tackle the issues we are now being presented with: how to manufacture in the most efficient, conservative ways possible, yet enable the economies of the world to progress. Both of these fields are ready for the innovative minds of tomorrow.

REFERENCES

Grace Communications Foundation 2016. Genetic engineering. http://www.sustainabletable.org/264/genetic-engineering (accessed October 15, 2016).

GMC. 2014. Genetically modified crops—Field research. *Economist.* (accessed November 8, 2014).

Lovley, D. R. 2003. Cleaning up with genomics: Applying molecular biology to bioremediation. *Nature Reviews Microbiology.* 1(1):35–44.

USDA. 2016. Organic production. United States Department of Agriculture Economic Research Service. http://www.ers.usda.gov/data-products/organic-production/organic-production/#National Tables (accessed October 23, 2016).

SUGGESTED READING

Day, J. W., and C. Hall. 2016. *America's Most Sustainable Cities and Regions: Surviving the 21st Century Megatrends.* New York: Copernicus, 348 p.

Jacobsen, J. 2011. *Sustainable Business and Industry: Designing and Operating for Social and Environmental Responsibility.* Milwaukee, Wis: Quality Press, 216 p.

Kleppel, G. 2014. *The Emergent Agriculture: Farming, Sustainability, and the Return of the Local Economy.* Gabriola Island, BC: New Society Publishers, 192 p.

Matheson, B. 2011. *DIY Projects for the Self-Sufficient Homeowner: 25 Ways to Build a Self-Reliant Lifestyle.* Minneapolis, USA:Cool Springs Press, 160 p.

Markham, B. L. 2010. *Mini Farming: Self-Sufficiency on ¼ Acre.* New York: Skyhorse Publishing, 240 p.

Sachs, C., M. Barbercheck, K. Braiser, A. Terman, and N. E. Kieman. 2016. *The Rise of Women Farmers and Sustainable Agriculture.* Iowa City, IA: University of Iowa Press, 202 p.

17 Green Careers and Ecotourism

OVERVIEW

This chapter focuses on the green job market and what careers are available in the field of green energy and sustainable living. It introduces advice on finding green jobs and provides information about employment and research opportunities. It also provides an overview for the most rapidly-growing green careers. From there, the chapter shifts gears just a bit and talks about a market trend that has emerged in the past few years: that of ecotourism. Not only does it provide a new and exciting way to take a vacation, it has some wonderful features and for those interested in green living, presents a great way to enjoy a vacation and see the world.

INTRODUCTION

Green jobs are not only the wave of the future but they are also good for strengthening our economy. Green energy is already boosting it by providing 2.7 million jobs in clean energy—more than what is employed in the heavily-subsidized fossil fuel industry. In fact, across the diversity of clean energy projects (including renewable energy, transit, and energy efficiency) for every million dollars spent, 16.7 green jobs are created. That is over three times the jobs that are created from the same spending in the fossil fuel industries.

The clean energy fields that are showing the most promise right now are solar thermal energy (expanding by 18.4 percent annually), solar photovoltaic power (by 10.7 percent), and bio-fuels (by 8.9 percent). Wind energy has had an average annual growth over the past five years of 35 percent. Add to that the clean car industry—they are currently employing more than 150,000. In conjunction, median wages are 13 percent higher in green energy careers than the economy average (DiPasquale, 2011). And also of note: green products are made in America with American ingenuity.

In the solar industry, the United States has a positive trade balance. Contrast that with the oil industry, where imports are over $250 billion in petroleum-related products. In addition, clean energy jobs are especially benefitting small businesses: especially those involved in construction and retrofits. There are also an abundance of manufacturing jobs with an upward career track. The bottom line—the green industry gives us a place in the forefront of innovation if we invest in it. It is the wave of the future. In these next few sections, we will find out why.

DEVELOPMENT

With the environmental push and going green movement starting to gain serious momentum, there is a new class of job emerging in the work force: the green-collar job. Green jobs have become the new opportunity when it comes to career options. These next sections will address this exploding job market and answer questions, such as: what is a green job, what are its sub-specialties, and why green jobs may be the wave of the future. From there, it shifts gears slightly and moves to ecotourism—a new way to travel that is visitor-friendly to the environment and leaves the landscape no worse for the wear; the newest trend in traveling.

WHAT IS A GREEN ECONOMY?

One sector that has shown constant growth in the job market is the green economy. The green economy is defined as an economy that aims at reducing environmental risks and ecological scarcities, while aiming for sustainable development without degrading the environment. It is an economy that simultaneously promotes sustainability and economic growth. In general, it can be thought of as an alternative vision for growth and development; one that can generate growth and improvements in people's lives in ways consistent with sustainable development. It is an economy that uniquely promotes a triple bottom line: sustaining and advancing economic, environmental, and social well-being.

The prevailing (traditional) economic growth model focuses on increasing GDP above everything else. While this system has improved incomes and reduced poverty for hundreds of millions, it comes with significant and potentially irreversible social, environmental, and economic costs. Issues it does not address are poverty and resource conservation. Poverty is a way of life for as many as two and a half billion people, and the natural wealth of the planet is rapidly being depleted—a completely unsustainable solution. Today, approximately 60 percent of the world's ecosystem services are either degraded or used unsustainably (Manish and Talberth, 2011). The gap between the rich and poor continues to increase.

It has been suggested that there are "missing markets" in the traditional model. This means that existing markets do not systematically account for the inherent value of services provided by nature, such as water filtration or coastal protection. A strict "market economy" alone cannot provide public goods, such as efficient electricity grids, sanitation, or public transportation (WRI, 2010). Because of this, the current market is not capturing the voice and perspectives of those most at risk. A Green Economy, on the other hand, attempts to remedy the problems through a variety of institutional reforms and regulatory-, tax-, and expenditure-based economic policies and tools. Two main components of its success in countries that are currently making it work, such as Mexico, Namibia,

China, and the Republic of Korea, are commitment and funding by the government and involvement and commitment by the public and private companies.

THE GREEN ECONOMY AND JOB MARKET

Carol McClelland of job-hunt.org noted: "I've seen authors refer to the shift into this new economic era as the Third Industrial Revolution (Jeremy Rifkin) and the Climate/Energy Era (Thomas Friedman, author of Hot, Flat and Crowded). Time will tell which term will rise to the top as the defining phrase."

Her observation is that the terms will probably keep evolving for a while. She also points out that with the current changes developing in business and economics because of the green revolution, we are also being handed some great opportunities: new, evolving career possibilities and trends. Because of it, she has divided the green job market up into four distinct employment sectors: (1) New green industry; (2) Existing companies rethinking their business practices; (3) Environmental science and natural resource management; and (4) Business promoting understanding and creating a demand for green lifestyle products and choices.

1. *New green industry*: These are the emerging industries in the renewable energy, Smart Grid, and clean tech fields. The principle companies involved are the new start-up companies and already well-known companies investing strongly in green/clean enterprises.
2. *Existing companies rethinking their business practices*: These companies are investing significant amounts of money into their business in order to reduce the impact their industry is having on the environment right now. This includes companies that are involved in green building, manufacturing, and alternative transportation. They are rethinking their operating practices and looking for a solution on how to lower their carbon footprint.
3. *Environmental science and natural resource management*: These are organizations that are striving to understand exactly how natural environmental systems work so that business systems will be designed to work in balance with the environment.
4. *Businesses promoting understanding and creating a demand for green lifestyle products and choices*: These include businesses that increase public awareness and focus on increasing public demand for greener, more sustainable actions. This involves businesses that specialize in communication, legal issues, education, and finance components (this book, helping to teach you about green living, is an example of this business).

So if a green job is of interest, it is never too late to take that first step, whether you are just entering the job market or are a veteran and would like to have work in the sustainable living,

or green, market. Some companies that are switching their vision for the future may already be helping you along that path. For others, the first step may be an educational program.

A GREEN EDUCATION

A need for green jobs has recently put into focus the need for advanced qualifications in sustainability and green business. While the corporate world focuses on devising strategies to address the business of sustainability, several business schools and other higher educational systems across the country are now incorporating this relatively new field of study into their curriculums.

Many universities are now offering green MBA programs. This degree emphasizes leadership in sustainability and focuses on the emerging needs of the green economy. Green MBA graduates receive a solid background in environmental issues along with traditional business training, which prepares them to become the leaders in the new green corporate world.

And this seems to be a growing area of demand. According to a report from Net Impact in 2010, a global nonprofit organization, the number of publicly advertised corporate social responsibility (CSR) jobs has gone up by 37 percent since 2004. This illustrates the genuine need in the market for managers and senior executives who are knowledgeable about the environment, which can lead the green future.

Today, there are several green MBA programs available through university curriculums. If the business aspect is not for you, and you are more into the technical, scientific, or social relations aspects, there are also several other avenues of higher education degrees, such as those in the following list:

- Environmental science
- Environmental studies
- Environmental policy and management
- Green law
- Environmental engineering
- Sustainable entrepreneurship
- Sustainable management
- Waste management
- Renewable or clean energy technology
- Ecology
- Ecotourism
- Geothermal development
- Green interior design
- Organic agriculture

In addition to on-campus, college-based education, there are also opportunities at tech and trade schools and online. No matter what area your specific interest may be in, with a little research you may be able to find your niche in a brand-new job market just now being defined and growing. You may be able to get in on the ground floor and be there right from a new career's inception. By the time your children and grandchildren are looking for jobs, these positions will probably be a way of life for them instead of the cutting-edge adventures they are right now. Or they may have a new spin-off of your new job.

SOME SOUND GREEN ADVICE

There are many job consultants willing to offer advice to green-job seekers. Douglas MacMillan in Bloomberg Businessweek advises people who are interested in switching to green-collar jobs to follow a few simple rules before filling out that résumé:

- Describe your personal skills in a green context: Look at the skills you already possess, see how they may fit into a green company, and apply for jobs that require those skills. For example, if you are in a human resources field, you may want to apply for a human resources position at an emerging green company.
- *Awake your inner passion*: For those who want to explore a new frontier and make a more radical life change, working at a green job may be your opportunity to do it. This may or may not require new training, depending on what you transition to.
- *Give yourself a new perspective*: Maybe this is the time to change the way you look at the world. As an example, one individual who previously focused on fast-paced city life and the great financial race of the stock market developed a new perspective and found a job raising funding for international projects in alternative energy and carbon trading. To that person, there was an additional bonus (other than making money): helping others obtain a better environment and a sustainable future.

Other advice from leading green companies follows similar lines of not only bringing skills and energy into the picture but also a genuine devotion to the green philosophy. The reason for this is that most green companies stand for more than just business. Most green business models are concerned not just about the bottom line, but about how their activities contribute to a better future. And that is the key to remember when seeking employment in the green world. So here are a few more bits of advice from green companies:

- *Keep your personal message consistent*: It is no longer possible to separate your business persona from your personal life. Employers are sensitive to their employees' true character and what they stand for. So, sure, your education and experience are part of it, but so is what you do with your free time (are you active in charities and causes that support the environment? Do you pay attention to your own carbon footprint?).
- *Your online image*: Yes, the world has gone digital, and your future employers will check out your digital connections if they are interested in you. So be consistent. If you have a Facebook page, do not post a picture of yourself racing down the road at 90 mph in your monster truck, blowing thick clouds of exhaust out the tailpipe (refer back to the part about needing to be genuinely involved in environmental causes).

First impressions speak volumes. Be sure you always portray yourself honestly—that is what this is all about.

- *Understand your future employers' concerns*: Do your homework first. If you know what is important to your potential employers, if you are familiar with each company's projects, concerns, and challenges, you will be able to relate to them better, and you will have more information to help you decide if the company is right for you.
- *Keep it real; passion is one thing, extremism is another*: Do not portray yourself as the world's biggest environmental guru. Companies will see right through that and question your real intentions—are you just trying to impress them, are you a con-artist? Be honest and realistic. If you truly stand for their green cause, your sincerity will prove that all on its own.

HOW CAN I FIND A GREEN JOB?

The good news is that there are plenty of green jobs to be found, and getting a green job is not much different from finding any other kind of job: you have to scout out job boards, get your résumé out, network, and convince the decision-makers that you are the only one they should hire. Be persistent and keep in mind that your best chances of getting a green job are in positions that utilize the skills you already have. After all, who knows how to do what you do better than you?

There are a couple of good sources to get you going. One is the Green Job List, which is an e-mail list for those seeking jobs that focus on environmental and social responsibility. Powered by Yahoo! Groups, you can subscribe or unsubscribe at any time at http://www.greenjoblist.com. You can also check out the green job Internet boards at Yahoo and Treehugger and many other sites you can find through an Internet search.

Checking through individual colleges and universities that offer green degrees is another option. Oftentimes, they have connections or employment opportunities through internships and other programs.

The Green Collar Blog is also a good resource for identifying the latest job postings in green energy, LEED (sustainable buildings), solar, wind power, green nonprofit, and many more. This blog lists job fairs and provides a comprehensive list of green job boards. Just know what you are looking for, search with those criteria in mind, make sure you do your homework and have obtained the proper training, and be persistent and thorough using the resources you have available to you.

EMPLOYMENT AND RESEARCH OPPORTUNITIES FOR GREEN LIVING

Green careers attempt to address a wide range of environmental concerns, so they cover a wide range of areas and specialties. Across every industry, new job possibilities are emerging for those with the skills to make that transition between the old fossil fuel[en]based economy and the new energy-efficient one. The good news here is that industries are now being held accountable for their effect on the environment. The new wave of thinking is that reducing their impact on the environment is as beneficial for future profits as for the environment[em]a win-win situation.

The Political Economy Research Institute has identified six areas that will require a specific focus in order to build an effective green economy[em]which also means more green jobs. They deal with the components necessary to effectively manage climate change and new environmental requirements. These areas include:

- Building retrofitting
- Mass transit restructuring
- Transition to energy-efficient automobiles
- Utilization of wind power
- Development of solar power technology
- Creation of cellulosic biomass fuels

The long-term goal of all new green careers is to have a resulting positive effect on the economic, environmental, and energy security of the future. Currently, renewable energy and clean technology is the fifth-largest market sector in the United States. According to Cleantech Venture Network, a green-oriented venture capitalist fund, there are roughly half a million green-collar jobs that have been recently created in the United States. And this number is growing, along with the range of job types and skill requirements.

New Technologies

Going green means switching to new technologies when it comes to energy sources and finding ways to use renewable energy. The focus will be on alternative fuels and finding ways to formulate them, and then on how to design and build the transportation systems that will be able to efficiently use these new fuels.

To go green, people need to look at city planning and the issue of sustainability, meaning that population centers will have to be rethought and redesigned, making new construction a key concept. It may involve using electronics and their components differently.

And as one thing changes, it affects another, like a chain reaction, until new systems are implemented that complement each other. For example, if a new energy-efficient fuel is developed, a new engine may be required to use it. This may necessitate a newly designed vehicle. Think right there of how many new jobs will have to be created from the initial finding of the new energy source to its manufacture to the development of everything else affected by it to the point the new fuel can actually be used.

These projects can take years and require dozens of new positions. So, it is not hard to see that many new jobs can potentially be created for each small component in the green sector. The possibilities are staggering.

Science

Science will continue to play a major role in going green. A slew of new or altered traditional careers will open up for those who want to major in going green. The fields in science, of course, will be greatly affected. Following are some of the fields and positions that will open up as going green becomes more of a major factor in our world:

- *Alternative energy*: This is a big one. There are so many different kinds of alternate types of energy and fuels that the needs for those people with a strong science and math background in several fields will be great for many years. Getting in on the ground level in any field of alternative energy will offer vast opportunities because it covers the discovery, conceptual stage, development, testing, and implementation stages of commodities such as fuels; cars; public transportation systems such as light rail, underground rail, renewable fuel buses; and alternative fuel cars (flex fuel, biofuel, electric, fuel cell, etc.). This also involves energy sources, such as wind, solar, hydropower, geothermal, biomass, and new sources of renewable energy currently in the research stages.
- *Atmospheric sciences*: With the issue of climate change so critical in every aspect of our lives, experts in research, development, and application positions in the field of atmospheric science will be increasingly in demand. Discoveries in the atmospheric sciences also have a direct bearing on public policy and national and international politics.

Carbon consulting and emissions trading: Understanding the science behind carbon emissions is going to become more critical as standards tighten up on carbon emissions and the United States joins the rest of the world and becomes a bigger player in the carbon trading field. This area will become even more important as the specific effects of carbon become better understood and appropriate controls and restrictions are set. These positions will involve both corporate and consulting positions. Emissions traders and brokers will also be needed to complete licensing requirements. In addition to CO_2, there are also expected to be controls and trading on nitrogen oxide (NO_x) and sulfur oxide (SO_x).

- *Carbon capture and sequestration*: This has been a hot topic lately. Carbon capture and sequestration involves finding places to store excess carbon out of the way of the environment. Projects currently being worked on include underground storage off the West Coast of the United States. This field will be very important as the CO_2 concentration rises and innovative ways are sought to get rid of it in order to keep climate change at bay.
- *Environmental scientists*: These professionals, found in both government and private enterprises, are the ones who complete critical environmental

investigations and write the environmental assessments that construction and industry use to move ahead. They also prepare assessments on climate change. Already utilized in situations such as oil spills, these professionals will be key players in the planning and implementation of anything that could negatively affect the environment, such as construction of a Smart Grid, any environmental[em]both short- and long-term[em]cleanup projects, restoration projects, and so forth.

- *Land-use planners*: These scientists, also found in both the public (government) and private sectors, are important because it is critical that when land is designated for a specific use, the long-term effects are analyzed for sustainability and possible conflicts with adjoining (and potentially conflicting) uses. This includes land uses such as recreation, forestry, urban interface, wildlife planning, and any other long-term use where sustainability is a critical component. Planners can help avoid land use conflicts that could occur later on and impact sustainability.
- *Landscape architecture and horticulture*: With climate change a key environmental issue, landscape architecture and horticulture is an important component. These fields will help conserve energy by affecting heating and cooling of homes, and water conservation with new practices like xeriscaping. Plants are also used to clean up oil spills, giving both research and applications horticulture specialists new opportunities.

- *Environmental engineering*: Environmental engineers are the professionals who design water and wastewater treatment systems. They also work on other large-scale projects, including both new and upgrades to existing systems. For example, some major cities in the United States, such as Portland and Oregon, have multibillion dollar water infrastructure upgrade programs currently in progress.
- *Electrical engineering*: Electrical engineers design and maintain the electrical grid and power plants among major renewable energy facilities. They will be major players in green buildings and research projects and continue to be critical as technology evolves.
- *Civil engineering*: This profession will have an active role in every infrastructure-related project, including airports, bridges, and storm water systems. It will continue to have a long growth period.
- *Architecture*: Architects will be in high demand as they are sought after to redesign buildings to fit new green standards. Just the government buildings alone that will be retrofitted with green equipment and features will necessitate the architect to become certified with LEED requirements (green building program) as part of their credentials.
- *Public transit*: One of the strongest focus in sustainable living is on the growth and utilization of public transit systems. This area will be in the spotlight as people look toward it as an alternate solution to rising fuel prices.

Engineering

Engineering will be affected as new fuels are developed and new technologies are invented to go along with new green practices. The following are some of the fields and positions that will be changing for the greener:

- *Structural engineering*: This field is critical as new energy sources are developed and green buildings and other structures are designed to be safe, efficient, practical, and cost effective.
- *Mechanical engineering*: Mechanical engineers will be important for a variety of areas. Indoors, they will be key players in keeping green buildings running; outdoors, they will be critical resources for various structures and devices, such as the new generation wind turbines, solar towers, and other newly emerging technology.
- *Geotechnical engineering*: Geotechnical engineers will be in great demand, needed to design and construct dams, foundations, levees, tunnels, and underground drainage systems. With the increasing implementation and use of alternative energy and the structures to support it, as well as the thousands of ongoing environmental projects, geotechnical engineering may become one of the hottest jobs around.

Technology

Technology will have to keep up with the green innovations of science and engineering. This will open up new fields and careers and alter some of the existing ones to work better in the green world. Following are some of the technologies and careers that will be important as we move toward a more sustainable way of life:

- *Computer-aided design*: A multitude of businesses will require these services. Any business that has any construction-related projects will need to have plans and diagrams prepared. These include architecture, engineering, and construction projects. This specialty is a good example of a crossover job from the nongreen to the green sector.
- *Environmental information systems*: This involves operating computer systems for facilities that are capable of monitoring equipment to ensure that employees, the public, and the facility are operating as planned. These types of systems will be in high demand.
- *Geographic information systems*: These systems involve the monitoring and analysis of spatial data, interaction of environmental components, and allow

the analysis of environmental decisions. This is another rapidly growing field in the green sector. Geographic information systems will be highly useful in network applications such as Smart Grids, carbon trading, and road networking impacts. It will serve as an integral part of any spatial business component with regional, national, and international jurisdiction.

- *Remote sensing*: This involves the use of imagery—both satellite and airplane platforms—that enable the user to analyze the digital product and interpret, classify, monitor, and perform analysis of impacts. Applications include change detection and determination of land degradation from activities like deforestation, impacts from energy operations, urban impacts on habitats and ecosystems, suitability determination of green sites, and many other green applications.
- *Waste management*: This involves waste management in industrial and hazardous waste industries. The recycling industry is one area where this position will be important. If a waste product is not recycled, it must be managed in some other way. This type of management is the green field that deals with the types of questions that involve typical household waste, nuclear waste, and other types of waste that must be disposed of in a special manner.
- *Recycling*: Recycling methods will be a profession in great demand. As more and more people practice sustainable living, new ways of reusing materials will need to be developed in order to make society more efficient.

Mathematics and Business

Math plays a role in all technology and business plays a part in nearly all fields. Both will be used in similar and new ways as we adjust to a greener way of doing things. Here are some examples:

- *Accounting*: Accountants will be critical in the business end of endeavors such as carbon offsets, carbon trading credits, and all other green businesses. This is another prime example of a green crossover career.
- *Regulation auditing and inspection*: As industries have to meet specific regulations concerning emissions, inspectors will play a crucial role in the green world. There will be no way around the fact that every facility that has a smokestack or produces greenhouse gas byproducts will be inspected and monitored for compliance because there will be both national and international regulations in place.
- *Insurance, underwriters, and risk management assessors*: These represent other great crossover fields and will be as necessary in green businesses as they are in nongreen ones.

- *Regulatory compliance*: Government agencies will monitor industries and other green businesses to ensure compliance toward determined standards. As new green businesses join the force, the demand in this field will grow. Examples of government agencies that will be involved in this area include the Environmental Protection Agency, the Department of Energy, the Bureau of Land Management, and the United States Forest Service.

Travel and Service

The tourism industry is a rapidly growing green industry. In fact, ecotourism is growing at three times the rate of the general tourism sector and requires a solid understanding of global sustainability issues. Hotels see themselves in a position to become educators, able to teach people from around the world about sustainability practices, such as water conservation, and issues particularly relevant to the area in which the traveler is visiting.

Ecotourism is responsible travel to fragile, pristine, usually protected areas that are at off-the-beaten-path locations (as opposed to mainstream tourist sites). Besides providing entertainment for the traveler, ecotourism's purpose is to educate the traveler and provide funds for ecological conservation of the area, directly benefit economic development of the area, assist in politically empowering the local communities, and foster respect for different world cultures and human rights.

Ecotourism teaches travelers about conservation and sustainable travel. It is expected that the traveler follow certain guidelines, including:

- Minimize impact to the area
- Build environmental and cultural awareness and respect
- Provide direct financial benefits for conservation
- Help provide empowerment opportunities for the local inhabitants
- Raise the sensitivity to the host countries' political, environmental, and social atmosphere
- Leave the destination a better place for having visited

It should be a win-win situation for both the visitor and visited by the end of the vacation, with both having learned and gained from the unique exchange of values and ideas.

The ecotourism industry has career opportunities in these areas:

- *Developing sustainable tourism destinations*: This would include becoming a tourism development specialist, having a career in ecosystem management, or becoming a sustainable development consultant.

- *Managing ecotourism destinations sustainably*: This could include becoming a park manager, a parks and recreation director, a wildlife visitor center advisor, conservation project manager, a conservation specialist, or learning one of many types of expedition field positions.
- *Education and marketing the benefits of sustainable travel*: This could involve jobs in the area of tourism marketing, guidebook writing, nature center research, becoming a specialist in biology and biodiversity, getting involved in humanitarian work, or a career in conservation science.

Social, Practical, and Other Sciences

There are many positions that fall under these categories. Many of these are also crossover careers. Here are a few of them:

- *Foreign language specialists and interpreters*: This will be a field in high demand as green businesses go international—many of which already have—and need to cross language barriers.
- *Planning and land use*: Local governments are now starting to look critically at their new development plans, account for projected carbon footprints, and devise ways to deal with solving issues before they become problems. This includes green planning for transportation and urban design, such as designing adequate green public transportation methods, carpooling incentives, and bicycle paths; adequate storm water management, such as efficient systems that will be maintained and not be inefficient or allowed to become clogged; construction of carbon-neutral buildings, such as those with solar power, automatic light sensors, automatic water sensors at sinks, and passive solar lighting; and wetlands restoration, such as planning around major flyways and maintaining critical wetland habitat. All of these steps are major steps in the right direction for a greener lifestyle.
- *Legal careers*: Legal positions are necessary in many aspects of the green economy. These are commonly involved in areas such as water, air, and land pollution issues, carbon trading, animal endangerment, and trespass issues. These careers can be with government agencies, private firms, and nonprofit organizations.
- *Green education*: These are the world's eco-educators and very important in the whole scheme of things. For positive action to take place, people have to understand what the issues are all about. Without eco-educators spreading the knowledge so that people can understand just how their actions contribute to the big picture, change would not happen.
- *Food and farming*: Agriculture is also a changing industry. The demand and increased revenue generated from organic products has begun to change the way farmers are doing business. The U.S. Department of Agriculture says that organic farming has been one of the fastest growing segments of U.S. agriculture for over a decade. As a result, this career path will offer not only jobs in agriculture and food production technology but also research in the biological sciences.
- *Corporate social responsibility*: To ensure that corporations become more responsive to environmental issues, human rights, and health issues, corporate responsibility is now advocating that new business ethnics move away from focusing primarily on profits to a new "triple bottom line," which is also referred to as the three Ps—people, planet, and profits. As business trends move this way, along with the green movement, this opens up new career options, as well.

According to Erica Dreisbach of Social Venture Network, a nonprofit organization designed to educate businesses on social responsibility, "The fact that corporations are starting to talk about reform means that corporate social responsibility is going to become more mainstream in the future."

This field does require knowledge of labor law and human resource management and the career path will become even more important as corporations begin to see the advantages of having an environmental focus.

- *Political science and geography*: As new joint ventures and regulations are formed between international corporations, careers in political science and geography will play a significant role. This is also a viable crossover career path that many will find practical.

International Business and Foreign Relations

As we go greener, we will be interacting more and more with foreign governments and businesses. This opens possibilities and options in the international realm. The following are some careers and fields that will be much needed in our greener world:

- *Carbon trading*: GHG emissions will get capped at certain levels by government or an international regulating body. Companies or other organizations are issued permits to emit a certain amount of carbon dioxide. Those that exceed their limit must buy allowances from those that emit less.

A carbon trader puts together financing plans for projects that can produce carbon credits to sell. For example, the project might be a wind farm in South America or a carbon sequestration project in Asia. The company that is emitting too much CO_2 finances the green project elsewhere, which buys them credits to offset the pollution they are causing.

Businesses can even finance projects to use as assets for the future. These assets today in the market are worth more than $6 billion, and as carbon trading expands throughout the world, the Deutsche Bank projects it could grow to

$500 billion by 2050, making this one of the jobs to have in the future.

- *Eco-investing*: As these transitions evolve over the next years, one thing is certain: millions of skilled workers across a wide range of familiar occupations and skill levels will benefit from transforming into a green economy. Investment opportunities will especially be important in businesses such as alternative energy, clean technology, computer technology, and research fields.
- *Green recruiting*: Recruiting firms will also play a major role in finding experienced employees to fill positions in the environmental sciences, in corporate social responsibility positions, in sustainable development fields, and in careers dealing with climate change issues, such as analysts and project managers.
- *Environmental bankers*: Environmental banking is going to play a key role because as new businesses emerge, they will need loans just as other emerging businesses do. Environmental banking will vary from the traditional banking approach in order to make it easier for emerging green businesses to get the financing they will require in order to be successful. This should open up many opportunities in the financial market.

Strategists and consultants: There will be a growing need for green professionals who can effectively examine the processes that comprise a company's business and look for practical ways to make them more energy efficient and green. This will incorporate all aspects of business processes, offering many new opportunities.

THE MOST RAPIDLY-GROWING GREEN CAREERS

A recent survey conducted by National Geographic identified 11 of the most rapidly growing green careers (National Geographic, 2016). They are:

- Urban growers
- Water quality technicians
- Clean car engineers
- Recyclers
- Natural scientists
- Green builders
- Solar cell technicians
- Green design professionals
- Wave energy producers
- Wind energy workers
- Biofuel jobs

Urban Growers

Urban growers involve rooftop orchards planted on high-rise buildings. Green rooftop gardens provide locally sourced foods that protect the environment because they minimize

FIGURE 17.1 Urban rooftop gardening on a high rise building provides a sustainable alternative as a food source. (From ecolonomics.org.)

the use of pesticides as well as fossil fuels because they do not have to be transported from commercial farms grown at distant locations. Green roofs also improve the urban environment by insulating buildings against energy loss, managing storm water, improving air quality, and providing recreation opportunities (Figure 17.1).

Careers in water quality are in high demand and currently undergoing rapid changes due to climate change, population growth, and urbanization. These factors are causing negative impacts on water resources which must be addressed in order to reduce them. The current shift in skills currently requires:

- Environmental competencies in analyzing and interpreting environmental samples and data
- Developing sustainable development indicators
- Developing plans and strategies
- Implementing and monitoring sustainable development strategies and programs
- Conducting environmental assessments
- Developing and implementing environmental communications and awareness programs

Water Quality Technicians

The demand for water quality professions is expected to increase, especially with employers of chemists, hydrogeologists, green building professionals, water utility engineers, and environmental engineers.

Clean Car Engineers

Clean car engineers will continue to be in increasing demand. Manufacturing accounts for the majority of U.S. green jobs. According to the U.S. Department of Labor's Bureau of Labor Statistics (BLS) report, more than 462,000 of the nation's 3.1 million green jobs are involved in manufacturing. Clean car engineering is on the forefront of the green industry as it becomes necessary to redesign vehicles that consume less fossil fuel and produce less pollution. Because transportation currently consumes approximately two-thirds of the United States's oil and produces about one-third of its GHG emissions, clean cars are a necessary alternative, making this career field an extremely lucrative one.

Recyclers

The recycling industry is extremely promising. The United States recycles more paper than all other materials combined (excluding steel). Data from the American Forest and Paper Association show that in 2011, two-thirds of all the paper consumed in the United States was recovered, rather than dumped into landfills. Data from the EPA indicate that more than one-third of the raw material used at U.S. paper mills is recovered paper. Based on this promising information, this not only makes a great impact on the environment but also creates many green jobs. EPA analysis shows that recycling a single ton of paper saves enough energy to power an American home for half a year, saves 26,500 liters of water, and reduces GHG emissions by a metric ton of carbon.

Natural Scientists

The role of natural scientists is to monitor and analyze human impacts on all aspects of the natural world and on the natural resources on which we all ultimately depend for daily living, health, and survival. This is a very broad field and includes specialties such as biology, chemistry, physics, zoology, botany, biochemistry, molecular biology, earth sciences, ecology, psychology, mathematics, neurobiology, pathology, computer science, evolution, materials science, genetics, history, physical science, plant sciences, and so forth (Figure 17.2). Examples of careers include:

- Geologist
- Marine biologist
- Chemist

FIGURE 17.2 Energy from abundant, renewable, domestic biomass can reduce U.S. dependence on oil, lower impacts on climate, and stimulate jobs and economic growth. (Courtesy of Pontafon, GNU Free Documentation License.)

- Geographer
- Oceanographer
- Microbiologist
- Geoscientist
- Environmental scientist
- Occupational health and safety specialist
- Biochemist
- Biophysicist
- Medical scientist
- Physicist
- Hydrologist
- Remote sensing specialist

Green Builders

Green building integrates natural resource, human health, and community concerns into building design, construction, and operation that significantly reduce or eliminate the negative impact on the environment and occupants by addressing energy efficiency, water conservation, indoor air quality, waste reduction, and occupant productivity and health.

Greener options are in high demand. With rising energy costs, tightening budgets, expanding populations, and diminishing resources, an increasing number of businesses are currently turning to green buildings, and cities and towns are implementing greener development strategies.

The change in mindset is being felt in the marketplace. The evolution of green building has now shifted from educating the public to keeping up with market demand for a greener way of doing things. This sector is beginning to see a revolution of its own. It seems to be going beyond sustainability in many cases and trending into innovation. Where traditional practice in green design focuses on minimizing damage to the environment and human health and using resources more efficiently, a school of thought that is evolving within the green building movement is *regenerative design*. Regenerative design shifts the frame of reference from minimal to positive impact with the idea that human activity should result in net benefits to both the ecological and social environments. What is seen as the "next stage of green development" is intended to repair and improve the building site.

Regenerative design is a concept based on process-oriented systems theory. The word "regenerate" means "to create again." A regenerative system makes no waste; its output is equal to or greater than its input; and part or all of this output goes toward creating further output—in other words, it uses as input what in other systems would become waste.

Although regenerative design is a part of sustainable living, it is not the same as sustainable design. Sustainability implies something that endures over time without degrading, but it does not regenerate itself or create anything new. A plastic bottle sustains; a plant regenerates. Sustainable design aims to provide for fundamental human needs; regenerative design goes further in that it plans for the future co-existence and co-evolution of humans and other species.

The regenerative approach uses biomimicry, or the study of ecological systems to find solutions to human problems, to model patterns for industry, agriculture, and human habitats.

Just as in nature, organic and synthetic materials are not only metabolized (used) but also metamorphosed (changed) into another vital element of a closed system.

One element of biomimicry is the use of all species within a system. Much of environmental action today focuses on preservation by segregating wild areas from human habitation. A better system, according to the regenerists, is conservation: recognizing that humans are a part of the ecosystem and need to be incorporated into it.

This is closely related to permaculture, another model for sustainable living that relies on synergy, or the idea of separate components forming a whole that is greater than the sum of its parts. It emphasizes patterns and groupings that occur naturally. Food crops are not only organically and sustainably grown, but part of the harvest can be used to support the next year's crop. A holistic farm model uses the outputs from one crop or animal to grow another; one crop does not necessarily need to be self-recreating, but the farm as a whole would require no additional input, and it would not generate any waste products. Permaculture relies on polyculture, or the use of multiple crops instead of a single crop, in imitation of natural biodiversity. It is permanent, or continually renewing, agriculture. In addition, the crops themselves and the materials used to build the farm would already exist in the area, so they would already be ideally suited to the climate.

Another regenerative concept is natural building, which, like permaculture, seeks to use natural materials and relies more on human labor than industry to produce living spaces that are ecologically and aesthetically harmonious. Natural building emphasizes sustainability and minimal environmental impact without sacrificing the health or comfort of the human inhabitants. It too makes use of the site's climate and conditions to reduce the amount of energy required for ventilation and temperature control. For example, native shade trees are planted next to buildings to cool the interior; windows are placed to take advantage of breezes. Simple wind turbines can provide the building's energy, and rainwater is collected for drinking and washing.

All these ideas combine to create patterns that mimic nature so that humans can take a symbiotic role in their environment rather than a destructive one. Obviously, a perfect closed system that regenerates itself 100 percent is not possible, so the current goal is 99.9 percent regeneration. Even that goal is a difficult one, but the process of attempting it is worth trying to obtain. Ultimately, what needs to be kept in mind is the end result of not adapting to a changing environment—extinction.

Solar Cell Technician

According to the European Commission, production of solar cells and photovoltaic systems has doubled every 2 years during the past decade. (ec.europa.eu). Costs have continued to decrease and financing options are attractive enough that converting a home to solar power as its primary electric supply source has finally become a competitive option for the majority of households.

In the solar cell production market, Germany leads the market, which has some 40,000 solar workers. Europe currently produces about 30 percent of the world's photovoltaic power. Japan and the United States are the other major users, though off-the-grid applications are now surging in the developing world, where they provide crucial power to those in poor and rural locations.

A solar energy technician may help construct and install active systems, which require solar collectors, concentrators, pumps, and fans. They may also install passive systems, which rely on the best use of windows and insulation to absorb and reflect solar radiation for heating and cooling.

Green Design Professionals

Green design professionals, also referred to as sustainable engineering or sustainable architecture, design the built environment to comply with the principles of social, economic, and ecological sustainability (Figures 17.3 and 17.4).

Sustainable engineering, in particular, is the process of designing or operating systems such that they use energy and resources sustainably, in other words, at a rate that does not compromise the natural environment, or the ability of future generations to meet their own needs. Sustainable engineering has several sub-specialties, including the following:

- Managing water supply
- Designing housing and shelter
- Energy systems development
- Transportation design
- Development of natural resources
- Planning projects to reduce environmental and social impacts
- Recommending the appropriate and innovative use of technology
- Improving industrial processes to eliminate waste and reduce consumption
- Providing medical care to those in need
- Restoring natural environments such as forests, lakes, streams, and wetlands

FIGURE 17.3 California academy of sciences in golden gate park, San Francisco, California, an example of green design. (From Public domain.)

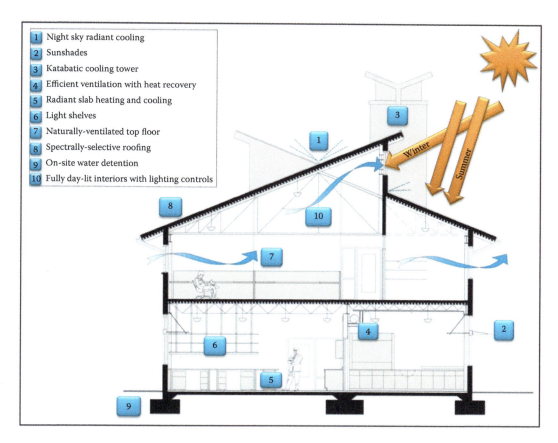

1 Night sky radiant cooling
2 Sunshades
3 Katabatic cooling tower
4 Efficient ventilation with heat recovery
5 Radiant slab heating and cooling
6 Light shelves
7 Naturally-ventilated top floor
8 Spectrally-selective roofing
9 On-site water detention
10 Fully day-lit interiors with lighting controls

FIGURE 17.4   Design for a sustainable living space. (Courtesy of Kyle Butler, Public domain.)

- Cleaning up polluted waste sites
- Industrial processing
- Sanitation and waste management
- Food production

Green engineering attempts to achieve four goals:

1. Waste reduction
2. Materials management
3. Pollution prevention
4. Product enhancement

Green engineering utilizes numerous ways to improve processes and products to make them more efficient from an environmental and sustainable standpoint. Every one of these approaches depends on viewing possible impacts in space and time. The ultimate design must consider both short- and long-term impacts. Those impacts beyond the near-term are the key to sustainable design (Figure 17.5). The effects may not manifest themselves for decades but still need to be anticipated. While practical applications may vary from project to project, there are common sustainable design principles that generally prevail in this field:

- *The use of low-impact materials*: Nontoxic, sustainably produced or recycled materials that require little energy to produce

FIGURE 17.5   Photovoltaic sunshade "SUDI" is an autonomous and mobile station that replenishes energy for electric vehicles using solar energy. (Courtesy of Tatmouss, Creative Commons, Toronto, CA.)

- *Design for reuse and recycling*: Systems designed to have a commercial "afterlife" (the ability to be repurposed eventually)
- *Designed with the total carbon footprint and life-cycle in mind*: Designed with the knowledge of the

full earth-impact (called a *whole-earth estimate*) of the building from inception to repurposing

- *Sustainable design standards*: Use of developed project design guidelines that have been developed and already proven to be sustainable
- *Biomimicry*: Enabling the constant reuse of materials in continuous closed cycles (the concept of redesigning industrial systems on biological lines)
- *Renewability*: Materials should come from local, sustainably-managed renewable sources to reduce fossil fuel involvement

Through the use of sustainable technologies that use less energy and fewer limited resources, engineers can design future settings that do not deplete natural resources, do not pollute the environment, and can be reused or recycled at the end of their useful life. As the public catches on to the green lifestyle, this career field will continue to increase in demand.

Wave Energy Producers

The ocean contains an abundance of perpetual motion, and devices called wave energy generators can capture it to produce power. In 2011, a 180-meter-long, 1,300-ton rig in Scotland's Orkney Islands became the first commercial-scale marine device to produce grid energy from offshore. It has a 750-kilowatt rating and currently generates enough power per year to meet the average electric needs of 500 homes.

Composed of five tube sections linked by joints that flex in two directions, waves bend the partially submerged tubes at the joints, which house hydraulic cylinders that resist the motion and pump fluid into pressurized accumulators to generate electricity. This is then transported ashore via an undersea cable. Though the potential of wave power is enormous, the industry is still in its infancy. Supporters are hopeful that it will catch on as a new technology and become a competitive option in the renewable energy arena (Figure 17.6).

FIGURE 17.6 Pelamis wave power device off the coast of Portugal. (Courtesy of Marine and Hydrokinetic Technologies Program. U.S. Department of Energy: Energy Efficiency and Renewable Energy.)

Wind Energy Workers

As of the end of 2015, the worldwide total cumulative installed electricity generation capacity from wind power amounted to 432,883 MW, an increase of 17 percent compared to the previous year. Global wind power installations increased by 63,330, 51,447, and 35,467 MW in 2015, 2014, and 2013, respectively.

Since 2010 more than half of all new wind power was added outside of the traditional markets of Europe and North America, mainly driven by the continuing boom in China and India. At the end of 2015, China had 145 GW of wind power installed. In 2015, China installed close to half of the world's added wind power capacity (Global Wind Energy Council, 2015).

Several countries have dedicated a relatively high percentage of their energy sources into wind energy, such as Denmark, with 39 percent of its stationary electricity production tied to wind energy. Portugal currently has 18 percent of its power from wind and Spain has 16 percent. Ireland claims 14 percent; Germany 9 percent. As of 2011, 83 countries worldwide were utilizing wind power on a commercial basis. Wind power's share of worldwide electricity usage at the end of 2014 was 3.1 percent (Ren21, 2015). This rate can only continue to increase, making this profession promising (Figure 17.7).

Biofuel Jobs

The Energy Independence and Security Act (2007) mandates that renewable fuel must increase from 34 billion liters of U.S. motor fuel in 2008 to 136 billion liters by 2022. The policy makes rapid growth in biofuel jobs highly likely, including the construction and operation of ethanol and other biofuel plants, production of feedstock, and creation of delivery infrastructure (Figures 17.8 and 17.9).

SAMPLE LIST OF GREEN CAREERS

There are jobs and career types continually being added to the field of green careers. The list below is a sampling of typical jobs that can currently be found in the fields of renewable energy, sustainable living, green careers, environmental stewardship, and green industry. Explore http://www.renewableenergyjobs.com/ for more ideas and options. As technology continues to develop, new career fields will continue to be added.

-A-
Account Manager/Executive
Account Manager–Building
Automation Systems Sales
Advisor for Management and Energy
Agency Energy and Environmental Manager
Application Engineer/Technician
Architect
Auditor
Automated Controls Technician

-B-
Building Automation Systems Engineer
Building Consultant/Engineer

FIGURE 17.7 Red Hills wind farm in Elk City, Oklahoma, covers 20 square kilometers. This wind farm production will offset 294,000 tons of CO_2 emissions annually and provide enough energy for 40,000 homes. This project is also a good example of the cohabitation of cattle grazing land and wind power production. The towers themselves are built in three sections: the base section, which is 20 meters tal, and the top section, or spike, which is 30.5 meters. The blades are each 36 meters long. (Courtesy of Todd Spink, NREL, Golden, CO.)

FIGURE 17.8 A biomass research farm operated by the State University of New York college of environmental science and forestry. A patchwork quilt of willow and poplar plots—different species of biomass are grown and tested for future energy potential. (Courtesy of National Renewable Energy Laboratory, Golden, CO.)

Building Control Specialist
Building Energy Engineer
Building Energy Analyst
Building Services Engineer
Business Development Manager
Business Services Engineer
Business Unit Energy Services

-C-
Campus Energy Manager
Carbon Consultant
Carbon Reduction Manager
Central Plant Engineering Manager
Certified Energy Manager
Certified Geothermal Designer
Chief of Energy Engineering
Certified Energy Auditor
Chief Electrical Engineer

FIGURE 17.9 Tree farms devoted to power generation. These hybrid poplar trees in Oregon are harvested for fiber and fuel. Once the trees are cut, they are chipped into smaller pieces and used to generate power. (Courtesy of Warren Gretz, National Renewable Energy Laboratory, Golden, CO.)

Chief Operations and Maintenance
Chiller Team Manager
Civil Engineer
Climate Change Consultant
Climate Initiative Program Director
Cogeneration Manager
Command Utility Energy Engineer
Commercial Engineering Manager
Commercial and Industrial Energy Services Engineer
Commercial Energy Management
Consultant
Commissioning Manager
Commissioning Engineer
Commissioning Technician
Commercial/Industrial Services Manager
Compressed Air Specialist
Conservation Engineer
Conservation Program Engineer
Construction Manager
Construction Manager/Designer
Construction Project Manager
Construction Services Manager
Construction/Facilities Manager
Consultant
Consulting Engineer
Controls and Commissioning Manager
Controls and Energy
Controls Engineer
Controls Manager
Controls Specialist
Coordinator of Environmental Services
Coordinator of Infrastructure
Coordinator of Energy Management
Coordinator of Residential Energy Programs
Corporate Director Engineering and
Construction
Corporate Director of Energy Management
Corporate Director of Energy and Environmental Services
Corporate Director Plant Operations
Corporate Energy Coordinator
Corporate Energy Director
Corporate Energy Management Leader
Corporate Energy Manager
Corporate Facilities/Energy manager
Corporate Facilities Manager
Corporate Facilities Engineer
Corporate Manager of Building Operations
Corporate Project Manager
Corporate Utilities Manager

-D-
Design Engineer
Design Build/Energy Engineer
Design Manager
Design/Energy Engineer
Development Engineer
Distributed Energy Manager

Distribution Standards Engineer
Demand-Side Management Coordinator
Demand-Side Management Engineer
Demand-Side Management P&E Specialist
Demand-Side Management Program Consultant

-E-
Electric Engineering Officer
Electric Meter Relay Technician
Electric Utility Analyst
Electric Utility Project Coordinator
Electrical & Energy Manager
Electrical & Instrument Engineer
Electrical and Controls Engineer
Electrical and Mechanical Engineer
Electrical Consultant Engineer
Electrical Coordinator
Electrical Design Engineer
Electrical Designer
Electrical and Automation Engineer
Electrical Engineer
Energy Engineer
Energy Manager
Energy and Sustainability Program Manager
Energy and Commissioning Engineer
Energy and Environment Engineer
Energy and Environmental Consultant
Energy and Environmental Manager
Energy and Sustainability Solutions Specialist
Energy and Utilities Manager
Energy and Water Resource Planner
Energy Advisor
Energy Analyst
Energy Auditor
Energy Balance Engineering
Energy Conservation Advisor
Energy Conservation Analyst
Energy Conservation Consultant
Energy Conservation Coordinator
Energy Conservation Engineer
Energy Conservation Manager
Energy Conservation Program Manager
Energy Conservation Technical Specialist
Energy Coordinator
Energy Education Specialist
Energy Efficiency Program Director/Engineer
Energy Efficiency Consultant
Energy Efficiency Coordinator
Energy Efficiency Engineer
Energy Efficiency Program Manager
Energy Efficiency Specialist
Energy Efficiency Technician Instructor
Energy Facility Manager
Energy Information Manager
Energy Management Analyst
Energy Management Consultant
Energy Management Engineer

Energy Management Quality Consultant
Energy Management Representative
Energy Management Specialist
Energy Management/Environmental System Consultant
Energy Manager
Energy Performance Analyst
Energy Plant Manager
Energy Procurement Manager
Energy Program Coordinator
Energy Program Engineer
Energy Program Manager
Energy Program Specialist
Energy Programming Technician
Energy Project Development
Energy Project Engineer
Energy Project Manager
Energy Projects Specialist
Energy Resource Conservation Specialist
Energy Resource Manager
Energy Sales Consultant
Energy Scientist
Energy Service Engineer
Energy Service Manager
Energy Service Specialist
Energy Services Account Executive
Energy Services Advisor
Energy Services Analyst
Energy Services Coordinator
Energy Services Program Consultant
Energy Services Project Developer
Energy Services Project Manager
Energy Services Representative
Energy Services Specialist
Energy Services Supervisor
Energy Solutions Analyst
Energy Solutions Consultant
Energy Solutions Manager
Energy Specialist
Energy Supply Manager
Energy Systems Engineer
Energy Technician
Energy Trader
Energy/Environmental Consultant
Energy/Mechanical Manager
Energy/Resource Conservation Program Manager
Energy/Utilities Engineer
Energy/Utilities Manager
Engineer
Engineering and Environmental Services
Engineering Analyst
Engineering and Design Supervisor
Engineering Consultant
Engineering Development Director
Engineering Director
Engineering Manager
Engineering Specialist
Engineering Technician

Engineer–Load Management
Environmental Health, Safety and Security Manager
Environmental Solutions Manager
Environmental and Utilities Manager
Environmental and Energy Manager
Environmental Consultant
Environmental Director
Environmental Engineer
Environmental Health and Safety Mgr.
Environmental Manager
Environmental Program Manager
Environmental Project Manager
Environmental Specialist
Equipment Engineer

-F-
Facilities and Compliance Analyst
Facilities and Maintenance Manager
Facilities and Construction Manager
Facilities Consultant
Facilities Coordinator
Facilities Design and Construction Supervisor
Facilities Design Engineer
Facilities Electrical Engineer
Facilities Electrical Systems Coordinator
Facilities Energy Engineer
Facilities Engineer
Facilities Environmental Manager
Facilities HVAC Engineer
Facilities Manager
Facilities Operations Manager
Facilities Performance Analyst
Facilities Plant Engineer
Facilities Project Manager
Facilities Systems Specialist
Facility Energy Conservation Manager
Field Engineer
Fuel Analyst

-G-
Gas and Energy Coordinator
Gas Management Service Analyst
Gas Supply and Technology Engineer
General Engineer
Geo-Exchange Designer
Geothermal Operations Manager
Geothermal Designer
Global Commodity Manager-Energy
Global Director Energy and Sustainability
Global Director–Energy and Plant Services
Global Energy and Climate Change Program
Global Energy Director
Global Energy Manager
Global Energy Program Manager
Global Energy Solutions Manager
Global Environmental Engineer
Global Energy Manager

Green Building Engineer
Greenhouse Gas Lead Assessor

-H-
Health and Safety Manager
Health, Safety, and Environmental Manager
HVAC Design Engineer
HVAC Designer
HVAC Engineer
HVAC Facilities Manager
HVAC Services Manager
HVAC Systems Engineer
HVAC Technician
Hydro Geologist

-I-
Indoor Environmental Consultant
Industrial Engineer
Industrial Gas Engineer
Infrastructure Engineer
Installation Energy Manager
Installation Manager
Intelligent Building Specialist

-L-
LEED AP
LEED Performance Assurance Engineer
Lenders Engineer
Lighting Consultant
Lighting Design Specialist
Lighting Engineer
Load Analyst

-M-
Maintenance Engineer
Maintenance Manager
Maintenance Planner
Manager of Energy Initiatives
Measurement and Verification Engineer
Measurement and Verification Specialist
Mechanical Design Engineer
Mechanical Designer
Mechanical Engineer
Mechanical Operations Manager
Mechanical Project Engineer
Mechanical Systems Engineer
Mechanical Utilities Engineer

-N-
National Energy Manager
National Energy Services Manager
National Energy Solutions Specialist

-O-
Operations and Energy Manager
Operations and Maintenance Engineer
Operations Analyst

Operations Engineer
Operations Manager
Owners Engineer

-P-
Performance Assurance Consultant
Performance Assurance Engineer
Performance Assurance Specialist
Performance Contract Engineer
Performance Contracting Energy Engineer
Performance Engineer
Performance Manager
Plant Engineer
Plant Manager
Pollution Prevention Engineer
Power Application Engineer
Power Management Applications Engineer
Power Optimization Engineer
Power Plant Manager
Power Quality Engineer
Power Quality Specialist
Power Resources Manager
Power Systems Engineer
Power Utilization Engineer
Power, Energy and Environmental Market Manager
Procurement Analyst
Procurement Manager (Electric and Gas) Product
    Manager
Product Manager—Energy
Production Engineer
Professional Engineer
Program Analyst
Program Director
Program Director—Energy
Program Manager
Program Manager—Corporate Energy Services
Program Manager—Energy Services
Program Manager—Renewable Energy
Project Analyst
Project Coordinator
Project Development Engineer
Project Developer
Project Director
Project Energy Engineer
Project Energy Manager
Project Engineer
Project Manager
Purchasing Director
Purchasing Manager
Purchasing Specialist

-Q-
Quality Control Manager Quality Engineer

-R-
Refrigeration Engineer
Reliability Engineer

Renewable Energy Business Manager
Renewable Energy Consultant
Resource Conservation Engineer
Resource Conservation Manager
Resource Efficiency Manager
Retro-Commissioning Engineer

-S-
Safety and Environmental Coordinator
Shipboard Resource Efficiency Manager
Smart Grid Engineer
Solar Energy Engineer
Solar Energy Systems Designer
Solution Design Specialist
Solution Development Engineer
Staff Engineer
Staff Research Engineer
State Energy Manager
Storm water Planning and Engineer
Supervisor
Strategic Energy Analyst
Surveillance Coordinator, Work Control
Sustainable Development Professional
Sustainability Coordinator
Sustainability Engineer
Sustainability Leader
Sustainability Manager
Sustainable Building Analyst
Sustainable Building Technologist
Sustainable Design Specialist
Sustainable Development Coordinator
Sustainable Solutions Engineer
System Applications Engineer
System Control Operator
Systems Analyst
Systems Engineer

-T-
Technical Advisor
Technical Analyst
Technical Consultant
Technical Director
Technical Engineer
Technical Facilities Assistant
Technical Manager
Technical Solutions Engineer
Technical Specialist
Test and Balance Engineer

-U-
Urban Planning
Utilities and Energy Director
Utilities Engineer
Utilities Manager
Utility Analyst
Utility Conservation Representative

-W-
Waste Reduction Consultant
Wastewater Engineer
Water and Energy Management Specialist
Wind Power Engineer

TURNING LEISURE TIME INTO GREEN TIME

Next, we will touch on how to turn leisure time into green time. Instead of the expected travel destination approach, we will take a bit different twist and focus on what seems to be transforming into the new adventure of the travel industry: the green experience of ecotourism. We will also focus on camping, hiking, and the green etiquette appreciated there.

Rather than taking a tour and being pampered, maybe taking on a new adventure and experiencing the life and culture of the destination might be the greener thing to try. A new approach and experience is often worth a lifetime of stories.

THE ECOTOURISM EXPERIENCE

You may have heard several new terms floating around lately and wondered what exactly each referred to: ecotourism, sustainable tourism, responsible tourism, nature-based tourism, and green tourism. Are they the same? If not, how do they relate? They definitely all sound green. Well, let us take a closer look and see how the travel industry defines them:

- Ecotourism is defined by the Ecotourism Society as "responsible travel to natural areas, which conserves the environment and improves the welfare of the local people." So the bottom line is that the visit must conserve and improve the place of destination.
- *Sustainable tourism*: This is any form of tourism that does not reduce the availability of resources and does not inhibit future travelers from enjoying the same experience. So what exactly is sustainable in the context of tourism? Here are some examples: If the presence of large numbers of tourists disturbs an animal's mating patterns so that there are fewer of that species in the future, then that visit was not sustainable. Taking a kayaking class on a free-flowing river where no fish species are disturbed in the process is an example of sustainable tourism. Big game hunting in Alaska is not sustainable. Whale watching in an area to the point where the whales stop visiting that spot is not sustainable.
- *Responsible tourism*: This is defined as tourism that operates in such a way as to minimize negative impacts on the environment. A wilderness camping trip using the Leave No Trace ethics would be considered responsible tourism, whereas dune buggy tours would not. The first example leaves the area the way it was initially found; the second example

disturbs the area so that the next visitor does not get the same quality of experience.

- *Nature-based tourism*: This is a more generic term for any activity or travel experience with a focus on nature. A cruise to Hawaii to see the whales would be a good example of this type of tourism. This type of trip is not necessarily environmentally sustainable or responsible; it just involves observing something nature-related.
- *Green tourism*: This is often used to describe any ecotourism or sustainable tourism activity and is thought of as any activity or facility operating in an environmentally friendly fashion. An example of this could be a lodge with composting toilets, a gray water system, or solar powered lighting. Obviously, there are various degrees of greenness. Overall, the goal is to promote an awareness of where resources are coming from and where wastes are going.

KEEPING CONSERVATION AND SUSTAINABILITY IN MIND

Now, with all these types of tourism defined, let us go back to the heart of the issue: lots of people traveling and lots of resources being stressed and used vs the need to protect, conserve, and respect those resources. Sounds like a juggling act; especially when a lot of people simply do not care about being green or understand the importance of it in the first place.

With that stated, this important fact cannot be minimized: as people become more mobile and tourism increases, associated activities must also be managed with conservation in mind to ensure sustainable landscapes. The key is to keep the landscape preserved for our enjoyment now and in the future for the enjoyment of others.

With millions of people traveling each year, tourism is a growing source of revenue for people living in areas that are rich in plant and animal life, and threatened with destruction. The double-edged sword is that many of these areas are in undeveloped countries, they have beautiful places to visit, and they desperately need the income. The other side of the sword is that tourism can lead to problems such as waste management, habitat destruction, and the displacement of local people and wildlife if it is not done sustainably (in a green manner).

So we are at a crossroads of sorts. The conservation-minded people of the world have a wonderful opportunity right now to help people in undeveloped countries with their economy by providing ecotourism, which, in turn, has the potential to provide incentives for education and conservation. This can bring the much-needed revenue to these economies, teach tourists about sustainability and the beauty of these areas, and also show the native inhabitants that they do not need to depend on land developers or deforesters as a source of income. And while tourists live sustainably at their vacation spot, they can participate in some once-in-a-lifetime experiences (Figure 17.10).

FIGURE 17.10   One example of ecotourism in West Bengal, India. The Jaldapara Wildlife Sanctuary is a protected park located at the foothills of the eastern Himalayas on the bank of the Torsa River. It was declared a sanctuary in 1941 to protect the vast variety of flora and fauna, particularly the endangered one-horned rhinos. (Courtesy of JKDs Creative Commons.)

ECO-HOTELS AND GUIDELINES

According to the Rainforest Alliance, many countries are now beginning to look at sustainable tourism as a preferred method of reducing negative impacts on their environment. Associated management includes the development of guidelines for sustainable tourism and providing training and information to those in the tourist industry on the fundamentals of environmentally sound management in order to obtain a healthy balance between tourism and nature.

As locations gear up to establish eco-hotels (also called green hotels), they must meet strict guidelines to qualify. In general, they must meet the following basic criteria:

- Be centered on the natural environment
- Be ecologically sustainable
- Contribute to conservation programs
- Be actively engaged in environmental training programs
- Incorporate cultural aspects into their business
- Provide a positive economic return to the community

Many people wonder what these eco-hotels look like—they may even have visions of a tent with a bucket of water outside the zip-up door—such as a place where they will really have to rough it. Not hardly. This typically is what they offer:

- Renewable energy sources, such as solar or wind energy
- Cleaning agents and laundry detergent that is nontoxic
- 100 percent organic cotton sheets, towels, and mattresses
- A strict, no-smoking rule
- Bulk organic soap and amenities rather than the individually wrapped personal packages common to mainstream hotels
- Guest room and hotel lobby recycling bins
- Towel and sheet reuse options
- Energy-efficient lighting
- Menus featuring organic and locally grown food
- No disposable items such as plates, eating utensils, etc.
- On-site transportation with green vehicles only
- Fresh-air exchange systems (flow-through air ventilation rather than standard air conditioning)
- Gray water recycling (reusing the kitchen, bath, and laundry water for garden and landscaping purposes)
- Newspaper recycling programs

In addition, when these hotels are constructed, the materials (such as wood, stone, etc.) are certified from sustainable sources, making them green from the construction phase onward.

THE REAL PURPOSE OF ECOTOURISM

Although many think ecotourism is just an adventure vacation—a way to mimic the TV show Survival, this is not the case. Ecotourism involves travel to fragile, pristine, and, usually, protected areas with the goal to be low-impact, small-scale destinations. The purpose is the defining key and includes the following:

- To educate the traveler
- To provide funds for direct ecological conservation of the destination
- To benefit the economic development and political empowerment of the local communities
- To foster respect for different cultures and for human rights

Behind these goals and philosophy, the net result is geared toward enabling future generations the same opportunities to partake in the same unspoiled destinations and experiences, promoting respect and appreciation for all humankind.

The goal of ecotourism is to foster a greater appreciation of the natural habitat. It also involves participating in this experience while promoting recycling activities, practicing energy efficiency, water conservation, and other environmental and socially responsible activities—kind of like green boot camp or college green living 101: the field version!

ROUGHING IT OUT ON THE LAND

Now let us talk about the "roughing it" experience on the land—the "being one with nature," such as camping, hiking,

backpacking, and so forth. An activity exploding in popularity in the United States for summer vacations is to go to national forests, parks, and other scenic destinations. Because of the subsequent stress of these lands due to the high volumes of visitors, it is critical that these areas are treated with extra care (sustainability) and we keep them in pristine condition. Because of this, treating the lands you briefly visit in a sustainable manner is deserving of mention.

Perhaps the best way to sum all this up is to refer to the Leave No Trace program set up by the Boy Scouts of America and sponsored by federal government agencies, such as the U.S. Forest Service, National Park Service, and Bureau of Land Management.

THE LEAVE NO TRACE PROGRAM

The Leave No Trace program, adopted by the federal government and private organizations, is designed to educate outdoor recreation enthusiasts and build awareness of the environment and land stewardship. Its goal is to avoid or minimize impacts to natural area resources and help create a positive recreational experience for all visitors.

This program is important because the United States's public lands are a finite resource with social and ecological values that are linked to the health of the land. Today, land managers face a constant struggle in their efforts to find an appropriate balance between programs designed to preserve the land's natural and cultural resources and provide high-quality recreational use.

The Leave No Trace educational program is designed to teach visitors low-impact care of the environment. If visitors follow the program's guidelines and act responsibly when using the land, then more direct regulations will not be necessary, such as restricting the number of people who can visit a particular area at a time or having to heavily police the public lands.

In summary, the Leave No Trace program stresses actions that maintain the beauty and integrity of the land. To follow the program, do the following before you go:

- Obtain information about area and use restrictions.
- Plan your trip for off season or nonholiday times. If this is not possible, go to less popular areas.
- Choose equipment in earth tone colors: blue, green, tan, etc.
- Repackage food in lightweight, burnable, or pack-out containers.

Follow these protocols on your way to your destination:

- Stay on designated trails.
- Do not cut across switchbacks.
- When traveling cross-country, hike in small groups and spread out.
- Do not get off muddy trails.
- Avoid hanging signs and ribbons or carving on trees to mark travel routes.
- When meeting horseback riders, step off on the lower side of the trail, stand still, and talk quietly.

It is also important to follow the Leave No Trace program while you are at your destination. At campsites, be sure you do the following:

- If you are in a high-use area, choose existing campsites.
- If you are in a remote area, choose sites that cannot be damaged by your stay.
- All campsites should be at least 75 paces, or 200 feet (61 meter) from water and trails.
- Hide your campsite from view.
- Do not dig trenches around tents.
- Avoid building camp structures. If temporary structures are built, dismantle completely before leaving.

If not tended to properly, campfires can mar pristine areas, spoiling the experience for others. You can minimize the potential effects of a campfire by doing the following:

- Use a lightweight gas stove rather than build a fire.
- In areas where fires are permitted, use existing fire rings.
- Do not build new fire rings.
- Do not build fires against large rocks.
- Learn and practice alternative fire-building methods that Leave No Trace.
- Use dead and down wood no larger than the size of your forearm.
- Do not break branches off trees.
- Put fires completely out (cold to the touch) before leaving.

Sanitation is important when enjoying the land. Be sure to do the following to keep nature clean:

- Deposit human waste and toilet paper in cat holes. Cat holes are 15 to 20.5 centimeter deep and should be located at least 75 paces from water or camp. Cover and disguise cat holes when finished.
- Wash dishes, clothes, and yourself away from natural water sources.
- Cover latrine and wash water holes thoroughly before breaking camp.
- Pick up all trash and pack it out (yours and others).

Part of leaving no trace is being courteous. Be sure to do the following:

- Avoid loud music, talking loudly, and other loud noises.
- Keep pets under control at all times. Better still, leave them home.
- Leave flowers, artifacts, and picturesque rocks and snags for others to enjoy.

Before you leave a site, take one last look at where you have been and do your best to Leave No Trace. (*Source: U.S. Bureau of Land Management, U.S. Department of Agriculture, the National Outdoor Leadership School, and the Boy Scouts of America. Use: Courtesy of Bureau of Land Management*).

Following this philosophy is becoming more critical each year as the visitor loads increase on the Nation's public lands. The three principal land-holding federal agencies manage a whopping 2.1 million square kilometers of public-accessible lands that need to be respected and protected (about 24 percent of the land mass of the United States). They include:

1. *The Bureau of land management*: 1 million square kilometers, or 13 percent of the United States
2. *The forest service*: 772,949.5 square kilometers, or 8 percent of the United States
3. *The national park service*: 34,195 square kilometers, or about 3 percent of the United States

The Park Service alone in 2010 received 281,303,769 visitors! Without sustainable land practices in place, can you imagine the stress to the landscape 5 years from now? What about 10 years from now? 20 years?

If you follow these simple rules and procedures, you really will Leave No Trace. And is not that what we all really expect to see when we go on a "back to nature," adventurous experience?

CONCLUSIONS

As you can see, there are many opportunities for growth and employment in the green economy and job market. As the environmental movement continues to grow and green living and employment practices become more common, opportunities in this sector should become even stronger. Innovations are being made at a rapid rate—and as they are, they come with a demand for job support.

Each new innovation requires support. That support is in the form of a cadre of new green jobs. As existing companies make the push to go green and move forward, they are looking for employees to fill those crossover positions. Even if someone's expertise is not science-related, it does not mean they are out of luck. Education, communication, business, and most other avenues of work all have jobs evolving from society's drive to become more eco-friendly. Science teachers and professors school the public at an early age about environmental well-being. Public health officials keep an eye out for health and environmental safety. Eco-friendly interior designers and architects create buildings and spaces that save energy without losing style. Housekeepers and dry-cleaners are transitioning from harsh chemicals and processes in favor of more energy- and air-friendly means of cleaning. Therefore, even if someone's interests and work history do not involve years of agricultural studies or water conservation, they will still be able to find a job allowing them to "go green." And it will only continue to get better.

REFERENCES

Bapna, M. and J. Talberth. April 5, 2011. What is a "Green Economy?" World Resources Institute. http://www.wri.org/blog/2011/04/qa-what-green-economy-0 (accessed May 22, 2017).

DiPasquale, C. and K. Gordon. September 7, 2011. Top ten reasons why green jobs are vital to our economy. *Economy.* https://www.americanprogress.org/issues/economy/news/2011/09/07/10333/top-10-reasons-why-green-jobs-are-vital-to-our-economy/ (accessed October 4, 2016).

Global Wind Energy Council. 2015. Global wind report: Annual market update. http://www.gwec.net/wp-content/uploads/vip/GWEC-Global-Wind-2015-Report_April-2016_22_04.pdf (accessed October 10, 2016).

National Geographic. 2016. 11 of the fastest growing green jobs. *National Geographic.* http://environment.nationalgeographic.com/environment/sustainable-earth/11-of-the-fastest-growing-green-jobs/#/rio-20-green-jobs-roof-top-garden_55050_600×450.jpg (accessed October 15, 2016).

Ren21. 2015. Renewables 2015: Global status report. Renewable energy policy network for the twenty first century. http://www.ren21.net/wp-content/uploads/2015/07/REN12-GSR2015_Onlinebook_low1.pdf (accessed October 15, 2016).

SUGGESTED READING

Cassio, J. 2009. *Green Careers: Choosing Work for a Sustainable Future.* Gabriola Island: New Society Publishers, p. 368.

Dakers, D. 2011. *Touring, Trekking, and Traveling Green.* Ottawa, Canada: Crabtree Publishing. Green-Collar Careers (Book 8), p. 64.

Deitche, S. M. 2010. *Green collar jobs: Careers for the 21st century.* Westport, CT: Praeger, p. 170.

Environmental Careers Organization. 2008. *The Eco Guide to Careers that Make a Difference.* Washington, D.C.: Island Press, p. 320.

Fennell, D. A. 2014. *Ecotourism,* 4th ed. London, NY: Routledge, p. 288.

Fletcher, R. 2014. *Romancing the Wild: Cultural Dimensions of Ecotourism.* Durham, NC: Duke University, p. 264.

Flint, A. May 19, 2015. Can regenerative design save the planet? *Atlantic City Lab.* http://www.citylab.com/design/2015/05/can-regenerative-design-save-the-planet/393626/ (accessed October 1, 2016).

18 Your Home and Backyard

OVERVIEW

In this chapter, we will take a look at how you can be a little greener at your home and in your backyard. You will be presented with many practical and functional ideas to get you inspired to take action. This chapter first takes a look at energy loss and keeping the temperature comfortable in your house without overburdening your wallet. It offers several practical suggestions on just how to do that. From there, it explores ways to go green in the kitchen and in keeping things clean and tidy without depending on chemicals. Next, it looks at the progress of the Energy Star program, its benefits, and the latest developments in fighting climate change. From there it goes to the backyard and a multitude of ways you can easily make it a little healthier as well. It also offers some practical advice on how to create your own compost and other practical applications.

INTRODUCTION

Going green at home is often one of the first places to make a change for a greener lifestyle. It is where we have the most control to make the most changes in our life, so it is usually the easiest place to do it. And the beauty of it is we can choose to make a series of small changes or we can make large changes all at once. We can caulk a few windows, add some insulation, replace some windows, or convert to solar power. Usually when people make positive changes like this and see the results, they keep going; adding changes as they can, continuing to improve their lifestyle and green it up. This chapter will provide you with many ideas on just how to make that possible.

DEVELOPMENT

One of the best areas to green up is your home. There are many choices we make each day at home that directly affect the environment. By making those choices greener, we can drastically cut our carbon footprint. These include choices in our energy use, how we use our water, how green we make our kitchen, how green we clean, and how eco-friendly we cook. So let us take a look at the home front.

Is Your Home an Energy Hog?

Not only is using a lot of electricity expensive, but it is a major source of GHG emissions due to the fact that most electricity is generated from coal-powered electrical generation. This means that any time the amount of electricity being used is cut back, the more the environment is being benefitted. The same applies to the use of natural gas when it is used to run the furnace and heat the water.

Wasted energy can originate from two basic sources:

- Waste you directly cause by your actions
- Waste that is indirectly caused by your house

You cause waste by not conserving energy. Almost every little gadgets and appliances in your home use some amount of electricity, from the digital clock on the stove and microwave to the TV to the water heater to the cell phone charger to the refrigerator, and so on. You might be surprised if you go from room to room and note every little gadget that is plugged into the grid. All of these electricity sippers add up to a good-sized electric load, which shows up on your electric bill at the end of the month. The pie chart illustrates where the typical energy usages occur in a home (Figure 18.1).

One of the most common problems of wasted energy are air leaks—small open areas in your home that allow cold air from outside to come into your home and heated air from inside your home to escape outside. Common areas where this can happen—areas around windows, air ducts, exterior vents—are shown in the illustration. Although each, individually, may not seem significant, together they can really add up (Figure 18.2).

According to the EPA, homes can cause twice the greenhouse gas (GHG) emissions of a car. One way to find out how much energy a home consumes, how to lessen a home's carbon footprint, and how energy and money can be saved is to have a home energy audit completed by a local utility or private company. An energy audit provides information about how much power a household uses and can supply specific strategies designed to help reduce energy consumption. It can be surprising how much energy a home uses and wastes. A home energy audit can help the members of an average household find simple ways to reduce their CO_2 emissions by 450 kilograms annually and lower utility bills at the same time.

You can have a thermal scan done of your home to detect areas where heat and air may be escaping during the winter or air conditioned air may be escaping during the summer, causing your home to use much more energy than necessary. In a thermal scan, areas are displayed in different

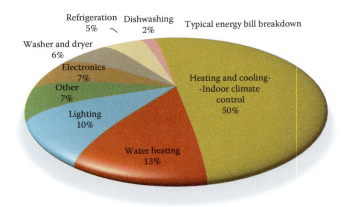

FIGURE 18.1 Typical energy use breakdown in an average home. (Courtesy of Kerr.)

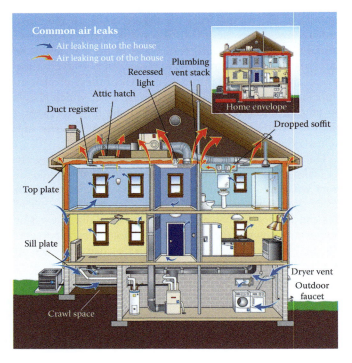

FIGURE 18.2 Common areas where air leaks can occur in a home, allowing cold air in and warm air to escape. (From EPA, Washington, DC.)

FIGURE 18.3 Heat escaping from a house can be detected through the use of thermal infrared scanners in a branch of science called thermography. The yellow, orange, and red (brighter areas) indicate the places where the most energy is escaping. In these homes in London, it is obvious which fireplaces are being used (bright areas above chimneys). The bright areas around the windows and doors show where heat is also escaping. (Courtesy of Wildgoose Education, Ltd., Coalville, UK; From Gines, J. K., *Climate Management Issues: Economics, Sociology, and Politics*, CRC Press, Boca Raton, FL, 2012.)

THE IMPORTANCE OF INSULATION

One of the best ways to green up and beat the high-energy prices is by insulating your home and getting rid of all the drafty cold spots. With some good home insulation, most homeowners can save 30 percent or more on their home heating costs. And this applies to both old and new homes. One of the most important places to add insulation is the attic. The attic is a major source of home heat loss and is usually one of the easiest places to get into. A good rule of thumb is that if the attic does not contain at least 30.5 centimeters of insulation, then more should be added on top of it. It is not difficult to lay paperless rolls of insulation on top of existing insulation such as cellulose or vermiculate insulation. These types of insulation compact over time, which reduces their efficiency.

The test of a well-insulated attic is that it should be cold inside. For example, icicles hanging from the eaves of the house is a sign that there is warm air in the attic (snow is melting and the runoff is forming as icicles), which is an indicator that the attic needs to be insulated better. If the insulation is doing its job, the attic will be cold (it is keeping the warm air inside the house where it belongs instead) and the snow on the roof will not be melting to form icicles.

The next thing to consider is how much additional insulation is necessary. That depends on what the house is made of and other factors. One of the biggest factors is the climate and how cold or warm it gets. The Energy Star program has developed

colors based on the temperature they are emitting. The darker blues and purples denote cooler areas; the brighter reds, yellows, and oranges signify the warmer areas. The hotspots (bright colors) are areas where energy is escaping (Figure 18.3). During the winter, this is especially critical because these are the areas where the expensive heated air is escaping and wasting energy. When areas of energy loss are detected—especially common in older homes—measures can be taken to fix the problems, such as adding better or thicker insulation (for energy loss from the roof) or more energy-efficient windows and doors.

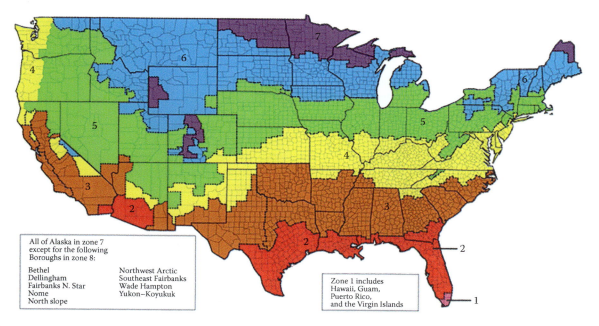

All of Alaska in zone 7
except for the following
Boroughs in zone 8:

Bethel   Northwest Arctic
Dellingham   Southeast Fairbanks
Fairbanks N. Star   Wade Hampton
Nome   Yukon–Koyukuk
North slope

Zone 1 includes
Hawaii, Guam,
Puerto Rico,
and the Virgin Islands

FIGURE 18.4 Recommended insulation levels for retrofitting existing wood-framed buildings. (From EnergyStar, DOE, Washington, DC.)

a guide that recommends the total amount of insulation needed based on an R-value. The R-value is a measurement of the amount of heat that can pass through the insulation. The higher the R-value, the better the thermal performance of the insulation. The map and table illustrate what levels of insulation are cost-effective for different climates in the United States (Figure 18.4).

Zone	Add Insulation to Attic		Floor
	Uninsulated Attic	Existing 3–4 inches (7.5–10 centimeters) of Insulation	
1	R30 to R49	R25 to R30	R13
2	R30 to R60	R25 to R38	R13 to R19
3	R30 to R60	R25 to R38	R19 to R25
4	R38 to R60	R38	R25 to R30
5 to 8	R49 to R60	R38 to R49	R25 to R30

How to Cut Back on Energy Use

Fortunately, there are several easy ways to cut back on energy use by greening up—and you will cut back on expenses at the same time. One of the easiest ways, as we mentioned before, is to just unplug the items that you are not using: unplug the TV when it is off, unplug your cell phone charger when it is not being used; and so forth. Again, using power strips reduces the number of cords you need to unplug.

Weather Strip for Windows and Doors

The edges of the windows and doors are another highly vulnerable spot for losing heated or air-conditioned air. To fix this problem, weather stripping material can be used to inexpensively and effectively minimize the loss of heat from a home. And even better—weather stripping is easy to install.

Expansion Foam

Any opening, no matter how small, allows heat to escape, and that can make a real difference in both your level of comfort and energy bills. These tiny spaces are behind electric switch plates on exterior walls, and where electrical wires, cables, Internet lines, and plumbing pipes come into the house. It can make a significant difference to fill in these small spaces with expansion foam.

Basements and Crawlspaces

Basements—especially unfinished basements—and crawlspaces can be notorious for letting out heated air and letting in cold air. Paper-faced insulation can be used on exposed ceiling joists in unfinished basements. Insulation may need to be stuffed where the basement walls meet the ceiling. It is also wise to check for small openings around the foundation. Here, expansion foam insulation can be pumped into eliminate drafts.

Window Coverings

One of the easiest ways to keep heat inside, especially at night, is to hang curtains or blinds on all windows and keep them closed. Just remember to open them during the day to let the sunshine in to naturally warm the house. The thicker the curtain, the better it will be at insulating. Many people purchase the double thickness theater curtains because they insulate even better.

In the summer, draw closed curtains and blinds during the day to block the sun and the summer heat so the air conditioner does not have to work any harder than necessary.

Turn Down the Hot Water Heater

Heating water can represent a huge portion of your entire electric bill—up to 20 percent. To green up and save here, turn your water heater temperature to no higher than 54.5°C. This still makes it comfortable for showers but costs less to run. Turning down the temperature of your water heater is safer for you and your family because it lowers the likelihood of scalding. Additional energy savings can be obtained if you install a thermal wrap for your water heater.

If your water heater storage tank has a low R-value, then adding insulation to it can reduce standby heat losses by 25 to 45 percent. Insulating the water storage tank translates into a savings of about 4 to 9 percent in water heating costs!

If you are not sure of your water heater tank's R-value, touch it. A tank that is warm to the touch needs additional insulation. Insulating your storage water heater tank is fairly simple and inexpensive, and it will pay for itself in about a year. Inexpensive precut jackets or blankets are available. Choose one with an insulating value of at least R-8. Some utilities sell them at low prices, offer rebates, and even install them at low or no cost.

Tankless Water Heaters

Another option is to install a tankless water heater. Tankless water heaters, also known as demand-type or instantaneous water heaters, provide hot water only as it is needed. They do not produce the standby energy losses associated with storage water heaters, which can save you money. As their name implies, tankless water heaters heat water directly without the use of a storage tank. When a hot water tap is turned on, cold water travels through a pipe into the unit. Either a gas burner or an electric element heats the water, delivering a constant supply of hot water. You do not need to wait for a storage tank to fill up with enough hot water. A tankless water heater's output limits the flow rate, however.

For homes that use 41 gallons or less of hot water daily, demand for water heaters can be 24 to 34 percent more energy efficient than conventional storage tank water heaters. They can be 8 to 14 percent more energy efficient for homes that use a lot of hot water—around 325 liters per day. Energy Star estimates that a typical family can save $100 or more per year with an Energy Star qualified tankless water heater.

The initial cost of a tankless water heater is greater than that of a conventional storage water heater, but tankless water heaters will typically last longer and have lower operating and energy costs, which could offset its higher purchase price. Most tankless water heaters have a life expectancy of more than 20 years.

Digital Thermostats

Programmable thermostats are a great way to save on your energy bill because they can be programmed to automatically reduce or increase the temperature (depending on the season) late at night when you are sleeping or not at home. You can then program them to bring the house's temperature back to comfortable levels in the morning or just before you get back home.

Home Improvement Tips

Here are some tips from the website CNN Living you can implement during home improvement projects:

- *Unplug your power tools:* Take an inventory of all your power tools and determine which ones you use a lot, then unplug all the rest. According to CNN Living, most of the cordless tools have nickel cadmium (NiCad) batteries, which will hold some charge for up to a year. They may lose up to 15 to 20 percent of their charge each month, but they take only a couple of hours to charge up again.

  Newer tools with lithium ion batteries lose only 2 to 5 percent of their charge each month, so they will still be ready to use even if they have not been charged for a long time.
- *Spread the sawdust around:* Take the superfine shavings gathered by your dust collection system, wet them down, and push them around with a stiff broom to sweep the concrete floor of your workshop or garage. This little trick works as well as an energy-guzzling shop vac but does not cloud the air.
- *Eat your leftover takeout:* Finish your partially filled containers of takeout food and hold onto those plastic containers with snap-on lids. They cannot be recycled in most municipal waste systems, so use them to organize your nails, screws, and leftover paints. Their tight seal helps preserve solvents and their see-through containers stack neatly and display the contents conveniently, helping you stay organized in your shop.

Replacing Older Appliances

When your budget allows, a great way to go green and save money and energy is to replace your older appliances and equipment: furnaces, hot water heaters, washers, dryers, refrigerators, dishwashers, and so on. A good rule of thumb is that appliances older than five years are not as efficient as new models.

With the big push to help the environment by going green and the major improvements in technology, appliances and equipment today save a significant amount of energy. For example, there are many models of water heaters available that are Energy Star certified. The tankless, on demand, water heaters offer the greatest energy efficiency. What makes these great is that unlike the traditional water heater, they do not store and heat water all the time—they heat it only as is needed. What this translates to is a double bonus: you never run out of hot water, and you use less energy.

And remember when you are replacing major items to always look for the Energy Star logo. That way, you know it is energy efficient and that it typically exceeds federal energy-efficiency guidelines. On top of that, you may even be eligible for some of those federal or state rebates or tax credits, making your wise decision even wiser!

Weatherizing Measures

In addition to insulating, there are the other measures to keep in mind that can also make a big difference. These include caulking, installing storm windows and doors, and placing draft guards in front of doors. And after the long summer months, if you have a window air conditioner, do not forget to weatherize it.

One of the best ways to weatherize is take a tour inside your house on a windy day. Sometimes, you can feel the drafty places when you walk by them. For the more subtle leaks, light a candle and hold it up to the doors and windows so that any tell-tale drafts will make the flame flicker—then weatherize the area.

Swap Out the Light Bulbs

Switch your light bulbs from the old-fashioned incandescent style to the new compact fluorescent bulbs (CFLs). They may cost a little more, but they last a lot longer and they light a room with only a quarter of the energy that the old-style bulbs required. In January 2014, the incandescent bulbs were no longer manufactured, but existing stock continued to be available for purchase until supplies were sold out; therefore, some areas do still have them available. Lighting often accounts for up to 10 percent of the overall energy use in a typical home, so switching the bulbs can really make a difference in the long run.

Change the Furnace Filter

This is one of the easiest things to overlook, but it is a simple task that can make a significant difference. Changing the filter reduces the load on the heating and cooling system, making it more energy efficient.

When trying to green up your home's energy consumption, one of the best things you can do is contact your local utility company and ask for an assessment. Many utilities offer free home energy audits in an effort to help educate their customers on ways to conserve energy.

Greening Up the Kitchen

One room that is easy to be eco-friendly in is the kitchen. If your family is like many others, you tend to spend a lot of time in the kitchen, so it makes sense that this is a great place to employ some eco-friendly skills.

At the Store

You can begin reducing your kitchen carbon footprint from the moment you enter the grocery store. The types of food you buy will determine how environmentally friendly your meals will be. You can make a significant impact by buying organically grown foods. Remember, this means that they are grown without the use of chemical fertilizers and pesticides. It also means that they do not contain any hormones, antibiotics, artificial ingredients, genetically engineered ingredients, and they have not been irradiated. The organically grown label gives you the peace of mind that the food's production helped the environment by fostering healthy soils and a diverse ecosystem. There is a double bonus here, not only are you keeping the chemicals away from your family but you are also keeping them away from the environment.

It greatly reduces the carbon footprint when you buy locally produced items. The bulk of the produce sold in the United States travels about 2,414 kilometers. If you buy locally, you eliminate this need for travel, thereby eliminating the GHGs produced to get the product to its faraway destination.

Another easy trick is to carry reusable shopping bags with you. The kind made out of durable canvas lasts for years—saving countless plastic or paper bags over months of shopping. And even better: many businesses give you a rebate of 5¢ per bag (Figure 18.5).

FIGURE 18.5 LED bulbs provide a convenient way to go green. (Courtesy of Kerr.)

Try to select items that come available in bulk and in as little packaging as possible. This causes less waste; and if you can reuse containers, that is even better. Also, for things you use on a regular basis, such as laundry detergent, you also help the environment when you buy the large sizes. Instead of buying a storage container over and over, just keep one container and purchase the detergent refills. Buying items in bulk also significantly cuts back on waste.

Once You Get to the Kitchen

In the kitchen, after all the groceries have been unpacked and stored, green up your cooking habits by using less energy. You can accomplish this by updating your appliances to more energy-efficient ones, and using less energy. One of the best ways to use less energy is to be efficient in the way you cook.

Have you ever noticed how much you open and close the refrigerator? Each time the door is opened, it lets precious cold air escape. Once the door is closed, the system must then work hard again to refrigerate the air. Another significant thing to note: when you are done cooking—whether on the stove or in the oven—turn off the appliance! Even letting it run an extra five minutes can add up on that energy meter. And speaking of the oven—if it is possible to cook with a microwave oven instead of a conventional oven—do that, too. Much less cooking time means much less energy use.

And for those very small cooking jobs you could try a toaster oven or a slow cooker. These are energy efficient and do not add a lot of extra heat to the kitchen to make you uncomfortable on a hot day.

Another energy-saving tip is to skip the oven preheating step that is often recommended. Most ovens today heat up in seconds, making it unnecessary to preheat ovens, as was necessary in the past.

If your oven has a self-clean option, it is best to clean it right after cooking in it because it is already warmed up, so you can save some energy there, as well.

Planning Ahead

Another energy saver is to plan your meals ahead and do the cooking once, eat twice strategy. Basically, this means cooking in larger portions so that you can have leftovers. That way, the big expenditure of energy to cook a meal is stretched over several meals, making both your time and the energy used more efficient. These meals include the traditional casseroles and pasta dishes that keep well refrigerated over the span of a few days.

Shopping for items only when they are in season is another important consideration. If you eat locally (avoiding those long transportation factors), you will probably also purchase only what is available in season. This is good because it means that growers did not have to use extra fertilizers or chemicals that could, in the long run, harm the environment. For some items, it also means they did not have to use additional heat and energy.

Here are some simple ways you can help prevent waste:

- Use regular washable cups, silverware, and plates instead of disposable ones.
- Use cloth napkins instead of paper ones.
- Use cloth diapers and cloth baby wipes.
- Start a compost heap using your kitchen scraps— it can be a great learning activity for kids.
- Donate items you have at home that you no longer use—clothes, furniture, toys, books—to thrift stores or charities so that others can use them.
- Recycle your old eyeglasses and cell phones.
- Use your own reusable cloth bags at the grocery store instead of paper or plastic ones.
- Purchase rechargeable batteries.
- Send digital holiday and birthday cards.

Avoid Disposables

Another biggie is to avoid using disposable items. This includes everything from paper plates, napkins, cups to plastic utensils, and anything else you use once and then throw away. Today's society is often referred to as the disposable society; but disposable and green do not go well together. It takes precious resources and energy to produce those goods, plus the energy to ship them to the store where you buy them.

When a product is used only once and then thrown away, it is wasteful. Try greening things up a bit and using standard dinnerware (yes, even if it has to be washed!), cloth napkins, silverware, and the like.

You will stay on track if you just keep these general rules of thumb in mind:

- Do not use too much of one item.
- Do not overdo it; use in moderation.
- Do not waste a commodity.
- Do not throw items away if they can be reused.

Preparing meals in a more environmentally friendly way is not hard to do—it just means making some adjustments on how you shop, cook, and use items. Once you get the hang of it and it becomes routine, eco-friendly cooking is a piece of cake.

THE BIG GREEN CLEAN

According to the New York Department of Environmental Conservation, there is an eco-friendly alternative to just about every household cleaning situation you will ever find yourself in. It is desirable to use green cleaning methods, of course, so that people and the environment do not have to be exposed to harmful or hazardous chemicals. The following table lists several alternative green cleaning options to spruce up the environment as well as our homes:

Household Hazardous Waste Disposal Symbols and Green Alternatives Chart Cleaners

Key to Disposal Codes

 Use according to the label directions

 Wrap and discard with other household trash

 Pour down drain slowly with plenty of water

 Save for a household hazardous waste collection program

 Dry out and then discard with other household trash

 Take to a recycling center

Product	Disposal Symbols	Green Alternatives
General purpose liquid		Vinegar or lemon juice diluted with water or three tablespoons washing soda in 1 quart water
Scouring powder		Baking soda, salt, or borax
Metal polishes		Use baking soda paste
Furniture polish		Dip cloth in olive, soybean, or raw linseed oil or mix 2 tablespoons vinegar and slowly stir into 1 quart water
Rug cleaners		Sprinkle baking soda on rug then vacuum; use club soda on stains
Spot removers		Clean spill quickly with club soda or use baking soda paste on stains
Toilet bowl cleaner		Flat cola, borax, or baking soda
Oven cleaner		Wipe up spills quickly; wash with baking soda using a scrubber or sprinkle with baking soda or salt, let sit, and then rinse

(Continued)

Home Maintenance

Product	Disposal Symbols	Green Alternatives
Latex paint		Use whitewash or milk paint
Oil-based paint		Use latex paint when possible
Paint thinner		Let paint settle out and then reuse
Glues/adhesives		Use white or yellow glue; let glue dry out if it is water based; save if solvent based
Drain opener		Prevent problems by using strainer and flushing pipes weekly with boiling water; use a plunger or snake; and put 1/2 cup each vinegar and baking soda down the drain, flushing 15 minutes later
Air freshener		Use flowers, herbs, or potpourri; place vinegar or vanilla in an open dish; or clean the source of the odor with baking soda
Degreasers		Detergents
Paint stripper		Sand or scrape paint

Auto Repair

Product	Disposal Symbols	Green Alternatives
Motor oil		None
Antifreeze		None
Gasoline		None; do not use as a cleaner
Vehicle batteries		None
Carburetor cleaner		None

(Continued)

Pesticides[a]

Product	Disposal Symbols	Green Alternatives
Bug sprays/insecticides		Handpick or trap pests, keep household clean and food covered, caulk or seal entryways
Weed killer		Maintain a healthy lawn by adjusting the pH to 6.5, mowing high with a sharp blade, and watering deeply when the soil is too dry
Flea killers		Use a flea comb, vacuum often, and wash pet's bedding
Wood preservatives		Use wood that is naturally resistant (cedar, honey locust, oak) and protect from dampness and insects; look for recycled plastic lumber
Disinfectant		Borax or pine oil with soap
Insect repellant		Do not wear scented products outdoors; burn citronella candles, punk, or incense

Hobby/Miscellaneous

Product	Disposal Symbols	Green Alternatives
Photographic chemicals		None
Artist paints		Use water-based products whenever possible
Swimming pool chemicals		None
Household batteries		Use rechargeable batteries
Aerosols		Use pump or liquid; if not hazardous, wrap it up and discard with other household trash; save for a household hazardous waste collection program if it contains hazardous materials
Mothballs		Keep garments clean; wrap in linen or seal in paper packages or cardboard boxes; use cedar chips; kill eggs by running dry garment through a warm dryer

Source: New York Department of Environmental Conservation.

[a] Do not use up pesticide products that have been banned or restricted.

Ultimately, the greenest you can get your home is to be completely self-sufficient. But that is difficult to obtain, and, quite simply, may not be for you. And that is all right. As long as we are striving to do the best we can with what we have, we are heading in the right direction.

THE ENERGY STAR PROGRAM

The Energy Star program has been an overwhelming success in the United States since its beginning in 1992. Established by the EPA for energy-efficient computers, the program has grown to encompass more than thirty-five different product categories for homes and offices—and it is still growing.

Realized Benefits

One of the largest benefits of the program is that using energy more efficiently avoids emissions from power plants, avoids the need to construct additional new power plants, and reduces energy bills. Because of this, the EPA has determined significant benefits have already been realized. These are just a few of the program's achievements:

- In one year, the Energy Star program prevented GHG emissions equivalent to those from 14 million vehicles and avoided using the power that fifty 300-MW power plants would have produced, while saving more than $7 billion.
- Since its beginning, thousands of organizations have joined forces and partnered with the federal government in order to protect the environment through energy efficiency.
- Americans have purchased more than one billion Energy Star—qualified products.
- Today, nearly half the American public recognizes Energy Star and is aware of the program.
- Thousands of buildings have already undergone effective energy-improvement projects.
- More than 1,100 buildings in the United States have earned the Energy Star label for superior energy performance.

Fighting Climate Change

As environmental awareness continues to increase, the popularity of Energy Star products continues to grow. As the program gains momentum and more Energy Star products are put into use, it helps the fight against climate change. According to the EPA, the Energy Star program has dramatically increased public use and preference of energy-efficient products and practices and is expected to continue to do so in the future.

The Superior Energy Management Program

As part of its program, the EPA has put in place what it calls its superior energy management criteria, which to date has proven highly successful. It offers the Energy Star partnership to organizations of all types and sizes.

As part of the program, senior-level executives make a commitment to the superior energy management of their buildings or facilities. This top-level organizational commitment has proven to be the catalyst for energy efficiency investments in many of the most successful partner organizations.

Almost 12,000 organizations have now partnered with EPA in the pursuit of superior energy management. Partners include the following:

- More than 425 public organizations such as state and local governments
- Schools and universities
- More than 880 businesses across the commercial and industrial sectors
- More than 8,000 small businesses

The EPA will continue to forge partnerships across the commercial and industrial sectors to create and ensure energy efficiency at the top management levels and to facilitate the development of best practices and information sharing. The EPA has already been able to help commercial real estate, public buildings, schools (K through 12), higher education, health care, hospitality, automobile manufacturing, cement manufacturing, wet corn milling, and others (EPA, 2009).

What about the Future?

The future also looks strong for Energy Star. The EPA expects the program to keep successfully expanding. In 2015 alone, millions of consumers and more than 16,000 business partners invested in Energy Star certified products and began conserving energy. More than 300 million certified products were purchased, and 82,000 new homes earned the Energy Star rating, which brought the total number of certified new homes to over 1.6 million nationwide since 1992 when the program began. Also since the program began, 5.5 billion Energy Star certified products have been purchased. Savings on utility bills in 2015 equaled approximately $430 billion and GHG emissions were reduced by 2.7 billion metric tons (www.energystar.gov). Programs like this are what help educate the public on the realities of environmental health as it relates to energy usage and the negative effects it can have on the earth due to global warming (Figure 18.6).

There are ways to do your part for the environment outside your home, such as by doing the following:

- Limiting pesticide use
- Conserving water
- Using low-maintenance plants
- Reducing and recycling yard waste
- Using techniques that conserve energy
- Providing for the wildlife that visits your yard every day, even when you are not looking

THE SMART HOME

The eventual implementation of the Smart Grid will also affect homes, as they are one of the endpoints in the network.

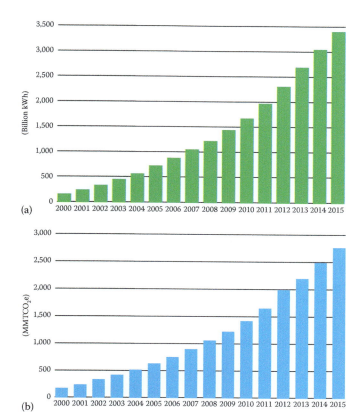

(a)

(b)

FIGURE 18.6 Energy Star benefits continue to grow. (a) Energy savings (2000–2015) and (b) GHG emission savings (2000–2015). (*Courtesy of EPA, Washington, DC.*)

Although it will not make homes look much different than they do now, there will be a lot going on behind the scenes. Even now, in several places across the United States, there is new equipment and software that is already using the emerging Smart Grid technologies to save energy, seek out the lowest rates, and contribute to the smooth, efficient, and effective functioning of the electric grid.

One of the key aspects that makes the Smart Grid so desirable is its ability to coordinate all the technologies that it is comprised of and enable them to all work in sync with each other in an interactive relationship between the grid operators, utilities, and individual homes. At the home level, computerized controls within the individual homes and appliances will be set up to respond to signals from the local energy provider to minimize their energy use at times when the power grid is under stress from high demand. The Smart Grid will also have the versatility to be able to shift some of its power use to times when power is available at a lower cost.

Smart Meters and Home Energy Management Systems

Smart meters will provide the Smart Grid with the interface between each home and the local energy provider. It will take the place of the old meter and operate digitally, allowing for automated and complex transfers of information between the home and energy provider. For example, smart meters will deliver signals from the energy provider that can help the homeowner cut energy costs. Smart meters will also provide utilities with greater information about how much electricity is being used throughout their service areas. This energy information that is sent to and from each home through each home's smart meter can then be run through a home Energy Management System (EMS), which will enable the homeowner to view it in a user-friendly format on either a computer or hand-held device. The home EMS enables each homeowner to track their energy use in detail to better save energy. For instance, it will be possible to see the energy impact of various appliances and electronic products by monitoring the EMS while switching the devices on and off.

The EMS will also allow the homeowner to monitor in real-time mode information and price signals from the utility company and create settings to automatically use power when prices are the lowest. The homeowner can then choose settings that allow specific appliances and equipment to turn off automatically when a large demand threatens to cause an outage. This will help the homeowner financially because it will avoid using electricity during peak demand rates. This then evens out the load in the neighborhood and prevents blackouts. Utility companies may even provide financial incentives to those who use that method to conserve energy.

Smart Appliances

Appliances within the home will be able to be added to the Smart Grid through the EMS. The EMS will provide the ability to turn on the heater or air conditioner from work just before the homeowner leaves the office to go home. It will even be possible to keep track of the energy use of specific appliances or equipment, such as seeing how much energy is saved with a new Energy Star appliance. Smart appliances will also be able to respond to signals from your energy provider and not use energy during peak energy demands—saving the homeowner money and keeping the power grid from blacking or browning out.

Home Power Generation

As consumers move toward home energy generation systems (generating their own power), the interactive capacity of the Smart Grid will become increasingly important. Rooftop solar electric systems and small wind turbines are already widely available, and people in rural areas could consider installing a small hydropower system on a nearby stream. Companies are also starting to introduce home fuel cell systems, which produce heat and power from natural gas.

The Smart Grid, with its system of controls and smart meters, will help to effectively connect all of these on-site generating systems to the grid, to provide data about their operation to utilities and owners, and to know what surplus energy is feeding back into the grid versus being used on-site. A potential feature of the Smart Grid will be to allow your community to use your solar array—and your neighbor's—to keep the lights on even when there is no power coming from a utility. A feature called *islanding*, it will allow a home to utilize power from "distributed resources," such as local rooftop solar, small hydropower, and wind projects, until utility workers can bring the grid back online.

MAKING YOUR BACKYARD ENVIRONMENTALLY FRIENDLY

Conservation practices in our own backyards can help increase food and shelter for birds and other wildlife, conserve water and improve water quality, control soil erosion, beautify the landscape, and inspire us to take better care of that special piece of land put under our care.

Urban revitalization is popular with inhabitants of cities and suburban areas. Many residents take pride in producing on their land, whether it is fruits and vegetables or beautiful flower gardens and landscaping. Many cities also strive to beautify the environment by creating large tracts of land devoted to parks, horticulture, and beautiful gardens. This appreciation of nature and the land is often referred to as backyard conservation and can be done by anyone to conserve and improve natural resources on the land and help the environment. And size does not matter here: whether you have acres of land in the country, an average-sized suburban yard, or a tiny plot within a busy city, you can help protect the environment and beautify your surroundings. Backyard conservation provides habitat for birds and other wildlife, healthier soil, erosion control, water conservation, and nutrient management. The following conservation practices are popular and easy for anyone to do.

Backyard Ponds and Wetlands

Backyard ponds and water gardens are not only beautiful but they also provide habitat for birds, butterflies, frogs, and fish as well. A backyard pond does not need to be big; it can be as small as one meter wide.

A pond is usually built where it can be seen from a deck or patio, and the landscaping around the pond can provide shelter for wildlife. The pond must be made with a protective liner to keep the water from seeping into the soil. Pumps and filters can be installed in a pond along with waterfalls. Different species of fish—such as the beautiful orange koi—can be added as well for additional habitat and aesthetic enjoyment. One of the benefits of keeping fish is that they keep insect populations under control (a form of integrated pest management) (Figure 18.7).

A properly located and maintained pond can reduce erosion and improve water quality. It also provides a good water source for wildlife, birds, and fish. It is common for people who live off the beaten path to have moose, elk, and deer drink from their pond.

Wetlands are areas where water covers the soil or keeps it saturated for at least 2–3 weeks during the growing season. They can be placed anywhere water accumulates faster than it drains away. Grasses, cattails, and other marshy vegetation grow in wetlands. How long the soil is kept wet at a time determines which plants can grow there. A mini-wetland in a backyard can provide many of the benefits that natural wetlands offer. A mini-wetland can also replace the natural function of the land that was in place before the ground was developed and houses were built. Backyard wetlands are advantageous because they temporarily store, filter, and clean runoff water from the house and lawn. They also provide habitat for many forms of life, such as toads, frogs, salamanders,

FIGURE 18.7 Having a backyard pond, such as this one in Cantril, Iowa, is like having your own little piece of nature. Ponds can be a rewarding way to practice conservation and provide habitat and biodiversity for the life around you. (Courtesy of Lynn Betts: U.S. Department of Agriculture/Natural Resource Conservation Service, USDA/NRCS, Des Moines, IA.)

butterflies, bees, and birds. Fortunately, many wetland plants do not require standing water to grow. Wetlands can be naturally built into a low or continually wet spot in the yard or an area can be converted into a wetland.

Backyard Mulching

Mulching is one of the simplest and most beneficial practices used in a garden. Mulch is simply a protective layer of material that is spread on top of the soil. Mulches can be organic such as grass clippings, straw, or bark chips, or they can consist of inorganic materials such as stones, brick chips, and plastic.

Mulching is beneficial in backyard gardens for many reasons. It protects the soil from erosion, reduces soil compaction from heavy rains, and prevents weed growth. It also helps conserve soil moisture so the garden does not have to be watered as often and helps the soil maintain a more even temperature. The mulch should not be more than 7.5 centimeters high around the plants or within 2.5 centimeters of tree trunks or plant stems.

Organic mulch improves the condition of the soil. When mulch decomposes in the soil, it provides the organic matter that helps keep the soil structure loose and aerated, not compacted. Because of this, plant roots grow better and water infiltrates the soil better, allowing the soil to hold more water. It is also a source of plant nutrients and provides a healthy environment for earthworms and other beneficial soil organisms.

Nutrient Management

There are many nutrients required by plants. Nitrogen, phosphorus, and potassium are required in the largest amounts. Nitrogen is responsible for lush vegetation, phosphorus for flowering and fruiting of plants, and potassium for improving resistance to disease. In addition, calcium, magnesium, and sulfur are also very important. These six nutrients are referred to as macronutrients. Soil also needs micronutrients—important nutrients required in only small amounts. These nutrients include zinc, iron, copper, and boron.

The level of nutrients can be a delicate balancing act. Not enough nutrients, and plants can't grow properly; too much and plants can be harmed and the extra nutrients can infiltrate and pollute groundwater or surface waters.

One way to manage nutrients in a backyard environment is to use a soil-testing kit. These kits test and determine the concentration of nutrients and soil pH levels. Once the results of the soil test are known, nutrients or other soil amendments such as lime can be added as needed. Compost can also be added to provide the essential nutrients.

The mow and grow concept is also important: do not cut the grass too short; mow more frequently instead. Be sure to leave the clippings on the grass to help you fertilize your lawn the natural way. And speaking of fertilizers—use natural, organic fertilizers, which as an additional bonus tend to last longer than the comparable bags of chemical-based fertilizers.

Your best bet is to grow native plants or plants from other areas that thrive in your climate. Overall, these plants require less maintenance on your part, adapt more easily to soil and weather changes, and are much more resilient.

Pest Management

Just like everywhere else, backyard gardens have pest problems. Yard pests include weeds, insects, and diseases such as fungi, bacteria, and viruses. Insects can damage plants in many ways. They can chew plant leaves and flowers. Some can suck out plant juices (such as aphids, mealy bugs, and mites); others cause damage by burrowing into stems, fruits, and leaves. Planting resistant varieties of plants can prevent many pest problems, such as disease. Rotating annual crops in a garden also prevents some diseases. Plants that have the correct amount of nutrients available in the soil are more resistant to disease, because they are in a healthier environment.

Backyard conservation methods can include some form of integrated pest management (IPM). IPM relies on several techniques to manage pests without the excessive use of chemical controls. IPM can include monitoring plants, determining tolerable injury levels, and applying appropriate pest management. IPM does not treat the entire area with a chemical; it looks at specific areas that need control and applies chemicals only in the affected areas, in the correct amount, and at the correct time. Spot spraying is an example of this type of management. It is cost-effective and limits damage to nontargeted species.

Management practices for weeds include hoeing, pulling, and mulching. Weeding is most important when plants are small and have a hard time competing for space and sunlight. Well-established plants can often tolerate competition from weeds.

FIGURE 18.8 Planting a backyard is an excellent way to learn about conservation. (Courtesy of Lynn Betts: USDA/NRCS, U.S. Department of Agriculture/Natural Resource Conservation Service, Des Moines, IA.)

Terracing

Terraces are used in backyards that consist of steep slopes, and they can create several backyard mini-gardens. They prevent erosion by shortening a long slope into a series of level steps, enabling heavy rains to soak into the soil rather than run off and erode the soil. This technique can transform a steep yard into a useable yard.

Planting Trees

When trees are planted in a backyard, they can provide valuable habitat for many types of wildlife. Trees can also help reduce home heating and cooling costs, help clean the air, and provide shelter from the wind (Figure 18.8).

Water Conservation

Wise use of water for gardens and lawns helps protect the environment and provides for optimum growing conditions. There are many practices available to promote backyard water conservation. Growing species of xerophytes (plants that can survive in dry conditions) is one way. Plants that use little water include yucca, California poppy, blanket flower, moss rose, juniper, sage, thyme, crocus, and primrose.

Besides using mulches and windbreaks, watering in the early morning is a beneficial conservation practice. If watering is done before the sun is intense enough to cause evaporation, more water will be utilized by the plant. Another method some gardeners use is drip, or trickle, irrigation—plastic tubing that supplies a slow, steady source of water from sprinkler heads suspended above the ground. Drip hoses can be designed to conserve water by delivering water only to the places where it is needed and in the amounts it is needed.

On farms, trickle irrigation is commonly used for high-value crops such as vegetables, grapes, and berries. High-efficiency irrigation systems for row crops use less energy to pump water. In addition, because they spray water downward, less water evaporates before it reaches the crop.

Growing native plants is also important because they naturally use less water than nonnative species. Native species have evolved under local conditions and usually have well-developed mechanisms for surviving extremes in the weather.

Wildlife Habitat

Habitat is a combination of food, water, shelter, and space that meets the needs of a species of wildlife. Even a small yard can be landscaped to attract birds, butterflies, small mammals, and insects. Trees, shrubs, and other plants provide shelter and food for wildlife. Nesting boxes, feeders, and watering sites can be added to improve the habitat (Figure 18.9).

Wildlife habitat covers the horizontal dimension (the size of the yard) as well as a vertical component from the ground to the treetops. Different wildlife species live in the different vertical zones, enabling several habitats to exist in a backyard setting. Trees and shrubs are also important sources of food for wildlife.

Birdhouses and other shelters are easy to add to a backyard habitat to increase the wildlife that visit the area. Plant species that birds enjoy can be grown to encourage them to visit. Clean, fresh water is also critical in a backyard habitat—birds, bats, butterflies, and other wildlife need a source of water. Water can be stored in a saucer, birdbath, or backyard pond.

Private landowners provide most of the habitat for wildlife on 70 percent of the land in the United States. Farmers accomplish this by installing grasses, trees, shrubs, riparian buffer strips, ponds, wetlands, and other types of wildlife habitat. Some farmers even plant food plots especially suited for wildlife, or put up structures that geese, ducks, and other birds can use as protected nests.

When private landowners care about conservation and ecosystems and strive to reduce, reuse, and recycle precious resources, they are taking a responsible and active role in land stewardship—and land stewardship is the key to a successful future.

XERISCAPING

It takes an enormous amount of water to successfully landscape a traditional yard in a dry, arid climate with lawns and flowerbeds. A green solution in this type of climate is xeriscaping—a low-maintenance landscape in harmony with the environment. By choosing plants that do not require heavy watering, you will conserve this precious resource, help the environment, and make your gardening chores much easier.

Going-Green Landscaping

Xeriscaping, also called going-green landscaping, is quickly becoming one of the most popular forms of landscaping because it is designed specifically to save water—a wise choice whether you live in a dry climate or not. Xeriscaping is a great way to create a water-wise garden that has continuous blooms on plants that naturally grow in the area from spring through fall.

These natural gardens also provide habitats for small wildlife, such as songbirds, hummingbirds, butterflies, bees, and beneficial insects. By grouping plants according to their water needs, using mulch and drought-tolerant plants, you can greatly cut back on your water usage. You will also be able to have a beautiful garden without using a lot of fertilizer or pesticides that could potentially contaminate runoff water. In fact, once these plants are established, they require much less water and do great with organic fertilizers (Figure 18.10).

The types of species you plant in your xeriscape are determined by where you live. The table lists some of the most common species that are frequently used.

FIGURE 18.9 Hummingbirds are frequent visitors in backyard gardens. Hanging out feeders is a great way to be a part of nature. (Courtesy of Kerr.)

FIGURE 18.10 Xeriscaping is a great option in arid climates. These gardens grow with little water. (Courtesy of U.S. Department of Agriculture, Des Moines, IA.)

Plant List for Xeriscapes

Latin Name	Common Name	Height/Width	Water	Hardiness	Sun Exposure	Seasonality	Flower Color	Flower Season
Achillea millefolium	Yarrow	2 ft/3 ft	Moderate	Up to 9,000	Sun-part shade	Evergreen	White	Spring-fall
Agastache cana	Double bubble mint	3 ft/2 ft	Moderate	Up to 6,000	Full sun	?	Pink	Summer-fall
Alyssoides utriculata	Bladderpod	2 ft/1 ft	Very low	Up to 8,500	Sun	?	Yellow	Spring
Anacyclus depressus	Atlas daisy	0.5 ft/1.5 ft	Low	Up to 10,000	Sun	?	White	Spring
Antennaria rosea	Pink puss toes	1 ft/1 ft	Very low	Up to 5,000?	Full sun	–	Pink	Summer
Aquilegia formosa	Red columbine	2 ft/2 ft	Moderate	Up to 11,000	Sun-full shade	–	Red	Spring-fall
Argemone platyceras	Prickly poppy	3 ft/2 ft	Very low	Hardy to 15°F	Sun	–	White	Spring-summer
Artemisia ludoviciana	White sage	2 ft/3 ft	Very low	Hardy to 0°F	Sun-part shade	Evergreen	Yellow/white	Summer-fall
Aurinia saxatile	Basket-of-gold	1.5 ft/1 ft	Low-mod	Up to 9,000	Full sun	–	Yellow	Spring
Baileya multiradiata	Desert marigold	1.5 ft/1.5 ft	Very low	Up to 5,000	Full sun-part shade	Evergreen	Yellow	Spring-fall
Berlandiera lyrata	Chocolate flower	2 ft/3 ft	Low-mod	Up to 5,000	Full sun	–	Yellow	Summer-fall
Calamintha grandiflora	Beautiful mint	2 ft/1.5 ft	Moderate	Up to 6,000	Sun-part shade		Pink	Summer
Callirhoe involucrata	Wine cup; poppy mallow	1 ft/3 ft	Low-mod	Up to 6,000	Full sun	–	Purple	Summer-fall
Campanula rotundifolia	Bluebells-of-Scotland	2 ft/3 ft	Low-mod	?	Full sun-full shade	–	Blue	Summer
Castilleja sp.	Indian paintbrush	1.5 ft/1 ft	Low	Hardy to -30	Sun	–	Red/orange	Spring
Centaurea cineraria	Dusty miller	1.5 ft/2 ft	Moderate	Up to 6,000	Sun-shade	Evergreen	Yellow	Summer
Centranthus ruber	Red valerian	3 ft/2.5 ft	Low	Up to 9,000	Sun-part shade	–	Red	Summer
Cerastium tomentosum	Snow-in-summer	1 ft/1.5 ft	Low-mod	Up to 7,500	Sun-part shade	–	White	Summer
Coreopsis lanceolata	Lanceleaf coreopsis	2 ft/2 ft	Moderate	Up to 5,500?	Sun	–	Yellow	Summer
Datura wrightii	Sacred datura	3 ft/6 ft	Low	Up to 6,500	Sun-part shade	Deciduous	White	Summer
Delosperma cooperi	Hardy pink ice plant	0.5/2 ft	Low	Up to 6,000	Sun	Evergreen	Pink	Spring-fall
Delosperma nubigenum	Hardy yellow ice plant	0.5 ft/1.5 ft	Low	Up to 7,500	Full sun	Evergreen	Yellow	Spring
Digitaria californica	Arizona cottontop	4 ft/5 ft	Very low	Up to 6,000	Sun	Deciduous	White	Fall
Echinacea purpurea	Purple coneflower	3 ft/2 ft	Moderate	Up to 5,500?	Sun	–	Purple	Summer
Erigeron divergens	Fleabane daisy	1 ft/1.5 ft	Very low	Up to 9,000	Sun-part shade	–	White	Summer
Fragaria vesca	California strawberry	0.5 ft/1 ft	Low	Up to 6,000	Shade-part shade	Evergreen	White	Summer?
Gaillardia aristata	Blanketflower; firewheel	3 ft/3 ft	Moderate	Up to 9,000	Sun	–	Red/yellow	Spring-fall
Glandularia gooddingii	Goodding's verbena	1.5 ft/1.5 ft	Low	Up to 6,000	Sun-part shade	–	Lavender	Summer-fall
Glandularia wrightii	Wright verbena	1 ft/1.5 ft	Very low	Hardy to 0°F	Sun	Evergreen	Pink/rose	Spring-fall
Gutierrezia sarothrae	Snakeweed	2 ft/2 ft	Very low	Up to 10,000	Sun	Deciduous	Yellow	Fall
Helianthus maximiliani	Maximilian sunflower	10 ft/6 ft	Moderate	Up to 8,000	Sun	–	Yellow	Fall
Hemerocallis species	Daylily	3 ft/2 ft	Low-mod	Up to 8,000	Sun-part shade	Evergreen	Yellow/various	Summer
Heuchera sanguinea	Coral bells; alum root	1.5 ft/1.5 ft	Low-high	Up to 9,000	Sun-part shade		Pink/white	Spring?
Iris hybrids	Bearded iris; Dutch iris	2 ft/1.5 ft	Low	Up to 8,500	Sun	Evergreen	Various	
Lessingia filaginifolia	California aster	3 ft/5 ft	Very low	Up to 8,500	Sun-part shade	–	Purple	Spring
Liatris punctata	Gayfeather	2 ft/1.5 ft	Low	Up to 8,000	Sun	–	Purple	Summer
Linum perenne var. lewisii	Blue flax	1.5 ft/1 ft	Low-mod	Up to 9,500	Sun-part shade	–	Blue	Spring
Melampodium leucanthum	Blackfoot daisy	1.5 ft/1.5 ft	Low	Up to 5,000	Sun-part shade	Evergreen	White	Summer
Mirabilis multiflora	Wild four o'clock	2 ft/4 ft	Very low	Up to 7,000	Sun-part shade	–	Purple	Summer
Nepeta x faassenii	Catmint	1.5 ft/3 ft	Low	Up to 6,000	Sun-part shade	–	Purple	Summer
Oenothera caespitosa	Tufted evening primrose	1 ft/2 ft	Very low	Up to 7,500	Sun-part shade	Evergreen	White/pink	Spring-fall

(Continued)

Plant List for Xeriscapes

Latin Name	Common Name	Height/ Width	Water	Hardiness	Sun Exposure	Seasonality	Flower Color	Flower Season
Oenothera missouriensis	Missouri evening primrose	1 ft/2 ft	Low	Up to 8,000	Sun-part shade	Evergreen?	Yellow	Summer
Oenothera speciosa	Mexican evening primrose	1.5 ft/2 ft	Mod	Hardy to 0°F	Sun-shade	Evergreen	Pink	Summer
Penstemon ambiguus	Bush penstemon	3 ft/3 ft	Low	Hardy to 15°F	Sun-part shade	Evergreen	White/pink	Summer
Penstemon eatonii	Eaton's penstemon	4 ft/3 ft	Very low	Up to 7,000?	Sun-part shade	Evergreen	Red	Spring-summer?
Penstemon palmeri	Palmer penstemon	3 ft/2 ft	Low	Up to 6,000	Sun-part shade	–	Pink	Spring-summer
Penstemon parryi	Parry's penstemon	4 ft/3 ft	Very low	Up to 5,000	Sun-part shade	Evergreen	Pink	Spring
Penstemon spectabilis	Royal penstemon	3 ft/2 ft	Very low	Up to 7,500	Sun-part shade	Evergreen	Purple	Spring
Penstemon thurberi	Thurber penstemon	1.5 ft/1 ft	Low	Up to 5,000	Sun-part shade	–	Lavender	Spring-summer
Perovskia atriplicifolia	Russian sage	5 ft/4 ft	Low	Up to 8,000	Sun	Deciduous	Lavender	Summer
Polygonum affine	Himalayan fleeceflower	1 ft/3 ft	Low-mod	Up to 8,000	Sun-part shade	Evergreen	Red/pink	Fall
Psilostrophe cooperi	Paperflower	1 ft/1.5 ft	Low	Hardy to 15°F	Sun	–	Yellow	Spring-summer
Pulsatilla vulgaris	European pasqueflower	1 ft/1 ft	Low-mod	Up to 9,500	Sun-part shade	–	Purple	Spring
Ratibida columnifera	Mexican hat; coneflower	2 ft/1.5 ft	Low	Up to 8,000	Sun		Red/yellow	Summer
Salvia farinacea	Mealycup sage	2 ft/1.5 ft	Moderate	Hardy to ?	Sun	Deciduous	Blue	Spring-summer
Salvia officinalis	Garden sage	2 ft/1.5 ft	Low-mod	Up to 9,000	Sun	Evergreen	Blue	Fall
Sedum spectabile	Showy stonecrop	2 ft/2 ft	Moderate	Up to 8,000	Sun-part shade	Evergreen	Pink/red/white	Summer-fall
Sedum spurium	Two-row stonecrop	0.5 ft/2 ft	Moderate	Up to 6,500	Sun-part shade	Evergreen	Red/yellow/ white	Summer?
Sempervivum species	Hens and chicks	0.5 / 1 ft	Moderate	Up to 8,000	Sun-part shade	Evergreen	Various	Summer
Senecio douglasii	Threadleaf groundsel	3 ft/2.5 ft	Very low	Hardy to 0°F	Sun	Evergreen	Yellow	Spring-fall
Senecio spartioides	Broom groundsel	2 ft/ 2 ft	Low	Up to 9,000	Sun	–	Yellow	Summer-fall
Sisyrinchium bellum	Blue-eyed grass	2 ft/2 ft	Very low	Up to 8,000	Sun-part sun	Deciduous	Blue	Spring
Stachys coccinea	Scarlet betony	3 ft/3 ft	Mod-high	Up to 8,000	Sun-shade	Evergreen	Red	Spring-fall
Stachys lanata	Lamb's ears	2 ft/3 ft	Low	Up to 8,000	Sun-part shade	Evergreen	Pink	Summer-fall
Tanacetum densum	Partridge feather	0.5 ft/1 ft	Low	Up to 6,500	Sun	Evergreen	Yellow	Summer
Teucrium laciniatum	Germander	3 ft/3 ft	Low	Up to 7,500	Sun-part shade	Evergreen	Purple	Summer
Thymus pseudolanuginosus	Woolly thyme	0.5 ft/1.5 ft	Low-mod	Up to 8,500	Sun-shade	Evergreen	Pink	Summer
Tritoma (Kniphofia) uvaria	Red hot poker	3 ft/2 ft	Moderate	Up to 7,500	Sun-part shade	Evergreen	Red/yellow	Summer
Verbena rigida	Sandpaper verbena	2 ft/ 3 ft	Moderate	Up to 4,500?	Sun	–	Purple	Summer
Veronica pectinata	Blue woolly speedwell	0.5 ft/2 ft	Low	Up to 8,500	Sun-part shade	Evergreen	Blue	Spring
Viguiera stenoloba	Skeletonleaf goldeneye	4 ft/4 ft	Very low	Up to 6,200	Sun-part shade	Deciduous	Yellow	Summer?
Zauschneria arizonica	Hummingbird trumpet	2 ft/4 ft	Low	Up to 6,000	Sun	–	Red	Fall
Zinnia acerosa	Dwarf white zinnia	0.5 ft/1.5 ft	Very low	Up to 5,000	Sun-part shade	Evergreen	White	Spring-fall
Zinnia grandiflora	Paper flower	1 ft/1 ft	Very low	Up to 6,000	Sun	–	Yellow	Summer-fall

Source: Arizona Cooperative Extension, 2005.

You can plant a drought-tolerant lawn, or alternative turf. Native grasses include buffalo grass and blue grama, which can survive with only a quarter of the water that typical grass varieties require. All grasses do require watering, so keeping those landscaped areas to a minimum is the wisest choice.

Xeriscaping is becoming more and more popular in regions not typically arid as climate patterns change and municipal water supplies become more restricted during drought years.

COMPOSTING—AN ECO-FRIENDLY WAY TO MANAGE ORGANIC WASTE

Composting is a practical and eco-friendly way to care for your vegetable garden, flowerbeds, and lawn. It may take a little bit of time to get the hang of it, but once you do and it becomes a habit, you will be surprised at all the uses and benefits it has.

Composting involves placing organic waste into a compost pile, where bacteria and other microorganisms break it down and turn it into dark and crumbly fertilizer (Figures 18.11 and 18.12). Compost is plant matter that has been decomposed and recycled as a fertilizer and soil amendment. A key ingredient in organic farming, it is easy to create—all that is involved is collecting scraps of organic matter, putting them in a bin, and waiting about a year for them to decompose into compost.

Of course, for the compost connoisseurs, there are more sophisticated processes involving closely monitored sequences of steps that require measured inputs of water, air, and carbon- and nitrogen-rich materials.

FIGURE 18.12 Compost can be combined with the soil and used as a fertilizer and soil amendment to enhance the soil's composition and fertility. (Courtesy of Normanack, USDA/NRCS.)

Soil amendments are components that can be added to soil to augment its natural properties. All soils are different, varying from sand to clay, from acidic to alkaline, and in nutrients. Plants also vary, requiring differing soil densities, pH, and amounts of nutrients. Each type of soil amendment is designed to remedy specific problems or tailor the soil to specific plants:

- Microbes: Enhances the soil food web
- Perlite: Improves aeration and drainage
- Organic peat moss: Acts as an overall soil conditioner and improves water retention
- Bone meal: Provides a good source of organic phosphorous
- Coco chips: Increases nutrient uptake
- Chicken manure: Supplies essential nutrients
- Agricultural gypsum: Raises the calcium content of the soil
- Soft rock phosphate: Adds essential minerals
- Organic cottonseed meal: Promotes green growth
- Organic fish meal: Provides a good source of nitrogen
- Glacial rock dust: Reintroduces trace minerals
- Ocean-floor greensand: Introduces potash
- Organic kelp meal: Enhances nitrogen, potash, and other minerals

These are just a few examples of soil amendments and what they are used for to increase the fertility, structure, and other properties of the soil, making it better to work with.

FIGURE 18.11 This is a compost bin in Escuela Barreales, Chili. Bins can range from very simple buckets like this one to custom-made, elaborate, home-built containers. (Courtesy of Diego Grez, USDA/NRCS.)

The Dirt on the Dirt

Composting may seem like a special process, but it is not. It goes on in nature all the time. If you have been hiking in the forest and noticed the mat of decomposing leaves, pine cones, and needles under your feet—that layer of dark, rich soil—then you have seen Mother Nature's compost in the making. When you make compost yourself, it just involves mixing yard and household organic waste in a pile or bin and providing the same conditions as Mother Nature does that encourage the concoction to decompose. The actual decomposition process is powered by millions of microscopic organisms—bacteria and fungi—that live inside the compost pile and continuously devour and recycle the material in it to eventually transform it into a rich organic fertilizer and soil amendment.

It helps to shred the plant matter, add water, and regularly stir the entire mixture, and expose it to air to help the decomposition process along. While this is being done, worms and fungi inside the mixture are further breaking it down. Aerobic bacteria are doing their part, as well. They are an important part of the chemical process and convert the processes into heat, carbon dioxide, and ammonium. Bacteria then take over the ammonium and further refine it into plant-nourishing nitrites and nitrates. And then, over time, it becomes compost.

There are many places on the Internet that provide step-by-step instructions on the finer points of setting up a compost operation, but basically, it boils down to a few simple steps:

1. Select the location in your yard for the compost bin. If it is outside, you will want it away from direct view, not right next to your patio, BBQ, or picnic table.
2. Place the bin where it can receive partial shade (compost gets very warm), good drainage, and enough air ventilation.
3. Place the bin as near the garden and a water source as possible. You may also want to put it near your kitchen for convenience.
4. Make your pile around $1 \times 1 \times 1$ meter. This way, the compost will "cook" (decompose) efficiently. It will also be manageable to churn and stir it up.
5. Add water to keep the compost moist (not wet). By adding green material (cut grass, weeds, leaves), it will cut down on the amount of water you have to add.
6. Make sure the bin has adequate ventilation so that the microorganisms can have the right conditions to decompose the materials.
7. Use a thermometer to monitor the temperature of the compost pile. A pile that is decomposing properly will produce temperatures of 60°C to 71°C.
8. As the compost cures, it shrinks. When you add more material to it, mix in the new material without packing down the existing material. Cover the entire pile with leaves or grass clippings to discourage animals from disturbing it.

What Goes In, What Stays Out

Compost helps the land in many ways. Finished compost can be very rich in nutrients. The browns (dead leaves, branches, and twigs) supply carbon; the greens (grass clippings, vegetable waste, fruit scraps, and coffee grounds) contribute nitrogen; and water helps break down organic products.

The following materials are good for composting (Compost Instructions, 2012):

Alfalfa
Aquarium plants
Bagasse (sugar cane residue)
Banana peels
Bird cage cleanings
Bread crusts
Brown paper bags
Burned toast
Cardboard cereal boxes (shredded)
Cattail reeds
Chicken manure
Chocolate cookies
Citrus peels
Clean paper and cotton products
Clover
Coconut hull fiber
Coffee grounds
Coffee grounds with filters
Cooked rice
Cover crops
Cow or horse manure
Crab shells
Date pits
Dead flower arrangements
Dead leaves
Dolomite lime
Expired yogurt
Flower petals
Freezer-burned fruit
Freezer-burned vegetables
Fruit salad
Fruits
Goat manure
Granite dust
Grass
Guinea pig cage cleanings
Hay
Healthy plant materials
Horse manure
Jell-O (gelatin)
Junk mail
Kitchen wastes
Kleenex tissues
Leather watchbands
Leaves
Limestone
Lobster shells

Macaroni and cheese
Matches (paper or wood)
Melted ice cream
Moldy cheese
Nut shells
Old or outdated seeds
Olive pits
Onion skins
Paper napkins
Paper towels
Peanut shells
Peat moss
Pencil shavings
Pet hair
Pie crust
Popcorn
Potash rock
Potato peelings
Produce trimmings from grocery store
Pumpkin seeds
Rabbit and hamster manure
River mud
Sawdust
Scrap paper
Shredded cardboard
Shredded newspapers
Shrimp shells
Soy milk
Spoiled canned fruits and vegetables
Stale bread
Stale breakfast cereal
Stale potato chips
Straw
Tea bags and grounds
Tree bark
Unwanted bills
Vacuum cleaner bag contents
Vegetables
Watermelon rinds
Wheat bran
Wheat straw
Wood ashes
Wood chips
Wood chips and dust
Wooden toothpicks

The following materials are not recommended for composting:

Black walnut leaves and twigs
Colored or glossy paper
Dairy products
Diseased plants
Fish or meat products
Grains and carbohydrates
Mayonnaise and salad dressing
Meat bones, meat scraps, and any fatty food

Pernicious weeds (such as morning glory that reproduces)
Pet wastes
Pine needles
Trimmings from a chemically treated yard
Weeds

At the home-gardening level, compost works great as a soil conditioner, a fertilizer, an addition of humus, and a natural pesticide for soil. At an ecosystem level, compost is useful for erosion control, land and stream reclamation, wetland construction, and as a landfill cover.

Conserving Energy with Plants

Greening up the yard can also save the homeowner more money. The landscaping method can have a drastic effect on possible energy savings. Through just a few simple rules, the climate around the home can be modified, which can be converted into immediate financial savings.

One method of saving money on energy bills is by creating windbreaks. By placing trees, shrubs, and other landscaping vegetation oriented in the right direction relative to the house, homeowners can reduce the energy required to keep homes heated and cooled during the winter and summer months. In fact, homeowners can reduce their winter heating bills by as much as 15 percent, and summer cooling energy needs may be cut by up to 50 percent, just by the way their yard is landscaped.

Because houses can lose heat when air escapes through cracks and around doors or through open windows and doors, the average home loses 20 to 30 percent of its heat in the winter by air infiltration. Heat conduction (the transfer of heat through a house's building materials) is another source of heat loss. The third source of heat loss is solar radiation—loss of heat through windows that do not have drapes or shutters.

Green landscaping choices can solve—or at least alleviate—many of these heat loss problems. There are three basic landscape applications that have proven to save energy:

• Shade trees
• Windbreaks
• Foundation plants

Shade trees can make a significant difference in the summer. Large trees shading the roof of a house from the afternoon sun can reduce temperatures inside the home by as much as 4.5°C to 5.5°C.

Deciduous trees are an excellent choice because they provide summer shade, and then drop their leaves in the fall. This enables the warmth of the sun to filter through the bare branches in the winter and help warm the home. If a home can be situated to take advantage of shade from existing trees on southeast and west exposures, energy expended to cool the house can be greatly reduced.

A windbreak is a row of trees planted close together to form a barrier. They are planted so that the wall they form is perpendicular to the wind direction so the wall of vegetation blocks the onslaught of cold air from reaching and entering the home. It has been shown that buildings, such as homes, protected by windbreaks use 10 to 20 percent less energy for heating and cooling compared to unsheltered homes.

Windbreak trees and shrubs also provide food and habitat for game birds and other wildlife, as well as assist in the carbon cycle. It is estimated that for each acre planted with windbreaks, more than 21,000 kilograms of carbon dioxide will be stored in the trees by age twenty (AFTA, 2016).

Foundation plants create a dead air space that has less cooling power than moving air, which can also decrease the loss of warm air through walls.

By following these relatively simple landscaping practices, it can make a serious difference on the resulting monthly energy bill.

CONCLUSIONS

This chapter has pointed out many ways to green up your home and yard. There are several benefits to living in a green home. Some of the biggest, and most direct benefits, include multiple reasons that benefit the environment, our health, natural resources, and the economy. For example, the environment is positively impacted because by using renewable and clean energy sources, it lessens our reliance on fossil fuels and other depleting resources. This lessens negative emissions into the environment, such as from GHGs. Our health is positively impacted because it enables us to have a healthier home environment through improved indoor air quality. Green homes also take advantage of nontoxic materials. By using renewable and clean energy sources, it lessens our reliance on fossil fuels and other depleting resources. The construction process of a traditional home alone emits much more construction waste than a green home. Recyclability of materials with a green home lessens negative emissions in the environment.

Green homes mean lower energy bills. When better, more efficient insulation is used; ductwork is tighter; doors and windows are sealed tighter; and heating/cooling and lighting/appliances are more efficient, the increased efficiency translates directly into savings for the home owner. Health is another issue. Green building means healthier indoor air quality. Green means paints, adhesives, sealants, and finishes are low VOC so they do not stink and cause respiratory illness. This means fresh air is brought into the house and nontoxic air filters are working to keep the air clean, fresh, and healthy.

Green homes are more durable. Green quality building lasts longer than a comparable nongreen product. Over time, the owner will see savings in maintenance and replacement costs. All of these advantages increase the value of the home.

REFERENCES

AFTA. 2016. What is agroforestry: Windbreaks. Agroforestry extension education for benefitting farmers and landowners. *Association for Temperate Agroforestry.* http://www.aftaweb.org/about/what-is-agroforestry/windbreaks/9-page.html (accessed October 22, 2016).

Arizona Cooperative Extension. 2005. Xeriscape plant list. *Yavapai County, ArizonaCooperative Extension.* http://www.ag.arizona.edu/yavapai/anr/hort/xeriscape/ (accessed October 15, 2016).

Compost Instructions. 2012. Composting instructions: How to xcompost at home. http://www.compostinstructions.com/what-you-can-and-cannot-compost/ (accessed September 15, 2016).

EPA. 2009. Celebrating a decade of energy star buildings. *United States Environmental Protection Agency.* https://www.energystar.gov/ia/business/downloads/Decade_of_Energy_Star.pdf (accessed October 10, 2016).

Gines, J. K. 2012. *Climate Management Issues: Economics, Sociology, and Politics.* Boca Raton, FL: CRC Press.

SUGGESTED READING

Cook, M. 2014. *Green Home Building: Money-Saving Strategies for an Affordable, Healthy, High-Performance Home.* Gabriola, BC: New Society Publishers, 432 p.

Maclay, B. 2014. *The New Net Zero: Leading-Edge Design and Construction of Homes and Buildings for a Renewable Energy Future.* White River Junction, VT: Chelsea Green Publishing, 576 p.

Madigan, C. 2009. *The Backyard Homestead: Produce all the Food You Need on Just a Quarter Acre!.* North Adams, MA: Storey Publishing, LLC. 14th ed., 368 p.

Rapinchuk, B. 2014. *He Organically Clean Home: 150 Everyday Organic Cleaning Products You Can Make Yourself—The Natural, Chemical-Free Way.* Holbrook, MA: Adams Media, 224 p.

New York Department of Environmental Conservation. 2017. Household hazardous waste (HHW). http://www.dec.ny.gov/chemical/8485.html (accessed May 22, 2017).

Toht, D. 2011. *Backyard Homesteading: A Back-to-Basics Guide to Self-Sufficiency (Gardening).* Petersburg, PA: Creative Homeowner, 256 p.

Trubble, T. 2016. *How to Build a Backyard Greenhouse: A DIY Practical Guide.* Boynton Beach, FL: CDI Publications LLC, 2162 KB.

19 The Sustainable Lifestyle
Low Impact and Healthy Living

OVERVIEW

This chapter explains what a sustainable lifestyle involves. It enlightens on low-impact living, on what it means, and on how it can impact one's life in a healthy way. It first explains the various health benefits of going green; why buying local is a much better option; the different choices of green household products available today; water conservation strategies; and available eco-friendly, energy-saving products. Next, it discusses the multiple benefits of recycling and how it helps not only the environment, but also every one of us, now and in the future.

INTRODUCTION

Low-impact living is more than just a new trend—it is an important way of thinking and taking action that has valuable effects on our environment. With all the harm being done to the environment today, many people believe that their actions alone cannot possibly make a real difference. But the fact is, if each person did their part, then collectively, it would make a significant difference.

Everyone is capable of making changes in a certain way to live a greener, more sustainable lifestyle. Even if each one of us just take small steps, cumulatively it can have a notable impact. Low-impact living is a lifestyle that incorporates sustainable habits into the choices we make and the way we live, which really do make a difference. Low-impact living is a way to learn that each one of us can take responsibility and live smarter in order to reduce the waste and improve the health of our environment. For example, a more eco-friendly lifestyle could include conserving your use of water, building a compost pile, recycling, buying local, and making your own earth-friendly household cleaners. All of these may seem like simple activities, but they do make a difference—this chapter will show you how.

DEVELOPMENT

Today, sustainable living is a hot topic. Many people may have heard about sustainable living but are not sure what it exactly means. At its most basic, sustainable living means living a lifestyle that uses as few resources as possible and causes the least amount of environmental damage for future generations to deal with. There are different forms of sustainable living, since the concept can apply to almost every part of a daily life.

In terms of housing, sustainable houses are built in such a way that they use few nonrenewable resources, do not require much energy to run, and cause little or no damage to the surrounding environment. A sustainable house should be constructed from materials that have been produced in an environmentally friendly manner. For example, a sustainable house might be built from straw bales, adobe, or reclaimed stone or brick. Many homeowners pursue sustainable living by making their houses as low energy as possible, either by making sure that they have very high energy efficiency or by producing their own power from the sun or wind.

Sustainable living must include a reliance on sustainable energy. The energy sources used must be renewable rather than limited in quantity. Energy sources such as fossil fuels cannot readily be renewed, so instead of these, use renewable energy sources such as solar, wind, water, or geothermal. The energy must also be captured and used in an environmentally friendly way that does not damage the environment for future generations. For sustainable living, a person must leave as little impact on the world as possible.

The diet of a person who is focused on sustainable living should contain foods that are at the base of the food chain. A vegetarian lifestyle is best suited to sustainable living because it requires fewest resources to produce and causes the least amount of environmental damage and degradation. The food should also be grown organically without the use of chemicals such as pesticides or herbicides that can pollute the environment and cause health problems for other animals or humans. It is best to eat food that has been grown locally to reduce the problems caused by transporting food over long distances. Many people try to grow their own produce in their yard or in a community garden near their home. Local farmers markets can also be a good source for sustainably grown fruits and vegetables.

Essentially, sustainable living involves living as lightly on earth as possible. Someone who succeeds at living a sustainable lifestyle will use very few resources and will leave the environment as untouched as possible so that future generations will be able to enjoy the same high quality of life that people do today.

Global warming, the greenhouse effect, pollution, and ever-growing landfills are all part of the big environmental picture. But many people have a belief that their actions, as just one person, have little impact on the world around us.

While one person's attempt at fixing the world's problems may not seem to make much of a difference, collectively, a lot can be done when everyone works together to slow down big issues. What is low-impact living? Low-impact living is a way for each of us to take responsibility and live smarter to reduce waste and give the environment a rest. You do not have to take drastic steps; even consistent small steps can have big impacts. The next sections will offer some advice, benefits, and how-tos of a low-impact, healthy lifestyle.

THE MANY HEALTH BENEFITS OF GOING GREEN

Working the fundamentals of green living into your busy life is much simpler than it might seem. Most of the changes that environmental scientists recommend are easy to make—and the benefits are definitely worth it. Not all the benefits are about the environment. Many of the eco-friendly habits you incorporate into your life can also have a positive impact on you and your health—your physical, emotional, and psychological well-being.

LIVING GREEN FOR A HEALTHIER LIFE

Here are a few benefits you will get from adopting a greener lifestyle:

- *A healthier heart*: One of the biggest benefits of a greener lifestyle is a healthier heart. You can achieve this by getting out and exercising more. If you can walk or ride a bike instead of drive a car for those short trips you need to make, you not only save gas and reduce your carbon footprint, but you also get some great cardio exercise. For a longer trip, try walking or riding your bike to and from the bus or train stop.
- *Healthier drinking water*: You may be surprised to know that bottled water is one of the least green items around. The chemicals that are used to produce the bottles mainly come from petroleum products, and when they are thrown away (millions of disposable bottles end up in landfills), they can leach into the water causing low-level contamination.
    - Many people drink bottled water because they claim it is purer than tap water, but that is not true. Unfortunately, it is just more expensive. No matter where you live, you are better off just filtering regular tap water if you want pure water. You can find a variety of water filters that remove common contaminants. Some drinking water purification systems utilize a reverse osmosis filtration. These systems include various filters to trap sediments and other impurities. River and ocean water can also be filtered with these.
    - Instead of a bottled water buy an environmentally friendly, reusable stainless-steel, canteen-like bottle to carry your filtered water in.

- *Healthier lighting for your eyes*: When you use natural sunlight instead of electricity in your rooms during the day, it not only saves your money, but also gives you several health benefits such as assisting your body in its production of vitamin D. And it has another benefit in the winter: It helps alleviate depression. With shorter daylight hours during winter months, many people suffer bouts of depression. But if they make it a point to be exposed to as much direct sunlight as possible, it often helps alleviate symptoms.
- *Healthier diet*: Purchasing locally produced food not only saves on the fuel it would normally cost to ship the food to you, but your local growers will also appreciate you. Local growers are more likely to have fresher, organically grown produce available than the stores that ship food from halfway across the country. Shopping at farmers markets is also a great way to eat healthier. That way, your food is most likely to be pesticide-free.
    - Another way to eat healthier is to eat less meat. You may not give it much thought when you sit down to a nice steak dinner, but a lot of energy, water, and resources are used to produce meat compared to vegetables. Eating less red meat is also good for your health.
- *Healthier remodeling*: If you are planning on remodeling your house and doing any painting, be sure to use a low-volatile organic compound (VOC) paint. Low-VOC paints are much healthier for you and the environment because they have a lower amount of airborne toxins.
- *Healthier clothes*: The World Wildlife Fund (WWF) has announced that perfluorinated chemicals (PFCs) are used on some clothes to help keep them wrinkle-free, but they are toxic to the environment and have also been linked with health concerns, including cancer. Hence, choosing more natural fabrics is a greener choice.
    - WWF also warns against some natural fabrics that may not be very eco-friendly. For instance, cotton may not be a good green choice because it can require heavy pesticide use. A quick online search can help you find out which stores sell eco-friendly fabrics, such as hemp, organic cotton, silk, or wool, produced using the least amount of harsh chemicals.
- *Healthier cleaners*: Rather than using cleaners that are made of harsh chemicals, which can harm the environment, it is best to make your own cleaners by using natural cleaning substances such as white vinegar or natural castile soap mixed with salt and baking soda to cut through mold or soap scum; natural products like washing soda and borax can be used to clean floors and remove grease stains. For example, one part vinegar and one part water in a spray bottle makes a glass, shower, floor, and toilet cleaner. Not only are the traditional chemical products harmful to

the environment, they can also cause skin rashes and asthma flare-ups.

- *Healthier dry-cleaning methods*: According to Crissy Trask, founder of Greenmatters.com, many dry cleaners use a chemical called perchloroethylene, or perc, which the International Agency for Research on Cancer (IARC) has listed as a probable human carcinogen. After your clothes are dry-cleaned, some of this chemical remains in the fabric. When you wear the clothes, the chemical is close to your skin and escapes into your home environment, which can be harmful. Trask recommends investigating dry cleaners in your area that offer wet-cleaning technologies and use water-based equipment to clean garments that previously were dry-cleaned.

- *Healthier food*: Another major boost for your health is to grow your own food by planting and maintaining an organic vegetable garden in your yard. Not only is home-grown food fun, but it also tastes much better than anything you can buy. Plus, you will have the peace of mind that it did not require fossil fuels to ship it halfway across the country to get it to you. And the benefits continue: It was farmed without harmful chemical pesticides.

Health Benefits of Alternative Energy

Energy use is one of the biggest issues we face today as we make choices between renewable forms of energy or continue to use fossil fuels and other forms of energy that is hard on the environment. But renewable, or alternative, energy sources not only benefit the environment, they also benefit each one of us. Here are some of those benefits:

- *Fewer respiratory problems*: By weaning ourselves off of fossil fuels, we can lower air pollution levels significantly. Pollution can aggravate asthma and allergies. Pollution is also linked to lung cancer. Therefore, by using nonpolluting, alternative energy sources, we can clean up the air and environment as well as improve our respiratory health.

- *A healthier heart*: Some of the particulates in the air that come from pollution caused by fossil fuels can actually contribute to heart disease and other cardiac issues. Reducing pollution with greener energy practices can actually help your heart stay strong and healthy.

- *Fewer infectious diseases*: As was mentioned before, climate change can increase the spread and range of infectious diseases. Therefore, if each one of us remembers to unplug all those electrical items we were not using, it can ultimately help curb the spread of infectious diseases.

Buying Local and Avoiding Prepackaged Food

Going green can also be accomplished through your dining habits. You can improve your health by selecting foods that are greener and by buying locally grown foods. Here are some green shopping advices:

- *Stay away from prepackaged foods*: Prepackaged foods use an extraordinary amount of materials and energy to produce. They are also often full of excess sodium and sugar and countless empty calories. So it should not be too hard to see that avoiding these types of items at the grocery store not only helps out the environment, but also helps keep you healthy. If you work on cutting prepackaged foods out of your diet, you can receive multiple benefits such as weight loss and lowering your chances of heart disease and cancer.

- *Avoid pesticides*: Foods that are not certified as having been grown organically have probably been grown with the use of pesticides. This can cause illness or other health issues, so it is better for your health to choose organically grown produce.

- *Avoid products from meat farms*: Fish caught in the wild and chickens and cattle that are raised as free-roaming range animals are farmed in a more environmentally sustainable manner. Another benefit is that the animals are not given hormone injections. When you avoid purchasing products from meat farms, you avoid the chemicals and hormones associated with them as well, and get a healthier, leaner meat overall.

Green Household Products

You use many supplies in your home every day. Without realizing it, some of them may not be healthy for you or the environment. How much thought have you given to the following?

- *Fertilizers*: The fertilizers you put on your lawn can have an effect on you and your family. Rainfall can wash the fertilizer off the grass and into the water supply, which can seep into the ground. You can eliminate this hazard, however, if you use an environmentally friendly fertilizer or an organic option, such as mulching, to keep your lawn safe and healthy.

- *Household paints*: Many household paints release compounds into the air that can be harmful to your health, causing headaches, dizziness, nausea, respiratory tract problems, and liver, kidney, and central nervous system problems. If you are painting inside your house, make sure you look for a paint with a low incidence of VOCs.

- *Diapers*: Disposable diapers may not be as much of a value as you think they are. Experimentation has shown that viruses can live in dirty disposable diapers for up to two weeks, making sanitation a health problem. Infections and diaper rash are more problematic with disposable diapers, too. The tried-and-true cloth diapers are still the choice for reducing environmental impact and keeping your baby healthier.

TRYING NEW TRANSPORTATION OPTIONS

Shifting from driving your own car every day to greener transportation options can help your health. Here are a few suggestions:

- *Ride a bike*: If you live close enough to your workplace, riding a bike can provide you with great exercise. If you do not live close enough, try riding public transportation part way and biking the rest of the way. Many cities provide bike racks on buses. Biking is an effective way to strengthen your cardiovascular system and to lose weight.
- *Take public transportation*: Automobile traffic contributes to air pollution, which in turn means more illnesses related to breathing problems such as asthma. Also, every additional car on the road can lead to increases in the numbers of injuries and deaths from car accidents, which already kill more than 40,000 people each year. If your community offers public transportation options—buses, light rail, subways—give that a try. They are going your way, anyway. Leave your car at home; this not only cuts down on air pollution, making your lungs healthier, but it also gives you some exercise as you walk or bike to and from your stops.
- *Telecommute*: This is a great option if your employer offers this choice. Working from home keeps you off the road, avoiding contributing to air pollution altogether. It also keeps you from being exposed to unhealthy air if you work in a congested city.

Here are some more advices on living green:

- *When you have to run errands*: If you have more than one car, use the one that is more energy efficient!
- *When you have to fly*: Like cars, planes burn large quantities of gasoline as fuel, which also adds to the greenhouse gas problem. Consider flying less to save energy and reduce those emissions. If you have to fly, consider buying carbon offset credits, which fund energy saving efforts to reduce the total impact of the greenhouse gases produced by your flights.
- *Give some time to the outdoors*: Do not forget some leisure time for yourself. Try taking a bike ride or walk just for the sake of enjoying the outdoors on a nice day. Being outside in the sunshine and enjoying nature can help your mental and emotional well-being, elevate your mood, reduce mental fatigue, and increase your concentration if you need a break from work.

Adopting a greener lifestyle can do wonders for your health in many ways. And the bottom line is if you feel better physically, emotionally, and mentally, you will probably feel those positive results financially, too!

PREVENTION AND PREPAREDNESS

The climate is already changing and these changes are likely to have an effect on human health. Individuals, families, and entire communities can make adjustments, along with green living, to prepare for these changes. Taking simple prevention and preparedness steps can help you stay healthy today and in the future.

Many of the potential health effects of climate change are related to threats we now face, including heat waves, extreme weather events, and the spread of new infectious diseases. These threats can seem overwhelming, but taking preparedness steps now can help you and your family to be safer and healthier when they do occur.

The Centers for Disease Control (CDC) and the Federal Emergency Management Agency (FEMA) advise preparing for a disaster by accessing FEMA's Ready America website (www.ready.gov); you can obtain preparedness plans designed to help families and businesses facing multitude of potential emergency situations. You can find out how to prepare an emergency kit, to make an emergency plan ahead of time, and to stay informed if an emergency does occur. The website offers information for situations such as:

- Winter storms and extreme cold
- Hurricanes
- Floods
- Extreme heat
- Biological threat
- Chemical threat
- Tornadoes
- Tsunamis
- Food safety in an emergency

It is always wise to be prepared for unexpected surprises in our changing environment.

WATER CONSERVATION CALCULATORS

Water conservation calculators can be used to determine ways to save water in the home. Once it can be determined where the majority of water use is focused, it may then be possible to make significant changes in water use in order to minimize waste. First, conserving water means using less water as well as reducing amounts of wastewater. Reducing water consumption also reduces energy consumption. In order to calculate water consumption, there are several online calculators available.

Using an online tool is a quick and easy way to calculate the total household water consumption. In addition, most tools compare individual household usage to the usage of the average American family or the average household in a specific region or county. A fun challenge—and a worthwhile objective—for the whole family is to make it a goal to use only 50 percent of the average household. Examples of calculators can be found at these websites:

Often significant volumes of water get wasted through inefficient household equipment or wasteful practices. A water conservation calculator determines where you currently use the most water and identifies areas of significant potential savings (Hanson, 2011).

ECO-FRIENDLY AND ENERGY-SAVING PRODUCTS

One of the best aspects of low-impact living is to use eco- or environmentally friendly products that would not have a major impact on the environment. There are many common items used on a daily basis that have a biodegradable or recycled version. If these are used instead of the nonbiodegradable version, the environment can be cared for at the same time. Today, there are a multitude of eco-friendly products available. For example, there are:

    Cleaners
    Clothes
    Compostable bags
    Cups
    Flooring
    Food packaging
    Furniture
    Grocery bags
    Light bulbs
    Mulch
    Paint
    Plastic bags
    Plates
    Shipping supplies
    Shopping bags
    Straws
    Sunscreen
    Textiles
    Tile adhesive remover
    Toys
    Water bottles

Those are just a few of the items. In addition to products that have a low impact when you dispose of them, there are also plenty of other products that will help save water, heat, and electricity when used. For example, one way to save water in the home is through the use of the Dual Flush System or low-flow toilets. As a reminder, one of the major uses of water in the home is flushing the toilet. A single flush can use several gallons of water, depending on the individual toilet. For toilets pre-1992, the average amount of water used is 3.5–5 gallons per flush. That can add up to an enormous amount of water in one day. Toilets newer than 1992 should use no more than 1.6 gallons of water per flush. Although this is a significant improvement, it also adds up. The Dual Flush Toilet System is a newer solution to the problem. This system offers a short

flush (designed to remove fluids) and a long flush (designed to remove solids). This helps significantly with water conservation in households. The Environmental Protection Agency (EPA) now offers a program called WaterSense, which is similar to their EnergyStar program. This program ensures that the product is a high-efficiency toilet that includes a dual-flush mechanism that works resourcefully (Pullen, 2007).

Low-flow toilets are another option for water-saving results. They only use 1.6 gallons per flush. Use of these can save a family of four up to 22,000 gallons of water a year, which equals to about a $100 in annual savings (Weber, 2007).

THE BEAUTY OF RECYCLING

One of the easiest ways to have a low-impact lifestyle is to recycle. It is easy, convenient, and the results (and satisfaction) are immediate. Recycling is finding alternate ways to reprocess and reuse items instead of throwing them away as garbage. Recycling is one of the most known aspects of the green movement. The recycling logo is one of the most recognized symbols around—it appears on items that can be recycled and on recycling bins that are located in many public locations; it is taught throughout schools at all levels of education, beginning at very young ages, even as young as kindergarteners. And if you take a brief trip down memory lane and remember the program, the progress it made has been remarkably successful.

A Brief Look Back

As recently as 20 years ago, everybody in this country threw everything away. We were a disposable society without shame. We tossed cans, plastics, and paper in ridiculous amounts without even giving it a second thought.

In the United States, the recycling movement actually began in the late 1960s as a result of pressure from environmental groups. But the market for recyclables was not well established at that point, so the movement did not begin with any real gusto. In fact, it was not until 1973 that the first curbside recyclable pickup service in the nation was established in California. The initial push was actually against littering.

Collecting Cans for a Little Cash

It was during the 1960s and 1970s when it was popular to collect cans and sell them to scrap markets. This was when the recycling programs began to emerge around the country. This turned out to be a bonus in more ways than one because its reward, in a sense, triggered some of the initial awareness of collecting and returning recyclable items rather than simply disposing of them.

The Environmental Movement Takes Hold

Once the public had embraced the idea of recycling cans, mindsets were finally to a point where a full-scale recycling program could be presented and successfully received by the general public.

Interestingly, it was during this time period that several things came about that began to make a difference from the public viewpoint and really got them thinking about the environment. One of those fortunate developments was the creation of Earth Day (April 22).

The Creation of Earth Day

Earth Day was the brainchild of Senator Gaylord Nelson of Wisconsin. The idea evolved over seven years beginning in 1962. Nelson created the special day because he was worried that national politics at the time did not seem to be concerned about the environment. Interestingly, President Kennedy was receptive to Nelson's environmental awareness push, and he also tried to promote interest and support in Washington, but without much luck.

Senator Nelson found, however, that the public was concerned about the environmental degradation issues he brought to their attention. So because of public interest, he continued to speak about and promote his cause.

In 1969, when he was out in the West promoting his environmental campaign, he had the foresight to organize a huge grassroots protest over what was happening to the environment. He patterned it after the anti-Vietnam War demonstrations (called teach-ins) that were popular at the time on college campuses. He thought that if he could tap into the environmental concerns of the general public, feed off the student antiwar energy, and then steer that energy into a demonstration that would force the environmental issue onto the political agenda, maybe he could finally get the attention of Washington, DC.

So, in 1970 he held a nationwide grassroots demonstration on behalf of the environment, and invited everyone he could to participate. The wire services helped him get the word out across the country. The event raised concerns nationwide about environmental issues and even grabbed attention away from Vietnam War protests that were common on college campuses.

Recycling Becomes Standard

Beginning 20 years ago, the curbside recycling program was set in place in cities across the United States, and the public began to actively participate in a serious recycling effort. Today, through various green campaigns, educational opportunities are readily available to the public to teach the importance of the three Rs: *Reduce, Reuse*, and *Recycle*. As a culminating result of the efforts and examples of the past three decades, many products today are recyclable and several recycling programs exist.

THE MULTIPLE BENEFITS OF RECYCLING

There are many recognized benefits of recycling. One of the biggest and most obvious benefits is that it helps control both consumer and commercial wastes, keeping it out of landfills. When waste is discarded in landfills and these landfills become too full, the waste must be removed or relocated. This takes, and wastes, energy. Recycling eliminates this need.

Recycling Saves Energy Resources

Recycling reduces the need to make new materials from scratch, which takes not only new resources (short changing future generations), but also energy to produce them. Recycling existing materials and reusing them saves energy resources.

Many materials, such as most metals (especially aluminum and steel), can be recycled indefinitely. In addition, the energy required to recycle materials is generally much less than the energy needed to produce a product from new materials. Here are some examples:

- Ninety-five percent more energy is needed to produce a new aluminum can than to recycle an existing one.
- Forty percent more energy is needed to produce a new glass item than to recycle an existing one.
- Sixty percent more energy is needed to produce new steel than to recycle it.
- Seventy percent more energy is needed to produce new plastic than to recycle it.
- Every ton of paper made from recycled materials saves 17 trees, about 1,703.5 liters of oil, and about 26.5 liters of water.
- Over 40,000 trees could be saved if all the morning newspapers produced in the United States for just one day were recycled.
- For every 100,000 people that put themselves on a "do not mail" list for junk mail, up to 150,000 trees could be saved.

Another major saving is a reduction on our reliance of imported oil. The EPA has estimated that people who opt to change their cars' oil at home throw away over 757 million liters of used motor oil. Oil can cause environmental damage when it is poured into the sewer system or onto the ground. The problem with motor oil is that it gets dirty, but it does not wear out. Used oil is recyclable; it can be cleaned and used again. If you are one who likes to change your car's oil, check with your local waste-management company to find out how you can recycle your used oil.

Recycling Reduces Pollution

Items that are thrown away instead of recycled can lead to water pollution through the leaching of chemicals into the soil. Once pollution enters the soil, some of it can eventually leach into the water table, contaminating supplies there.

Air pollution is another resultant problem. In landfills, it is common for methane and other gases to be released. Again, if the items were recycled, that would eliminate this problem.

Job Security

Another benefit of recycling is that it contributes to job security. Because recycling is more labor-intensive than landfill management, there are employment opportunities involved with recycling positions. This can help out job markets in rural areas, where a lot of recycling centers are located.

Recycling also helps revitalize other manufacturing industries because many items can be made from recycled materials. This process involves collecting a wide range of recyclable materials, breaking them down into raw materials, and then manufacturing the raw materials into new products.

Moreover, as recycling technology continues to improve and become more sophisticated in its ability to cost effectively collect, clean, and remanufacture recycled goods, it will have the increased ability to make more items from recycled materials.

Easing the Newspaper Burden at Landfills

Currently, about 14 percent of landfills are occupied by newspapers. This represents several problems—the biggest being the waste of paper resources. Recycling is obviously helpful here; the resource can be reused over and over.

There are several benefits of recycling paper. One is that it saves energy from having to be expended to produce new paper from virgin materials. Recycled paper usually does not have to be re-bleached, which means that fewer harmful chemicals are released into the environment. If bleaching recycled stock is required, oxygen is typically used instead of chlorine. This is also beneficial because it reduces the amount of dioxins produced as a by-product of the chlorine bleaching process.

It needs to be stressed that along with recycling products, there should be a market demanding them. Therefore, when you purchase paper, always make sure you purchase recycled paper so that you help create a demand for the recycled product in the first place.

WASTE AND RECYCLING FACTS

Different recycling centers have different types of systems, which can determine what types of items they can accept. Therefore, it is always a good idea to do your homework before going to a recycling center to make sure exactly what types of items it accepts.

The following is a list of items that you can recycle:

- Acids
- Adhesives
- Aerosol cans (empty)
- Aluminum: beverage cans, foil, food trays, and pie plates
- Antifreeze
- Appliances (freon)
- Appliances (nonfreon)
- Bags, brown paper, and pet feed
- Bags, plastic
- Batteries
- Books
- Boxes: cereal boxes, egg cartons, frozen food boxes (if not wax coated), milk cartons (if not wax coated), laundry detergent boxes (clean), orange juice cartons, shoe boxes, and tissue boxes
- Brake fluid
- Brochures, glossy
- Brown paper bags
- Cans: aluminum (beverage) and steel (food/juices)

- Cardboard boxes
- Cardboard rolls: gift wrap, paper towels, and toilet paper
- Cards, greeting
- Catalogs
- Cell phones
- Cleaners, household
- Coffee cans
- Electronics
- Floor polish
- Furniture
- Furniture polish
- Glass containers
- Grass clippings
- Junk mail
- Kerosene
- Leaves
- Newspapers
- Oil
- Paint
- Paint-related products
- Paper
- Pallets, wooden
- Pesticides
- Phone books
- Pipes
- Plastic bags
- Plastic bottles
- Post cards
- Printer Inkjet Cartridges
- Rolls, cardboard
- Sinks
- Steel cans (food/juice)
- Stumps
- Tables
- Telephone books
- Tires
- Toilets
- Trunks
- Windows
- Wood chips/shavings
- Wrapping paper

The following is a list of items that you cannot recycle:

- Carbon paper
- Wet cardboard
- Carpet
- Ceramics
- Mirrors
- Lightbulbs (incandescent)
- Styrofoam
- Laminated paper
- Thermal fax paper

The best method of recycling electronic items that are still useable is to donate them to a school or charity. If the item is broken and not fixable, it usually can be recycled. Recyclable electronics include the following:

- Answering machines
- Cables
- Camcorders
- CD players
- Cell phones
- Cell phone batteries
- Computers
- Computer batteries
- Fax machines
- Keyboards
- Laptops
- Monitors
- Computer mice
- Printers
- PDAs
- Remote controls
- Televisions

THE DILEMMA OF PLASTIC WATER BOTTLES

One of the biggest problems being faced today is the waste produced by the bottled water industry. The sad truth is that the recycling rate is extremely low for water bottles—most consumers just toss them. In fact, bottled water is the single largest growth area among all beverages, and that includes alcohol, juices, and soft drinks. Over the past 10 years, average consumption per person has more than doubled (Figure 19.1).

FIGURE 19.1 Bottled water is one of the most popular—and most tossed—drinks in America, causing mountains of landfill waste. (Courtesy of Brett Weinstein; From Gines, J. K., *Climate Management Issues: Economics, Sociology, and Politics*, CRC Press, Boca Raton, FL, 2012.)

Here are some facts to consider about plastic:

- Each year 2.4 million tons of plastic bottles are thrown away.
- Of 60 million single-use drink containers that are purchased, 75 percent are thrown away directly after use.
- Plastic bottles are the number one source of pollution found on American beaches today.
- Every 2.5 square kilometers (square mile) of the ocean has 46,000 pieces of floating plastic in it (Source: United Nations Educational, Scientific and Cultural Organization, 2017).
- Of the plastic produced every year worldwide, 10 percent ends up in the ocean. Of that, 70 percent ends up on the ocean floor, where it will probably never degrade (Source: United Nations).
- Plastic bottles take 700 years to begin decomposing.
- Americans toss about 38 million plastic bottles each year (not including soda), and 80 percent of plastic bottles are never recycled.
- It takes 91 million liters of oil to produce 1 billion plastic bottles.
- The average American consumes 167 bottles of water a year.
- Bottling and shipping water is the least energy-efficient method ever used to supply water.
- Tap water is distributed through an energy-efficient infrastructure, but transporting bottled water over long distances involves burning massive quantities of fossil fuels.
- Bottled water companies do not have to release their water-testing results to the public, whereas municipalities do.

The issues surrounding plastics in general, and plastic water bottles in particular, is a good illustration of how critical green living and the value of education are in making of a meaningful difference in the lives of others and in the health of the environment.

CONCLUSIONS

This chapter has explained the idea of sustainable living and provided several suggestions on how to achieve a low-impact lifestyle. The next step is to go out and do it. The most difficult part is making the commitment, choosing to do it, and sticking with it. But once it becomes a habit, and you see the rewards, it becomes an enjoyable part of life because you can see the fruits of your labor. Once you see the benefits, you can be an inspiration to others, and one step at a time, we can make this a successfully sustainable world.

REFERENCES

Gines, J. K. 2012. *Climate Management Issues: Economics, Sociology, and Politics*. Boca Raton, FL: CRC Press.

Hanson, R. 2011. Water conservation calculator. *Ilovetonow, Home & Garden*. http://greenliving.lovetoknow.com/Water_Conservation_Calculator (accessed October 20, 2016).

Pullen, K. 2007. Dual flush toilet. *LovetoKnow*. http://greenliving.lovetoknow.com/Dual_Flush_Toilet (accessed October 5, 2016).

United Nations Educational, Scientific and Cultural Organization (UNESCO). 2017. Facts and figures on marine pollution. http://www.unesco.org/new/en/natural-sciences/ioc-oceans/focus-areas/rio-20-ocean/blueprint-for-the-future-we-want/marine-pollution/facts-and-figures-on-marine-pollution/ (accessed May 22, 2017).

Weber, S. 2007. Low flow toilets. *Lovetoknow*. http://greenliving.lovetoknow.com/Low_Flow_Toilets (accessed October 5, 2016).

SUGGESTED READING

Bowe, A. 2011. *High-Impact, Low-Carbon Gardening: 1001 Ways to Garden Sustainably*. Portland, OR: Timber Press, 264 p.

Chatterton, P. 2014. *Low Impact Living: A Field Guide to Ecological, Affordable Community Building*. London: Routledge, 248 p.

Environmental Science.org. 2016. What is sustainability and why is it important?. *Environmental Science.org*. http://www.environ mentalscience.org/sustainability (accessed October 23, 2016).

Thorpe, D. 2015. *The 'One Planet' Life: A Blueprint for Low Impact Development*. London: Routledge, 476 p.

20 Making Changes for the Benefit of Future Generations

OVERVIEW

In this final chapter, we will be looking at what will be required to work together as a world community to effectively deal with the problems of climate change and jointly achieve global sustainability. We will examine past global attempts at working through issues for the good of humanity to see what worked and what did not in an effort to try and understand what we may be facing in the road immediately ahead of us. We will also look at what programs the U.S. government currently has in progress to get us on our way toward a zero-carbon future. Next, we will highlight those cities in the United States and worldwide that have taken measures to cut greenhouse gases (GHGs) and reach toward a zero-carbon balance and applaud their outstanding achievements. In conclusion, we will look at why the final choice is ultimately up to each one of us to decide whether or not to choose a sustainable lifestyle and strive to make a difference in the future.

INTRODUCTION

Currently, 7.5 billion people inhabit earth; and every one of them matters. Every person has a right to clean air, clean water, healthy food, and adequate resources for shelter and living. Today, we are faced with some of the most trying problems of all time: warming temperatures, heavy pollution, rising tides, drought, depletion of the ozone layer, wildfires, changing seasons, shifting growing zones, and resource depletion. We do know that by adopting a sustainable lifestyle, however, through the actions we take—reducing energy consumption, using renewable resources, becoming more eco-friendly—we can remedy some of these issues by reducing our environmental impact and make this planet a cleaner, safer place. All it requires is making the commitment and doing our part—one action at a time.

DEVELOPMENT

City, state, and federal governments are doing much more today in an attempt to promote sustainable living, as are many countries around the world. The tide seems to be changing as populations are becoming more educated about climate change and are seeing the effects of it firsthand. Ultimately, it will take the power of government leadership worldwide to make the large-scale changes necessary to make societies, as a whole, greener because it requires sustainable infrastructure as well as both national and international leadership and cooperation. That, however, in no way minimizes the power of the individual. Oftentimes, it is through the power of individuals that these movements become trends—which catch on in society as a whole—leading to changes in lifestyle and then leadership. As we have seen in events like Earth Day, these movements can change the mindset of a nation.

TAKING THE LEAD

There have already been several excellent examples of leadership in taking successful assertive action to go green. Two of the best past examples have been the areas of ozone damage and climate change. It is worth mentioning them here to illustrate the power of public involvement—including local, regional, national, and international efforts—and how collective action really can make a difference.

Ozone and the Montreal Protocol

In 1985, the world was shocked when British Antarctic Survey scientists published the proof there was a "hole" in the ozone layer. Without hesitation, 20 nations—including most of the major chlorofluorocarbon producers (the cause of the ozone hole)—signed the Vienna Convention, a document that established a framework for negotiating international regulations concerning ozone-depleting substances. From this, the Montreal Protocol was born.

The protocol is an international treaty designed to protect the ozone layer by phasing out the production of many of the substances believed to be the cause of the ozone depletion. This treaty evolved rapidly. It was opened for signature on September 16, 1987, and entered into force on January 1, 1989. It has undergone seven revisions since that time to keep it updated and workable so that participating nations can cooperate and make progress. It is believed that if the international agreement is adhered to, the ozone layer will fully recover by 2050.

Because of its widespread acceptance, adoption, and successful international agreement after all these years, it is considered to be one of the most successful international agreements that have ever been put into place to solve an environmental issue. It is a landmark example of exceptional international cooperation. So far, it has been ratified by 196 nations.

Climate Change, the Kyoto Protocol, and the IPCC

Climate change and the Kyoto Protocol is another example of international cooperation, although it has not run quite as smoothly as the Montreal Protocol has. There have been several measures to cooperatively deal with climate change issues among and between nations, but the Kyoto Protocol is probably one of the most well known.

It is a protocol to the United Nations Framework Convention on Climate Change (UNFCCC) and was created as an international environmental treaty with the goal of stabilizing GHG concentrations in the atmosphere below dangerous levels. Initially adopted in December 1997 at a convention held in Kyoto, Japan; it was entered into force in February 2005. So far, 192 countries have signed and ratified it. Different countries fall under different groups with specific conforming regulations that they have agreed to follow to meet specific GHG concentration reduction levels. Unfortunately, the United States is not one of the ratifying countries in this landmark document.

The Intergovernmental Panel on Climate Change (IPCC) is another great example of countries working together toward a common goal. The IPCC is a scientific organization that was established by the United Nations Environment Programme (UNEP) and the World Meteorological Organization (WMO) in 1988. It comprised the world's top scientists in all relevant fields who review and analyze scientific studies of climate change and provide authoritative assessments of the state of knowledge regarding climate change; it was established to provide decision makers and others interested in climate change with an objective source of information.

Reports are produced at regular intervals and have been instrumental in providing both the public and the decision makers with relevant, reliable, and up-to-date information. The IPCC has been a driving factor in finding an international, cooperative solution to the problem.

PRACTICAL EXAMPLES OF GOVERNMENT GREEN PROGRAMS

The federal government is on board today with green technology and research to support renewable technology. There are several ongoing research programs in existence currently promoting the sustainable lifestyle, such as:

- The Biomass Program
- The Building Technologies Program
- The FreedomCAR and Vehicle Technologies Program
- The Geothermal Technologies Program
- The Solar Energy Technologies Program

The Biomass Program

A primary goal of the National Energy Policy is to increase energy supplies using a more diverse mix of existing resources available in the country and to reduce the dependence on

FIGURE 20.1 A fermentation tank with pumps and piping. This system converts animal waste into methane gas for use as a fuel and energy source. (Courtesy of NREL, Golden, CO.)

imported oil. The U.S. Department of Energy's Biomass Program develops technology for conversion of biomass to valuable fuels, chemicals, materials, and power to reduce the United States' dependence on foreign oil, cut back on emissions that contribute to pollution, and encourage the growth of biorefineries, which provides jobs.

Biomass is one of the United States' most important resources—it has been the largest U.S. renewable energy source since 2000. It also provides the only renewable alternative for liquid transportation fuel. Today's biomass uses include ethanol, biodiesel, biomass power, and industrial process energy.

In the future, biorefineries will use advanced technology such as hydrolysis of cellulosic biomass to sugars and lignin, and thermochemical conversion of biomass to synthesis gas for fermentation and catalysis of these platform chemicals to produce biopolymers and fuels. To expand the role of biomass in America's future, the Biomass Program's extensive and ongoing research and development helps biomass technologies advance (Figures 20.1 through 20.5).

The main goal of the government's energy program is to increase the nation's energy supplies using a more diverse mix of domestic resources. Its goal is to create a new bio-industry and reduce U.S. dependence on foreign oil by supplementing the use of petroleum for fuels and chemicals.

The Building Technologies Program

The federal government's Building Technologies Program focuses on analyzing the components inside different types of buildings and determining the most efficient forms of the major features that compose them. For example, the program looks at building types, such as homes, multifamily dwellings (apartments), offices, retail stores, health care facilities, hotels, schools, government buildings, and laboratories.

FIGURE 20.2 The internal workings of a gas production module that converts wood chips into producer gas. (Courtesy of NREL, Golden, CO.)

FIGURE 20.4 Sludge from paper mills is being used to produce levulinic acid. In the future, the acid may be used to make automotive fuel. (Courtesy of NREL, Golden, CO.)

FIGURE 20.3 An ethanol-powered snowplow in Hennepin County, Minnesota. (Courtesy of NREL, Golden, CO.)

FIGURE 20.5 Examples of plastic products that can be made from pyrolysis oil. (Courtesy of NREL, Golden, CO.)

It researches the elements inside a building, including appliances, ducts, heating and cooling, insulation, lighting, water and water heating, and windows, to find the most energy-efficient options. The program also develops energy-efficient strategies when doing a remodel of an existing building. For those who rent living space, it provides energy-saving tips for responsible energy management.

The program offers help in incorporating energy-saving features into new home construction right from the start, as well as for those who remodel. It also provides resources and information on solar panels, solar water heaters, and ground source heat pumps.

The FreedomCAR and Vehicle Technologies Program

The FreedomCAR and Vehicle Technologies Program is developing more energy-efficient and environmentally friendly highway transportation technologies that will enable Americans to use less petroleum. The goal of the program is to develop emission- and petroleum-free cars and light trucks. It is conducting the research necessary to develop new technologies such as fuel cells and advanced hybrid propulsion systems.

The program's hybrid and vehicle systems research is done with industry partners. Automobile manufactures and scientists from the program work together to design and test cutting-edge technologies. Energy storage technologies, especially batteries, are also part of the program. The battery is a technology critical to the development of advanced, fuel-efficient, light-duty, and heavy-duty vehicles. The program is in the process of developing durable and affordable batteries that cover many applications in a car's design, from starting and stopping to full-power hybrid electric, electric, and fuel cell vehicles. New batteries are being developed to be affordable, perform well, and be durable.

Advanced internal combustion engines are also being developed to be more efficient in light-, medium-, and heavy-duty vehicles. Along with efficiency in mind, they are being developed to meet future federal and state emissions regulations. Scientists believe that this technology will lead to an overall improvement of the energy efficiency of vehicles. Advanced internal combustion engines may also serve as an important element in the transition to hydrogen fuel cells.

New fuel types and lubricants are also being developed as part of the energy management program. The program's goal is to identify advanced petroleum- and nonpetroleum-based fuels and lubricants for more energy-efficient and environmentally friendly highway transportation vehicles. Nonpetroleum fuel components will come from nonfossil fuel sources, such as biomass, vegetable oils, and waste animal fats.

Research into materials technologies is another important component in the program. Advanced materials, including metals, polymers, composites, and intermetallic compounds, can play an important role in improving the efficiency of transportation engines and vehicles. Weight reduction is one of the most practical ways to improve fuel efficiency. The use of lightweight, high-performance materials will contribute to the development of vehicles that provide better fuel economy but are still comparable in size, comfort, and safety to today's vehicles.

The Geothermal Technologies Program

The Geothermal Technologies Program works as a partner with industries to establish geothermal energy as an economically competitive contributor to the U.S. energy supply. The Department of Energy works with individual power companies, industrial and residential consumers, and federal, state, and local officials to provide technical industrial support and cost-shared funding. The federal government even offers tax credits to businesses that use geothermal energy. Geothermal power is most commonly found in the western portions of the United States.

The Solar Energy Technologies Program

The Solar Energy Technologies Program is designed to develop solar energy technologies that supply clean, renewable power to the United States. This program focuses on five different types of energy management:

- Low-grade thermal energy for heating homes and businesses
- Medium-grade thermal energy for industrial processes
- High-grade thermal energy for driving turbines to generate electricity
- Electrical energy converted directly from sunlight to provide electricity for homes and other buildings
- Chemical energy in hydrogen for use in fuel cells and many electrical, heating, and transportation applications

Like other clean, renewable energy sources, solar energy technologies have great potential to benefit the nation. They can diversify our energy supply, reduce the dependence on imported fuels, improve air quality, reduce the amount of GHGs generated from other energy sources, and provide jobs for many people. Heat energy created from the sun's energy can be used to generate electricity in a steam generator. Because solar energy is fairly inexpensive and has the ability to provide power when and where it is needed, it can be a major contributor to the nation's future needs for energy.

This program's goal is to ensure that solar thermal technologies make an important contribution to the world's growing need for energy. An exciting new development in solar technology that the program is working on is hybrid solar lighting, which collects sunlight and routes it through optical fibers into buildings where it is combined with electric light in "hybrid" light fixtures. Automatic sensors keep the room at a steady lighting level by adjusting the electric lights based on the sunlight available.

EPA's Regulatory Initiatives

We discussed in Chapter 6 the work EPA is involved in with the mandatory GHG reporting rules, as well as the U.S. Climate Action Plan and the Clean Power Plan from the White House. The EPA is also in the process of developing common-sense regulatory initiatives to reduce GHG emissions and increase efficiency. Part of their work along this line includes their vehicle GHG rules, which are slated to save consumers $1.7 billion at the pump by 2025 and eliminate 6 billion metric tons of GHG pollution. The regulatory initiatives involve stationary sources with their standards on municipal solid waste (MSW) landfill air pollution. On July 14, 2016, EPA presented final updates to its new source performance standards designed to reduce emissions of methane-rich landfill gas (LFG) from

MSW landfills (to include new, modified, and reconstructed). In another action concerning stationary sources, they also issued guidelines for reducing emissions from existing MSW landfills. Both rules combined are expected to reduce methane emissions by an estimated 334,000 metric tons by 2025. This is roughly the equivalent of reducing 8.2 million metric tons of CO_2. These rules also cut CO_2 emissions directly, which yield an estimated 303,000 metric tons of additional CO_2.

On May 12, 2016, EPA issued three final rules that together will cut emissions of methane, smog-forming VOCs, and toxic air pollutants (e.g., benzene) from new, reconstructed, and modified oil and natural gas sources. This provides greater clarity about the Clean Air Act permitting requirements for the industry. The rules are expected to reduce methane emissions by 510,000 short tons of methane in 2025, which is the equivalent of reducing 11 million metric tons of CO_2.

The EPA also partners with the private sector. Through voluntary energy and climate programs, EPA's partners reduced more than 345 million metric tons of GHGs in 2010 alone, which is equivalent to the emissions from 81 million vehicles. They also saved consumers and businesses approximately $21 billion. The programs they offer include the following:

- Center for Corporate Climate Leadership
- The Combined Heat and Power (CHP) Partnership
- Energy Star
- Greenchill
- Green Power Partnership
- High Global Warming Potential Gases Voluntary Programs
- Methane Reduction Voluntary Programs
- Transportation and Air Quality Voluntary Programs
- Responsible Appliance Disposal Program
- WasteWise
- WaterSense
- Climate Ready Water Utilities
- Climate Ready Estuaries
- State and Local Climate and Energy Program
- Smart Growth Program

The *Center for Corporate Climate Leadership* is a resource center for any company that is interested in expanding their work in the area of GHG measurement and management. The purpose is to reduce GHGs through cost-effective partnerships to advance clean energy and energy efficiency across the U.S. economy. The center was created in 2012 to establish and coordinate climate leadership by encouraging organizations that have climate-related objectives to identify and achieve cost-effective GHG emission reductions and also to help more advanced organizations spearhead innovations to reduce their GHG impacts. The center serves as a central resource to assist organizations of all sizes measure and manage their GHG emissions. They provide the technical tools needed, assist with ground-tested guidance, provide educational resources, and offer opportunities for information sharing and peer exchange among various organizations that are interested in reducing the environmental impacts associated with climate change.

The *Combined Heat and Power (CHP) Partnership* is a voluntary program designed to reduce the environmental impact of power generation by promoting the use of CHP, which is seen as an efficient, clean, and reliable approach to generating power and thermal energy from a single-fuel source. The goal of the CHP Partnership is to reduce the air pollution and water usage that is associated with electric power generation by promoting the use of CHP. The EPA's goal is to remove the policy barriers and facilitate the development of new projects in the United States and its territories by promoting the economic and environmental benefits of CHP. They provide the tools, policy information, and other resources to energy users; the CHP industry; clean air officials; and others that are involved in promoting clean energy to promote sustainability.

As discussed previously, *Energy Star®* is a joint program of EPA and the U.S. Department of Energy, helping us all save money and protect the environment through energy-efficient products. It is the most successful voluntary energy conservation movement in history. Not only has the program helped consumers and businesses save money and energy, but it also has had a significant impact on climate change mitigation. Because energy use in homes, buildings, and industry account for two-thirds of GHG emissions in the United States, the use of Energy Star® products has been key in reducing energy use, which has significantly reduced GHG emissions. Because the program has grown steadily over time—nearly tripling in the past decade—the benefits to health and the environment have also increased. The program has also strengthened the economy.

GreenChill is a voluntary partnership between EPA and the supermarket and refrigeration industry to promote green technologies, strategies, and practices that protect the stratospheric ozone layer, reduce GHGs, and save money. The program reduces refrigerant emissions and decreases their impact on the ozone layer. The program accomplishes this by helping food retailers' transition to environmentally friendlier refrigerants; lower refrigerant charge sizes and eliminate leaks; adopt green refrigeration technologies, and employ the best environmental practices available.

The *Green Power Partnership* is a voluntary partnership between EPA and organizations that are interested in using green power, which is electricity produced from a subset of renewable resources—such as wind, solar, geothermal, biomass, and low-impact hydropower. This program, launched in 2001, was designed to increase the use of renewable electricity in the United States. In return for technical assistance to organizations that want to utilize green power as a way to reduce their environmental impacts, the green power partners commit to use green power for all, or a portion, of their annual electricity consumption. The EPA provides technical assistance and education on renewable power procurement. Currently, the partnership has hundreds of partner organizations using billions of kilowatt-hours of green power annually.

The *High Global Warming Potential Gases Voluntary Programs*, also referred to as the Fluorinated Gas Partnership Programs, are voluntary industry partnerships that are considerably reducing U.S. emissions of the highly potent GHGs. These are gases emitted from a variety of industrial processes. Unlike many other GHGs, fluorinated gases (also called F-gases) have no natural sources—they originate only from human-related activities. They are emitted through several industrial processes such as aluminum and semiconductor manufacturing. Many F-gases have very high global warming potentials relative to other GHGs, which means that even small atmospheric concentrations can have significant effects on global temperatures. They can also have long atmospheric lifetimes—in some cases, lasting thousands of years. Like other long-lived GHGs, F-gases are well mixed in the atmosphere, spreading around the world after they are emitted. F-gases are removed from the atmosphere only when they are destroyed by sunlight in the far upper atmosphere. In general, F-gases are the most potent and longest lasting type of GHGs emitted by human activities. That is why these partnerships are so important. The reduction of F-gases is critical in fighting climate change and EPA and their partners are actively participating in their reduction.

The *Methane Reduction Voluntary Programs* work with U.S. industries, as well as state and local governments, to promote the reduction of methane (CH_4) emissions. They do this through the following programs:

- AgSTAR
- Natural Gas STAR
- Global Methane Initiative
- Coalbed Methane Outreach Program
- Landfill Methane Outreach Program

AgSTAR encourages the use of biogas recovery systems to reduce methane emissions from livestock waste. They also promote the use of anaerobic digestion systems, which is a process through which bacteria break down organic matter without oxygen. As the bacteria process the organic matter, they generate biogas. The biogas is composed mainly of methane, which is the primary component of natural gas. The nonmethane components of the biogas are then removed, so that the methane can be used as an energy source.

The *Natural Gas STAR* program involves oil and gas operations and works to reduce methane emissions. This is geared to improve the performance of the oil and gas operations, increase the natural gas supply, help the company save money, and protect the environment. It also enables partners to participate in Technology Transfer Workshops, Annual Implementation Workshops, and Web-based communications to build networks with fellow industry peers, so that they can keep up with the latest technologies.

The *Global Methane Initiative* involves 38 governments (national partners), the European Commission, the Asian Development Bank, and the Inter-American Development Bank, and it is designed to fight climate change while developing clean energy and strengthening economies. The initiative is intended to build on the existing Methane to Markets Partnership (a global partnership led by the United States

and Mexico in 2004 to focus attention on the importance of reducing methane emissions) to reduce emissions of methane, while enhancing and expanding those efforts and encouraging new resource commitments from each country. This initiative includes the following elements:

- *Expanded scope*—Including methane abatement and avoidance from existing and new sectors, such as municipal waste water.
- *Methane action plans*—All partner countries will develop actions plans to coordinate methane reduction efforts at home and abroad. Developed countries would provide coordinated assistance to developing country partners.
- *New resource commitments*—From developed country partners, and others in a position to do so, as well as the broader international community, so as to accelerate global methane abatement.

The Coalbed Methane Outreach Program (CMOP) has been involved with the coal-mining industry in the United States and other major coal-producing countries since 1994, in an effort to reduce coal mine methane (CMM) emissions. In their efforts to help identify and implement strategies to recover and utilize CMM instead of emitting it directly to the atmosphere as GHG emissions, the Coalbed Methane Outreach Program has played a significant role in the United States' efforts to significantly reduce GHG emissions in an effort to fight climate change. In 2014, the program reduced CMM emissions by more than 10 million metric tons of CO_2 equivalent. In addition to projects in the United States, the program has also helped promote CMM recovery in China, India, Kazakhstan, Mongolia, Poland, the Russian Federation, Turkey, and Ukraine.

The Landfill Methane Outreach Program (LMOP) is a technical assistance program that helps reduce methane emissions from landfills by encouraging the recovery and use of LFG as a renewable energy resource. LFG contains methane, which when captured can be used to fuel power plants, manufacturing facilities, vehicles, and more. In order to do this, the LMOP forms partnerships with communities, landfill wners, utilities, power marketers, states, project developers, tribes, and nonprofit organizations. They help their Partners with project development by assisting them with determining project feasibility, procuring financing, and marketing the benefits of the finished project to the community to get the project underway and completed.

The *Transportation and Air Quality Voluntary Programs* are designed to reduce pollution and improve air quality through the formation of partnerships with both small and large businesses, citizen groups, industrial interests, manufacturers, trade associations, and state and local government. This is done primarily through three programs:

- The National Clean Diesel Campaign
- The SmartWay Transport Partnership
- The Clean School Bus USA

The National Clean Diesel Campaign offers funding in the form of grants and rebates in addition to other forms of support for projects that protect human health and improve air quality by reducing harmful emissions from diesel engines. The SmartWay Transport Partnership is designed to help freight shippers, carriers, logistics companies, and others measure, benchmark, and improve their logistics operations so that they can reduce their carbon footprint. It enables partners to ship freight smarter and more sustainably. Launched in 2004, the public–private program

- Provides a smart system for tracking, documenting, and sharing information about fuel use and freight emissions across supply chains.
- Helps companies identify and select more efficient freight carriers, transportation modes, equipment, and operational strategies to improve supply chain sustainability and lower costs.
- Supports global energy security and offsets environmental risks.
- Reduces freight transportation-related climate change and air pollutant emissions.
- Is supported by major transportation industry associations, environmental groups, state and local governments, international agencies, and the corporate community.

The *Clean School Bus USA* program is a national program designed to help communities reduce emissions from older diesel school buses. EPA offers funding, as appropriated annually by Congress, for projects that reduce emissions from existing diesel engines. EPA also provides information on strategies for reducing emissions from older school buses. One of the easiest ways to reduce school bus emissions and save money is to reduce idling. Another effective method is to replace the oldest school buses in the fleet. This program monitors these issues.

The *Responsible Appliance Disposal (RAD) Program* is a partnership program established to help protect the ozone layer and reduce emissions of GHGs by recovering ozone-depleting chemicals from old refrigerators, freezers, window air conditioners, and dehumidifiers. Under this program, the RAD partners ensure that:

- Refrigerant is recovered and reclaimed or destroyed.
- Foam is recovered and destroyed, or the blowing agent is recovered and reclaimed.
- Metals, plastic, and glass are recycled.
- PCBs, mercury, and used oil are recovered and properly disposed.

EPA serves as a technical advisor and provides partner recognition for achievement and awards. RAD partners include utilities, retailers, local governments, manufacturers, universities, and other organizations.

WasteWise eliminates costly municipal solid and select industrial wastes, benefiting their bottom line and reducing the amount of waste deposited in landfills and GHG emissions. This is part of EPA's sustainable materials management effort, which promotes the use and reuse of materials over their entire lifecycle. A good approach to environmental stewardship, this is seen as a comprehensive approach to reducing waste. Benefits of this program include reductions in purchasing new materials by repurposing used materials, lowering input into landfills, and lowering wasted resources by eliminating the need for new manufacturing. The program participants can search databases, such as the Sustainable Materials Management (SMM) Data Management System for materials and recycling information.

WaterSense is a partnership program that aims to protect the future of the nation's water supply by offering the public a simple way to use less water with water-efficient products, new homes, and services, which reduces energy use and GHG emissions in the process. The program provides resources to learn about water efficiency and educational tools to learn how to become more water smart and conserve water through daily activities. It also educates the public about products that bear the WaterSense label, which are designed to be more water conservative.

The *Climate Ready Water Utilities (CRWU) Initiative* is a partnership effort with the public sector (states, tribes, localities, and resource managers). This provides resources for the water sector to adapt to climate change by promoting a clear understanding of climate science and adaptation/mitigation options. They also provide education concerning integrated water resources management planning in the water sector. This is a program largely designed to help water utilities plan for climate change impacts. CRWU works with drinking water utilities, wastewater utilities, and stormwater utilities, providing them with the practical tools, training, and technical assistance necessary to adapt to future climate changes by promoting a clear understand of climate science and what adaptation options are available to use, why they will be necessary, and when.

The *Climate Ready Estuaries* program works with the National Estuary programs to help coastal managers understand and determine climate change vulnerabilities, implement adaptation strategies, educate others involved, and be able to share what they have learned with others. This will be important as sea levels rise and coasts become affected and inundated. Adaptation will become necessary and by working together through this program, adaptation options will run smoother, be easier to access, and a data communication base will be available for coordination and emergencies.

The *State and Local Climate and Energy Program* is an effort to offer expertise to local governing agencies about energy efficiency, renewable energy, and climate change policies and programs that can help communities at these levels. For example, it helps local areas understand where their sources of GHG emissions are coming from and helps them implement plans of action to cut back on emissions. It looks at specific community impacts and helps them cut back and also implement plans where necessary that can deal with climate change adaptation and mitigation.

The *Smart Growth Program* covers several development and conservation strategies that help protect our health and natural environment and make our communities more attractive, economically stronger, and more socially diverse. Designed to achieve a well-balanced society, "smart growth" is applied to communities of all sizes across the country, utilizing creative strategies to develop in ways that preserve natural lands and critical environmental areas, protect water and air quality, and reuse already developed land. They adhere to the following principles:

- They conserve resources by reinvesting in existing infrastructure and rehabilitating historic buildings.
- By designing neighborhoods that have homes near shops, offices, schools, houses of worship, parks, and other amenities; communities give residents and visitors the option of walking, bicycling, taking public transportation, or driving as they go about their business.
- By offering a range of different housing types, it makes it possible for senior citizens to stay in their neighborhoods as they age; young people to afford their first home; and families at all stages in between to find a safe, attractive home they can afford.
- By using smart growth approaches that directly involve residents in the development decisions, these communities are creating vibrant places to live, work, and stay involved.
- By creating a high quality lifestyle, these communities become economically competitive, which strengthens the local tax base.

The concept has created what EPA calls their Smart Growth Network, which has proven to be very successful and faces a positive future.

The Paris agreement on climate change entered into force on Friday, November 4, 2016, marking the first time that governments have agreed to legally binding limits on global temperature rises. Seen as the fruit of more than two decades of often tortuous international negotiations on combating climate change, it was hailed by nations and observers around the world. Under the agreement, all governments that have ratified the accord, which includes the United States, China, India, and the European Union, now have an obligation to hold global warming to no more than 2°C above preindustrial levels. That is what scientists regard as the limit of safety—also called the tipping point—beyond which climate change is likely to become catastrophic and irreversible. Countries have put forward commitments on curbing carbon emissions under the agreement, but a report on Thursday found those pledges would see temperature rises significantly overshoot the threshold, with 3°C of warming. Environmental groups urged governments to do more. (Source: Harvey, 2016).

OTHERS SETTING A GOOD EXAMPLE

Several cities and states across America, as well as countries worldwide, are taking measures to live a greener lifestyle and are achieving noteworthy results.

America's Green Cities and States

There are quite a few notable cities and states that are leading the way to a greener country. Here are some examples:

- Texas has added more than 4,000 MW of wind power-generating capacity in the past decade. Wind power now provides 3 percent of Texas' electricity, which is enough to keep about 8 million metric tons of GHGs out of the atmosphere each year.
- New Jersey has doubled its solar power-generating capacity within the past few years through public policies that promote solar panels on rooftops.
- California uses 20 percent of less energy per capita than it did in 1973 thanks to strong energy efficiency policies for buildings and appliances.
- Wisconsin has adopted several environmental policies to promote energy efficiencies in industry. These programs have not only been able to save businesses money and create new jobs within the state, but they have also kept 200,000 metric tons of CO_2 out of the atmosphere.
- Portland, Oregon, has more than doubled the number of bicyclists in the past decade by making the city bicycle friendly.
- Improvements to the mass transit systems in Rosslyn and Ballston, Virginia, have encouraged about 40 percent of the residents to take mass transit.
- Southeastern Pennsylvania has a 20 percent increase in the number of passengers on the trains that travel to Harrisburg and Philadelphia because the travel speeds were increased, making the trains more efficient, reliable, and attractive to use (Figure 20.6).

A highlight being reported from Seattle, Washington (one of America's greenest cities), is that their Seattle Climate Action

FIGURE 20.6 Portland, Oregon, has been voted one of the greenest cities of America. (Courtesy of Truflip99, Creative Commons.)

Plan will lay out a roadmap for how they can become a carbon neutral city by the year 2050 and be prepared for the impacts of climate change. The plan will include strategies identifying how they can reduce their GHGs in the transportation, building, energy, and waste sectors. Some of the initiatives in the action plan include:

- Construction of green buildings and green site development.
- Creation of bike- and pedestrian-friendly neighborhoods.
- Enhancement of public transportation, such as increases in local bus service, construction of light rail, and completion of streetcars.
- Energy efficiency and conservation programs with city residents.
- City recycling programs.
- Water conservation measures (current measures have kept water consumption levels in Seattle equivalent to what they were in 1975, even though there are 400,000 more residents living there now).
- Planting 649,000 trees over the 30 years. This will increase canopy cover from 18 percent to 30 percent.

According to *EcoWatch*, who completed a 2014 assessment on America's cities based on the following criteria:

- *Environmental Quality: Healthy air*: defined as "2014 median of the U.S. Environmental Protection Agency's daily Air Quality Index—a measure of sulfur dioxide, carbon monoxide, nitrogen dioxide, and ozone levels that guide the assessment of acute health effects. The higher the index number, the more polluted an area"
- *Transportation: Green transportation*: defined as "data from the U.S. Census Bureau's American Community Survey on commuting methods to find out how many workers are walking, biking, carpooling and taking public transit"
- *Energy sources: Green energy*: defined as "data from the American Community Survey on heating in homes. We were specifically interested in the number of homes using coal and wood because these fuels are particularly harmful. We also examined the use of solar energy because of its sustainability"
- *Housing density: Urban concentration*: defined as "data from the American Community Survey on the percentage of residential buildings with 10 or more residences in each city" (Figures 20.7 through 20.9).

FIGURE 20.7   Honolulu, Hawaii. (Courtesy of Cristo Vlahos, Creative Commons.)

FIGURE 20.8   Salt Lake City, Utah. (Courtesy of Garrett, Creative Commons, Mountain View, CA.)

FIGURE 20.9   Seattle, Washington. (Courtesy of Daniel Schwen, Creative Commons, Mountain View, CA.)

They identified the top 25 of the 2014 Greenest Cities in America as (EcoWatch, 2015):

America's 25 Greenest Cities–2014

Honolulu, Hawaii
Washington, DC
Arlington, VA
San Francisco, CA
Miami, FL
New York City, NY
Boston, MA
Orlando, FL
Seattle, WA
Jersey City, NJ
Fort Lauderdale, FL
Minneapolis, MN
Austin, TX
Madison, WI
Hialeah, FL
Atlanta, GA
Portland, OR
Rochester, NY
Durham, NC
Irving, TX
St. Paul, MN
Salt Lake City, UT
Lincoln, NE
Denver, CO
Oakland, CA

The following list highlights some of the achievements of the green cities:

- Portland, Oregon, came in first place as America's greenest city. More than half of Portland's energy comes from renewable sources. Portland has in place the Clean Energy Works, a first-in-the-nation program that gives homeowners free energy assessments and provides $2,000 rebates and loans for home retrofitting. They also have a curbside composting program that has resulted in a 38 percent drop in the city's trash output.
- San Francisco, California, has a mandatory recycling and composting ordinance, which requires all people within the city to separate their recyclables from their trash, and also separate out compostable food and packaging. As a result, by October 2012, 80 percent of the city's waste was going to recycling and composting facilities, instead of to landfills. San Francisco now leads America in sustainable waste disposal. They also have approximately 700 LEED-certified building projects underway.
- Seattle, Washington's utility company became the first in the nation to go carbon neutral in 2005, thanks to their hydroelectric dams. They no longer invest their money in fossil fuel companies.
- Minneapolis, Minnesota, has more than 160 miles of bikeways, more than half of them are not alongside a road. They are advocates of the city's tap water, discouraging bottled water.

- Austin, Texas has committed to becoming carbon neutral by 2020. Citizens there can purchase their energy from green sources instead of fossil sources; and it has been highly successful for nine consecutive years. The city has more than 19,000 acres of parkland (more than 30 square miles), covering more than 10 percent of the city.
- New York City, New York, has one of the lowest levels of emission per capita and is also one of the most walkable cities in the country. Many initiatives have been launched that deal with sustainability and fighting climate change, such as reducing GHG emissions by more than 30 percent, achieving the cleanest air quality of any large U.S. city, diverting 75 percent of solid waste from landfills, improving their green building codes, preparing coastal areas for the effects of climate change, and planning for a warmer climate in the future.
- Salt Lake City, Utah, has made it a goal to reduce carbon emissions produced by city government buildings by 21 percent over the next decade. The decade is not over yet and the city has already reduced emissions by 31 percent. In addition, the city has installed solar-powered parking meters throughout the downtown area, is using alternative fuel vehicles, installed electric vehicle charging stations in the city to prepare for a "new wave of affordable, electric-only vehicles" (Light, 2013).

The World's Role Models

Other countries are ahead of the United States in their efforts to green up the world. The environmental movement has been around much longer in other countries, so they have had much more time to green up and be an excellent role model. They have also been much more amenable to living sustainable lifestyles, which has put them ahead of the United States.

Freiburg, Germany, for example, is a quaint town that has been greening itself up for decades—since the 1940s, to be exact. Freiburg offers its car-free sector of Vauban as well as its Solar Village (Figure 20.10). Zermatt, Switzerland, another example, is a quaint alpine village town at the base of the world famous Matterhorn. Only pedestrians and bicycles use its green streets. Deliveries of merchandise are pulled through the town either on hand carts, horse carts, or on manually steered electric carts. The small electric freight vehicles that are allowed to travel the streets require special permits (emergency vehicles are exempt).

The cities of the world are also rated for their level of sustainability by the Centre for Economic and Business Research in the Sustainable Cities Index. They are rated on three characteristics: *people*, *planet*, and *profit*. The people factor is a measure of social performance, which looks at health aspects, education levels and opportunities, income inequality, work–life balance, dependency ratio, crime, housing, and living costs. Together, these factors provide a *quality of life* measure. The planet factor ranks cities on energy consumption and renewable energy share, green space within cities, recycling efforts, GHG emissions, natural catastrophe risk, drinking water quality, sanitation, and air pollution. These are considered the city's *green factors*. The profit factor looks at business perspectives, transportation infrastructure, ease of doing business, tourism, GDP per capita, the city's importance in global economic networks, communication connectivity, and employment rates. This is referred to as the *economic health* of the city. The top 10 cities for 2016 include the following (Figures 20.11 and 20.12):

---

10 Greenest Cities in the World 2016

Zürich, Switzerland

Singapore

Stockholm, Sweden

Vienna, Austria

London, UK

Frankfurt, Germany

Seoul, South Korea

Hamburg, Germany

Prague, Czech Republic

Munich, Germany

FIGURE 20.10 Freiburg's solar village or solar settlement. (Courtesy of Andrew Glaser, Creative Commons.)

FIGURE 20.11 Zürich, Switzerland and Lake Zürich. (Courtesy of MadGeographer, Creative Commons, Mountain View, CA.)

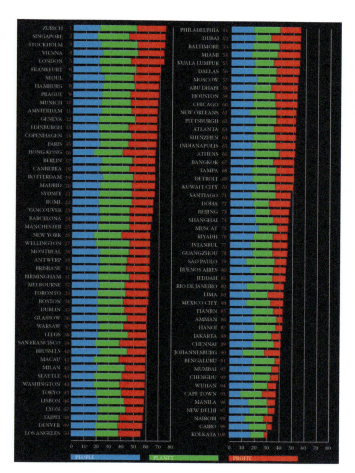

FIGURE 20.12 Sustainable Cities Index 2016. (Courtesy of Arcadis, Amsterdam, the Netherlands; http://ecourbanlab.com/.)

The index report emphasizes that there were no cities that did equally well in all three areas. Many did well in two, but very few cities even came close to being somewhat balanced in three. What the authors felt this indicated was that it was a challenge for cities to balance all three needs effectively to ensure long-term sustainability. Developing sustainability takes effort, time, and trial; but with practice and cities being willing to learn from each other, they will eventually figure out the solutions.

Consider these worldly green facts:

- Germany has recycled 60 percent of its municipal waste for the past 20 years. It has enacted policies that put the responsibility of recycling on product manufacturers instead of individual consumers and taxpayers.
- In Israel, more than 90 percent of the homes use solar water heaters, which have greatly reduced the need for natural gas and electricity for water heating. Israel requires that all new homes are equipped with solar water heaters (Figure 20.13).
- In Copenhagen, Denmark, pedestrians and bicyclists are given preference over cars in its downtown city center section. Currently, about 40 percent of the population walks or rides bicycles as a principal form of transportation.

FIGURE 20.13 Solar boiler on a rooftop in Israel. (Courtesy of Gilabrand, GNU Free Documentation License, Albany, NY.)

- Spain is now third in the world for wind farms and wind power capacity and is the world's fourth leading market for solar photovoltaics.

THE FINAL CHOICE IS YOURS

There are just a few things left to cover. With everything we have covered in this book—natural resources, the climate system, GHGs, ecosystems, human psychology, geopolitics, energy basics, nonrenewable and renewable energy, sustainability, urban planning, smart technology, alternate and new fuel technology, sustainable agriculture and industry, green careers, making your own home green, and living a green lifestyle—you are now aware of the challenges we face as a global community; but of the enormous payoffs if we face them, adapt, and succeed. You understand the situation we are in, are aware of the options we have to do something about the situation—and the very real consequences if we do not. You are now prepared to make educated choices about your future and make a meaningful difference. You are fully prepared to adopt a healthier lifestyle, lead by example, and teach others. So, you may wonder: what is next? That is the final key. And the answer is you must make the choice to do it. You must make the commitment. Anyone can have the knowledge; but knowledge in itself, unfortunately, is not enough. Action is required. You much choose to act.

PUTTING IT ALL TOGETHER: PERSONAL CHOICES

Look at this chapter as the last piece of the big picture. This is what puts everything else that has been discussed into perspective.

What we have talked about is critical—the tools you need to succeed in going green. Now you need to put them to use.

Again, you have heard all the great reasons to go green: for the earth, for all its scarce and precious resources, and for its atmosphere; for your health, your family's health, and for the future; for new job opportunities, for tax breaks and rebates, and because it is simply the right thing to do. You know we need to conserve, preserve, and reserve. You know how to go green at home, at school, at work, on vacation, while driving, during your vacation, and anywhere else in between. You know you can get into it a little at a time or jump right in with gusto. You can be a little green or a lot. You have the ability to succeed, to be empowered, to influence others, to even be a leader, or to be a role model if you choose. You have learned a little bit about human nature—what makes us tick, why we make the decisions we do, why we worry about certain things, what it takes to make us worry and finally take action, why we sometimes procrastinate and ignore serious issues, and why we often resist making lifestyle changes, even when they might be good for us, our families, and future generations. And because you know these little idiosyncrasies about human nature, you are now empowered to see past that and make wiser decisions because of it.

You know the importance of following a few common sense rules—and the far-reaching ramifications if you do not. You know how to recycle, to renew, to recharge. You are now armed with all kinds of tips and tricks about treating the environment around you with care. But when it comes down to it, the big question is what are you going to do with it? You can know everything there is to know about sustainability, but it does not matter until you take that first step toward living it. To get positive results, your actions must cause reactions.

The Future Is Now

The time to look toward the future is now. Decisions we make, systems we create, facilities we build, and laws we enact will all affect us and the generations of the future. It is critical that society makes sustainable, intelligent choices now. It also comes down to electing politicians who will do the same and act responsibly; and we are the ones who put them there. While scientists are engaged in advanced research, the time to boldly move ahead to help the populations of tomorrow is today.

Everyone Is Somebody's Neighbor

As population increases and environmental degradation occurs (such as air pollution, deforestation, soil erosion, water pollution, overcrowded landfills, waste of natural resources, etc.), it is becoming an increasing struggle to maintain a healthy, well-balanced environment—and that responsibility is for everyone. Everyone is somebody's neighbor, and everyone's actions have the potential to hurt or help the land.

Everyone Can Get Involved

It does not matter where you live—in a city, on a farm, or in a suburban neighborhood. You can practice responsible conservation techniques in your own backyard. You can do your part to promote water quality and biodiversity. Your responsible lifestyle practices can benefit the environment no matter where you live because there are programs and plans available that fit every lifestyle and living arrangement.

For example, if you are lucky enough to own land, you can practice conservation techniques on your own land; if you live in a subdivision with a backyard, there are many backyard conservation opportunities for gardening, composting, feeding wild birds, soil conservation activities, and many other activities you can get involved in.

If you live in an inner city environment, there are community programs available you can get involved in that enable you to participate in various conservation activities such as gardening, recycling, and other conservation initiatives.

Everyone can get involved with conservation organizations, community volunteer organizations, and government environmental initiatives. There are even international environmental volunteer opportunities available for anyone interested in participating. With a little research and initiative, there is no limit to how much you can become involved.

Make Every Day Earth Day

If truth be told, everyone would like to have a safe and healthy environment to live in and be assured their families and future generations would have the same. So why not work together and strive for every day to be an Earth Day? If you have never participated in any Earth Day celebration or related activities, pencil the next April 22 onto your calendar and give it a try so you know what all the hubbub is about.

This is a great opportunity to contact an environmental organization in your community and get involved. Simply working together greatly increases the chance of improving the health of the environment on a long-term, lasting basis. Joining forces with environmental organizations makes it easy for you to get involved and find out how you can contribute to going green and really make a difference.

Every year on Earth Day volunteers around the world plant trees, restore trails and wildlife habitat, clean up beaches, parklands, and rivers, and contribute to making our planet a healthier and more beautiful place to live in. Why keep the celebration to just one day a year? Why not take action to improve the earth more often? You can have as many Earth Days as you like. You can get every member of your family involved for a rewarding experience.

A small step you can take is to put aside one Saturday each month to get involved in an Earth Day activity. It could be something as simple as cleaning up a playground or taking a trip around the neighborhood or grocery store parking lot to pick up litter.

Let me take a moment to interject a personal story here on just how powerful these little parent teaching moments can be. Years ago when my kids were in elementary school, the day after school was out for the summer, someone had toilet papered the school grounds and dumped what looked like multiple backpacks of old school papers and assignments all over the playground. We lived close to the school and I saw what had happened and (yes, in my greenie, tree hugger way) was mortified. So I grabbed some plastic garbage bags and told the kids we were going on a little field trip, and off we went.

When we walked to the school and they realized what I had in mind, at first, they were reluctant—after all, they protested, they did not make that mess! So I started cleaning up the grounds by myself while my kids watched and explained to them that I had not make the mess either, but someone has to step up to the plate because the garbage was not going to clean itself up. It only took a few minutes before my three elementary-age kids each slowly picked up a garbage bag and began to pick up trash alongside me. Between the four of us, we had the grounds cleaned within an hour.

And that one-hour investment taught my kids a lesson for life. Since that time, we have been many places—hiking, in a parking lot, at a national park, you name it—where they have seen litter on the ground and cleaned it up without any prompting. I have even seen them pick up trash in front of friends. When the friends ask, my kids explain why they are picking up the trash. Then the friends follow suit and pick up the litter! That right there makes it all worthwhile. Those kids are our future, and with examples like that, we are in good hands.

Never underestimate the potential of youth; and as parents, never underestimate the effect one of those little impromptu teaching moments just might have.

FIGURE 20.14 Elbert Wells, a Natural Resources Conservation Service (NRCS) Project Leader, and students from the Hartranft Elementary School in Philadelphia, collect seed heads from marigolds. The students plant, maintain, and harvest flowers and vegetables in a garden; they built themselves. (Courtesy of NRCS, Washington, DC.)

Now back to Earth Day—if you want to join in, here is a sampling of some organizations you might want to start with:

- *Campaign Earth, http://www.campaignearth.org*: This national organization works to make local communities aware of how they can help prevent serious environmental threats like climate change. It sponsors a program called the Monthly Challenge, which is a one step at a time approach that shows you how to take those baby steps to make the small changes necessary to create a greener environment.
- *Earth Day Network (EDN), http://www.earthday. org*: Organized as a result of the first Earth Day back in 1970, EDN currently works with over 22,000 partners in 192 different countries to broaden, diversify, and mobilize the global environmental movement. More than 1 billion people now participate in Earth Day activities each year, making it the largest civic observance in the world. EDN works with several partner organizations at many different levels and scales—local, regional, national, and international.
- *EarthCorps, http://www.earthcorps.org/*: This is a nonprofit volunteer-based organization that works to improve environmental health by getting young people involved. It focuses on school-aged children because it sees them as the strength of the future.
- *Envirolink Resource Guide, http://earthday.envirolink.org/*: The EnviroLink network has served as an online clearinghouse for environmental information since 1991.

These suggestions should get you started. You can also check your local schools, libraries, and county and community parks and recreation departments. They often sponsor activities of their own. Also check your newspapers, call your local radio stations, and check with your local news stations for local information. Finding activities to attend on Earth Day these days is not hard; sometimes it is finding enough hours in the day to attend all the events you would like to. Do a little bit of exploring and see what you can find—you may be surprised by how much fun you have while doing so much good (Figure 20.14).

A LONG-TERM COMMITMENT

Our willingness to recycle, reduce, and reuse will help protect the environment now and for generations to come. To conserve our resources and promote land stewardship, we must make a long-term commitment. Actions we take now can affect the environment long into the future. Everyone has a role in maintaining healthy ecosystems and keeping the land productive.

You Do Make a Difference!

Perhaps one of the best ways to make the transition to going green and for it to be meaningful is to get directly involved and hold a personal stake in the outcome. If you make a personal commitment, then you have a stake in how making the transition to going green turns out because it directly affects you, your children, grandchildren, and all future generations. If you own that piece of the future, you are responsible for its outcome, and you have a direct interest in it turning out successfully. Perhaps the entire situation is best summed up by these immortal words from one very remarkable person that set such a good example for us all to follow:

You must be the change you wish to see in the world.

– Mahatma Gandhi

CONCLUSIONS

You now know that sustainability is a very broad subject, dealing with the environment, physical science, social science, business practices, politics, and more—an entire spectrum of important interactions between different components. The world today needs these skills to solve the most pressing problems humankind is now facing—how to solve global problems like climate change; dwindling nonrenewable energy resources and replacing those with renewable ones; replacing dirty energy with clean energy and fighting a warming climate. These are the next big problems the world must solve if we are to advance, or even survive. It all comes down to the concepts you have just learned in this book. It all depends on how you react to what is going on around you and how you choose to deal with it. This is not a problem that will go away on its own, not is it a problem that can be left for someone else alone to fix. It is a community problem that will require the participation and commitment of every single human being because each person really does count—really does make a difference. The future is, quite literally, in your hands.

REFERENCES

EcoWatch. April 15, 2015. Top 25 greenest cities in America. *EcoWatch*. http://www.ecowatch.com/top-25-greenest-cities-in-america-1882033203.html (accessed November 4, 2016.)

Harvey, F. November 3, 2016. Paris climate change agreement enters into force. *TheGuardian*. https://www.theguardian.com/environment/2016/nov/04/paris-climate-change-agreement-enters-into-force (accessed November 6, 2016.)

Light, J. 2013. 12 cities leading the way in sustainability. *Moyers and Company*. http://billmoyers.com/search-results/?q=5+cities+with+game-changing+sustainability+projects#gsc.tab=0&gsc.q=5%20cities%20with%20game-changing%20sustainability%20projects&gsc.page=1 (accessed November 3, 2016.)

SUGGESTED READING

Gifford, D. 2015. *Sustainability Starts at Home: How to Save Money While Saving the Planet*. San Diego, CA: Small Footprint Family, 346 p.

Jacques, P. 2014. *Sustainability: The Basics*. London, UK: Routledge, 250 p.

Johnson, B. 2013. *Zero Waste Home*. New York, NY: Scribner, 304 p.

Roderick, J. 2014. *Recycled Thoughts: Just how Green is Green?*. Seattle, WA: Amazon Digital Services, 65 p.

Glossary

acid rain: rainfall made sufficiently acidic by atmospheric pollution that causes environmental harm, typically to forests and lakes. The main cause is the industrial burning of coal and other fossil fuels; the waste gases emitted contain sulfur and nitrogen oxides that combine with atmospheric water to form acid rain

aerosol: a colloid of fine solid particles or liquid droplets in air or another gas. Aerosols can be natural or artificial. An example of a natural aerosol is fog. Examples of artificial aerosols are haze, dust, particulate air pollutants, and smoke

albedo: the relative reflectivity of a surface. A surface with high albedo reflects most of the light that shines on it and absorbs very little energy; a surface with a low albedo absorbs most of the light energy that shines on it and reflects very little energy

anthropogenic: made by people or resulting from human activities. This term is usually used in the context of emissions that are produced as a result of human activities

astroturf group (astroturfing): the practice of masking sponsors of a message or organization (e.g., political, advertising, religious, or public relations) to make it appear as though it originates from and is supported by a grassroots participant(s)

biodegradable: an object that is capable of being decomposed by bacteria or other living organisms

biodiversity: a contraction of "biological diversity," it refers to the variety and variability of life on earth. It refers to the variability within species, between species, and between ecosystems

biofuels: converting biomass into liquid fuels for transportation

biomass feedstock: any renewable, biological material that can be used directly as a fuel or converted to another form of fuel or energy product

biopower: burning biomass directly or converting it into a gaseous fuel or oil to generate electricity

bioproducts: converting biomass into chemicals for making products that are typically made from petroleum

bioremediation: a waste-management technique that utilizes organisms to remove or neutralize pollutants from a contaminated site

biosphere: the global sum of all ecosystems. It is also referred to as the zone of life on earth, a closed system (apart from solar and cosmic radiation and heat from the interior of the Earth), and mostly self-regulating

carbon sequestration: the uptake and storage of carbon. Trees and plants absorb carbon dioxide, release the oxygen, and store the carbon

chemical energy: energy stored in chemicals that can be released when a chemical takes part in a chemical reaction

clerestory: the upper part of an interior ceiling rising above adjacent rooftops and having windows allow daylight to the interior. Often used in solar heating applications

climate justice: a term used for framing climate change as an ethical and political issue, rather than one that is purely environmental or physical in nature

climate: the usual pattern of weather that is averaged over a long period of time

Coriolis force: a representative artifact of the Earth's rotation. In physics, it is an apparent deflection of moving objects when they are viewed from a rotating reference frame. In a reference frame with clockwise rotation, the deflection is to the left of the motion of the object; in one with anti-clockwise rotation, the deflection is to the right

desiccant cooling: open cycle cooling systems using water as a refrigerant in direct contact with air

direct solar gain: the heat from the Sun being collected and contained in an occupied space. This heat can be retained by the building's thermal mass or can be avoided with reflective materials

distillate fuel: a petroleum fraction typically obtained through a conventional distillation process. It includes diesel fuels, which are normally used by diesel engines in cars and trucks, and fuel oils, which are used for space heating and to generate electric power

ecosystem: a community of living organisms called producers, consumers, and decomposers. These biotic and abiotic components are linked together through nutrient cycles and energy flows. Ecosystems exist at multiple scales and levels

electrical energy: energy carried by an electric current flowing around a circuit

electromagnetic spectrum: a family of waves, including light, radio waves, microwaves, infrared rays, and X-rays

environmental pollution: the contamination of the physical and biological components of the earth/atmosphere system to such an extent that normal environmental processes are adversely affected

eutrophication: in a lake, this is characterized by an abundant accumulation of nutrients that support a dense growth of algae and other organisms, the decay of which depletes the shallow waters of oxygen in summer

front group: an organization that purports to represent one agenda; while in reality, it serves some other party or interest whose sponsorship is hidden or rarely mentioned

genetically modified crops: also called GMCs, GM crops, or biotech crops, these are plants used in agriculture, the DNA of which has been modified using genetic engineering techniques. In most cases, the aim is to introduce a new trait to the plant that does not occur naturally in the species

geopolitics: the battle for space and power played out in a geographical setting

green building: also called green construction or sustainable building, green building refers to both a structure and the use of processes that are environmentally responsible and resource-efficient throughout a building's life cycle: from siting to design, construction, operation, maintenance, renovation, and demolition

green collar worker: green-collar worker is a worker who is employed in the environmental sector of the economy. Environmental green-collar workers (or green jobs) satisfy the demand for green development. Generally, they implement environmentally conscious design, policy, and technology to improve conservation and sustainability

green economy: an economy that aims at reducing environmental risks and ecological scarcities and that aims for sustainable development without degrading the environment

heat energy: energy that an object has because of its temperature. Hotter, larger objects have more heat energy

hybridization: a controlled method of pollination in which the pollen of two different species or varieties is crossed by human intervention. Hybridization can occur naturally through random crosses, but commercially available hybridized seed, often labeled as F1, is deliberately created to breed a desired trait

hypolimnion: the layer of water below the thermocline in some deep lakes

industrial revolution: the time in history of the transition to new manufacturing processes, from abo ut 1760 to 1840. This became the age of machine power and affected nearly every aspect of daily life. This was also the turning point when population began its unprecedented growth

kinetic energy: energy that an object has because it is moving or spinning around

low-carbon fuel standard (LCFS): a rule enacted to reduce carbon intensity in transportation fuels as compared to conventional petroleum fuels, such as gasoline and diesel. The most common low-carbon fuels are alternative fuels and cleaner fossil fuels, such as natural gas (CNG and LPG)

mechanical energy: energy transferred to an object

monoculture: the agricultural practice of only growing a single crop, plant, or livestock species in an area at a time. This can cause buildup of pests and diseases. The opposite situation, with several crops over time rotated through an area, is called polyculture

nuclear energy: energy stored inside an atom, which is released when the atom's nucleus splits up or combines with another nucleus

particulate: also called particulate matter, particulates are microscopic solid or liquid matter suspended in the Earth's atmosphere

phonon: a definite discrete unit or quantum of vibrational mechanical energy, just as a photon is a quantum of electromagnetic or light energy. Phonons and electrons are the two main types of elementary particles or excitations in solids

photon: the fundamental particle of visible light. In some ways, visible light behaves like a wave phenomenon, but in other respects, it acts like a stream of high-speed, submicroscopic particles. Isaac Newton was one of the first scientists to theorize that light consists of particles—making it both waves and particles

potential energy: energy that can be released by an object or a substance, such as gravitational energy, elastic energy, or chemical energy

primary energy source: an energy source that can be used directly, as it appears in the natural environment: coal, oil, natural gas and wood, nuclear fuels (uranium), the Sun, the wind, tides, mountain lakes, the rivers (from which hydroelectric energy can be obtained), and the heat from the earth's interior that supplies geothermal energy

pseudoscience: a collection of beliefs or practices mistakenly regarded as being based on scientific method

radiate: one of the ways that heat is transferred. The Sun's heat is radiated through space in the form of photons. The atmosphere, the land, and the oceans then absorb these photons

replacement level fertility: the fertility rate at which a population exactly replaces itself from one generation to the next without migration. In other words, as many die as are being born. This keeps the population constant

secondary energy source: an energy resource that is made using a primary resource. Electricity is a secondary resource and can be generated by a number of different primary resources, such as coal, which is a primary energy source

smart grid: an electrical grid that includes a variety of operational and energy measures including smart meters, smart appliances, renewable energy resources, and energy efficiency resources

solar power tower: also called a central tower or heliostat power tower, it is a type of solar furnace using a tower to receive focused sunlight. It uses an array of flat, movable mirrors (called heliostats) to focus the Sun's rays upon a collector tower (the target)

suburban redevelopment: reinventing infrastructure for compact development, thereby removing the necessity to travel extensive distances outside residential areas

sunspace: a room or area in a building having a glass roof and walls and intended to maximize the power of the Sun's rays

Trombe wall: a passive solar building design where a wall is built on the winter Sun side of a building with a glass external layer and a high heat capacity internal layer separated by a layer of air. Light close to UV in the electromagnetic spectrum passes through the glass almost unhindered and then is absorbed by the wall that re-radiates in the far infrared spectrum that does not pass back through the glass easily, instead heating the inside of the building

urban heat island effect: The term "heat island" describes built-up areas that are hotter than nearby rural areas. The annual mean air temperature of a city with 1 million people or more can be 1°C–3°C warmer than its surroundings. In the evening, the difference can be as high as 12°C. Heat islands can affect communities by increasing summertime peak energy demand, air conditioning costs, air pollution and greenhouse gas emissions, heat-related illness and mortality, and water quality

urban in-fill: new development that is sited on vacant or undeveloped land within an existing community and that is enclosed by other types of development. The term "urban in-fill" itself implies that the existing land is mostly built-out and what is being built is in effect "filling in" the gaps

Periodic Table

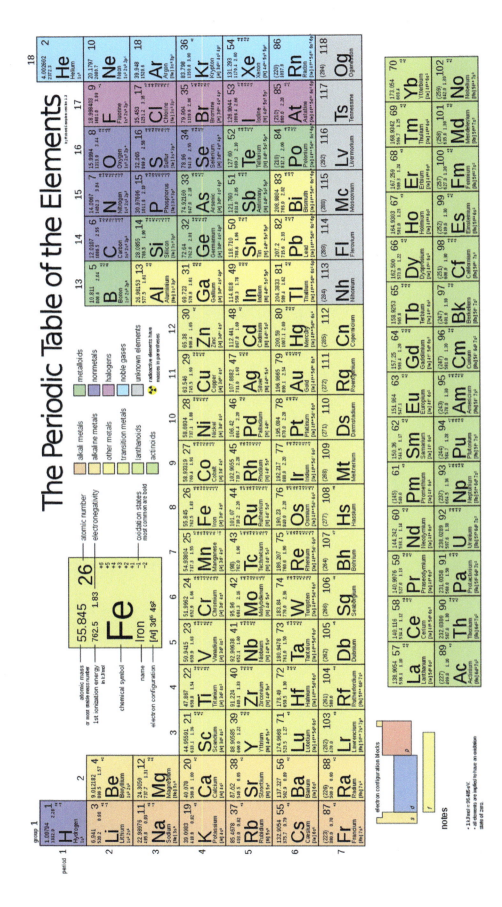

Index

Note: Page numbers followed by f and t refer to figures and tables, respectively.